T0180526

Transactions on Computational Science and Computational Intelligence

Series Editor
Hamid Arabnia
Department of Computer Science
The University of Georgia
Athens, Georgia
USA

Computational Science (CS) and Computational Intelligence (CI) both share the same objective: finding solutions to difficult problems. However, the methods to the solutions are different. The main objective of this book series, "Transactions on Computational Science and Computational Intelligence", is to facilitate increased opportunities for cross-fertilization across CS and CI. This book series will publish monographs, professional books, contributed volumes, and textbooks in Computational Science and Computational Intelligence. Book proposals are solicited for consideration in all topics in CS and CI including, but not limited to, Pattern recognition applications; Machine vision; Brain-machine interface; Embodied robotics; Biometrics; Computational biology; Bioinformatics; Image and signal processing; Information mining and forecasting; Sensor networks; Information processing; Internet and multimedia; DNA computing; Machine learning applications; Multi-agent systems applications; Telecommunications; Transportation systems; Intrusion detection and fault diagnosis; Game technologies; Material sciences; Space, weather, climate systems, and global changes; Computational ocean and earth sciences; Combustion system simulation; Computational chemistry and biochemistry; Computational physics; Medical applications; Transportation systems and simulations; Structural engineering; Computational electro-magnetic; Computer graphics and multimedia; Face recognition; Semiconductor technology, electronic circuits, and system design; Dynamic systems; Computational finance; Information mining and applications; Astrophysics; Biometric modeling; Geology and geophysics; Nuclear physics; Computational journalism; Geographical Information Systems (GIS) and remote sensing; Military and defense related applications; Ubiquitous computing; Virtual reality; Agent-based modeling; Computational psychometrics; Affective computing; Computational economics; Computational statistics; and Emerging applications. For further information, please contact Mary James, Senior Editor, Springer, mary.james@springer.com.

More information about this series at http://www.springer.com/series/11769

Transactions on Computational Science and Computational Intelligence

Series Editor
Hamid Arabnia
Department of Computer Science
The University of Georgia
Athens, Georgia
USA

Computational Science (CS) and Computational Intelligence (CI) both share the same objective: finding solutions to difficult problems. However, the methods to the solutions are different. The main objective of this book series, "Transactions on Computational Science and Computational Intelligence", is to facilitate increased opportunities for cross-fertilization across CS and CI. This book series will publish monographs, professional books, contributed volumes, and textbooks in Computational Science and Computational Intelligence. Book proposals are solicited for consideration in all topics in CS and CI including, but not limited to, Pattern recognition applications; Machine vision; Brain-machine interface; Embodied robotics; Biometrics; Computational biology; Bioinformatics; Image and signal processing; Information mining and forecasting; Sensor networks; Information processing; Internet and multimedia; DNA computing; Machine learning applications; Multi-agent systems applications; Telecommunications; Transportation systems; Intrusion detection and fault diagnosis; Game technologies; Material sciences; Space, weather, climate systems, and global changes; Computational ocean and earth sciences; Combustion system simulation; Computational chemistry and biochemistry; Computational physics; Medical applications; Transportation systems and simulations; Structural engineering; Computational electro-magnetic; Computer graphics and multimedia; Face recognition; Semiconductor technology, electronic circuits, and system design; Dynamic systems; Computational finance; Information mining and applications; Astrophysics; Biometric modeling; Geology and geophysics; Nuclear physics; Computational journalism; Geographical Information Systems (GIS) and remote sensing; Military and defense related applications; Ubiquitous computing; Virtual reality; Agent-based modeling; Computational psychometrics; Affective computing; Computational economics; Computational statistics; and Emerging applications. For further information, please contact Mary James, Senior Editor, Springer, mary.james@springer.com.

More information about this series at http://www.springer.com/series/11769

Robert Stahlbock • Gary M. Weiss
Mahmoud Abou-Nasr • Cheng-Ying Yang
Hamid R. Arabnia • Leonidas Deligiannidis
Editors

Advances in Data Science and Information Engineering

Proceedings from ICDATA 2020 and IKE 2020

 Springer

Editors

Robert Stahlbock
HBS – Hamburg Business School, Institute
of Information Systems
University of Hamburg
Hamburg, Hamburg, Germany

Gary M. Weiss
Department of Computer & Information
Science
Fordham University
New York, NY, USA

Mahmoud Abou-Nasr
College of Engineering & Computer
Science
University of Michigan-Dearborn
Dearborn, MI, USA

Cheng-Ying Yang
Department of Computer Science
University of Taipei
Taipei City, Taiwan

Leonidas Deligiannidis
School of Computing and Data Sciences
Wentworth Institute of Technology
Boston, MA, USA

Hamid R. Arabnia
Department of Computer Science
University of Georgia
Athens, GA, USA

ISSN 2569-7072   ISSN 2569-7080   (electronic)
Transactions on Computational Science and Computational Intelligence
ISBN 978-3-030-71706-3   ISBN 978-3-030-71704-9   (eBook)
https://doi.org/10.1007/978-3-030-71704-9

This Springer imprint is published by the registered company Springer Nature Switzerland AG
The registered company address is: Gewerbestrasse 11, 6330 Cham, Switzerland

Preface

It gives us great pleasure to introduce this collection of papers that were presented at the following international conferences: Data Science (ICDATA 2020) and Information & Knowledge Engineering (IKE 2020). These two conferences were held simultaneously (same location and dates) at Luxor Hotel (MGM Resorts International), Las Vegas, USA, July 27–30, 2020. This international event was held using a hybrid approach, that is, "in-person" and "virtual/online" presentations and discussions.

This book is composed of nine Parts. Parts I through V (composed of 46 chapters) include chapters that address emerging trends in data science (ICDATA). Parts VI through IX (composed of 25 chapters) include a collection of chapters in the areas of information and knowledge engineering (IKE).

An important mission of the World Congress in Computer Science, Computer Engineering, and Applied Computing, CSCE (a federated congress to which this event is affiliated with), includes *"Providing a unique platform for a diverse community of constituents composed of scholars, researchers, developers, educators, and practitioners. The Congress makes concerted effort to reach out to participants affiliated with diverse entities (such as: universities, institutions, corporations, government agencies, and research centers/labs) from all over the world. The congress also attempts to connect participants from institutions that have **teaching** as their main mission with those who are affiliated with institutions that have **research** as their main mission. The congress uses a quota system to achieve its institution and geography diversity objectives."* By any definition of diversity, this congress is among the most diverse scientific meeting in the USA. We are proud to report that this federated congress had authors and participants from 54 different nations representing variety of personal and scientific experiences that arise from differences in culture and values.

The program committees (refer to subsequent pages for the list of the members of committees) would like to thank all those who submitted papers for consideration. About 50% of the submissions were from outside the USA. Each submitted paper was peer reviewed by two experts in the field for originality, significance, clarity, impact, and soundness. In cases of contradictory recommendations, a member of the conference program committee was charged to make the final decision, often this involved seeking help from additional referees. In addition, papers whose authors included a member of the conference program committee were evaluated using the double-blind review process. One exception to the above evaluation process was for papers that were submitted directly to chairs/organizers of pre-approved sessions/workshops; in these cases, the chairs/organizers were responsible for the evaluation of such submissions. The overall paper acceptance rate for regular papers was 20%; 18% of the remaining papers were accepted as short and/or poster papers.

We are grateful to the many colleagues who offered their services in preparing this book. In particular, we would like to thank the members of the program committees of individual research tracks as well as the members of the steering committees of ICDATA 2020 and IKE 2020; their names appear in the subsequent pages. We would also like to extend our appreciation to over 500 referees.

As sponsors-at-large, partners, and/or organizers, each of the following (separated by semicolons) provided help for at least one research track: Computer Science Research, Education, and Applications (CSREA); US Chapter of World Academy of Science; American Council on Science and Education & Federated Research Council; and Colorado Engineering Inc. In addition, a number of university faculty members and their staff, several publishers of computer science and computer engineering books and journals, chapters and/or task forces of computer science associations/organizations from three regions, and developers of high-performance machines and systems provided significant help in organizing the event as well as providing some resources. We are grateful to them all.

We express our gratitude to all authors of the articles published in this book and the speakers who delivered their research results at the congress. We would also like to thank the following: UCMSS (Universal Conference Management Systems & Support, California, USA) for managing all aspects of the conference; Dr. Tim Field of APC for coordinating and managing the printing of the programs; the staff at Luxor Hotel (MGM Convention) for the professional service they provided; and Ashu M. G. Solo for his help in publicizing the congress. Last but not least, we would like to thank Ms. Mary James (Springer Senior Editor in New York) and

Arun Pandian KJ (Springer Production Editor) for the excellent professional service they provided for this book project.

Hamburg, Germany	Robert Stahlbock
New York, NY, USA	Gary M. Weiss
Dearborn, MI, USA	Mahmoud Abou-Nasr
Taipei City, Taiwan	Cheng-Ying Yang
Athens, GA, USA	Hamid R. Arabnia
Boston, MA, USA	Leonidas Deligiannidis

Book Co-editors and Chapter Co-editors:
Advances in Data Science and Information Engineering + ICDATA 2020 & IKE 2020

Preface

It gives us great pleasure to introduce this collection of papers that were submitted and accepted for the 16th International Conference on Data Science 2020, ICDATA'20 (https://icdata.org), July 27–30, 2020, at Luxor Hotel, Las Vegas, USA. Obviously, the year 2020 is very different from others due to the Covid-19 pandemic that had severe impact on all our lives. That was not at the horizon when planning the conference. The conference was held, but almost all authors were not allowed to travel during the summer, and even if it would have been allowed, it would have been wise to stay at home instead of travelling if possible. As a consequence, the typical communication, face to face, during sessions, in front of the conference rooms and during social events, was replaced by the opportunity to give talks via the web, either as pre-recorded talk or "live." All organizers and presenters did their best in that situation. Thank you very much for all your effort!

Some words about ICDATA and data mining: data mining or machine learning is critically important if we want to effectively learn from the tremendous amounts of data that are routinely being generated in science, engineering, medicine (take Covid-19 and the search for better understanding of the disease as well as for medicine and better treatment as an example), business, sports and e-sports, and other areas. The aim is gaining insight into processes and transactions, extract knowledge, make better decisions, and deliver value to users or organizations. This is even more important and challenging in an era in which scientists and practitioners are faced with numerous challenges caused by exponential expansion of digital data, its diversity, and complexity. The scale and growth of data considerably outpace technological capacities of organizations to process and manage it. During the last decade, we all observed new, more glorious, and promising concepts or labels emerging and slowly but steadily displacing "data mining" from the agenda of CTOs. It was and still is the time, more than ever before, of data science, big data, advanced-/business-/customer-/data-/predictive-/prescriptive-/ . . . /risk-analytics, to name only a few terms that dominate websites, trade journals, and the general press – although there is even a rebirth of terms such as artificial intelligence (AI) and (machine) learning (e.g., deep learning) in academia, companies, and even on the agenda of political decision makers.

All the concepts of data science aim at leveraging data for a better understanding of complex real-world phenomena. They all pursue this objective using some formal, often algorithmic, procedures, at least to some extent. This is what data miners have been doing for decades. The very idea of all those similar or identical concepts with different labels; the idea to think of massive, omnipresent amounts of data as strategic assets; and the aim to capitalize on these assets by means of analytic procedures is, indeed, more relevant and topical than ever before. Although there are very helpful advances in hardware and software, there are still many challenges to be tackled in order to leverage the promises of data analytics. Obviously, technological change is never ending and appears to be accelerating. The world is especially focused on machine learning and data mining (not contradictory but similar or even equivalent to data science), as these disciplines are making an ever-increasing impact on our society. Large multinational corporations are expanding their efforts in these areas, small startups are founded, and students are flocking to computer science and related disciplines in order to learn about these disciplines and take advantage of the many lucrative job opportunities. Many industries, even conservative ones like, for example, the port industry, are working towards "Version 4.0" (e.g., "Port 4.0"), with digitization, digitalization, and even digital transformation of traditional processes resulting in improved workflows, new concepts, and new business plans. Their goal usually includes data analytics, automation, autonomization, robotics, and AI. The industry is interested in feasibility studies and results of scientific research. Data science is popular like never before. Data scientists are rare on the job market and, therefore, very well compensated.

The growth in all these areas has been dramatic enough to require changes in nomenclature. Most of these "hot" technologies and methods are increasingly considered part of the broad field of data science, and there are benefits to viewing this field as a unified whole, rather than a collection of disparate sub-disciplines. ICDATA, the former data mining conference DMIN merged with the big data conference ABDA, is much broader than just data mining and big data. It includes all of the following main topics: all aspects of data mining and machine learning (tasks, algorithms, tools, applications, etc.), all aspects of big data (algorithms, tools, infrastructure, and applications), data privacy issues, and data management. The conference is designed to be of equal interest to researchers and practitioners, academics and members of industry, computer scientists, physical and social scientists, and business analysts.

ICDATA'20 attracted submissions of theoretical research papers as well as industrial reports, application case studies, and, in a second phase, late breaking papers, position papers, and abstract/poster papers. The program committee would like to thank all those who submitted papers for consideration. We strived to establish a review process of high quality. To ensure a fair, objective, and transparent review process, all review criteria are published on the website. Papers were evaluated regarding their relevance to ICDATA, originality, significance, information content, clarity, and soundness on an international level. Each aspect was objectively evaluated, with alternative aspects finding consideration for application papers. Each paper was refereed by at least two researchers in the topical area, taking the reviewers' expertise and confidence into consideration, with most of the papers receiving three reviews. The review process was competitive. The overall acceptance rate for submissions was 47%.

We are very grateful to the colleagues who helped in organizing the conference. In particular, we would like to thank the members of the program committee of ICDATA'20 and the members of the congress steering committee. The continuing support of the ICDATA program committee has been essential to further improve the quality of accepted submissions and the resulting success of the conference. The ICDATA'20 program committee members are (in alphabetical order): Mahmoud Abou-Nasr, Ruhul Amin, Jérôme Azé, Kai Brüssau, Paulo Cortez, Zahid Halim, Tzung-Pei Hong, Wei-Chiang Hong, Andrew Johnston, Madjid Khalilian, Robert Stahlbock, Chamont Wang, Gary M. Weiss, Yijun Zhao, and Zijiang Yang. They all did a fantastic job in evaluating a lot of submissions in very short time. We are aware that their workload was particularly high due to the Covid-19 situation, so we are grateful for their support of ICDATA'20. The conference's quality depends on reliable and good reviewers. We would also like to thank Mahmoud Abou-Nasr for organizing the special session on "Real-World Data Mining & Data Science Applications, Challenges, and Perspectives" for more than a decade. We would like to thank our publicity co-chair Ashu M. G. Solo (Fellow of British Computer Society, Principal/R&D Engineer, Maverick Technologies America Inc.) for circulating information on the conference, as well as www.KDnuggets.com, a platform for analytics, data mining, and data science resources, for listing ICDATA'20. We are also grateful for support by the Institute of Information Systems at Hamburg University, Germany and would like to thank all supporters and sponsors of CSCE. Last but not least, we wish to express again our sincere gratitude and utmost respect towards our colleague and friend Prof. Hamid R. Arabnia (Professor, Department of Computer Science, University of Georgia, USA; Editor-in-Chief, *Journal of Supercomputing* [Springer]), General Chair and Coordinator of the federated congress, and also Associate Editor of ICDATA'20, for his excellent, tireless, and continuous support, organization, and coordination of all affiliated events, particularly in these hard and difficult times of Covid-19. His exemplary and professional effort in 2020 and all the earlier years in the steering committee of the congress make these events possible. We are grateful to continue our data science conference as ICDATA'20 under the umbrella of the CSCE congress.

Thank you all for your contribution to ICDATA'20! We hope to see you at ICDATA'21. Stay safe and healthy!

We present the proceedings of ICDATA'20.

ICDATA'20 General Conference Chair   Robert Stahlbock

Steering Committee ICDATA'20
https://icdata.org

Data Science: ICDATA 2020 – Organizing Committee (Leadership)

- *Dr. Robert Stahlbock (ICDATA 2020 Chair); University of Hamburg, Germany*
- *Dr. Gary M. Weiss; Fordham University, New York, USA*
- *Dr. Sven F. Crone; Lancaster University, UK*
- *Dr. Mahmoud Abou-Nasr, USA*
- *Dr. Hamid R. Arabnia, USA*

For the complete list of program committee refer to: https://icdatascience.org/

Information & Knowledge Engineering: IKE 2020 – Program Committee

- *Prof. Emeritus Nizar Al-Holou (Congress Steering Committee); Vice Chair, IEEE/SEM-Computer Chapter; University of Detroit Mercy, Detroit, Michigan, USA*
- *Prof. Emeritus Hamid R. Arabnia (Congress Steering Committee); The University of Georgia, USA; Editor-in-Chief, Journal of Supercomputing (Springer); Fellow, Center of Excellence in Terrorism, Resilience, Intelligence & Organized Crime Research (CENTRIC).*
- *Dr. Travis Atkison; Director, Digital Forensics and Control Systems Security Lab, Department of Computer Science, College of Engineering, The University of Alabama, Tuscaloosa, Alabama, USA*
- *Dr. Arianna D'Ulizia; Institute of Research on Population and Social Policies, National Research Council of Italy (IRPPS), Rome, Italy*
- *Prof. Emeritus Kevin Daimi (Congress Steering Committee); Department of Mathematics, Computer Science and Software Engineering, University of Detroit Mercy, Detroit, Michigan, USA*
- *Prof. Zhangisina Gulnur Davletzhanovna; Vice-rector of the Science, Central-Asian University, Kazakhstan, Almaty, Republic of Kazakhstan; Vice President of International Academy of Informatization, Kazskhstan, Almaty, Republic of Kazakhstan*
- *Prof. Leonidas Deligiannidis (Congress Steering Committee); Department of Computer Information Systems, Wentworth Institute of Technology, Boston, Massachusetts, USA*
- *Prof. Mary Mehrnoosh Eshaghian-Wilner (Congress Steering Committee); Professor of Engineering Practice, University of Southern California, California, USA; Adjunct Professor, Electrical Engineering, University of California Los Angeles, Los Angeles (UCLA), California, USA*
- *Prof. Ray Hashemi (Session Chair, IKE & Steering Committee member); Professor of Computer Science and Information Technology, Armstrong Atlantic State University, Savannah, Georgia, USA*

- *Prof. Dr. Abdeldjalil Khelassi; Computer Science Department, Abou beker Belkaid University of Tlemcen, Algeria; Editor-in-Chief, Medical Technologies Journal; Associate Editor, Electronic Physician Journal (EPJ) - Pub Med Central*
- *Prof. Louie Lolong Lacatan; Chairperson, Computer Engineerig Department, College of Engineering, Adamson University, Manila, Philippines; Senior Member, International Association of Computer Science and Information Technology (IACSIT), Singapore; Member, International Association of Online Engineering (IAOE), Austria*
- *Dr. Andrew Marsh (Congress Steering Committee); CEO, HoIP Telecom Ltd (Healthcare over Internet Protocol), UK; Secretary General of World Academy of BioMedical Sciences and Technologies (WABT) a UNESCO NGO, The United Nations*
- *Dr. Somya D. Mohanty; Department of CS, University of North Carolina - Greensboro, North Carolina, USA*
- *Dr. Ali Mostafaeipour; Industrial Engineering Department, Yazd University, Yazd, Iran*
- *Dr. Houssem Eddine Nouri; Informatics Applied in Management, Institut Superieur de Gestion de Tunis, University of Tunis, Tunisia*
- *Prof. Dr., Eng. Robert Ehimen Okonigene (Congress Steering Committee); Department of Electrical & Electronics Engineering, Faculty of Engineering and Technology, Ambrose Alli University, Nigeria*
- *Prof. James J. (Jong Hyuk) Park (Congress Steering Committee); Department of Computer Science and Engineering (DCSE), SeoulTech, Korea; President, FTRA, EiC, HCIS Springer, JoC, IJITCC; Head of DCSE, SeoulTech, Korea*
- *Dr. Prantosh K. Paul; Department of CIS, Raiganj University, Raiganj, West Bengal, India*
- *Dr. Xuewei Qi; Research Faculty & PI, Center for Environmental Research and Technology, University of California, Riverside, California, USA*
- *Dr. Akash Singh (Congress Steering Committee); IBM Corporation, Sacramento, California, USA; Chartered Scientist, Science Council, UK; Fellow, British Computer Society; Member, Senior IEEE, AACR, AAAS, and AAAI; IBM Corporation, USA*
- *Chiranjibi Sitaula; Head, Department of Computer Science and IT, Ambition College, Kathmandu, Nepal*
- *Ashu M. G. Solo (Publicity), Fellow of British Computer Society, Principal/R&D Engineer, Maverick Technologies America Inc.*
- *Prof. Fernando G. Tinetti (Congress Steering Committee); School of CS, Universidad Nacional de La Plata, La Plata, Argentina; also at Comision Investigaciones Cientificas de la Prov. de Bs. As., Argentina*
- *Varun Vohra; Certified Information Security Manager (CISM); Certified Information Systems Auditor (CISA); Associate Director (IT Audit), Merck, New Jersey, USA*
- *Dr. Haoxiang Harry Wang (CSCE); Cornell University, Ithaca, New York, USA; Founder and Director, GoPerception Laboratory, New York, USA*

- *Prof. Shiuh-Jeng Wang (Congress Steering Committee); Director of Information Cryptology and Construction Laboratory (ICCL) and Director of Chinese Cryptology and Information Security Association (CCISA); Department of Information Management, Central Police University, Taoyuan, Taiwan; Guest Ed., IEEE Journal on Selected Areas in Communications.*
- *Prof. Layne T. Watson (Congress Steering Committee); Fellow of IEEE; Fellow of The National Institute of Aerospace; Professor of Computer Science, Mathematics, and Aerospace and Ocean Engineering, Virginia Polytechnic Institute & State University, Blacksburg, Virginia, USA*
- *Prof. Jane You (Congress Steering Committee); Associate Head, Department of Computing, The Hong Kong Polytechnic University, Kowloon, Hong Kong*

Contents

Part I
Graph Algorithms, Clustering, and Applications

Phoenix: A Scalable Streaming Hypergraph Analysis Framework

Kuldeep Kurte, Neena Imam, S. M. Shamimul Hasan, and Ramakrishnan Kannan

1 Introduction

Over the last few years, we have witnessed the explosive growth of data due to the technological advancements in the fields of social networking, e-commerce, smart mobile devices, etc. This necessitates the development of novel data mining/analysis approaches to address the various analytical challenges posed by the massive growth in data. Some examples of data analytics include live tracking in the transportation sector, fraud management in insurance, product recommendations in the retail industry, and predictive analysis in health care. These analyses study the relations, dynamics, and behavior at an individual level (entity level) as well as at the group level. The graph representation, $G = (V, E)$, in which entities are represented by vertices ($V = \{v_1, v_2, .., v_n\}$) and relations among entities are represented by edges ($E = \{e_1, e_2, \ldots, e_m\}$), is a natural way to model such relational information. For instance, in an e-commerce system, customers and products are modeled as vertices, and customer-product relations are represented by edges.

This manuscript has been authored in part by UT-Battelle, LLC, under contract no. DE-AC05-00OR22725 with the US Department of Energy (DOE). The US government retains and the publisher, by accepting the article for publication, acknowledges that the US government retains a nonexclusive, paid-up, irrevocable, worldwide license to publish or reproduce the published form of this manuscript, or allow others to do so, for US government purposes. DOE will provide public access to these results of federally sponsored research in accordance with the DOE Public Access Plan (http://energy.gov/downloads/doe-public-access-plan).

K. Kurte (✉) · N. Imam · S. M. S. Hasan · R. Kannan
Computing and Computational Sciences Directorate, Oak Ridge National Laboratory, Oak Ridge, TN, USA
e-mail: kurtekr@ornl.gov; imamn@ornl.gov; hasans@ornl.gov; kannanr@ornl.gov

© Springer Nature Switzerland AG 2021　　　　　　　　　　　　　　　　　　　　3
R. Stahlbock et al. (eds.), *Advances in Data Science and Information Engineering*,
Transactions on Computational Science and Computational Intelligence,
https://doi.org/10.1007/978-3-030-71704-9_1

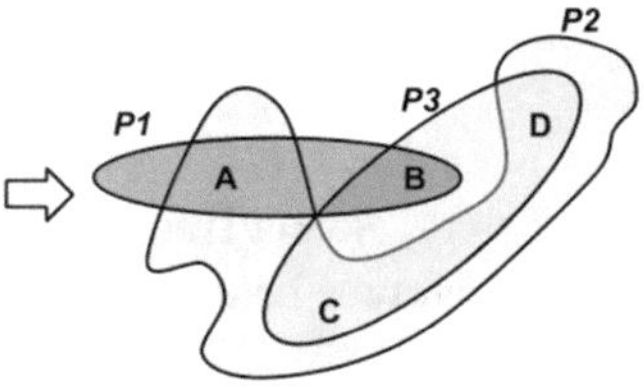

Fig. 1 Example hypergraph showing social media users (rows) and three social media posts (columns). Each post P_i is represented as an hyperedge, and those users who interacted with that post are the hypergraph vertices incident on that hyperedge

The graph representation of the information is able to capture the dyadic relations, i.e., relations between two entities, but fails to model the group-level interactions. Due to the fact that the individual's behavior is mainly influenced by the group-level interactions, modeling group-level dynamics is important. Hypergraphs—the generalization of graphs—provide an excellent way to model the group-level interactions [6, 9, 28]. A hypergraph $HG = (V, H)$ is an ordered pair of "n" vertices, i.e., $V = \{v_1, v_2, v_3, \ldots, v_n\}$, and H is a set of "m" hyperedges, i.e., $H = \{H_1, H_2, H_3, \ldots, H_m\}$. Each hyperedge H_i is a vector of incident vertices such that $V \equiv h_1 \cup h_2 \cup h_3 \cup \ldots \cup h_m$. Figure 1 shows an example hypergraph which includes four social network users, A, B, C, D, and three social media posts, P_1, P_2, P_3. Each post P_i represents a hyperedge, and its incident vertices are the users who interacted with the content, say, shared, liked, or commented on the post (represented by "X"). From this example, it is evident that such hypergraph-based representation is useful to understand the information propagation among entities and the categorization of groups according to specific interests over the social network.

Although the efficacy of hypergraphs for modeling group dynamics is well documented [1], efficient hypergraph analytics must overcome challenges associated with accurate hypergraph representation and scalable computation models that can deal with very high data ingestion rates without creating bottlenecks. While several large-scale graph processing software are available such as [5, 7, 18, 26], only a limited number of options are available for *hypergraph* analysis frameworks [28]. Very-large-scale hypergraph analysis requires scalable and distributed computing systems which present novel challenges as well as opportunities. The situation becomes more challenging when streaming data need to be incorporated in the framework. Some challenges posed by the streaming scenario include variability in the streaming rates from various external hypergraph sources, heterogeneity in representing the hypergraph, and efficient hypergraph representation at a system level to sustain the streaming scenario.

Little research has been done for methodical performance evaluation of large-scale hypergraph analysis frameworks in a streaming scenario. The leadership class high-performance computing facilities, such as hosted at Oak Ridge National Laboratory, provide petascale to exascale computing powers, large amounts of per node memory, efficient storage, and high-speed interconnects. Such leadership class computing facilities can meet the computational requirements of large-scale streaming hypergraph analysis. As such, researchers at Oak Ridge National Laboratory developed *Phoenix*, a high-performance, hybrid system enabling concurrent utilization of online and offline analysis worlds. Phoenix architecture is distributed for scalability of problem size and performance. In addition, Phoenix is designed for fast and scalable ingest of streaming data sources. Phoenix also incorporates fast online (CRUD) operations and has dynamic (and fixed) schema. Using Phoenix, researchers are able to perform fast decoupled offline global analytics with in-memory snapshots and commit logs. Phoenix was deployed on Oak Ridge National Lab's Titan (ranked number one on top500[1] list in 2012) and showed good performance. Originally designed for simple graph analytics, we recently enhanced Phoenix to handle *hypergraphs*. The performance of Phoenix for streaming datasets is the subject of this paper.

In the following sections, we present our approach to scalable streaming hypergraph analysis as implemented in Phoenix. Section 2 presents an overview of the various hypergraph analysis tools. Section 3 presents the Phoenix framework for streaming hypergraph analysis and describes various technical aspects of Phoenix. Section 4 presents results of the numerical experiments we performed to evaluate metrics such as streaming performance, ingestion performance, and hypergraph clustering efficiency. Section 5 summarizes our observations and discusses few future extensions of this work.

2   Related Work

Many hypergraph analysis tools are available. However, none of these tools presents the scalability and flexibility associated with Phoenix. In addition, Phoenix incorporates scalable hypergraph generators. Most other hypergraph analytics software tools do not have this attribute. In the following paragraphs, we present an overview of the various hypergraph analysis tools and the advantages and disadvantages of each.

HyperNetX is a Python library that supports hypergraph creation, hypergraph-connected component computation, sub-hypergraph construction, hypergraph statistics computation (e.g., node degree distribution, edge size distribution, toplex size computation for hypergraphs), and hypergraph visualization (e.g., draw hyper-graphs, color nodes, and edges). *HyperNetX* was released in 2018 under the Battelle Memorial Institute license [21]. *HyperNetX* library does not support high-

---

[1] https://www.top500.org/system/177975.

performance computing (HPC)-based parallel processing. Also, *HyperNetX* library documentation does not provide any scalability information.

Chapel HyperGraph Library (CHGL) was developed in the Chapel programming language by the Pacific Northwest National Laboratory. In the CHGL, users can use both shared and distributed memory systems for the storage of hypergraphs. The CHGL is not well documented and requires knowledge of the Chapel programming language, which is Partitioned Global Address Space (PGAS) language. PGAS languages are not as widely used as the C or C++ programming language. However, CHGL does offer valuable functionality within the context of parallel computations [2, 4].

HyperX offers a scalable framework for hypergraph processing and learning algorithms, which is developed on top of Apache Spark. It replicates the design model that is utilized within GraphX. *HyperX* directly processes the hypergraph rather than converting the hypergraph to a bipartite graph and employs GraphX to do the processing [2, 15]. Apache Spark programming paradigm cannot match the scalability offered by a leadership class computing platform.

HyperGraphLib package was developed in the C++ programming language, which supports k-uniform, k-regular, simple, linear, path search, and isomorphism algorithms. *HyperGraphLib* employs both OpenMP and Boost libraries. *HyperGraphLib* cannot represent a hypergraph as a bipartite graph or a two-section graph. Moreover, *HyperGraphLib* is not integrated with any graph libraries for advanced analytics [2, 14].

Halp is a Python library that provides both directed and undirected hypergraph implementations as well as a range of algorithms. These include a variety of hypergraph algorithms–for instance, k-shortest hyperpaths as well as random walk and directed paths [2, 13]. However, *Halp* does not provide parallel implementation of the algorithms.

SAGE hypergraph generator was developed in the Python language and supports the creation of complete random, uniform, and binomial random uniform hypergraphs. Nevertheless, large-scale hypergraph generation is not possible in *SAGE*. Besides, *SAGE* does not support parallel hypergraph generation.

Karlsruhe Hypergraph Partitioning (KaHyPar) was developed in C++ and is a multilevel hypergraph partitioning framework. It supports hypergraph partitioning with variable block weights and fixed vertices. Although *KaHyPar* is a useful tool, it does not support the hypergraph generation facility [16, 24].

The Julia programming language was used to develop the *SimpleHypergraphs.jl* hypergraph analysis framework. It is an efficient hypergraph analysis tool that supports distributed computing. However, *SimpleHypergraphs.jl* is heavily dependent on the *HyperNetX* library, specifically for hypergraph visualization. Moreover, *SimpleHypergraphs.jl* tool provides limited hypergraph analysis functionalities and is not highly scalable [2].

networkR was developed in the R programming language, which supports hypergraphs' projection into graphs. networkR also supports degree distribution, diameter, centrality, and network density computation. One of the limitations of the *networkR* is that it needs to project hypergraph into graph structure for analysis.

Moreover, vertices and hyperedge-related meta-information is not available in networkR [2, 20].

Gspbox provides hypergraph modeling capability. Although in Gspbox, one can manipulate the hypergraph by transforming a model into a regular graph, it does not provide specific solutions or optimizations for hypergraphs [2, 8].

BalancedGo software was developed in the Go programming language. BalancedGo supports generalized hypertree decompositions via balanced separators. *BalancedGo* supports a limited number of algorithms mainly focused on hypertree decompositions. Moreover, *BalancedGo* supports only HyperBench format or PACE Challenge 2019 format [3] as input.

Pygraph was released under the MIT license and is a Python library that can be used to process graphs. It includes hypergraph support along with standard graph functionalities. However, Pygraph does not offer any hypergraph optimization feature [2, 22].

Yadati et al. developed *HyperGCN*, a new graph convolutional network (GCN) training approach for semi-supervised learning (SSL) on hypergraphs [30]. The Python implementation of the tool is available in [12]. The quality of the hypergraph approximation heavily depends on weight initialization, which is a limitation of *HyperGCN* [30].

Multihypergraph is a Python package that provides support for multi-edges, hyperedges, and looped edges. The main focus of the *Multihypergraph* package is the mathematical understanding of graph than algorithmic efficiency. Moreover, the *Multihypergraph* package is limited with graph model memory definition and isomorphism functionalities and does not provide any other functionalities for hypergraphs [2, 19].

d3-hypergraph is a hypergraph visualization tool developed on top of the D3 JavaScript library. Another example of the hypergraph visualization tool is *visualsc,* which is similar to the open-source graph visualization tool Graphviz. *d3-hypergraph* and *visualsc* tools are solely used for hypergraph visualization.

3   Framework for Analyzing Streaming Hypergraphs

This section describes the overall Phoenix framework and its various components which enable the analysis of the streaming hypergraph. Figure 2 shows Phoenix's end-to-end framework which is composed of various essential modules for analyzing the streaming hypergraphs in a distributed and scalable fashion.

3.1   Hypergraph Sources and Generation

Phoenix is capable of utilizing a diverse set of graph generators as inputs to the framework. One of the candidates is a distributed hypergraph generator called

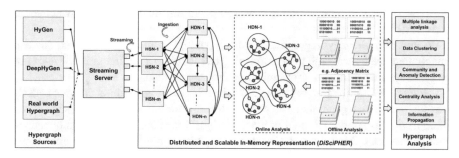

Fig. 2 Phoenix's end-to-end framework for scalable and distributed hypergraph analysis. Streaming server acts as a gateway where various hypergraph generators/external sources can connect. Next the streaming server streams the hypergraph in the form of hyperedge or incidences to the graph service nodes (GSNs). GSNs handle the communication with the streaming server and consume the hypergraph and send it to the graph data nodes (GDNs) where GDNs store the ingested hypergraph as its in-memory representation

HyGen, which is capable of generating synthetic hypergraphs. HyGen is another high-performance graph analytics project at Oak Ridge National Laboratory and was incorporated in the Phoenix architecture. HyGen takes input parameters such as number of clusters, number of vertices, and number of hyperedges to generate the corresponding hypergraph. For instance, if we have a rough understanding about the number of the clusters in the real-world hypergraph (e.g., communities), HyGen will enable the rapid production of the different sizes of hypergraphs which can be further consumed (by HSNs) and stored in-memory (by HDNs) in a distributed fashion. Refer to Fig. 2 and Sect. 3.3 for more information on HSNs and HDNs. Further, various online and offline analyses can be performed on this generated hypergraph. Similarly, the external hypergraph sources can also connect to the streaming server. More detailed discussions on graph generators can be found in references [17, 27, 32].

3.2 Hypergraph Streaming and Consumption

A streaming server is developed to facilitate the streaming of hypergraphs generated by hypergraph generators and from external sources to the internal core component called *DiSciPHER* (refer Sect. 3.3) which is responsible for hypergraph consumption and in-memory storage. The three advantages of having this layer of streaming server are as follows:

1. **Decoupling**: Streaming server acts as a gateway and prevents hypergraph generators and external sources from directly accessing the *DiSciPHER* which is a core internal module of Phoenix. This provides the flexibility to make changes in the *DiSciPHER* module without impacting the accessibility of the hypergraph

sources. Moreover, syntactic changes made by hypergraph sources do not have any impact on the *DiSciPHER*'s representation.

2. **Standardization**: Streaming servers can acquire data either as a bipartite representation or as a hyperedge representation. It is unlikely that all external sources comply with a unified syntax even though the data follow the semantics of bipartite or hyperedge representation. The streaming server can implement various methods for data translation to address this syntactic heterogeneity problem.

3. **Intermediate caching**: The rate of streaming from different external sources of hypergraphs can be different. At the system level, the heterogeneity in the streaming rates could cause data loss in case of extremely high data streaming and longer wait time for HSN processes in case of slow data streaming. We believe that the intermediate layer of the streaming server can stabilize the rate of streaming hypergraph from various external sources to HSN. The streaming server can provide a temporary storage capability to store the acquired hypergraph data before sending it to the HSNs of *DiSciPHER* module. This way streaming servers can stabilize the streaming rate.

The streaming server can acquire hypergraphs in one of two ways: (1) bipartite representation, a list of incidences, and (2) hyperedge representation, a list of hyperedges. Each incidence in a bipartite representation is a two-dimensional vector $\langle i, j \rangle$, such that $v_i \in H_j$, i.e., vertex v_i incident upon hyperedge H_j. On the other hand, the hyperedge representation constitutes a set of hyperedges (H) in which each hyperedge is a vector of incident vertices, i.e., $H_k = \langle v_{k1}, v_{k2}, v_{k3}, \ldots, v_{kp} \rangle$ and "p" is the total number of incident vertices on hyperedge k.

The streaming server opens multiple communication ports where several hypergraph service node (HSN) processes of *DiSciPHER* module, which is responsible for the consumption of the hypergraph, can connect and consume the hypergraph. In the case of bipartite representation, the streaming server performs streaming of incidences in a batched fashion. The batch size represents the maximum number of hypergraph incidences that can be packed in a batch. The batch size in case of hyperedge representation is the maximum number of hypergraphs per batch. Due to the variable size of hyperedges in a batch, the batch creation is not as straightforward as in the bipartite representation. Here, each hyperedge is reformatted as $\langle h_{id}, p, v_1, v_2, v_3, \ldots, v_p, -1 \rangle$ by appending hyperedge identifier h_{id}, its length in the beginning p, followed by a list of incident vertices, i.e., v_i and "-1" at the end to indicate the termination of the hyperedge. In this way, the hypergraphs are packed to form a batch such that each element in the batch represents either hypergraph identifier, length of hypergraph, vertex identifier, or "-1."

As mentioned in the paragraph above, the hypergraph service node processes (HSNs) connect to the communication ports of the streaming server and consume a hypergraph either as a batched incidences or as hyperedges. We implemented a handshaking and communication protocol to enable the streaming and consumption of the hypergraphs. Figure 3 shows a sequence of commands and data exchanges

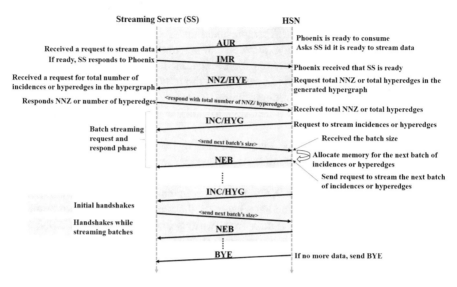

Fig. 3 Sequences of messages and data exchanges take place between the streaming server and hypergraph service node (HSN) process. The green portion depicts the message and data exchanges during the initial handshake. The blue portion depicts messages and data exchanges take place during streaming the batches of incidences (or hyperedges)

occurring while streaming the hypergraph. Initial handshake includes HSN process sending a message "AUR" asking if the streaming server is ready to stream the hypergraph. HSN waits until it receives "IMR" from the streaming server which indicates that the server is ready to stream the hypergraph. Next, depending on the format in which HSN wants to consume the hypergraph, it either sends "NNZ" to ask for the number of incidences (non-zeros) in case of bipartite representation or sends "HYG" to ask for the number of hyperedges in the hypergraph. In response to this message, streaming server sends total number of incidences (non-zeros) or total number of hyperedges to HSN.

After this initial handshake (as depicted in the green portion of Fig. 3), the streaming server sends "INC" (for bipartite) or "HYG" (for hyperedge) message to the streaming server. After receiving this message, the streaming server prepares the batch of incidences (or hyperedges) and sends the batch size to the HSN so that HSN can reserve sufficient memory to consume the upcoming batch of incidences (or hyperedges). Further, HSN sends a "NEB" message to the streaming server to indicate that it is ready to consume the batch of incidences (or hyperedges). Upon receiving "NEB," the streaming server sends the prepared batch of incidences (or hyperedges) to HSN. This communication between HSN and the streaming server (as depicted by the blue portion in Fig. 3) continues until the entire hypergraph is consumed by HSN.

3.3 Distributed and Scalable in-Memory rePresentation of HypERgraph (DiSciPHER)

Phoenix's *DiSciPHER* module is responsible for the efficient in-memory representation of the consumed hypergraph such that it enables both offline and online analysis on the streaming hypergraphs. Here, we describe, (1) how the *DiSciPHER* module represents hypergraphs and (2) two essential components of *DiSciPHER* which enable the efficient in-memory representation, i.e., hypergraph service node (HSN) processes and hypergraph data node (HDN) processes.

3.3.1 In-Memory Representation of Hypergraph

In this subsection, we describe the in-memory representation of hypergraphs at a system level which enables the online and offline analysis of hypergraphs in a streaming scenario inside Phoenix framework. In a streaming scenario, it is highly likely that only the part of hyperedge is available which is being streamed at any given time. In other words, the complete knowledge of the incident vertices of an hyperedge is not available at the time when that hyperedge is being streamed. The remaining (partial) hyperedge can arrive later. This characteristic of the streaming scenario, along with the variable sized nature of the hyperedge, increases overhead of the hypergraph ingestion process. For instance, every time the partial hyperedge arrives, the hyperedge insertion involves updating and redistributing the vertices (or hypergraphs) among the compute nodes (i.e., HDNs in Phoenix).

To alleviate the hyperedge ingestion overhead problem in a streaming scenario, we adopted a bipartite representation to represent the hypergraph at a system level. Figure 4 shows the bipartite representation of the hypergraph shown in Fig. 1. In this representation (refer to Fig. 4), users are vertices represented as green circles on the top, and each social media post is a hyperedge shown as a blue circle at the bottom. The additional edges (E_{vh}) are inserted from the nodes on the top side (vertices) to the hyperedge nodes on the bottom side to represent the incident vertices of the

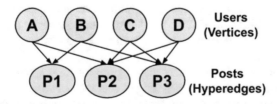

Fig. 4 Bipartite representation used in Phoenix to represent hypergraph. Example hypergraph is represented in Fig. 1. Each social media post is an hyperedge, and users who share, like, or comment on a common post are the vertices incident upon that hyperedge. In bipartite representation, both vertices and hyperedges are represented by nodes, and additional edges are added from incident vertices to their hyperedge to preserve the semantics of the hyperedge

hyperedge. This approach translates to hyperedge insertions with multiple edges. Such a representation in streaming scenarios can accommodate hyperedge with partial information, where the complete vertex set is still unknown. However, this approach increases the number of edge insertions for a given hyperedge. So there is a trade-off between update operations to accommodate the partial hyperedge issue and increased number of edge insertions in case of bipartite representation. In our opinion, the scalable and distributed nature of Phoenix can handle the increased number of edge insertions without impacting the streaming performance.

3.3.2 Hypergraph Service Nodes (HSNs) and Hypergraph Data Nodes (HDNs)

DiSciPHER is an essential component of the Phoenix's ecosystem consisting of hypergraph service nodes (HSNs) and back-end data storage and processing nodes (HDNs). The responsibilities of HSNs include communicating with the streaming server, consuming the hypergraph, redirecting the consumed hypergraph to the HDNs in a load-balanced fashion, and keeping track of the progress of the system by broadcasting messages to HSNs and HDNs. HDN's main task is to process and store the consumed hypergraph in a distributed in-memory fashion.

DiSciPHER's implementation is rather memory intensive (i.e., best suited for large clusters and supercomputers with very large memory). *DiSciPHER*'s deployment typically contains a number of HSNs (service nodes) that could be load balanced and a larger number of HDNs (data nodes)—depending on the size of the graph to be processed. A balanced shared distributed memory-multi-processing and multi-threading design approaches are used to achieve the best concurrent performance.

Figure 5 shows the cluster view of the HSNs and HDNs inside *DiSciPHER*. The number of HSNs and HDNs to be used (or deployed) depends on the size of the hypergraph and the availability of the compute resources, i.e., size of the cluster and memory per compute node. The *DiSciPHER* module makes sure that the HSNs and HDNs are deployed before initiating the communication. The process begins by allocating the sufficient numbers of compute nodes. Once the sufficient nodes are allocated, they are divided into two groups. One set of nodes are grouped as HSNs and the remaining as HDNs. We designed this cluster as a multi-processing and multi-threaded architecture in which each node has a main process and several worker and I/O threads.

Figure 6 shows the programmers view of the main components of HSNs and HDNs. The master process of HSN-"0" collects the network addresses of all the nodes in the cluster and groups them as HSNs and HDNs. Each node has dedicated queues for input and output messages. The input message queue is connected to the input socket which is set up to receive the incoming messages from all other nodes. Similarly, multiple output sockets are set up, each is dedicated to one node in the cluster. The output message queue is connected with these output sockets. The message (incoming or outgoing) contains *command*, indicates task to be performed;

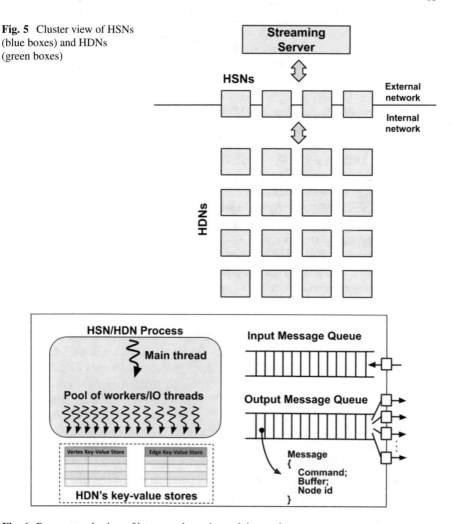

Fig. 5 Cluster view of HSNs (blue boxes) and HDNs (green boxes)

Fig. 6 Programmer's view of hypergraph service and data nodes

buffer, contains data to operate on; and *node id*, identifier of the destination HSN or HDN node. The messaging scheme supports peer-to-peer and broadcasting of the messages. The HDNs have additional in-memory distributed local key-value storage to store the local vertices (or hypergraph node) and edges (connecting vertex and hypergraph node) and their mapping with the corresponding global identifiers.

The input and output message queues are thread-safe and accessed by both workers and I/O threads within the node. The I/O threads are responsible for dispatching the outgoing messages, queued in the output message queue, to the appropriate socket based on the destination node. The I/O threads are also responsible for receiving the incoming messages from the input socket and putting them in the input message queue. The worker threads, depending on the requirements, form messages and put them in the output message queue and retrieve messages from input message queue and further perform the required task.

Once the cluster of HSNs and HDNs is deployed and ready to consume the hypergraph, HSN connects to the streaming server and requests for a batch of either incidences or hyperedges. Since, at the system level, *DiSciPHER* represents hypergraph in a bipartite form, the HSN creates message for inserting edge, i.e., E_{vh} (representing incidences connecting hypergraph's vertex v and hyperedge, h and directs it to the appropriate HDN based on the source vertex identifier. The HSN balances the load among HDNs by distributing the E_{vh} to HDNs in a round-robin fashion based on the identifier of the source vertex v.

The message for inserting E_{vh} includes *command*, inserting edge (E_{vh}); *buffer*, containing *source id* representing hypergraph vertex identifier; *destination id* representing hyperedge identifier; and *edge id* an identifier representing E_{vh} and *node id* representing the identifier of destination HDN. The worker threads of HSN are responsible for creating these messages and putting them in the output message queues. The I/O threads periodically take these messages out of the output message queue and send them to the appropriate HDNs through their dedicated communication sockets. The I/O threads of HDNs receive the input massages from the receiver socket and put them in the input message queue. The worker threads of HDNs further remove the messages from input queue and perform the required tasks, in this case storing the vertex (v) and the edge connecting them, i.e., E_{vh}.

If the destination vertex representing the hyperedge identifier belongs to the different HDN, then the worker thread of the current HDN forwards the edge insertion message to the appropriate HDN based on the hyperedge identifier. In this case, both the HDNs store the edge E_{vh} such that the destination field of the edge in the first HDN's points to the second HDN's identifier and the source field of the edge in the second HDN points to the identifier of the first HDN. The current implementation of Phoenix requires hypergraph generators (and external sources of hypergraph) to differentiate the vertex and hyperedge identifiers, in order to avoid potential conflicts between the vertex and hyperedge.

3.4 Hypergraph Clustering

The graph clustering problem involves partitioning the vertices, such that the similarity of vertices within a cluster is higher than the inter-cluster similarity. While most approaches on graph clustering assume edges as pairwise relationships between vertices, many real-world applications participate in multi-way relations represented as hyperedges in hypergraphs. Analogous to the graph clustering task, hypergraph clustering seeks to find partitions among vertices using hyperedges [25].

Within the ML community, the seminal work of Zhou, Huang, and Schölkopf [33] looked at learning on hypergraphs. They sought to support spectral clustering methods on hypergraphs and defined a suitable hypergraph Laplacian. This effort, like many other existing methods for hypergraph learning, makes use of a reduction of the hypergraph to a graph [1]. We apply similar techniques on this paper.

Formally, in this paper, given a hypergraph $HG = (V, H)$, we determine k partitions on V, $\pi_V = \pi_1, \pi_2, \cdots, \pi_k$, where $\pi_i \subset V$, $\cup_i \pi_i = V$, and $\cap_i \pi_i = \emptyset$.

Let us consider there are $|V| = m$ vertices in the hypergraph and $|E| = n$ hyperedges of the hypergraph are represented as a $P \in \mathbb{R}^{m \times n}$. We can perform the spectral cut on P as follows.

Say, $D_v \in \mathbb{R}^{m \times m}$ and $D_e \in \mathbb{R}^{n \times n}$ be diagonal matrices of row and column sums of H. Determining the top-k eigenvectors of Laplacian matrix $S = D_v^{-\frac{1}{2}} H D_e^{-1} H^T D_v^{-\frac{1}{2}}$ will provide the soft clustering on V. Obtaining the highest cluster membership of each v_i will provide us k clusters of the hypergraph clustering π_V.

In our case, S is a sparse symmetric case, and we are determining k leading eigenvector as value decomposition $S = U D U^T$, where $U \in \mathbb{R}^{n \times k}$. Algorithm 1 presents the listing of the high-performance spectral clustering for hypergraphs.

Algorithm 1 Hypergraph clustering

Input: $H \in \mathbb{R}^{m \times n}$, k clusters
Output: Vertex cluster π_V
1   Compute $D_v = rowsum(H)$;
2   Compute $D_e = columnsum(H)$;
3   Compute $S = D_v^{-\frac{1}{2}} H D_e^{-1} H^T D_v^{-\frac{1}{2}}$;
4   /* eigsh is eigen value decomposition for symmetric square
      matrix.   */
5
6   Compute eigen vectors $U = eigsh(S, k)$;
7   Compute $\pi_V = $ argmax; U

In this paper, we used Scalable Library for Eigenvalue Problem Computations (SLEPC) [10] for computing the eigenvalue decomposition problem in Step 6 of the above algorithm for scaling to very large hypergraphs in distributed MPI environment.

The output of the above algorithm for a generated hypergraph HG is shown in Fig. 7. We generated a ground truth graph as shown in Fig. 7a and permuted the

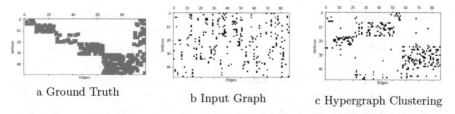

a Ground Truth   b Input Graph   c Hypergraph Clustering

Fig. 7 Hypergraph clustering demonstration. (**a**) Ground truth. (**b**) Input graph. (**c**) Hypergraph clustering

rows and columns as in Fig. 7b. We took this sparse random hypergraph HG and determined five clusters. The output is shown in Fig. 7c.

4   Performance

In this section, we describe the performance evaluation of the various components of the Phoenix framework. We first discuss the dataset, computation environment, software environment, and performance metrics considered for this performance evaluation process and further present the performance results.

4.1   Approach for Evaluating the Performance

As a proof of concept, we evaluated the performance of various components of the Phoenix framework for streaming hypergraphs. Here, we specifically focused on the hypergraph ingestion performance. We evaluated the streaming performance with varying batch sizes. We also investigated ingestion performance with varying numbers of HSNs and HDNs. Finally, we evaluated the performance of the distributed hypergraph clustering approach.

4.1.1   Dataset

We used the synthetic hypergraphs generated by our distributed hypergraph generator, i.e., HyGen, by varying the parameters such as #clusters, #vertices, and #hyperedges. Tables 2 and 3 show various synthetic hypergraphs (generated using HyGen) used for this performance evaluation.

4.1.2   Computational Environment

We used Oak Ridge Leadership Computing Facility (OLCF) called Rhea cluster. It is a 521-node commodity-type Linux® cluster. Each node of Rhea contains two 8-core 2.0 GHz Intel Xeon processors with Intel's Hyper-Threading (HT) Technology and 128GB of main memory. Rhea also has nine GPU nodes, and each node is equipped with 1TB of main memory and two NVIDIA K80 GPUs with two 14-core 2.30 GHz Intel Xeon processors with HT Technology. Rhea is connected to the OLCF's high-performance Lustre® filesystem, Atlas, through a high-speed interconnect 4X FDR Infiniband with maximum data transfer rate of 56 GB/s. More information on the specification of Rhea can be found at [23].

4.1.3 Software Environment

The codebase of Phoenix is developed in C++ (specifically C++11 standards). Inter-node communication is implemented using binary message structures over a message-oriented middleware. Currently, ZeroMQ over Transmission Control Protocol (TCP) is implemented [11], and MPI is planned for the future developments. ZeroMQ is a high-performance asynchronous messaging library that supports common messaging patterns (pub/sub, request/reply, client/server, and others) over a variety of transports (TCP, in-process, inter-process, multicast, WebSocket, and more). Intel®'s Thread building blocks, version 4.3+, is used to develop a scalable implementation of the concurrent queues. Further, we used the Scalable Library for Eigenvalue Problem Computations (SLEPC) [10] for computing the eigenvalue decomposition.

4.1.4 Performance Metrics

We mainly measured the performance in terms of streaming rate in a batched streaming scenario, ingestion rate with different settings, and scaling performance of the developed hypergraph clustering method.

4.2 Results

First, we present the time performance of the streaming hypergraph from streaming server to HSN of *DiSciPHER*. Although the additional layer of a streaming server provides few architectural benefits (refer to Sect. 3.2), this experiment is necessary to understand its overall overhead. Once the hypergraph data is acquired at streaming server from hypergraph generators and external sources, the streaming server further streams the data in the batches of incidences (or hyperedges) instead of streaming only one incidence at a time (refer to Sect. 3.2 for streaming strategies). The overhead includes batch preparation time followed by the time to stream those batches. Table 1 presents the total batch preparation and streaming timings for different sizes of batches. One more motivation behind this experiment was to understand the ideal batch size for streaming hypergraph data. From the timings shown in Table 1, it is clear that, although the batch size is varying, the overall batch preparation and streaming timings are roughly the same for all the scenarios, i.e., $\approx$8.2 s and $\approx$2.1 s for batch preparation and streaming respectively.

As mentioned in Sect. 3.2, the external sources of hypergraph can stream data in varying rates and formats. At the system level, such heterogeneity in the streaming rates could cause the loss of data in case of extremely high data streaming rates and longer wait times for HSN processes in case of slow data streaming rates. Based on the results of the batched streaming experiments in Table 1, we argue that the intermediate layer of the streaming server can stabilize the rate of streaming hypergraph from various external sources to HSN.

Next, we describe the time performance of the scaling experiments in two different scenarios: (1) weak-scaling, i.e., increasing the number of incidences (increasing hypergraph size) with number of HDNs, and (2) strong-scaling, i.e., adding more HDNs for a fixed number of incidences (i.e., fixed sized hypergraph). We want to mention that, in both settings, one compute node was used as HSN; each HSN has 12 worker threads and 2 I/O threads. Each HDN has eight worker threads and two I/O threads. As described earlier, HDNs are responsible for storing hypergraphs in memory and perform necessary computations for its consumption. We carried out both weak- and strong-scaling experiments in two different settings. In the first setting, various hypergraphs were generated in which #vertices<#hyperedges and in the second setting various hypergraphs were generated in which #vertices>#hyperedges. The motivation behind these experiments was to analyze the ingestion performance for the streaming data on a leadership class computing platform.

In the weak-scaling experiment, one compute node was used as HSN, and the number of compute nodes used for HDNs was increased along with the hypergraph size. Table 2 shows two different settings which were used to generate hypergraphs for weak-scaling experiments. #Incidences indicate the size of the hypergraph. Ideally, a constant ingestion time is expected in this weak-scaling experiment as the workload of hypergraph consumption per HDN roughly remains the same with the growing size of hypergraph and number of HDNs. In our case, one HSN is used, and the total ingestion time includes the time that HSN takes to prepare messages and send it to the respective HDNs. For the same reason, as the hypergraph size increases, we expect some linear growth in the HSN's contribution to the total ingestion time; however, in an ideal scenario, the HDN's consumption timing should be constant. Along with the ingestion time, we also measured the ingestion rate, i.e., number of incidences ingested per second. In an ideal scenario, the ingestion rate should grow as we increase the hypergraph size and number of HDNs.

Except the first column in Table 2, the other columns represent different scenarios. The main goal of this experiment is to understand the variations in the ingestion times and ingestion rates while increasing the hypergraph size and number of HDNs. We measured total ingestion time and derived the ingestion rates. Table 2 shows that the total ingestion time is increasing with increasing hypergraph size and

Table 1 Timing for batched streaming of hypergraph data from streaming server to HSN. ≈2.5M hyperedges and ≈208M incidences (NNZ) used	Streaming batch size (#n batches)	Total batch preparation overhead (s)	Total streaming time (s, excluding batch preparation)
	300M (1 batch)	8.265	2.108
	200M (2 batches)	8.268	2.08
	100M (3 batches)	8.265	2.097
	50M (5 batches)	8.255	2.086
	10M (21 batches)	8.271	2.126

Table 2 Setting for weak-scaling and ingestion timings

Weak scaling1 (#Vertices <#Hyperedges)					
#Clusters	1000	3000	6000	12,000	24,000
#Vertices	60,000	200,000	400,000	800,000	1,600,000
#Hyperedges	200,000	600,000	1,200,000	2,400,000	4,800,000
#Incidences	2,388,362	16,042,887	56,189,594	208,488,077	786,869,455
#HDN	1	2	4	16	64
Ingestion time (s)	3.5	6.6	17.1	63.1	248.0
Ingestion rate (#ing/s)	≈682K	≈2.4M	≈3.3M	≈3.3M	≈3.2M
Weak scaling2 (#Vertices >#Hyperedges)					
#Clusters	300	1000	2000	4000	8000
#Vertices	200,000	600,000	1,200,000	2,400,000	4,800,000
#Hyperedges	60,000	200,000	400,000	800,000	1,600,000
#Incidences	5,437,372	23,855,054	71,276,112	239,667,974	859,300,720
#HDN	1	2	4	16	64
Ingestion time (s)	4.7	8.7	21.7	72.2	269.9
Ingestion rate (#ing/s)	≈1.1M	≈2.7M	≈3.3M	≈3.3M	≈3.2M

HDNs. Figure 8 shows the variations in the ingestion timings for different scenarios for both settings. It can be seen that the ingestion rate increased for the first three scenarios and after that it remained stable, i.e., ≈3.3M ingestion per second. The potential reasons for the increase in the ingestion time and the ingestion rate are use of single HSN and background network traffic created by other jobs executing on Rhea cluster. However, we would like to emphasize the fact that we observed the stable ingestion performance in both settings, i.e., one with #vertices<#hyperedges and the other with #vertices>#hyperedges which represent to different hypergraph structures.

In the strong-scaling experiment, we kept the hypergraph size the same and increased the number of HDNs used for consumption. The intent behind this experiment is to understand the workload sharing ability of the HDNs when the hypergraph is fixed and we add more HDNs. We measured the total ingestion time for these scenarios and derived the ingestion rate (refer to Table 3). Table 3 shows the ingestion timings and rates for two strong-scaling settings, each with different number of NNZ (incidences) with increasing number of HDNs, and Fig. 9 shows the ingestion timings for two strong-scaling scenarios. Ideally, we expect a decreasing trend in the ingestion time as workload of hypergraph consumption per HDN decreases with the increase in the number of HDNs. From Table 3 and Fig. 9, we can observe that the total ingestion time for the strong-scaling settings decreased when two HDNs were used; however, the ingestion time remained roughly constant for all of the subsequent scenarios. Similarly, in ideal cases, the ingestion rate in a strong-scaling setting should increase with the addition of more HDNs due to the decrease in the ingestion time for the constant workload. The ingestion rate showed some increase for the first two scenarios but remained nearly constant for other

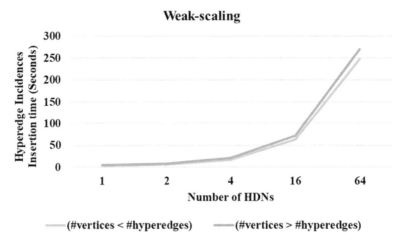

Fig. 8 Ingestion time for different scenarios for both weak-scaling settings with varying numbers of HDNs. The ingestion time/rate and overall scaling of the hypergraph ingestion in a streaming scenario is largely determined by the number of HSNs

Table 3 Settings for strong-scaling experiment and ingestion timing

	#HDN	Ingestion time (s)	Ingestion rate (#ing./s)
Strong scaling1	1	151.3	1.3M
#clusters:12,000;	2	58.6	3.5M
#vertices:800,000;	3	58.1	3.5M
#hyperedges:2,400,000	4	58.7	3.5M
NNZ = 208,486,247	6	58.4	3.5M
	8	58.6	3.5M
	16	63.1	3.3M
Strong scaling2	1	786.7	1.1M
#clusters:8,000;	2	373.2	2.3M
#vertices:4,800,000;	3	251.6	3.4M
#hyperedges:1,600,000	4	252.8	3.4M
NNZ=859,300,720	6	250.6	3.4M
	8	248.6	3.5M
	16	263.9	3.3M

settings where the number of HDNs is increasing. We emphasize that the role of the HSN is to formulate messages and distribute the hypergraph data to the HDNs. The HDNs are responsible for the necessary computation and communications with other HDNs to store the consumed hypergraph in memory. The potential reasons for the deviation from the ideal strong-scaling behavior could be attributed to the use of a single HSN resulting and the increased message communication among HDNs with the increase in the number of HDNs.

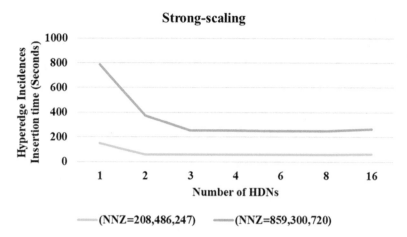

Fig. 9 Ingestion time for different scenarios for both strong-scaling settings with fixed hypergraph size and varying number of HDNs. The ingestion time/rate and overall scaling of the hypergraph ingestion in a streaming scenario is largely determined by the number of HSNs

As mentioned above, in the strong-scaling scenario, one can expect an increased ingestion rate with increase in the number of HDNs when the hypergraph size is kept constant; however, the results show some deviation from this ideal behavior. It should be noted that both strong- and weak-scaling experiments were performed with one HSN only which could be a potential reason for this performance deviation. Therefore, to understand the impact of varying number of HSNs, we fixed the number of HDNs to 512 and varied the numbers of HSNs from 1 to 8. Figure 10 shows the variation in the ingestion timing with the increasing number of HSNs. The results are favorable, and it can be clearly observed that the ingestion time decreases with increase in the number of HSNs for a fixed problem size. Therefore, we expect to see improved weak and scaling experiments by increasing the number of HSNs. Further analysis to understand the optimal number of HSNs is one of the objectives of our future research. In the future, we intend to perform similar scaling experiments on even larger-scale systems such as Oak Ridge National Laboratory's Summit supercomputer, which currently holds the number one spot on the top500[2] list [29].

Figure 11 shows the strong-scaling performance of the distributed hypergraph clustering algorithm (refer to Sect. 3.4). In an ideal setting for strong-scaling, the total execution timing of hypergraph clustering and an average MPI message length should be decreased with the increasing number of MPI processes. We observed that both the job execution time and message length decreased exponentially with the increase in the MPI processes. The results showed an $\approx 38\times$ speedup when 64 MPI processes were used for hypergraph clustering.

---

[2]https://www.top500.org/system/179397.

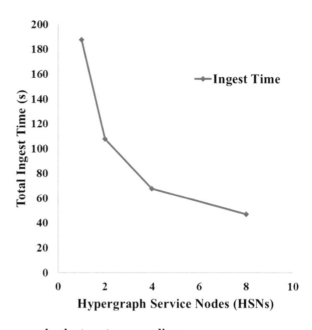

Fig. 10 Ingestion time variation with increase in HSNs with fixed 512 HDNs. The hypergraph ingestion rate increases significantly by increasing HSNs

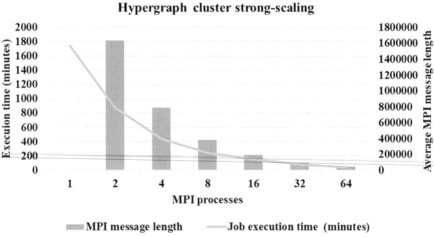

Fig. 11 Strong-scaling performance of the distributed hypergraph clustering analysis algorithm. Observed 38 speedup when 64 MPI processes were used

4.3 Observations

From the experiments performed to analyze the performance of the Phoenix framework, we draw following key observations which should inform future research and development in hypergraph analysis:

- The additional layer of the streaming server is important in the streaming hypergraph scenario to stabilize the streaming process.
- The ingestion time/rate and overall scaling of the hypergraph ingestion in a streaming scenario is largely determined by the number of HSNs.
- The hypergraph ingestion rate increases significantly with the increasing number of HSNs. However, further performance analysis is required to obtain the optimal number of HSNs for a given size of the hypergraph.
- The distributed hypergraph clustering algorithm showed $\approx 38\times$ speedup when 64 MPI processes were used. More experiments are needed with larger hypergraphs to further validate the usefulness of the algorithm.

5   Conclusion

Graphs are becoming ubiquitous and growing in volume. From social networks to language modeling, the growing scale and importance of graph data have driven the development of numerous graph analytic systems. While graph analytic systems have many applications, they are not able to model group-level interactions with high fidelity. In this paper, we present our approach to hypergraph analysis to better capture the nuances of complex multilateral relations in group interactions. Although other hypergraphh analytic tools exist, they are not well suited to tasks such as generating the hypergraphs, modifying hypergraph structures, or expressing computation that spans multiple graphs and compute nodes.

In this paper, we present Phoenix, a scalable hypergraph analytics framework that was implemented on the leadership class computing platforms at Oak Ridge National Laboratory. Our software framework is implemented in a distributed fashion. Phoenix has the capability to utilize diverse hypergraph generators, including HyGen, a very-large-scale hypergraph generator developed by Oak Ridge National Laboratory. Phoenix also incorporates specific algorithms for efficient data representation by exploiting hidden structures of the hypergraphs. We presented experimental results that demonstrate Phoenix's scalable and stable performance on massively parallel computing platforms. In the future, we will optimize our load balancing techniques for better strong- and weak-scaling performances. Also, we plan to implement 2-D partitioning techniques to improve the scalability of HyGen [31]. Other future directions include the development of machine learning-based hypergraph generators, which will learn structures of real-world hypergraphs, and based on that information, the hypergraph generator will generate massive-scale hypergraphs.

Acknowledgments Support for this work was provided by the US Department of Defense. We used resources of the Computational Research and Development Programs and the Oak Ridge Leadership Computing Facility at the Oak Ridge National Laboratory, which is supported by the Office of Science of the US Department of Energy under contract no. DE-AC05-00OR22725.

References

1. S. Agarwal, K. Branson, S. Belongie, Higher order learning with graphs, in *Proceedings of the 23rd international conference on Machine learning* (2006), pp. 17–24
2. A. Antelmi et al., SimpleHypergraphs.jl–novel software framework for modelling and analysis of hypergraphs, in *International Workshop on Algorithms and Models for the Web-Graph* (Springer, Cham, 2019), pp. 115–129
3. *BalancedGo* (2020). https://github.com/cem-okulmus/BalancedGo. Online. Accessed 03 Apr 2020
4. *Chapel Hypergraph Library* (2020). https://github.com/pnnl/chgl. Online. Accessed 03 Apr 2020
5. D. Ediger et al., STINGER: High performance data structure for streaming graphs, in *2012 IEEE Conference on High Performance Extreme Computing* (2012), pp. 1–5
6. E. Estrada, J. Rodriguez-Velazquez, Complex networks as hypergraphs (2005) . Arxiv preprint physics 0505137
7. G. Feng, X. Meng, K. Ammar, DISTINGER: A distributed graph data structure for massive dynamic graph processing, in *2015 IEEE International Conference on Big Data (Big Data)* (2015), pp. 1814–1822
8. *Graph Signal Processing Toolbox (GSPBox)* (2020). https://github.com/epfl-lts2/gspbox. Online. Accessed 03 Apr 2020
9. B. Heintz, A. Chandra, Beyond graphs: toward scalable hypergraph analysis systems. SIG-METRICS Perform. Eval. Rev. **41**(4), 94–97 (2014). ISSN: 0163-5999. https://doi.org/10.1145/2627534.2627563
10. V. Hernandez, J.E. Roman, V. Vidal, SLEPc: A scalable and flexible toolkit for the solution of eigenvalue problems. ACM Trans. Math. Softw. (TOMS) **31**(3), 351–362 (2005)
11. P. Hintjens, *ZeroMQ: messaging for many applications* (O'Reilly Media, Newton, 2013)
12. *HyperGCN: A New Method of Training Graph Convolutional Networks on Hypergraphs* (2020). https://github.com/malllabiisc/HyperGCN. Online. Accessed 03 Apr 2020
13. *Hypergraph Algorithms Package* (2020). https://murali-group.github.io/halp/. Online. Accessed 03 Apr 2020
14. *HyperGraphLib* (2020). https://github.com/alex87/HyperGraphLib. Online. Accessed 03 Apr 2020
15. W. Jiang et al., HyperX: A scalable hypergraph framework. IEEE Trans. Knowl. Data Eng. **31**(5), 909–922 (2019)
16. *KaHyPar - Karlsruhe Hypergraph Partitioning* (2020). https://github.com/kahypar/kahypar. Online. Accessed 03 Apr 2020
17. J. Leskovec et al., Kronecker graphs: An approach to modeling networks. J. Mach. Learn. Res. **11**, 985–1042 (2010)
18. A. Lugowski et al., Scalable complex graph analysis with the knowledge discovery toolbox, in *2012 IEEE International Conference on Acoustics, Speech and Signal Processing (ICASSP)* (2012), pp. 5345–5348
19. *multihypergraph* (2020). https://github.com/vaibhavkarve/multihypergraph. Online. Accessed 03 Apr 2020
20. *networkR - An R package for analysing social and economic networks* (2020). https://github.com/O1sims/networkR. Online. Accessed 03 Apr 2020
21. *pnnl/HyperNetX* (2020). https://pnnl.github.io/HyperNetX/build/index.html. Online. Accessed 03 Apr 2020
22. *Pygraph* (2020). https://github.com/jciskey/pygraph. Online. Accessed 03 Apr 2020
23. *RHEA: A conduit for large-scale scientific discovery by pre- and post- processing and analysis of simulation data.* https://www.olcf.ornl.gov/olcf-resources/compute-systems/rhea/. Accessed 14 Apr 2020

24. S. Schlag et al., K-way hypergraph partitioning via n-level recursive bisection, in *2016 Proceedings of the Eighteenth Workshop on Algorithm Engineering and Experiments (ALENEX)* (SIAM, Philadelphia, 2016), pp. 53–67
25. P. Sen et al., Collective classification in network data. AI Mag.**29**(3), 93–93 (2008)
26. N. Sundaram et al., GraphMat: high performance graph analytics made productive. Proc. VLDB Endow. **8**(11), 1214–1225 (2015). ISSN: 2150-8097. https://doi.org/10.14778/2809974. 2809983
27. J. Winick, S. Jamin, Inet-3.0: Internet topology generator. Technical Report CSE-TR-456-02, University of Michigan, 2002
28. M.M. Wolf, A.M. Klinvex, D.M. Dunlavy, Advantages to modeling relational data using hypergraphs versus graphs, in *2016 IEEE High Performance Extreme Computing Conference (HPEC)* (2016), pp. 1–7
29. D.E. Womble et al., Early experiences on summit: data analytics and AI applications. IBM J. Res. Devel. **63**(6), 2:1–2:9 (2019)
30. N. Yadati et al., HyperGCN: a new method for training graph convolutional networks on hypergraphs, in *Advances in Neural Information Processing Systems* (2019), pp. 1509–1520
31. A. Yoo, A.H. Baker, R. Pearce, A scalable eigensolver for large scale-free graphs using 2D graph partitioning, in *Proceedings of 2011 International Conference for High Performance Computing, Networking, Storage and Analysis* (2011), pp. 1–11
32. J. You et al., Graphrnn: Generating realistic graphs with deep autoregressive models (2018). Preprint. arXiv:1802.08773
33. D. Zhou, J. Huang, B. Schölkopf, Learning with hypergraphs: Clustering, classification, and embedding, in *Advances in Neural Information Processing Systems* (2007), pp. 1601–1608

Revealing the Relation Between Students' Reading Notes and Scores Examination with NLP Features

Zhenyu Pan, Yang Gao, and Tingjian Ge

1 Introduction

Predicting individual or group performance in the coursework or school has been studied in the past many years [1]. Accurate predictions of students' exam grades or course performance have the potential to make schools and teachers flexible enough to adjust their teaching methods according to the needs of students. Students' performance prediction can be used to detect students who have difficulties with the course materials and to help at-risk students to keep them retained.

In the literature, there are many efforts devoted to predicting student performance with two kinds of data: student behavior data and exercise content data. The authors in [2] use general student behavior records to predict students' final scores, while the authors in [3] focus on course-related behaviors, such as watching video records, quiz answering, and assignment completion. The approach in [1] builds a social network based on student collaboration. Some other approaches resort to modeling the exercise records for exam score prediction. However, the work of labeling knowledge concepts for each exercise may be labor-intensive. The method in [4] learns an exercise representation from the exercise content data and then combines both exercise records and exercise content data to precisely predict student performance of the online course.

Student behavior data and exercise content data are both popular in educational psychology and data mining areas for student exam score prediction. Our previous work [5] uses student behaviors of taking the course that require textbook reading notes and peer evaluations. In this paper, we explore the relationship between the content of students' reading notes and their midterm or final exam scores. We use

Z. Pan (✉) · Y. Gao · T. Ge
University of Massachusetts Lowell, Lowell, MA, USA
e-mail: Zhenyu_Pan@student.uml.edu; Yang_Gao@student.uml.edu; Tingjian_Ge@uml.edu

© Springer Nature Switzerland AG 2021   27
R. Stahlbock et al. (eds.), *Advances in Data Science and Information Engineering*,
Transactions on Computational Science and Computational Intelligence,
https://doi.org/10.1007/978-3-030-71704-9_2

information retrieval methods to analyze the reading notes' contents and learn their context information.

One popular method is called the *Bag of Words*, in which each word is assumed to be independent. With Bag of Words assumption, since each word is associated with a parameter, the topic selection will be a problem. TFIDF has been used to filter out stop words, which are the words shown in most documents.

Latent semantic indexing (LSI) [6] is a vector-based indexing and retrieval method that uses the singular value decomposition (SVD) to discover hidden concepts in document data. Instead of using terms or documents as the orthogonal basis of semantic space, LSI expresses each term and document in a low-rank subspace, as a vector with elements corresponding to these concepts. It is able to extract the document or term similarity or semantic relationship by representing documents and terms in a uniform way and then computing the semantic similarity based on the proximity between the vectors [7]. By replacing the independence hypothesis with an exchangeable hypothesis, Blei comes up with the LDA model [8].

The language models that we mention above usually focus on maximizing the probability of observing the entire corpus. However, in order to predict the exam score of a student with the given reading notes, we need to find out the distribution of the conditional probability of the exam score given the student's reading notes. We denote the conditional probability as $Pr\{S|W, \alpha, \beta\}$; the goal of our model is to maximize the probability of the observed score for the given reading notes. Thus, we firstly implement the Two-Step LDA model as the baseline to have a view of the challenges. We use LDA as our topic model to extract the topics for each reading note and train a three-layer neural network to predict students' examination scores. Then, we introduce a Topic-Based Latent Variable Model and achieve more accurate prediction results. For further exploration, we also propose a knowledge graph-based graph embedding model to predict the exam score for the student.

The rest of this paper is organized as follows. In Sect. 2, we discuss related work. We describe the experimental design and data collection in Sect. 3. We discuss our language model and prediction model in Sect. 4 and conclude in Sect. 5.

2   Related Work

2.1   Students' Performance Prediction with Learning Behaviors

There has been previous work that focuses on discovering the relation between learning behaviors and students' performance. In [2], Andrew et al. built a prediction model to identify the students at risk using the students' behavior data collected from an online learning initiative platform. Cheng et al. intended to explore the behavior facts related to low completion rates and dropout prediction in the massive open online course [3]. They tracked the lecture video watching behaviors, lecture-embedded quiz answering behaviors, and the completion of peer-graded assignments.

The research from Michael Fire et al. introduced a method for the prediction of student exam scores by building a social network of students in a course [1]. The course social network was created by analyzing the collaboration among the students. This work demonstrated that the best friend's grade has a direct impact on a student's exam grade, and other social parameters also have some influence on the grade.

Our previous research [5] focused on the students' behaviors in reading chapters of the textbook and taking reading notes. We presented a regression model to predict students' midterm and final examinations by using student interaction history.

2.2 Students' Performance Prediction with Exercise Records

The authors in [4] proposed an Exercise-Enhanced Recurrent Neural Network (EERNN) framework to predict student performance in an online learning system. This framework takes full advantage of both exercise records and exercise text contents with its LSTM architecture. They used EERNNM with the Markov property and EERNNA with the attention mechanism to achieve effectiveness in student performance prediction.

By contrast, in this paper, we study students' performance prediction with exercise records of an offline course. We use students' textbook reading notes to predict the exam scores of the course.

3 Experimental Design and Data Collection

Our prediction model was built on the text data collected from the peer-based online homework of the students enrolled in an undergraduate-level Algorithms and Data Structures course at the University of Massachusetts Lowell. The online homework was distributed on a web-based peer-assessment platform which we built for this project. Since the goal of our work is to predict students' midterm and final exam scores, we designed our experiment as follows:

3.1 Homework Design

The homework activities included two main tasks: write a reading note and rate the reading notes written by other students. Textbook reading notes could tell us what the knowledge points that students had learned, and they could also gave a sense of which part of the class made them felt difficult. According to the reading notes, we could easily figure out what content of the course students pay more attention to, and what important point was easy to be missed. Thus, we chose reading notes as weekly homework for the course. In each week, the instructor gave a lecture for a new chapter, and the students were asked to write a reading note for this chapter independently. They were not allowed to seek help from classmates. Besides, anonymity was preserved as students did not know whose notes they were rating. The time of completion of tasks was also logged by the system for reference.

3.2   Examination Design

There were two examinations for the course, namely, the midterm exam and the final exam. In both of the exams, we set up like that (1) all the questions have the same score, (2) each question only belongs to one chapter, and (3) all the chapters have the same number of questions appearing in the exam.

3.3   Data Collection

In total, there were 36 students writing reading notes for all 9 chapters in the textbook. We required the students to make at least five notes for each chapter. Thus, we collected 1645 reading notes at the end of the experiment period.

4   Language Model and Prediction Model

In this section, we first use the Two-Step LDA model as our baseline to solve the exam score prediction problem. To address the defects in the Two-Step LDA, we introduce the combination of the Topic model and the Latent Variable Model, which is named as the Topic-Based Latent Variable Model. After that, we use the topics extracted by the Topic-Based Latent Variable Model (TB-LVM) to construct a knowledge graph. Finally, we propose a graph embedding model for the knowledge graph as another approach.

4.1   Two-Step LDA Model

In natural language processing, latent Dirichlet allocation (LDA) [9, 10] is a topic model used to identify topics in a set of documents. It provides the probability distribution of the topics for each document. At the same time, it is an unsupervised learning algorithm, which only requires the document set and the number of specified topics rather than a manually labeled training set during training. For our purpose, we first used the LDA models to extract the topic features for each reading

Fig. 1 The pipeline of the
Two-Step LDA

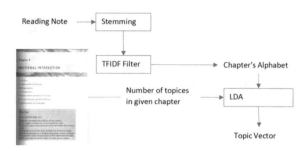

note. All the topic feature vectors were used to train a three-layer neural network.
The observed examination grades were applied as the ground truth of the neural
network. Then, we used the three-layer neural network to predict the examination
score of the test set.

4.1.1 Pipeline of Two-Step LDA

As Fig. 1 shows, we first use the Stemming method to reduce inflected and derived
words to their word stem. Then, we used the TFIDF filter to remove the unimportant
words for reading notes of each chapter, which is not relevant to the core knowledge
points of the chapter. In a textbook, the topic of each chapter should be different.
Therefore, if a word appears in each chapter, the word must be irrelevant to the
topics of the chapter. We use the TFIDF value to represent the relevance between a
word and a chapter. We use $\vec{w}_{si}$ to denote the reading notes of student s for chapter
i; and d_i is the collection of all students' reading notes for chapter i, which is $d_i =
\{\vec{w}_{si}|s \in Student\}$. The TFIDF value of the word w in chapter j will be

$$TF_j(w) = \frac{C(w, d_j)}{C(d_j)} \ln\left(\frac{N}{|\{d|C(w, d_i) \neq 0 \& i = 1 \dots N\}|}\right)$$

where $C(w, d_j)$ is the number of occurrences of word w in a collection d of reading
notes for chapter j and $C(d_i)$ is the total number of words in d_i. The TFIDF module
will filter out the words with a TFIDF value 0 from the word list of the chapter.

We trained LDA models for the reading notes of each chapter. For the training
of the LDA model, we determine the number of the topics for each chapter with the
number of concepts listed in the chapter. In the next step, we use our LDA model to
convert the student's reading note $\vec{w}_{si}$ into a topic vector $\vec{t}_{si}$. The kth component in
this vector corresponds to the frequency of the kth Topic appearing in the Reading
Note.

The relationship between the student's exam score and the topic vector of each
chapter is as follows:

$$f \sim N(f | f(T_s; \theta, \vec{\alpha}), \beta^{-1}) = N(f | \sigma(\sum_c a_c \sigma(\vec{\theta}_c^T \cdot \vec{T}_{sc}), \beta^{-1}))$$

where $\sigma(x) = \frac{1}{1+exp\{-x\}}$. The subscript c of the parameter θ represents the weight of each topic in chapter c. The subscript c of the parameter represents the weight of each topic in chapter c. And the c component in parameter $\vec{a}$ represents the weight of chapter c. The maximum likelihood estimation results of the corresponding parameters can be calculated using the least square method.

4.1.2 Cross-Validation Result

We run the leave-one-person-out (LOPO) evaluations for our Two-Step LDA model. Table 1 shows the prediction results from the LOPO evaluations. Each row corresponds to one LOPO evaluation. The average difference between the prediction and the truth of the midterm exam is 12.03; and the standard deviation of the midterm exam score is 14.38. For the final exam, the average difference is 13.69 and the standard deviation is 16.56.

4.1.3 Discussion

The problem of Two-Step LDA is that the original model of LDA is searching for a Latent structure that maximizes the overall probability, but what we need our task is a latent structure that can highlight the difference in student performance. In other words, since the objective function $Pr\{F|W\} = Pr\{F, W\}/Pr\{W\}$, the objective function of the feature extraction $max\, Pr\{W\}$ will hold the training result back.

4.2   Topic-Based Latent Variable Model

In this subsection, we combine the modeling assumptions of LDA to build a Probabilistic Generative model. After appropriately modifying the modeling assumptions, we use the Probabilistic Generative model to derive the corresponding discriminative model and then obtain the evaluation function for our training process. We named the model with the Topic-Based Latent Variable Model (TB-LVM). Since our model contains latent variables, we will next introduce the Expectation Maximum Algorithm for optimal parameters searching. Then, we will analyze the cross-validation result of the TB-LVM.

4.2.1   Probabilistic Generative Model

As the core of LDA, the De Finetti theorem [11] provides a way to weaken the modeling assumptions. By introducing Latent variables, the original exchangeable sequence is transformed into a conditional independence sequence. Following this idea, the modeling assumptions of our TB-LVM are the following:

– Assume that the grades of the students are independent of each other. Ignore the cooperation among the students.
– Assume that the knowledge points of chapter i and chapter j are independent. Ignore the interdependence between knowledge points, that is, the knowledge points are context-independent.
– The words in the student's reading notes are exchangeable. The order in which the characters appear does not affect the reading.
– Each reading note corresponds to one topic. Let g_c devote to the topic of chapter c; the topics for all chapter is $\vec{g} =< g_1, \ldots, g_c >$.
– The final grade f is only related to the subject in the reading notes, that is, $Pr\{f|g, W\} = Pr\{f|g\}$.
– The topic of chapter i has an influence on the distribution of knowledge points in chapter i, $\vec{\theta}_i =< \theta_{i1}, \ldots, \theta_{ik} >$; θ_{ik} is the probability of observing the kth knowledge point in chapter i.
– Knowledge point z affects word distribution in notes.

With these modeling assumptions, we built a Bayesian network (Fig. 2). The algebraic expression of the Bayesian network is

$$Pr\{f, \vec{g}, \theta, Z, W\} = Pr\{f|\vec{g}\} \cdot Pr\{\vec{g}, \theta, Z, W\}$$

Derived from modeling the fourth assumption:

$$Pr\{f, \vec{g}, \theta, Z, W\} = Pr\{f|\vec{g}\} \cdot \prod_i Pr\{g_i, \vec{\theta}_i, \vec{z}_i, \vec{w}_i\}$$

Derived from modeling the second assumption:

$$Pr\{f, \vec{g}, \theta, Z, W\} = Pr\{f|\vec{g}\} \cdot \prod_i \left(Pr\{g_i\} \cdot Pr\{\vec{\theta}_i|g_i\} \cdot Pr\{\vec{z}_i, \vec{w}_i|\vec{\theta}_i\} \right)$$

Derived from modeling the sixth assumption:

$$Pr\{f, \vec{g}, \theta, Z, W\} = Pr\{f|\vec{g}\} \cdot \prod_i \left(Pr\{g_i\} \cdot Pr\{\vec{\theta}_i|g_i\} \cdot \prod_j (Pr\{w_{ij}|z_{ij}\} \cdot Pr\{z_{ij}|\vec{\theta}_i\}) \right)$$

Derived from modeling the De Finetti theorem and the exchangeable property:

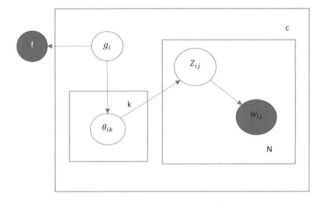

Fig. 2 The Bayesian network

4.2.2 Probabilistic Discriminative Model

The probability of observing the reading Node W for a given student is

$$Pr\{W\} = \sum_f \sum_{\vec{g}} \sum_\theta \sum_Z Pr\{f, \vec{g}, \theta, Z, W\} = \sum_{\vec{g}} \sum_\theta \sum_Z \left(Pr\{\vec{g}, \theta, Z, W\} \cdot \left(\sum_f Pr\{f|\vec{g}\} \right) \right)$$

Since $\sum_f Pr\{f|\vec{g}\} = 1$,

$$Pr\{W\} = \sum_{\vec{g}} \sum_\theta \sum_Z Pr\{\vec{g}, \theta, Z, W\} = \prod_i \left(\sum_{\vec{\theta}_i} Pr\{\vec{\theta}_i\} \cdot \prod_j \left(\sum_Z Pr\{w_{ij}|z\} \cdot Pr\{z|\vec{\theta}_i\} \right) \right)$$

Also, we have

$$Pr\{f, W\} = \sum_{\vec{g}} \sum_\theta \sum_Z Pr\{f, \vec{g}, \theta, Z, W\}$$

$$= \sum_{\vec{g}} \left(Pr\{f|\vec{g}\} \cdot \prod_i \left(\sum_{\vec{\theta}_i} Pr\{g_i, \vec{\theta}_i\} \cdot \prod_j \left(\sum_Z Pr\{w_{ij}|z\} \cdot Pr\{z|\vec{\theta}_i\} \right) \right) \right)$$

Thus,

$$Pr\{f|W\} = \frac{Pr\{f, W\}}{Pr\{W\}}$$

$$= \sum_{\vec{g}} \left(Pr\{f|\vec{g}\} \cdot \prod_i \frac{\sum_{\vec{\theta}_i} Pr\{g_i, \vec{\theta}_i\} \cdot \prod_j \left(\sum_Z Pr\{w_{ij}|z\} \cdot Pr\{z|\vec{\theta}_i\} \right)}{\sum_{\vec{\theta}_i} Pr\{\vec{\theta}_i\} \cdot \prod_j \left(\sum_Z Pr\{w_{ij}|z\} \cdot Pr\{z|\vec{\theta}_i\} \right)} \right)$$

In order to simplify the model as much as possible, we assume the student's final score distribution is a mixed Gaussian model $Pr\{f|\vec{g}\} \sim N(f|\mu_{\vec{g}}, \sigma_{\vec{g}})$.

Each category corresponds to a permutation and combination of a topic, and the probability of the corresponding permutation and combination depends on the content of the notes submitted by the student.

Assuming there is a one-to-one correspondence between knowledge points and topics. We can get

$$Pr\{\vec{\theta}_i, g_i\} = Pr\{\vec{\theta}_i|g_i\} * Pr\{g_i\} = \begin{cases} Pr\{g_i\}, & \text{if } g_i \text{ mapping with } \theta_i \\ 0, & \text{otherwise} \end{cases}$$

With the assumption that student performance is independent of each other, the probability of the entire data set is

$$\ln\left\{\prod_{s \in student} Pr\{f^{(s)}|W^{(s)}\}\right\} = \sum_{s \in student} \ln\{Pr\{f^{(s)}|W^{(s)}\}$$

$$= \sum_{s \in student} \ln\left\{\sum_{\vec{g}} Pr\{f^{(s)}|\vec{g}, \mu, \Sigma\} \cdot \prod_i Pr\{g_i = t|\vec{w}_i, \beta\}\right\}$$

Although the maximum likelihood estimation expression of the entire model has been obtained, considering the existence of the latent variable, directly using the steepest descent to calculate the optimal parameter will encounter problems such as Untraceable [12]. Therefore, we use Expectation Maximum to transform the problem into $\text{argmax}_{\theta_{new} in \Theta} \sum P(z|x, \theta_{old}) \ln\{P(x, z|\theta_{new})\}$ problem.

4.2.3 Cross-Validation Result

When the set of reading notes of the student W is observed, the student's grade f is subject to the distribution $Pr\{f|W\}$. We have two ways to predict the exam score of the student, maximum likelihood estimation (MLE) $f_{MLE} = \text{argmax}_{f \in Z^+} P(\{f|W\})$ and moment estimation (ME). $E_{f|W} = \sum Pr\{\vec{g}|W\} \cdot \mu_{\vec{g}}$ Table 1 lays out the results of the LOPO evaluation of the MLF approach and the ME approach with the assumption that there are only two types of knowledge points. For midterm exams, the average difference and the standard deviation of the MLE approach are 5.60 and 7.79, respectively. The prediction results of the ME approach give an average difference of 5.84 and a standard deviation of 7.63. Compared with the Two-Step LDA model, the prediction error is reduced by half. The prediction results of the final exam also indicate that the TB-LVM model has the same improvement. The MLF approach gets a 6.17 average difference and an 8.68 standard deviation. Likewise, the average difference and the standard deviation of the EM model are 5.89 and 7.72, respectively.

Table 1 The results of LOPO evaluations for the Two-Step LDA and the TB-LVM

Midterm truth	Mid 2S-LDA	Midterm MLE	Midterm ME	Final truth	2s-LDA	Final MLE	Final ME
85	83	85	81.80	78	76	81	79.82
75	100	83	84.96	85	92	80	78.29
81	87	87	87.20	68	54	86	82.49
83	88	85	84.85	83	100	81	80.34
85	79	80	79.03	84	94	82	76.80
90	100	87	86.38	82	100	82	80.98
85	99	86	85.44	82	98	84	84.07
86	79	84	82.90	83	100	82	81.52
89	100	84	83.62	82	92	82	81.11
78	100	84	83.51	88	84	80	76.38
87	95	84	84.35	83	89	86	86.02
87	79	86	85.99	79	90	84	83.11
89	77	83	76.59	90	50	81	75.92
93	63	81	79.08	86	100	87	83.56
83	86	82	78.86	81	99	81	80.08
78	83	86	86.19	82	50	63	71.44
74	72	85	84.53	75	89	85	84.83
100	100	85	83.33	96	84	83	82.44
70	100	95	89.52	77	86	82	79.51
87	87	83	84.00	85	64	84	84.21
80	96	86	84.72	74	100	64	74.08
89	87	83	82.79	91	96	83	83.05
78	93	85	84.35	76	79	80	78.66
87	80	85	84.51	80	79	84	81.58
76	81	85	84.00	60	74	85	83.43
89	71	86	85.84	80	100	84	83.58
58	82	78	75.06	79	89	83	83.12
83	100	81	80.84	82	92	87	86.24
79	78	82	81.31	73	86	83	83.55
86	86	88	89.35	87	100	83	81.50
84	50	78	76.86	75	83	78	77.66
82	86	80	80.11	84	82	83	83.07
88	96	87	86.93	90	100	82	82.01
81	69	83	80.59	60	100	79	78.60
84	53	85	84.86	84	98	84	83.58

4.3   Knowledge Graph Model

In previous research [5], we use learning behaviors to predict student performance. With TB-LVM, we predict student performance by analyzing the reading notes.

Fig. 3 The knowledge graph

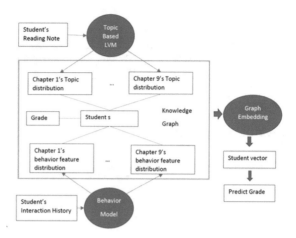

For further exploration, we studied how to combine these two kinds of data in one prediction model. One straightforward approach is to build a knowledge graph (Fig. 3) to combine the learning behavior information and the reading notes topic information. Then, the graph embedding method can be applied to this knowledge graph to obtain a general description vector of the student. With the description vector, we can determine the grade of the student.

Let g_i denote the topic of chapter i, and $W_i(s)$ is the reading note of the student S for chapter i. We use TB-LVM to get the topic distribution for each chapter for the student. The probability that the student s has mentioned Topic g_i is $Pr\{g_i|W_s\}$. Let $h_{ij}^{(s)}$ be the intensity of the relevant learning behavior feature recorded by student s when completing chapter j reading notes. The probability that student s exhibits behavior B_{ij} in the homework of chapter j is $Pr\{B_{ij}|h_{ij}^{(s)}\}$.

4.3.1 Leaning Behavior LVM

First of all, we need to make appropriate adjustments to the previous work [5]. In the previous research, we focused on the relationship between the learning behavior features of each chapter and the final student performance. In this work, we need to get the probability that the student has a specific Learning Behavior based on the strength of each learning behavior feature (Table 2) by a learning behavior LVM. Thus, we use a new Bayesian network (Fig. 4) to construct the learning behavior LVM. The variables of the new model are listed in Table 3.

Table 2 Learning behavior and behavior feature

Learning behavior	Behavior feature
Willing to ask question	The total number of questions issued in his reading note
Submit in time	The number of late days
Hard working	The total time spend on the reading note
Perfectionism	The total number of edit and delete operations

Fig. 4 The Bayesian network for the learning behavior LVM

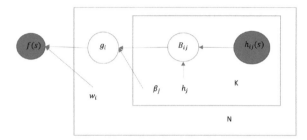

Table 3 The variables of the behavior LVM

Notation	Type	Meaning
$H(s)$	Observable vector	The interaction history of the student s
$\vec{H}_i(s)$	Observable vector	The interaction history of the student s for the chapter i
$h_{ij}(s)$	Observable vector	The strength of $j^t h$ behavior feature of the student s for the chapter i
h_{ij}	Parameter	The threshold of $j^t h$ behavior for chapter i Because different chapter has different difficulty and importance
β_j	Parameter	The weight of $j^t h$ behavior (Different chapter use the same weight)
B_j	Latent variable	The $j^t h$ behavior
$\vec{B}$	Latent variable	The learning behavior for the chapter i
g_i	Latent variable	The grade for chapter i
$\vec{g}$	Latent variable	The grade of each chapter covered by examination
w_i	Parameter	The importance of grade from chapter i
$f(s)$	Target vector	The examination score associated with student s

4.3.2 Graph Embedding

After training topic LVM and behavior LVM with the training data set and the existing model parameters, we obtained the adjacency matrix of student s and the topic g

$$(T_{ij})_{|s| \times |g|} = (Pr\{g_i | S_j\})_{|s| \times |g|}.$$

The adjacency matrix of student s and learning habits b is

$$(B_{ij})_{|s|\times|B|} = (Pr\{b_i|S_j\})_{|s|\times|B|}.$$

The adjacency matrix of students s and their results in the training data set is

$$(F_{ij})_{|s|\times|f|} = (Pr\{f_i|S_j\})_{|s|\times|f|}.$$

Student scores are divided into grade levels (A+, A, B+, B, C+, C, D). Above grade level D, every 5 is a new level. For example, level D corresponds to [0, 70], level C level corresponds to (70, 75], and A+ corresponds to (95, 100].

In order to obtain the score of the object s, we used the probability of the occurrence of each edge to randomly generate 2000 graphs (denoted as $G_1, \ldots, G_{2000}$). Then, we applied TransH on each graph and got the prediction results of the score in the graph. Let $f_i(s)$ denote the prediction result of the test object s in the i-th graph. The mathematical expectations is $E(s) = \frac{\sum_{i=1}^{N} f_i(s)}{N}$.

4.3.3 Cross-Validation Result

Table 4 gives the results of leave-one-person-out evaluations of the knowledge graph embedding model. The accuracy of the midterm exam performance prediction is 41.7%, and the accuracy of the final exam performance prediction is 33.3%. The results are not as good as some of our previous methods.

5 Conclusions

In this paper, we use the Latent Variable method to construct a TB-LVM, which achieves acceptable prediction results for examination scores. Compared to the Two-Step LDA model, the TB-LVM doubles the accuracy. We also explore a graph embedding method to combine the topic LVM and behavior LVM, in order to obtain a comprehensive description of student performance. However, the result of the knowledge graph embedding does not reach the desired accuracy. The main reason is that behavior features do not distinguish the students well, which essentially results in doing graph embedding over topic LVM. As future work, we plan to compare against more machine learning prediction models.

Table 4 LOPO evaluations for the knowledge graph embedding model

Mid truth	Mid prediction	Final truth	Final prediction	Mid truth	Mid prediction	Final truth	Final prediction
85(B+)	B+	78(C+)	B+	70(C)	B+	77(C+)	B
75(C+)	B+	85(B+)	A	87(B+)	B+	85(B+)	A
81(B)	B+	68(D)	B+	80(B)	B+	74(C)	B
89(B+)	B+	91(A)	A	84(B)	B+	81(B)	B
85(B+)	B+	84(B)	A	89(B+)	B+	91(A)	A
90(A)	B+	82(B)	B	78(C+)	B+	76(C+)	B
85(B+)	B+	82(B)	B	87(B+)	B+	80(B)	B
86(B+)	B+	83(B)	A	76(C+)	B+	60(D)	B
89(B+)	B+	82(B)	A	89(B+)	B+	80(B)	B
78(C+)	B+	88(B+)	B	58(D)	B+	79(C+)	A
87(B+)	B+	83(B)	B	83(B)	B+	82(B)	A
87(B+)	B+	79(C+)	A	79(C+)	B+	73(C)	B
89(B+)	B+	90(A)	A	86(B+)	B+	87(B+)	B
93(A)	B+	86(B+)	B	84(B)	B+	75(C+)	A
83(B)	B+	81(B)	B	82(B)	B+	84(B)	B
78(C+)	B+	82(B)	A	88(B+)	B+	90(A)	A
74(C)	B+	75(C+)	B	81(B)	B+	60(D)	A
100(A+)	B+	96(A+)	B	84(B)	B+	84(B)	A

References

1. M. Fire, G. Katz, Y. Elovici, B. Shapira, L. Rokach, Predicting student exam's scores by analyzing social network data, in *International Conference on Active Media Technology* (Springer, Berlin, 2012), pp. 584–595
2. A.E. Krumm, C. D'Angelo, T.E. Podkul, M. Feng, H. Yamada, R. Beattie, H. Hough, C. Thorn, Practical measures of learning behaviors, in *Proceedings of the Second (2015) ACM Conference on Learning @ Scale*, ser. L@S '15 (Association for Computing Machinery, New York, 2015), pp. 327–330. [Online]. Available: https://doi.org/10.1145/2724660.2728685
3. C. Ye, J.S. Kinnebrew, G. Biswas, B.J. Evans, D.H. Fisher, G. Narasimham, K.A. Brady, Behavior prediction in moocs using higher granularity temporal information, in *Proceedings of the second (2015) ACM conference on Learning@ Scale* (2015), pp. 335–338
4. Y. Su, Q. Liu, Q. Liu, Z. Huang, Y. Yin, E. Chen, C. Ding, S. Wei, G. Hu, Exercise-enhanced sequential modeling for student performance prediction, in *Thirty-Second AAAI Conference on Artificial Intelligence* (2018)
5. Z. Pan, J. Xue, Y. Gao, H. Wang, G. Chen, Revealing the relations between learning behaviors and examination scores via a prediction system, in *Proceedings of the 2018 2nd International Conference on Computer Science and Artificial Intelligence* (2018), pp. 414–419
6. M.W. Berry, Z. Drmac, E.R. Jessup, Matrices, vector spaces, and information retrieval. SIAM Rev. **41**(2), 335–362 (1999). [Online]. Available: https://doi.org/10.1137/S0036144598347035
7. Y.X.K. Umemura, Very low-dimensional latent semantic indexing for local query regions, in *Proceedings of the Sixth International Workshop on Information Retrieval with Asian Languages - Volume 11*, ser. AsianIR '03 (Association for Computational Linguistics, New York, 2003), pp. 84–91. [Online]. Available: https://doi.org/10.3115/1118935.1118946
8. D.M. Blei, A.Y. Ng, M.I. Jordan, Latent Dirichlet allocation. J. Mach. Learn. Res. **3**, 993–1022 (2003)
9. J.K. Pritchard, M. Stephens, P. Donnelly, Inference of population structure using multilocus genotype data. Genetics **155**(2), 945–959 (2000). [Online]. Available: https://www.genetics.org/content/155/2/945
10. D. Falush, M. Stephens, J.K. Pritchard, Inference of population structure using multilocus genotype data: Linked loci and correlated allele frequencies. Genetics **164**(4), 1567–1587 (2003). [Online]. Available: https://www.genetics.org/content/164/4/1567
11. L. Accardi, De finetti theorem, in *Encyclopaedia of Mathematics*, ed. by M. Hazewinkel (Kluwer Academic Publishers, Cham, 2001)
12. C.M. Bishop, Periodic variables, in *Pattern Recognition and Machine Learning*, vol. 1 (Elsevier, New York, 2006)

Deep Metric Similarity Clustering

Shuanglu Dai, Pengyu Su, and Hong Man

1 Introduction

Data clustering based on similarity measures has been studied as a fundamental topic in data mining and machine learning for decades. Well-known clustering methods such as spectral clustering [26], affinity propagation [6], minimum spanning trees [23], and normalized cut [30] are mainly based on certain data similarities. While it is relatively easy to partition the type of data possessing consistent similarities, a wider range of real-world data with implicit non-linear similarities is more challenging for clustering.

Conventional clustering methods often construct certain non-linear similarity functions explicitly with special geometric and subspace properties. Typical methods include manifold clustering, subspace clustering [8], spectral clustering, etc. Manifold clustering identifies a low-dimensional manifold to represent the structure of the data while maintaining linear similarities of the data. Methods using manifold embedding have shown promising performance on clustering data with implicit shape structures such as sphere, spiral, trefoils, and Mobius bands [7]. Subspace clustering optimizes a collection of subspaces to represent the observed data. Data sparsity can be easily introduced in these methods for clustering high-dimensional data. Spectral clustering introduces low-rank subspaces of the affinity matrix to encode the cluster information. Compared to subspace clustering and manifold

This work was supported in part by US Army Research, Development and Engineering Command (RDECOM) under contract W15QKN-18-D-0040.

S. Dai (✉) · P. Su · H. Man
Stevens Institute of Technology, Hoboken, NJ, USA
e-mail: sdai1@stevens.edu; psu2@stevens.edu; hman@stevens.edu

© Springer Nature Switzerland AG 2021
R. Stahlbock et al. (eds.), *Advances in Data Science and Information Engineering*,
Transactions on Computational Science and Computational Intelligence,
https://doi.org/10.1007/978-3-030-71704-9_3

43

clustering, spectral clustering can be adapted to various types of data with properly defined kernel functions. In addition, graph cut provides a different view of using data similarities for clustering. Besides pair-wise potentials, unary potentials are also modeled as energy terms in these graph cut-based methods [20, 22].

Recent studies on clustering have mainly focused on specific extensions to conventional methods for improved performance. The approaches include exploring multi-view of data, similarity learning, as well as co-training. Kumar et al. proposed a co-training method to obtain a semi-supervised similarity matrix from multi-view of data, which extends spectral clustering to various vision tasks [21]. Nie et al. proposed to estimate multi-view similarities on projected data spaces. For efficiency, similarities were estimated from a linearly projected space. Unfortunately, such setting restricts the application of their method [27, 28]. Gao et al. proposed a multi-view subspace method for clustering collections of visual features [9].

While many existing clustering methods have been introduced using multi-view feature collections, there lacks of effective end-to-end methods for unsupervised similarity learning and clustering. Hershey et al. proposed a deep clustering method to separate multi-speakers from a single channel [12]. However, this method can only be applied to acoustic sequences.

This paper proposes a deep metric clustering method with simultaneous non-linear similarity learning and clustering. First of all, a deep metric network is proposed to approximate implicit non-linear data similarities. The graph Laplacian matrix is introduced for cluster assignment. A co-training approach of the deep metric network is adopted to further improve visual data clustering. A stochastic optimization method and a sparse similarity learning are introduced to achieve high-level algorithm efficiency for similarity learning and deep metric network updating on large-scale data sets. Finally, the theoretical connection between the proposed method and spectral clustering is discussed. The proposed method is evaluated on both synthetic and benchmark machine learning data sets. Real-world visual data sets are employed to evaluate the clustering performance with co-trained optimal deep metric networks. Experimental results show that the proposed method outperforms the benchmark methods and related state-of-the-art methods in clustering various data types.

The major contributions of this paper are summarized as follows. First, this paper proposes an end-to-end model for simultaneous data similarity learning and clustering. Second, non-linear data similarities are approximated by the proposed deep metric network for clustering various data types. Third, a stochastic optimization method is proposed, which improves the efficiency of the learning algorithm on large-scale data sets. Fourth, the co-training method introduced in this paper further improves the clustering performance. A residual self-expressive metric network (ResEM-Net) is introduced in this paper for deep metric co-training and clustering on visual data. Figure 1 illustrates the network structure of ResEM-Net. This paper constructs the non-linear mapping f_W using residual convolution blocks [11] with an implicit self-expressive layer [17]. The proposed deep metric $d_{f_W}(Z_1, Z_2)$ measures the closeness of pairs of samples (X_1, X_2) in their self-expressive subspaces (Z_1, Z_2).

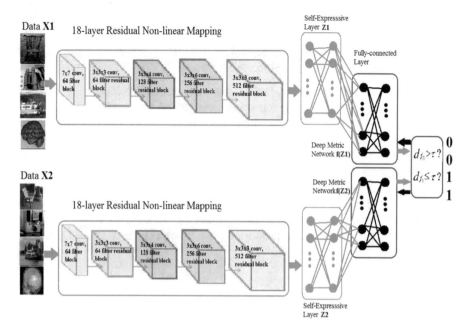

Fig. 1 Residual self-expressive metric network (ResEM-Net) for visual clustering

2 Related Works

2.1 Clustering Based on Affinity Space

Data affinities have been explored as a useful property for clustering. Clustering methods using data affinities such as spectral clustering and affinity propagation have been widely applied to data mining and visual clustering [6]. Dueck et al. proposed an affinity propagation method for unsupervised image categorization. Their work is applied to dynamically clustering visual data with increasing categories. In addition to pairwise affinities, graph cut-based methods are also able to model unary energy induced by data samples. Ng et al. discovered that the subspace of the graph Laplacian matrix could properly encode cluster information [26]. Although affinity transformations have been well explored, the construction of data affinity space is still limited to pre-defined kernels. Recent attempts to remedy this issue can be mainly summarized in the following.

Multi-View Subspace Affinity Nie et al. proposed that data similarities from multi-view features could be properly approximated by projected and weighted Euclidean distances [27, 28]. Their method was applied to various types of data by introducing data locality. However, the linear projection prevents the embedding of non-linear similarities. Moreover, exhaustive similarity estimations for all data pairs make their algorithm inefficient on large-scale data sets. Gao et al. proposed to learn data affinities from sparse subspaces for clustering high-dimensional visual data [9]. Their work showed that learning subspaces from multi-view visual features could improve the targeted clustering. However, it is nontrivial to identify and explore visual feature combinations for optimal performance.

Deep Subspace Affinity Besides multi-view features, convolutional auto-encoders have also been explored as an effective approach for representative and discriminative feature learning [13]. Subspace affinities from deep auto-encoders have been recently studied for clustering analysis. Xie et al. proposed a deep subspace embedding to simultaneously address deep embedding and clustering. A representative deep subspace was obtained by their method [34]. Ji et al. explored a deep subspace clustering network (DSC-Net) with a self-expressive layer as sparse subspace [17]. However, their method did not optimize clustering in an end-to-end fashion. Huang et al. improved the deep subspace embedding with a similarity constraint [15]. However, the selection of pre-defined similarity function in their method is nontrivial.

2.2 Deep Metric Learning

Learning discriminative distance metric has been studied for decades, and non-linear distance metrics have been well explored. Information theoretic distance measures such as Mahalanobis distance and manifold distances have shown great effectiveness in clustering and classification [3, 4, 32]. These distances can be viewed as similarity measures for distributions of data samples or sets [16].

However, estimating large-scale metrics for high-dimensional data is often intractable. For symmetric distance metrics such as Mahalanobis distance, low-rank factorization can be adopted to mitigate such a problem. With a low-rank square root matrix factorization, Hu et al. transformed the Mahalanobis distance to an Euclidean distance on a projected space, where the non-linear projection can be approximated by deep networks [14]. The deep metric learning can be further transferred to different data sets for face verification and person reidentification [18].

3 Deep Metric Similarity Clustering

Clustering high-dimensional and multichannel data is generally a challenging problem. Information theoretic distances such as Mahalanobis distances have been studied as effective measures for such data [4]. Given N high-dimensional and multichannel data samples $X=\{x_i \in \mathbb{R}^d\}_{i=1,\dots,N}$, Mahalanobis distance for $(x_i, x_j) \in X$ is defined by

$$d_A(x_i, x_j) = \sqrt{(x_i - x_j)^T A(x_i - x_j)} \tag{1}$$

where the distance metric A is symmetric positive definite (SPD). With a low-rank factorization $A = WW^T$, $W \in \mathbb{R}^{d' \times d}$, $d' < d$, the Mahalanobis distance can be relaxed to

$$\begin{aligned} d_W(x_i, x_j) &= \sqrt{(x_i - x_j)^T W^T W(x_i - x_j)} \\ &= \sqrt{(Wx_i - Wx_j)^T (Wx_i - Wx_j)} \\ &= ||Wx_i - Wx_j||_2, \quad W \in \mathbb{R}^{d' \times d}, d' < d \end{aligned} \tag{2}$$

However, a linearly projected space is not effective for non-linearly related data samples. Therefore, this paper introduces a non-linear mapping $g : \mathbb{R}^d \rightarrow \mathbb{R}^{d''}$ as a generalized distance metric. The non-linear function g in this paper is approximated by a multilayer convolutional mapping followed by a linear mapping W at the last stage, which is in the form of a deep neural network. Introducing function $f_W(x) = Wg(x)$ where mapping $f_W : \mathbb{R}^d \rightarrow \mathbb{R}^{d''} \rightarrow \mathbb{R}^{d'}$, the Mahalanobis distance can be generalized by

$$\begin{aligned} d_{f_W}(x_i, x_j) &= ||f_W(x_i) - f_W(x_j)||_2 \\ &= d_W(g(x_i), g(x_j)), \quad W \in \mathbb{R}^{d'' \times d'}, d' < d \end{aligned} \tag{3}$$

An implicit subspace $g(x)$ can be approximated by learning the network f_W. Distance $d_W(g(x_i), g(x_j))$ measures the non-linear data similarity in domain $\mathbb{R}^+$, where the upper bound of the similarity cannot be well defined. Therefore, a probability measure is introduced to remedy such an issue. Let a density distribution $p(x_i, x_j) = s_{ij}, \sum_j s_{ij} = 1, \forall i, 0 \le s_{ij} \le 1$ measure how likely the distance relationship between x_i and x_j exists. The similarity measure s_{ij} can be estimated by the following max-margin optimization.

$$\max_{s_{ij}} \sum_{i,j} (d_{f_W}^2(x_i, x_j) - \tau)(1 - s_{ij}), \; s.t. \sum_j s_{ij} = 1, \forall i, 0 \le s_{ij} \le 1, \forall j \tag{4}$$

where $\tau > 0$ is the similarity margin. The i-th sample x_i and the j-th sample x_j are dissimilar when $d_{f_W}^2(x_i, x_j) > \tau$, and they are similar when $d_{f_W}^2(x_i, x_j) \le \tau$. Correspondingly in Eq. 4, $s_{ij} \rightarrow 0$ maximizes the objective function when x_i and x_j are dissimilar; $s_{ij} \rightarrow 1$ maximizes the objective function when x_i and x_j are similar. Compared to the distance measure, assigning clusters based on the similarity affinity s_{ij} incorporates finite similarity boundaries defined by a probability measure. However, Eq. 4 is a linear programming and a trivial solution can be obtained by the extreme points of the convex set $\{s_{ij} | \sum_j s_{ij} = 1, \forall i, 0 \le s_{ij} \le 1, \forall j\}$, where for the i-th sample, only the entity with minimum $d_{f_W}^2$ is 1 and the other entities are zeros. Such an issue can be avoided by introducing sparsity penalties $s_{ij}^2, \forall i, j$ to the objective function.

Based on the similarity affinity, the computational framework of spectral clustering is further introduced for cluster assignment. Denoting an SPD matrix $S = [s_1^T, \ldots, s_N^T]$ formed by $s_i, \forall i$, where vector $s_i = [s_{i1}, \ldots, s_{iN}]$ parameterizes the similarity probability measure on the i-th sample, the Laplacian matrix of S is computed by $L_S = D_S - (S + S^T)/2$ where the degree matrix $D_S = diag(\sum_j (s_{1j} + s_{j1}), \ldots, \sum_j (s_{Nj} + s_{jN}))/2$. As proposed by Ng et al. [26], minimizing the spectral $tr(Y_c L_S Y_c)$ assigns the clusters on the subspace Y_c, where $Y_c \in \mathbb{R}^{N \times c'}, Y_c Y_c^T = I$ is introduced as a cluster assignment matrix for c clusters encoded by the c'-dimensional subspace. Together with Eq. 4, an end-to-end optimization problem for simultaneous metric learning, similarity learning, and clustering can be formulated as

$$\min_{f_W, s_{ij}, Y_c} \gamma tr(Y_c L_S Y_c) + \sum_{i,j} [(d_{f_W}^2(x_i, x_j) - \tau)(s_{ij} - 1) + \beta s_{ij}^2]$$

$$s.t. \sum_j s_{ij} = 1, \forall i, 0 \leq s_{ij} \leq 1, \forall j, Y_c Y_c^T = I, Y_c \in \mathbb{R}^{n \times c} \tag{5}$$

where s_{ij}^2 is introduced as a penalty term to avoid trivial solution of s_{ij}. Parameters γ and β are introduced to control the objective function values induced by the spectral cluster assignment term and the penalty term.

4 Optimizations

The alternating directional method of multipliers (ADMM) is adopted to solve the optimized metric network f_W, similarity s_{ij}, and cluster assignment Y_c [1]. The optimization w.r.t. Y_c with fixed s_{ij} and f_W is a typical spectral clustering $\min_{Y_c} tr(Y_c L_S Y_c), s.t. Y_c Y_c^T = I, Y_c \in \mathbb{R}^{n \times c}$, where Y_c^* is solved by the eigenvectors of L_S with smallest c' eigenvalues. The optimization w.r.t. s_{ij} with fixed Y_c and f_W can be formulated as

$$\min_{s_{ij}} \sum_{i,j} (\gamma ||y_i - y_j||_2^2 s_{ij} + (d_{f_W}^2(x_i, x_j) - \tau)s_{ij} + \beta s_{ij}^2)$$

$$s.t. s_i^T 1 = 1, \forall i, 0 \leq s_i \leq 1, y_i, y_j \in \mathbb{R}^{1 \times c'} \tag{6}$$

where $1 \in \mathbb{R}^{N \times 1}$ denotes an all-one vectors; y_i, y_j are the i-th and j-th row vectors of Y_c. Introducing a cluster-sample distance $d_{ij}^{y,f} = \gamma ||y_i - y_j||_2^2 + (d_{f_W}^2(x_i, x_j) - \tau)$ with vector denotation $d_i^{y,f} = [d_{i1}^{y,f}, \ldots, d_{iN}^{y,f}]$, Eq. 6 can be reformulated as the following problem.

$$\min_{s_i} ||s_i - \frac{1}{2\beta} d_i^{y,f}||_2^2, s.t.s_i^T \mathbf{1} = 1, 0 \leq s_i \leq 1, \forall i \tag{7}$$

In practice, a locally connected graph may achieve better performance. Given data with only k related nearest neighbors ($k < N$), the sub-optimal of s_{ij} with fixed Y_c and f_W can be solved by KKT condition as follows.

$$s_i = (-\frac{d_i^{y,f}}{2\beta} + u)_+, \quad u = \frac{1}{k} + \frac{1}{2k\beta}(d_i^{y,f} \mathbf{1}_k) \tag{8}$$

where $u = [u_1, \ldots, u_N]$ denotes the Lagrangian multipliers w.r.t s_i; $\mathbf{1}_k \in \mathbb{R}^{N \times 1}$ is introduced to denote a sparse vector with k ones and N-k zeros. Introducing locality benefits the proposed clustering method in two aspects. First, the computation of the SPD matrix S can be largely reduced. Second, the following inequalities can be derived to determine the parameter value of β.

$$\frac{k}{2}d_{i,k}^{y,f} - \frac{1}{2}\sum_{j=1}^{k} d_{ij}^{y,f} < \beta \leq \frac{k}{2}d_{i,k+1}^{y,f} - \frac{1}{2}\sum_{j=1}^{k} d_{ij}^{y,f}, \forall i \tag{9}$$

Accordingly, $\beta = \frac{1}{N}\sum_{i=1}^{N}(\frac{k}{2}d_{i,k+1}^{y,f} - \frac{1}{2}\sum_{j=1}^{k} d_{i,k}^{y,f})$ is set in this paper.

With fixed similarity S and cluster assignment Y_c, the deep metric network f_W can be approximated by solving

$$\min_{f_W} J(f_W) = \sum_{i,j} ||f_W(x_i) - f_W(x_j)||_2^2 (s_{ij} - 1) \tag{10}$$

In each iteration, the deep non-linear distance measure d_{f_W} is consistently updated to approach the data similarity S. The proposed loss function is in the form of a soft-margin loss. Adam optimizer is used to estimate f_W [19].

Two challenging issues exist when learning the metric network on a data set with N instances. First, N^2 pairs of instances are necessary to be constructed for optimizing Eq. 10, which exponentially increases training and test computations. Second, the number of dissimilar pairs is c-1 times of the number of the similar pairs for data evenly labeled in c classes. Except for c=2, the constructed pairs are unbalanced for training in terms of similarity. Therefore, a stochastic method is introduced to remedy these challenging issues. Dividing N^2 to the loss function, the objective function $J(f_W)$ can be derived by

$$\frac{1}{N^2}J(f_W) = \frac{1}{N}\sum_{i}\frac{1}{N}\sum_{j} ||f_W(x_i) - f_W(x_j)||_2^2 (s_{ij} - 1) \tag{11}$$

$$\lim_{N \to \infty} \frac{1}{N^2}J(f_W) = \mathcal{E}_{x_i}\mathcal{E}_{x_j|x_i}[||f_W(x_i) - f_W(x_j)||_2^2 s_{ij}]$$

According to the Monte Carlo method, $J(f_W)$ can be approximated as the expectation over x_j and x_i given x_j using large epochs but small randomly sampled batches. Hence, we propose to first randomly sample M_2 samples from X and then sample M_1 samples from X to pair with them. $M_1 \times M_2 \ll N^2$ pairs of data are constructed and fed back to update f_W in each epoch. The deep metric learning can also be balanced in terms of similarity by the proposed sampling. For simplicity, $M_1 = M_2 = M$ is set in this paper. The detailed procedure for deep metric similarity clustering is given in Algorithm 1.

Algorithm 1 Deep metric similarity clustering (DMSC)

procedure DMSC(Data set $\{x_i \in \mathbb{R}^d\}_{i=1,\dots,N}$, initialized metric network f_W, similarity margin τ, spectral penalty γ, connection sparsity k, encoding subspace dimension c', cluster number c, metric learning epochs N, batch size M)

Initialize f_W

Initialize s_{ij} by Eq. 8, with $d_{ij}^{y,f} \leftarrow d_{f_W}^2(x_i, x_j) - \tau$, $\forall i, j$

while not converge **do**

$\quad L_S \leftarrow D_S + \frac{S + S^T}{2}$

$\quad Y_c \leftarrow$ Eigenvector(L_S) with c' smallest eigenvalue

$\quad d_{f_W}^2 \leftarrow$ Metric-Learning(X,S,f_W,N,M)

$\quad d_{ij}^{y,f} \leftarrow \gamma ||y_i - y_j||_2^2 + (d_{f_W}^2(x_i, x_j) - \tau)$, $y_i, y_j \in Y_c$, $\forall i, j$

$\quad$ Update $S = [s_1^T, \dots, s_N^T]$ by Eq. 8

end while

clusters $\leftarrow$ K-means(Y_c,c)

return clusters, Y_c, S

procedure Metric-Learning(Data set X, similarity matrix S, Network f_W, epochs N, sample size M, similarity margin τ)

Initialize $d_{f_W}^2(x_p, x_q)$, $\forall x_p, x_q \in X$

i=1,...,N

$\quad \{(z_{j1}, z_{j2}), j \in [1, M]\} \leftarrow$ Sampling(X,M,$d_{f_W}^2$,τ)

$\quad f_W \leftarrow$ Backward loss $\frac{1}{M^2} \sum_{j1 < j2} d_{f_W}^2(X[z_{j1}], X[z_{j2}]) S[z_{j1}, z_{j2}]$

$\quad$ Update $d_{f_W}^2(x_p, x_q)$, $\forall x_p, x_q \in X$

end for

return $d_{f_W}^2(x_p, x_q)$, $\forall x_p, x_q \in X$

procedure Sampling(Data set X, sample size M, distance d, similarity margin τ)

$\quad$ X indices $\{z_{1,j}, j \in [1, M]\} \leftarrow$ Random sample$\{x \in X\}$

$\quad$ X indices $\{z_{2,j}, j \in [1, M/2]\} \leftarrow$ Random sample$\{x \in X, d(z_{1,j}, x) > \tau\}$, X indices $\{z_{2,j}, j \in [M/2 + 1, M]\} \leftarrow$ Random sample$\{x \in X, d(z_{1,j}, x) \leq \tau\}$

$\quad$ **return** Pairs of indices $\{(z_{1,j}, z_{2,j}), j \in [1, M]\}$

4.1 A Large-Scale Network Co-training Approach

The $DMSC$ procedure in Algorithm 1 provides an end-to-end solution for deep metric similarity learning and clustering. However, the procedure $Metric - Learning$ is called multiple times in the outer loop for updating the deep network. Let $O(N_1)$ denote the complexity of the outer loop and $O(N_2)$ denote the complexity of the inner loop. The computational complexity of Algorithm 1 is $O(N_1 N_2)$. Large-scale networks have large N_2, which leads to inefficient clustering. Therefore, a co-training approach is further introduced to remedy this issue by reducing the complexity to $O(N_1 + N_2)$.

Given a subset of data $X_s \in X$ with labeled similarities $\{l(x_i, x_j) \in X_s \forall i, j\}$, a near-optimal deep metric f_W can be pre-trained as distance measure. Let $l(x_i, x_j) = 1$ if x_i, x_j are similar and 0 otherwise, an expected minimum square error (MSE) loss can be constructed by

$$\min_{f_W} \mathcal{E}_{x_i} \mathcal{E}_{x_j | x_i} (\tau(l(x_i, x_j) - 1) + d_{f_W}(x_i, x_j)))^2 \tag{12}$$

to optimize the deep metric. With an optimal distance $d_{f_W^*}$, samples x_i, x_j are dissimilar if $d^2_{f_W^*}(x_i, x_j) > \frac{\tau}{2}$; otherwise similar.

Given an optimized network f_W^*, the proposed model in Eq. 5 can be rewritten as

$$\min_{s_{ij}, Y_c} \gamma tr(Y_c L_S Y_c) + \sum_{i,j} (d^2_{f_W^*}(x_i, x_j) s_{ij} + \beta s_{ij}^2)$$

$$s.t. \sum_j s_{ij} = 1, \forall i, 0 \leq s_{ij} \leq 1, \forall j, Y_c Y_c^T = I, Y \in \mathbb{R}^{n \times c} \tag{13}$$

Let $d_{ij}^{y, f^*} = \gamma ||y_i - y_j||_2^2 + d^2_{f_W^*}(x_i, x_j), \forall i, j$ with vector $d_i^{y, f^*} = [d_{i1}^{y, f^*}, \ldots, d_{iN}^{y, f^*}]$. The solution to s_i with fixed Y_c becomes

$$s_i = (-\frac{d_i^{y, f^*}}{2\beta} + u^*)_+, u^* = \frac{1}{k} + \frac{1}{2k\beta} (d_i^{y, f^*} 1_k) \tag{14}$$

The detailed procedures for deep metric co-training and clustering with optimized distance measure are described in Algorithm 2. The $Sampling$ procedure proposed in Algorithm 1 is reused by Algorithm 2 for labeled data sampling, where the data label L is considered as a true distance with similarity margin of 0.5.

Algorithm 2 Similarity clustering with optimized metric (Optim-MSC)

---

procedure Optimal-MSC(Data Set X, Similarity labeled data X_s, similarity label L=$\{l(x_i, x_j), \forall x_i, x_j \in X_s\}$, learning epochs N, batch size M, initialized metric network f_W, similarity margin τ, spectral penalty γ, connection sparsity k, encoding subspace dimension c', cluster number c)

$\quad f_W^* \leftarrow$ Metric-Learner(X_s,L,N,M, f_W,τ)

$\quad$ Compute $d^2_{f_W^*}(x_p, x_q), \forall x_p, x_q \in X$

$\quad$ **while** not converge **do**

$\qquad L_S \leftarrow D_S + \frac{S+S^T}{2}$

$\qquad Y_c \leftarrow$ Eigenvector(L_S) with c' smallest eigenvalue

$\qquad d_{ij}^{y,f^*} \leftarrow \gamma \|y_i - y_j\|_2^2 + d^2_{f_W^*}(x_i, x_j), y_i, y_j \in Y_c, \forall i, j$

$\qquad$ Update $S = [s_1^T, ..., s_N^T]$ by Eq. 14

$\quad$ **end while**

$\quad$ clusters $\leftarrow$ K-means(Y_c,c)

$\quad$ **return** clusters, Y_c, S

procedure Metric-Learner(Similarity labeled data X_s, similarity label L=$\{l(x_i, x_j), \forall x_i, x_j \in X_s\}$, learning epochs N, sample size M, initialized metric network f_W, similarity margin τ)

$\quad$ **For** i=1, ..., N

$\qquad \{(z_{j1}, z_{j2}), j \in [1, M]\} \leftarrow$ Sampling(X_s,M,L,0.5)

$\qquad L_z \leftarrow \{L(z_{1,j}, z_{2,j}), j \in [1, M]\}$

$\qquad f_W \leftarrow$ Backward loss $\frac{1}{M^2} \sum_{j1<j2}(\tau(L[z_{1,j}, z_{2,j}] - 1) + d_{f_W}(X_s[z_{j1}], X_s[z_{j2}]))^2$

$\quad$ **end for**

$\quad$ **return** f_W

---

5 Connection to Spectral Clustering

Given an optimized network f_W^*, this section shows the proposed model is a variation of spectral clustering with optimal similarity affinity and deep metric. Let matrix $D^* = [(d^2_{f_W^*,1})^T, \ldots, (d^2_{f_W^*,N})^T]$ where $d^2_{f_W^*,i} = [d^2_{f_W^*}(x_i, x_1), \ldots, d^2_{f_W^*}(x_i, x_N)]$. The objective function in Eq. 13 can be rewritten as

$$J(S, Y_c) = \gamma tr(Y_c Y_c^T L_S) + tr(S^T D^*)$$
$$= tr(\gamma L_S + S^T D) = tr(Y_c(\gamma L_S + S^T D^*)Y_c^T) \tag{15}$$

where $Y_c \in \mathbb{R}^{N \times c'}$, $Y_c Y_c^T = I$. Introducing a metric similarity modified Laplacian matrix $L_{S,D^*} = \gamma L_S + S^T D^*$, the proposed method is in the form of a spectral clustering.

$$\min_{Y_c} tr(Y_c L_{S,D^*} Y_c^T), s.t. Y_c Y_c^T = I \tag{16}$$

Equation 16 can be degenerated to a spectral clustering with a pre-defined distance measure $d_{f_W^*}(x_i, x_j) = e^{||x_i - x_j||_2^2}$ and uniformly distributed similarity relationship $s_{ij} = p(x_i, x_j) = \frac{1}{N}, \forall i$, where $L_{S,D^*} = \gamma L_S + S^T D^* = \gamma L_S + L_S = (\gamma + 1)L_S$. This degeneration shows that the spectral clustering models data affinity with an explicitly defined distance and uniform similarity density, which is a special case of the proposed method. With a generalized form in Eq. 16, DMSC and Optim-MSC are able to achieve optimized affinity for clustering. This advantage makes the proposed method more effective on complex real-world data with challenging non-linear relationships.

Besides similarity learning, the learned affinity subspace Y_c shows another difference from spectral clustering. In the proposed method, the learned cluster assignment Y_c lies in the subspace of an optimized deep non-linear affinity d_{f_W}. With the optimized affinity, the number of dimensions used to encode the cluster in $Y_c \in \mathbb{R}^{c' \times c}$ can be reduced to as small as possible with a perfect clustering, i.e., $c' = \lceil log_2 c \rceil$ can be set in general as a binary cluster encoder for c clusters. This property shows that the proposed method can be applied as a minimum cluster encoder.

6 Experiments

6.1 Experiment Setups

Three properties of the proposed method are evaluated in the experiments. They are clustering, subspace learning, and cluster encoding. Data visualization is introduced to evaluate the capability of subspace learning. Clustering accuracy (ACC) and normalized mutual information (NMI) are introduced to evaluate the clustering performance. ACC$=\frac{\sum_\ell n(C,\ell)}{N}$, where $n(C, \ell)$ denotes the largest number of samples in some cluster $C \in [1, c]$ that matching the true label $\ell \in [1, |\ell|]$, $|\ell|$ is the number of the labels. NMI measures the distribution difference between the cluster and label space. The dimensions of the cluster assignment subspace are recorded to evaluate the cluster encoder.

Synthetic data, benchmark machine learning data, and visual recognition data are used to comprehensively evaluate the proposed clustering method. The benchmark machine learning data sets used in the experiment include Stock, Pathbased, Movements, Wine, Compound, Spiral, Yeast, Class, Ecoli, Umist, COIL20, JAFFE, and USPS, where three sets are shape data, six sets are biological data from UCI Machine Learning Repository, and the last four sets are image data sets for hand-written digits recognition, facial expression recognition, and 20-class object recognition. K-means, spectral clustering (SPCL), ratio-cut [10], normalized-cut [30], and nonnegative matrix factorization (NMF) [5] are evaluated on these data sets as benchmark methods. The clustering and projected clustering with adaptive neighbors (CAN, PCAN) proposed by Nie et al. is evaluated as the state-of-the-art method [27].

Benchmark visual recognition databases including Extended Yale B, ORL, COIL20, COIL100, Caltech101, MSRCV1, and ETH80 are introduced in the experiment. For a fair comparison, the same experimental setup in [17] is adopted on Extended Yale B, ORL, COIL20, and COIL100 databases. Seven classes from Caltech101 and 20 classes from Caltech101 are selected as two data sets for evaluation. The seven classes include "tree," "building," "airplane," "cow," "face," "car," and "bicycle," and 30 images are sampled for each class. The 20 classes include "binocular," "brain," "camera," "carside," "faces," "ferry," "garfield," "hedgehog," "leopards," "motobike," "pagoda," "rhino," "snoopy," "stapler," "stop-sign," "waterlily," "windsor," "chair," "wrench," and "yin-yang," and 60 images are sampled for each class. Seven classes in MSRCV1 including "tree," "building," "airplane," "cow," "face," "car," and "bicycle" are used, and 30 images per class are sampled for evaluation. All 8 classes in ETH80 are used and 50 images per class are sampled in the evaluation. Low-rank representation (LRR) [24], multi-view sub-space clustering (MSC) [9], co-regularized multi-view spectral clustering (CMSC) [21], low-rank subspace clustering (LRSC) [31], sparse subspace clustering (SSC) [8], sparse subspace clustering with orthogonal matching pursuit (SSC-OMP) [35], efficient dense subspace clustering (EDSC) [35], and deep subspace clustering network (DSC-Net) [17] are evaluated as comparative state-of-the-art methods.

A general deep structure is introduced to evaluate the synthetic data and benchmark machine learning data. The structure of the network f_W follows $input - Linear - Tanh - Linear$. For data with N samples, c channels, and d dimensions $d < 20$, the size of the linear layers is chosen by $(d \times c \times 10) + (10 \times c \times 10)$. For data with N samples, c channels, and d dimensions $d < 256$, the size of the network is chosen by $(d \times c \times 256) + (256 \times c \times 256)$. For data with N samples, c channels, and d dimensions $d \geq 256$, the size of the network is chosen by $(d \times c \times 1024) + (1024 \times c \times 1024)$. The end-to-end model proposed by Eq. 5 and Algorithm 1 is used on the synthetic data and benchmark machine learning data. The co-training approach proposed in Algorithm 2 is evaluated on the visual data. For improved visual clustering, the state-of-the-art deep networks for visual recognition are adopted to implement Algorithm 2. The code is available at https://github.com/daishuanglu/dmClustering.

6.2   Evaluations on Synthetic Data

First of all, shape- and distribution-based synthetic data sets are generated to evaluate the subspace learning and cluster encoding performance obtained by the proposed clustering Algorithm 1. The dimension of the affinity subspace is strictly set to $c' = \lceil log_2 c \rceil$ as the minimum cluster encoder in this experiment, e.g., c'=1 for two clusters and c'=2 for two and three clusters. As a related baseline method, the affinity subspaces obtained by spectral clustering are compared to those obtained by the proposed method. Note DMSC with optimized affinity always achieves the

Table 1 Accuracies (%) obtained by spectral clustering and our method for synthetic data clustering

Methods	Crescent moon	Circle (3:6:10)	Blobs3.0
SPCL	87	66	66
Ours	100	100	89

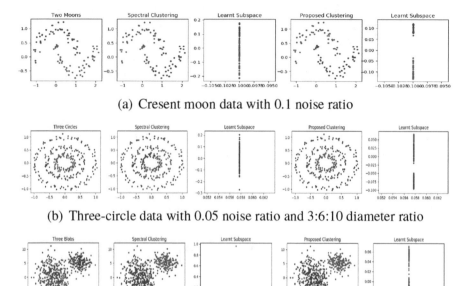

(a) Cresent moon data with 0.1 noise ratio

(b) Three-circle data with 0.05 noise ratio and 3:6:10 diameter ratio

(c) Three-blobs data with 3.0, 2.5 and 2.0 noises

Fig. 2 Comparative methods for synthetic data clustering. From left to right: the generated data; results obtained by spectral clustering; the minimum subspace learned by spectral clustering; results obtained by our method; the minimum subspace learned by our method. (**a**) Crescent moon data with 0.1 noise ratio. (**b**) Three-circle data with 0.05 noise ratio and 3:6:10 diameter ratio. (**c**) Three-blobs data with 3.0, 2.5, and 2.0 noises

same cluster space on the synthetic data. We repeat K-means 100 times for spectral clustering and compare its best possible performances with DMSC.

Three toy data sets are randomly generated with different levels of noise for evaluation. The data sets are in three different shapes: "crescent moon," "circle," and "Gaussian blob." The clustering results obtained by the comparative methods with minimum-dimension cluster encoder are visualized in Fig. 2. The accuracies obtained by spectral clustering and our method are recorded in Table 1. As shown in Fig. 2, spectral clustering fails to encode the clusters using minimum dimensions with 100% cluster separability. Compared to spectral clustering, the proposed method effectively encodes shapes and distribution-based data on a one-dimensional line. As presented by Table 1 and Fig. 2c, the proposed method becomes ineffective when the noisy data samples from different classes become inseparable. Adding prior data distributions to the proposed model may be able to remedy this issue.

6.3 Evaluations on Real-World Benchmark Data

Secondly, the real-world benchmark data is used to evaluate the clustering and cluster encoding performance obtained by the proposed clustering Algorithm 1. The information of the data sets including the sizes, dimensions, and number of classes is summarized in Table 2. All the comparative tests are repeated 100 times. The average ACCs and NMIs are recorded in Table 4. The minimum encoded cluster assignment dimensions obtained by the comparative methods are recorded in Table 3.

As presented in Table 4, the proposed method generally outperforms other comparative methods in terms of ACC and NMI, except that PCAN marginally outperforms the proposed method on JAFFE data set. This shows that the linear mapping in PCAN is more suitable to model the data relationships in JAFFE data

Table 2 Summarization of the 13 benchmark data sets

Data	# of instances	Dimensions	Classes
Umist	165	3456	15
COIL20	1440	1024	20
JAFFE	213	1024	7
USPS	1854	256	10
Stock	950	85	5
Pathbased	300	2	3
Movements	360	90	15
Spiral	312	2	3
Wine	178	13	3
Compound	399	2	6
Yeast	1484	8	10
Glass	219	9	6
Ecoli	327	7	4

Table 3 Dimensions of the minimum encoded cluster assignment obtained by the comparative methods

Data sets	Classes	SPCL	NMF	PCAN	Ours
Umist	15	50	15	15	**10**
COIL20	20	80	20	20	**5**
JAFFE	7	20	7	7	**3**
USPS	10	50	10	10	**4**
Stock	5	10	5	5	**3**
Pathbased	3	10	3	3	**2**
Movements	15	50	15	15	**5**
Spiral	3	10	3	3	**2**
Wine	3	10	3	3	**2**
Compound	6	20	6	6	**3**
Yeast	10	15	10	10	**4**
Glass	6	20	6	6	**3**
Ecoli	4	20	4	4	**2**

Table 4 Average NMIs (%) and ACCs (%) obtained by the comparative methods on real-world benchmark data sets

Data	K-Means		SPCL		Ratio cut		Normalized cut		NMF		CAN		PCAN		Ours	
	ACC	NMI	ACC	NMI	ACC	NMI	ACC	NMI	ACC	NMI	ACC	NMI	ACC	NMI	ACC	NMI
Umist	43.31	64.44	63.39	63.9	60.44	77.78	59.51	77.26	62.26	78.97	77.57	88.52	68.17	85.6	**79.65**	**88.40**
COIL20	56.54	73.45	71.6	63.88	69.42	84.01	70.3	84.42	70.42	81.22	90.14	94.6	83.33	89.1	**99.93**	**89.48**
JAFFE	73.7	82.44	82.54	84.19	84.78	90.24	81.63	89.43	96.71	96.23	96.17	96.23	**100**	**100**	99.53	98.94
USPS	64.27	62.07	68.64	58.34	67.59	73.95	68.43	73.97	67.37	74.15	78.96	80.46	63.81	68.93	**87.73**	**83.3**
Stock	74.02	76.35	50.0	57.72	55.23	56.56	55.39	56.2	56.21	60.29	67.79	74.89	77.16	68.3	**80.0**	**86.67**
Path	74.34	51.3	86.33	66.31	77.67	55.16	77.67	55.16	78.0	52.51	87.0	75.63	87.0	75.63	**87.67**	**72.58**
Move	44.04	57.2	42.78	54.95	46.01	61.89	44.49	59.85	43.33	58.95	49.17	64.07	49.17	60.84	**56.94**	**70.36**
Spiral	34.56	0.05	51.92	26.82	96.92	94.98	96.03	94.19	91.03	75.95	**100**	**100**	**100**	**100**	**100**	**100**
Wine	94.65	82.41	82.47	80.12	95.44	84.37	94.99	84.02	94.94	83.24	97.19	88.97	**100**	**100**	**100**	**100**
Comp	65.49	69.6	70.18	60.11	52.94	70.71	52.67	70.29	52.38	73.26	80.2	79.27	79.7	78.65	**90.48**	**94.48**
Yeast	38.0	25.19	55.32	28.62	38.11	24.94	36.99	23.88	35.65	23.66	50.27	30.3	50.07	30.55	**51.48**	25.62
Glass	45.57	33.13	48.13	30.89	38.28	29.1	38.26	28.58	37.85	28.7	50.0	26.91	49.53	33.82	**71.83**	**56.45**
Ecoli	55.1	51.04	77.37	59.69	49.08	43.96	48.1	44.78	49.17	42.12	78.04	67.2	78.33	67.44	**79.82**	**70.98**

Bold values indicate the best results

sets, while the non-linear mappings in the proposed method can only approximate them to a certain degree. Besides competitive ACC and NMI, Table 3 further shows that the proposed method encodes the clusters with their minimum possible dimensions. In NMF and PCAN, the cluster assignment dimension is modeled equivalent to the number of clusters. Large-scale matrix factorization will be required by these methods when the number of clusters is increased. However, the proposed method is able to avoid this expensive computation by encoding c clusters to $c' = \lceil log_2 c \rceil$ dimension.

6.4 Evaluations on Visual Data

The last experiment evaluates the proposed method with large-scale networks for visual clustering. The proposed residual self-expressive metric network is used in this experiment for real-world visual clustering. The proposed co-training Algorithm 2 is evaluated. In this experiment, the residual convolution blocks and self-expressive layers of ResEM-Net are adopted to compute the non-linear mapping f_W. The deep metric network d_{f_W} is co-trained based on data similarity. Network parameters of ResNet-18 trained on ILSVRC15 [11] are used to initialize the residual convolution blocks. The whole network d_{f_W} is then retrained on target data sets.

First, the state-of-the-art subspace clustering methods are evaluated on the ORL, COIL20, COIL100, and Extended Yale-B (EYB) databases. 10, 20, and 30 subjects are obtained from EYB for evaluation. Average ACCs of 20 trials are recorded in Table 5. As shown by Table 5, the proposed method outperforms the comparative methods in terms of ACC. This shows the subspace and similarity measure obtained by ResEM-Net and Optim-MSC are more discriminative than the comparative methods. Second, the comparative multi-view subspace clustering methods are evaluated on MSRCV1, ETH80, and Caltech101 (Cal) databases. The results obtained by these methods are recorded in Table 6. HOG [2], LBP [29], SIFT [25], color moment [36], and CENTRIST [33] are used as multi-view visual features by the comparative methods. The subspaces of these features are summed

Table 5 Average ACC (%) obtained by the comparative subspace clustering methods

Methods	ORL	COIL20	COIL100	EYB10	EYB20	EYB30
LRR	61.75	68.99	40.18	77.78	69.77	62.02
LRSC	67.5	68.25	49.33	69.05	71.24	69.36
SSC	67.5	85.14	55.0	89.88	80.25	71.24
SSC-OMP	64.0	64.1	44.8	87.92	84.84	79.25
EDSC	72.25	85.14	61.87	94.36	90.7	88.76
DSC-Net	86.0	94.86	69.2	98.41	98.20	97.93
Ours	**100**	**100**	**85.83**	**100**	**98.27**	**98.15**

Table 6 Average ACC (%) and NMI (%) obtained by the comparative multi-view subspace clustering methods

Methods	Cal7		MSRCV1		ETH80		Cal20	
	ACC	NMI	ACC	NMI	ACC	NMI	ACC	NMI
LRR	71.68	59.26	62.52	53.84	51.27	52.0	46.17	44.57
SSC	71.02	56.45	63.43	54.15	52.47	51.43	55.27	51.51
CMSC	71.88	67.68	66.67	57.45	55.56	**61.47**	57.14	61.64
MVSC	74.15	71.53	56.5	58.41	**64.59**	61.3	58.79	65.32
Ours	**97.56**	**96.52**	**80.48**	**81.85**	55.22	58.22	**95.03**	**96.33**

Bold values indicate the best results

up as input to SSC and LRR. CMSC is evaluated as a baseline method, and MVSC is evaluated as the state-of-the-art method for multi-view similarity clustering. The average ACCs over 20 trials are recorded in Table 6. As observed from Table 6, the proposed method outperforms other comparative methods on Caltech and MSRCV1 databases in terms of ACC and NMI. The multi-view features adopted by MVSC may be more representative on ETH80 than the subspace learned by the residual blocks in ResEM-Net and thus lead to a better performance. In summary, the proposed method generally outperforms the related state-of-the-art multi-view and subspace clustering methods in terms of ACC, NMI, and the dimension of the cluster encoder.

7 Discussions

This paper proposed a deep metric similarity clustering method for end-to-end similarity learning and clustering. With the learned deep similarity, the proposed method performs effectively on real-world and visual data with non-linear relationships. The proposed co-training method improves the clustering with targeted purposes. The stochastic optimization method introduced in co-training reduces the computation cost. The learned affinity subspaces encode the clusters with approximated minimum dimensions. Promising results show that the proposed method outperforms benchmark and state-of-the-art clustering methods in terms of clustering performances, subspace learning, and cluster encoding.

8 Theoretical Analysis

This section explains the derivation of the sparsity penalty terms $\beta s_{ij}, \forall i, j$ and provides detailed solutions to the optimization problem for deep metric similarity clustering.

8.1 Optimization Analysis of the Sparsity Penalty

Given the linear programming problem w.r.t. the similarity affinity s_{ij} as the following.

$$\max_{s_{ij}} \sum_{i,j} (d_{fw}^2(x_i, x_j) - \tau')(1 - s_{ij}), \ s.t. \ \sum_{j} s_{ij} = 1, \forall i, 0 \leq s_{ij} \leq 1, \forall j \tag{17}$$

where τ' is a constant, the optimal to Eq. 17 is directly located at the extreme points of the convex set $\{s_{ij} | \sum_j s_{ij} = 1, \forall i, 0 \leq s_{ij} \leq 1, \forall j\}$, which are a group of one-hot index vectors given by $[argmin_j\{d_{1j}, \forall j\}, \ldots, argmin_j\{d_{Nj}, \forall j\}]$. d_{ij} is introduced to denote $d_{fw}^2(x_i, x_j)$. This solution only shows the relationships within the instances. However, we hope that the optimal similarity affine space s_{ij} of d_{ij} also explores between-instance similarities. A sparsity constraint on the similarity vectors $s_i, \forall i$ is introduced to achieve such a purpose, where the ℓ-1 norm is constrained by sparsity constants $||s_i||_1 \geq \epsilon_i, \forall i, s_i = [s_{i1}, \ldots, s_{iN}]$. Therefore, the optimization problem becomes

$$\min_{s_{ij}} \sum_{i,j} (d_{ij} - \tau) s_{ij}, \ s.t. ||s_i||_1 = \epsilon_i, \ \sum_{j} s_{ij} = 1, \forall i, 0 \leq s_{ij} \leq 1, \forall j \tag{18}$$

For simplicity, the constraint $||s_i||_1 = \epsilon_i$ with sparsity constants $\epsilon_i > 1 \forall i$ is formulated as a penalty term in the objective function, where

$$\min_{s_{ij}} \sum_{i,j} d_{ij} s_{ij} + \sum_{i} \beta_i \left(\sum_{j} ||s_{ij}||_1 - \epsilon_i \right)^2, \ s.t. \ \sum_{j} s_{ij} = 1, \forall i, 0 \leq s_{ij} \leq 1, \forall j \tag{19}$$

Let $\beta = \min\{\beta_i, \forall i\} = \beta_1 = \beta_2 = \ldots = \beta_N, \beta > 0$ as a penalty parameter. Large β will enforce the solution to the constraint boundary. With the inequalities $(\sum_j ||s_{ij}||_1 - \epsilon_i)^2 \geq (\sum_j ||s_{ij}||_1)^2 - 2\epsilon_i \sum_j s_{ij} \geq \sum_j s_{ij}^2 - 2\epsilon_i \sum_j s_{ij}$ and $0 \leq s_{ij} \leq 1$, the optimization problem given by Eq. 18 is equivalent to

$$\min_{s_{ij}} \sum_{i,j} (d_{ij} - \tau' + \beta \epsilon_i) s_{ij} + \beta \sum_{ij} s_{ij}^2, \ s.t. \ \sum_{j} s_{ij} = 1, \forall i, 0 \leq s_{ij} \leq 1, \forall j \tag{20}$$

As $\tau', \beta, \epsilon_i, \forall i$ are all constant parameters, we introduce two constant parameters $\tau = \tau' + \beta \epsilon_i > 0$ and $\beta > 0$ for simplicity. The optimization problem is thus derived as

$$\min_{s_{ij}} \sum_{i,j} (d_{ij} - \tau) s_{ij} + \beta \sum_{ij} s_{ij}^2, \ s.t. \ \sum_{j} s_{ij} = 1, \forall i, 0 \leq s_{ij} \leq 1, \forall j \tag{21}$$

Equation 21 shows the soft-margin parameter τ encloses both the margin of the deep metric distance and the sparsity of data similarity.

8.2 Solution to the Optimizations

Given the proposed optimization problem w.r.t. f_W, s_{ij}, Y_c

$$\min_{f_W, s_{ij}, Y_c} \gamma tr(Y_c L_S Y_c) + \sum_{i,j} [(d_{f_W}^2(x_i, x_j) - \tau)(s_{ij} - 1) + \beta s_{ij}^2],$$

$$s.t. \sum_j s_{ij} = 1, \forall i, 0 \le s_{ij} \le 1, \forall j, Y_c Y_c^T = I, Y_c \in \mathbb{R}^{n \times c} \tag{22}$$

where $L_S = D_S - S \in \mathbb{R}^{N \times N}$, $D_S = \begin{bmatrix} \sum_j s_{1j} \cdots & 0 \\ \vdots & \ddots & \vdots \\ 0 & \cdots \sum_j s_{Nj} \end{bmatrix}$. Introduce vector

denotation $Y_c = [y_1^T, \ldots, y_N^T]$. The first term of Eq. 17 can be derived as

$$tr(Y_c L_S Y_c) = tr(Y_c Y_c^T L_S) = tr(Y_c Y_c^T D_S) - tr(Y_c Y_c^T S)$$

$$= \sum_i y_i y_i^T \sum_j s_{ij} - \sum_{i,j} y_i y_j^T s_{ij}$$

$$= \frac{1}{2} \sum_{i,j} (y_i y_i^T s_{ij} - 2 y_i y_j^T s_{ij} + y_j y_j^T s_{ij}) \tag{23}$$

$$= \frac{1}{2} \sum_{i,j} ||y_i - y_j||^2 s_{ij}$$

Therefore, an empirical value of gamma can be 0.5. The Lagrangian function of 22 is written as

$$\mathcal{L}(f_W, s_{ij}, Y_c, u_1, \ldots, u_N, \lambda, \eta)$$

$$= \sum_{i,j} (\gamma ||y_i - y_j||^2 s_{ij} + (d_{f_W}^2(x_i, x_j) - \tau)(s_{ij} - 1) + \beta s_{ij}^2$$

$$+ \eta_i (1 - s_{ij})) + \sum_i u_i (1 - \sum_j s_{ij}) + \lambda \cdot (I - Y_c Y_c^T) \tag{24}$$

where $s_{ij} \le 0$. $\lambda = [\lambda_1, .., \lambda_N] \in \mathbb{R}^{N \times 1}$ denotes a vector of Lagrangian multipliers. Introduce vector denotation $s_i = [s_{i1}, \ldots, s_{iN}] \in \mathbb{R}^N$ and an all-one vector $\mathbf{1}$, $\sum_j s_{ij} = s_i \mathbf{1}$, the optimal of 22 can be derived and approximated in the following ADMM form.

$$s_i^{t+1} = \min_{s_i^t \geq 0} \mathcal{L}_i(s_i^t, \eta_i, u_i) = s_i^t d_i^{y,f} + \beta ||s_i^t||_2^2 + \eta_i^T(s_i^t - \mathbf{1})$$

$$+ u_i \cdot (1 - s_i^t \mathbf{1}), \forall i = 1, \ldots, N$$

$$Y_c^{t+1} = \min_{Y_c^t} \mathcal{L}_Y(Y_c^t, \lambda) = \sum_{i,j}(\gamma ||y_i^t - y_j^t||^2 s_{ij}^{t+1}) + \lambda(I - Y_c^t(Y_c^t)^T) \tag{25}$$

$$= \gamma tr(Y_c^t L_{S^{t+1}}(Y_c^t)^T) + \lambda(I - Y_c^t(Y_c^t)^T)$$

$$f_W^{t+1} = \min_{f_W^t} \mathcal{L}_f(f_W^t) = \sum_{i,j}(d_{f_W^t}^2(x_i, x_j)(s_{ij}^{t+1} - 1)$$

where t indicates the t-th step of iteration; vector $\mathbf{d}_i^{y,f} = [d_{i1}^{y,f}, \ldots, d_{iN}^{y,f}]$ with $d_{ij}^{y,f} = \gamma ||y_i - y_j||^2 + d_{f_W}^2(x_i, x_j) - \tau$, $\eta \in \mathbb{R}^{N \times 1}$ and $u_i \in \mathbb{R}$, $\forall i$ are Lagrangian multipliers, $\cdot$ denotes the dot product. In each step of the sub-minimization given by Eq. 25, the solutions of the non-linear mapping network f_W and the cluster assignment subspace Y_c are conditionally independent given the similarity affinity S, where $f_W \perp Y_c|S$. Therefore, S must be computed first, but the computation orders of Y_c and f_W are not sensitive in the algorithm.

8.2.1 Solution to S

The sub-minimization w.r.t. s_i can be formulated as

$$\min_{s_i \geq 0} \mathcal{L}_i(s_i, \eta_i, u_i) = \frac{1}{2}||s_i - \frac{1}{2\beta}d_i^{y,f}||_2^2 + \eta_i^T(s_i - \mathbf{1}) + u_i(1 - s_i\mathbf{1}) \tag{26}$$

According to KKT condition, the following linear system equations can be derived.

$$s_i + \frac{d_i^{y,f}}{2\beta} + \eta_i^T - (u_i \cdot \mathbf{1}) = 0$$

$$s_i^T \mathbf{1} = 1 \tag{27}$$

$$\eta_i \cdot (s_i - \mathbf{1}) = 0$$

As $s_i - \mathbf{1} = \mathbf{1}$ contradicts to the boundary condition $s_i^T \mathbf{1} = 1$, valid solution can only be achieved by $\eta_i = 0$, $s_i - \mathbf{1} \neq 0$, where $s_i = (-\frac{d_i}{2\beta} + (\frac{1}{N} + \frac{1}{2\beta N}d_i^{y,f}) \cdot \mathbf{1})_+$ and $s_i < \mathbf{1}$ holds. Given only k connected nearest neighbors (k<N), the solution becomes $s_i = (-\frac{d_i}{2\beta} + (\frac{1}{k} + \frac{1}{2\beta k}d_i^{y,f}) \cdot \mathbf{1}_k)_+$, where $\mathbf{1}_k \in \mathbb{R}^N$ is introduced to denote a sparse vector with k 1s and N-k 0s. The range of parameter β is identical given s_i with $0 \leq s_i < \mathbf{1}$.

8.2.2 Solution to Y_c

The sub-minimization w.r.t. Y_c can be formulated as

$$\min_{Y_c, \lambda} \mathcal{L}_Y(Y_c, \lambda) = \gamma tr(Y_c L_S Y_c^T) + \lambda(I - Y_c Y_c^T) \tag{28}$$

The Euler-Lagrangian equation is

$$(\gamma L_S - \Lambda)Y_c = 0 \tag{29}$$

where matrix $\Lambda = \begin{bmatrix} \lambda_1 \dots 0 \\ \vdots \ddots \vdots \\ 0 \dots \lambda_N \end{bmatrix}$ is constructed by the Lagrangian multipliers λ. Y_c

and λ can be solved by the c-th smallest eigenvector and eigenvalue of γL_S.

References

1. S. Boyd, N. Parikh, E. Chu, B. Peleato, J. Eckstein et al., Distributed optimization and statistical learning via the alternating direction method of multipliers. Found. Trends® Mach. Learn. **3**(1), 1–122 (2011)
2. N. Dalal, B. Triggs, Histograms of oriented gradients for human detection, in *IEEE Computer Society Conference on Computer Vision and Pattern Recognition, 2005. CVPR 2005*, vol. 1 (IEEE, Piscataway, 2005), pp. 886–893
3. J.V. Davis, I.S. Dhillon, Differential entropic clustering of multivariate gaussians, in *Advances in Neural Information Processing Systems* (2007), pp. 337–344
4. J.V. Davis, B. Kulis, P. Jain, S. Sra, I.S. Dhillon, Information-theoretic metric learning, in *Proceedings of the 24th International Conference on Machine Learning* (ACM, New York, 2007), pp. 209–216
5. C.H. Ding, T. Li, M.I. Jordan, Convex and semi-nonnegative matrix factorizations. IEEE Trans. Pattern Anal. Mach. Intell. **32**(1), 45–55 (2010)
6. D. Dueck, B.J. Frey, Non-metric affinity propagation for unsupervised image categorization, in *IEEE 11th International Conference on Computer Vision 2007. ICCV 2007* (IEEE, Piscataway, 2007), pp. 1–8
7. E. Elhamifar, R. Vidal, Sparse manifold clustering and embedding, in *Advances in Neural Information Processing Systems* (2011), pp. 55–63
8. E. Elhamifar, R. Vidal, Sparse subspace clustering: Algorithm, theory, and applications. IEEE Trans. Pattern Anal. Mach. Intell. **35**(11), 2765–2781 (2013)
9. H. Gao, F. Nie, X. Li, H. Huang, Multi-view subspace clustering, in *Proceedings of the IEEE International Conference on Computer Vision* (2015), pp. 4238–4246
10. L. Hagen, A.B. Kahng, New spectral methods for ratio cut partitioning and clustering. IEEE Trans. Comput. Aided Des. Integr. Circ. Syst. **11**(9), 1074–1085 (1992)
11. K. He, X. Zhang, S. Ren, J. Sun, Deep residual learning for image recognition, in *Proceedings of the IEEE Conference on Computer Vision and Pattern Recognition* (2016), pp. 770–778
12. J.R. Hershey, Z. Chen, J. Le Roux, S. Watanabe, Deep clustering: Discriminative embeddings for segmentation and separation, in *2016 IEEE International Conference on Acoustics, Speech and Signal Processing (ICASSP)* (IEEE, Piscataway, 2016), pp. 31–35
13. G.E. Hinton, R.R. Salakhutdinov, Reducing the dimensionality of data with neural networks. Science **313**(5786), 504–507 (2006)

14. J. Hu, J. Lu, Y.P. Tan, Discriminative deep metric learning for face verification in the wild, in *Proceedings of the IEEE Conference on Computer Vision and Pattern Recognition* (2014), pp. 1875–1882
15. P. Huang, Y. Huang, W. Wang, L. Wang, Deep embedding network for clustering, in *22nd International Conference on Pattern Recognition (ICPR)* (IEEE, Piscataway, 2014), pp. 1532–1537
16. P. Jain, B. Kulis, I.S. Dhillon, K. Grauman, Online metric learning and fast similarity search, in *Advances in Neural Information Processing Systems* (2009), pp. 761–768
17. P. Ji, T. Zhang, H. Li, M. Salzmann, I. Reid, Deep subspace clustering networks, in *Advances in Neural Information Processing Systems* (2017), pp. 24–33
18. H. Junlin, L. Jiwen, T. Yap-Peng, Z. Jie, Deep transfer metric learning. IEEE Trans. Image Process. 25(12), 5576–5588 (2016)
19. D.P. Kingma, J. Ba, Adam: A method for stochastic optimization, in *International Conference on Learning Representations (ICLR)* (2015)
20. D. Koller, N. Friedman, F. Bach, *Probabilistic Graphical Models: Principles and Techniques* (MIT Press, Cambridge, 2009)
21. A. Kumar, H. Daumé, A co-training approach for multi-view spectral clustering, in *Proceedings of the 28th International Conference on Machine Learning (ICML-11)* (2011), pp. 393–400
22. L. Ladicky, C. Russell, P. Kohli, P.H. Torr, Graph cut based inference with co-occurrence statistics, in *European Conference on Computer Vision* (Springer, Berlin, 2010), pp. 239–253
23. M. Laszlo, S. Mukherjee, Minimum spanning tree partitioning algorithm for microaggregation. IEEE Trans. Knowl. Data Eng. 17(7), 902–911 (2005)
24. G. Liu, Z. Lin, S. Yan, J. Sun, Y. Yu, Y. Ma, Robust recovery of subspace structures by low-rank representation. IEEE Trans. Pattern Anal. Mach. Intell. 35(1), 171–184 (2013)
25. D.G. Lowe, Distinctive image features from scale-invariant keypoints. Int. J. Comput. Vis. 60(2), 91–110 (2004)
26. A.Y. Ng, M.I. Jordan, Y. Weiss, On spectral clustering: Analysis and an algorithm, in *Advances in Neural Information Processing Systems* (2002), pp. 849–856
27. F. Nie, X. Wang, H. Huang, Clustering and projected clustering with adaptive neighbors, in *Proceedings of the 20th ACM SIGKDD International Conference on Knowledge Discovery and Data Mining* (ACM, New York, 2014), pp. 977–986
28. F. Nie, G. Cai, X. Li, Multi-view clustering and semi-supervised classification with adaptive neighbours, in *AAAI* (2017), pp. 2408–2414
29. T. Ojala, M. Pietikainen, T. Maenpaa, Multiresolution gray-scale and rotation invariant texture classification with local binary patterns. IEEE Trans. Pattern Anal. Mach. Intell. 24(7), 971–987 (2002)
30. J. Shi, J. Malik, Normalized cuts and image segmentation. IEEE Trans. Pattern Anal. Mach. Intell. 22(8), 888–905 (2000)
31. R. Vidal, P. Favaro, Low rank subspace clustering (lrsc). Pattern Recogn. Lett. 43, 47–61 (2014)
32. R. Wang, S. Shan, X. Chen, Q. Dai, W. Gao, Manifold-manifold distance and its application to face recognition with image sets. IEEE Trans. Image Processing 21(10), 4466–4479 (2012)
33. J. Wu, J.M. Rehg, Centrist: A visual descriptor for scene categorization. IEEE Trans. Pattern Anal. Mach. Intell. 33(8), 1489–1501 (2011)
34. J. Xie, R. Girshick, A. Farhadi, Unsupervised deep embedding for clustering analysis, in *International Conference on Machine Learning* (2016), pp. 478–487
35. C. You, D. Robinson, R. Vidal, Scalable sparse subspace clustering by orthogonal matching pursuit, in *Proceedings of the IEEE Conference on Computer Vision and Pattern Recognition* (2016), pp. 3918–3927
36. H. Yu, M. Li, H.J. Zhang, J. Feng, Color texture moments for content-based image retrieval, in *2002 International Conference on Image Processing. Proceedings*, vol. 3 (IEEE, Piscataway, 2002), pp. 929–932

Estimating the Effective Topics of Articles and Journals Abstract Using LDA and K-Means Clustering Algorithm

Shadikur Rahman, Umme Ayman Koana, Aras M. Ismael, and Karmand Hussein Abdalla

1 Introduction

Topic modeling and text clustering perform very significant roles in many research areas, such as text mining, text labels, natural language processing, text classification, and information retrieval. Recently, the ever increasing electronic documents which is becoming a challenging task to extract out the exact topics without reading the entire text documents. It provides a researcher or reader an advantage to find out what is going on through the journals and articles text documents. However, for multidimensional issues, it is a bit tedious for the reader or researcher to understand the point. If it is labeled correctly, it is better to understand it. There are various algorithms for topic modeling and text clustering. The most popular are LDA [3], LSA [5], NMF [11], pLSA [7], and HDP [17] and K-means cluster [12]. We have chosen popular topic modeling and text clustering technique, i.e., latent Dirichlet allocation (LDA) and K-means cluster respectively are handled to extract latent topics and clusters of documents. These techniques can automatically identify the

S. Rahman (✉)
Department of Software Engineering, Daffodil International University, Dhaka, Bangladesh
e-mail: shadikur35-988@diu.edu.bd

U. A. Koana
Department of Electronics and Communication Engineering, Khulna University of Engineering and Technology, Khulna, Bangladesh

A. M. Ismael
Department of Information Technology, Sulaimani Polytechnic University, Kurdistan, Iraq
e-mail: Aras.masoud@usa.com

K. H. Abdalla
Department of Software Engineering, University of Raparin, Ranya, Iraq
e-mail: karmand.hussein@uor.edu.krd

© Springer Nature Switzerland AG 2021   65
R. Stahlbock et al. (eds.), *Advances in Data Science and Information Engineering*,
Transactions on Computational Science and Computational Intelligence,
https://doi.org/10.1007/978-3-030-71704-9_4

abstract topics that occur in a collection of text documents. Topic modeling and text clustering are two widely studied problems that have many applications. Text clustering strives to organize similar documents into groups, which is essential for text document organization, text labels, summarization, classification, and retrieval. Topic modeling develops probabilistic generative models to discover the latent semantics embedded in document collection and has illustrated widespread success in modeling and analyzing text documents.

In our previous studies [8] and [16], we proposed a model to find generic labels for the polynomial topics over short text documents, in which we have used only LDA model to generate topic models and also proposed to find out automated labels with the description for the perennial topics in short online text documents. We have used the same model for generating the generic label and used LDA, LSI, and NMF for training the model. We have operated these models over the short text documents and measure the WUP similarities of each label. By comparing the WUP similarities for each model, we come to the conclusion that we can find the exact labels by using the LDA model. Though we have done this experiment in short text documents, it can be done on large texts or documents. Correspondingly, we have used a unique technique of topic modeling and text clustering to solve text document problems for automated labeling. We have used LDA and K-means cluster for training the model and also used Word2vec [6] to classify the vectors of similar words together in vector space. We have operated these models over the large-scale journals and articles text documents and measure the WUP similarities of each label. By comparing the WUP similarities for each model, we have come to the conclusion that using the technology, we can label properly.

The rest of the paper is organized as follows: Related Work is discussed in Sect. 2 followed by Research Methodology and Result and Discussion in Sects. 3 and 4, respectively. And finally, Sect. 5 concludes the summary.

2 Related Work

The efficiency of topic labeling to automatically assign labels to topics is described by Mei et al. [14] for LDA topics which were an unsupervised approach. In their approach, they showed potential ways to automatically label multinomial models in objective ways. In [16] the topic model is used to create generic labels and LDA, LSI and NMF to train the model. They conducted this procedure on short text documents and measured the WUP similarity of each label. By comparing the WUP matches for each model, we conclude that we can make the correct labels using the LDA model. In [19], the proposed multi-grain clustering topic model (MGCTM) which integrates text clustering and topic modeling together performs the two tasks to achieve the overall best performance. In [10] is presented an approach for topic labeling. Primarily, they created a set of candidate labels from the top-ranking topic terms, titles of Wikipedia articles including the top-ranking topic terms, and subphrases extracted from the Wikipedia article titles. Then finally, they ranked the label candidates using a sequence of association measures, lexical features, and an information retrieval feature. Natural language processing (NLP) is also used for topic modeling and labeling method. Using Word2vec word embedding, [1] labeled online news considering the article terms and comments. In this work, [13] proposed a method for labeling topics induced by a hierarchical topic model. Their label candidate set is the Google Directory service (gDir) hierarchy, and label selection takes the form of ontological alignment with gDir. In a recent work, [9] proposed to approach topic labeling via best term selection, selecting one of the top ten topic terms to label the overall topic. In [2], the author presented an approach to select the most relevant labels for topics by computing neural embedding for documents and words.

The phrase-based labels in the above works are still pretty short and are sometimes not sufficient for interpreting the topics. Though several approaches had been proposed in various studies, as per our knowledge, no proof is present for assessing the models for generating topic labels.

3   Research Methodology

In this section, the entire process of our research activities has been described. In the first instance, we have selected our journals and articles datasets.[1] For the dataset, we have selected some online journals and articles documents to perform our research process. As preparation, we have to take several steps to complete the process such as text preprocessing, chunking and N-gram, Word2vec, training model, WordNet process, noun phrase selection, and label processing with the help of WordNet synset. Then we receive topic labeling based on our topic modeling and text clustering. We transfer a retrieve responsibility to compare three topic representations: (1) clustering result, (2) topic labels, and (3) keyphrases extraction. Figure 1 shows an overview of our research experiment.

Fig. 1 Research methodology

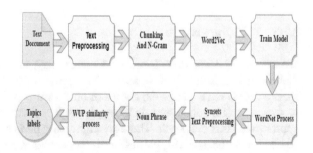

---

[1] https://github.com/sadirahman/Estimating-the-Effective-Topics-of-Articles-and-journals/tree/master/Datasets.

Fig. 2 Text preprocessing
process

3.1   Text Preprossessing

To analyze our text document datasets, we preprocess unsupervised journals and articles text documents. In this research, we have used online journals and articles. In most of the cases, we realize that text data is not effectively cleaned. For cleaning the text data, text preprocessing is required. For doing the preprocessing, we need to follow several steps like tokenization, stop words removing, POS tag, lemmatizing, and removing punctuation. Figure 2 shows an overview of our text document preprocessing process.

3.1.1   Text Tokenization Process

We create a bag of words in our text documents and divide them into punctuation marks. However, we make sure that small words such as "don't," "I'll," and "I'd" remain as one word. Splitting the provided text into smaller portions is called a token.

3.1.2   Stop Words Removing

Stop words are common words in a language like "almost," "an," "the," "is," and "in." These words have no significant meaning, and we usually remove words from our text documents.

3.1.3   POS Tag Process

We used the NLTK word tokenizer to parse each text into a list of words. After that text tokenization, then we used POS tagging in NLP using NLTK. The parts of speech (POS) tag explains the corpus of how a word is used in a sentence. POS tag is separated into subclasses. POS tagging solely means labeling words with their appropriate parts of speech. There are eight main parts of speech: nouns, pronouns, adjectives, verbs, adverbs, prepositions, conjunctions, and interjections.

3.1.4 Lemmatizing Process

We used the lemmatizing process of decreasing words to their word lemma, base, or root form, for example, worked–work, loved–love, roads–road, etc. c

3.2 Chunking and N-Gram Process

After preprocessing, we use the N-gram and chunking process to sort the sentence sequence of N words and extracting phrases from our text documents.

3.2.1 Chunking Process

We used the chunking process of extracting phrases from unstructured text documents. Chunking works on top of POS tagging, it uses POS tags as input and provides chunks as output. We search for chunks corresponding to an individual some POS tag phrase. The journals and articles include many parts of speech that are irrelevant to detect semantic orientation in our case. We consider only adjectives (JJ), verb (VB), adverb (ADV), and noun (NN) parts of speech (POS) tag from Penn Treebank annotation. For example, "I think the experience is wrong now." Figure 3 shows an overview of our chunking process.

3.2.2 N-Gram Model Process

An N-gram is a sequence of N words, which computes p(w–h), the probability of a word w* [4]. We have used the N-gram model using the journals and articles text documents held-out data as we used for the word-based natural language process model we discussed above in our research. The purpose of using this is to maintain the sequence of candidate labels in our text document. For example, if we consider

Fig. 3 Overview of chunking process

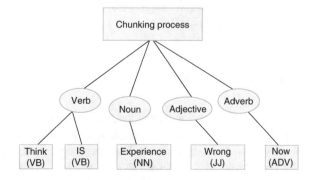

Table 1 Process of an N-gram	2-gram (bigram)	3-gram (trigram)
	"Please enter"	"Please enter your"
	"Enter your"	"Enter Your abstract"
	"Your abstract"	"Your abstract text"
	"Abstract text"	——

Table 2 Process of Word2vec	Similar words (cluster)	Numerical representations
	Corruption	0.82
	Wheel	0.82
	Weapons	0.82
	Ministration	0.81
	Blast	0.81
	Sinners	0.80
	Covert	0.80
	Bird	0.80
	Necessity	0.80
	Transgression	0.79

the text "Please enter your abstract text." For this text, the 2-gram and 3-gram words are presented in Table 1.

3.3 Word2vec Model Process

After the previous process, we used the Word2vec method in our research work. Word2vec gives direct access to vector representations of journals and articles text words. Word2vec [6] is used to classify the vectors of similar words together in vector space. It recognizes similarities mathematically. Word2vec produces vectors that are distributed numerical representations of word features, features such as the context of individual words. Word2vec model is based on journals and articles text documents and gensim Word2vec module. We used Word2vec for the most similar word-finding parameter "topn = 10." For example, the word "cluster" is a vector representation in Table 2

3.4 Train Model

In this research, we used the most popular text clustering and topic modeling algorithms K-means cluster and LDA. In order to train our models, we must first need journal and article documents. Each topic model and text cluster is based on the same basic role: each document contains a mixture of topics, and each cluster

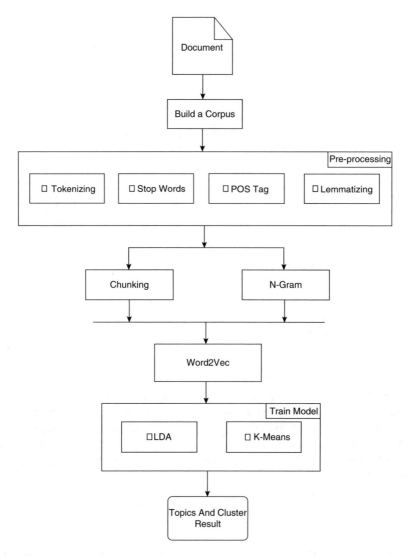

Fig. 4 Train model

is composed of a set of words. As a result, the purpose of the topic modeling and text clustering algorithms is to uncover these latent topics and cluster variables that shape the meaning of our text documents and training text documents. Figure 4 shows an overview of our research training models.

Fig. 5 Process of LDA model

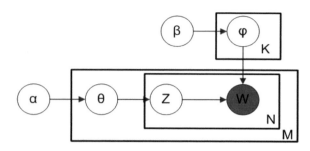

Table 3 Result of an example Document S1 in LDA model

Topic 1	Topic 2	Topic 3
0.071*"topic"	0.048*"nmf"	0.065*"document"
0.046*"algorithm"	0.047*"used"	0.045*"topic"
0.034*"document"	0.046*"lsi"	0.037*"modeling"

3.4.1 Latent Dirichlet Allocation (LDA)

Latent Dirichlet analysis is a probabilistic model to obtain cluster assignments. LDA utilizes Dirichlet priors for the document–topic and word–topic distributions, and it utilizes two probability values: P (word—topics) and P (topics—documents) [3]. Figure 5 shows an overview of our LDA models where K denotes the number of topics, M denotes the number of documents, N is the number of words in a given document, and w is the specific word.

To perform the LDA model works by defining the number of "topics" that are begun in our set of training text documents. Now, we will present the model output below: here we have chosen the number of topics = 3 and number of words = 3. We have also set the random state = 1 which is enough for our journals and articles text documents. Table 3 presents the result of training LDA model after executing Document S1.

3.4.2 K-Means Cluster

K-means clustering [12] is a method usually used to automatically partition a data set into k groups. It is an unsupervised learning algorithm. K-means is to minimize the total sum of the squared distance of every point to its corresponding cluster centroid. K-means clustering aims to partition n observations into k clusters in which each observation belongs to the cluster with the nearest mean, serving as a prototype of the cluster. Figure 6 shows process flow of K-means clustering algorithm.

We train the clustering algorithm for over our text document to split it into word groups. To perform the K-means cluster algorithm begins by defining the number of "clusters" that is begun in our set of training text documents. Now we will present the model output below: here we have chosen the number K = 3 and number of words = 3. We have also set the n-init = 1 and max-iter = 100 which is enough for

Fig. 6 Process flow of
K-means cluster

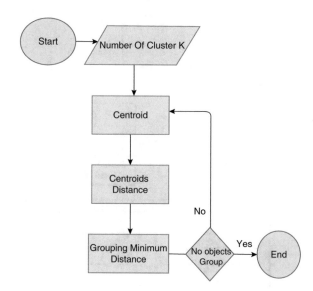

Table 4 Result of an
example Document S1 in
K-means cluster

Cluster 1	Cluster 2	Cluster 3
Lda	Documents	Topics
Nmf	Popular	Meaning
Used	Lsi	Select

our journals and articles text documents. Table 4 presents the result of training K-means cluster algorithm after executing Document S1.

3.5   WordNet Process

WordNet [15] is a comprehensive lexical database of English. Nouns, verbs, adjectives, and adverbs are classified into sets of cognitional synonyms (synsets), each meaning a distinct idea. It partly compares a dictionary, in that it classifies words respectively based on their suggestions. However, there are any significant variations. First, it interlinks not just word that makes sequences of words but special functions of words. As a result, words that are seen in near concurrence to one extra in the system are semantically disambiguated. Second, it specifies the semantic relationships between words, whereas the classification of words in a dictionary does not match any specific design other than discovering the identity. For doing the WordNet process, we need to follow several steps such as synsets text preprocessing, noun phrase process, and WUP similarity process. Figure 7 shows an overview of our text document WordNet process for keyphrases extraction.

In topic modeling and text clustering results in most comprehensive relevant words and next words in WordNet term. This WordNet term gives a word definition

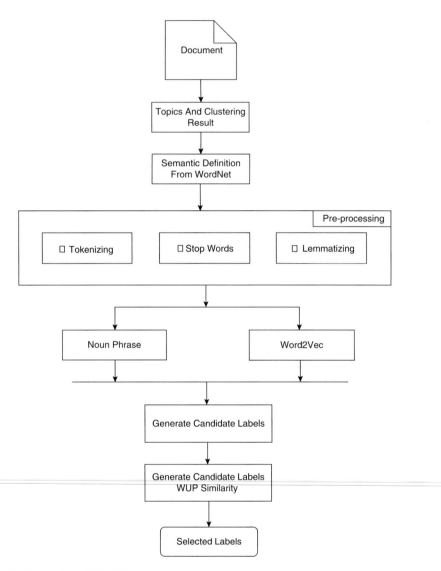

Fig. 7 Process flow of WordNet process

in our selected top-weighted word. Suppose our top-weighted selected word is "algorithm," then WordNet synset gives a definition "S: (n) algorithm, algorithmic rule, algorithmic program (a precise rule (or set of rules) specifying how to solve some problem)." Then we started again preprocessing in our WordNet definition and also pick up the noun and proper noun phrase.

Table 5 Result of noun phrase words

POS tagging words	Only noun words
LDA, K-means, Used,	LDA, K-means, Topic,
Topic, Select, Transportation	Transportation

Table 6 WUP similarity labels process

Document (S1)	Words	Top weighted word(s)	Candidate labels	Labels
Topic 1	Topic	Topic	Matter, Conversation,	Conversation
	Algorithm		Discussion, Situation,	
	Document		Event	

3.5.1 Noun Phrase Process

After synsets text preprocessing, we have only picked the noun and proper noun from the preprocessed result. Applying this approach, the topic is taken by top noun words with the largest frequency in the text corpus. For noun phrase choosing, first, the tokenization of text is executed to lemma out the words. The tokenized text is then tagged with parts of speech NN (nouns), NNP (proper nouns), VB (verbs), JJ (adjectives), etc. Before lemmatizing and stop words removing, parts of speech (POS) tagging is done. The stop words are removed after POS tagging. In the final stage, words including their tags and rounds are put in a hash table, and most solid nouns are obtained from these to create a heading for a text. The results of noun phrase words are presented in Table 5.

3.5.2 WUP Similarity Labels Process

After the noun phrase process, we take topics and clusters from the documents we only get top-weighted words from the topic and cluster because its result is well for its topic set and cluster set. Then we explore the semantic definition from the lexical database for English parts of speeches which are termed WordNet. We choose the definition description because within its phrases we find the candidate labels. After preprocessing the definition, we prepare the candidate labels, and from those candidate labels, we estimate the candidate labels with the main topic word for conceptual semantic relatedness measure by WUP similarity [18]. Then we begin in our WUP similarity process for labeling. The results of WUP similarity labels process are presented in Table 6.

4   Result and Discussion

We estimate the text clustering and topic modeling performance of our model based on journals and articles text documents. In this section, we consider the overall results of our research work. We have picked the top three words for each model and obtained the top word considering the highest weighted value. Then we obtain the definition of top-weighed word using the WordNet synsets. After that, we preprocess the found definition and choose the candidate labels of each definition by using noun phrase and Word2vec.

Finally, we estimate the keyphrases (label) comparing the WUP similarities accuracy between each candidate label and the top-weighted word. Tables 7 and 8 are showing the details of selecting the labels of three journals and articles text

Table 7 Label of chosen three documents with LDA

Document(s)	Topic(s) Topic	word(s)	Top weighted word(s)	Candidate label(s)	Label
S1	Topic 1	Topic, Algorithm, Document	Topic	Matter, Conversation, Discussion, Situation, Event	Conversation
	Topic 2	Use, Nmf, Lsi	Use	Employ, Service, Purpose, Act, Put	Service
	Topic 3	Document, Topic, Modeling	Document	Representation Information, Instruction, File, Obligation	Information
S2	Topic 1	News Misconception Avoiding	News	Information, Event Magazine, Newspaper Commentary	Information
	Topic 2	News, Headline, Increasing	News	Information, Event Magazine, Newspaper Commentary	Information
	Topic 3	Headline, News Sentiment	Headline	Caption, Newspaper Action, Story, page	Story
S3	Topic 1	Topic, word, Reproduced	Topic	Matter, Conversation Discussion, Situation, Event	Conversation
	Topic 2	Label, Topic labeling	Label	Identification, Description, Name, Mechanism Reaction	Identification
	Topic 3	Topic, Labeling, Idea	Topic	Matter, Conversation Discussion, Situation, Event	Conversation

Table 8 Label of chosen three documents with K-means cluster

Document(s)	Cluster(s)	Cluster word(s)	Top weighted word(s)	Candidate label(s)	Label
S1	Cluster 1	Topic, Study, Labeling	Topic	Matter, Conversation, Discussion, Situation, Event	Conversation
	Cluster 2	Documents Popular, Everyday	Document	Representation, Information, Instruction File, Obligation	Information
	Cluster 3	Algorithm, Lda, Topic	Algorithms	Rule, Solve, Problem, Set	Rule
S2	Cluster 1	News, Headlines Interactivity	News	Information, Event Magazine, Newspaper, Commentary	Information
	Cluster 2	News, Sentiment, Headlines	News	Information, Event Magazine, Newspaper, Commentary	Information
	Cluster 3	Sentiment, Accuracy, Achieves	Sentiment	Felling, Emotion Judgment, Tender, Belief	Felling (0.90)
S3	Cluster 1	Topic, Labels, Words	Topic	Matter, Conversation Discussion, Situation, Event	Conversation
	Cluster 2	User, Cognitional Understanding	User	Person, Someone, Use Thing, Drug	Person
	Cluster 3	Noun, Phrase, Proposed	Noun	Content, Person, Word, Preposition, Object	Preposition

documents by using topic modeling algorithm LDA and text clustering K-means algorithm, respectively.

Here in choosing the final label of each topic and cluster, WUP similarities values are used. The values of WUP similarities accuracy are shown in Tables 9 and 10 for models LDA and K-means cluster, respectively.

After choosing the topic and cluster labels, and then getting the WUP similarities values of each label, we average the WU similarities values. We can see in Table 11 that LDA model shows an accuracy of 63% and the k-means cluster model shows

Table 9 WUP similarity between topic and label with LDA

Document(s)	Top-weighted word(s)	Labels	WUP similarity	Average WUP
S1	Topic	Conversation	0.54	0.62
	Use	Service	0.80	
	Document	Information	0.54	
S2	News	Information	0.90	0.73
	News	Information	0.90	
	Headline	Story	0.40	
S3	Topic	Conversation	0.54	0.56
	Label	Identification	0.62	
	Topic	Conversation	0.54	

Table 10 WUP similarity between cluster and label with K-means

Document(s)	Top-weighted word(s)	Labels	WUP similarity	Average WUP
S1	Topic	Conversation	0.54	0.47
	Document	Information	0.54	
	Algorithm	Rule	0.35	
S2	News	Information	0.90	0.90
	News	Information	0.90	
	Sentiment	Feeling	0.90	
S3	Topic	Conversation	0.54	0.69
	User	Person	0.80	
	Noun	Preposition	0.75	

Table 11 WUP similarities difference among models

Models	Document sets	Average WUP	Total average
LDA	Document S1	0.62	0.63
	Document S2	0.73	
	Document S3	0.56	
K-means	Document S1	0.47	0.68
	Document S2	0.90	
	Document S3	0.69	

an accuracy of 68%. Both our models work together to give us the best results in our keyphrases extraction based on our journals and articles text documents.

5 Conclusion

In this research, we presented a unique technique of topic modeling and text clustering to solve text document problems for automated labeling. This technique observed that our proposed method is an application to find the relevant topic and cluster label for the polynomial topic in the text documents of our journals and articles. We have used topic modeling LDA and text clustering K-means algorithm to train our model. For each model, we have determined the top three words and picked the top-weighted word to obtain the definition of chosen top words and generate the candidate labels for each of the words using the WordNet synset of the lexical database. We have compared the WUP similarities between candidate labels and top words, and we chose the most accurate label for topics. This research can be broadly expanded into topics and labels for large-scale text documents, and a company text document can effectively solved the problem.

References

1. A. Aker, M. Paramita, E. Kurtic, A. Funk, E. Barker, M. Hepple, R. Gaizauskas, Automatic label generation for news comment clusters, in *Proceedings of the 9th International Natural Language Generation Conference* (2016), pp. 61–69
2. S. Bhatia, J.H. Lau, T. Baldwin. Automatic labelling of topicswith neural embeddings (2016). Preprint. arXiv:1612.05340
3. D.M. Blei, A.Y. Ng, M.I. Jordan, Latent Dirichlet allocation. J. Mach. Learn. Res. **3**, 99–1022 (2003)
4. P.F. Brown, P.V. Desouza, R.L. Mercer, V.J. Della Pietra, J.C. Lai, Class-based n-gram models of natural language. Comput. Linguist. **18**(4), 467–479 (1992)
5. S.T. Dumais, Latent semantic analysis. Annu. Rev. Inf. Sci. Technol. **38**(1), 188–230 (2004)
6. Y. Goldberg, O. Levy, word2vec explained: deriving Mikolov et al.'s negative-sampling word-embedding method (2014). Preprint. arXiv:1402.3722
7. T. Hofmann, Probabilistic latent semantic analysis, in *Proceedings of the Fifteenth Conference on Uncertainty in Artificial Intelligence* (Morgan Kaufmann Publishers, San Francisco, 1999), pp. 289–296
8. S.S. Hossain, M.R. Ul-Hassan, S. Rahman, Polynomial topic distribution with topic modeling for generic labeling, in *International Conference on Advances in Computing and Data Sciences* (Springer, Singapore, 2019), pp. 409–419
9. J.H. Lau, D. Newman, S. Karimi, T. Baldwin, Best topicword selection for topic labelling, in *Coling 2010: Posters* (2010), pp. 605–613
10. J.H. Lau, K. Grieser, D. Newman, T. Baldwin, Automatic labelling of topic models, in *Proceedings of the 49th Annual Meeting of the Association for Computational Linguistics: Human Language Technologies*, vol. 1 (Association for Computational Linguistics, Strouds-burg, 2011), pp. 1536–1545
11. D.D. Lee, H.S. Seung, Learning the parts of objects by non-negative matrix factorization. Nature **401**(6755), 788 (1999)
12. J. MacQueen et al., Some methods for classification and analysis of multivariate observations, in *Proceedings of the Fifth Berkeley Symposium on Mathematical Statistics and Probability, Oakland, CA*, vol. 1 (1967), pp. 281–297
13. D. Magatti, S. Calegari, D. Ciucci, F. Stella, Automatic labeling of topics, in *2009 Ninth International Conference on Intelligent Systems Design and Applications* (IEEE, Piscataway, 2009), pp. 1227–1232
14. Q. Mei, X. Shen, C.X. Zhai, Automatic labeling of multinomial topic models, in *Proceedings of the 13th ACM SIGKDD International Conference on Knowledge Discovery and Data Mining* (2007), pp. 490–499
15. G.A. Miller, Wordnet: a lexical database for English. Commun. ACM **38**(11), 39–41 (1995)
16. S. Rahman, S.S. Hossain, M.S. Arman, L. Rawshan, T.R. Toma, F.B. Rafiq, K.B.M. Badruz-zaman, Assessing the effectiveness of topic modeling algorithms in discovering generic label

with description, in *Future of Information and Communication Conference* (Springer, Cham, 2020), pp. 224–236

17. Y.W. Teh, M.I. Jordan, M.J. Beal, D.M. Blei, Sharing clusters among related groups: Hierarchical Dirichlet processes, in *Advances in Neural Information Processing Systems* (2005), pp. 1385–1392
18. Z. Wu, M. Palmer, Verbs semantics and lexical selection, in *Proceedings of the 32nd Annual Meeting on Association for Computational Linguistics* (Association for Computational Linguistics, Stroudsburg, 1994), pp. 133–138
19. P. Xie, E.P. Xing, Integrating document clustering and topic modeling (2013). Preprint. arXiv:1309.6874

Part II
Data Science, Social Science, Social Media, and Social Networks

Modelling and Analysis of Network Information Data for Product Purchasing Decisions

Md Asaduzzaman, Uchitha Jayawickrama, and Samanthika Gallage

1 Introduction

Due to the prodigious advancement of technologies and increasing competition, the life cycle of products is shortened. An important source of pre-purchase information is online media for many buyers nowadays. Online reviews, feedback and comments have been playing a significant role for customers to make a purchase decision [1, 2]. In a survey by PWC [3], it is found that 50% customers have used online media before making their purchases. Although there are multiple studies discussing the influence of online information on consumer purchase decision, the literature contradicts in establishing relationships.

Marketing literature suggests that consumers usually go through five steps while making a purchase decision. In the first step, a customer recognises the problem, the second step is the search of information, the third step is the customer evaluates the alternatives, the fourth step is the customer takes the product purchase decision, and the fifth step is the post-purchase behaviour [4, 5]. Therefore, a customer's purchase decision can be defined as the purchase intention of a product and loyalty to the product in the post-purchase behaviour. However, the discipline of operational research suggests a slightly different approach. Holtzman [6] introduced a decision-making model which consists of three stages: (1) formulation, (2) evaluation and (3) appraisal.

Md. Asaduzzaman (✉) · S. Gallage
Staffordshire University, Stoke on Trent, UK
e-mail: md.asaduzzaman@staffs.ac.uk, http://www.mdasad.com/

U. Jayawickrama
University of Nottingham, Nottingham, UK

© Springer Nature Switzerland AG 2021
R. Stahlbock et al. (eds.), *Advances in Data Science and Information Engineering*,
Transactions on Computational Science and Computational Intelligence,
https://doi.org/10.1007/978-3-030-71704-9_5

In each of the consumer decision-making process, digital media and information influence significantly to the process of purchase [4, 7], where customers start with their intention to purchase a product. In contrast, it is natural that loyal customers eliminate competitors from their consideration while making a purchase decision. "True loyalty" is a cognitive behaviour that leads to positive word of mouth (WOM) and repeats purchases [8, 9]. More diverse communication allows effective purchasing process. Recent studies show that social media helps organisations and customers to communicate dynamically and more effectively [10, 11].

Electronic word of mouth (eWOM), for instance, social media information, online product rating and reviews, plays a prominent role in customers' product choice [2, 12]. Online product reviews on the Internet and product information and rating posted on social media are found to be useful sources of online WOM communication [13]. Due to the great advancement of wireless communication and Internet of things (IoT) product information, reviews and rating on the Internet are found to be extremely useful when gathering pre-purchase product information and making purchase intentions by the customers [14]. eWOM communication provides consumers opinion, which is easily and quickly accessible [10]. Due to higher accessibility, the eWOM has been proven to be more effective than offline WOM [15].

As there has been increased availability of information, their accessibility and higher knowledge on consumer psychology, as a consequence, the market competition has also been growing exponentially. Nonetheless, these give more opportunities than ever to be exploited by the organisations to increase product performance, profit margins, etc. More specifically, organisations can easily identify factors affecting the product choice, product features customers likes the most, etc. Hence, the main purpose of this paper is to model these factors and analyse the consumers' purchase decision for product choices and quantify the effect of consumers' attitudinal factors. Specifically, we are mainly concerned with the following questions: (1) how to model and analyse the consumers' purchase decision for product choices and (2) how to quantify the effect of consumers' attitudinal factors, social factors, demographic factors and economic factors.

There are many model-based and algorithm-based analytical and machine learning techniques to extract social media information [16–18]. See the book [17], the review paper by Ferrara et al. [16] and references therein for the details of the methods and analysis. The influence of social media on the product purchase decision is a well-established fact [19–22]. The online product reviews are also known to have a significant impact on product purchase decisions [23–25]. Social media and online product reviews are the most popular and important information sources in digital age product purchase decisions [22–25]. However, consumers express their perception and views on social media and product reviews in many different forms for various products. The expressions can be in nominal, ordinal, ratio or interval scale of measurements, while some of the reviews are in text formats. Although many works have been done in the past, it yet remains unclear how we can model the consumers' complex attitude expressed in the social media in different forms, scales and dimensions.

Our aim in this paper is to demonstrate the modelling and analysis of the complexities of the consumers' purchase decision and quantify the effects of many factors related to consumer behaviour. We propose a novel framework to model and analyse the consumers' purchase decision for product choices using a network model and a utility-based discrete choice model (DCM) to quantify the effect of various factors related to consumers including attitudinal factors, which has received little attention in the literature. We consider two central information sources, i.e. social networks and online product reviews also referred to as eWOM along with some socio-economic and demographic factors. Our detailed simulation experiments show that the complexities of data structure measured in many dimensions and scales can be captured through a network model and convert them into some meaningful factors through our latent class analysis.

The rest of this paper is organised as follows. In Sect. 2, we review the different streams of literature relevant to our research. Section 3 presents the network model formulation, discrete choice model and parameter estimation. Section 4 provides some simulation results and discussion. In Sect. 5, we discussed some management insights and concluding remarks of the paper.

2 Literature Review

Our study builds on literature including marketing and operational research. We link these two to develop a model to build a relationship between digital information and consumer product purchase decision. According to marketing literature, decision-making is a complex cognitive process that involved multiple steps. Engel et al. [26] presented a recognised model of consumer purchase decision-making with five stages: (1) problem recognition, (2) information search, (3) evaluation of alternatives, (4) purchase decision and (5) post-purchase behaviour. According to operational research literature, widely used [6] model suggests three different stages such as formulation, evaluation and appraisal [27]. Information search (in marketing model) or appraisal stage (operational research) is the first step consumers actively seek for information. In these stages, consumers are motivated to activate their knowledge from the memory or acquire information from external sources [26, 28]. Memory-related information is the fundamental source of decision-making when a consumer is planning to purchase a product. However, when no such information is available from the memory or a customer is unable to reacquire the information from memory or if the information is not acquired previously, the customer relies upon obtaining information from the external sources [29].

In the external environment, online and offline information sources influence to a greater extent. While offline information sources are peers, family, friends and the company-generated information, online information sources are Internet-based platforms such as social media, web sites, chat rooms, blogs, etc. [30]. The motivation for external information search depends on many factors, for example, involvement, the need of cognition and the stage of decision-making process they are in, etc. Especially when it comes to high involvement products (such as electric cars), consumers would put more efforts in searching and processing information from various sources [31]. In the digital age, online information sources have become more significant and widely accessible than offline information sources [1, 2]. Although information search using internal and external sources has been properly established by researchers over the past years [32–34], majority of these studies have either focused on online or offline information sources only to a limited depth from the marketing point of view for product promotions. Thus, in this study, we focus on online information sources in-depth and to enable manufacturers to forecast their demand.

Social media information regarding products and services is perceived to be more trustworthy by consumers than corporate-sponsored communications transmitted via traditional information sources [35]. Therefore, the importance of social media marketing rapidly grew over the past few years. Social media technologies enabled both consumers and producers to create and distribute information easily. There are many of such established technologies, for example, *Wikipedia* for collaborative writing, *IoT* devices for content sharing (text, video and images) and *Facebook, Twitter, Delicious, LinkedIn and YouTube* for social networking, social bookmarking and syndication (e.g. ratings, tagging, RSS feeds). These are popular Internet-based applications built on the Web 2.0 technological platform and allow to generate and share user-generated contents [36]. Therefore, social media marketing includes both user-generated content and firm-generated content. It is a general understanding that the user-generated contents are stored online by users themselves for relatively reliable sites other than a few exceptions. The firm-generated content includes page publishing, stories, apps and advertisements by marketing organisations. These online information sources and offline sources are working as hybrid models to provide information to consumers.

The offline and online media activity can be categorised into three categories: paid media, owned media and earned media [37]. Paid media can be referred to as a media activity that a company and its agents generate either in online or offline channels. Owned media can be referred to as the media activity that a company and its agents generate in channels it controls such as press releases, brochures, posts, etc. Earned media are referred to the media activity that is generated by consumers as electronic word of mouth (eWOM) in online platforms. It is challenging for researchers and practitioners as how to use these factors to model consumer purchase decision-making. Yet, this is crucial in the current context.

The Internet penetration in the UK is 95% and social media penetration is 66% [38]. This has enabled the eWOM as a powerful communication source for consumer decision-making [39]. eWOM is defined as the statement (positive or negative) made by customers (potential, actual or former) about a product or an organisation, which is available via the Internet to the mass people and institutions [10]. Further, it is also known as the informal person-to-person communication done through online channels [39] and has become many-to-many communications between consumers via online platforms [40]. Consumers usually join online groups to seek for advice and information and exchange ideas specific to their interest, and this information has made a significant influence in their decision-making process. However, the level of interaction in these platforms depends on many factors, and based on the level of interaction, the speed and amount of eWOM exchange vary.

Having known the different sources of information on consumers' perception, views and attitudes toward a specific product alternative, it is a complex task to establish connections among those information and extract the actual significant factors of purchase decisions. This is also difficult given that modelling has received less attention in understanding consumer purchase decision [27]. Extensive investigations have been performed, and more efforts are continuing to gather and to make the best use of those optimised factors and other extracted information. Big data modellers, data analytics and marketing analytics experts have been making consistent efforts to model and make valid conclusions from the network-based information. Some notable efforts have been made recently [41–44]. Acemoglu et al. [41] proposed a utility-based analytical model to model the network connections and information exchange, while [42] performed a multidimensional network analysis for consumer preference modelling. Recently, a Bayesian social learning model has been proposed by Ifrach et al. [44] based on binary product reviews.

Although there have been some efforts on information extract from the web or social media and modelling and analysis of that information, to the best of our knowledge, there is no study to capture such social media data and convert them into suitable factors with consistent dimensions, scales and layers for further modelling and analysis in the context of consumers' purchase decisions. In the next section, we propose an information network model to gather attitudinal or factor information and consumers' purchase decision analysis using a discrete choice model.

3 Methodology

3.1 The Information Network Model Formulation

Consider a network $\mathcal{G}$, where a set of individuals $\{1, 2, \ldots, n\}$ are connected by a set of direct and indirect nodes. The jth individual in the network is identified by a vector consisting of a set of p exogenous characteristics $X_j = \{X_{j1}, X_{j2}, \ldots, X_{jp}\}$. The collection of characteristics of n individual is the matrix $\mathbf{X} = \{X_1, X_2, \ldots, X_n\}$. The network $G \in \mathcal{G}$ can be represented as an $n \times n$ matrix with elements being either 0 or 1. The entry of the matrix $g_{jk} = 1$ if individual j forms a connection with individual k and $g_{jk} = 0$ if not. We assume $g_{jj} = 0$ for any j, by convention. As the network is directed, $g_{jk} = 1$ does not necessarily implies $g_{kj} = 1$. Network links among the potential customers are depicted in Figs. 1 and 2. However, a potential customer gets information on his product information not necessarily from all his links.

Suppose that the total utility obtained by an individual j for the product alternative i from the network g with the population attribute $\mathbf{X} = \{X_1, X_2, \ldots, X_n\}$

Fig. 1 A typical network
with direct and indirect links

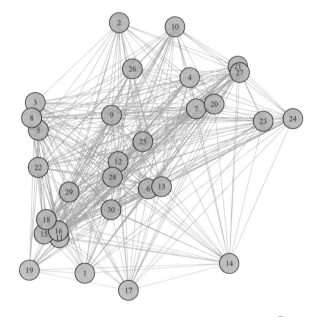

Fig. 2 Network connections
of the ith person

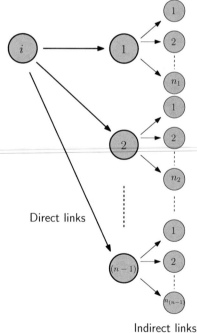

can be composed into two components: (1) a social network $\mathcal{G}^{(s)}$ and (2) a product
information network $\mathcal{G}^{(p)}$. Define $g_{jk}^{(h)}$ be the connection that individual j forms

with individual k for the network h, which takes the values $\{s, p\}$ corresponding to a social and product information network with parameter vectors (θ_u^s, θ_v^s) and (θ_u^p, θ_v^p), respectively. The amount of utility of the jth person for the ith product alternative obtain from a network h is denoted by $u_{ij}(g^{(h)}, \theta)$ and can be defined as

$$
u_{ij}(g^{(h)}; \boldsymbol{\theta}) = \begin{cases} \underbrace{\sum_{k=1}^{n} g_{jk}^{(s)} \cdot f(\theta_u^s)}_{\text{directlinks}} + \underbrace{\sum_{k=1}^{n} g_{jk}^{(s)} \sum_{\substack{l=1 \\ l \neq j,k}}^{n} g_{kl}^{(s)} \cdot f(\theta_v^s)}_{\text{indirectlinks}}; \ g^{(s)} \in \mathcal{G}^{(s)} \\[4ex] \underbrace{\sum_{k=1}^{n} g_{jk}^{(p)} \cdot f(\theta_u^p)}_{\text{directlinks}} + \underbrace{\sum_{k=1}^{n} g_{jk}^{(p)} \sum_{\substack{l=1 \\ l \neq j,k}}^{n} g_{kl}^{(p)} \cdot f(\theta_v^p)}_{\text{indirectlinks}}; \ g^{(p)} \in \mathcal{G}^{(p)} \end{cases}
$$

(1)

For the product alternative i, the total utility obtained from the two information networks—social network and product information network—by an individual j is assumed to be additive and can be written as

$$
\mathbf{u}_{ij}(g, \mathbf{Z}; \boldsymbol{\theta}) = \sum_{h \in \{s, p\}} u_{ij}(g^{(h)}, \mathbf{Z}; \boldsymbol{\theta})
$$

$$
= u_{ij}(g^{(s)}, \mathbf{Z}; \boldsymbol{\theta}) + u_{ij}(g^{(p)}, \mathbf{Z}; \boldsymbol{\theta}).
$$

(2)

However, the utility values from the social and product network cannot be observed directly and are, in general, latent variables. Therefore, they are described by the measurement model equations as

$$
\mathbf{u}_{ij}(g^{(s)}, \mathbf{Z}; \boldsymbol{\theta}) = \Lambda_s \boldsymbol{\xi}_s + \boldsymbol{\epsilon}_s
$$

(3)

and

$$
\mathbf{u}_{ij}(g^{(p)}, \mathbf{Z}; \boldsymbol{\theta}) = \Lambda_p \boldsymbol{\xi}_p + \boldsymbol{\epsilon}_p,
$$

(4)

where Eqs. (3) and (4) are latent factor model measurement equations with parameters Λ_s and Λ_p. The variables $\boldsymbol{\xi}_s$ and $\boldsymbol{\xi}_p$ are two sets of measured variables from social and product networks, respectively. These variables can be categorical and can be measured in nominal and ordinal scale and also can be interval and ratio scale.

3.2 The Discrete Choice Model

A discrete choice model is a utility-based choice model composed of two main components, namely, the observed exogenous variables (demographic and socio-economic) and accumulated utility based on the information obtained from the social networks and product information networks. The choice of a particular product alternative i for an individual j can be generated by the utility obtained from different networks and effect of the exogenous factors. We assume that an individual's choice is a single nominal indicator, and we express the choice as a function of the effect from the exogenous variables and total utility from the individual's information network. The model can be expressed as

$$
v_{ijl} = \sum_{j=1}^{n} \sum_{l=0}^{q} \beta_{jl} \, x_{ijl} + \sum_{h \in \{s, p\}} \beta_{ij}^{(h)} \, u_{ij}(g^{(h)}, \mathbf{Z}; \boldsymbol{\theta}) + \epsilon_{ijl}, \quad
\begin{cases}
i = 1, 2, \ldots m \\
j = 1, 2, \ldots n \\
l = 1, 2, \ldots q
\end{cases}
\tag{5}
$$

where v_{ijl} is the random utility of an individual j for the product alternative i, x_{ijl} denotes the lth endogenous variable of the individual j for the product alternative i, $u_{ij}(\cdot)$ is the expected utility of an individual j for the product alternative i from the hth information network, β_{jl} denotes the effects of the lth endogenous variables for the ith product alternative and $\beta_{ij}^{(h)}$ the effect of the hth information network of the jth individual for the product alternative i and ϵ_{ijl} is the error term distributed as $\sim$ Gumbel $(0, \sigma_{ij})$. The discrete choice model is developed under the assumption of an individual's utility-maximisation behaviour and can be written as

$$
Y_{ij} =
\begin{cases}
1, & \text{if } v_{ij} = \max_{i,j} \{v_{ij}\} \\
0, & \text{otherwise.}
\end{cases}
\tag{6}
$$

In matrix notation, the model for ith product alternative is written as

$$
\mathbf{P}(Y_i = 1) = \frac{\exp\left(\mathbf{X}'\boldsymbol{\beta} + \mathbf{u}'\boldsymbol{\beta}^h\right)}{1 + \exp\left(\mathbf{X}'\boldsymbol{\beta} + \mathbf{u}'\boldsymbol{\beta}^h\right)}.
\tag{7}
$$

The likelihood of the model can be written as

$$
\mathcal{L}(\mathbf{y}_n) = \int_{\mathbf{u}_{ij}} f\left(\mathbf{y}_{ij} \mid \mathbf{X}_{ij}, \mathbf{u}_{ij}^{(h)}\right) f(\mathbf{u}_{ij} \mid \mathbf{Z}_{ij}) \, d\mathbf{u}_{ij}
\tag{8}
$$

where $f\left(\mathbf{y}_{ijk} \mid \mathbf{X}_{ij}, \mathbf{u}_{ij}^{(h)}\right)$ is the likelihood of the n sample observation under the set of covariates $\mathbf{X}_{ij}$ and expected utility $\mathbf{u}_{ij}^{(h)}$. The function $f(\mathbf{u}_{ij} \mid \mathbf{Z}_{ij},)$ is the joint density of the latent utility function given the observed set of covariates $\mathbf{Z}_{ij}$.

3.2.1 Model Fitting and Parameter Estimation

To estimate the parameters of the model, we perform a two-stage estimation procedure. In stage 1, we estimate the latent utility variables with covariates $\mathbf{Z}_{ij}$. The method is analogous to factor score estimation with a mixture of categorical and continuous variables. At the end of stage 1, utility scores are calculated using the estimated coefficients. These utility scores are then passed to stage 2 with the assumption that these are fixed value explanatory variables for the full regression model. The rest of the parameters of the full model are estimated by the maximum likelihood method in stage 2.

4 Simulation Experiment and Discussion

To establish the model framework, we have taken the "electric car purchase decision" for our simulation experiment. Electric cars are becoming popular around the world despite their high price. Many countries are imposing laws and providing incentives to their citizens to popularise electric cars to reduce carbon emissions. The "electric car purchase decision" has received significant attention recently, and many factors are found to be involved in electric vehicle purchase decision [45–47]. Among the wide range of factors involved in the purchase decisions of an electric car, some crucial factors are highly influenced by social network information and online reviews (product information). Therefore, the "electric car purchase decision" is considered as one of the most suitable examples to check the performance of our proposed model framework.

In our model framework, we considered that a consumer's decision is influenced by social network data, in which their online reviews come along with their socio-economic and demographic variables. Previous studies [45–47] have reported that purchase price, cruising range, charging time, max. speed, charging point distance, battery life, environmental aspect and innovation aspect are the key influential variables for purchasing decisions. We have considered these variables for our study given in Table 1. However, the data structure is complex in types and characteristics. They can be nominal, ordinal, ratio and interval scaled variables due to the type of entries in the social media. We considered that the social media data, purchase price, cruising range, charging time, max. speed and charging point distance, are of the ordinal type and battery life, environmental aspect and innovation aspect are of the nominal type.

Table 1 List of variables for simulation experiment

Variable list		
Social network	Product network	DemographicSocio-economic variables
Purchase price	Purchase price	Age
Cruising range	Cruising range	Gender
Charging time	Charging time	Marital status
Max. speed	Max. speed	Educational level
Charing point distance	Charing point distance	Income level
Battery life	Battery life	Educational level
Environmental aspect		
Innovation aspect		

In our simulation, we have used parameter values that are consistent with previous studies. In the first step, 5 ordinal variables to represent purchase price, cruising range, charging time, max. speed and charging point distance and 3 nominal variables for battery life, environmental aspect and innovation aspect for $n = 100$ observations are generated. These data assumed to form the latent class measurement of social influence. Another set of 6 ordinal variables to represent purchase price, cruising range, charging time, max. speed, charging point distance and battery life are generated to obtain the latent class factor online product network. Finally, we have generated 5 categorical variables: age (> 30 or ≤ 30), gender (M or F), marital status (married or single), educational level (lower than/equal to A level or higher than A level) and income level ($\leq 30K$ or $> 30K$).

The full set of data are generated in such a way that a more likely response (favouring to purchase an electric car) is positive or given a higher rating (in ordinal scale) for all 8 items in set 1 (social network influence) and positive for all 6 items in set 2 (online product network). The socio-economic and demographic variable set is also generated according to having a positive and consistent impact as found in the literature. The correlation structures for variables considered in social network latent factor and online product information are displayed in Figs. 3 and 4, respectively. At the first stage of estimation, we calculate the scores of the latent social network factor and the score of latent online product information, which is considered as expected utility from these two factors using Eqs. (3) and (4), respectively. The direction of those variables on the two latent factors (social network factor and online product information) are shown in Figs. 5 and 6, respectively. In the second step, the discrete choice model in Eq. (5) is fitted using these two sets of utility scores along with the socio-economic and demographics factors. Higher values of the generated variables mean that the association between purchase decision and social network and online product information influence is stronger. The estimated parameter values in Table 2 indicate a strong positive impact of the considered variables in the study.

Fig. 3 Correlation circle of the numeric data (five variables) in the principal component axes for social network data

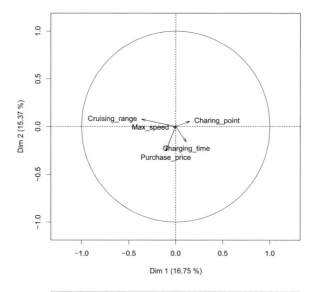

Fig. 4 Correlation circle of the numeric data in the principal component axes for product information data

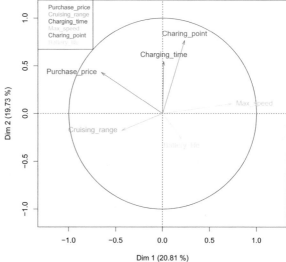

5 Conclusion

In this study, we proposed a framework to model and analyse the impact of network information on consumer product purchase decision: social media and online product information using a latent class approach. The nature of data obtained in social networks is highly complex and measured in several measurement scales. We proposed to model the expected random utility measure for each customer on their purchasing decisions based on data obtained from social networks. The unstructured

Fig. 5 Direction of variables in the principal component axes for social network data

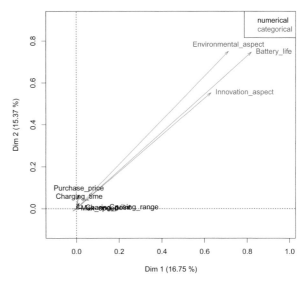

Fig. 6 Direction of variables in the principal component axes for product information data

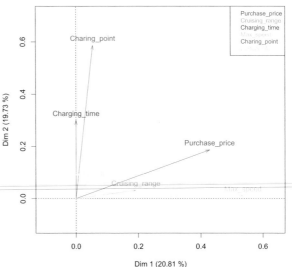

product preference data are converted into structured format along with their socio-economic and demographic characteristics. The measured expected utility for two latent factors social network influence and online product information with the socio-economic and demographic is then regressed with their purchasing decisions in the discrete choice modelling framework. The likelihood function method is used to fit the full model. Using the electric car purchase decision as a case study with simulated data, we showed that the model framework can find the significant factor explicitly. The model will be useful for decision-makers to forecast the demand by understanding how consumers make decisions [16, 18]. It also provides a yardstick

Table 2 Estimated
parameters, std. error and
p-values

Variables/factors	Estimate	SE	*p*-values
Intercept	−5.651	1.252	<0.0001
Age	2.176	0.699	0.0018
Gender	1.679	0.688	0.0147
Marital status	2.563	0.760	0.0008
Educational level	2.520	0.759	0.0009
Income level	3.373	0.843	<0.0001
Social network	0.063	0.023	0.0073
Product network	0.110	0.027	<0.0001

for marketers who make communication decisions. In the future, a more generalised approach of complex data extraction will be adopted, and each separate variable will be taken as the latent factor for analysis.

References

1. F.R. Jiménez, N.A. Mendoza, Too popular to ignore: The influence of online reviews on purchase intentions of search and experience products. J. Int. Mark. **27**(3), 226–235 (2013)
2. S. Prasad, A. Garg, S. Prasad, Purchase decision of generation Y in an online environment. Mark. Intell. Plan. **37**(4), 372–385 (2019). https://doi.org/10.1108/MIP-02-2018-0070
3. PWC: They say they want a revolution, total retail 2016 (2016); PWC (PricewaterhouseCoopers) company technical reort – Published by PWC, USA
4. P. Kotler et al., *Marketing management: The Millennium Edition*, vol. 199 (Prentice Hall, Upper Saddle River, 1999)
5. V. Zwass, K. Kendall, Structure and macro-level impacts of electronic commerce, in *Emerging Information Technologies: Improving Decisions, Cooperation, and Infrastructure* (Sage, Beverly Hills, 1999), pp. 289–315
6. S. Holtzman, *Intelligent Decision Systems* (Addison Wesley, Reading, 1989)
7. V.A. Zeithaml, L.L. Berry, A. Parasuraman, The behavioral consequences of service quality. J. Mark. **60**(2), 31–46 (1996)
8. V. Shankar, A.K. Smith, A. Rangaswamy, Customer satisfaction and loyalty in online and offline environments. Int. J. Res. Mark. **20**(2), 153–175 (2003)
9. L.C. Wang, J. Baker, J.A. Wagner, K. Wakefield, Can a retail web site be social? J. Mark. **71**(3), 143–157 (2007)
10. T. Hennig-Thurau, K.P. Gwinner, G. Walsh, D.D. Gremler, Electronic word-of-mouth via consumer-opinion platforms: what motivates consumers to articulate themselves on the internet? J. Int. Mark. **18**(1), 38–52 (2004)
11. C.H. Ho, K.H. Chiu, H. Chen, A. Papazafeiropoulou, Can internet blogs be used as an effective advertising tool? The role of product blog type and brand awareness. J. Enterp. Inf. Manag. **28**(3), 346–362 (2015). https://doi.org/10.1108/JEIM-03-2014-0021
12. M. Kulmala, N. Mesiranta, P. Tuominen, Organic and amplified eWOM in consumer fashion blogs. J. Fash. Mark. Manag. **17**(1), 20–37 (2013). https://doi.org/10.1108/13612021311305119
13. S. Sen, D. Lerman, Why are you telling me this? An examination into negative consumer reviews on the web. J. Int. Mark. **21**(4), 76–94 (2007)
14. F. Zhu, X. Zhang, Impact of online consumer reviews on sales: The moderating role of product and consumer characteristics. J. Mark. **74**(2), 133–148 (2010)

15. Chatterjee, Patrali, Online Reviews: Do Consumers Use Them?. ACR 2001 PROCEED-INGDS, M. C. Gilly, J. Myers-Levy, eds., pp. 129–134, Association for Consumer Research, 2001, Available at SSRN: https://ssrn.com/abstract=900158

16. E. Ferrara, P. De Meo, G. Fiumara, R. Baumgartner, Web data extraction, applications and techniques: A survey. Knowl. Based Syst. **70**, 301–323 (2014)

17. S. Hai-Jew, *Social Media Data Extraction and Content Analysis* (IGI Global, Hershey, 2016)

18. D. Gupta, S. Tripathi, A. Ekbal, P. Bhattacharyya, A hybrid approach for entity extraction in code-mixed social media data. Money **25**, 66 (2016)

19. X. Wang, C. Yu, Y. Wei, Social media peer communication and impacts on purchase intentions: A consumer socialization framework. J. Int. Mark. **26**(4), 198–208 (2012)

20. T. Powers, D. Advincula, M.S. Austin, S. Graiko, J. Snyder, Digital and social media in the purchase decision process: A special report from the advertising research foundation. J. Advert. Res. **52**(4), 479–489 (2012)

21. K. Goodrich, M. De Mooij, How social are social media? a cross-cultural comparison of online and offline purchase decision influences. J. Mark. Commun. **20**(1–2), 103–116 (2014)

22. C.M. Cheung, B.S. Xiao, I.L. Liu, Do actions speak louder than voices? The signaling role of social information cues in influencing consumer purchase decisions. Decis. Support. Syst. **65**, 50–58 (2014)

23. Y. Chen, J. Xie, Third-party product review and firm marketing strategy. Mark. Sci. **24**(2), 218–240 (2005)

24. J. Pickett-Baker, R. Ozaki, Pro-environmental products: marketing influence on consumer purchase decision. J. Consum. Mark. **25**(5), 281–293 (2008). https://doi.org/10.1108/07363760810890516

25. S. Jang, A. Prasad, B.T. Ratchford, How consumers use product reviews in the purchase decision process. Mark. Lett. **23**(3), 825–838 (2012)

26. J. Engel, R.D. Blackwell, P. Miniard, Consumer behavior, Chicago. Dry Den **12**, 251–264 (1995)

27. S. Karimi, K.N. Papamichail, C.P. Holland, The effect of prior knowledge and decision-making style on the online purchase decision-making process: A typology of consumer shopping behaviour. Decis. Support Syst. **77**, 137–147 (2015)

28. J.B. Schmidt, R.A. Spreng, A proposed model of external consumer information search. J. Acad. Market Sci. **24**(3), 246–256 (1996)

29. D. Gursoy, K.W. McCleary, Travelers prior knowledge and its impact on their information search behavior. J. Hosp. Tour. Res. **28**(1), 66–94 (2004)

30. L.R. Klein, G.T. Ford, Consumer search for information in the digital age: An empirical study of prepurchase search for automobiles. J. Int. Mark. **17**(3), 29–49 (2003)

31. T. Zhang, D. Zhang, Agent-based simulation of consumer purchase decision-making and the decoy effect. J. Bus. Res. **60**(8), 912–922 (2007)

32. D.H. Furse, G.N. Punj, D.W. Stewart, A typology of individual search strategies among purchasers of new automobiles. J. Consum. Res. **10**(4), 417–431 (1984)

33. S.E. Beatty, S.M. Smith, External search effort: An investigation across several product categories. J. Consum. Res. **14**(1), 83–95 (1987)

34. S. Moorthy, B.T. Ratchford, D. Talukdar, Consumer information search revisited: Theory and empirical analysis. J. Consum. Res. **23**(4), 263–277 (1997)

35. G. Foux, Consumer-generated media: Get your customers involved. Brand Strategy **8**(202), 38–39 (2006)

36. A.M. Kaplan, M. Haenlein, Users of the world, unite! The challenges and opportunities of social media. Bus. Horiz. **53**(1), 59–68 (2010)

37. M.J. Lovett, R. Staelin, The role of paid, earned, and owned media in building entertainment brands: Reminding, informing, and enhancing enjoyment. Mark. Sci. **35**(1), 142–157 (2016)

38. Number of daily internet users in Great Britain from 2006 to 2019. https://www.statista.com/statistics/275786/daily-internet-users-in-great-britain/. Accessed 18 Mar 2020

39. Y.Y.Y. Chan, E.W.T. Ngai, Conceptualising electronic word of mouth activity. Mark. Intell. Plan. **29**(5), 488–516 (2011). https://doi.org/10.1108/02634501111153692

40. M.R. Jalilvand, N. Samiei, The effect of electronic word of mouth on brand image and purchase intention. Mark. Intell. Plan. **30**(4), 460–476 (2012). https://doi.org/10.1108/02634501211231946
41. D. Acemoglu, K. Bimpikis, A. Ozdaglar, Dynamics of information exchange in endogenous social networks. Theor. Econ. **9**(1), 41–97 (2014)
42. M. Wang, W. Chen, Y. Huang, N.S. Contractor, Y. Fu, Modeling customer preferences using multidimensional network analysis in engineering design. Des. Sci. **2**, 1–28 (2016)
43. O. Besbes, M. Scarsini, On information distortions in online ratings. Oper. Res. **66**(3), 597–610 (2018)
44. B. Ifrach, C. Maglaras, M. Scarsini, A. Zseleva, Bayesian social learning from consumer reviews. Oper. Res. **67**(5), 1209–1221 (2019)
45. W. Sierzchula, S. Bakker, K. Maat, B. Van Wee, The influence of financial incentives and other socio-economic factors on electric vehicle adoption. Energy Policy **68**, 183–194 (2014)
46. J. Kim, S. Rasouli, H. Timmermans, A hybrid choice model with a nonlinear utility function and bounded distribution for latent variables: application to purchase intention decisions of electric cars. Transportmetrica A Transp. Sci. **12**(10), 909–932 (2016)
47. B. Junquera, B. Moreno, R. Alvarez, Analyzing consumer attitudes towards electric vehicle purchasing intentions in Spain: technological limitations and vehicle confidence. Technol. Forecast. Soc. Chang. **109**, 6–14 (2016)

Novel Community Detection and Ranking Approaches for Social Network Analysis

Pujitha Reddy and Matin Pirouz

1 Introduction

Real-world networks such as information, transportation, and social networks generate a voluminous amount of data in today's world [1–3]. Network analysis that involves community detection and ranking has become prominent in predicting the primary features and topology of these kinds of networks. Community is an important parameter that leads to identifying connections deep down the network. Ranking can also help in identifying the patterns that influence and cluster similar members of a network into a community. The ranking determines how important a node is in a network [4]. There are many ranking measures to find the importance of a node in a network such as degree centrality, closeness centrality, betweenness centrality, eigenvector centrality, Katz centrality, and PageRank centrality. All these centrality measures have their importance as well as limitations. Each works well only for probing certain phenomena. However, in most cases, none of them is complete enough to find the overall importance and influence [5]. Degree centrality is the simplest measure that reflects the number of immediate connections and ignores the influence of those direct links. Betweenness and closeness centrality are based on the shortest path [6, 7]. However, information flow is not always through the shortest path in real-world networks. Eigenvector centrality overcomes the degree centrality to a certain extent by capturing the influence of neighbors. However, it fails if the influence is passed to its neighbors. Katz centrality gives constant importance to all its neighbors but unnecessarily distributes its importance over the network if there are many outgoing edges to many neighbors [8]. In the case of undirected graphs, the page rank algorithm divides by out-degree, which

P. Reddy · M. Pirouz (✉)
Department of Computer Science, California State University, Fresno, CA, USA
e-mail: pujithareddys@mail.fresnostate.edu; mpirouz@csufresno.edu

© Springer Nature Switzerland AG 2021
R. Stahlbock et al. (eds.), *Advances in Data Science and Information Engineering*,
Transactions on Computational Science and Computational Intelligence,
https://doi.org/10.1007/978-3-030-71704-9_6

is equal to the number of neighbors which makes the node less important even if it is important [9]. Eigenvector centrality also makes nodes centrality zero even it has many neighbors and if a neighbor's degree is zero in sparse graphs [10]. A community (cluster) is a densely connected group of vertices, with only sparser connections to other groups [11]. A network is said to have community structure if the nodes of the network can be meaningfully grouped into sets of nodes such that each set of nodes is densely connected internally [12]. We need a numeric criterion of inside-ness and outside-ness. Something we can try to maximize. A quantity called modularity is often used to decide the best number of communities [6]. The main aim of this research is to create a new clustering and community ranking method that involves in identifying efficient clusters and most central nodes. Community detection is key to understand the structure of complex networks and extract useful information from network, and this project could be used to find optimal clusters [13].

The rest of this paper is organized as follows: "Methodology" introduces the proposed method and its applications. "Analysis" depicts the application of the proposed method to real-world datasets and comparing them with ground truths. Conclusion and future work discuss the outline of the proposed algorithm, its applications, and improvements that can be made in the future.

2   Literature Review

In this paper [14], a new type of network community in online social networks (OSNs) is identified using the association between network nodes. [author] Proposes a virtual community detection algorithm that uses the basics of link prediction to detect communities. Li at [2] says that link prediction works on a concept called information diffusion. Prime nodes: If a pair of nodes share two or more consistent neighbor sets, such nodes are called prime nodes. Virtual communities: The consistent set of neighbors shared by prime nodes along with prime nodes are identified as virtual communities. The main contribution of this research is to identify communities, by comparing the similar nodes connected with prime nodes. Link prediction completes the incomplete OSNs using association algorithms. This research is validated using centrality methods page rank and k-core before and after link predictions in OSNs. The research proposed here in [15] is mainly based on the graph theory. Community is defined as a group with more connections inside than connections outside. There is a possibility that a node belongs to more than one community at the same time. This is referred as community overlapping, which we often see in real-world networks such as social networks. Community detection starts with the properties of chordal graphs. First, the maximal cliques are identified and it is extended with nodes and edges. Later, it is optimized greedily by local clustering function. The proposed algorithm is bench marked using the ground truths of the networks experimented. K-means clustering is a classical algorithm proposed long back [16]. It always starts with choosing appropriate K, which always does not give efficient clustering. It works well for global clusters as it always assigns the nearest items in the radius. If the size of the network is huge, K-means is always sensitive to initialization. Affinity propagation is a new type of algorithm that was introduced in this research. Communities are clustered based on nearest neighbor influences a node has. A threshold of nearest neighbors delta is set and influence of a node is calculated. The list is sorted based on the influence and more influenced neighbors are clustered. A semi-supervised community detection algorithm is proposed [17] based on the concept of must-links and cannot link. The must-link nodes are started as a seed, and it is extended based on the transitive property. Nodes that are similar to must-link nodes make a cluster. Transitive property does not get hold here in cannot links. In the social network, the centrality plays an import role in the graph. Finding the centrality could help to improve the accuracy in classification data mining techniques. Hui [8] studied the correlation of degrees and betweenness centrality to investigate BBS reply networks. And the result showed central nodes with high degree or high betweenness centrality do have high influence and power in online social networks. On the other hand, when central nodes with highest degree and highest betweenness centrality are removed, network centralizations decreased typically. Hajibagheri et al. [9] provided an idea to use centrality metric such as degree, eigenvector, betweenness, and closeness to define the most central state within a country and use this information to design the road/rail transportation networks. Girvan and Newman [18] presented to use a new metric, namely, cross-closeness centrality, for measuring the multiplex social network and simple network. The datasets were the families which were from two different areas (Danio Rerio and Florentine). After analyzing, the data showed multiplex network offers much valuable and concrete information compared to the simple network. Prantik et al. [5] observed the influence users from Twitter use degree centrality and eigenvector centrality to collect the data. The result showed that indegree and eigenvector centrality should both be considered when finding users who are influential.

Sudhakaran and Renjith [19] presented a new centrality which combined betweenness centrality and Katz centrality to measure the importance of node. This new centrality not only reduced the problem of betweenness centrality which only focused on the shortest path but also solve the problem of Katz centrality which focused on the adjacent nodes. Lingjie [12] gave a new centrality which depended on the betweenness changes caused by the removal of the largest node in the network. This method was useful to identify the functional and structural importance of the nodes in a network.

3   Methodology

Clustering coefficient explains how connected the neighbors are to each other. Mathematically, clustering coefficient of a node i is

$$CC_i = \frac{2|e_{jk} : V_j, V_k \in N_i, e_{jk} \in E|}{K_i * (K_i - 1)} \qquad (1)$$

Local clustering coefficient of a node in a graph describes how connected its neighbors are where e_{jk} represents the edges among neighbor set V_j, V_k, N_i is the neighborhood of vertex i, K_i represents the degree of node i, and E represents edge set of graph G. If an important node points to many less important nodes unnecessarily, other nodes also get importance, so page rank divides it by out-degree which in case of undirected network is not justified. So, we use clustering coefficient and influence as measures and propose a new centrality method.

Algorithm 1: Influence and clustering centrality

---

 1 Input: *Directed Graph , InfluenceLimit*
 2 Output: *Centrality Matrix*
 3 **for** $i \leftarrow 1$ **to** n **do**
 4 | find cc(i)
 5 | Set influence weight of each node to cc[i]
 6 **end**
 7 **for** $i \leftarrow 1$ **to** n **do**
 8 | Empty open/close list
 9 | Set all nodes to be unexplored
10 | OpenList-PushStack (Node(i))
11 | Set Node(i).depth = 0
12 | **while** *OpenList is not empty* **do**
13 | | currNode = OpenList-PopStack ( )
14 | | Node(i).influence Vector(currNode) += (currNode.depth)-2
15 | | Pop all nodes in CloseList with depth $\geq$ currNode.depth
16 | | Set those nodes to be unexplored;
17 | | **if** *currNode.depth < influenceLimit* **then**
18 | | | **for each** out-link neighbor j of the currNode **do**
19 | | | | **if** *Node(j) is unexplored* **then**
20 | | | | | OpenList-PushStack(Node(j))
21 | | | | | Set Node(j).depth = currNode.depth + 1
22 | | | | **end**
23 | | | **end**
24 | | Set currNode.isExplored = True
25 | | CloseList-PushStack (currNode)
26 | **end**
27 | **end**
28 | M(i) = Node(i).influence Vector
29 **end**
30 return M

Here we propose community detection based on the common neighbor and influence they have in the network [13]. This influence centrality indicates if two nodes are close to a common set of neighbors, then they both are closer to each other. This makes the nodes stay in one community. To find the common neighborhood,

Dataset	Sparse/dense	Weighted/unweighted
Karate Club	Dense	Unweighted
Dolphins	Sparse	Unweighted
Les Miserables	Dense	Unweighted
Football	Sparse	Unweighted
Political blogs	Dense	Unweighted
Neural networks	Dense	Unweighted

Fig. 1 Community ranking
for Karate Club

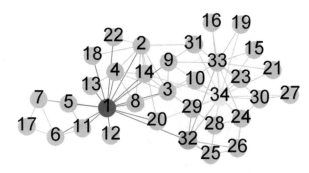

we can make use of the centrality measures. An influence matrix will be generated. First, the influence weight for all the nodes is set to 1. Every level updates the influence weight with the clustering coefficient.

4 Results

Karate Club:

In the Karate Club dataset, Node 8 has CC = 1.0, which means all the neighbor nodes are connected and it forms a clique, where any information easily spreads at this part of the network. Node 8 has important nodes in terms of other centralities 1,3 connected to it, which means it can easily pass information to other nodes as it forms a clique around it. It is also pointing to only four nodes, and it justifies the problem that if an important node is pointing many other nodes, the other is getting importance unnecessarily. It is also close to many important nodes such as instructor which has high connectedness. The new centrality measure is defined by taking some centralities and ranking methods, so the central nodes of this ranking include the nodes in order which have a high clustering coefficient and neighbors with high connectedness, and this ranking will not give importance which it has unnecessarily by pointing to many nodes (Fig. 1 and Table 1).

In Dolphin Community, there are no nodes with clustering coefficient 1, which means connectedness among neighbors in the Dolphin Community seems less compared to other datasets taken. Notch has CC = 0.6 which forms a partially

Table 1 Clustering
coefficients of top three nodes
in Karate Club

Node	CC
8	1.0
13	1.0
15	1.0

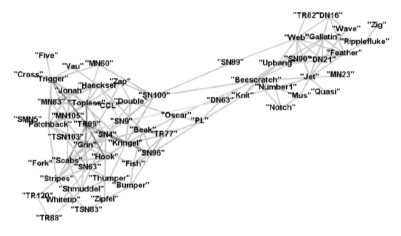

Fig. 2 Community ranking for Dolphin Community

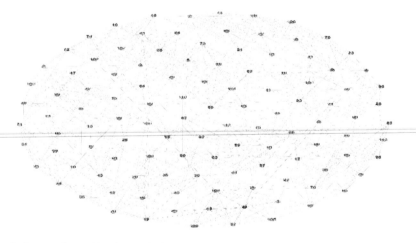

Fig. 3 Community ranking for Football dataset

connected graph around it. The neighbors of the Notch are Grin and SN4 which are highly connected nodes. The hook is close to the connected nodes, has a high clustering coefficient, and is not making less important nodes unnecessarily important (Figs. 2 and 3).

In football matches conducted, almost all teams that played with Wake Forest have played with each other. So Wake Forest team has a high clustering coefficient.

Table 2 Clustering coefficients of top three nodes in Dolphin Community

Node	CC
Mus	0.66
Notch	0.63
Hook	0.6

Table 3 Clustering coefficients of top three nodes in Football dataset

Node	CC
Wake Forest	0.66
Virginia	0.64
Clemson	0.62

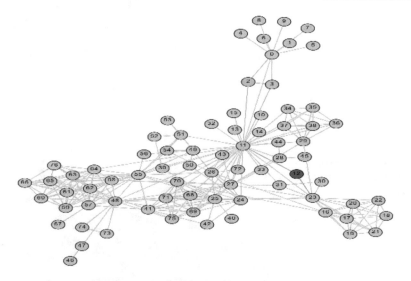

Fig. 4 Community ranking for Les Miserables dataset

Wake Forest played with teams that played the highest games in the tournament. It played with very few less popular teams which mean a team that played a very few times (Tables 2 and 3).

In Les Miserables' novel, Marguerite is the central character according to new centrality measure. That is, Marguerite made good connections, i.e., clique around it. It also appeared with the most popular character of the novel Valjean. It also appeared with Enjolras, Courfeyrac, and Fantine which also appeared with important characters (Figs. 4 and 5).

In Fig. 6 The clusters are identified with different colors. The nodes with high degree (1, 33, 34), high closeness (1, 3, 34), and eigen (1, 3, 34) are one cluster away (1 is one cluster away from 33, 34). The two central nodes 33, 34 are in the same cluster, so the fourth cluster might be the important community. Homophily seems to be the centrality and closeness (Tables 4, 5, and 6).

Fig. 5 Community ranking
for political blogs

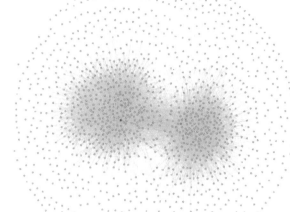

Fig. 6 Visualization of
Communities in Karate Club
dataset

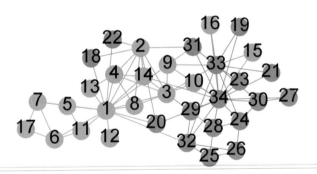

Table 4 Clustering
coefficients of top three nodes
in Les Miserables play

Node	CC
MlleBaptistine	1.0
MmeMagloire	1.0
Marguerite	0.90

Table 5 Clustering
coefficients of top three nodes
in pol blogs

Node	CC
angryhomo.blogspot.com	1.0
arkansastonight.com	1.0
battlegroundstates.blogspot.com	1.0

In Dolphins, community homophily might be centrality. Most of the central
nodes discovered by centrality methods are in clusters 3, 4, and 2. This says those
three clusters are the most important communities in the network as shown in Fig. 7

Table 6 Modularity values of detected communities

Dataset	Max modularity
Karate Club	0.256
Dolphins	0.334
Lesmis	0.421
Football	0.597
Celegan neural networks	0.781
Pol blogs	0.672

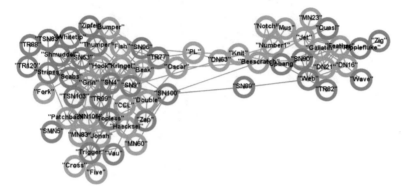

Fig. 7 Visualization of communities in Dolphin dataset

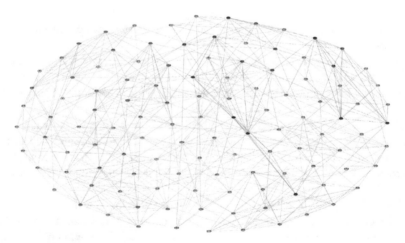

Fig. 8 Visualization of communities in Football dataset

In Fig. 8, the homophily of clusters formed might be geographical location because cluster 1 (Nevada, California, Washington) have all the teams from western states and cluster 2 (Florida, Virginia, Carolina) have all the eastern states, cluster 3 (Michigan, Illinois, Minnesota) have all the northeastern states, and cluster 4 (Texas,

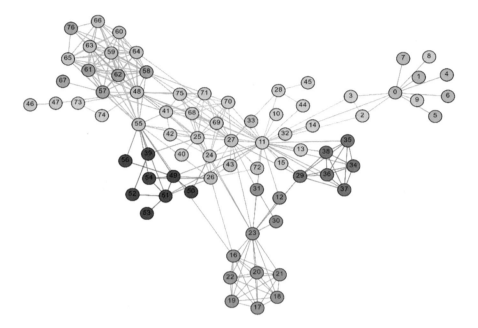

Fig. 9 Visualization of communities in Les Miserables dataset

Oklahoma) have all the southeastern states There are ten clusters in the set with maximum modularity.

In Fig. 9, in this novel, all the important characters, Gavroche, Enjolras, Courfeyrac, and Marius, belong to the same community which makes the community more central in the entire network. Here also homophily is centrality. There are four clusters in the set with maximum modularity.

Figure 10 explains the connectivity of neurons of nematode Caenorhabditis elegans. All of these neurons are clustered based on their connections and the functions they perform. The clusters are identified as sensory receptors, interneurons, motor neurons, and amphids. Homophily here is the local transmission of signals in the network.

Figure 11 depicts the network of links between US political web blogs. However, the clusters are not clear, the reason being it is densely connected. It is clear from the picture that all the blogs are connected in the center. The communities explain that there is high betweenness centrality.

5   Conclusion

The outcome of this research is to improve ranking and detection algorithms in communities. We addressed the issue by explaining the importance of local

Fig. 10 Visualization of
communities in neural
network dataset

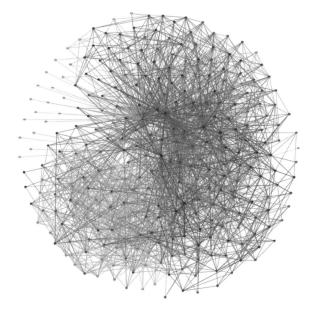

Fig. 11 Visualization of
communities in political
blogs dataset

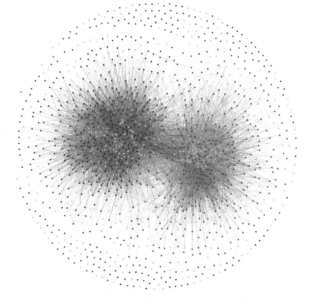

influence and the number of cliques. Modularity is used to measure the efficiency
of the communities detected. The modularities of identified clusters are positive
values and varied between 0.25 and 0.63. This shows that the communities identified
are optimal. In the future, we will apply these algorithms on highly dynamic and
complex datasets to find the efficiency and load of the proposed model.

References

1. M. Pirouz, J. Zhan, S. Tayeb, An optimized approach for community detection and ranking. J. Big Data **3**, 22 (2016)
2. L. Zhou, Y. Zeng, Y. He, Z. Jiang, J. Ma, Multi-hop based centrality of a path in complex network, in *Proceedings of the 2017 13th International Conference on Computational Intelligence and Security (CIS)* (IEEE, New York, 2017), pp. 292–296
3. Z. Song, H. Duan, Y. Ge, X. Qiu, A novel measure of centrality based on betweenness, in *Proceedings of the 2015 Chinese Automation Congress (CAC)* (IEEE, New York, 2015), pp. 174–178
4. U. Kang, S. Papadimitriou, J. Sun, H. Tong, Centralities in large networks: Algorithms and observations, in *Proceedings of the 2011 SIAM International Conference on Data Mining* (SIAM, New York, 2011), pp. 119–130
5. A.P. Naik, S. Bojewar, Tweet analytics and tweet summarization using graph mining, in *Proceedings of the 2017 International Conference of Electronics, Communication and Aerospace Technology (ICECA)*, vol. 1 (IEEE, New York, 2017), pp. 17–21
6. P. Howlader, K. Sudeep, Degree centrality, eigenvector centrality and the relation between them in twitter, in *Proceedings of the 2016 IEEE International Conference on Recent Trends in Electronics, Information and Communication Technology (RTEICT)* (IEEE, New York, 2016), pp. 678–682
7. R. Mittal, M. Bhatia, Cross-layer closeness centrality in multiplex social networks, in *Proceedings of the 2018 9th International Conference on Computing, Communication and Networking Technologies (ICCCNT)* (IEEE, New York, 2018), pp. 1–5
8. N. Meghanathan, R. Lawrence, *Centrality Analysis of the United States Network Graph* (2016)
9. A. Hajibagheri, A. Hamzeh, G. Sukthankar, Modeling information diffusion and community membership using stochastic optimization, in *Proceedings of the 2013 IEEE/ACM International Conference on Advances in Social Networks Analysis and Mining* (ACM, New York, 2013), pp. 175–182
10. C. Wang, W. Tang, B. Sun, J. Fang, Y. Wang, Review on community detection algorithms in social networks, in *Proceedings of the 2015 IEEE International Conference on Progress in Informatics and Computing (PIC)* (IEEE, New York, 2015), pp. 551–555
11. Y. Zhang, Y. Bao, S. Zhao, J. Chen, J. Tang, Identifying node importance by combining betweenness centrality and katz centrality, in *Proceedings of the 2015 International Conference on Cloud Computing and Big Data (CCBD)* (IEEE, New York, 2015), pp. 354–357
12. Z. Yang, R. Algesheimer, C.J. Tessone, A comparative analysis of community detection algorithms on artificial networks. Sci. Rep. **6**, 30750 (2016)
13. W. Wang, W.N. Street, A novel algorithm for community detection and influence ranking in social networks, in *Proceedings of the 2014 IEEE/ACM International Conference on Advances in Social Networks Analysis and Mining (ASONAM 2014)* (IEEE, New York, 2014), pp. 555–560
14. M.S. Khan et al., Virtual community detection through the association between prime nodes in online social networks and its application to ranking algorithms. IEEE Access **4**, 9614–9624 (2016)
15. C. Lee, F. Reid, A. McDaid, N. Hurley, Detecting highly overlapping community structure by greedy clique expansion (2010). arXiv preprint arXiv:1002.1827
16. X. Chen, A new clustering algorithm based on near neighbor influence. Expert Syst. Appl. **42**, 7746–7758 (2015)
17. J. Cheng, M. Leng, L. Li, H. Zhou, X. Chen, Active semi-supervised community detection based on must-link and cannot-link constraints. PloS One **9**, e110088 (2014)
18. M. Girvan, M.E. Newman, Community structure in social and biological networks. Proc. Natl. Acad. Sci. **99**, 7821–7826 (2002)
19. D. Sudhakaran, S. Renjith, *Survey of Community Detection Algori MS to Identify e Best Community in Real-time Networks*, vol. 2(1) (2016), pp. 529–533

How Is Twitter Talking About COVID-19?

Jesus L. Llano, Héctor G. Ceballos, and Francisco J. Cantú

1 Introduction

Social networks offer us the possibility for sharing information, news, interests and most important opinions about specific topics that currently impact the society [12]. People take to social media to comment about practically anything that happens in their lives, from trending topics to the current social crisis that COVID-19 possesses. This commentary could offer an insight into the overall opinion that exists related to the current health crisis the world is facing.

The ability to understand society and how it sees things is something that has eluded experts for years. As an attempt to achieve this, helped invariably by the boom that technology had over the last few years, many techniques for analysing natural language and the relationship that exists between words, phrases and speech have arisen. We intend to use these techniques to explore a more social-driven aspect of the current public health crisis.

Our primary interest is to perform sentiment and emotional analysis on social media and how it impacts the general opinion towards this particular topic. In this case, the results could shine a light on the opinion of the users of Twitter. We followed the Cross-Industry Standard Process for Data Mining (CRISP-DM) as a way to provide structure during the process of this project.

J. L. Llano (✉) · H. G. Ceballos · F. J. Cantú
Tecnológico de Monterrey, Monterrey, Mexico
e-mail: jesus_llg@me.com; A01748867@itesm.mx; ceballos@tec.mx; fcantu@tec.mx

© Springer Nature Switzerland AG 2021   111
R. Stahlbock et al. (eds.), *Advances in Data Science and Information Engineering*,
Transactions on Computational Science and Computational Intelligence,
https://doi.org/10.1007/978-3-030-71704-9_7

2   Methodology

We extracted information regarding the overall emotions and sentiments that exist towards the current pandemic as well as the correlation that exists between different repeating words related to COVID-19. To do this we used a dataset composed of three samples, each spanning over a month, obtained from a repository containing a collection of tweetIDs associated with COVID-19 made public by Chen et al. in [3] and applying different natural language processing (NLP) techniques.

2.1   Business Understanding

Social networks, as we have stated before, offer more than a glimpse of the behaviour, response and opinion that society as a whole develops towards practically any affair that concerns it. Studying and extracting this information is known as sentiment analysis or opinion mining. In recent years, sentiment analysis has become a popular research area in data science, decision-making and computational linguistics. This analysis allows us to assess the general opinion of people, by classifying it into three categories: positive, negative and neutral. Obtaining this information allows different organisms from the private sector to the government to study the opinion of society from yet another angle with aims to take measures to better the opinion of people [4, 8, 10, 14].

Furthermore, in many cases, it is vital not only to understand the polarity of opinions but to assess the emotions that give shape to it. By analysing the word composition of texts, experts can identify up to a certain degree the emotions of the author. Several studies have demonstrated that words contain an inherent emotion or set of emotions tied to them. The extraction of emotions from texts would allow us to see how society is reacting to the pandemic with increase granularity and how it is affecting their emotional state [6, 7].

We propose topic discovery as an attempt to understand the structure and patterns of the *discussion* that exists around COVID-19. By topics, we refer to mixtures of words that are more commonly found within the texts and are more probable of appearing in texts belonging to a particular topic. Extracting this information, we should be able to obtain various sets of words that are more closely related and used within the same context of COVID-19.

2.2   Data Understanding and Preparation

Tweet IDs were collected over 3 months, from January to March 2020, using a set of keywords to try and obtain only those related to the COVID-19 health crisis. Using a tool to extract the information tied to the IDs, we formed a sample of 150,000 tweets written in English and 120,000 in Spanish to assure that the obtained results reflect those of target population with a precision of 95% ± 1%. While each tweet has an extensive amount of information for this particular work, we will focus on extracting the following information:

- hashtags: A list of the existing metadata tags inside the tweet.
- text: The content of the tweet, the message the author sent out.
- URLs: Any reference to web resources that may exist inside.

These features are essential for our study as we intend to analyse the contents of each tweet. To do this, we have to move on to the phase of data preparation. As it may sound simple to analyse text for us, a computer needs to decompose and transform phrases into more easily understandable pieces. This process is somewhat tricky, and it is entirely dependent on the task and model that is to be applied over the data.

For this project, we begin by normalising to remove any particular pieces of irrelevant text such as style retweet text (RT), URLs, punctuation, stop words, extra white spaces and hyphens found in the content of the tweet. Then using the WordNetLemmatizer defined in the Natural Language Toolkit (NLTK) library, we applied lemmatisation to our text. This standardises text depending on the grammatical type of word returning it to its root. Finally, we tokenised using NLTK Tweettokenizer, especially design for this. It splits the text into words, which facilitates the creation of a corpus the computer can handle easily.

Having the text cleaned and split into tokens, it is trivial to create a bag of words to have as a corpus. This simplifies the process of analysing the data both for the LDA model and the sentiment/emotional analysis.

2.3 Modelling

As it has been mentioned in previous sections, this project is concerned with three main tasks to be performed: sentiment analysis, emotional analysis and topic discovery. A series of experiments were performed for each, and the results will be described in this section.

Starting with the sentiment analysis, not having any data that has been labelled makes the idea of applying pre-existing methods of machine learning quite challenging in particular for the proper assignation of an opinion polarity [5]. The fact that we are working with raw data obtained from Twitter motivated us to explore the use of the Valence Aware Dictionary for sentiment Reasoner (VADER), a sentiment analysis tool that is specifically designed for social media.

This tool allows us not only to identify if a tweet is positive or negative but also to determine the intensity of the polarity for each. The results of analysing the entirety of the tweets datasets can be seen in Figs. 1 and 2. In general, the polarity of both corpora leans towards a negative opinion.

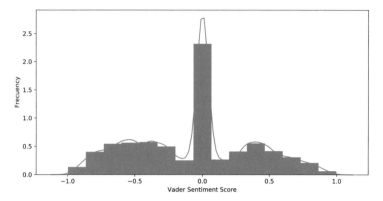

Fig. 1 Results obtained by performing VADER's sentiment analysis over the English tweets

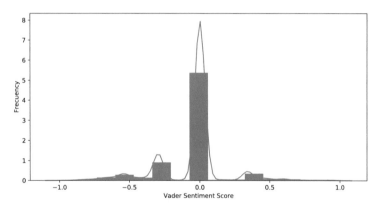

Fig. 2 Scores obtained by VADER for the Spanish tweets

From just studying this data, it is not possible to accurately identify the underlying reason for this negative feeling. In the following phases of experimentation, one of which focuses on topic discovery, we could gain more insight into the cause of this negativity.

As we can see, the majority of the tweets had a neutral score, yet the amount of negative tweets seems to be slightly larger than the number of positive tweets, especially for the Spanish tweets (Fig. 2), a fact that could signify that there is a negative perception or opinion towards the COVID-19, which would be understandable.

For the second phase of experiments, we continued mining the data, this time aiming to extract the emotions associated with the content of the tweets as well as confirm the previous analysis, and to do this, we used the NRC-Emotion-Lexicon (NRC) proposed in [9]. This particular lexicon describes eight basic emotions: joy, anger, sadness, surprise, trust, anticipation, disgust, fear and the polarity associated with them, negative or positive. Using this lexicon, we were able to separate the

Table 1 Average TF-IDF for each emotion related to the tweets, according to the NRC lexicon

Emotion	Avg. TD-IDF
Fear	0.122933
Trust	0.120723
Sadness	0.089662
Anticipation	0.087558
Anger	0.066408
Surprise	0.060779
Disgust	0.055382
Joy	0.050357
Negative	0.195718
Positive	0.173378

As we can see, fear, trust and sadness are the top three emotions. We can also observe that the majority of the tweets had a negative connotation

words in our corpus according to the emotions associated with them. However, this lexicon is exclusive for English words which stopped us from being able to analyse the Spanish written tweets.

By calculating the term frequency-inverse document frequency (TF-IDF) as described by Sammut and Webb in [11], only using words inside the NRC we can approximate the average percentage of emotions of the entire corpus. The results obtained are quite interesting; *fear* is the emotion that appeared most frequently followed by *trust* with a minimal difference. The TF-IDF for each emotion is described in Table 1 and visually represented in Fig. 3.

As a way to further test the polarity of the tweets, NCR allowed us to calculate the TF-IDF of words associated with positive and negative opinions; again it is by the smallest amount that the analysed tweets seem to relate more to a negative opinion towards the COVID-19. These results are also described in Table 1 and in Fig. 3b.

Lastly, for the third model, we decided to apply a latent Dirichlet allocation (LDA), a generative model presented by Blei et al. [1], to extract topics from the tweets. This model should provide a deeper understanding of the way people refer to COVID-19 and the context in which they do. By finding relations using Bayesian probabilities, an LDA finds sets of words that better serve to separate documents, in this case, tweets, into groups. These groups can then be used to find trends about what is being discussed and even to create directed content [2, 13]. We extracted five topics using an LDA model that was initialised using a dictionary composed of all the words in our original corpus transformed using TF-IDF. The extraction was performed independently for both language groups of tweets. We can see the cluster maps that describe the topics in Figs. 4 and 5.

From the topics, we extracted the five most representative words and the assigned weights of each which are depicted in Table 2. We can observe that three common words that appear in several of the found topics are *corona*, *coronavirus* and *China*.

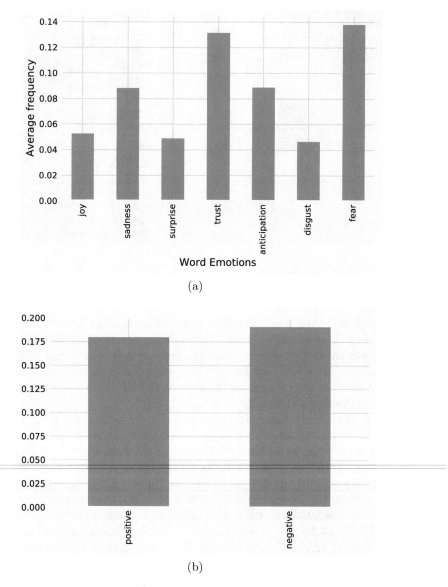

(a)

(b)

Fig. 3 After calculating the average percentage of emotions and sentiments appearing in the English tweets corpus using NCR, *fear* appeared the most. (**a**) Emotion analysis using NRC-Emotion-Lexicon. (**b**) Sentiment analysis using NRC-Emotion-Lexicon

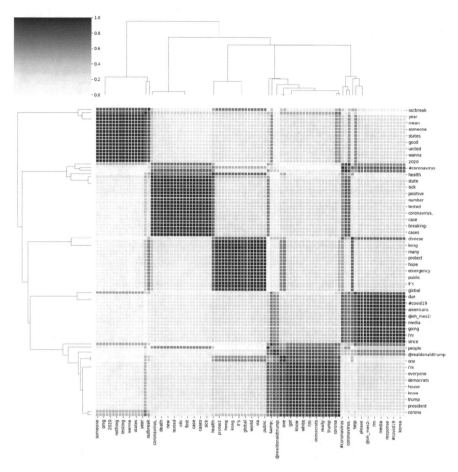

Fig. 4 Cluster map obtained after processing the corpus composed of English written tweets using LDA. There are three clear clusters of words (topics) depicted in the darker areas of the map. With words like *outbreak*, *corona*, *coronavirus* and *testing* appearing on them

In the case of English tweets, it seems that words directly related to the government like *democrats*, *states* and *president* appeared with enough frequency to be relevant. This could indicate some relation with the sentiments found in the other two analyses.

Finally, just as a way to visually represent the word frequency analysis performed over content of the entire corpus of tweets, we present a word cloud depiction in Fig. 6. Some of the most relevant words in both were also found during the topic discovery phase, natural since they appear with more frequency through the corpora.

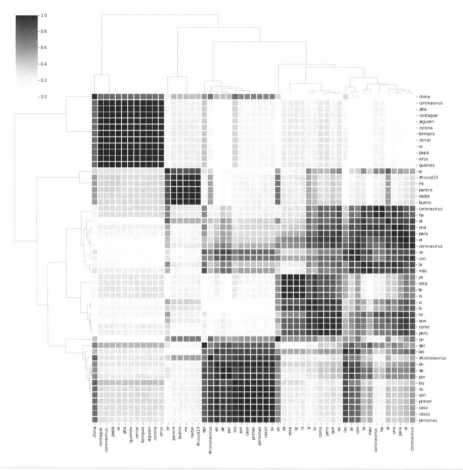

Fig. 5 Cluster map obtained after processing the corpus composed of Spanish written tweets using LDA. In this case, the topics are considerably less discriminated from one another as there seems to be certain overlapping of terms. Words like *coronavirus*, *contagion*, *people* and *cases* seem to be within these groups

3 Discussion

The principal goal of this project is to provide an insight into the public opinion towards the novel coronavirus (COVID-19 or SARS-CoV-2) virus. Thanks to the diffusion channels that social media platforms like Twitter provide, it is now easier than ever to access sentimental information regarding particular topics of interest. We show that it is possible to model the sentiment of people towards a pandemic such as the one sweeping the globe as of the day this document was written.

Sentiment and emotional analysis of social networks have gained the interest of experts since it provides a way to gauge public opinion without interference.

Table 2 Five most representative words of the topics extracted from the English (left) and the Spanish (right) tweets corpus

Topic	Words	Weights	Topic	Words	Weights
	Outbreak	0.047		Italia	0.046
	Testing	0.040		Hoy	0.040
1	Coronavirus	0.020	1	Hospital	0.034
	Death	0.016		Hora	0.032
	Someone	0.014		Chino	0.031
	Coronavirus	0.019		Si	0.128
	Kong	0.014		País	0.073
2	Flu	0.012	2	Vamos	0.031
	Hope	0.010		Solo	0.026
	Die	0.009		Confirma	0.026
	Corona	0.015		Coronavirus	0.565
	Said	0.014		Caso	0.045
3	Virus	0.013	3	Persona	0.037
	Coronavirus	0.012		España	0.031
	Democrats	0.011		Oms	0.013
	Coronavirus	0.027		China	0.211
	Health	0.023		Casos	0.100
4	President	0.018	4	Wuhan	0.035
	Public	0.014		Gripe	0.034
	Global	0.014		Vez	0.022
	Hands	0.020		Virus	0.099
	Wash	0.016		Salud	0.078
5	Don't	0.015	5	Nuevo	0.068
	Cases	0.010		Madrid	0.026
	Workers	0.007		Medios	0.020

Mining key words through topic discovery could help in the scope of public health as valuable, early clues about disease outbreaks, providing feedback on public health policies and response measures. For this is necessary to continue exploring the possibilities provided by the vast amount of information social platforms generate every second.

4   Conclusion

The different results presented in this document point towards a negative opinion in regard to the current situation. Furthermore, the polarisation on the opinion for the data-based of Spanish written tweets seems to have a more substantial lean towards negativity than its English counterpart. At this moment, we have demonstrated that it is possible to extract enough information that provides a glimpse of the general

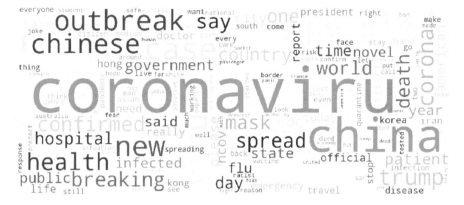

Fig. 6 Word cloud of the entire dataset

opinion using only simple natural language processing techniques. Further, more sophisticated analyses need to be performed to provide more precious insights as it would be the introduction of more sophisticated word embeddings extracted through deep learning.

As a last comment, social media is now ingrained within society deeper than any other technological advancement. We, as a community, use it to share interests, media, news and information. It would be a misuse not to seize this information ethically towards the benefit of society. For which an essential aspect of it needs to be addressed, even when it was not actively discussed through the project, is the importance of ethics in data science appears as a necessity.

References

1. D.M. Blei, A.Y. Ng, M.I. Jordan, Latent dirichlet allocation. J. Mach. Learn. Res. **3**(null), 993–1022 (2003)
2. D. Bracewell, J. Yan, F. Ren, S. Kuroiwa, Category classification and topic discovery of japanese and english news articles. Electr. Notes Theor. Comput. Sci. **225**, 51–65 (2009). DOI 10.1016/j.entcs.2008.12.066
3. E. Chen, K. Lerman, E. Ferrara, *Covid-19: The First Public Coronavirus Twitter Dataset* (2020)
4. R. Feldman, Techniques and applications for sentiment analysis. Commun. ACM **56**(4), 82–89 (2013). DOI 10.1145/2436256.2436274. https://doi.org/10.1145/2436256.2436274
5. A. Joshi, P. Bhattacharyya, S. Ahire, *Sentiment Resources: Lexicons and Datasets* (Springer, Cham, 2017), pp. 85–106. https://doi.org/10.1007/978-3-319-55394-8_5
6. J. Li, F. Ren, Emotion recognition from blog articles, in *Proceedings of the 2008 International Conference on Natural Language Processing and Knowledge Engineering* (2008), pp. 1–8
7. K.H. Lin, C. Yang, H. Chen, Emotion classification of online news articles from the reader's perspective, in *Proceedings of the 2008 IEEE/WIC/ACM International Conference on Web Intelligence and Intelligent Agent Technology*, vol. 1 (2008)

8. S.M. Mohammad, P.D. Turney, Emotions evoked by common words and phrases: using mechanical turk to create an emotion lexicon, in *Proceedings of the NAACL HLT 2010 Workshop on Computational Approaches to Analysis and Generation of Emotion in Text (CAAGET '10)* (Association for Computational Linguistics, USA, 2010), pp. 26–34

9. S.M. Mohammad, P.D. Turney, *NRC Emotion Lexicon* (National Research Council, Canada, 2013)

10. R. Narayanan, B. Liu, A. Choudhary, Sentiment analysis of conditional sentences, in *Proceedings of the 2009 Conference on Empirical Methods in Natural Language Processing* (Association for Computational Linguistics, Singapore, 2009), pp. 180–189. https://www.aclweb.org/anthology/D09-1019

11. C. Sammut, G.I. Webb (eds.), *TF–IDF* (Springer, Boston, 2010), pp. 986–987. DOI 10.1007/978-0-387-30164-8_832. https://doi.org/10.1007/978-0-387-30164-8_832

12. V. Varma, L.J. Kurisinkel, P. Radhakrishnan, *Social Media Summarization* (Springer, Cham, 2017), pp. 135–153. DOI 10.1007/978-3-319-55394-8_7. https://doi.org/10.1007/978-3-319-55394-8_7

13. X. Wang, N. Mohanty, A. Mccallum, *Group and Topic Discovery from Relations and their Attributes* (2005)

14. D.D. Zeng, H.C. Chen, R. Lusch, S.H. Li, Social media analytics and intelligence. IEEE Intell. Syst. **25**, 13–16 (2011). DOI 10.1109/MIS.2010.151

Detecting Asian Values in Asian News via Machine Learning Text Classification

Li-jing Arthur Chang

1 Introduction

Past research has used content analysis to locate Asian values of harmony and support by hand [1]. In the research, harmony is classified as either "harmony" or "conflict," with "harmony" defined as a void of conflict [1]. Also, support is classified as either "supportive" or "critical," with "supportive" defined as supporting economic, political, or societal strength at home [1].

This study is designed to develop machine learning models to classify Asian news articles as "harmony" or "conflict" (for the harmony variable) and as "supportive" or "critical" (for the support variable). This approach differs from past news classification studies, which often classified news as having "positive" or "negative" sentiments [2–3].

2 Literature Review

This study used supervised machine learning algorithms for text classification. Supervised machine learning starts from pre-labeled data and uses the feature vector and labels to build a prediction model [4]. The model is used to predict the labels for the test dataset. The predictions are compared with actual labels to measure performance [4].

Sentiment analysis is a form of opinion mining that explores the viewpoints of people [5]. The text analytical skills have been used to analyze news articles [2–3,

L.-j. A. Chang (✉)
Jackson State University, Jackson, MS, USA
e-mail: j00093959@jsums.edu; li-jing.a.chang@jsums.edu

© Springer Nature Switzerland AG 2021
R. Stahlbock et al. (eds.), *Advances in Data Science and Information Engineering*,
Transactions on Computational Science and Computational Intelligence,
https://doi.org/10.1007/978-3-030-71704-9_8

6–10].While focusing on the "positive" and "negative" sentiments of news, studies on news sentiment analysis differs in their approaches and focuses. For example, one study proposed a news sentiment-based trading strategy after analyzing the relationship between company-related sentiment and investment returns [8]. The results showed that trading based on a "neutral" sentiment gives good market returns [8].

A couple of studies used sentiment lexicon to categorize news sentiments [6, 10]. The lexicon-based sentiment approach uses a lexicon with words or a multi-word group labeled as positive, negative, or neutral to analyze text sentiment [4].

More recent news sentiment analysis adopted the machine learning approach [2–3]. The research used machine learning algorithms such as support vector machine, naïve bayes, k-nearest neighbors, maximum entropy (or logistic regression), random forest, decision tree, gradient boosting, AdaBoost, and artificial neural network [2–3, 6, 11–19]. The present study is aimed at using the algorithms to classify the news text for the target variables of harmony and support. Based on the above review, two research questions are generated:

RQ 1: *How do the machine learning algorithms perform in predicting the Asian value of harmony in Asian newspaper stories?*
RQ 2: *How do the machine learning algorithms perform in predicting the Asian value of support in Asian newspaper stories?*

3   Methodology

The sample consisted of 3062 articles published between late July 2018 and the first half of August 2018 from English-language newspapers in seven Asian countries.[1] The articles were accessed from a database at a local library. Only hard news articles with a focus on issues related to home news are picked. Home news is events occurring in the newspaper's home country or abroad involving the nation's citizens, governments, or residents. The rationale for the choice is that home news will better reflect the Asian values of harmony and support [1].

To build the machine learning model for harmony, the dataset of 3062 articles was processed to derive an input feature vector and a harmony target variable vector. The support model used the same input vector and a support target variable vector.

To derive labels for the target variables, 269 of the articles were first manually labeled with intercoder reliabilities of over 0.85 for both variables. The rest of the articles were labeled with a machine learning model built from the initial hand-labeled sample. The machine-assigned labels were verified manually. The manual verification showed that the machine-coded labels are at least 95% accurate for

---

[1]The seven Asian nations are China, India, Malaysia, Pakistan, Singapore, Thailand, and The Philippines. Of the 3062 articles, 25.9% came from China, 30.1% were from India, 9.3% were from Malaysia, 13.5% were from Pakistan, 9.3% were from Singapore, 3.3% were from Thailand, and 8.6% were from the Philippines.

both target variables. The mislabeled machine labels were corrected with manual labels. The harmony labels (either "harmony" or "conflict") derived from the above process formed the harmony target variable vector. The support labels ("supportive" or "critical") derived formed the support target variable vector.

To obtain the input vector, the articles were processed to remove stop words, special characters, punctuations, overly frequent words, and rare words because these words/symbols did not add much meanings and the removal would ease data complexity [20]. The text was also lemmatized to linguistic roots based on the assumptions that words of the same root mostly describe the same or relatively close concepts [20–21].

At this point, there are 9769 unique words in the dataset. They were converted to a vector of 9769 features using TFIDF to give more weight to frequent words in each article while offsetting the influence of overly frequent words in the whole sample [22].

After the TFIDF transformation, the number of input features remained at 9769. To ease model complexity, principal component analysis was used. Principal component analysis, an unsupervised method, is chosen because it does not require labeled data and is more suitable for model deployment for real-world unseen data such as news where no labels exist [23]. After the dimensionality reduction using principal component analysis, the input vector had 3062 features.

At this point, the input vector and the target variable harmony vector is the dataset for the harmony model. The same input features vector and the target variable support vector is the dataset for the support model. Each dataset was randomly split into 80% train dataset and 20% test dataset. Each algorithm was tuned via a grid search with tenfold cross validations on the train dataset. The test dataset is used to compare the performance of the tuned algorithms using metrics such as accuracy, precision, recall, and F1 score [24].[2]

4 Results

The results of testing RQ 1 showed logistic regression is the top performer in terms of accuracy and F1 score (see Table 1). Support vector machine leads the performance in precision. Also, artificial neural network is the top performer in the recall measure.

The testing of RQ 2 showed logistic regression is the top performer in accuracy, recall, and F1 score (see Table 2). Support vector machine leads the performance in precision. The results are similar to those from the testing of RQ 1.

---

[2] $Accuracy = \frac{(TP+TN)}{(TP+TN+FP+FN)}$; $precision = \frac{TP}{(TP+FP)}$; $recall = \frac{TP}{(TP+FN)}$; $F1 = \frac{(2 \times Precision \times Recall)}{(Precision+Recall)}$ where TP is true positive, TN is true negative, FP is false positive, and FN is false negative.

Table 1 RQ 1 – ML models (target variable harmony)

Algorithm	Accuracy	Precision	Recall	F1
Logistic regression	93.0%	91.0%	92.1%	91.6%
Support vector machine	90.9%	94.5%	85.2%	89.6%
Gradient boost	88.3%	84.0%	87.4%	85.7%
Artificial neural network	88.3%	78.1%	92.6%	84.7%
K nearest neighbors	83.7%	78.9%	81.5%	80.2%
Adaboost	83.2%	78.5%	80.7%	79.6%
Decision tree	78.8%	71.5%	76.3%	73.8%
Random forest	74.4%	55.5%	76.8%	64.4%
Naive Bayes	62.8%	46.9%	56.6%	51.3%

Notes: Results ranked in a descending order by accuracy; $n = 3062$

Table 2 RQ 2 – ML models (target variable support)

Algorithm	Accuracy	Precision	Recall	F1
Logistic regression	91.2%	81.9%	90.1%	85.8%
Support vector machine	90.4%	82.9%	86.8%	84.8%
Artificial neural network	89.7%	81.9%	85.8%	83.8%
Gradient boost	85.5%	71.9%	81.3%	76.3%
K nearest neighbors	82.2%	69.3%	74.2%	71.7%
Adaboost	81.9%	68.8%	73.7%	71.2%
Decision tree	73.4%	57.3%	59.4%	58.3%
Random forest	71.1%	11.6%	95.8%	20.6%
Naive Bayes	67.2%	34.7%	49.3%	40.7%

Notes: Results ranked in a descending order by accuracy; $n = 3062$

To apply the trained logistic regression model for harmony, this study deployed the model to a website. Similar procedure was done to deploy the trained logistic regression model for support. A news story was picked from the publicly accessible Reuters news corpus (Reuters-21578, Distribution 1.0) of Python's nltk library [25]. Describing Asian exporters' concerns about the US-Japan trade conflict, the story was classified as "conflict" by the harmony-conflict predictor website. Also, the story was labeled as "critical" by the supportive-critical predictor website. This showed the utility of the models.

5 Conclusion

Whether it is the harmony model or the support model, the results are quite similar. In both situations, logistic regression performs best in the measures of accuracy and F1 score, while support vector machine performs best in the measure of precision. Beyond accuracy, logistic regression is a fast algorithm compared to most other algorithms tested in this study. In addition, the deployment of the trained logistic algorithm models demonstrated their practical utilities.

Lastly, the present study used the manual coding of an initial sample to build a machine learning model to label the remaining sample. This procedure is a time-saver compared to the complete manual coding of the entire dataset.

Bibliography

1. B.L. Massey, L.-j.A. Chang, Locating Asian values in Asian journalism: A content analysis of web newspapers. J. Commun. **52**, 987–1003 (2002)
2. V. Bobichev, O. Kanishcheva, O. Cherednichenko, Sentiment analysis in the Ukrainian and Russian news, in *Proceedings of 2017 IEEE First Ukraine Conference on Electrical and Computer Engineering (UKRCON)*. IEEE, USA (multi-national) (2017), pp. 1050–1055
3. D. Nguyen, K. Vo, D. Pham, M. Nguyen, T. Quan, A deep architecture for sentiment analysis of news articles, in *Proceedings of International Conference on Computer Science, Applied Mathematics and Applications*, (Springer, Cham, 2017), pp. 129–140
4. C.S.G. Khoo, S.B. Johnkhan, Lexicon-based sentiment analysis: Comparative evaluation of six sentiment lexicons. J. Inf. Sci. **44**, 491–511 (2018)
5. B. Liu, Sentiment analysis and subjectivity, in *Handbook of Natural Language Processing*, (CRC Press, Boca Raton, 2010), pp. 627–666
6. S. Krishnamoorthy, Sentiment analysis of financial news articles using performance indicators. Knowl. Inf. Syst. **56**, 373–394 (2018)
7. A. Balahur, R. Steinberger, M. Kabadjov, V. Zavarella, E. van der Goot, M. Halkia, B. Pouliquen, J. Belyaeva, Sentiment analysis in the news, in *Proceedings of the 7th International Conference on Language Resources and Evaluation (LREC' 2010)*, Cornell University Repository, USA (2010), pp. 2216–2220 https://arxiv.org/abs/1309.6202
8. W. Zhang, S. Skiena, Trading strategies to exploit blog and news sentiment, in *Proceedings of the Fourth International AAAI Conference on Weblogs and Social Media*, (AAAI Press, Menlo Park, 2010), pp. 375–378
9. M. Bautin, L. Vijayarenu, S. Skiena, International sentiment analysis for news and blogs, in *ICWSM*, Association for the Advancement of Artificial Intelligence, USA (multi-national), (2008) https://www.aaai.org/Papers/ICWSM/2008/ICWSM08-010.pdf
10. N. Godbole, M. Srinivasaiah, S. Skiena, Large-scale sentiment analysis for news and blogs, in *ICWSM/W3*, Boulder, Colorado, USA (2007) https://pdodds.w3.uvm.edu/files/papers/others/2007/godbole2007a.pdf
11. J. Kalyani, H.N. Bharathi, R. Jyothi, Stock trend prediction using news sentiment analysis, in *arXiv preprint arXiv:1607.01958* (2016)
12. P.D. Azar, in *Sentiment Analysis in Financial News*, Bachelor Thesis, Harvard College, Cambridge, Massachusetts, USA, 2009
13. S. Fong, Y. Zhuang, J. Li, R. Khoury, Sentiment anlaysis of online news using MALLET, in *Proceedings of 2013 International Symposium on Computational and Business Intelligence*, IEEE, USA (multi-national) (2013), pp. 301–304
14. K. Cortis, A. Freitas, T. Daudert, M. Hurlimann, M. Zarrouk, S. Handschuh, B. Davis, SemEval-2017 Task 5: Fine-grained sentiment analysis on financial microblogs and news, in *Proceedings of the 11th International Workshop on Semantic Evaluations (SemEval-2017)*, (Association for Computational Linguistics (ACL), Stroudsburg, 2017), pp. 519–535
15. Y. Gao, L. Zhou, Y. Zhang, C. Xing, Y. Sun, X. Zhu, Sentiment classification for stock news, in *Proceedings of International Conference on Pervasive Computing and Applications (ICPCA)*, IEEE, USA (multi-national) (2010), pp. 99–104

16. A.E. Khedr, S.E. Salama, N. Yaseen, Predicting stock market behavior using data mining technique and news sentiment analysis. Int. J. Intell. Syst. Appl. (IJISA) **9**, 22–30 (2017)
17. W. Zhang, S. Skiena, Improving movie gross prediction through news analysis, in *Proceedings of 2009 IEEE/WIC/ACM International Joint Conference on Web Intelligence and Intelligent Agent Technology*, IEEE, USA (multi-national) (2009), pp. 301–304
18. K.-Y. Ho, W.W. Wang, Predicting stock price movements with news sentiment: An artificial neural network approach, in *Artificial Neural Network Modelling*, (Springer, Cham, 2016), pp. 395–403
19. S.A. Haider, R. Mehrotra, Corporate news classification and valence prediction: A supervised approach, in *Proceedings of the 2nd Workshop on Computational Approaches to Subjectivity and Sentiment Analysis*, Association for Computational Linguistics; Portland, Oregon, USA (2011), pp. 175–181 https://aclanthology.org/W11-1723.pdf
20. V. Srividhya, R. Anitha, Evaluating preprocessing techniques in text categorization. Int. J. Comput. Sci. Appl. **47**(11), 49–51 (2010)
21. D. Sarkar, *Text Analytics with Python*, Springer, USA (multi-national) 2016 https://link.springer.com/content/pdf/10.1007/978-1-4842-4354-1.pdf
22. B. Trstenjaka, S. Mikacb, D. Donkoc, KNN with TF-IDF based framework for text categorization. Proc. Eng. **69**, 1356–1364 (2014)
23. P. Cunningham, Dimension reduction, in *Machine Learning Techniques for Multimedia*, (Springer, Berlin, 2008), pp. 91–112
24. V.A. Kharde, S.S. Sonawane, Sentiment analysis of twitter data: A survey of techniques. Int. J. Comput. Appl. **139**(11), 5–15 (2016)
25. F. Debole, F. Sebastiani, An analysis of the relative hardness of Reuters-21578 subsets. J. Am. Soc. Inf. Sci. Technol. **56**(6), 584–596 (2005)

Part III
Recommendation Systems, Prediction Methods, and Applications

The Evaluation of Rating Systems in Online Free-for-All Games

Arman Dehpanah, Muheeb Faizan Ghori, Jonathan Gemmell, and Bamshad Mobasher

1 Introduction

Online competitive games pit players against one another in player-versus-player (PvP) matches. A common goal of PvP games is to match players based on their skills. When a new player is matched against an experienced player, neither is likely to enjoy the competition. Therefore, competitive games often use rating algorithms to match players with similar skills.

Rating systems often represent players with a single number, describing the player's skills. For example, Elo [1] considers 1500 as the default skill rating for new players and updates this value based on the outcome of the matches they played. Rating systems leverage skill ratings to predict ranks. While researchers have made numerous efforts in improving rank prediction, less attention has been given to how predicted ranks are evaluated.

There are several metrics commonly used to evaluate ratings. However, these metrics often do not capture important characteristics of the ratings. They might give equal weighting to low-tier and top-tier players, even when matching top-tier players is more important for the goals of the system. They might also be hampered by the inclusion of new players since the system does not possess any knowledge of these players.

In this paper, we consider six evaluation metrics. We include traditional metrics such as accuracy, mean absolute error, and Kendall's rank correlation coefficient. We further include metrics adapted from the domain of information retrieval, including mean reciprocal rank (MRR), average precision (AP), and normalized

A. Dehpanah (✉) · M. F. Ghori · J. Gemmell · B. Mobasher
School of Computing, DePaul University, Chicago, IL, USA
e-mail: adehpana@depaul.edu; mghori2@depaul.edu; jgemmell@cdm.depaul.edu;
mobasher@cs.depaul.edu

© Springer Nature Switzerland AG 2021
R. Stahlbock et al. (eds.), *Advances in Data Science and Information Engineering*,
Transactions on Computational Science and Computational Intelligence,
https://doi.org/10.1007/978-3-030-71704-9_9

discounted cumulative gain (NDCG). We analyze the ability of these metrics to capture meaningful insights when they are used to evaluate the performance of three popular rating systems: Elo, Glicko, and TrueSkill.

To perform this analysis, we limit our experimentation to free-for-all matches. Free-for-all is a widely used game-play mode where several players simultaneously compete against one another in the same match. The winner is the "last man standing"; this mode of game-play is more commonly referred to as battle royale. Our real-world dataset includes over 100,000 matches and over 2,000,000 unique players from PlayerUnknown's Battlegrounds (PUBG).

Our evaluation shows that in free-for-all matches, the metrics adapted from information retrieval can better evaluate the rating systems while being more resistant to the influence of new players. NDCG, in particular, could more precisely capture the predictive power of these systems. NDCG distinguishes between the prediction errors for top-tier players with higher ranks and those for low-tier players with lower ranks by applying a weight to positions. This is particularly important for companies whose business model is based on user interaction and engagement at the top levels of play. Top-tier players are the ones who most probably stay in the system and play more games.

The rest of this paper is organized as follows: In Sect. 2, the related works are reviewed. Rank prediction and its application in rating systems are discussed in Sect. 3. In Sect. 4, the evaluation metrics are introduced. In Sect. 5, the dataset and the experiments are explained, and then we discuss the results in detail in Sect. 6. Finally, we conclude the paper and mention future works in Sect. 7.

2   Related Work

The popularity of online competitive games has exploded in the last decade. There are more than 800 million users playing online games, and this number is expected to grow to over 1 billion by the year 2024 [2].

An important mode of game-play is the free-for-all which can be divided into deathmatch and battle royale. In deathmatch games, many players are pitted against one another; the winner is the one with the most points at the end of the game. In battle royale, players eliminate one another; the winner is the last one standing.

A critical component of these zero-sum games is the ability to rate a player's skill. An accurate skill rating enables the system to generate balanced matches [3, 4]. It allows for the assignment of teams with similar skill levels [3, 5–8]. It serves as feedback for the players so that they can track their performance over time [4, 8, 9].

The most prominent examples of such systems are Elo [1], Glicko [10], and TrueSkill [4]. Elo and Glicko are designed for head-to-head matches with only two players. TrueSkill, developed by Microsoft, extends upon these approaches to handle multiplayer games with multiple teams.

A rating system like those above can be used to predict the outcome of games. Those predictions can then be evaluated to judge the quality of the rating system.

Accuracy is often used for evaluating rating systems. It has been used to evaluate first-person shooter games [4–7, 11–14], real-time strategy games [5, 12, 15, 16], as well as tennis [15, 17, 18], soccer [14, 18], football [14], and board games [19–21].

Other metrics have been used for rank prediction. Log-likelihood has been used to evaluate first-person shooters [6], real-time strategies [15], and board games [21]. Information gain has been used to evaluate first-person shooters [14]. Mean squared error has been used to evaluate real-time strategy games [22] and soccer [23]. Mean absolute error (MAE) has been used to evaluate first-person shooters [14] and real-time strategies [22]. Root mean squared error has been used to evaluate chess games [20]. Spearman's rank correlation coefficient has been used to evaluate first-person shooter games [24]. However, most of these examples are head-to-head games; there are only two sides such as in chess or squad-vs-squad first-person shooters. When there are more than two sides, these metrics are less appropriate.

Information retrieval, the process of obtaining relevant resources from document collections [25], includes several metrics for evaluating the ranking of search results. These metrics can be adapted to the evaluation of predictions in online competitive games. Mean reciprocal rank considers the rank positions of relevant documents in the results list to compute relevance scores [26]. Average precision considers the number of relevant documents among retrieved documents along with the number of documents retrieved out of all available relevant documents to evaluate the performance of the system [27]. NDCG considers a graded relevance for determining the gain obtained by each retrieved document [28].

Our work differs from previous efforts. We focus on free-for-all games, one of the most popular game-play modes in which several players compete against one another in a single match. We extend Elo and Glicko to free-for-all games. Whereas research often attempts to improve rank prediction, we seek to better evaluate the predicted ranks. We explore six evaluation metrics, including traditional metrics and those drawn from the domain of information retrieval. We analyze the explanatory power of these metrics using a large real-world dataset and consider how the metrics describe different populations such as the top-tier and the most frequent players.

3   Predicting Rank

Predicting rank in online competitive games is important for several reasons. It can be used to evaluate the performance of the players. It can be used to assign players to teams. It can be used to create balanced matches. Often, rank is predicted by first evaluating the skills of the players.

The skill level of a player p can be described as a single number: μ_p. In a head-to-head match between two players, p_1 and p_2, with skill ratings, μ_{p_1} and μ_{p_2}, the player with the higher rating can be predicted to win that match. In a free-for-all match with several players $p_1, p_2, p_3, \ldots, p_n$, their skill ratings $\mu_{p_1}, \mu_{p_2}, \ldots, \mu_{p_n}$ can be used to create a rank ordering of the players R^{pred}. After the match, the observed rankings R^{obs} can be used to update the skill levels.

There are several ways to update a player's skill rating. In the remainder of this section, we describe three common algorithms: Elo, Glicko, and TrueSkill.

3.1 Elo

Originally developed for ranking chess players, Elo has been used for ranking players in many competitive environments [1]. Elo calculates the skill level of players by appraising a set of historical match results. Elo assumes that players' skill follows a Gaussian distribution with the same standard deviation for all players.

As a convention, the default rating for new players is set to 1500. The rating is updated after each match. The winner is awarded points and the loser surrenders points after each match. The amount of these points is dependent on the probability of the outcome of the match based on the two players' initial ratings.

The probability that player p_i wins the match against player p_j can be calculated by:

$$Pr(p_i \ wins, p_j) = \left(1 + e^{\frac{\mu_j - \mu_i}{D}}\right)^{-1}$$

where μ_i and μ_j are the ratings of the two players. D represents the weight given to ratings when determining players' estimated scores. Using higher values for D decreases the influence of the difference between ratings and vice versa. Conventionally, D is set to 400.

After the match, the rating of player p_i is updated by:

$$\mu_i' = \mu_i + K[R - Pr(p_i \ wins, p_j)]$$

where R is 1 if player p_i wins the game, 0.5 if the match is a draw, and 0 if it is a loss. K is a scaling factor determining the magnitude of the change to players' ratings after each match. Using higher values for K leads to greater changes in players' ratings. The value of K should be tuned based on the nature of the game and the players' characteristics. For example, the World Chess Federation (FIDE) considers several tiers for the value of K. It uses $K = 40$ for new players until they participate in at least 30 matches, $K = 20$ as long as their ratings remain under 2400, and $K = 10$ once their ratings reach 2400.

We can extend Elo skill ratings from head-to-head matches to free-for-all matches that include many players. Several possibilities exist. One way is to consider the match as a set of head-to-head matches.

We recalculate the probability of winning for each player by summing the probability of winning values in all their pairwise matches versus the other players. Assuming N players competing against each other in a field F, the overall probability of winning for player p_i can be calculated as:

$$Pr(p_i \ wins, F) = \frac{\sum\limits_{1 \le j \le N, i \ne j} \left(1 + e^{\frac{(\mu_j - \mu_i)}{D}}\right)^{-1}}{\binom{N}{2}}$$

where $\binom{N}{2}$, the total number of pairwise comparisons, is used to normalize the probability values to sum up to 1.

Because free-for-all matches include several players, we normalize the sum of the observed outcomes to 1 in order to conform to Elo's design that the total number of points awarded is equal to the total number of points deducted. For player p_i, we transform the observed rank R_i^{obs} into a normalized result, R_i', calculated as:

$$R_i' = \frac{N - R_i^{obs}}{\binom{N}{2}}$$

The player's Elo rating in this multiplayer environment can then be updated as:

$$\mu_i' = \mu_i + K \left[R_i' - Pr(p_i \ wins, F) \right]$$

One criticism of Elo is that it assumes a fixed skill variance for all players and may not handle uncertainty well. This could result in reliability issues [10]. Glicko addresses this problem.

3.2 Glicko

The Glicko rating system [10] extended Elo by introducing a dynamic skill deviation σ for each player. Players are characterized by a distribution with a mean μ representing their skill and a deviation σ representing the uncertainty about their skill. The frequency that a player competes in the game is used to modify their skill deviation σ. New players are assigned $\mu = 1500$ and $\sigma = 350$. Both these numbers are updated after each match.

The probability that player p_i wins the match against player p_j can be calculated by:

$$Pr(p_i \ wins, p_j) = \left(1 + 10^{\frac{-g\left(\sqrt{\sigma_i^2 + \sigma_j^2}\right)(\mu_i - \mu_j)}{400}}\right)^{-1}$$

where μ_i, μ_j, σ_i, and σ_j represent the skill ratings and skill deviations of the two players. The function g takes the sum of the square of the two skill deviations and uses them to weight the deviation in the players' skills. It is defined as:

$$g(\sigma) = \left(\sqrt{\frac{1 + 3q^2\sigma^2}{\pi^2}} \right)^{-1}$$

where Glicko sets q as a constant equal to 0.0057565. After the match, the skill rating and deviation of player p_i are updated:

$$\mu_i' = \mu_i + \frac{q}{\frac{1}{\sigma_i^2} + \frac{1}{d^2}} \left[g(\sigma_j)(R - Pr(p_i \ wins, \ p_j)) \right]$$

$$\sigma_i' = \sqrt{\left(\frac{1}{\sigma_i^2} + \frac{1}{d^2} \right)^{-1}}$$

where R is 1 if player p_i wins the game, 0.5 if the match is a draw, and 0 if it is a loss. The variable d^2 is minus of inverse of Hessian of the log marginal likelihood and is calculated as:

$$d^2 = \left[q^2 g(\sigma_j)^2 Pr(p_i \ wins, \ p_j)(1 - Pr(p_i \ wins, \ p_j)) \right]^{-1}$$

We can extend Glicko to free-for-all matches as we did with Elo. We consider each free-for-all match with N players as $\binom{N}{2}$ separate matches between each pair of players. The probability of winning for player p_i can be calculated as:

$$Pr(p_i \ wins, \ F) = \frac{\displaystyle\sum_{1 \le j \le N, i \ne j} \left(1 + 10^{\frac{-g\left(\sqrt{\sigma_i^2 + \sigma_j^2} \right)(\mu_i - \mu_j)}{400}} \right)^{-1}}{\binom{N}{2}}$$

Again we normalize the function with $\binom{N}{2}$ so that the probability values sum up to 1. The variable d^2 can be updated for this scenario as:

$$d^2 = \left[q^2 g(\sigma_j)^2 Pr(p_i \ wins, \ F)(1 - Pr(p_i \ wins, \ F)) \right]^{-1}$$

Similar to Elo, Glicko is a zero-sum rating system; an equal number of points is awarded and deducted. In order to achieve such a balance in a multiplayer free-for-all match, we once again normalize the match results as before. Finally, the rating of player p_i can be updated as:

$$\mu_i' = \mu_i + \frac{q}{\frac{1}{\sigma_i^2} + \frac{1}{d^2}} \left[g(\sigma_j) \left(R_i' - Pr(p_i \ wins, \ F) \right) \right]$$

Glicko has been proven to be successful at rating players. However, it does have some drawbacks. Glicko requires an average of five to ten matches for each player in order to accurately describe a player's skill [29]. Moreover, when players compete very frequently, their skill deviation σ becomes very small, and there are no noticeable changes in their ratings, even when they are truly improving. Finally, while we have extended Elo and Glicko from head-to-head matches to large multiplayer free-for-all matches, they were not initially designed to do so. TrueSkill was designed for this purpose.

3.3 TrueSkill

TrueSkill [4] is a Bayesian ranking system developed by Microsoft Research for Xbox Live that can be applied to any type of game-play mode with any number of players or teams. TrueSkill derives individual skill levels from the outcome of matches between players by leveraging factor graphs [30] and expectation propagation algorithm [31].

Similar to Glicko, TrueSkill assumes that the performance of players follows a Gaussian distribution with mean μ and standard deviation σ representing their skills and skill deviations. New players are assigned $\mu = 25$ and $\sigma = 8.333$. These values are updated after each match.

TrueSkill follows different update methods depending on whether a draw is possible. For a non-draw case, if μ_i, μ_j, σ_i, and σ_j represent skill ratings and deviations of players p_i and p_j, assuming player p_i wins the match against player p_j, his skill rating is updated by:

$$\mu_i' = \mu_i + \frac{\sigma_i^2}{c} \left[\frac{N(\frac{t}{c})}{\Phi(\frac{t}{c})} \right]$$

where $t = \mu_i - \mu_j$ and $c = \sqrt{2\beta^2 + \sigma_i^2 + \sigma_j^2}$. N and Φ represent the probability density function and cumulative distribution function of a standard normal distribution. The parameter β is the scaling factor determining the magnitude of changes to ratings. Skill deviations for both players are updated by:

$$\sigma' = \sigma - \sigma \left(\frac{\sigma^2}{c^2} \left[\frac{N(\frac{t}{c})}{\Phi(\frac{t}{c})} \right] \left[\frac{N(\frac{t}{c})}{\Phi(\frac{t}{c})} + t \right] \right)$$

TrueSkill has been used in online games [32], sports [18], education [33], recommender systems [34], and click prediction for online advertisements [35]. Despite the popularity of TrueSkill, it suffers from a conceptual issue. It ignores interactions of players within a team and assumes their performance is independent of one another. This issue was addressed by several following works through which many different algorithms and extensions were introduced [6, 11, 12, 14].

3.4   *PreviousRank*

Elo, Glicko, and TrueSkill have several similarities in how they model and update a player's skill. To provide a naive baseline, we describe PreviousRank.

PreviousRank simply assumes that a player's predicted rank is equal to their observed rank in their previous match. If a player is new to the system, we assume that their predicted rank is equal to $\frac{N}{2}$ where N is the number of players competing in the match. We use PreviousRank as our naive baseline to achieve a better understanding of the predictive power of other mainstream models.

3.5   *Calculating Predicted Ranks*

Elo, Glicko, TrueSkill, and our naive baseline all maintain a number that can be interpreted as the skill level of a player. This estimation of the player's skill can be updated after every match. These matches might be head-to-head or larger free-for-all matches. Given such ratings, we can predict the ranking of a player in a field of other players.

For an upcoming match, we collect the players in the match. For each player, we retrieve their rating. If the player is new, we use the default rating value. Players are sorted by their ratings, thereby producing a rank prediction for the list of players. Ties in the ratings are randomly broken. By comparing this pre-match predicted ranking to the post-match observed ranking, we can evaluate the performance of the rating systems.

4   Metrics

As shown in the previous section, a rating system can be used to produce a ranking for a field of players in a match. Given the predicted rankings R^{pred} and observed rankings R^{obs} after the match is finished, several metrics can be used to evaluate the performance of a rating algorithm.

Traditional metrics such as accuracy, mean absolute error, and rank correlation coefficients are commonly used for evaluating head-to-head games, but may not be as appropriate in free-for-all games. We describe these three metrics and leverage three additional metrics taken from the field of information retrieval: average precision, mean reciprocal rank, and normalized discounted cumulative gain. These metrics may yield different insights into the rating systems they evaluate. In the remainder of this section, we present these metrics.

4.1 Accuracy

The problem of ranking may be viewed as a classification problem. In a head-to-head match, accuracy can be used as the evaluation metric for a ranking problem with only two outputs or labels (three if a draw is possible). In this case, each rank could be assumed as a nominal or categorical value.

Free-for-all matches can have many more players. We could use the same assumption and treat the problem as a multi-label classification problem, evaluating the rankings based on how accurately they classify players into their observed ranks. As such, accuracy is calculated as the ratio of correctly classified ranks to the total number of players.

Accuracy is not generally suited to ranking problems because it treats all the ranks as labels. If a player was predicted to achieve rank 5 and earns rank 5, that is a hit. But if he earns rank 6 or rank 96, it is a miss even when these two scenarios differ greatly.

4.2 Mean Absolute Error

Mean absolute error (MAE) is one of the most common measures to evaluate the similarity of two sets of values. Assuming two rankings, the predicted ranks R^{pred} and the observed ranks R^{obs} of players, MAE is the average of the absolute errors:

$$MAE = \frac{1}{N} \sum_{i=1}^{N} \left| R_i^{pred} - R_i^{obs} \right|$$

where N is the total number of players competing in a match and R_i^{pred} and R_i^{obs} are the predicted and observed rankings for a player p_i. An MAE of zero means that the two rankings are identical. A higher MAE suggests higher dissimilarities between the two rankings.

MAE is more suited than accuracy to compare two sets of non-ordinal values. Unlike accuracy, the case of a player having the predicted rank of 5 while earning the 96th rank will have a much larger impact on the metric than if the player earned the sixth rank.

However, MAE misses potentially useful information. It does not distinguish between the prediction errors in higher ranks and those in lower ranks. For example, MAE treats the difference between rank 1 and rank 6 the same as the difference between rank 90 and rank 95. Accuracy suffers from the same issue.

4.3 Kendall's Rank Correlation Coefficient

Kendall's rank correlation coefficient, referred to as Kendall's tau and denoted by τ, is a common statistic to measure the ordinal association between two variables [36]. Kendall's tau leverages a more interpretable approach compared to other rank correlation coefficients by looking at the number of concordant and discordant pairs of observations.

Assume two rankings of R^{pred} and R^{obs} as the predicted rank and observed rank of players in a match between N players. For two players, p_i and p_j, any pair of observations (R_i^{pred}, R_i^{obs}) and (R_j^{pred}, R_j^{obs}) are concordant if $R_i^{pred} > R_j^{pred}$ and $R_i^{obs} > R_j^{obs}$, or $R_i^{pred} < R_j^{pred}$ and $R_i^{obs} < R_j^{obs}$. Otherwise, they are considered discordant.

Kendall's tau can be calculated as:

$$\tau = \frac{n_c - n_d}{\binom{N}{2}}$$

where n_c and n_d are the number of concordant and discordant pairs. The denominator is the total number of pair combinations. Tau is equal to 1 if the predicted and observed rankings completely agree, is equal to -1 if they completely disagree, and is 0 if there is no correlation between the two rankings.

Kendall's tau, unlike accuracy and MAE, does not consider the deviation in predicted and observed rankings, but instead considers the pairwise agreement between two rankings. For example, if two players were predicted to have ranks 5 and 10 and achieved ranks 3 and 12, tau considers this a concordant pair without regard to the deviations in predicted versus observed ranks. However, like accuracy and MAE, it does not distinguish between higher rank and lower rank errors.

4.4 Mean Reciprocal Rank

As the first metric adapted from the field of information retrieval, we leveraged mean reciprocal rank (MRR) [26]. It is often used to evaluate the performance of a query-response system that returns a ranked list based on a query.

We extend MRR to the evaluation of rank prediction in online competitive games. Given the predicted ranks, R^{pred}, and observed ranks, R^{obs}, we compute the error for each player as the absolute difference between his predicted rank and observed rank. In a free-for-all match with N players, MRR can be calculated as:

$$MRR = \frac{1}{N} \sum_{i=1}^{N} \frac{1}{1 + error_i}$$

where $error_i$ is the error in prediction. The fraction of $\frac{1}{1+error_i}$ may be considered as the relevance of the prediction for player p_i. When the prediction is perfect, the fraction is 1; the worse the prediction, the closer to 0 it becomes. Therefore, MRR can be considered as a summation of relevance scores.

While the modified MRR applies a different penalty function than MAE, it is similar in that it considers higher penalties for higher differences between predicted and observed ranks. However, like the above metrics, it considers the deviation between rank 1 and rank 6 to be the same as the difference between rank 90 and rank 95.

4.5 Average Precision

Average precision (AP) is the second metric we borrow from the field of information retrieval. Given a ranked list of generated responses for a query, AP uses list-wise precision and relevance scores to evaluate the system [27].

We extend AP to the evaluation of rank prediction in online competitive games. Similar to MRR, we consider $\frac{1}{1+error_i}$ as the relevance score of each prediction. In a free-for-all match with N players, AP can be calculated as:

$$AP = \frac{1}{N} \sum_{i=1}^{N} P(i) \times \frac{1}{1 + error_i}$$

where $P(i)$ is the overall precision value up to the ith position and $error_i$ is the error in prediction.

AP works exactly like MRR if all predictions are correct. However, AP is generally more strict since it weights relevance scores for each position with the overall precision value up to that position. Using precision values as weights causes AP to distinguish between prediction errors in higher ranks and lower ranks. However, it puts a higher concentration on hits, especially in higher ranks. This may have negative impacts on the evaluation of a model whose overall performance is great but its first few incorrect predictions occur in higher ranks, even when the prediction error is as small as 1 rank. AP may be an appropriate metric for evaluating systems whose main focus is on high-rank or top-tier players.

4.6 Normalized Discounted Cumulative Gain

Normalized discounted cumulative gain (NDCG) is the third metric we adapt from the field of information retrieval. Given a ranked list of responses for a query, NDCG evaluates the quality of the generated responses based on their relevance score and weighted position in the list [28]. The overall score is accumulated from individual scores at each level from the top of the list to the bottom.

We extend NDCG to the evaluation of rank prediction in online competitive games. Similar to MRR and AP, we consider $\frac{1}{1+error_i}$ as the relevance score of predictions. In a free-for-all match with N players, NDCG can be computed as:

$$NDCG = \frac{\sum_{i=1}^{N} \frac{1}{log_2(i+1)} \times \frac{1}{1+error_i}}{IDCG}$$

where $error_i$ is the prediction error, $\frac{1}{log(i+1)}$ is the weight assigned to the ith position, and $IDCG$ is a normalizing factor.

Similar to AP, NDCG distinguishes between errors in higher ranks and lower ranks by weighting the evaluation of each position. However, in NDCG, we can adjust the weights based on the evaluation goals. For example, by increasing the weights, the model may direct its attention to evaluating good or regular players who often appear in higher ranks.

5 Methodology

In this section, we introduce the dataset used to perform our experiments. We detail our methodology along with the parameters we used to implement the rating systems. Finally, we explain how we performed our evaluations.

Player Unknown's Battlegrounds (PUBG) is a popular free-for-all online multi-player video game developed and published by PUBG Corporation. The game pits up to 100 players in a battle royale match against each other. The last player or team standing wins the match. PUBG is played in teams of four, teams of two, or singletons. The dataset is publicly available on *www.kaggle.com*, a public data platform.

For this research, we only considered solo matches in the dataset, free-for-all matches where each player competes against every other player at the same time. In this paper, we are concerned with how to evaluate a single player, a necessary step before we can evaluate a team. The dataset provides in-game statistics such as the number of kills, distance walked, and rank for over 100,000 unique matches and 2,260,000 unique players.

We sorted the matches by their timestamps. For each match, in order, we retrieved the list of players. New players—those that have yet to appear in a match—were assigned default ratings, 1500 for Elo and Glicko and 25 for TrueSkill. We also retrieved the skill rating of returning players.

Based on these skill ratings, we sorted the players and used this sorting as the rank prediction for the match. The ratings for the players were updated after each match by comparing their predicted ranks with their observed ranks. The parameters we used for each rating system include $k = 10$ and $D = 400$ for Elo and $\beta = 4.16$ and $\tau = 0.833$ for TrueSkill as suggested in its official documentation.

For each match, the metrics (accuracy, MAE, etc.) were computed creating a time-series reporting the performance of the rating systems. To aid in our exploration of the evaluation metrics, we implemented four different set-ups.

First, we evaluated the performance of the rating systems for all players and all unique matches in the dataset sorted by date. In this set-up, every player is treated equally regardless of their skill or how often they play.

Second, we evaluated the performance of the rating systems for the best players in the system. To identify the best players, we selected 1000 players with the highest ratings who had played more than 10 games. Since these players did not compete at the same time, we evaluated the predictive performance of the rating systems on their first ten games.

Third, we evaluated the performance of the rating systems for the most frequent players in the system. To identify the most frequent players, we randomly selected 1000 players who played more than 100 games. We evaluated the predictive performance of the rating systems on their first 100 games.

Finally, we evaluated the performance of the rating systems for binned ranks—a grouping of players based on their observed ranks. For each individual match, we divided the competing players into five different bins. The top 20% players of each match may be considered as skilled players, while the last 20% may be viewed as novice players. The three middle bins may contain seasonal players or players who are still learning the game and advancing their skills. New players may negatively influence the system's performance in this set-up. Almost half of the players in the dataset only played one game. The rating systems do not have any knowledge about these players and yet they can be placed in any of the five bins. For each evaluation metric, we averaged the score of each rating system for each bin over all matches.

6   Results and Discussions

In this section, we discuss the results of four experiments on all players, best players, most frequent players, and binned ranks. We compare the performance of the competing models using six evaluation metrics discussed earlier. Finally, we analyze the ability of these metrics in capturing prediction patterns of the rating systems.

The results of these experiments are given in Fig. 1. In this figure, the rows correspond to evaluation metrics and the columns are associated with experimental set-ups. The performance of the models is represented by trend lines with different colors. For example, in accuracy plot for the best players, the accuracy of TrueSkill, shown by the red trend line, starts from 0.5% and increases up to 8.5% after ten games.

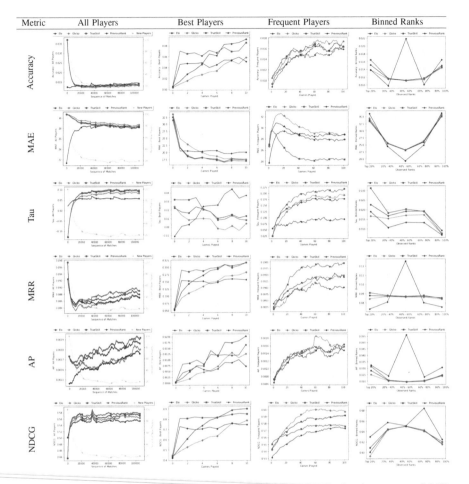

Fig. 1 The results of evaluating Elo, Glicko, TrueSkill, and PreviousRank using accuracy, MAE, Kendall's tau, MRR, AP, and NDCG on four different experimental set-ups: all players, best players, frequent players, and binned ranks

6.1 Accuracy

Previous works demonstrated that Elo, Glicko, and TrueSkill achieve high accuracy for predicting ranks in head-to-head games. However, the results shown in Fig. 1 indicate that it does not hold true for free-for-all games.

The observed accuracy values are fairly small in all set-ups. The highest value is around 10% in the case of Elo for the best players. In addition, the results display different patterns. For example, accuracy suggests that PreviousRank, Elo, and Glicko have relatively better performance for evaluating all players, best players, and frequent players, respectively.

Accuracy seems to be influenced by factors such as the number of new players, players' behavior, and their frequency of play. We expect the rating systems to achieve a better knowledge of players' skills over time by observing more games. For all players, Elo, Glicko, and TrueSkill achieved higher accuracy than PreviousRank at the early stages of the sequence of matches where most of the players are newly added to the system (showed by the gray trend line). These matches contain all or many ties in ratings that are randomly broken by the rating systems. The patterns show that the rating systems cannot outperform random predictions of the early stages of the sequence. Accuracy values significantly drop as the number of known players to the system increases. However, the patterns observed for the best and frequent players suggest the opposite when the influence of new players is excluded. The ratings converge faster for the best players—the players whom we expect to show consistent behaviors.

Results of binned ranks reveal that the models can correctly distinguish between different types of players based on accuracy. Elo, Glicko, and TrueSkill achieved higher accuracy for players in the first and last bins who are assumed to demonstrate consistent behavior (highly skilled players and novice players). On the other hand, the models achieved less accuracy for the middle bins that presumably contain players with inconsistent behavior (less known or seasonal players).

Accuracy is a reliable metric for evaluating head-to-head games. However, as the results suggest, it is not suited to evaluating free-for-all games. Accuracy highly exaggerates the negative influence of new players on the performance of rating systems. It also treats all the ranks as labels and thus is unable to explain the real differences between two ranks. Finally, since accuracy only considers hits as good predictions, it is unable to capture a large part of the predictive behavior of rating systems.

6.2 MAE

The results of evaluating rating systems using MAE suggest significant errors for predicting rank in free-for-all games. For example, the lowest MAE value observed was 17 for TrueSkill in the case of the best players.

The results display fairly similar patterns for Elo, Glicko, and TrueSkill in all, best, and frequent players set-ups. For example, TrueSkill achieved the lowest MAE values, while Glicko experienced higher errors in all these three set-ups.

While accuracy was mostly influenced by the number of new players, MAE seems fairly resilient in comparison. Elo, Glicko, and TrueSkill achieved a better knowledge of players improving with a slow rate as the number of new players in each match decreases over time. On the other hand, MAE seems to be highly influenced by players' behavior. The models demonstrate considerable improvements by observing more games from the best players. However, for frequent players set-up that includes players with different skills, the patterns are not as clear.

Finally, the MAE plot for binned ranks suggests that the rating systems achieved the lowest errors for the three middle bins (i.e., players with inconsistent behaviors or less-known players) while achieving higher errors for the first and last bins (i.e., players with consistent behaviors or known players).

While MAE is more suited than accuracy for evaluating rank predictions, the results suggest that it is not suited for evaluating free-for-all games. MAE presents a global evaluation of the model's performance. However, since MAE does not consider rank positions, it cannot explain the real differences between players and fails to capture the evaluation details.

6.3   Kendall's Tau

Similar to accuracy, Kendall's tau seems to amplify the influence of new players. Elo, Glicko, and TrueSkill show negative correlations at the early stages of the sequence in all players set-up. However, the correlations increase with a fast rate as the number of new players in each match decreases.

The patterns observed for the best players suggest that Kendall's tau is unable to capture the learning ability of the rating systems from players with consistent behavior. Elo, Glicko, and TrueSkill demonstrate decreasing correlation values as they observe more games, while PreviousRank shows an upward trend. On the other hand, the frequency of play seems to be an important factor influencing the evaluations of Kendall's tau. The results of the frequent players set-up suggest that, regardless of players' skill levels, Kendall's tau captures the ability of rating systems in achieving a better knowledge of players by observing more games. The learning demonstrated by Kendall's tau for frequent players happens much faster than what we observed for accuracy and MAE. For example, TrueSkill starts from 2.5% correlation while constantly increasing to reach 22% correlation at the end of the 100th game.

Finally, the patterns observed in the binned ranks plot imply that based on Kendall's tau, Elo, TrueSkill, and Previous Rank perform best for the first bin which consists of highly skilled players. This is inconsistent with the patterns Kendall's tau showed for the best players set-up.

The results suggest that Kendall's tau is not suited to evaluating rank predictions in free-for-all games. It cannot represent the real predictive power of rating systems under the influence of new players. It also fails to capture the learning patterns of the systems for top-tier players over time. Kendall's tau only considers the pairwise agreement between predicted and observed ranks without regard to their deviations. Therefore, it cannot capture the real differences between players.

6.4 MRR

MRR correctly captured the ability of rating systems in achieving more knowledge of players by observing more games based on their playing behavior and frequency of play. The models show different learning patterns. The fastest learner is TrueSkill in both best and frequent players set-ups. MRR also suggests that TrueSkill is the best performer overall.

Results of MRR show fairly similar patterns to accuracy. However, MRR values are higher than accuracy for all set-ups. In addition, MRR seems to be more resilient than accuracy to the influence of new players. Accuracy required at least 80% of players in the match to be known to the system in order to start its upward trend after the early drop, while this value for MRR is around 60%. All models also learn faster based on MRR compared to what we observed for accuracy.

Finally, based on the results of MRR for binned ranks, Elo and TrueSkill show a fairly similar pattern to that of Kendall's tau. Elo and TrueSkill performed their best predictions for the first bin while showing higher errors for the last bin. On the other hand, Glicko and PreviousRank show the exact opposite of what we expected, higher MRR values for the middle bins and lower MRR values for the first and last bins.

Although MRR is a generic rank-based evaluation metric (like Kendall's tau), it relatively captured the ability of rating systems to learn the player's behavior over time. However, it is highly influenced by the number of new players in matches. It also does not distinguish between higher rank and lower rank prediction errors. Using metrics that adjust their evaluation scores based on the prediction error's position may better benefit the evaluation by giving us the flexibility to adjust the evaluations based on our goals.

6.5 Average Precision

AP patterns suggest that the models learn more about the players over time by observing more games. This is particularly true for players with consistent behavior and players who play frequently. However, AP suggests a fairly slow learning process for all models.

While accuracy, Kendall's tau, and MRR were substantially influenced by the number of new players and changed dramatically over the first few matches in all players set-up, the models experience a small drop in AP values for the same matches and start improving with a fairly fast rate afterward. This pattern demonstrates that AP is more resistant to the influence of new players in the system. For example, TrueSkill corrected its early downward trend when at least 45% of the players in a match are known to the system.

Finally, AP results for binned ranks show that Elo, Glicko, and TrueSkill achieved higher AP for players who are assumed to be known to the system (the first and last bins) while displaying more uncertainties for less-known players (the middle bins). This pattern is fairly similar to that for accuracy.

AP demonstrated three main benefits over previous metrics. First, it is more resistant to the influence of new players. Second, it considers rank positions when evaluating predictions and thus can capture the differences between players. Finally, it more accurately makes predictions for the four experimental set-ups. However, AP values are extremely small. AP provides strict evaluations of the models by using precision values as weights for the relevance of each prediction, corresponding to the relative importance of each position. These weights may over-penalize trivial errors that are not considered bad predictions in other metrics.

6.6   NDCG

NDCG accentuates the high learning ability of the rating systems based on the consistent behavior of players and their frequency of play. For example, in the best players set-up, TrueSkill shows a significant increase from 45% to 80% after observing one game. The patterns show that Elo is the best performer for the best players achieving NDCG of 90% at the end of the tenth game. On the other hand, Glicko shows the best performance for the most frequent players.

NDCG seems to be the metric least affected by the influence of new players in the system. It starts increasing from the start and improves with a highly fast rate as the number of new players in matches decreases. The learning process gradually becomes slower as the number of new players in each match does not change much.

Finally, although NDCG showed good performance in capturing the predictive performance of the rating systems, the patterns observed for binned ranks contradict our expectations. The models achieved their best performance for the middle bins while achieving lower scores for the first and last bins. As mentioned before, this could well be the result of the large number of new players who may be placed in either of bins.

NDCG alleviates all the challenges faced by other metrics. The weighting factor used in NDCG is separate from the predictions and is directly based on the positions. Therefore, before evaluating the predictions, we can adjust the weights based on the goals of the system. The results suggest that NDCG correctly captures the learning ability of the rating systems for both players' behavior and frequency of play. It can also represent the predictive power of the rating systems even when a large number of players in a match are new to the system. Finally, NDCG seems to be the best metric for evaluating rank predictions in free-for-all games.

7 Conclusion and Future Works

In this paper, we evaluated the predictive performance of three popular rating systems in free-for-all games. We performed our experiments on four different groups of data to paint a clear picture of the evaluations.

The results indicated that many metrics were negatively influenced by the number of new players in each match. Some metrics captured the ability of rating systems to learn more about the behavior of players by observing more games. Others correctly captured the differences between players. Some metrics while being well suited to evaluate the rating systems on a certain group of players may not be appropriate for other groups of players. Achieving better predictions for top-tier players is particularly more important since these players often stay in the system and play more games.

Among all metrics tested, NDCG best represented the predictive power of rating systems while resolving all challenges faced by other metrics. It was more resistant than other metrics to the influence of new players. It also correctly captured the learning patterns of these systems based on both player's behavior and frequency of play.

Our experimentation was limited to free-for-all games. Evaluating other modes of game-play is part of our future work. This work is a part of more comprehensive research on group assignment in online competitive games. Evaluating rank predictions is the first step in building a framework for predicting rank. We plan to extend rating systems by incorporating players' behavioral features to achieve better predictions. We will extend rank prediction to building a more comprehensive framework for predicting the success of proposed teams and making assignments.

References

1. A.E. Elo, *The Rating of Chessplayers, Past and Present* (Arco, Portugal, 1978)
2. J. Ilic, *Number of Online Gamers to Hit 1 Billion by 2024*. https://leagueofbetting.com/number-of-online-gamers-to-hit-1-billion-by-2024/. Accessed 2020-04-22
3. M. Myślak, D. Deja, Developing game-structure sensitive matchmaking system for massive-multiplayer online games, in *International Conference on Social Informatics* (Springer, Berlin, 2014), pp. 200–208
4. R. Herbrich, T. Minka, T. Graepel, Trueskill^{TM}: a bayesian skill rating system, in *Advances in Neural Information Processing Systems* (2007), pp. 569–576
5. L. Zhang, J. Wu, Z.-C. Wang, C.-J. Wang, A factor-based model for context-sensitive skill rating systems, in *Proceedings of the 2010 22nd IEEE International Conference on Tools with Artificial Intelligence*, vol. 2 (IEEE, New York, 2010), pp. 249–255
6. O. Delalleau, E. Contal, E. Thibodeau-Laufer, R.C. Ferrari, Y. Bengio, F. Zhang, Beyond skill rating: Advanced matchmaking in ghost recon online. IEEE Trans. Comput. Intell. AI in Games **4**(3), 167–177 (2012)
7. J.E. Menke, T.R. Martinez, A bradley–terry artificial neural network model for individual ratings in group competitions. Neural Comput. Appl. **17**(2), 175–186 (2008)

8. D. Buckley, K. Chen, J. Knowles, Predicting skill from gameplay input to a first-person shooter, in *Proceedings of the 2013 IEEE Conference on Computational Inteligence in Games (CIG)* (IEEE, New York, 2013), pp. 1–8
9. J.R. López-Arcos, F. Gutiérrez, N. Padilla-Zea, N.M. Medina, P. Paderewski, Continuous assessment in educational video games: a roleplaying approach, in *Proceedings of XV International Conference on Human Computer Interaction* (2014), pp. 1–8
10. M.E. Glickman, Parameter estimation in large dynamic paired comparison experiments. J. R. Stat. Soc. Series C (Appl. Stat.) **48**(3), 377–394 (1999)
11. C. DeLong, N. Pathak, K. Erickson, E. Perrino, K. Shim, J. Srivastava, Teamskill: modeling team chemistry in online multi-player games, in *Pacific-Asia Conference on Knowledge Discovery and Data Mining* (Springer, Berlin, 2011), pp. 519–531
12. I. Makarov, D. Savostyanov, B. Litvyakov, D.I. Ignatov, Predicting winning team and probabilistic ratings in "dota 2" and "counter-strike: global offensive" video games, in *Proceedings of the International Conference on Analysis of Images, Social Networks and Texts* (Springer, Berlin, 2017), pp. 183–196
13. R.C. Weng, C.-J. Lin, A bayesian approximation method for online ranking, J. Mach. Learn. Res. **12**(Jan), 267–300 (2011)
14. S. Guo, S. Sanner, T. Graepel, W. Buntine, Score-based bayesian skill learning, in *Joint European Conference on Machine Learning and Knowledge Discovery in Databases* (Springer, Berlin, 2012), pp. 106–121
15. S. Chen, T. Joachims, Predicting matchups and preferences in context, in *Proceedings of the 22nd ACM SIGKDD International Conference on Knowledge Discovery and Data Mining* (ACM, New York, 2016), pp. 775–784
16. Y. Li, M. Cheng, K. Fujii, F. Hsieh, C.-J. Hsieh, Learning from group comparisons: exploiting higher order interactions, in *Advances in Neural Information Processing Systems* (2018), pp. 4981–4990
17. S. Motegi, N. Masuda, A network-based dynamical ranking system for competitive sports. Sci. Rep. **2**, 904 (2012)
18. J. Ibstedt, E. Rådahl, E. Turesson, et al., *Application and Further Development of Trueskill*[TM] *Ranking in Sports* (2019)
19. S. Cooper, C.S. Deterding, T. Tsapakos, Player rating systems for balancing human computation games: testing the effect of bipartiteness, in *Proceedings of the 1st International Joint Conference of DiGRA and FDG*. DIGRA Digital Games and Research Association (2016)
20. B. Morrison, *Comparing ELO, Glicko, IRT, and Bayesian IRT Statistical Models for Educational and Gaming Data* (2019)
21. M. Stanescu, Rating systems with multiple factors, in *Master's thesis, School of Informatics, University of Edinburgh, Edinburgh, UK* (2011)
22. L. Yu, D. Zhang, X. Chen, X. Xie, Moba-slice: a time slice based evaluation framework of relative advantage between teams in moba games (2018). arXiv preprint arXiv:1807.08360
23. J. Lasek, Z. Szlávik, S. Bhulai, The predictive power of ranking systems in association football. Int. J. Appl. Pattern Recognit. **1**(1), 27–46 (2013)
24. D. Buckley, K. Chen, J. Knowles, Rapid skill capture in a first-person shooter. IEEE Trans. Comput. Intell. AI Games **9**(1), 63–75 (2015)
25. C.D. Manning, P. Raghavan, H. Schütze, *Introduction to Information Retrieval* (Cambridge University, Cambridge, 2008)
26. E.M. Voorhees, The trec-8 question answering track report, in *Proceedings of the Trec*, vol. 99 (Citeseer, New York, 1999), pp. 77–82
27. G. Salton, Developments in automatic text retrieval. Science **253**(5023), 974–980 (1991)
28. K. Järvelin, J. Kekäläinen, Cumulated gain-based evaluation of ir techniques. ACM Trans. Inf. Syst. (TOIS) **20**(4), 422–446 (2002)
29. M.E. Glickman, *The Glicko System*, vol. 16 (Boston University, Boston, 1995)
30. F.R. Kschischang, B.J. Frey, H.-A. Loeliger et al., Factor graphs and the sum-product algorithm. IEEE Trans. Inf. Theory **47**(2), 498–519 (2001)

31. T.P. Minka, *A Family of Algorithms for Approximate Bayesian Inference*. Ph.D. dissertation (Massachusetts Institute of Technology, Cambridge, 2001)
32. J. Huang, T. Zimmermann, N. Nagapan, C. Harrison, B.C. Phillips, Mastering the art of war: how patterns of gameplay influence skill in halo, in *Proceedings of the SIGCHI Conference on Human Factors in Computing Systems* (ACM, New York, 2013), pp. 695–704
33. C. Kawatsu, R. Hubal, R.P. Marinier, Predicting students' decisions in a training simulation: a novel application of trueskill. IEEE Trans. Games **10**(1), 97–100 (2017)
34. L.C. Quispe, J.E.O. Luna, A content-based recommendation system using trueskill, in *Proceedings of the 2015 Fourteenth Mexican International Conference on Artificial Intelligence (MICAI)* (IEEE, New York, 2015), pp. 203–207
35. T. Graepel, J.Q. Candela, T. Borchert, R. Herbrich, Web-scale Bayesian click-through rate prediction for sponsored search advertising in microsoft's bing search engine (Omnipress, Madison, 2010)
36. M.G. Kendall, *Rank Correlation Methods* (1948)

A Holistic Analytics Approach for Determining Effective Promotional Product Groupings

Mehul Zawar, Siddharth Harisankar, Xuanming Hu, Rahul Raj,
Vinitha Ravindran, and Matthew A. Lanham

1 Introduction

The pricing strategy adopted by different companies is a crucial deciding factor that impacts their revenue and margin. Selecting the right pricing tactics and applying them in the right manner can have a major impact on the bottom line. Promotional product grouping (or PPGs) is a technique where multiple products are grouped and sold as a single unit for one price. Grouping or bundling is a way to get customers to buy multiple products. The discounted price of such bundles attracts customers in droves. This could be advantageous to a company not only because of the increased sales but also because bundling enables consumers to discover new products that are grouped with goods that are commonly purchased. As with any strategy, improper management of bundling could hurt sales. As per a research report published by The Boston Consulting Group, 20–50% of promotions generate no noticeable lift in sales. Another 20–30% dilute margins, where the increase in sales does not offset the cost of promotion [1].

A survey of marketing professionals from leading companies was conducted to analyze the problems faced by different companies while creating a good promotion plan. Some of the issues identified from the survey include not having a track of their effective promotions that help build profits. Hence, the companies are hesitant to cut back the sales volume, leading them to run all the promotions year long. This not only results in monetary losses but is also driving customers toward better opportunities elsewhere. The major reason for the failure of the promotions is due to the lack of a clear goal. Companies do not consider whether the promotions are

M. Zawar (✉) · S. Harisankar · X. Hu · R. Raj · V. Ravindran · M. A. Lanham
Department of Management, Purdue University, West Lafayette, IN, USA
e-mail: mzawar@purdue.edu; sharisan@purdue.edu; hu661@purdue.edu; rajr@purdue.edu;
ravindr1@purdue.edu; lanhamm@purdue.edu

© Springer Nature Switzerland AG 2021
R. Stahlbock et al. (eds.), *Advances in Data Science and Information Engineering*,
Transactions on Computational Science and Computational Intelligence,
https://doi.org/10.1007/978-3-030-71704-9_10

being held to improve margins, to reach out to new customers, or to improve some other metrics.

This chapter looks at the promotional grouping strategy employed by a global consumer products firm (hereafter referred to as "Company"). The research conducts a competitive analysis of similar groupings sold across the industry. This analysis could help identify the areas where the Company is finding success with its strategy and where it is surpassed by its competitors. Several parameters, such as total sales and quantity sold, are taken to generate a score, which is then compared across the various competitors to identify which grouping had the highest "success." Building on this analysis, the study uses machine learning models to estimate the importance of various drivers behind a successful promotional bundle. The model created considers a systematic analysis of historical performance to determine whether specific promotions are meeting the Company's strategic objectives. It learns from past PPG groupings, sales patterns, and the PPGs of the competitors and will then recommend what PPG label to place at each SKU level. For example, a particular flavor of a shampoo grouped with other personal care products could prove to have traction with customers. This could be successfully adopted by the Company to boost sales. Finally, the study develops an optimization model to generate promotional groupings that can be adopted by the Company to maximize revenue.

An advantage of using machine learning to study and generate promotional groupings is that it would identify trends in customer preferences. The model constantly learns from these patterns and suggests groupings that capture customer requirements of today. This is a far cry from employing a manual approach to identifying product bundles, which could fail to identify the changing trends in customer preferences. In a cut-throat industry, any lapses or delays in adapting to market requirements could lead to the Company losing significant market share.

The remainder of this chapter is organized as follows: A review of the literature on various criteria and methods used for PPG selection is presented in the next section. In Sect. 3, the proposed methodology is presented, and the criteria formulation is discussed. In Sect. 4, various models are formulated and tested. Section 5 outlines the performance of our models. Section 6 concludes the paper with a discussion of the implications of this study, future research directions, and concluding remarks.

2   Literature Review

Prior research on product bundling encompasses various types of product groupings, across various industries. Though there are fundamental differences in the approaches, a majority of the studies agree upon the fact that bundling can be viewed similar to volume discount where the volume is based on aggregate sales across products [1]. Currently, bundling is approved and reviewed manually, which is time-

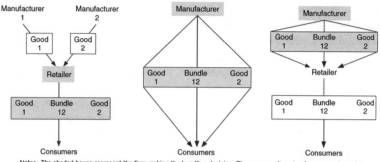

Notes. The shaded boxes represent the firm making the bundling decision. The c_is are unit costs of component goods.

Fig. 1 Different distribution structures for product bundling

consuming and could result in human errors. The focus is to utilize modeling to help consumer product companies identify these drivers and how their performance can be improved with respect to their competitors.

Bundling can be done at various levels. Manufacturers could group the products before passing the bundles to retailers or sell directly to consumers. Bhargava [2] details a third type of bundling wherein retailers sell a bundle of component goods from independent manufacturers, a practice common in travel, technology, and media industries. He depicts the various types of bundling as shown in Fig. 1.

The type of bundling is also studied by Cao et al. [3], where they examine how bundling decisions affect wholesale price and profit.

However, this approach would not apply to the problem in hand as it would not make sense for a consumer products company to sell their products alongside a competitor's. Hence, the main point of focus would be on the third method, where the bundling decision is made by the manufacturer before passing the products on to the retailer, as depicted in the diagram. Several factors, such as marketing and economic aspects, govern the decision to bundle products. However, there are other interesting aspects which could be considered as well. Sheikhzadeh and Elahi [4] examine the impact of heterogeneity, such as the difference in average prices, in the products to be bundled. Perhaps a more important feature would be the firm's stance on taking a risk. Some firms might prefer to experiment with novel bundling options, with the aim of increasing the expected profit, while reducing the profit variance.

While analyzing product groupings, studying the characteristics of the bundle would not yield much information as purchase patterns of the component products could vary significantly. Hence, it would be important to analyze the features of the individual products and identify characteristics that enable them to be grouped correctly.

In order to develop a model that helps companies automatically assign bundles, the study examined related papers detailing multi-classification models. Tang et al.

[5] analyze different approaches of support vector machines (1 v. 1 and 1 v. rest) to develop new SVM methods based on a binary tree, which resolve unclassifiable region problems common in a conventional multiclass SVM. Their methods help reduce training time while maintaining acceptable accuracy. Fabio Aiolli and Alessandro Sperduti [6] also introduced a novel approach of SVM Single- and Multi-Prototype SVM. This allows us to combine several vectors in a principled way to obtain large margin decision functions, and the model reduces overall problem for a series of more straightforward convex problems, with faster processing speed. Cheng and Hüllermeier [7] conducted ML-KNN (Multi-label K-Nearest Neighbor), considering the correlations among the different labels. Thus, they are not only considering the independent variables resulting in binary classifications but also how does classification vary for different labels. Valizadegan et al. [8] consider the information loss while building a multi-classification model using the binary tree approach. Valizadegan and his team proposed a semi-supervised boosting framework model named Multi-Class Semi-Supervised Boosting (MCSSB). This model exploits both classification confidence and similarities among examples when deciding the pseudo-labels for unlabeled examples.

While the above research methods have employed sophisticated models that improve the accuracy, a lot of the interpretability of the results is lost. As a result, they cannot accurately identify what the best grouping for a specific product is. Furthermore, multiclassification methods are hard to provide accurate results when the target variable has a large number of levels, as is the case in this study. Therefore, this research aims to leverage logistic regression to identify the key drivers of a successful grouping under each sub-category. Subsequently, the research develops a new approach to analyze data from consumer products companies and provide solutions for product bundling, namely, using optimization techniques to find out what the PPG bundle can maximize the overall revenue for the company. Many researchers have used standard optimization software to analyze the relationships between different products in the vendor. While the goal of that study is different, it provides a methodology that is taken as a barometer on how to utilize optimization techniques in product bundling. Ye et al. [9] conducted a more comparable study, where they develop a computationally efficient optimization algorithm to approximate the optimal product bundle and identifies conditions that would generate highly profitable bundling.

3 Data

The data for this study was provided by the Company. The data consists of two files – "Transaction data" and "Promotions data."

Transaction data contains 553,651 observations and 21 variables over 4 years, from January 2016 to December 2019. The datasets consist of product hierarchy, description, weekly sales amount and quantity, and number of stores the product was in display for every SKU. All the products under transaction data are from

Table 1 Summary of transactional data by category

Category	Count of transactions
Home care	90,860
Oral care	109,766
Personal care	353,025
Grand total	**553,651**

Table 2 Transactional data dictionary

Particulars	Data-type	Description
Category	String	Highest level of product hierarchy
Product-category	String	Child of 'Category'
Sub-category	String	Child of 'Product-category'
Manufacturer	String	Company who produces/ sells products
Brand	String	Child of 'Product-category'
Sub-brand	String	Child of 'Brand'
Variant	String	Child of 'Sub-brand'
Pack type	String	Packaging type
Size	String	Number of L, mls, gms, KGs, pks that a product contains per consumer unit sold
Segment	String	Shopper positioning of the product
Product name	String	Product name
Size range	String	Size range
Unit of measure	String	Measurement unit
Sub-segment	String	Child of 'Segment'
UPC/SKU	Integer	Unique product code
Sales value	Float	The sales amount of the product for the 'Customer' by 'Week'
Store count	Integer	Number of stores product is displayed at for 'Customer' by 'Week'
Sales quantity	Integer	The sales quantity of the product for the 'Customer' by 'Week'
Average sale price	Float	Average selling price of product for a 'Customer' in a given week
customer number	Integer	Customer ID
Week	Timestamp	7 day period of sales for the product within the customer

three categories: home care, oral care, and personal care, and 21 sub-categories, and sub-categories are crucial in later analysis (Table 1).

Data dictionary for transaction data is as follows (Table 2):

Promotion data contains 559,023 records and 8 variables. Similar to transaction data, promotion data includes records from January 2016 to December 2019. The one variable that different is example PPG, which is the PPG bundle that been signed by Company (Table 3).

Table 3 Promotional data dictionary

UPC/SKU	Integer	Unique product code
Sales value	Float	The sales amount of the product for the 'Customer' by 'Week'
Store count	Integer	Number of stores product is displayed at for 'Customer' by 'Week'
Sales quantity	Integer	The sales quantity of the product for the 'Customer' by 'Week'
Average sale price	Float	Average selling price of product for a 'Customer' in a given week
Customer number	Integer	Customer ID
Week	Timestamp	7day period of sales for the product within the customer
Example PPG	String	PPG bundle signed to individual product in a given week

4   Methodology

The primary research of the chapter focuses on analytically examining the promotions and identifying the significant factors that drive those promotions. Once the factors that affect the promotions are identified, the optimal product bundles, which will considerably boost the sales, are recommended in a more systematic, data-driven way that utilizes machine learning to bring hidden insights to life.

The study identifies the following questions that will help the business take an analytical approach to maximize the impact of their bundles:

- What are the drivers of promotional groupings?
- How can better predictions of promotional groupings provide essential insights, decision-support, cost-savings, etc., to the business?

The study uses Python so that the codes can be easily reproduced, and the workflow can be integrated with the company's existing capabilities, solving the aforementioned problems.

4.1   Process Flow of Methodology (Fig. 2)

4.2   Data Preprocessing

Data cleaning steps were performed to make the data suitable for analysis. In the Quantity column, there were some rounding-off errors in the data obtained from the system. Since quantities cannot have any decimal points, they are rounded off to the nearest 0. Furthermore, some numbers in the sales column appeared to be negative. These transactions were dropped from the dataset after receiving confirmation from the client.

Typically, one UPC code is assigned to a product. However, in the dataset, some UPC linked to more than one product. On closer inspection of the cases, this case was identified as a system issue where new names were stored in the system

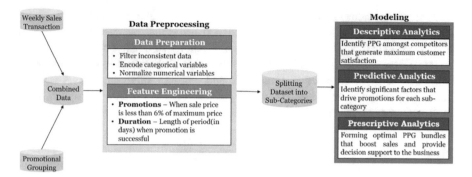

Fig. 2 Abridged methodology process flow

along with old names. Such inconsistencies have been corrected by aggregating the transactions.

4.3 Feature Engineering

Further information is required to obtain the solution for the research questions defined above. The dataset has details of promotion groupings at a transactional level. However, identifying promotion success is still uncertain. The first thought process was to identify for each sub-category what promotions were successful.

After conducting some detailed analysis, the study decided a promotional success period to be defined as from the starting date to the date when the store count for that bundle falls below 100. If the store count falls below 100, then it is considered that the promotion is no longer active. Thus, the duration of promotion is defined below:

$$\text{Duration} = T_{i\_below\_100} - T_{i\_start}$$

$T_{i\_below\_100}$: the first day that promotion goes below to 100 after the promotion. start for product i.
T_{i_start}: the first day that promotion start for product i.

To identify whether the product was sold at its base price or a discount, a Promotions variable was created. If the product was sold at a value lower than 6% of the base price, then the Promotions is 1, and is 0 otherwise.

$$\text{Promotions} = \text{Price}_{xt} \le 0.94 * \max(\text{Price}_x)$$

$\max(\text{Price}_x)$: Max price of Product x during the given period,
Price_{xt}: Price of Product x sold at time t.

In order to incorporate the factors and find the product bundle with the high success of the promotion, a score is calculated using the revenue, quantities, and duration values from the data. The values are standardized using a min-max scaler so that all the values are relative to each other and fall between 0 and 1. An equal weightage is given to revenue, quantities, and duration value, and the average is taken to form the score value. This score gives a value to the bundles, which has a high consumer preference:

$$\text{Score} = \text{Average}\left(\frac{x - \min(x)}{\max(x) - \min(x)} + \frac{y - \min(y)}{\max(y) - \min(y)} + \frac{z - \min(z)}{\max(z) - \min(z)}\right)$$

where

x: Revenue,
y: Sales quantity,
z: Duration.

The PPG with the highest score across sub-category and size range is called a 'Model PPG' which is compared with the corresponding bundles of each manufacturer at the same level. The target variable is created, which is 1 if the bundle is the same as the Model PPG, and 0 otherwise.

5 Modeling and Results

5.1 Logistic Regression

The objective of this modeling is twofold. First, to understand the factors that drive the success of the promotion across each sub-category. Second, to quantify and estimate the probability of success of a promotion using the significant factors across each sub-category.

This study employs logistic regression as it is one of the most popular methods to solve this binary classification problem. It uses a Sigmoid distribution, and it is easier for organizations to interpret variables and implement this algorithm for decision-making (Fig. 3).

Promotion success is the target variable to be analyzed. The data is filtered for the specific manufacturer who excels in promotions of that sub-category. The independent variables used to understand the important attributes are: Variant, Pack Type, Segment, Size range, Sub-segment, and Promotions. Since all the independent variables are categorical, one-hot encoding is implemented. The data is then oversampled to balance the classes and modeled using "Statsmodels" in Python for logistic regression.

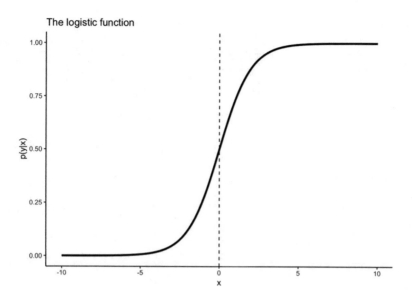

Fig. 3 Sigmoid function for logistic regression

5.1.1 Results

Identifying Important Attributes (Table 4)

The above results depict the modeling for shower gels. The levels Men's and Milk/Skincare under Segment, the levels 251–400 ml, 401–500 ml and 501–750 ml under Size Range, and Promotions have a significant impact on the success of a promotion as the p values are less than the 95% confidence interval. Although other levels within the variables are not significant in the model, due to the hierarchy principle, all levels are included when one of the classes is significant.

A similar approach is executed to observe the important features for other sub-categories. The results are shown as follows (Table 5):

Estimating Probability of Success

To interpret the coefficients, the logistic regression uses the equation:

$$p(x) = \frac{e^{\beta_0 + \beta_1 x}}{1 + e^{\beta_0 + \beta_1 x}}$$

From the above results, the probability of success of a promotion is estimated for one variable holding other factors constant.

Table 4 Identifying important attributes for a PPG at sub-category level

	coef	std err	z	P>| z |	[0.025	0.975]
Promotions_1_0	−0.9849	0.104	−9.477	0.000	−1.189	−0.781
Segment_EPOS_Sgel Men's	2.0342	0.148	13.716	0.000	1.743	2.325
Segment_EPOS_Sgel Milk/Skincare	4.6818	0.366	12.789	0.000	3.964	5.399
Size_Range_EPOS_Sgel 251-400ml	−14.6447	4.431	−3.305	0.001	−23.329	−5.960
Size_Range_EPOS_Sgel 401-500ml	1.0792	0.096	11.282	0.000	0.892	1.267
Size_Range_EPOS_Sgel 501-750ml	−9.2551	0.678	−13.645	0.000	−10.584	−7.926
Size_Range_EPOS_Sgel 751-1500ml	11.9239	7.705	1.547	0.122	−3.178	27.026
Size_Range_EPOS_Sgel <100ml	3.9743	3.115	1.276	0.202	−2.131	10.080
Size_Range_EPOS_Sgel >1501ml+	0.7517	0.566	1.329	0.184	−0.357	1.860

Table 5 Summary of important drivers of PPGs for each sub-category

Sub category	Variant	Segment	Size range	Sub segment	Promotions
Shower gels		✓ (Men's) (Milk/skincare)	✓ (251–400 ml) (001–500 ml) (501–750 ml)		✓
Cleaning wipes	✓ (Other fragrance)	✓ (General purpose)			✓
Electric TB			✓ (4pk+)	✓ (Rfl)	
Manual TB			✓ (2pk)		✓
Mouthwash botte	✓ (Ice)				✓
Sprays		✓ (General purpose) (Glass) (Kitchen)	✓ (376–300 ml)		✓

If Men's segment is used in the promotional bundling, keeping other factors constant, there is an 88% probability of success of promotion indicating that when shower gels are bundled with Men's segment, there is an 88% chance that the promotion is successful.

Similarly, the probability is estimated for other variables across each sub-category (Fig. 4).

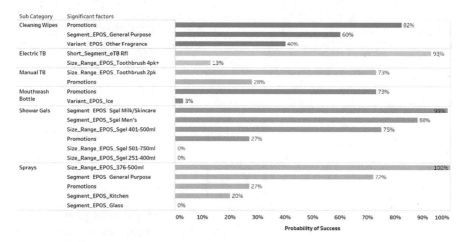

Fig. 4 Probability of success for important attributes in each sub-category

5.2 Optimization

Product bundling plays a vital role in enhancing both profits and the company's growth. Bundling is based upon the idea that customers usually save more on the value of the grouped package than the individual items when purchased separately. Customers generally compare product prices and love choices before making a purchase.

In this section, the optimization technique was used to form new promotion bundles.

5.2.1 Goal

The goal of the model is to identify new promotions groupings that can boost the Company's sales.

5.2.2 Solving Method

Since the revenue function is non-linear and non-smooth, an evolutionary algorithm is preferred as the solving method.

Process flow for the evolutionary algorithm is as follows (Fig. 5):

Evolutionary algorithms perform well-approximating solutions to all types of problems because they do not make any assumptions about the underlying fitness landscape.

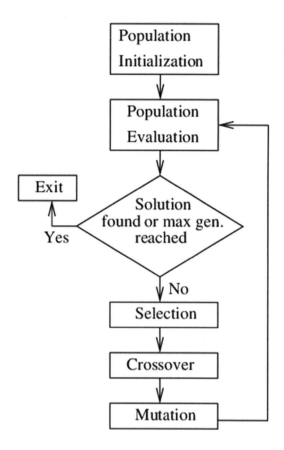

Fig. 5 Process flow of evolutionary algorithm

5.2.3 Advantages of Evolutionary Algorithm

Evolutionary algorithm optimizers are global optimization methods and scale well to higher-dimensional problems. They are robust to noisy evaluation functions, and the handling of evaluation functions that do not yield a sensible result in each period is straightforward. The algorithms can be easily adjusted to the problem at hand.

Evolutionary algorithm calculates the global optimum, and not only the local optima, giving the most optimal results (Fig. 6).

5.2.4 Disadvantages of Evolutionary Algorithm

While research has been done on which algorithm is best suited for a given problem, this question has not been answered satisfactorily yet. While the standard values usually provide reasonably good performance, different configurations may give better results. Furthermore, premature convergence to a local extremum may result from adverse configuration and not yield (a point near) the global extremum.

Fig. 6 Difference between local optima and global maxima

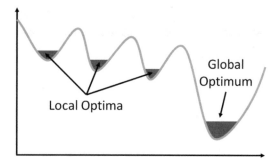

Furthermore, as evolutionary algorithm considers all possible combinations, it is time-consuming.

5.2.5 Objective Function

The objective of the optimization solver is to maximize revenue.

5.2.6 Decision Variables

Optimization model was developed for each sub-category. Each unique product (SKU) is considered as an individual decision variable. Our aim is to find a combination of products that can be combined together to form a bundle for promotion.

5.2.7 Constraints

Number of Products in a Bundle

$$2 \leq n\,(X_i) \leq 6$$

There will be at least 2 distinct products in a PPG and at-most 6 distinct products in a PPG.

Decision Variables Can Take Only Binary Non-Negative Values

$$X_i = \{0, 1\}$$

A product can be included in the bundle only once. This constraint ensures that no same products can be included in the bundle more than once.

Price of the Bundle

$$\sum_{i=1}^{n} P(X_i) \leq \frac{\sum_{j=1}^{N} P(X_j)}{N} * n(X_i)$$

The price of the bundle should not be too high. The sum of the price of individual products in the bundle should be less than the average price of all the products in the bundle multiplied by the total number of products in the bundle.

Quantity Sold

$$\sum_{i=1}^{n} Q(X_i) * \text{Corr}(X_i, n) \geq \frac{\sum_{j=1}^{N} Q(X_j)}{N} * 2n(X_i)$$

Corr $(X_{i,n})$ refers to the selling correlation between two products. For example, if A & B are sold separately, quantities sold are x & y, respectively. However, if they are sold together, then the total quantity sold will be $1.1 * (x + y)$. The factor of 1.1 suggests that the products complement each other.

Due to the limited availability of data, the study has considered 10% as the correlation factor between any two products. A more accurate correlation factor can be computed using market basket analysis if associations between products are known using the transaction data.

As per this constraint, the quantity sold for the formed PPG bundle should be more than the average sold quantities of products in the sub-category times the number of products in the bundle times 2. This will ensure that the quantities sold are higher through bundling.

Revenue Earned

$$\sum_{i=1}^{n} R(X_i) \geq \frac{\sum_{j=1}^{N} R(X_j)}{N} * 2n(X_i)$$

Total revenue after optimization should be higher than the current cumulative revenue earned by all the products in the sub-category.

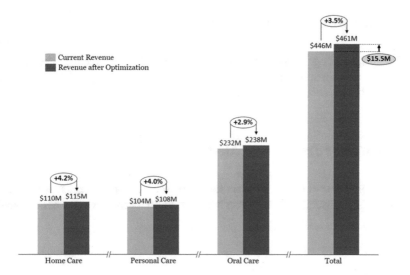

Fig. 7 Revenue generated for each category before and after optimization

5.2.8   Tool Used for Optimization

Microsoft Excel Solver was used to run optimization and obtain results. The parameters used are as follows:

5.2.9   Results (Fig. 7)

The Company has three categories, and the chart reflects the revenue generated before and after optimization. Using optimization, the study was able to increase the revenue of the Company by 3.5% per annum, which is tantamount to a $15.5 million growth in revenue.

6   Conclusions and Future Scope

6.1   *Manufacturing Cost of Product*

The product margin can be calculated if manufacturing cost of the products is provided. This can then be used in the analysis and optimization models. This will determine successful promotion campaigns with a positive and increased margin.

6.2   Store-Wise Analysis

By obtaining stores-wise transactions, data can be analyzed to identify demographic information and consumer preference. This data can be used to provide personalized recommendations to increase customer loyalty.

6.3   Customer Transaction Analysis

If transactional data of customer spending is obtained, market basket analysis can be used to uncover associations between items. It will help improve promotion bundle formation.

References

1. B. Nalebuff, Bundling as an entry barrier. Q. J. Econ. **119**(1), 159–187 (2004)
2. H.K. Bhargava, Retailer-driven product bundling in a distribution channel. Mark. Sci. **31**(6), 1014–1021 (2012). https://doi.org/10.1287/mksc.1120.0725
3. Q. Cao, X. Geng, J. Zhang, Strategic role of retailer bundling in a distribution channel. J. Retail. **91**(1), 50–67 (2015). https://doi.org/10.1016/j.jretai.2014.10.005
4. M. Sheikhzadeh, E. Elahi, Product bundling: Impacts of product heterogeneity and risk considerations. Int. J. Prod. Econ. **144**(1), 209–222 (2013). https://doi.org/10.1016/j.ijpe.2013.02.006
5. F.-m. Tang, Z.-d. Wang, M.-y. Chen, On multiclass classification methods for support vector machines. Control Decis. **20**(7), 743–754 (2005). https://doi.org/10.13195/j.cd.2005.07.29.tangfm.006
6. F. Aiolli, A. Sperduti, Multiclass classification with multi-prototype support vector machines. J. Mach. Learn. Res. **6**, 817–850 (2006)
7. W. Cheng, E. Hüllermeier, Combining instance-based learning and logistic regression for multilabel classification. Mach. Learn. **76**(2–3), 211–225 (2009). https://doi.org/10.1007/s10994-009-5127-5
8. H. Valizadegan, R. Jin, A.K. Jain, Semi-supervised boosting for multi-class classification, in *Machine Learning and Knowledge Discovery in Databases Lecture Notes in Computer Science*, (Springer, Berlin, Heidelberg, 2008), pp. 522–537. https://doi.org/10.1007/978-3-540-87481-2_34
9. L. Ye, H. Xie, W. Wu, J.C. Lui, Mining customer valuations to optimize product bundling strategy, in *2017 IEEE International Conference on Data Mining (ICDM)* (2017). https://doi.org/10.1109/icdm.2017.65

Hierarchical POI Attention Model for Successive POI Recommendation

Lishan Li

1 Introduction

In recent years, location-based social networks (LBSNs) have developed rapidly, which attract users to share their check-in behaviors and write comments on point-of-interests (POIs). The generated textual information and sequential check-ins are very useful for recommending where a user will visit next. Successive POI recommendation can not only improve user experience but also help business owners launch advertisements to targeted customers [1, 2].

POI recommendation task is achieved by modeling check-in context, which contains a variety of information, including sequence pattern, temporal characteristics, and POIs' textual contents.

To date, many proposals have been proposed to solve POI recommendation issue. Markov chain-based models are proposed to exploit a user's check-in sequence information [3–5]. Recurrent neural network (RNN)-based method [6] is designed to better model spatial information. POI embedding has also been used in POI recommendation to capture relationships among POIs [7, 8]. In addition, there have been some studies that further utilize POI's characteristics to improve model performance, including temporal characteristic [9, 10] or POIs' text contents [11, 12].

L. Li (✉)

Department of Computer Science and Technology, Tsinghua University, Beijing, China
e-mail: ls-li14@mails.tsinghua.edu.cn

© Springer Nature Switzerland AG 2021
R. Stahlbock et al. (eds.), *Advances in Data Science and Information Engineering*,
Transactions on Computational Science and Computational Intelligence,
https://doi.org/10.1007/978-3-030-71704-9_11

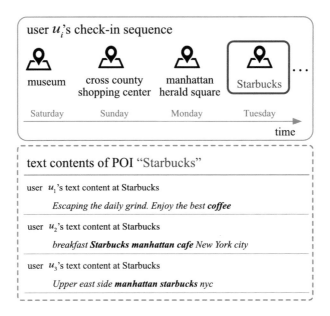

Fig. 1 A check-in sequence example from location-based social network. Each POI contains multiple text contents written by users

However, these proposals have limitations as follows:

(1) They do not jointly study the influence of sequential pattern and temporal and textual content information on POI recommendation task.
(2) They do not strengthen crucial context POIs. As shown in Fig. 1, POIs around office area, such as Manhattan area's Starbucks, are often visited on weekdays. Conversely, POIs around recreation area (e.g., museum and shopping mall) are always visited on weekends. Therefore, a good POI recommendation model should highlight POIs with different attention weights with temporal information.
(3) They do not highlight important words in the text content. As shown in Fig. 1, a long text content of a POI also contains many irreverent words. Considering this example, a user written the following text content "Escaping the daily grind. Enjoy the best coffee" for the POI "Starbucks." Intuitively, the words "enjoy" and "coffee" should attract more attention to indicate core meaning. However, the simple averaged vector of the long content might lose significant information of the POI.

Attention mechanism could enhance the influence of crucial components on final results, which has made impressive performance improvement in many NLP issues [13–16]. Inspired by it, we propose hierarchical POI attention model (HPAM), a novel hierarchical POI attention model that jointly learns check-in sequence, temporal characteristic, and textual content.

Specifically, HPAM proposes a lower-level POI representation layer to capture textual context with a word-level attention mechanism, and a higher-level contextual sequence layer with a temporal-level attention mechanism. The work-level attention mechanism enables it to attend differentially to more and less important content when constructing the POI representation. The temporal-level attention mechanism weighs a long POI sequence to build an attentive context that outputs the crucial POI with a high probability.

We conduct performance evaluation on a public data and compare HPAM with a variety of baseline models. The experimental results show that HPAM outperforms all baseline models. Furthermore, we also conduct performance comparison between HPAM variants. The evaluation results demonstrate that HPAM could efficiently capture the temporal influence and exploit POIs' characteristics from text contents.

2 Related Work

2.1 Successive POI Recommendation

There have been many successive POI recommendation approaches to the model sequential pattern of check-ins. Some previous work is designed based on Markov chain model [3–5] or hidden Markov chain model [17] to exploit a user's check-in sequence information. ST-RNN [6] employs the RNN model to find the sequential correlations among POIs. Feng [8] proposes a POI latent representation model to incorporate geographical influence by embedding techniques. It develops a hierarchical binary tree based on the physical distance between POIs to reflect the geographical influence. However, these methods do not utilize POIs' specific characteristics, such as temporal characteristics and text content information.

2.2 Temporal Characteristic Modeling

There has been some research work that takes advantage of temporal characteristics on POI recommendation task. The proposed method [7] considers temporal influence by training a time latent representation vector. STELLAR [9] proposes a spatial-temporal latent ranking method to capture the temporal influence. The work [10] designs a temporal POI sequential embedding model to capture the contextual check-in information and temporal characteristics as well. However, they could not consider POIs' text content information properly, which could further improve performance.

2.3 Textual Content Influence Modeling

The textual content information of a POI can also provide very useful information in POI recommendation task. The method [11] studies the relationships between a user's check-in actions and various types of content information in terms of POI properties, user interests, and sentiment indications. CAPE [12] designs a content-aware embedding model, which utilizes a user's check-in sequence information and text content for POI embedding. However, the above work does not jointly study temporal characteristic modeling. Besides, a long check-in sequence or text content often contains many irrelevant items to core meaning, which may confuse the overall representation learning for the final target POI recommendation.

2.4 Hybrid Characteristic Modeling

There are few studies that use both temporal characteristic and text content for POI recommendation. Previous work [18] simultaneously considers the textual and temporal factors. However, this method in capturing the temporal information is very arbitrary with some manually defined rules, such as simply dividing the datasets into different time intervals. Therefore, it destroys the valuable information about the ordering of check-ins and cannot take into account other temporal aspects such as seasonality, which is especially useful in the POI recommendation domain.

3   Model Design

We propose a hierarchical POI attention model (HPAM) on the successive POI recommendation task. The proposed approach not only captures the basic sequence correlations between POIs but also depicts POIs' textual and temporal characteristics.

We observed that a long check-in sequence often contains many items irreverent to the next choice, which tends to overwhelm the influence of a few relevant items. Attention mechanisms have proved to be successful in the NLP field to enhance the influence of crucial components on final results. Inspired by those approaches, we propose specific attention mechanisms in the POI recommendation task for performance improvement. Therefore, HPAM highlights important words when learning POI representation from text content through word attention mechanism and also strengthens the influences of critical POIs in the user's check-in sequence by temporal attention strategy. In this way, important words and POIs can contribute more to the prediction of successive POI.

In the following sections, we first formally define the POI recommendation task. Then, we briefly introduce the overall architecture of HPAM. Besides, we also present the low-level POI representation learning with word attention, and the high-level POI sequence representation with temporal attention. At last, we introduce the model training process in detail.

3.1 Task Definition

To better describe the model, we present some basic concepts as follows:

Definition 1 (Check-in) Each check-in is a quadruple $\langle u, l, t, s \rangle$ that depicts a user u visiting POI l at time t and writing text content s.

Definition 2 (Check-in Sequence) A check-in sequence is a set of check-ins of user u sorted by check-in time and grouped by day-hour. The check-in sequence of the user $u \in U$ is denoted as $C_u = \{\langle t_1, \{l_{1,1}, l_{1,2}, \dots\}\rangle, \langle t_2, \{l_{2,1}, l_{2,2}, \dots\}\rangle, \dots\}$, where U is the set of users and t_i is i-th time in day-hour granularity. Here we define day set as {Monday, Tuesday, ..., Sunday} and hour set as {0,1,2,...,23}.

Definition 3 (Context and Target POI) In a check-in sequence C_u, the last POI l in last day-hour t is the target POI, and other check-ins in C_u are context check-ins.

Definition 4 (Text Content) Each text content s consists of words, denoted as $s = \{w_1, w_2, \dots\}$, where w indicates a word. Each user u writes content s about his visited POI l_i.

Given the user check-in sequence and text contents of user visited POI, the successive POI recommendation task aims to recommend the next POI based on the context POIs and text contents.

We present the overall HPAM architecture in Fig. 2, which consists of a lower-level POI representation layer with word attention and a high-level contextual sequence layer with temporal attention. In the following sections, we first present each layer in detail.

3.2 POI Representation Layer

As shown in Fig. 3, each POI representation p consists of two parts, where one is learned from text content representation with word attention and the other comes from POI embedding layer.

Text Content Representation Firstly, the input word embedding layer maps each word w_i among the text content to a high-dimensional vector space. We employ the Glove method [19] to initialize the embedding matrix. Specially, we denote the embedding lookup matrix as $\mathbb{L}_w \in \mathbb{R}^{d_w \times |V_w|}$, where d_w is the word vector dimension and $|V_w|$ is the word vocabulary size.

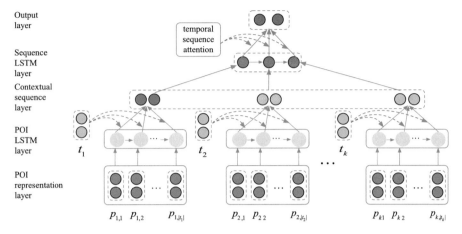

Fig. 2 The overall architecture of the proposed hierarchical POI attention model

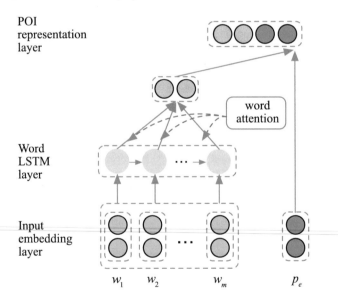

Fig. 3 POI representation layer to capture both sequence and text content characteristic of each POI

We treat its text content as a sentence and employ a long short-term memory network (LSTM) [20] to capture the correlations among words. Specially, given the input word embedding w, the update process of LSTM network at time t can be formalized as follows:

$$i_t = \sigma(W_i \cdot [h_{t-1}, w_t] + b_i), \qquad (1)$$

$$f_t = \sigma(W_f \cdot [h_{t-1}, w_t] + b_f), \tag{2}$$

$$o_t = \sigma(W_o \cdot [h_{t-1}, w_t] + b_o), \tag{3}$$

$$g_t = tanh(W_g \cdot [h_{t-1}, w_t] + b_g), \tag{4}$$

$$c_t = f_t * c_{t-1} + i_t * g_t, \tag{5}$$

$$h_t = o_t * tanh(c_t), \tag{6}$$

where σ is the sigmoid activation function and i_t, f_t, and o_t indicate the input gate, forget gate and output gate, respectively. $W_i, W_f, W_o, W_g \in \mathbb{R}^{d*(d+d_w)}$, $b_i, b_f, b_o, b_g \in \mathbb{R}^d$, and d is the hidden dimension size. We can obtain the final outputs of context words $H = \{h_1, h_2, \cdots, h_m\}$. Consider that all words do not contribute equally to the representation of POI text content, especially for case that contains noisy irrelevant words. We introduce attention mechanism to capture crucial words that are important to the meaning of text content and aggregate these representations of informative words to form the POI content vector. We employ the word attention mechanism on the hidden state sequence $\{h_1, h_2, \cdots, h_m\}$ to compose an aggregated vector p_c as the content representation. Formally, the POI content representation vector $p_c \in \mathcal{R}^d$ is learned by:

$$u_t = tanh(W_w h_t + b_w), \tag{7}$$

$$a_t = \frac{exp\left(u_t^T u_w\right)}{\sum_k exp\left(u_k^T u_w\right)}, \tag{8}$$

$$p_c = \sum_t a_t h_t, \tag{9}$$

where $W_w \in \mathcal{R}^{d \times d}$, $b \in \mathcal{R}^d$, and $u_w \in \mathcal{R}^{d \times 1}$ are the attention mechanism parameters.

Specifically, we first feed word annotation h_t through one-layer MLP to get u_t as hidden representation of h_t. After that, we measure the word importance as the similarity of u_t and word context vector u_w. Then we can obtain a normalized attention weight a_t through the softmax function. At last, we compute the POI content representation vector p_c as the weighted sum of word annotation h_t based on the weights. The word context vector u_w can be seen as high-level representation of characterizing importance of words. The word context vector u_w is randomly initiated and jointly learned during the training process.

POI Embedding Layer From another perspective, each POI could be visited by multiple users; the plain POI sequence also contains useful information to represent each POI. Similar to input word embedding layer, we pre-train the POI embedding matrix $\mathbb{L}_p \in \mathbb{R}^{d_p \times |V_p|}$ based on each user's raw POI sequence using the Glove method [19], where d_p is the POI embedding dimension and $|V_p|$ is the POI set size.

At last, we concatenate both the POI embedding vector and POI content representation vector as the final POI representation vector. The process can be formalized as follows:

$$p = [p_e; p_c]. \tag{10}$$

3.3 POI LSTM Layer

Given the concatenated POI representations as input, we employ the POI LSTM layer to capture the sequence correlations among POIs. Specifically, POI LSTM layer is similar to the word LSTM layer described in Sect. 3.2. Besides, for each user u, each POI sequence $\{l_{k,1}, l_{k,2}, \dots\}$ is the check-ins occurred in day-hour t_k. Assuming user u has $|T|$ day-hours check-ins, then user u will have $|T|$ POI sequences. After the POI LSTM layer, we could obtain the hidden sequence vectors $Q = \{q_{k,1}, q_{k,2}, \dots\}$ for day-hour t_k, which are fed to higher-level contextual sequence layer.

3.4 Contextual Sequence Layer

The check-in sequences in different days naturally show different temporal characteristics. Previous work [9, 10, 21] shows that check-in sequences exhibit different patterns on different days, especially weekday and weekend. For example, check-ins around offices on weekday are more frequent compared with others, while check-ins around shopping malls on weekend would attract more users' attention. The temporal influence is further verified in Sect. 6.3. Hence, we introduce the temporal attention mechanism to capture the various temporal characteristics among POI sequences.

Temporal Embedding Layer POI sequences in different users could have same day-hour input. Similar to POI embedding layer, we employ a temporal embedding layer to map the day-hour to high-dimensional vector space. Formally, the temporal embedding matrix $\mathbb{L}_t \in \mathbb{R}^{d_t \times |V_t|}$ is randomly initialized and jointly trained during the training process.

Temporal POI Attention For each hidden POI sequence vector $Q = \{q_{k,1}, q_{k,2}, \dots\}$ at day-hour time t_k, we first look up the time t_k through the temporal embedding layer and get temporal embedding vector u_k^s. The temporal vector u_k^s is regarded as the higher representation of measuring how important of each POI vector $q_{k,i}$ in sequence Q. The important POIs should contribute more in generating the contextual sequence vector, which depicts the overall representation for POIs visited in t_k. Formally, the contextual POI sequence representation vector $p_s \in \mathcal{R}^d$ is updated by:

$$u_{k,t} = tanh(W_s q_{k,t} + b_s), \tag{11}$$

$$a_{k,t} = \frac{exp\left(u_{k,t}^T u_k^s\right)}{\sum_k exp\left(u_{k,t}^T u_k^s\right)}, \tag{12}$$

$$p_s = \sum_t a_{k,t} q_{k,t}, \tag{13}$$

where $W_s \in \mathcal{R}^{d \times d}$, $b_s \in \mathcal{R}^d$, and temporal embedding vector $u_k^s \in \mathcal{R}^{d \times 1}$ are the temporal POI attention parameters.

Similarly, then we feed the contextual POI sequence representation to **sequence LSTM layer**. We employ **temporal sequence attention** vector to extract the important sequence representation to contribute more on the final successive POI prediction. Here the temporal sequence attention vector is randomly initialized and jointly learned during training. After that, we feed the weighted POI sequence vector m to the output layer.

3.5 Output Layer

The obtained representation m will be fed to a softmax layer for generating the target POI:

$$p = softmax(W_p * m + b_p), \tag{14}$$

where $p \in \mathbb{R}^C$ is the probability distribution for all possible POIs and $W_p \in \mathbb{R}^{C \times d}$ and $b_p \in \mathbb{R}^C$ are the weight matrix and bias, respectively. Here C indicates the number of all possible POIs.

3.6 Model Training

For training the hierarchical POI attention model (HPAM), we should optimize all the parameters Θ from the word and check-in LSTM networks, word, POI and temporal embeddings, word and temporal attention mechanisms, and softmax parameters. The final loss function is consisting of the cross-entropy loss and regularization item as follows:

$$\mathcal{L} = -\sum_{i=1}^{C} y_i log(p_i) + \lambda \|\Theta\|^2, \tag{15}$$

where $\lambda \geq 0$ controls the influence of the L_2 regularization item. We employ the stochastic gradient descent (SGD) optimizer to compute and update the training parameters. In addition, we utilize dropout strategy before softmax layer in training to avoid overfitting.

4 Experimental Setting

4.1 Performance Metric

In this work, we evaluate the model performance through two metrics: Recall@k and mean reciprocal rank (MRR). Recall@k is a generally used metric for ranking task, which is computed when the correct successive POI is in the top-k recommended POI list. In the following evaluation, we show the Recall@k for k = 1, 5, and 10. MRR measures the mean reciprocal rank of the true predicted target in ranked lists.

4.2 Dataset Preparation

We evaluate the baselines and proposed method on the public dataset in CAPE [12], which has large percentages of text content for POIs. This dataset is very suitable for content involved POI recommendation task, as analyzed in CAPE [12]. The dataset statistics are shown in Table 1.

To make our model satisfactory to the recommendation task, we divide the dataset into training set, validation set, and test set, following previous methods [6, 11, 12]. The most recent check-ins that compose each user's 20% check-ins are

Table 1 The statistics of the dataset

Number of check-ins	2,216,631
Number of POIs	13,187
Number of users	78,233
Number of words	958,386
Average number of check-ins per user	28.3
Average number of POIs per user	15.2
Average number of users per POI	90
Average number of text word per POI	22.7

used as test set. And the less recent check-ins that compose 10% check-ins work as the validation set. And then, the remaining 70% check-ins are used as the training set.

4.3   Compared Methods

To evaluate the proposed model, we compare it with the following methods:

GRU Utilizes one basic GRU network and uses averaged hidden vector to predict successive POI.

LSTM Utilizes one basic LSTM network to learn the hidden states and obtain the average vector to predict the successive POI.

ST-RNN [6] Employs RNN model and incorporates the time interval information between check-ins and distance information between POIs.

STELLAR [9] Proposes a spatial-temporal latent ranking method to capture the impact of time on successive POI recommendation. This method trains pairs of consecutive POIs to predict the successive POI.

Geo-Teaser [10] Is a geo-temporal sequential embedding rank model, which captures the contextual check-in information in sequences and the various temporal characteristics of different days.

CAPE [12] Is a content-aware hierarchical POI embedding model, where check-in context layer captures the geographical influence of POIs from check-in sequence and the text content layer captures the POI characteristics from text content.

4.4   Variants of HPAM Model

We also list the variants of HPAM model, which are used to analyze the effects of textual content and temporal patterns, respectively.

- **HPAM-C** utilizes the characteristics of a POI from text content.
- **HPAM-CA** further employs word-level attention model based on HPAM-C to strengthen the influence of important words to its core meaning.
- **HPAM-T** exploits the temporal patterns with temporal attention mechanisms and does not take advantage of text contents.
- **HPAM** is the complete hierarchical POI attention model.

4.5   Hyper-parameter Set

In our experiments, word embeddings for these methods are initialized by Glove [19]. The dimension size of word embedding and hidden state d are set to 300. The weight and bias are initialize by sampling from a uniform distribution $U(-0.01, 0.01)$. The coefficient λ of L_2 regularization item is 10^{-5}, the initial learning rate is set to 0.01, and the dropout rate is set to 0.5.

5   Evaluation

In this section, we conduct experiments to evaluate the performance of HPAM in the following three aspects:

(1) How does HPAM perform compared with state-of-the-art POI recommendation models? We present the performance comparison results in Sect. 5.1.
(2) How does each key idea of HPAM contribute to the final results? We show the performance of HPAM variants in Sect. 5.2.

5.1   Performance Comparison

Table 2 shows the performance comparison results of HPAM with other baseline methods. We can have the following observations:

(1) **ST-RNN**, **GRU**, and **LSTM** model the sequential pattern of check-ins to predict successive POI. They achieve good performance, which demonstrates the effectiveness of contextual sequential information. **HPAM** outperforms the above methods by considering POI characteristic influence, including temporal pattern and text content.

Table 2   The performance comparisons between HPAM and other baseline methods

Model	Recall@1	Recall@5	Recall@10	MRR
GRU	0.1197	0.2207	0.2726	0.1792
LSTM	0.1207	0.2225	0.2751	0.1805
ST-RNN	0.1185	0.2142	0.2529	0.1721
STELLA	0.1308	0.2251	0.2923	0.1857
Geo-Teaser	0.1291	0.2334	0.2980	0.1850
CAPE	0.1390	0.2433	0.3079	0.1953
HPAM	**0.1513**	**0.2608**	**0.3281**	**0.2154**

The results of baseline methods are retrieved from published papers. The best performances are marked in bold

(2) **STELLAR** and **Geo-Teaser** obtain better results than above methods by further considering the various temporal characteristics in different days, while the simple strategy could not depict the various temporal patterns properly. **HPAM** performs better than them by utilizing textual content and more reasonable temporal characteristic modeling.

(3) **CAPE** gains better performance compared with previous methods, which utilizes both textual content and geographical sequential influence. **HPAM** performs better than CAPE, since we also exploit temporal influence and apply hierarchical attention mechanisms to capture crucial components.

Our proposed model consistently outperforms state-of-the-art POI recommendation models due to the following considerations. On the one hand, **HPAM** jointly depicts POIs' temporal characteristics and textual contents. On the other hand, **HPAM** employs hierarchical attention mechanisms to better model POI representation and depict various temporal check-in sequence patterns. In this way, the model could strengthen important words for learning POI content representation and highlight crucial POI representations when predicting the final successive POI.

5.2   Performance Analysis

Table 3 shows the performance comparison results among the variants of HPAM model. We can have the following observations:

(1) **HPAM-C** achieves competitive performance compared with **CAPE** and outperforms **Geo-Teaser**, which demonstrates that text content can bring more useful information for POI representation learning. In addition, **HPAM-CA** performs better than **HPAM-C**, which indicates the effectiveness of the word attention mechanism by strengthening the contributions of important words to core text meaning.

(2) **HPAM-T** gains better performance compared with **HPAM-CA**, which demonstrates the benefits of temporal characteristics and the effectiveness of temporal attention mechanisms. The temporal attention could help to depict the various temporal patterns in user check-in sequences by learning the POI attention weights based on the temporal embedding vector.

(3) Compared with **HPAM-CA** and **HPAM-T**, the complete version **HPAM** achieves further performance improvement, which demonstrated that jointly depicting text content and capturing temporal characteristics could improve the performance from different perspectives.

Table 3 The performance comparisons between HPAM variants

Method	Recall@1	Recall@5	Recall@10	MRR
HPAM-C	0.1372	0.2467	0.3096	0.1928
HPAM-CA	0.1429	0.2518	0.3102	0.1973
HPAM-T	0.1462	0.2565	0.3228	0.2071
HPAM	**0.1513**	**0.2608**	**0.3281**	**0.2154**

POI sequence	blue water grill -> fresco by scotto -> brandy piano bar -> charming charlie -> redemption -> calexico -> American academy of dramatic ->rosa mexicano -> mulberry project -> the king and i -> **museum of modern art moma**
HPAM results	sheep meadow central park; **museum of modern art moma**; playstation theater

Fig. 4 POI recommendation results generated by HPAM given POI sequence. The target POI is highlighted in bold

6   Discuss

6.1   Case Study

Figure 4 shows the recommended POIs by HPAM which has given a user's preceding visited POIs. By jointly incorporating both textual content and temporal characteristic into POI recommendation, HPAM could achieve good successive POI recommendation. For example, HPAM can predict the next POI "museum of modern art moma" in the top three recommendations. In addition, the other two POI recommendations are also latently similar to the target POI.

6.2   Hyper-parameter Optimization

We change the embedding and hidden dimension of HPAM and observe the corresponding performance results, which are shown in Fig. 5. Here, we set the range of model dimensions to 50–450. We then observe the performance changes from increasing the dimension in increments of 50. The best performance of HPAM is obtained when the dimension is set to 300. After the dimension increases to 300, the performance becomes relatively steady since the embedding and hidden vectors already have enough capacity to represent.

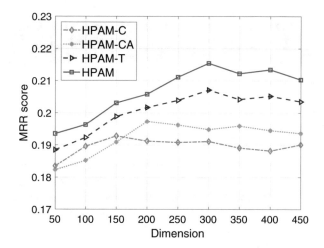

Fig. 5 Effects of hyper-parameters

Fig. 6 Various check-in
temporal patterns of different
day-hours

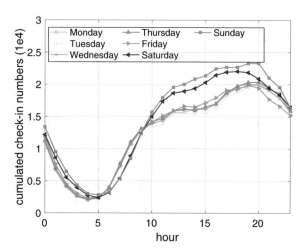

6.3 Effect of Temporal Characteristic

In order to present the effect of temporal characteristic, we collect the check-ins distribution of different day-hours from temporal pattern perspective, which is shown in Fig. 6. The figure could depict the various check-in temporal patterns of different day-hours, We could observe that the cumulated check-in numbers changed a lot in different day-hours. For example, "Saturday" and "Sunday" have a similar temporal pattern, which is obviously different with the pattern in weekday, such as larger check-in numbers between hours 11 to 19.

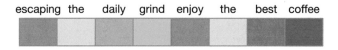

Fig. 7 The attention visualizations on content information

6.4 Effect of Word Attention

In order to demonstrate the effect of attention mechanism on text content, we visualize the attention weights of words. Figure 7 shows the attention weights of text content on POI "Starbucks." We can observe that the attention mechanism can enforce the model to pay more attentions on the important words with respect to the POI. For example, the words "best" and "coffee" have higher attention weights compared with other words.

7 Conclusion

In this paper, we propose HPAM, a novel hierarchical POI attention model for successive POI recommendation. More specially, we jointly take advantage of textual content influence and sequential and temporal characteristics. We propose a POI representation layer that consists of POI embedding vector and POI content representation layer, which captures the POI sequential context by highlighting crucial words. Furthermore, we design a contextual sequence layer with temporal attention mechanisms to strengthen the influences of crucial POIs in final successive POI recommendation. Experiment results show that the HPAM model outperforms the state-of-the-art POI recommendation models.

References

1. H. Yin, W. Wang, H. Wang, L. Chen, X. Zhou, Spatial-aware hierarchical collaborative deep learning for poi recommendation. IEEE Trans. Knowl. Data Eng. **29**(11), 2537–2551 (2017)
2. Q. Yuan, G. Cong, Z. Ma, A. Sun, N.M. Thalmann, Time-aware point-of-interest recommendation, in *Proceedings of the 36th international ACM SIGIR conference on Research and development in information retrieval* (ACM, New York, 2013), pp. 363–372
3. M. Chen, Y. Liu, X. Yu, Nlpmm: a next location predictor with markov modeling, in *Pacific-Asia Conference on Knowledge Discovery and Data Mining* (Springer, Berlin, 2014), pp. 186–197
4. C. Cheng, H. Yang, M.R. Lyu, I. King, Where you like to go next: successive point-of-interest recommendation, in *Proceedings of the Twenty-Third International Joint Conference on Artificial Intelligence* (2013)

5. J.-D. Zhang, C.-Y. Chow, Y. Li, Lore: exploiting sequential influence for location recommendations, in *Proceedings of the 22nd ACM SIGSPATIAL International Conference on Advances in Geographic Information Systems* (ACM, New York, 2014), pp. 103–112

6. Q. Liu, S. Wu, L. Wang, T. Tan, Predicting the next location: a recurrent model with spatial and temporal contexts, in *Proceedings of the Thirtieth AAAI Conference on Artificial Intelligence* (2016)

7. X. Liu, Y. Liu, X. Li, Exploring the context of locations for personalized location recommendations, in *IJCAI* (2016), pp. 1188–1194

8. S. Feng, G. Cong, B. An, Y.M. Chee, Poi2vec: geographical latent representation for predicting future visitors, in *Proceedings of the Thirty-First AAAI Conference on Artificial Intelligence* (2017)

9. S. Zhao, T. Zhao, H. Yang, M.R. Lyu, I. King, Stellar: spatial-temporal latent ranking for successive point-of-interest recommendation, in *Proceedings of the Thirtieth AAAI Conference on Artificial Intelligence* (2016)

10. S. Zhao, T. Zhao, I. King, M.R. Lyu, Geo-teaser: geo-temporal sequential embedding rank for point-of-interest recommendation, in *Proceedings of the 26th International Conference on World Wide Web Companion*. International World Wide Web Conferences Steering Committee (2017), pp. 153–162

11. H. Gao, J. Tang, X. Hu, H. Liu, Content-aware point of interest recommendation on location-based social networks, in *Proceedings of the Twenty-Ninth AAAI Conference on Artificial Intelligence* (2015)

12. B. Chang, Y. Park, D. Park, S. Kim, J. Kang, Content-aware hierarchical point-of-interest embedding model for successive poi recommendation, in *IJCAI* (2018), pp. 3301–3307

13. Y. Wang, M. Huang, L. Zhao, et al. Attention-based LSTM for aspect-level sentiment classification, in *Proceedings of the 2016 Conference on Empirical Methods in Natural Language Processing* (2016), pp. 606–615

14. T. Lei, R. Barzilay, T. Jaakkola, Rationalizing neural predictions, in *Proceedings of the 2016 Conference on Empirical Methods in Natural Language Processing* (2016), pp. 107–117

15. C. Li, X. Guo, Q. Mei, Deep memory networks for attitude identification, in *Proceedings of the Tenth ACM International Conference on Web Search and Data Mining* (ACM, New York, 2017), pp. 671–680

16. Z. Yang, D. Yang, C. Dyer, X. He, A. Smola, E. Hovy, Hierarchical attention networks for document classification, in *Proceedings of the 2016 Conference of the North American Chapter of the Association for Computational Linguistics: Human Language Technologies* (2016), pp. 1480–1489

17. J. Ye, Z. Zhu, H. Cheng, What's your next move: user activity prediction in location-based social networks, in *Proceedings of the 2013 SIAM International Conference on Data Mining* (2013), pp. 171–179

18. P. Kefalas, Y. Manolopoulos, A time-aware spatio-textual recommender system. Expert Syst. Appl. **78**, 396–406 (2017)

19. J. Pennington, R. Socher, C. Manning, Glove: global vectors for word representation, in *Proceedings of the 2014 Conference on Empirical Methods in Natural Language Processing (EMNLP)* (2014), pp. 1532–1543

20. S. Hochreiter, J. Schmidhuber, Long short-term memory. Neural Comput. **9**(8), 1735–1780 (1997)

21. S. Zhao, M. Lyu, I. King, Aggregated temporal tensor factorization for point-of-interest recommendation, in *ICONIP* (Springer, Berlin, 2016)

A Comparison of Important Features for Predicting Polish and Chinese Corporate Bankruptcies

Yifan Ren and Gary M. Weiss

1 Introduction

Corporate bankruptcy is an important topic in both the accounting and finance disciplines. The advent of data mining has provided a different set of methods for predicting these corporate bankruptcies. However, much of the relevant research in this area has focused more on the predictive value of the models than on which features they utilize to achieve this performance [1]. Although there have been attempts to explore the features associated with bankruptcy, these attempts generally rely on a statistical and econometrical approach rather than a data mining approach [2]. Furthermore, most work on studying bankruptcies is limited to a single country, and there is very rarely any comparison between the importance of different financial features between dissimilar financial systems. This study utilizes data mining methods to build predictive models of corporate bankruptcy in Poland and China, and then compares the importance of the financial features utilized in these models.

This study utilizes corporate bankruptcy data from Poland and China that indicates the bankruptcy status after 1 year. The predictive models are built using classification algorithms that report feature importance. We only choose those algorithms which can report feature importance. Thus powerful algorithms like neural nnetworks are not utilized since they cannot easily determine feature

Y. Ren
Information, Technology, and Operations Department, Fordham University, New York, NY, USA
e-mail: yren50@fordham.edu

G. M. Weiss (✉)
Department of Computer & Information Science, Fordham University, New York, NY, USA
e-mail: gaweiss@fordham.edu

© Springer Nature Switzerland AG 2021
R. Stahlbock et al. (eds.), *Advances in Data Science and Information Engineering*,
Transactions on Computational Science and Computational Intelligence,
https://doi.org/10.1007/978-3-030-71704-9_12

importance [3]. Researchers have utilized feature selection and ranking methods in the context of data mining, but rarely in the financial field [4].

2   Background

This section provides necessary background information about bankruptcies in Poland and China. We start with a definition of bankruptcy. Bankruptcy is a legal process through which people or entities who cannot repay debts to creditors seek relief from some or all of their debts. It is a legal definition where the details vary between nations. However, within the financial field, the definition is simpler: bankruptcy occurs when insolvency appears, which means the entity is no longer able to repay its liabilities. This definition can be used to conduct studies that span nations despite the variations in the legal processes and definitions associated with bankruptcy [5].

The key financial indicators that impact bankruptcy include assets, liabilities, and available cash. Nonetheless, even with these financial indicators, a bankruptcy may not be seen ahead of time, which is quite problematic. This study, therefore, can make a practical contribution by identifying a better set of financial indicators associated with bankruptcy. Any financial warning would be invaluable to investors and shareholders.

3   Predicting Corporate Bankruptcies in Poland

This section describes the data, experiments, results, and features associated with corporate bankruptcies in Poland.

3.1   Data Description

The data set used in this study is "Polish Companies Bankruptcy Data Set," which is available from the UCI Machine Learning Repository [1, 6]. The data was collected from the Emerging Markets Information Service, which is a database containing information on emerging markets around the world. The bankrupt companies are from the period 2000–2012, while the still operating companies were evaluated from 2007 to 2013. In order to predict whether a company will be bankrupt in the next year, only the fifth dataset is used for our task, which contains company financial information in the $t - 1$ year and the associated bankrupt status in the t year. This dataset is entitled "5thYear" in original format on the UCI Machine Learning Repository, which contains financial rates from the fifth year of the forecasting period, and corresponding class label that indicates bankruptcy status after 1 year.

The data set contains 5910 total instances (i.e., financial statements), of which 410 (6.9%) represent bankrupted companies and 5500 (93.1%) represent companies that were not bankrupted during the forecasting period. The data set contains the 64 attributes that are enumerated in Table 1. The missing values in the data set were imputed using the mean value. The one exception is feature P37, which had missing values for almost half of all records; this attribute was converted into a binary attribute that indicated if the features was missing or present. The target class has two values: "bankrupted" and "not bankrupted" (i.e., solvent). Feature engineering and dimensionality reduction techniques like principal components analysis were not used since they would obscure the importance of the original features.

3.2   Modeling

The classification models were induced from the data set using the Python-based scikit-learn data mining toolkit [7]. Since the primary goal of this study is to identify the features most responsible for predicting bankruptcy, only classification algorithms that measure/rank feature importance were utilized. This study used the following four classification methods:

- Logistic regression (LR).
- The C4.5 decision tree model (DT) [8].
- Extreme gradient boosting (XGBoost) [9].
- Random forest (RF) [10].

The XGBoost algorithm is not included in scikit-learn, so the XGBoost library was used for this [9]. The experiments used a training/test set split of 70%/30%, where the instances were randomly selected. Unless otherwise specified default parameters were used for all experiments. Because mistakenly predicting a company that will become bankrupt as a solvent is much worse than predicting a solvent company will become bankrupt, instances belonging to the "bankrupt" class are weighted five times more than instances belonging to the "solvent" class. In the DT, XGBoost, and RF models, when the algorithm calculated the gain ratio, it would regard one instance of the minor class as five instances in a node, so that adjusted the weighting. This weighting was not a default parameter in those models provided, so it was input manually. Meanwhile, we used to tenfold cross-validation method to avoid possible over-fitting.

3.3   Prediction Results

The classification results for the four algorithms are provided in Table 2. The evaluation metrics include precision, recall, F1-score, and accuracy. Note that precision, recall, and F1-score are with respect to the "bankrupt" class value.

Table 1 Description of features in Polish data set

ID	Description	ID	Description
P1	Net profit / total assets	P33	Operating expenses / short-term liabilities
P2	Total liabilities / total assets	P34	Operating expenses / total liabilities
P3	Working capital / total assets	P35	Profit on sales / total assets
P4	Current assets / short-term liabilities	P36	Total sales / total assets
P5	[(Cash + short-term securities + receivables − short-term liabilities) / (operating expenses − depreciation)] * 365	P37	(Current assets − inventories) / long-term liabilities
P6	Retained earnings / total assets	P38	Constant capital / total assets
P7	EBIT / total assets	P39	Profit on sales / sales
P8	Book value of equity / total liabilities	P40	(Current assets − inventory − receivables) / short-term liabilities
P9	Sales / total assets	P41	Total liabilities / ((profit on operating activities + depreciation) * (12/365))
P10	equity / total assets	P42	Profit on operating activities / sales
P11	(Gross profit + extraordinary items + financial expenses) / total assets	P43	Rotation receivables + inventory turnover in days
P12	Gross profit / short-term liabilities	P44	(Receivables * 365) / sales
P13	(Gross profit + depreciation) / sales	P45	Net profit / inventory
P14	(Gross profit + interest) / total assets	P46	(Current assets − inventory) / short-term liabilities
P15	(Total liabilities * 365) / (gross profit + depreciation)	P47	(Inventory * 365) / cost of products sold (COGS)
P16	(Gross profit + depreciation) / total liabilities	P48	EBITDA (profit on operating activities − depreciation) / total assets
P17	Total assets / total liabilities	P49	EBITDA / sales
P18	Gross profit / total assets	P50	Current assets / total liabilities
P19	Gross profit / sales	P51	Short-term liabilities / total assets
P20	(Inventory * 365) / sales	P52	(Short-term liabilities * 365) / COGS
P21	Sales (n) / sales (n − 1)	P53	Equity / fixed assets
P22	Profit on operating activities / total assets	P54	Constant capital / fixed assets
P23	Net profit/sales	P55	Working capital
P24	Gross profit (in 3 years) / total assets	P56	(Sales − COGS) / sales
P25	(Equity − share capital) / total assets	P57	(Current assets − inventory − short-term liabilities) / (sales − gross profit − depreciation)
P26	(Net profit + depreciation) / total liabilities	P58	Total costs /total sales
P27	Profit on operating activities / financial expenses	P59	Long-term liabilities / equity
P28	Working capital / fixed assets	P60	Sales / inventory
P29	Logarithm of total assets	P61	Sales / receivables
P30	(Total liabilities − cash) / sales	P62	(Short-term liabilities *365) / sales
P31	(Gross profit + interest) / sales	P63	Sales / short-term liabilities
P32	(Current liabilities * 365) / COGS	P64	Sales / fixed assets

Table 2 Evaluation tables (Poland)

Measure	LR	DT	RF	XGBoost
Precision	0.56	0.60	0.80	0.93
Recall	0.12	0.58	0.32	0.56
F1-Score	0.20	0.59	0.46	0.70
Accuracy	0.94	0.95	0.95	0.97

Table 3 Confusion Matrix of XGBoost Model (Poland)

		Predicted Labels	
Actual Labels		Bankrupt	Solvent
	Bankrupt	64	50
	Solvent	5	1654

The results clearly demonstrate that XGBoost performs best overall since it has the highest F1-score and accuracy. Table 3 shows more detailed results by providing the confusion matrix results for XGBoost when evaluated on the test set. The biggest issue with the results is that only 56% of the bankrupted companies are identified, but given the relatively severe level of class imbalance (1:13.5), the results are nonetheless impressive.

3.4   Feature Importance

This section describes the importance of the features with respect to classifying the Polish corporations. Logistic regression naturally generates feature importance based on the coefficient associated with each feature. The other three algorithms are all based on decision trees, since random forest is an ensemble of decision trees and XGBoost is based on boosted decision trees. Feature importance in decision trees can be calculated, although it is not as straightforward as for logistic regression. In decision trees, the closer the feature is to the root node, and the more often it appears in a boosted decision tree, the greater the weight and importance. Table 4 provides the top-10 features for XGBoost, the best-performing model.

4   Predicting Chinese Corporate Bankruptcies

This section describes the data, experiments, results, and feature importance associated with corporate bankruptcies in China.

Table 4 XGBOOST feature importance scores (Poland)

Rank	ID	Importance score	Description
1	P22	0.1052	Profit on operating activities / total assets
2	P35	0.0752	Profit on sales / total assets
3	P41	0.0482	Total liabilities / ((profit on operating activities + depreciation) * (12/365))
4	P34	0.0412	Operating expenses / total liabilities
5	P26	0.0393	(Net profit + depreciation) / total liabilities
6	P5	0.0389	[(Cash + short-term securities + receivables − short-term liabilities) / (operating expenses − depreciation)] * 365
7	P46	0.0321	(Current assets − inventory) / short-term liabilities
8	P21	0.0302	Sales (n) / sales (n − 1)
9	P39	0.0294	Profit on sales / sales
10	P6	0.0286	Retained earnings / total assets

4.1 Data Description

The data set of Chinese corporations was formed by merging three datasets, each manually collected from the Wind Financial Database. The data for each record was obtained by querying the financial information one company at a time from WIND financial terminal interface. A total of 61 bankrupted companies were obtained; each was listed on the Chinese stock market and had gone bankrupt since 2006. The 620 solvent companies were collected from SH380 and SZ300 index companies. Hence, the data set contained 681 entries, of which 9.0% represents bankrupted companies and 91.0% represent solvent companies. The feature value information is from the last year of their bankruptcy, or the newest data if solvent.

Table 5 describes the 84 features that were collected. Missing values were imputed using the mean value. The C43 and C84 features were dropped because there often was no net debt, which led to a zero denominator. The C1 and C2 features were removed since they are identifiers and do not provide useful information.

4.2 Modeling

The experiments for the Chinese data set are similar to those for the Polish data set, and the same four classification algorithms were used. The training and test sets were again partitioned using random sampling, but this time using a train/test split of 50%/50%, since the Chinese data set is smaller and has very few bankrupt instances, which makes accurate evaluation more difficult. As before, the bankrupt companies were given a weight of five times that of the solvent companies.

Table 5 Description of features in Chinese data set

ID	Description	ID	Description
C1	Company code	C43	Tangible assets / net debt
C2	Company name	C44	Capital expenditure / depreciation and amortization
C3	Return on equity (ROE) (average)	C45	Cash received for sales of goods and services / operating income (TTM)
C4	ROE (deducted / average)	C46	Net cash flow from operating activities / operating income (TTM)
C5	Return on total assets (ROA)	C47	Net cash flow from operating activities / operating profit (TTM)
C6	ROA (net)	C48	Proportion of net cash flow from operating activities
C7	Return on human investment (RHI)	C49	Proportion of net cash flow from investment
C8	ROE (annualized)	C50	Proportion of net cash flow from fundraising
C9	ROA (annualized)	C51	Net operating cash flow / total operating income
C10	ROA (net) (annualized)	C52	Cash operation index
C11	Sales margin	C53	Cash recovery rate of all assets
C12	Gross profit margin	C54	Asset-liability ratio
C13	Cost of sales ratio	C55	Asset-liability ratio (excluding advance receipts)
C14	Sales period expense rate	C56	Asset-Liability Ratio (excluding advance receipts) (announcement based)
C15	Main business ratio	C57	Long-term debt ratio
C16	Net profit / total operating income	C58	Long-term asset fit ratio
C17	Operating profit / total operating income	C59	Tangible assets / total assets
C18	EBIT / total operating income	C60	Non-current debt ratio
C19	Total operating cost / total operating income	C61	Current assets /short-term liabilities
C20	Management expenses / total operating income	C62	Current liabilities / total liability
C21	Financial costs / total operating income	C63	Capitalization ratio
C22	Asset impairment loss / total operating income	C64	Quick ratio
C23	ROA (trailing twelve months (TTM))	C65	Conservative quick ratio
C24	ROA (net) (TTM) − excluding minority shareholder profit and loss	C66	Cash ratio
C25	Return on invested capital (ROIC)	C67	Net asset-liability ratio
C26	ROIC (TTM)	C68	Net debt ratio
C27	EBIT/ total assets (TTM)	C69	Total equity attributable to shareholders of the parent company / liabilities
C28	Net sales margin (TTM)	C70	EBITDA / total liability

(continued)

Table 5 (continued)

ID	Description	ID	Description
C29	Gross profit margin (TTM)	C71	Net cash flow from operating activities / total liability
C30	Sales period expense rate (TTM)	C72	Net cash flow from operating activities / current liabilities
C31	Operating profit / total operating income (TTM)	C73	Net cash flow from operating activities / non-current liabilities
C32	Total operating cost / total operating revenue (TTM)	C74	Non-financing net cash flow / non-current liabilities
C33	Operating profit / operating income (TTM)	C75	Non-financing net cash flow / total liability
C34	Tax / total Profit (TTM)	C76	Proportion of long-term debt
C35	Net profit attributable to shareholders of the parent company / operating income (TTM)	C77	Working capital / total assets
C36	Asset impairment loss / total operating income (TTM)	C78	Tangible net worth debt ratio
C37	Asset impairment loss / operating profit	C79	Retained earnings / total assets
C39	Operating profit / total profit (TTM)	C80	EBIT (TTM) / total assets
C39	Total profit / operating income (TTM)	C81	Total market value / liabilities for the day
C40	Cash received from sales of goods and services provided/operating income	C82	Total shareholders' equity (including minority) / total liabilities
C41	Net cash flow from operating activities / operating income	C83	Operating income / total assets
C42	Net profit cash rate	C84	Net cash flow from operating activities / net debt

Table 6 Evaluation tables (China)

Measure	LR	DT	RF	XGBoost
Precision	0.65	0.74	1.0	0.96
Recall	0.79	1.0	0.90	0.93
F1-Score	0.71	0.85	0.95	0.95
Accuracy	0.95	0.97	0.99	0.99

4.3 Prediction Results

The classification results are provided in Table 6, using the same format that was used previously for the Polish results. Random forest and XGBoost both perform much better than logistic regression and decision trees. Random forest and XGBoost perform equally well for both accuracy and F1-Score, with the only real difference being that random forest has a higher precision while XGBoost has a higher recall. In this case we give slight preference to precision over recall and choose random forest for the remainder of our analysis. Table 7 shows the confusion matrix for random forest.

Table 7 Confusion matrix of
RF model (China)

Actual labels		Predicted labels	
		Bankrupt	Solvent
	Bankrupt	26	3
	Solvent	0	312

Table 8 Feature importance scores (China)

Rank	ID	Importance score	Description	Adjusted rank
1	C66	0.0630	Cash ratio	1
2	C77	0.0477	Working capital / total assets	2
3	C79	0.0445	Retained earnings / total assets	3
4	C6	0.0432	ROA (net)	4
5	C69	0.0426	Total equity attributable to shareholders of the parent company/liabilities	5
6	C54	0.0415	Asset-liability ratio	6
7	C24	0.0398	ROA (net) (TTM) – excluding minority shareholder profit and loss	–
8	C55	0.0378	Asset-liability ratio (excluding advance receipts)	–
9	C82	0.0375	Total shareholders' equity (including minority) / total liabilities	7
10	C56	0.0319	Asset-liability ratio (excluding advance receipts) (announcement based)	–
11	C17	0.0299	Operating profit / total operating income	8
12	C61	0.0292	Current assets /short-term liabilities	9
13	C8	0.0244	ROE (annualized)	10

4.4 Feature Importance

Random forest will generate different feature importance values for each run. The model is trained and tested many times and a specified number of the top features is extracted from each run. Ultimately the top features aggregated over all of the runs are collected. However, several features have a strong correlation with each other. For instance, C55, asset-liability ratio (excluding advance receipts) is very similar to C54, asset-liability ratio. In cases like this, where features are just slight variations of each other, we kept only the higher-ranked one and adjusted the ranks. Table 8 shows the top-ranked features for the Chinese companies. The features that were removed for being variations of other features are denoted in Table by "–".

5 Comparison of Important Features

This section compares the features that are important for identifying companies that are going to become bankrupt in Poland and China. Table 9 shows similar features,

Table 9 Pairs of relevant features

Poland		China	
Description	*ID*	*ID*	*Description*
Retained earnings / total assets	P6	C79	Retained earnings / total assets
(Current assets – inventory) / short-term liabilities	P46	C61	Current assets/short-term liabilities
Profit on operating activities / total assets	P22	C7	EBIT / total assets
Working capital / fixed assets	P28	C77	Working capital / total assets

which are ranked in the top 15 most important features, for both cases. If we only discussed perfect feature matches, then that would leave us with only P6 and C79. Each of these was one of the top 10 features in their respective lists. However, other pairs were quite similar. The main difference between P46 and C61 was that the inventory was subtracted in the numerator in P46. As far as the pair of P22 and C7, EBIT was earnings before interest and taxes, containing profit on not only the operating activities but also other profitable activities (this does not commonly make much difference). In terms of the pair of P28 and C77, there was little difference in the denominator, with one using fixed assets and the other total assets. Thus, we see that essentially there are four important features in common.

There are also some important differences between the important features for the best classification models for the two countries. Asset- and equity-related features appeared eight times in the ranking list for China but only four times in the list for Poland. Meanwhile operation-related features, like profit and income, appeared five times in the ranking list for Poland but only one time in the list for China. In China, assets played an important role in bankruptcy prediction, because there are many state-owned enterprises (SOE) with enormous assets (which comprise a large part of the Chinese market. Moreover, operating-related attributes were not a big factor in predicting bankruptcy. SOEs generally have more assets and lower costs to get financing [11], which have crucial influence in financing like loans or bonds, so whatever they obtained high profits or not, they always had more possibility to get a loan to maintain operation to avoid bankruptcy. On the contrary, in Poland, the capacity to have an excellent operating situation may become an important factor affecting bankruptcy.

6 Related Work

Even though there are many studies on bankruptcy in both the finance and machine learning areas, those studies tend not to focus on identifying the key financial characteristics. More importantly, there do not appear to be studies that compare the importance of features for predicting bankruptcy using data mining methods for multiple countries.

Table 10 Feature importance scores (Poland) (Zieba) [1]

Rank	ID	Importance score
1	P25	0.0627
2	P22	0.0480
3	P27	0.0379
4	P15	0.0356
5	P52	0.0326
6	P53	0.0284
7	P14	0.0248
8	P40	0.0247
9	P42	0.0238
10	P36	0.0236

In earlier research, a logistic regression model was used to predict financial distress in China [2]. Although it only utilized six features of 139 records, the excellent performance was achieved with an accuracy of 0.94 for overall and an F1-score of 0.93 for class "distress." However, the researchers did not explore features. They chose to conduct a univariate analysis and found the performance of ROA (net) was best. Several of the features in their study, including ROA and working capital to total assets ratio, were also important in our models as C6 and C77. Moreover, their asset turnover is very similar to P22 and P35. And the current ratio was similar to the P46.

Our study can also be compared to the prior research by Zieba et al. [1] that was conducted on the Polish data set. Table 10 shows the ranked list of important features found in that study.

The P22 feature, profit on operating activities / total assets, which had the highest importance score in our prediction (see Table 4), was still ranked very high. Furthermore, it appeared in the logistic regression model we introduced as well. However, except for P22, there were no features in Zieba's [1] list that also appeared in our list. The difference was probably caused by that Zieba calculating the importance score on multiple models, while we relied only on the best performing model.

7 Conclusion

The study described in this chapter involved generating and evaluating classification models for predicting bankruptcy in companies in Poland and China, identifying the most important features, and then comparing and contrasting the features for the two nations. This comparison demonstrated that while there are some commonalities in the models for Polish and Chinese corporate bankruptcies, there are some significant differences. The common indicators include ROA and retained earnings to total assets ratio (RE/TA), but in the Chinese markets the asset-related features were much more important than for the Polish markets, while the operations-related features were much more important in the Polish markets than in the Chinese markets. These differences were explained based on the role of large state-owned enterprises in China. This study is quite unusual in that it analyzed and compared the role of financial features in two very different markets.

There are several ways in which this research can be extended and improved. First, we only considered the financial index of the year before the bankruptcy. There was no longer-term or periodic consideration. Meanwhile, the study was based on the pre-existing list of companies, which may have introduced a bias. In particular, the Chinese data set is quite small, which made it more difficult to reliably evaluate, and which limits the generality of the results. Analysis using a much larger data set would be quite beneficial.

We hope that this study and its novel perspective will inspire future research. Data mining has not been widely adopted in the asset pricing field or in corporate finance, and we hope that this study will help to change that.

Acknowledgments The authors would like to thank Yi Wang for her assistance in obtaining the data for China.

References

1. M. Zieba, S. Tomczak, J. Tomczak, Ensemble boosted trees with synthetic features generation in application to bankruptcy prediction. Expert Syst. Appl. **58**, 93–101 (2016 April)
2. S. Wu, X. Lu, A study of models for predicting financial distress in China's listed companies. Econ. Res. J. **36**, 4 (2001 June)
3. R. Setiono, H. Liu, Neural-network feature selector. IEEE Trans. Neural Netw. **8**, 654–662 (1997 May)
4. R. Genuer, J. Poggi, C. Tuleau-Malot, Variable selection using random forests. Pattern Recogn. Lett. **31**, 2225–2236 (2010 October)
5. C. Lu, L. Xu, L. Zhou, Comparative analysis of corporate financial distress and financial bankruptcy. Econ. Res. J. **39**, 64–73 (2004 August)
6. D. Dua, C. Graff, in *UCI Machine Learning Repository* [http://archive.ics.uci.edu/ml] (Irvine)
7. F. Pedregosa et al., Scikit-learn: Machine learning in Python. J. Mach. Learn. Res. **12**, 2825–2830 (2011 October)
8. J.R. Quinlan, *C4. 5: Programs for Machine Learning*, 1st edn. (Morgan Kaufmann, San Mateo, 2014)
9. T. Chen, C. Guestrin, XGBoost, in *Proceedings of the 22nd ACM SIGKDD International Conference on Knowledge Discovery and Data Mining – KDD 16, San Francisco, CA, USA, August 2016*, (2016), pp. 758–794
10. T.K. Ho, Random decision forests, in *Proceedings of 3rd International Conference on Document Analysis and Recognition, IEEE, Montreal, QC, Canada*, (1995), pp. 278–282
11. G. Ferri, L. Liu, Honor thy creditors beforan thy shareholders: Are the profits of Chinese state-owned enterprises real? Asian Econ. Pap. **9**, 50–71 (2010 October)

Using Matrix Factorization and Evolutionary Strategy to Develop a Latent Factor Recommendation System for an Offline Retailer

Y. Y. Chang, S. M. Horng, and C. L. Chao

1 Introduction

E-commerce and online entertainment are becoming widespread applications of the internet. For example, people buy products from Amazon.com, view videos on Youtube.com, and watch movies from Netflix.com. These online companies provide various services to worldwide customers 24 h a day and collect a large number of customer behavior data. Most online users get only a fleeting glimpse of a single online service because they feel overwhelmed by information on the internet. Therefore, getting customers' attention became a critical issue for online service providers, and recommender systems play a crucial role in the internet business [2, 4, 5, 10, 19, 21].

When shopping in brick and mortar retailers is still a primary way for people to buy goods and a place for social activities, recommendation systems can help the retailers to improve their operational efficiencies. For retailing companies, launching target-marketing campaigns to attract customers is a tremendous momentum to enhance competitiveness. Although it was commonly observed that retailers have applied statistical techniques to identify more promising customers for target marketing, the recommendation system based on big data has not been widely used. In addition, practitioners have known that past purchasing experience would influence future behavior. A study has indicated that the impact of prior experience

Y. Y. Chang · S. M. Horng (✉)
Department of Business Administration, National Chengchi University, Taipei City, Taiwan
e-mail: 104363050@nccu.edu.tw; shorng@nccu.edu.tw

C. L. Chao
Department of Accounting & Information Technology, National Chung Cheng University, Chiayi, Taiwan
e-mail: actact@ccu.edu.tw

© Springer Nature Switzerland AG 2021
R. Stahlbock et al. (eds.), *Advances in Data Science and Information Engineering*,
Transactions on Computational Science and Computational Intelligence,
https://doi.org/10.1007/978-3-030-71704-9_13

on future behavior will be gradually reduced by its time away from the target period [16]. It is very likely that this timely effect will show different degrees of impact dependent on various product types in retailing. Based on customers' shopping records of a retailing company in Taiwan, this study developed a recommendation system integrating a matrix factorization with an evolutionary strategy to efficiently predict the customers' preferences toward products. A matrix factorization was used to uncover the latent factors while an evolutionary program was applied to optimize parameters of the duration adjustment functions that were applied to assign weights over time. In addition, the recommendation system developed in this study also discovered that the duration adjustment functions depend on product categories. That is, customers' preferences toward new products rely on past experience at different degrees. For some products, customers' preferences could be predicted primarily by their past experience of weeks, and by their past experience of months for other products. The findings of this study will have significant academic and practical contributions.

This chapter is structured as follows. The next section will review the past literature on recommender system technology and evolutionary strategy. Next, an introduction of methodologies will be presented, followed by the results of data analysis, and further investigation for the duration adjustment functions of different categories. The last section will discuss the conclusions, academic as well as practical contributions, and provide some directions for future research.

2   Literature Review

In this section, we will first discuss the literature regarding recommendation systems, and then evolutionary strategy.

2.1   Recommendation System

Recognizing the correlation of components is the primary task of developing a recommender system, and the key techniques are collaborative filtering. Two approaches to collaborative filtering have been developed. One is the nearest neighborhood model that calculates similarities between two users or two items, and the other one is the latent factor model that uncovers latent factors that explain observed ratings and calculate similarities through these factors between users and items [20]. These two approaches will be introduced in detail below.

The nearest neighborhood model was first developed as user-oriented [14]. The user-oriented method estimated unknown ratings based on recorded data of similar users. Taking the same approach, an item-oriented method was developed to calculate the ratings based on data of similar items [19, 22]. The latter became popular in many cases because it had better scalability and provided results with superior performance [4, 5, 22, 23].

A primary task of the nearest neighborhood model is to measure the similarity between items or users. SIM (i_a, i_b) denotes the similarity of item a (i_a) and item b (i_b). The conventional methods of calculating similarities are Pearson correlation coefficient and cosine similarity [1]. In Eq. (1), $\hat{r}_{u,i_a}$ is an unobserved rating by user u on an item i_a:

$$\hat{r}_{u,i_a} = \bar{r}_{i_a} + \frac{\sum_{i_b \in N(i_a)} SIM(i_a, i_b) * (r_{u,i_b} - \bar{r}_{i_b})}{\sum_{i_b \in N(i_a)} SIM(i_a, i_b)} \tag{1}$$

where $N(i_a)$ is a set of ratings which are most similar to i_a, $SIM(i_a, i_b)$ is the similarity of i_a, and i_b, $\bar{r}_{i_a}$ is an average rating of item a, and $\bar{r}_{i_b}$ is an average rating of item b. The predicted rating of $\hat{r}_{u,i_a}$ is measured as a weighted average of the ratings for neighboring items [22, 24].

The latent factor model measures similarities through latent features that are extracted from observed ratings [18]. Several approaches have been developed including matrix factorization, deep learning, and principal component analysis [15]. Consider a matrix factorization of the user-item observations matrix R ($R \in \mathbb{R}^{M \times N}$), which can be decomposed into user matrix U ($U \in \mathbb{R}^{M \times F}$) and item matrix V ($V \in \mathbb{R}^{F \times N}$). The user matrix is composed of the user-factors vector u_u, and the item matrix consists of the item-factors vector v_i. The unobserved rating, i.e., a rating by user u on item i_a, can be calculated by taking an inner product, $\hat{r}_{u,i_a} = u_u^T v_{i_a}$ [17]. The goal of training a matrix factorization is to minimize Eq. (2), which is used to minimize the sum of squared errors between actual rating, $r_{u,i}$ and predicted rating, $u_u^T v_i$:

$$\min_{U,V} \sum_{(u,i) \in R} \left(r_{u,i} - u_u^T v_i \right)^2 + \lambda \left(\|u_u\|^2 + \|v_i\|^2 \right) \tag{2}$$

where u_u is the u-th column of user matrix U, and v_i is the i-th column of item matrix V. The second term in Eq. (2), $\lambda(\|u_u\|^2 + \|v_i\|^2)$, is added in order to prevent overfitting. The parameter λ controls the strength of regularization [6]. Usually, a stochastic gradient descent [13] can be applied to train this dataset in order to achieve the minimum sum of squared errors. These two techniques, latent factor model of matrix factorization and item-based nearest neighborhood model, will be tested and compared in this study.

2.2 Evolutionary Strategy

Evolutionary strategy is inspired by the evolutionary processes of biological populations. An evolutionary algorithm uses recombination, mutation, and selection to create candidates from parental features [3]. By testing the candidates of their fitness, the algorithm will evolve with better solutions iteratively. Evolutionary strategy is an all-purpose algorithm, and can be applied in various scenarios that optimize discrete as well as continuous variables. Many versions of evolutionary strategy were found [8, 12], and the differences between the specific algorithms are approaches to implement recombination, mutation, and selection [8].

A typical evolutionary algorithm starts from generating an initial population with a pre-defined number of individuals. Next, a set of offspring or individual is produced by selecting parents based on their fitness, and recombining the selected parents to create different individuals. Last, individuals are mutated according to Gaussian distribution to increase the diversity of offspring. In each iteration, the individuals with better fitness scores will survive into the next generation. This process will iterate until the termination criterion, the number of total iterations or number of iterations without improvement, is met [8].

3   Methodology

Figure 1 shows the flowchart of the data analysis. Due to the complexity of the dataset, this study adopts several processes in order to analyze the dataset in a logical way. Data are sampled and split after collection. The next step is to convert the data based on the duration adjustment function so that the data are available for further analysis. Following the data conversion is to identify a measure to represent customers' preferences and decompose the user-item matrix to find the users' preferences. Last, an evolutionary algorithm is introduced to find the final solution. The following sections will introduce these processes in detail.

3.1   Data Collection

This study was based on the members' one-year purchasing records of a retailer in Taiwan. This retailer is an international company specialized in fast-moving consumer goods (FMCG) and has many stores across Taiwan. By utilizing a clustering analysis on members' purchasing records of 1 year, all of the items sold by this retailer in Taiwan were grouped into 19 categories, including alcohol & cigarettes, bacon & cheeses, bakery, bath & body, beverages, breakfasts, candies & cookies, canned goods, frozen foods, fruit, hot meals, household essentials, pets, seafood & meat, seasoning, soft drinks, Taiwan snacks, Taiwanese cuisine, and vegetables.

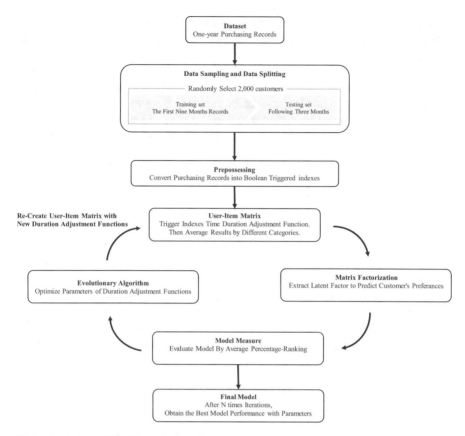

Fig. 1 Flowchart of the data analysis

3.2   Data Sampling and Data Splitting

Approximately one hundred thousand members and ten thousand items are available in the database. Because of the computation limit, records of two thousand members within one year were randomly selected as the research dataset. In addition, this study used the data of the first 9 months as a training dataset and data of the following 3 months as a testing dataset. The purpose of this arrangement is to have adequate data in both training and testing datasets [7].

3.3   Data Preparation and Duration Adjustment Function

In this study, $r(u,i)$ indicates how many times the user u bought item i within a day. $r(u,i)$ is set to zero, if user u did not buy the item i. This setting is based on related researches and pre-training results, and is different from the previously developed recommendation system in which the unrated items were neglected and excluded from further analysis [14].

Purchasing records represent customers' preferences, and this study did not use numerical values to express the level of preferences. Instead, triggered index $p(u,i)$ is defined to represent that the interest of user u is triggered by item i. If a user u bought an item i twice on a specific period, $r(u,i)$ is equal to 2. When $r(u,i)$ is greater than zero, the triggered index $p(u,i)$ will be set as one indicating that user u was interested in item i within a specific period. $p(u,i)$ will be zero when the user u did not show any interest in item i within the period. The values of $p(u,i)$ are derived by the following:

$$p_{u,i} = \begin{cases} 1, & r_{u,i} > 0 \\ 0, & r_{u,i} = 0 \end{cases}$$

Customers' preferences might not be constant over a long time. In the case of training dataset with 9 months, this study assumed that newer behavior represents stronger interests in items. To lower weights on the past records in the period further away, this study used duration adjustment functions, as shown in Eq. (3) [14], to control how the variable T, a maximum of 270 in this study, influences the customers' preferences along with two other variables, a and b, for the shapes of the function:

$$\frac{e^{-(a*T-b)}}{1 + e^{-(a*T-b)}} \tag{3}$$

In general, different combinations of parameter setting will produce various weights on the records further away, and weight on the records at time t is always no less than the weight on the records at time $t-1$. Extremes of the function are either having the same weights across the entire training period or having the highest weight at the time right before the testing period and zero elsewhere.

3.4 Model Measure

It is difficult to identify a measure representing customers' preferences. In this study, average percentage-ranking of items was used to evaluate models' performance [14]. First, an ordered list of items for each customer was generated. The list was sorted by models' predictions with the highest chance of being purchased on the top. For example, 〖rank〗_(u,i) represents item i within the ordered list that is derived for user u. If a list contains one thousand items, and the purchased item i is predicted to be the tenth item, then 〖rank〗_(u,i) will be 1% (10/1000=1%).

Table 1 An example of user-item matrix

User\item	A	b	c	D
1	3	1	4	4
2	5	3	1	2
3	5	4	3	2
4	1	2	1	5
5	4	?	4	1

On the other hand, ⟦rank ⟦_(u,i)=100% represented that the purchased item i is predicted to be the least possible item for user u.

To explore customers' hidden preferences, the models developed and tested in this study aimed to predict the items that had not been purchased by the member during the training period of 9 months. To measure the performance of the models tested, percentage-ranking of items for each member will be computed first, and then the average of percentage ranking of items for all of the two thousand members will be calculated to evaluate the models. The lower averaged percentage-ranking would present a better prediction ability of the model.

3.5 Matrix Factorization

This study decomposed the user-item observations matrix by a singular value decomposition. After the decomposition process, the user-item matrix is composed of user-factors vector u_u and the item-factors vector v_i. The prediction is calculated by taking an inner product of user-factors vector u_u and item-factors vector v_i. For example, the user-item observations can be represented in a feedback matrix as shown in Table 1:

The numbers in Table 1 are the users' preferences toward items. The intuition behind using matrix factorization to predict the missing value is that there should be some latent factors that determine how user 5 rates item b. Firstly, assume that the matrix factorization model would like to extract two latent factors. The task of the model is to find user matrix U ($U \in \mathbb{R}^{5 \times 2}$) and item matrix V ($V \in \mathbb{R}^{2 \times 4}$) such that their product approximates the user-item matrix in Table 1. In this way, the rows of U represent the strength of the associations between users and two factors. Similarly, the columns of V represent the strength of the associations between items and two factors.

For obtaining user matrix U and item matrix V. The typical approach is to initialize two matrixes with some values, and then, try to minimize the difference between the product of two matrixes and user-item matrix in Table 1 by stochastic gradient descent iteratively. This process is the same as minimizing Eq. (2). After converging to the minimum difference, these two matrixes can represent user matrix U and item matrix V.

Table 2 Evolutionary program

1. Initialize parent populations of parameter sets $P_\mu = \{s_1, \ s_2, \ \ldots, s_\mu\}$.
2. Generate offspring population $\widetilde{P}_\mu = \left\{\widetilde{s_1}, \ \widetilde{s_2}, \ldots, \widetilde{s_\mu}\right\}$, each $\widetilde{s}$ is generated by the following:
– Select ρ parents from P_μ randomly according to their fitness. Each pair of parents will generate two individuals.
– Recombine ρ parents, based on pre-defined probability, to generate the same number of individuals.
– Mutate each value of parameters of offspring $\widetilde{s}$ by Gaussian distribution $N(0, \sigma)$.
3. Select new parent population from the offspring $\widetilde{P}_\mu$ and the parent P_μ by their fitness values. Those with lower averaged percentage-ranking will remain as the new population for the next iteration.
4. End the algorithm when termination condition is fulfilled, otherwise go to step 2.

Through the above approach, results of matrix factorization are user-factors vector u_5 as (0.62, 1.73) and item-factor vector v_b as (0.72, 1.59). By taking an inner product of u_5 and v_b, the prediction of missing value is 3.19.

3.6 Evolutionary Algorithm

This study adopted an evolutionary algorithm to optimize the parameters of duration adjustment functions, T, a, and b in order to minimize the average percentage-ranking of items. Items were classified into 19 categories as mentioned previously, and each category has its own duration adjustment function. Therefore, 19 sets of parameters of T, a, and b will be derived. The evolutionary program will obtain the near-optimal combination of parameter setting [9, 11], and the detail is shown in Table 2.

4 Results

4.1 Duration Adjustment Function Results

Out of 19 item categories, six patterns based on their duration adjustment functions were identified. Figure 2 shows the curves of duration adjustment functions for the six patterns after two hundred generations, and Table 3 shows the value of parameters T, a, and b for each of the six patterns.

The category of fruit has a flat line across the 100% of weight indicating that customers' preference toward fruit is almost constant over the training period. It implies that the past purchasing experience within 9 months in this retailer has equal influence on the customers' purchasing behavior in the test period of 3 months.

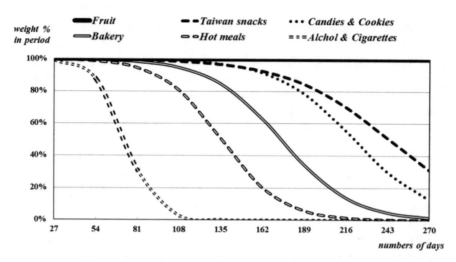

Fig. 2 Duration adjustment function of fruit, Taiwan snacks, candies & cookies, bakery, hot meals and alcohol & cigarettes

Table 3 Parameters T, a, and b for fruit, Taiwan snacks, candies & cookies, bakery, hot meals and alcohol & cigarettes

Category\parameter	T	a	b
Fruit	1.86620	0.18188	7.56427
Taiwan snacks	3.18732	2.58972	7.46066
Candies & cookies	4.22185	2.54015	8.79425
Bakery	3.72996	3.15518	7.58203
Hot meals	3.80535	3.69314	7.05311
Alcohol & cigarettes	6.74123	4.03122	7.37033

Out of the 19 categories, customers show the highest loyalty when purchasing new items of fruit. On the opposite, alcohol and cigarettes have the steepest curve compared to others. Past purchasing records over approximately 110 days have the weight close to zero indicating that only the past records, the purchasing experience in the training period, within approximately 3 months, can influence customers' purchasing behavior, in the testing period, for the new items in this category. Customers purchasing new items for alcohol and cigarettes have the least loyalty, compared to other categories, toward this retailer. Buying alcohol and cigarettes is convenient in Taiwan from thousands of convenience stores, and this retailer is not the primary channel for this category. Therefore, customers' purchasing behavior of this retailer for this category was possibly affected by its seasonal promotion campaigns. The curves of the other three categories, hot meals, bakery and candies & cookies lie between these two extremes. In general, curves toward the right, or having a smaller slope, shows higher loyalty toward this channel.

4.2   Model Evolution Results

In addition to the matrix factorization model developed in this research and item-based nearest neighborhood approach used as a competing model, a popular approach in practice, the popularity model is also applied to compare with others. The popularity model sorts all items based on their popularity, that is, the number of transactions. Items with more transactions are predicted to have a higher chance of being purchased. It is easy to understand and implement this model in practice, and therefore is widely used in the retailing industry of Taiwan. The item-based nearest neighborhood model predicts unknown preferences based on recorded observations of similar items. According to preliminary results, cosine similarity showed better results than Pearson correlation and was used in this study to calculate similarities between items.

Figure 3 shows the results of average percentage-ranking for the three models, popularity, item-based nearest neighborhood, and matrix factorization. Only the measures of the matrix factorization model evolved over generations, while the other two models remained the same. The popularity model has the highest average of percentage-ranking of 40.75% and the item-based nearest neighborhood model has a lower measure of 36.16%. The matrix factorization model reached the best results out of the three models at 35.73% after two hundred generations. Table 4 shows the values of parameters T, a, and b for the nineteen categories after two hundred generations.

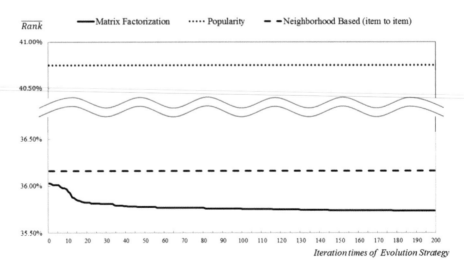

Fig. 3 Results of popularity, item-based nearest neighborhood and matrix factorization recommendation systems

Table 4 Parameters T, a, and b for whole nineteen categories after two hundred generations

Category\parameter	T	A	b
Alcohol & cigarettes	6.74123	4.03122	7.37033
Bacon & cheeses	2.37456	3.55500	6.68644
Bakery	3.72996	3.15518	7.58203
Bath & body	3.59946	3.04752	7.50740
Beverages	3.49104	3.16544	6.49497
Breakfasts	3.95482	2.71484	7.57106
Candies & cookies	4.22185	2.54015	8.79425
Canned goods	3.79674	2.54696	6.80136
Frozen foods	1.95966	0.99546	9.47475
Fruit	1.86620	0.18188	7.56427
Hot meals	3.80535	3.69314	7.05311
Household essentials	3.91813	2.84353	7.32999
Pets	3.76021	3.00918	7.31041
Seafood & meat	3.88733	2.52668	8.07790
Seasoning	4.13416	2.99593	9.19544
Soft drinks	4.24225	2.88737	6.57402
Taiwan snacks	3.18732	2.58972	7.46066
Taiwanese cuisine	3.88084	3.77489	7.71500
Vegetables	4.66090	3.29381	7.17516

5 Conclusions, Contributions, and Future Work

5.1 Conclusions

Recommendation systems have been developed for the online channel with a simple product mix. This study extended its applications to an offline retailer with a more complex product mix. Purchasing records of two thousand members of this retailer in one year were used as the dataset for this study. Datasets of the first 9 months were used for the purpose of training and models were tested by the records of the last 3 months. Duration adjustment functions were implemented to adjust the weights on the past purchasing records over time. Three models for the recommendation system, popularity, item-based nearest neighborhood, and the matrix factorization were tested in this study. The model developed in this study, the matrix factorization model with weights controlled by duration adjustment functions, showed the best performance out of the three models tested.

5.2 Contributions

The recommendation system was originally developed for online services with a simple product mix. This study successfully extended it into an offline retailer with a more complex product mix. In addition, this study integrated two approaches, a matrix factorization to uncover latent factors from the user-item matrix, and an evolutionary algorithm to optimize parameters of duration adjustment functions for assigning weights over time. The recommendation system developed by this innovative approach outperformed two other models in predicting customers' preferences when purchasing new items.

The results of this study also have significant practical contributions. First, the superior performance of the recommendation system in this study can help the retailer promote new items more efficiently. Compared to the widely used approach, popularity model, this system can improve the prediction percentage by 5%. In a highly competitive FMCG market, the retailer could transfer this improvement into a competitive advantage over competitors. Second, the duration adjustment functions for different categories can help the retailer identify the item groups having a higher potential to show customers' loyalty toward the retailer. By targeting these item groups, the retailer could have a better chance of retaining customers, and therefore increase the sales. Third, for the categories with the curves of duration adjustment function close to the left in Fig. 2, the retailer could target the customers with only purchasing records within a certain period determined by the curves. With the higher accuracy of targeting customers, the cost saved is equivalent to higher profitability.

5.3   Limitations and Future Work

Although the recommendation system developed in this study have shown better performance over the other two systems. The dataset used only contains 1 year of data and seasonal effect was not captured during model training and testing. A dataset covering a longer period of time will further enhance the validity of the results. Second, due to the computation complexity, the parameter settings of duration adjustment functions for 19 item categories were integrated into one task for the evolutionary algorithm. Separating these functions and running them individually by evolutionary programs could get parameter settings with better performance.

References

1. G. Adomavicius, A. Tuzhilin, Toward the next generation of recommender systems: A survey of the state-of-the-art and possible extensions. IEEE Trans. Knowl. Data Eng. **17**(6), 734–749 (2005)
2. K. Ali, W. Van Stam, TiVo: Making show recommendations using a distributed collaborative filtering architecture, in *Proceedings of the tenth ACM SIGKDD International Conference on Knowledge Discovery and Data Mining*, (ACM, Seattle, 2004), pp. 394–401
3. T. Bäck, H.P. Schwefel, An overview of evolutionary algorithms for parameter optimization. Evol. Comput. **1**(1), 1–23 (1993)

4. R.M. Bell, Y. Koren, Lessons from the Netflix prize challenge. ACM Sigkdd Explor. Newsl. **9**(2), 75–79 (2007a)
5. R.M. Bell, Y. Koren, Scalable collaborative filtering with jointly derived neighborhood interpolation weights, in *Proceedings of the 2007 Seventh IEEE International Conference on Data Mining*, (IEEE Computer Society, Omaha, 2007b), pp. 43–52
6. Y. Bengio, Practical recommendations for gradient-based training of deep architectures, in *Neural networks: Tricks of the trade*, (Springer, Berlin, Heidelberg, 2012), pp. 437–478
7. J. Bennett, S. Lanning, The Netflix Prize. ACM SIGKDD Explor. Newsl. – Spec. Issue Vis. Anal. **2007**, 3–6 (2007)
8. H.G. Beyer, H.P. Schwefel, Evolution strategies–A comprehensive introduction. Nat. Comput. **1**(1), 3–52 (2002)
9. L. Butcher, L. Owen, S. Harris, in *Decoupling Evolutionary Programming from Component-Based Software*. Recent advances in software engineering and computer science, 5(1) (2020)
10. J. Davidson, B. Liebald, J. Liu, P. Nandy, T. Van Vleet, U. Gargi, S. Gupta, Y. He, M. Lambert, B. Livingston, D. Sampath, The YouTube video recommendation system, in *Proceedings of the fourth ACM conference on Recommender Systems*, (ACM, Barcelona, 2010), pp. 293–296
11. D.B. Fogel, An introduction to simulated evolutionary optimization. IEEE Trans. Neural Netw. **5**(1), 3–14 (1994)
12. C.M. Fonseca, P.J. Fleming, An overview of evolutionary algorithms in multiobjective optimization. Evol. Comput. **3**(1), 1–16 (1995)
13. R. Gemulla, E. Nijkamp, P.J. Haas, Y. Sismanis, Large-scale matrix factorization with distributed stochastic gradient descent, in *Proceedings of the 17th ACM SIGKDD International Conference on Knowledge Discovery and Data Mining*, (ACM, San Diego, 2011), pp. 69–77
14. Y. Hu, Y. Koren, C. Volinsky, Collaborative filtering for implicit feedback datasets, in *Proceedings of the 2008 Eighth IEEE International Conference on Data Mining*, (IEEE Computer Society, Pisa, 2008), pp. 263–272
15. Y. Koren, Factorization meets the neighborhood: A multifaceted collaborative filtering model, in *Proceedings of the 14th ACM SIGKDD international conference on Knowledge Discovery and Data Mining*, (ACM, Las Vegas, 2008), pp. 426–434
16. Y. Koren, The BellKor solution to the Netflix Grand Prize. Netflix Prize Doc. **81**, 1–10 (2009)
17. Y. Koren, Factor in the neighbors: Scalable and accurate collaborative filtering. ACM Trans. Knowl. Discov. Data **4**(1), 1–24 (2010)
18. Y. Koren, R. Bell, C. Volinsky, Matrix factorization techniques for recommender systems. Computer **42**(8), 30–37 (2009)
19. G. Linden, B. Smith, J. York, Amazon.com recommendations: Item-to-item collaborative filtering. IEEE Internet Comput. **7**(1), 76–80 (2003)
20. M. Papagelis, D. Plexousakis, Qualitative analysis of user-based and item-based prediction algorithms for recommendation agents. Eng. Appl. Artif. Intell. **18**(7), 781–789 (2005)
21. B. Sarwar, G. Karypis, J. Konstan, J. Riedl, Analysis of recommendation algorithms for e-commerce, in *Proceedings of the 2nd ACM Conference on Electronic Commerce*, (ACM, Minneapolis, 2000), pp. 158–167
22. B. Sarwar, G. Karypis, J. Konstan, J. Riedl, Item-based collaborative filtering recommendation algorithms, in *Proceedings of the 10th International Conference on World Wide Web*, (ACM, Hong Kong, 2001), pp. 285–295
23. G. Takács, I. Pilászy, B. Németh, D. Tikk, Major components of the gravity recommendation system. ACM SIGKDD Explor. Newsl. **9**(2), 80–83 (2007)
24. J. Wang, A.P. De Vries, M.J. Reinders, Unifying user-based and item-based collaborative filtering approaches by similarity fusion, in *Proceedings of the 29th annual international ACM SIGIR conference on Research and Development in Information Retrieval*, (ACM, Seattle, 2006), pp. 501–508

Dynamic Pricing for Sports Tickets

Ziyun Huang, Wenying Huang, Wei-Cheng Chen, and Matthew A. Lanham

1 Introduction

Nowadays, in the world that is flooded by digital data, a one price for all strategy is no longer optimized for products like sports tickets. With continuously increasing costs in operation, maintenance for the venue, and contracts for players, sports team management are in need of a modern strategy for pricing to optimize their revenue and profits. Dynamic pricing strategies are widely adopted in the hotel industry in which the demand for products is very flexible. According to Hoisington [1], the hotel industry in the US reports the highest revenue in the budget hotel market and revenue for budget hotels increased 3.5% over year thanks to dynamic pricing strategy adapted in the industry. The demand for sports tickets could also change dynamically with time, participants in the events, and other factors. With its similarity to the hotel industry, it is believed that the sports industry could also be greatly benefited from a dynamic pricing strategy. In fact, some sports teams have already employed data analytics in their pricing strategy to maximize their ticket revenue. According to SAS [2], the Orlando Magic from the NBA league has accomplished a game ticket revenue growth of 91% since the 2013–2014 season with its data-driven strategy in pricing. 91% is a huge number, and not to mention that the baseline revenue from an NBA team is tremendous by itself. According to Khan [3], the benefits of dynamic pricing include (1) greater control over pricing strategy; (2) brand value with flexibility; (3) cost efficiency in the long run; and (4) efficiency in management.

Z. Huang · W. Huang · W.-C. Chen (✉) · M. A. Lanham
Department of Management, Purdue University, West Lafayette, IN, USA
e-mail: huang747@purdue.edu; huang814@purdue.edu; chen1614@purdue.edu;
lanahamm@purdue.edu

© Springer Nature Switzerland AG 2021
R. Stahlbock et al. (eds.), *Advances in Data Science and Information Engineering*,
Transactions on Computational Science and Computational Intelligence,
https://doi.org/10.1007/978-3-030-71704-9_14

With the benefits and potential growth in profits that could be brought by dynamic pricing, this paper aims at discussing how to apply dynamic pricing strategy to ticket price of individual sports teams in practice. A deep understanding of the methodologies of dynamic pricing in the sports industry is necessary to maximize the benefits of dynamic pricing in the industry. The scale of the sports industry is one of a kind in the US economy and the growth brought to the sports industry by dynamic pricing would be a strong boost to the overall economy.

It is impossible to develop an effective price optimization model without an accurate prediction of the demand under different circumstances. Therefore, after data cleaning, the first and the most important part of constructing the optimization model is to develop a predictive model for the demand. In the process, multiple machine learning methods were applied to the cleaned data, and the model that performs the best among the non-overfit candidate models was chosen as the foundation to build the dynamic pricing model. With predictable demands, economic rules are utilized to develop a dynamic pricing model. With the developed model, the optimized price was generated based on historical data to obtain a potential optimized revenue compared to the real historical revenue. The dynamic pricing model which builds upon this process provides sport teams insights and methods to capture the lost revenue due to ineffective pricing. The potential of this model was quantified by the historical data to prove its effectiveness.

In the remainder of this paper, Sect. 2 provides a literature review that discusses current studies in the field of dynamic pricing. Section 3 gives the sources and context of the data used. The next section provides detailed explanations of methodologies to develop the predictive model and the optimization model. The evaluation and performance demonstration of the developed model is the topic of the fifth section. Finally, the paper concludes with the consideration of future application and study direction of this research.

2   Literature Review

Dynamic pricing is becoming a prevailing method in the sporting event industry; hence, numerous studies have been performed to discuss the implementation of this approach. Dittmer and Carbaugh [4] have mentioned the frequency of game and size of the stadium in the sports industry provide an ideal condition for the adoption of dynamic pricing. The discussion mainly focused on applying microeconomic theory to the model. A dynamic model reduces the cost of re-pricing the ticket every time there is a gap in the price and consumer's perceived value because it will automatically match the expected value based on selected factors. Sweeting's paper [5] discussed dynamic pricing behavior in the secondary markets for Major League Baseball tickets and concluded that sellers cut prices by more than 40% as event day approaches and adopting dynamic pricing raises the average seller's expected payoff by 16%. Our work is similar to that of Sweeting's paper, but we also take into account data from the primary market in addition to the secondary market.

Before getting in touch with dynamic pricing, assessing consumer demand is a critical element before building a price optimization model. Strnad and Nerrat's paper [6] examined the accuracy of using different artificial neural network (ANN) models to capture soccer match attendance. Their paper suggests that all neural networks outperform linear predictor by a large margin and can capture attendance patterns. A more recent paper by Sahin and Erol [7] also evaluated the performance of neural network models on predicting soccer match demand. The paper used data from three sports teams and discovered that the ANN model with Elman network using one hidden layer and 20 neurons has the most accurate result in demand prediction. Even though neural network models are difficult to interpret the impact of each feature brings to the result, our paper also designed ANN models for comparison with other models.

Xu et al. [8, 9] evaluated the effect of dynamic pricing policy by developing a demand model for single-game ticket sales that were used to predict the revenue associated with a pricing strategy over the course of an MLB franchise in a season. The original dynamic pricing policies of this franchise results in a revenue decrease of 0.79% compared to static pricing. Hence, the authors proposed alternative pricing policies that helped the franchise find an optimized dynamic pricing policy that improved revenue by 14.3% compared to fixed pricing policy. A similar approach was proposed by Kemper and Breuer [10] where they applied mathematical theory to estimate a demand function before building a price optimization model. Their paper discover that the consumers' willingness to pay is significantly higher than the original ticket price. The average willingness to pay ranged between € 70 and € 178, depending on the seat and price category. Our paper also first developed a demand function for a sport event to obtain the probability that a ticket would be sold, then built a dynamic pricing model that reoptimizes prices on a daily or weekly basis.

Diehl and Maxcy [11] investigated elasticity of demand in the secondary market for NFL tickets and how elasticity varies across different seat types. The authors used the standard inverse elasticity rule to maximize profit when elasticity exceeds the unit elastic point. Price elasticity of demand was estimated for the entire venue and then location-specifically. Their research indicates that demand in the secondary market is price elastic and that the demand for higher quality seats is more price elastic than the demand for lower quality seats. An important factor that their research did not incorporate is that their data lack the timing of the sales relative to the game.

Shapiro and Drayer [12] assessed factors that influenced the ticket price for a sports team in both the primary and secondary markets. Correlation designs were first used to observe the relationships between other factors and ticket prices. Their paper is different from our paper in that they devised two regression models which include a 2-stage least square regression (2SLS) using season ticket prices and secondary market price as the independent variable and an OLS model excluding these variables. The main findings of their paper suggest that season ticket price is an important feature that affects secondary market price and that the secondary market affects dynamic ticket pricing. Our paper includes only single-game ticket data which could potentially reduce the effect of our dynamic price optimization model.

3   Data

To estimate the ticket demand for sports events and train all the necessary machine learning models, data is required to include all ticket transactions from both the primary and secondary markets. Primary market is where transactions happen directly between the team to customers, and secondary market is where a ticket owner resale his/her tickets on platform like Ticketmaster or StubHub. The datasets used in this study were retrieved from the internal database of a primary NFL team and used with the permission of the team.

The original datasets describe ticket pricings, online transactions, and event-related information in four separate tables: primary, secondary, unsold inventory, and opponent. The data covers all events in the team's main stadium from 2012 to 2019 seasons, and the total number of games played during the period was 79 after removing pre-season and post-season events. These four tables were cleaned, merged, and standardized before predictive modeling to acquire only significant measures for better forecasting results. Variables directly related to ticket pricing (e.g., seat location in the stadium, sale time, final sale price) are the primary internal measures for demand prediction and revenue management study. In addition to the internal factors, external factors that affect customer purchasing behaviors such as weather measures (e.g. temperature, snow/rain) or competitor measures (e.g. ranking, win probability) were also considered in the modeling. The data dictionary of the pre-processed input variables is given in Table 1.

4   Methodology

There are two phases to the dynamic pricing model, the first of which is to compute the demand probability of each ticket since this is an essential element in formulating the revenue function which we will discuss later. According to fundamental economic theories, the demand for a product increases as the price drops and decreases as the price rises. In many economic studies, product demand is modeled in a simple linear function between the quantity and the price. However, a more sophisticated model that considers other factors is needed to assess the demand for each ticket for this problem. The demand forecasting process in this phase is considered a binomial classification problem of whether a ticket is sold. We then train several predictive models using varying algorithms to obtain the demand probabilities of each sold ticket. The predictive model also calculates the probability

Table 1 Data dictionary of the pre-processed input variables

Variable	Type	Description
Target	Factor	The seat sold status, 1 = yes, 0 = no
Event name	Factor	The event code of game
Section name	Factor	The name of section
Row name	Factor	The row number
Seat num	Factor	The seat number on ticket
Team	Factor	The opponent team state and name
Season	Factor	The year of game
Event date	Date	The date of game
Sale date	Date	Ticket purchase's final transaction date (used the date time from the final transaction if resold in the secondary market)
Final price	Numeric	Ticket's final price (used the final transaction price if resold)
Original purchase price	Numeric	Ticket's original purchase price from the primary market
Sales channel	Factor	Ticket's sales channel
Opp win	Numeric	The winning probability of opponent team
Opp LS win	Numeric	Last season's win probability of opponent team
Team win	Numeric	The winning probability of the team
Team LS win	Numeric	Last season's win probability of the team
Road attendance	Numeric	Attendance percentage for opponent team. This is an indicator if a team has a big draw attendance wise or not
FB fans	Numeric	The Facebook fans number of opponent team
Distance	Numeric	Distance between the team's city and opponent city (in miles)
Home opener	Factor	Whether the team was home opener of the season, 1 = yes, 0 = no
Temp at kick	Numeric	Temperature at the first kick of the game
Rain/snow	Factor	Weather condition during the game, 1 = rain/snow, 0 = no
Team contention	Factor	Whether the team was out of contention, 1 = yes, 0 = no
Last visit years	Numeric	Number of years since the opponent's last visit
Opp scored LY	Numeric	Opponents' points scored last year (previous year due to at the time we pull this data it is February for the next year and wanted to remain consistent). Regular season only
Opp def GivenLY	Numeric	Opponents' points given up last year (previous year due to at the time we pull this data it is February for the next year and wanted to remain consistent). Regular season only
Opp playoff	Factor	Whether opponent had playoff game, 1 = yes, 0 = no
Off MVP	Numeric	Offensive NFL MVP votes the year before
Def MVP	Numeric	Defensive NFL MVP votes the year before
Odds f	Numeric	Super bowl odds from February of that year
GA_indy_L10	Numeric	Google trends index for opponent team over the past 10 years

of a ticket sold at any point in time, given the time information is also available in the data. We assume that a ticket is sold if the probability is greater than or equal to 0.5, and the sold tickets are then used in calculating the expected revenue.

We chose one specific game from the 2019 season and used it as the testing set. A random partition of 2:8 was conducted to ensure random performance testing. The selected testing set from the 2019 season is more recent and is in more proximity to the current and future situations. Leaving this game out of the training set guarantees the precision testing statistics and business performance analysis. The Area Under Curve (AUC) was the criterion used to evaluate and determine the performance of varying predictive models, the model with the highest AUC was chosen for the optimization process.

The second phase is the optimization process which involves obtaining the optimal price and calculating the expected revenue. It is necessary to find the balance point between the price and the effects the price has on the demand. We apply the predictive model with the highest accuracy to the optimization process using the recurring method to find an optimal price point at which the expected revenue is maximized for each ticket available at any given point of time. The expected revenue of the tickets is the product of the probability of a ticket demand given other factors and the optimal price. The formula is given as the following:

$$E \text{ (Revenue)} = P \text{ (Demand | Price\&other factors)} \times \text{Price}$$

For the game that was used for demonstration, we chose six different seat sections from the team's stadium to observe the price impact on the expected revenue using the optimization method and compared them to the actual historical revenue of the same sections to demonstrate the performance of the optimization model. The recommended price of each section at different points of time is the range from the lower 1.5 IQR bound to the maximum of the optimized ticket price in the section at that time. The overall process is summarized in Fig. 1:

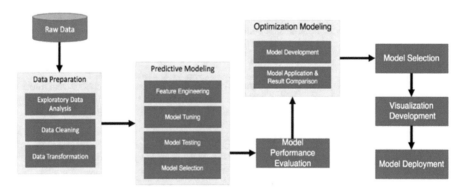

Fig. 1 Process flow chart

4.1 Data Cleaning

(a) *Primary Data:* This table includes the information (e.g., opponent, event date, sale date, section, ryow) of single-game tickets bought by customers directly from the team's website from 2012 to 2019. An individual with a ticket account ID could purchase multiple tickets at once; hence, a column named "num_seats" would indicate the number of tickets bought by the customer. We decomposed the rows with more than one ticket possession into separate rows. We also added a target column with all 1 s to this table indicating every ticket in the table is sold.

(b) *Secondary Data:* This table consists of tickets being sold on the secondary market. The table has over 600,000 rows of data because it includes different activities (e.g., update price, cancel posting, expired, successful resold) of the same ticket by the same sellers. To simplify the data, we subset only the resold, expired records. Then we assigned 1 s to those whose activity name is resold, and 0 s to the records that are expired. In this table, there were a lot of extreme posting price; therefore, we set a condition to only include prices that are below $1000. Event dates for this table also deviate from the actual event date by 1 day; we added 1 day to every event date in this table.

(c) *Unsold Inventory Data:* This table includes all the unsold single game tickets of the team's website from 2012 to 2019. A target column of 0 s was added indicating every ticket in the table is unsold. All the records in this table were combined with the primary table.

(d) *Opponent Data:* Opponent information including last season win rate, weather conditions, Facebook fans, etc., attributed to this table will be merged to combine ticket information and opponent information.

4.2 Feature Engineering

Time is an important factor in a dynamic pricing model; Ticket buyers usually purchase tickets days or weeks before the event occurs. Hence, we inserted a new column that took the difference between the sales and game dates which shows how many days in advance people purchase the tickets. In addition, we extracted the day from the event date column to observe whether a particular day of the week could affect the demand.

Categorical encoding was also necessary before moving into model building. Ticket information contains the section names, which generally have hundreds within a sports stadium. Therefore, performing one-hot encoding would impede the efficiency of the models. We used target encoding for the sections, which label the variable by the ratio of the target value occurrences. The remaining categorical features that have low numbers of levels were one-hot encoded.

Numerical features such as Facebook fans and distance were standardized using min-max scaling to ensure the model does not bias to the feature that has larger numbers. This method rescales every numeric value to lie in the range between 0 and 1 and puts equal weight among all of them.

4.3 Data Imputation

After combining the value of the day in advance column, it would appear as null for those tickets that were unsold. We used the mice package to impute the missing values by using the cart (classification/regression tree) method to replace missing values. However, it is crucial that we state the assumption that the distribution of the missing values follows the same distribution of the existing values. A density plot of the imputed and existing values was created to compare the distributions.

5 Models

To obtain the most accurate demand forecasting models, four different supervised learning algorithms were compared.

5.1 Logistic Regression

Logistic regression is a generalized linear regression for binary classification problems. The probability of both positive and negative event outcomes could be given by logistics. This suits our need for the predictive model. Nevertheless, a regression method can provide specific insights on how each factor affects the change of the probability of outcomes. However, as we will discuss in the following section, the performance of the logistic regression on this data set was not ideal. Therefore, more robust algorithms were tested.

5.2 Gradient Boosting/XGBoosting

Gradient boosting algorithm is suitable for both regression and classification problems. The algorithm is based upon the classical decision tree algorithm in an iterative stage-wise method. Gradient boosting seeks to optimize the mean squared error (MSE) of the model by performing gradient descent.

XGBoosting is the abbreviation for eXtreme Gradient Boosting. Indicating that XGBoosting is a derivative of gradient boosting. According to Tianqi Chen [13] who is the author of XGBoosting, the engineering goal for the model was to push the limit of computations resources of boosted tree algorithm. He also stated that "XGBoost used a more regularized model formalization to control over-fitting, which gives it better performance."

The performance of these two algorithms was challenged by the random forest algorithm which is adopted for the development of the optimization model in the next phase.

5.3 Random Forest

Similar to the gradient boosting algorithm, random forest operates by building a multitude of the decision tree. The algorithm is closely related to another algorithm called bootstrap aggregating (also known as bagging). The bootstrap aggregating algorithm repeatedly chooses random samples from the data with replacement to train decision trees. This method helps to reduce overfitting that is commonly seen in simple decision trees algorithm. The random forest takes one step further from bagging by limiting the candidate features that could be chosen at each split of a tree. This further reduces the risk of overfitting by decreasing the correlations between individual trees generated in the model training process.

5.4 Optimization

Having used the random forest algorithm to develop a predictive model, a function that optimizes the price of individual tickets at different point of time was then developed using an iterative method. The process was summarized in Fig. 2. Given any other features of a ticket equal, the price of the ticket was set at 0 initially and is increased by 1 dollar at each step of the iteration. The expected revenue of the ticket was calculated by:

$$E\ (\text{Revenue}) = P\ (\text{Demand} \mid \text{Price\&other factors}) \times \text{Price}$$

at each step of the iteration. The demand at each step is

$$P\ (\text{Demand} \mid \text{Price\& ...})$$

which was obtained from the predictive model. The expected revenue at each step is $E(\text{Revene})_t$ which was compared to the expected revenue at the last step $E(\text{Revenue})_{t-1}$. The iteration would stop when the increase in revenue between the two steps is less than 0.4.

Fig. 2 Optimization process

Fig. 2 Optimization process

6   Results

The results are explained in detail in the following two sections:

6.1   Demand Forecasting Results from Predictive Models

In the predictive modeling phase, all the season tickets were excluded while all the final ticket transaction records, i.e., tickets that are listed as "resold" and "expired" from the secondary market, the primary market tickets from the NFL team's internal ticket transaction records, and unsold inventory, were used in training the models. To demonstrate the prediction accuracy for future use, one event in the 2019 season was selected as the testing dataset while all other events from 2012 to 2019 seasons were used as the training set to train models. Here, a game with event code "CLT19DEN" was used to exemplify the prediction and optimization results. Since the testing and training data partition is very specific on the individual event, sophisticated machine learning algorithms such as random forest and gradient boosting were used to improve prediction results.

In the binary classification problem of this study, the model performance evaluation metric is the Area Under Curve (AUC). As mentioned in Sect. 4, the model with the highest AUC was selected for the optimization phase. Based on the AUC results, the best performing model is the random forest model with an AUC of 0.78. It is also worth mentioning the log loss measures, which quantifies the accuracy of a classifier, a lower log loss represents a better model. In our case, random forest has the second-lowest log loss value. Hence, a further evaluation of the model is needed to determine the best model (Tables 2 and 3).

To further assess the prediction accuracy and compare the results of existing machine learning models, the false-positive rate and the false-negative rate were calculated as above. The calculated false-positive rate and false-negative rate of the random forest model, as shown in Table 4, are 43.63% and 16.14%, respectively. If we compare the error rate to that of Tables 3, 5, and 6, we can observe that the classifier generated by those models perform relatively poorer. Based on these results, it can be concluded that random forest generates the best classification outcome of demand forecasting and was carried on to the optimization modeling in the next section.

6.2   Dynamic Pricing Results from Optimization Models

The exploratory data analysis results indicated that seats from section 600 to section 699, which were low in price yet had fairly good views, only had 40.65% average sold rate per section and a large number of unsold seats. Seats within sections 400 to 499 also have a sold rate of 43.26% and a lot of unsold seats. Therefore, the primary pricing optimization objective is to increase the sales and revenue over these sections. Meanwhile, an average of 25% of tickets from sections 100 to 399 did not have any tickets sold since 2012 season. Due to the high price and low demand, better price adjustment is required over these sections. Through optimizing pricing in sections described above, the NFL team can significantly boost sales and maximize revenue. We then selected 6 representative sections to perform optimization, which are section 640, 609, 625, 450, 437, and 117.

Additionally, the NFL team required updating prices to align with forecasted sales. The model was designed to conduct dynamic pricing simulation at the 30 days, 14 days, and 3 days in advance of the event day simultaneously. As discussed in the Sect. 4, recommended price ranged from the lower 1.5 IQR bound of all optimized prices of one single section as the minimum and the highest optimized ticket price in the section as the maximum. The minimum and the maximum prices recommendations at the 30 days, 14 days, and 3 days in advance of the event for each section are illustrated in Table 7. There is no significant pattern of optimized pricing change across the three time points.

Based on the optimization results, five of the six sections displayed different levels of need to adjust average ticket pricing; for example, the average ticket price of section 609 increased by 157.42% to reach the optimized price and maximize revenue. The comparison of original revenues and stimulated revenues of the six sections were summarized in Table 8:

The overall results showed that the forecasted demand at the three different periods fluctuated irregularly. With the dynamic pricing model, the optimal ticket price for a specific section is constantly changing at different time points to increase

Table 2 Model performance metric result

Model	AUC	Log loss	AUCPR	Mean per-class error	RMSE	MSE	Gini
Random forest	0.7848318	0.5944028	0.6987082	0.2989711	0.4516365	0.2039755	0.5696636
Gradient boosting	0.7686897	0.5707387	0.6674262	0.3144083	0.4406208	0.1941467	0.5373793
XGBoosting	0.7536417	0.6030912	0.6672684	0.3303482	0.4591871	0.2108528	0.5072833
Logistic regression	0.6959825	0.6069658	0.5354505	0.3493138	0.4588363	0.2105307	0.3919649

Table 3 XGBoosting model confusion matrix

	0	1	Error	Rate
0	3819	4613	0.547083	4613/ 8432
1	575	4486	0.113614	575/5061
Totals	4394	9099	0.384496	5188/13493

Table 4 Random forest model confusion matrix

	0	1	Error	Rate
0	4753	3679	0.436314	3679/8432
1	818	4243	0.161628	818/5061
Totals	4394	9099	0.384496	5188/13493

Table 5 Gradient boosting model confusion matrix

	0	1	Error	Rate
0	4491	3941	0.467386	3941/8432
1	817	4244	0.161431	817/5061
Totals	5571	7922	0.333284	4497/13493

Table 6 Logistic regression model confusion matrix

	0	1	Error	Rate
0	3964	4468	0.529886	4468/8432
1	854	4207	0.168741	854/5061
Totals	4818	8675	0.394427	5322/13493

Table 7 Change in optimized prices over three time points

Section	Days in advance	Min. price	Max. price
640	30	31	98
	14	Sold	Sold
	3	Sold	Sold
609	30	30	135
	14	48	169
	3	51	120
437	30	26	62
	14	Sold	Sold
	3	Sold	Sold
625	30	26	55
	14	37	50
	3	37	50
450	30	21	54
	14	33	61
	3	40	80
117	30	32	61
	14	51	59
	3	51	51

revenue opportunities. Comparing to the traditional approach of using static pricing strategy, the total unsold tickets were reduced, and the total revenue resulted in a significant increase after deploying dynamic pricing optimization.

Table 8 Revenue increase after optimization

Section	Original revenue ($)	Optimized revenue ($)	% Increase
640	4042	5550	37%
609	4556	11,728	157%
625	946	5669	499%
450	1199	1502	25%
437	150	656	337%
117	2470	2017	−18%

7 Conclusion

The sports ticket market has a giant potential increase in profit with the introduction of dynamic pricing to the market. A dynamic pricing model was done in two phases. First, the predictive model that targets whether a ticket is sold at a given price point and time portrays the demand for sports tickets as well as provides the probability of a ticket being sold. The probability is then used in the revenue function to calculate the expected revenue of the newly optimized price. The price is computed using an iterative method to find the point at which the expected revenue of the sale of a certain ticket is maximized. The result from the price optimizing model shows an immediate increase in expected revenue, which is caused by effective pricing leading to an increase in quantity demanded.

7.1 Limitation

Despite the good performance of the optimization on sections that are lower in price, the optimization result is not ideal for premium tickets like those in section 117. It is possible that the demand for premium tickets is not as elastic as the demand for cheaper tickets. Therefore, the increase in demand given by lower pricing could not compensate for the revenue because of the lowering of price. However, revenue increase by dynamic pricing for cheaper tickets like the upper-deck tickets is still huge. According to Overby [14], San Francisco Giants from the MLB was the first professional sports team to adopt dynamic pricing in their upper-deck seats which increased revenue of $500,000. It is reasonable to believe that optimization for cheaper tickets only could also bring enormous revenue increase to NFL teams.

The iterative method for optimization is time consuming and requires strong computing power resources. This is the reason why the research team only used a few sections for performance demonstration instead of using a whole game. With limited computing power, it is difficult to achieve the goal of daily updating for ticket prices.

7.2 Future Study Direction

7.2.1 Parallel Computing

Parallel computing could be a solution to the issue of slow computing with the iterative method. Future studies could focus on parallel computing to increase the speed of the optimization process. Other methods to solve the issue of slow computing include but are not limited to using environments other than R, which is used in this research.

7.2.2 Database Management

The research team would also suggest a better data management system or method to be introduced to the team. Typo and inconsistency are common in the data set and might become the obstacle for future implementation of the optimization method as well as any other potential future studies based on the data set. A potential future research topic could be finding a suitable data management method and/or system for sports teams in the NFL. This would not only benefit individual sports team in terms of their information management but also create a more organized data structure for follow up study in the optimization method discussed in this article.

7.2.3 Attendance Rates

When it is essential for the team to optimize ticket prices to maximize potential revenue, it is also important to optimize the attendance rate of games to maximize the experience of fans to maintain fans' loyalty for long-term revenue. New technologies like the Internet of Things (IoT) are available to keep track of attendance rate of game and other important information like when an attendant of the game left. Other information about fans and customers like posts from fan page on social media could also help the research in this field. With more information that could be provided by new technologies like IoT, it is possible to conduct studies to optimize attendance rates for better fans' experience.

Acknowledgments This study is carried out under the Industry Practicum Project of Purdue's Business Analytics and Information Management Program. We thank Professor Matthew Lanham (Purdue University, Krannert School of Management) for support and guidance. We thank Christopher Grecco and Alexander Romano (NFL Partner) for providing internal ticket sales data and additional information.

References

1. A. Hoisington, Research: U.S. reports highest revenue in budget sector worldwide (2020, January 28). Retrieved from https://www.hotelmanagement.net/development/research-u-s-reports-highest-revenue-budget-sector-worldwide

2. SAS, Predictive analytics and AI deliver a winning fan experience (2020). Retrieved from https://www.sas.com/en_us/customers/orlando-magic.html

3. J. Khan. The concept of dynamic pricing & how does if affect ecommerce (2015, March 26). Retrieved from https://www.business.com/articles/what-is-dynamic-pricing-and-how-does-it-affect-ecommerce/

4. T. Dittmer, B. Carbaugh, Major league baseball: Dynamic ticket pricing and measurement costs. J. Econ. Educ. **14**(1), 44–57 (2014) ISSN 2688-5956

5. A. Sweeting, Dynamic pricing behavior in perishable goods markets: Evidence from secondary markets for major league baseball tickets. J. Polit. Econ. **120**(6), 1133–1172 (2012). https://doi.org/10.1086/669254

6. D. Strnad, A. Nerat, Š. Kohek, Neural network models for group behavior prediction: A case of soccer match attendance. Neural Comput. & Applic. **28**(2), 287–300 (2015). https://doi.org/10.1007/s00521-015-2056-z

7. M. Şahin, R. Erol, Prediction of attendance demand in European football games: Comparison of ANFIS, fuzzy logic, and ANN. Comput. Intell. Neurosci. **2018**, 1–14 (2018). https://doi.org/10.1155/2018/5714872

8. J.J. Xu, P.S. Fader, S. Veeraraghavan, Designing and evaluating dynamic pricing policies for major league baseball tickets. Manuf. Serv. Oper. Manag. **21**(1), 121–138 (2019). https://doi.org/10.1287/msom.2018.0760

9. J.J. Xu, P.S. Fader, S. Veeraraghavan, Evaluating the effectiveness of dynamic pricing strategies on MLB single-game ticket revenue. Manuf. Serv. Oper. Manag. **21**(1) (2014). https://doi.org/10.1287/msom.2018.0760

10. C. Kemper, C. Breuer, How efficient is dynamic pricing for sport events? Designing a dynamic pricing model for bayern Munich. Int. J. Sport Financ. **11**(1), 4–25 (2016)

11. M.A. Diehl, J.G. Maxcy, J. Drayer, Price elasticity of demand in the secondary market: Evidence from the national football league. J. Sports Econ. **16**(6), 557–575 (2015). https://doi.org/10.1177/1527002515580927

12. S.L. Shapiro, J. Drayer, An examination of dynamic ticket pricing and secondary market price determinants in major league baseball. Sport Manag. Rev. **17**(2), 145–159 (2014). https://doi.org/10.1016/j.smr.2013.05.002

13. T. Chen, What is the difference between the R gbm (gradient boosting machine) and xgboost (extreme gradient boosting)? Quora (2015, Septermber 30). Retrieved from https://www.quora.com/What-is-the-difference-between-the-R-gbm-gradient-boosting-machine-and-xgboost-extreme-gradient-boosting

14. S. Overby, For san francisco giants, dynamic pricing software hits a home run (2011, June 29). Retrieved from https://www.cio.com/article/2406673/for-san-francisco-giants%2D%2Ddynamic-pricing-software-hits-a-home-run.html

Virtual Machine Performance Prediction Based on Transfer Learning of Bayesian Network

Wang Bobo ⓘ

1 Introduction

Cloud computing has provided the hardware foundation and some software plat-forms for the development of industries such as big data [5, 6, 11, 14] and the Internet of things [3, 18, 22]. The considerable benefits of cloud computing come from the effective allocation and utilization of resources [20]. Therefore, the reasonable dynamic allocation of cloud resources is the key to ensure their efficient utilization. The prediction of virtual machine (VM) performance is a prerequisite to ensure the reasonable and dynamic allocation of cloud resources.

However, the inevitable interference caused by resource competition among multiple co-located VMs will cause the fluctuation of VMs' performance, thus leading to the difficulties of accurate performance prediction. Moreover, it is hard to quantify the relationships among the features affecting the performance of VMs, which also increases the uncertainty during the prediction process. Then, many solutions have been proposed to solve the abovementioned difficulties.

Reference [21] proposed to use a linear model to describe the relationships among VM features and performance. But, the complex relationships among VM features and performances cannot be described by the simple linear relationships. In Reference [9], a neuromorphic system based on cogent confabulation is built to predict the resource usages from statistics of historical records in comprehensive dimensions. The system exploits the correlations between observations in multiple dimensions and then builds a probability network model to support the prediction. However, the method in this paper is not complete enough to characterize the relationship between VM features and performances. In Reference [7], the central

W. Bobo (✉)
School of Information Science and Engineering, Yunnan University, Kunming, China
e-mail: wangbobochn@gmail.com

© Springer Nature Switzerland AG 2021
R. Stahlbock et al. (eds.), *Advances in Data Science and Information Engineering*,
Transactions on Computational Science and Computational Intelligence,
https://doi.org/10.1007/978-3-030-71704-9_15

idea is to use low-level performance information to augment the process of Bayesian optimization. The prediction work in this paper is only to find the configuration scheme from the historical resources of VM configuration and give the historical VM performance corresponding to this configuration scheme type. This approach is limited by the search space of VM configuration information, and the feature-performance combination is not flexible enough.

Reference [17] proposed a Bayesian model to determine short- and long-term virtual resource requirement of the CPU/memory intensive applications on the basis of workload patterns at several data centers in the cloud during several time intervals. But, there are many types of cloud computing services, which cannot be limited to the performance prediction of CPU/memory intensive applications.

To make up the limitations of the above works, we propose to use the Bayesian network (BN) model to predict VM performance according to its resource utilization. BN uses the directed acyclic graph (DAG) and the conditional probability table (CPT) to represent the uncertain relationships among complex factors.

However, the hardware and software in cloud might be changed by system maintaining and upgrading, and the hardware and software configurations in different clouds might be different. Therefore, a VM performance prediction model trained in a period or on a certain cloud may get a lower precision rate when the model is used in another time period or to another cloud. We refer to this situation as the model's inability to multiplex in both the temporal and spatial dimensions. However, rebuilding a performance prediction model for each cloud is not only time consuming, but the collecting and processing of the large amount of training data are also full of challenges [8].

These problems can be solved through transfer learning. Reference [15] pointed out that transfer learning can integrate the knowledge learned from other related works, thereby improving the learning effects of the new task. Specifically, when the training data in the target task is not sufficient to build a reliable model, transfer learning allows the knowledge learned from the source task to be transferred to the target task to enrich the data features. It is equivalent to obtaining more feature data in the target task, which improves the training effectiveness of reliable target models.

Reference [13] investigate an approach for re-training neural network models, which is based on transfer learning. Using this approach, a limited number of neural network layers are re-trained, while others remain unchanged. In this study, good prediction of VM performance is achieved, but in this study, data on features in the cloud that could change at any time is used for modeling, and data collection and model updating are done continuously. This way of maintaining a dynamic model comes at the expense of increasing the load on the cloud. In a large-scale cloud, the system will be busy with updating the model due to the changing cloud environment at any time. And the prediction effect will be greatly reduced.

In this paper, we first build a VM's performance prediction BN (i.e., VMP-PBN) to address the difficulty of accurately predicting the performance of VMs. To solve the problem that a VM performance prediction model cannot be multiplexed in either the temporal or spatial dimensions, we propose a BN transfer learning algorithm based on the method proposed by Yue et al. [23]. The transfer learning approach in this paper is based on Markov equivalence theory and uses an improved scoring function with penalties as a model selection criterion for BN transfer learning. In transfer learning, we consider VMP-PBN constructed in the source domain as the source model (i.e., S-VMP-PBN), and a target model (i.e., T-VMP-PBN) applicable to the target domain is gained. Experiments have proven that our approach to transfer learning is accurate and effective.

The rest of the paper is organized as follows. Section 2 introduces the construction of VMP-PBN. Section 3 introduces the transfer learning method of S-VMP-PBN. Section 4 presents experimental results and performance analysis. Section 5 concludes this paper.

2 VMP-PBN Construction

2.1 Definition and Constraints

In this paper, we use the Bayesian network as a VM performance prediction model (VMP-PBN) in a cloud. We define VMP-PBN as Definition 1:

Definition 1 We appoint that G is the network structure of VMP-PBN, and then a VMP-PBN is denoted as $N = (V, E, \theta)$, $(E \cap C = \varnothing)$, where

- $V = U \cup R$ is the set of nodes in G. U denote the set of n VM-related features, among which the uncertain dependencies might exist. R denote the set of n VM-related performances.
- E is the set of directed edges representing the dependencies among node set V. $(V_i \rightarrow V_j)(i \neq j)$ or $e(V_i, V_j)(i \neq j)$ represents an edge in E which starts from V_i and ends at V_j $(V_i, V_j \in V)$.
- C is a set of constrained edges defined in Definition 2.
- θ is the set of parameters consisting of the probability distribution tables of all the nodes in V.

In our work, we use the execution time of applications as the measurement of VM's performance. In a cloud, the performance of a VM might be influenced by features, such as the average memory usage (*mem_avg*) and the average CPU usage, but the performance of a VM does not influence its average memory usage (*mem_avg*) reversely. That is, $e(mem\_avg, duration)$ may exist in VMP-PBN, but $e(duration, mem\_avg)$ is not allowed. For another example, the average CPU usage (*cpu_avg*) and the average memory usage (*mem_avg*) are independent; thus, the edges $e(cpu\_avg, mem\_avg)$ and $e(mem\_avg, cpu\_avg)$ will not exist. So we define the constraints of edges in VMP-PBN as Definition 2:

Definition 2 A constrained edges set is denoted as C. It is impossible that V_k impacts V_m and then $e(V_k, V_m) \in C$, $(k \neq m)$, where

- $V = U \cup R$ is the set of nodes in G. U denote the set of n VM-related features, among which the uncertain dependencies might exist. R denote the set of n VM-related performances.

2.2 Parameter Learning and Structure Learning

The process of constructing BN from data consists of two parts, parameter learning and structure learning. In structure learning, there is often a need for quantitative structural evaluation and selection. In this paper, a scoring function is used for the quantitative analysis of BN, which is calculated based on the network parameters and network structure of BN, and the parameter calculation relies only on the sample data and the determined network structure. Therefore, we first describe in this paper how to calculate the network parameters of a given network based on its structure, and then we describe how to make structure selection.

Parameter Learning We calculate its probability distribution table based on the sample data and use it as the parameter for the network. Specifically, for nodes with no parent in G, calculate its probability distribution table as its parameters, and for nodes with one or more parents, calculate its conditional probability table as its parameters. In this paper, the probability distribution table for all nodes is collectively referred to as CPT and is denoted by θ. The following is a full text example of how the network parameters of VMP-PBN are calculated in this paper.

We use G to denote the structure of a VMP-PBN and D to denote the sample dataset. There is a VMP-PBN structure G with three nodes in Fig. 1. Table 1 is an example of a sample dataset D. The columns *cpu*, *mem*, and *du* in Table 1 correspond to the nodes *cpu_avg*, *mem_avg*, and *duration* in G, respectively, and column Count indicates the number of times this combination of values appears in the sample dataset, and the total sample data in D is 180.

For node *cpu_avg*, it does not have a parent; its parameter is its probability distribution table; for the value 0, its probability is $(10+15+20+27+23+9)/180 = 0.58$, and for the value 1, its probability is $(8+34+5+6+16+7)/180 = 0.42$. For node *duration*, there are two parents; its parameter is its conditional probability table, e.g., when *cpu_avg* takes a value of 0, *mem_avg* takes a value of 1 and *duration* takes a value of 2; the conditional probability of the combination of values is $9/(15 + 27 + 9) = 0.18$.

Fig. 1 A simple structure G of VMP-PBN with only three nodes

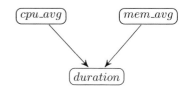

Table 1 Sample dataset:D

cpu	mem	du	Count	cpu	mem	du	Count
0	0	0	10	0	1	1	27
1	0	0	8	1	1	1	6
0	1	0	15	0	0	2	23
1	1	0	34	1	0	2	16
0	0	1	20	0	1	2	9
1	0	1	5	1	1	2	7

Structure Learning We use the hill-climbing method for BN structure learning. In order to avoid the algorithm falling into a local maximum, we use a random restart hill-climbing method for BN structure selection. The algorithm randomly generates an initial structure each time, and then randomly adds, deletes, and reverses the edges on the current structure to obtain a set of candidate structures, from which the structure with the highest BIC score is selected. After several random restarts, an approximately global optimal structure is found as the final BN structure. During the structure selecting, if a candidate structure contains any edge in the edge constraint set in Definition 2 or form cycles, then it will be discarded. Then, we use the BIC scoring function to calculate the scores of the candidate BN structure, which is shown in (1).

$$BIC(D|G) = \sum_{i=1}^{m} \log_2 P(D_i) + \frac{|\theta|}{2} \log_2 m, \ D_i \in D \tag{1}$$

where

- G is the network structure of the BN N
- D is the sample dataset
- D_i is a sample data in D
- m is the number of sample data
- $|\theta|$ is the number of all parameters in N

Algorithm 1 shows how to perform a VMP-PBN construction task using random restart hill-climbing. We take the training data as the input of Algorithm 1, the scoring function uses (1), and the output of the algorithm is VMP-PBN.

3 VMP-PBN Transfer Learning

The transfer learning approach proposed in this paper initially takes the network structure of the source VMP-PBN (S-VMP-PBN) tentatively as a commonality between the source and target domains and then calculates the network parameters with new data from the target domain based on node dependencies in this network structure to obtain a provisional BN as a transition model (i.e., T-BN). In transfer

---

Algorithm 1 VMP-PBN's construction

Input:
D: dataset.
Output:
N_s: VMP-PBN
Local variables:
G_r: a randomly generated network structure
G: current optimal VMP-PBN structure, initially an empty graph, contains only nodes and no edges
G_s: a set of candidate graphs
G': an element in G_s
S: the score of current optimal structure G
S_o: G's BIC score, initially negative infinity
S_n: the highest score in G_s
$Flag$: a flag
i: the controller to terminate the loop
Algorithm start:
$S \leftarrow -\infty$
$i \leftarrow 0$
while $i \leqslant 10$ **do**
  $i \leftarrow i + 1$
  $G_r \leftarrow$ generate a network structure randomly
  $G_s \leftarrow$ randomly add, reverse or delete an edge in G_r
  $S_o \leftarrow S, S_n \leftarrow -\infty, Flag \leftarrow true$
  while $Flag$ **do**
    $Flag \leftarrow false$
    while G_s is not empty **do**
      $G' \leftarrow$ pop an element from G_s.
      if G' is DAG and satisfy Definition 2 **then**
        $S_n \leftarrow BIC(D|G')$
        if $S_o < S_n$ **then**
          $G \leftarrow G', S_o \leftarrow S_n, Flag \leftarrow true, i \leftarrow 0$
        end if
      end if
    end while
    $S \leftarrow S_o$
    $G_s \leftarrow$ randomly add, reverse or delete an edge in G
  end while
end while
$N_s \leftarrow (G, \theta)$
return N_s

---

learning, if T-BN is not suitable for the target domain, we will modify the structure of T-BN node by node according to the descending sequence of node variation degree. After modification and updating, the final result will be a target VMP-PBN (i.e., T-VMP-PBN) that is applicable to the target domain.

3.1 Node Variation Degree and BIC Average Score

The degree of node variation reflects a measure of the variation about the information in the node between the source and target domains. The feature-performance relationship of the target domain is embedded in the new data of the target domain. Thus, the node variation is related not only to the respective CPTs in S-VMP-PBN and T-BN but also to the amount of new data in the target domain. In the following, we will still use the example in 2.2 to demonstrate how to perform the calculation of node variation.

We use G_t and G_o to denote the network structure of T-BN and S-VMP-PBN, respectively; D_t and D_o to denote the samples dataset in target domain and source domain, respectively; X_{ij} to denote the node X in the sample data, satisfying that itself has a value of x_i and its parent nodes $\Pi(X)$ have a combination of π_j ($X_{ij} \in (D_t \cup D_o)$). θ^t; and θ^o denote the network parameters of T-BN and S-VMP-PBN, respectively. Then, the function for calculating the variation degree of each node in T-BN is shown in (2).

$$F_{xv}(X) = \frac{1}{h} \sum_{\theta_{ij} \in \theta} \frac{m_{ij}}{\theta^o_{ij}} |\theta^o_{ij} - \theta^t_{ij}| \tag{2}$$

where

- X is the node in G_t
- $\theta = \theta^t \cup \theta^o$
- h is the size of θ ($h = |\theta|$)
- $\theta_{ij} \in \theta$ denotes the conditional probability of X_{ij}
- θ^t_{ij} and θ^o_{ij} are the parameters of T-BN and S-VMP-PBN, respectively
- m_{ij} is the number of samples X_{ij} in D_t

In (2), it must be noted, when $X_{ij} \in D_t$ and $X_{ij} \notin D_o$, the denominator (θ^o_{ij}) should be 1, but the difference of $|\theta^o_{ij} - \theta^t_{ij}|$ should take on the absolute value of θ^t_{ij} (i.e. $|\theta^t_{ij}|$).

We take dataset D in Table 1 as the dataset in source domain and rename it to D_o. T-BN and S-VMP-PBN both use G in Fig. 1 as the network structure. The nodes *cpu_avg* and *mem_avg* represent the average CPU occupation quantity and average memory occupation quantity (feature) of the VM during the running process, and the node *duration* represents the VM execution time (performance). Edges *cpu_avg* $\rightarrow$ *duration* and *mem_avg* $\rightarrow$ *duration* in the figure indicate that the performance of the VM depends on the CPU and memory occupation quantity. Table 2 lists the sample data D_t in the target domain, where Count denotes the number of the same sample.

According to the calculation method of CPT in 2.2, for S-VMP-PBN and T-BN, the CPTs of node *duration* are calculated as shown in Tables 3 and 4 (θ_t and θ_o), using Tables 1 and 2 as sample data, respectively. In this two tables, the cells show the probability that the VM running time is du_i when the combination of the VM's

Table 2 Sample dataset D_t in target domain

cpu	mem	du	Count	cpu	mem	du	Count
0	0	0	15	0	0	1	14
0	1	0	12	0	1	1	8
1	0	0	5	1	0	1	23
1	1	0	9	1	1	1	33
2	0	0	18	2	0	1	18
2	1	0	18	2	1	1	27

Table 3 θ^o: the parameters set of node *duration* in S-VMP-PBN

$P(du \vert cpu, mem)$	du_1	du_2	du_3
$\pi_1 = cpu_1, mem_1$	0.19	0.38	0.43
$\pi_2 = cpu_1, mem_2$	0.29	0.53	0.18
$\pi_3 = cpu_2, mem_1$	0.28	0.17	0.55
$\pi_4 = cpu_2, mem_2$	0.72	0.13	0.15

Table 4 θ^t: the parameters set of node *duration* in T-BN

$P(du \vert cpu, mem)$	du_1	du_2
$\pi_1 = cpu_1, mem_1$	0.52	0.48
$\pi_2 = cpu_1, mem_2$	0.6	0.4
$\pi_3 = cpu_2, mem_1$	0.18	0.82
$\pi_4 = cpu_2, mem_2$	0.22	0.78
$\pi_5 = cpu_3, mem_1$	0.5	0.5
$\pi_6 = cpu_3, mem_2$	0.3	0.7

CPU and memory usage is π_j. Obviously, $\theta^t \not\subseteq \theta^o$ and $\theta^o \not\subseteq \theta^t$. Then, according to (2), the variation degree of *duration* is:

$$h = |\theta| = |\theta^t \cup \theta^o| = 16$$

$$F_{xv}(duration) = \frac{1}{h} \sum_{\theta_{ij} \in \theta} \frac{m_{ij}}{\theta_{ij}^o} |\theta_{ij}^o - \theta_{ij}^t|$$

$$= \frac{1}{16} \Big(\frac{15}{0.19} |0.19 - 0.52| + \frac{14}{0.38} |0.38 - 0.48|$$

$$+ \frac{12}{0.29} |0.29 - 0.6| + \frac{8}{0.53} |0.53 - 0.4|$$

$$+ \frac{5}{0.28} |0.28 - 0.18| + \frac{23}{0.17} |0.17 - 0.82|$$

$$+ \frac{9}{0.72} |0.72 - 0.22| + \frac{33}{0.13} |0.13 - 0.78|$$

$$+ \frac{18}{1} |0.5| + \frac{18}{1} |0.5| + \frac{18}{1} |0.3|$$

$$+ \frac{27}{1} |0.7| + \frac{0}{0.43} |0.43| + \frac{0}{0.18} |0.18|$$

$$+ \frac{0}{0.55}|0.55| + \frac{0}{0.15}|0.15|)$$

$$\approx 21.74$$

In transfer learning, we modified the BIC scoring function in (1) to obtain (3) to eliminate the impact of sample data volume on the BIC score. Thus, using (3) to calculate the score of a model, we can obtain the contribution of each sample data to the quantitative score of the current model, which ensures that the quantitative comparison of T-BN and S-VMP-PBN is not affected by the amount of the new or old data.

$$BIC_{avg}(D|G) = \frac{\sum_{i=1}^{m} \log_2 P(D_i) + \frac{|\theta|}{2} \log_2 m}{|D|}, D_i \in D \tag{3}$$

where $|D|$ denotes the number of sample data in dataset D and the other symbols have the same meaning as in (1).

3.2 Get T-VMP-PBN with Transfer Learning

We set a ratio ξ_t, and if the value of the score $BIC_{avg}(D_t|G_o)$ of T-BN is greater than or equal to the product of ξ_t and $BIC_{avg}(D_o|G_o)$ (D_o and D_t is the dataset in the source domain and the target domain, respectively), it indicates that the BN structure of the T-BN and the target domain are consistent. At this time, we consider the T-BN as the final T-VMP-PBN. Otherwise, it indicates that the T-BN needs to be modified.

In combination with Reference [10], we will consider all nodes in the T-BN one by one in a descending order by variation degree, using the Markov blanket [16] of each node as the modification radius. First, find all nodes in the Markov blanket of node X_i (denoted as MB(X_i)) from T-BN. We use G_{ts} and G_{os} to denote the subgraph in T-BN and S-VMP-PBN, respectively, and both the set of nodes in G_{ts} and the set of nodes in G_{os} are equal to MB(X_i) $\cup$ X_i. We define the ξ_s as the proportional threshold when comparing the subgraphs. Next, we consider the modification of T-BN in three cases.

- If $BIC_{avg}(D_t|G_{ts}) \geq \xi_s \times BIC_{avg}(D_o|G_{os})$, then the subgraph with MB(X_i) as the node does not need to be modified, and the next node is considered at this time.
- If $BIC_{avg}(D_t|G_{ts} < \xi_s \times BIC_{avg}(D_o|G_{os})$, but there is a consistent extension structure G_{es} of G_{ts}, so that $BIC_{avg}(D_t|G_{es}) \geq \xi_s \times BIC_{avg}(D_o|G_{os})$. Reference [19] explained that the consistent extension structures of directed acyclic graphs are actually Markov equivalents. Therefore, we only need to replace G_{ts} with G_{es}. According to Reference [4, 12], it is possible to calculate the consistent extended structure G_{es} of G_{ts}.

– If $BIC_{avg}(D_t|G_{ts}) < \xi_s \times BIC_{avg}(D_o|G_{os})$, and, for all consistent extension structures G_{es} of G_{ts}, there are $BIC_{avg}(D_t|G_{es}) < \xi_s \times BIC_{avg}(D_o|G_{os})$, it indicates that new data in the target domain changes the node dependencies in the source BN. In this case, it is necessary to use a heuristic algorithm to find the most suitable subgraph G_{ns} from all the subgraphs generated by reversing, adding, or deleting edges in G_{ts} one by one and replace the subgraph G_{ts} in T-BN with G_{ns}. Here, we still use the random restart hill-climbing method for heuristic search.

In the above steps, for each newly generated network structure, the constraints defined in Definition 2 must be met. The idea of transfer learning in this paper can be explained by Algorithm 2. Based on several experiments, we find that the thresholds ξ_t and ξ_s in the algorithm are set to $\xi_t = 0.25$ and $\xi_s = 0.3$, transfer learning can achieve good results, and the transfer time and computational resource cost overhead are not very high.

4   Experimental Results

In this experiment, the main program is written in Python 3.6, and the data preprocessing is written in C. The computer we used is configured as 4 single-core CPUs(Intel(R) 2.30 GHz), 8 GB physical memory, the entire experimental program runs in Docker, and the operating system is CentOS 7.7.

4.1   Dataset

The Alibaba Group announced the cluster data cluster-trace-v2017 [1] and cluster-trace-v2018 [2] collected from production clusters in Alibaba in 2017 and 2018, respectively. We took the cluster-trace-v2017 as the source domain data and the cluster-trace-v2018 data as the target domain data due to the changes of hardware and software.

4.2   Data Pre-processing

In this paper, we extracted the data items which have been successfully terminated from the batch instance dataset and then selected four dimensions as VM-related features, namely *real_cpu_max (cpu_max)*, *real_cpu_avg (cpu_avg)*, *real_mem_max (mem_max)*, and *real_mem_avg (mem_avg)*. Furthermore, the results of *end_timestamp* minus *start_timestamp* are recorded as *duration* and taken as the corresponding VM performance. Thus, the feature-performance data

---

Algorithm 2 VMP-PBN's transfer learning

Input:

G_o: the structure of S-VMP-PBN

D_o: the set of old data

D_t: the set of new data

ξ_t: the proportional threshold of BIC_{avg} for T-BN

ξ_s: the proportional threshold of BIC_{avg} for subgraph

Output:

N_t: T-VMP-PBN

Local variables:

G_t: the structure of T-BN

ID_v: a sequence of node variation degree

$MB(X_i)$: the Markov blanket of X_i

S: the subset of nodes in G_o

S': the subset of nodes in $MB(S) \cup S$.

G_{os}: the subgraph of G_o on the nodes in S

G_{ts}: the subgraph of G_t on the nodes in S

G_{es}: a consistent extension structure of G_{ts}

Algorithm start:

if $BIC_{avg}(D_t|G_o) \leqslant \xi_t \times BIC_{avg}(D_o|G_o)$ **then**

    $ID_v \leftarrow$: calculate the variation degree of each nodes

    while ID_v is not empty **do**

        $X_i \leftarrow$ pop a node from ID_v, $S \leftarrow MB(X_i) \cup X_i$

        $G_{ts} \leftarrow$ the subgraph of G_t on the nodes in S

        $G_{os} \leftarrow$ the subgraph of G_o on the nodes in S

        $S' \leftarrow MB(S) \cup S$

        while $BIC_{avg}(D_t|G_{ts}) \leqslant \xi_s \times BIC_{avg}(D_o|G_{os})$ **do**

            $G_{es} \leftarrow$ a consistent extension structure of G_{ts}

            while $BIC_{avg}(D_t|G_{es}) \leqslant \xi_s \times BIC_{avg}(D_o|G_{os})$ **do**

                $G_{ts} \leftarrow$ searching the best structure

            end while

            Replace the subgraph G_{ts} in G_t by G_{es}.

            $S \leftarrow S'$, $S' \leftarrow MB(S) \cup S$

            $G_{ts} \leftarrow$ the subgraph of G_t on the nodes in S

            $G_{os} \leftarrow$ the subgraph of G_o on the nodes in S

        end while

    end while

end if

$N_t \leftarrow (G_t, \theta)$

return N_t

---

with five dimensions consists of the training set. We used K-means to discretize each feature into 10 values and then obtained 10^5 feature combinations in the training set.

We use the 2 times and 3-fold cross-validation to perform the experiments. We adopt simple non-replacement random sampling method to extract 1.9 million pieces of data from the new data in the target domain, and divide these new data and the old data from the source domain into six training-testing pairs according to the idea of 2 times and 3-fold cross-validation, and denoted by C_{o1}–C_{o6} and C_{t1}–C_{t6}, respectively.

In order to determine the minimum sample size required for transfer learning, we performed 15 times simple random sampling without replacement from the target domain sample data, and the sample sizes are 1K, 5K, 10K, 20K, 50K, 100K, 200K, 500K, 1M, 1.5M, 2M, 5M, 10M, 15M, and 50M separately (denoted by C_{s1}–C_{s15}). For each subset of sample, according to the idea of 2 times and 3-fold cross-validation, these subsets were divided into six training-test set pairs. There are a total of 90 training-test set pairs.

4.3   The Constraints of VMP-PBN in Cloud

In Definition 2, we give a formal definition of VMP-PBN constraints. In this experiment, based on the specific analysis of the source of the dataset, Table 5 shows 12 illegal edges that are not allowed to appear in the BN structure. For example, the edge ($duration \rightarrow cpu\_max$) is not allowed in VMP-PBN, because it is impossible that the performance of the VMP-PBN influences performance. Edge ($cpu\_max \rightarrow mem\_max$) is also not allowed, because in ordinary tasks (non-memory intensive calculations), the CPU will not influence the memory usage.

4.4   Performance Prediction of VM Based on S-VMP-PBN

We use six sets of training-test set pairs (C_{o1}–C_{o6}) from the source domain to perform model training and obtain six VM performance prediction models, which are named S-VMP-PBN. At the same time, as a comparison, the decision tree (DT), random forest (RF), and support vector machine (SVM) are used to train the model and obtain 18 VM performance prediction models accordingly. For the four groups of VM performance prediction models based on different principles, BN, DT, RF and SVM, we test their performance by the same test dataset which come from C_{o1}–C_{o6}, and the results are shown in Fig. 2.

In Fig. 2a–c respectively show the macro average precision, macro average recall, and macro average F1-score of the four groups of model. In the figure, BN represents

Table 5 The set of edges which is not allowed to appear in VMP-PBN's structure

Tail node	Edge	Head node	Tail node	Edge	Head node
duration	→	*cpu_max*	*mem_max*	→	*cpu_max*
duration	→	*cpu_avg*	*mem_max*	→	*cpu_avg*
duration	→	*mem_max*	*mem_avg*	→	*cpu_max*
duration	→	*mem_avg*	*mem_avg*	→	*cpu_avg*
cpu_max	→	*mem_max*	*cpu_avg*	→	*mem_max*
cpu_max	→	*mem_avg*	*cpu_avg*	→	*mem_avg*

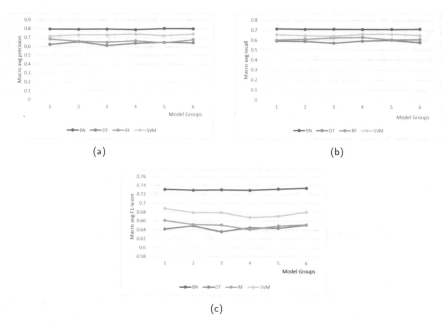

Fig. 2 Performance comparison of models based on four machine learning methods BN, DT, RF, and SVM, respectively. (**a**) Macro average precision. (**b**) Macro average recall. (**c**) Macro average F1-score

the performance of S-VMP-PBN based on Bayesian network. It is obvious that the performance of DT and RF is always similar. But in all six models, although RF is better than DT, there are no obvious performance gap. The performance of the six models based on SVM is significantly better than that of DT and RF, but compared with the VMP-PBN based on BN, the performance of all models of SVM is lower. It shows that compared with traditional machine learning methods, the BN can learn the knowledge better than between the features and performances of the VM embedded in the sample data.

To test whether S-VMP-PBN can be used directly to predict the performance of the VM in the target domain without transfer learning, we use the datasets C_{t1}–C_{t6} as training data and obtain six VMP-PBN-cons (the VMP-PBN which we called VMP-PBN-con is trained from the new data of target domain by BN learning instead of transfer learning).

For comparison, we compile the performance data of S-VMP-PBN and VMP-PBN-con in Table 6. As can be seen, the S-VMP-PBN performs poorly on all three performance metrics, only around 0.25. And the VMP-PBN-con is even better, with twice or even three times the performance of the S-VMP-PBN. It shows a big difference between the target domain and the source domain, and the source BN cannot be directly used in the target domain. To obtain the performance prediction BN of the target domain with good performance, we must use BN reconstruction or transfer learning.

Table 6 Performance comparison of S-VMP-PBN and VMP-PBN-con in the target domain

VMP–PBN	1	2	3	4	5	6
(a) Macro average precision						
S–VMP–PBN	0.307	0.324	0.278	0.285	0.31	0.288
VMP–PBN–rec	0.889	0.854	0.837	0.89	0.869	0.852
(b) Macro average recall						
S–VMP–PBN	0.246	0.259	0.221	0.243	0.227	0.261
VMP–PBN–rec	0.744	0.716	0.773	0.798	0.734	0.775
(c) Macro average F1-score						
S–VMP–PBN	0.278	0.302	0.305	0.285	0.276	0.297
VMP–PBN–rec	0.813	0.738	0.795	0.807	0.778	0.792

4.5   Performance Prediction of VM Based on T-VMP-PBN

In 4.2, each subset of sample from C_{s1}–C_{s15} is divided into six training-test set pairs, so that there are a total of $15 \times 6 = 90$ training-test set pairs (15 sample size levels) involved in transfer learning. We used the six S-VMP-PBNs obtained in Sect. 4.4 as the source BN and performed BN transfer learning using the dataset C_{s1}–C_{s15} and Algorithm 2. A total of 540 ($90 \times 6 = 540$) target VMP-PBN (i.e., T-VMP-PBN) and 540 sets of test results are obtained. And each test result data consists of four dimensions: macro average precision, macro average recall, macro average F1-score, and running time.

To facilitate the analysis of the results and to eliminate errors, we divide all 540 test sets into 15 groups based on the size of the training set, and there are 36 test result data in each group. Then, the test result data in each group are averaged according to the four dimensions above and regard them as performance data for transfer learning at that training set scale. A total of 15 sets of test data reflecting the effects of the transfer learning approach were obtained.

To compare with the performance of transfer learning, we reconstructed the VMP-PBN (VMP-PBN-rec) with subsets of sample C_{s1}–C_{s15} and tested its performance, respectively. A total of $15 \times 6 = 90$ VMP-PBN-rec were obtained by the reconstruction, and using the size of the sample dataset as a taxonomy, these models and test data can be divided into 15 groups, each containing 6 test result data. Similarly, we averaged the six result data in each group according to the four dimensions and regard them as performance data for what the model reconstruction could achieve at that data scale. A total of 15 sets of performance data of model reconstruction were obtained.

Figure 3 show the performance comparison between T-VMP-PBN and VMP-PBN-rec. Among them, (a), (b), and (c) show the macro average precision, macro average recall, and macro average F1-score, respectively. As can be seen, while all three performance metrics of VMP-PBN-rec and T-VMP-PBN end up at roughly equivalent levels as the sample data volume increases, the former grows significantly slower than the latter.

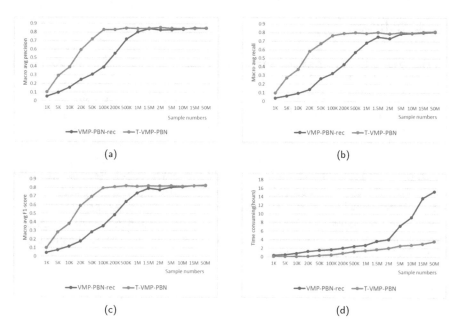

Fig. 3 Performance comparison of T-VMP-PBN and VMP-PBN-rec in the target domain. (**a**) Macro average precision. (**b**) Macro average recall. (**c**) Macro average F1-score. (**d**) Time consuming (hours)

It can be seen that the performance of the two BNs is only approximately equal at 1M data volume, before which the performance of T-VMP-PBN is always better than that of VMP-PBN-rec. In addition, at a sample size of 100K, the macro average precision, macro average recall, and macro average F1-score of the T-VMP-PBN are close to the best performance that can be achieved; however, it is not until a data volume of 1M that the VMP-PBN-rec can achieve this level of performance.

Figure 3d shows the time consumption of T-VMP-PBN and VMP-PBN-rec with various sample sizes. It can be seen that the time consumption of both methods will increase with the increase of the amount of data. However, the time consumption of transfer learning is always less than reconstruction, and as the amount of data increases, the time consumption of transfer learning increases faster than the time required for reconstruction.

As shown in Fig. 3, T-VMP-PBN can reach almost the best prediction performance when the amount of data is 100 K; at this time, the time consumption of transfer learning is only 0.49h. But VMP-PBN-rec can reach the same prediction performance when the amount of data is 1M, and it takes 2.75h.

5 Conclusion

In this paper, the experimental results show that the performance of the model based on BN is better than the models based on traditional machine learning methods, and compared with model reconstruction, while maintaining 79% macro average precision, our transfer learning method reduces the data requirements during modeling to 1/10 of the original and the time consumption to 1/4.

In the future work, we will explore the transfer learning method of BN when the source and target domains have different data dimensions, and by adding or deleting nodes in the source model to obtain the target model. Because when the hosts in cloud are maintained and updated, or the two clouds are compared, not only the hardware model and architecture are changed, sometimes new hardware may be installed or uninstalled, such as GPU. This will cause the dimensions of the feature-performance data of the source and target domains to change, that is, the number of nodes in the VMP-PBN will change.

References

1. I. Alibaba, *Alibaba Production Cluster Data*, vol. 2017 (2017). website
2. I. Alibaba, *Alibaba Production Cluster Data*, vol. 2018 (2018). website
3. A. Botta, W. De Donato, V. Persico, A. Pescapé, Integration of cloud computing and internet of things: a survey. Future Gener. Comput. Syst. **56**, 684–700 (2016)
4. D.M. Chickering, A transformational characterization of equivalent bayesian network structures (2013). arXiv preprint arXiv:1302.4938
5. F.J. Clemente-Castelló, B. Nicolae, M.M. Rafique, R. Mayo, J.C. Fernández, Evaluation of data locality strategies for hybrid cloud bursting of iterative mapreduce, in *Proceedings of the 2017 17th IEEE/ACM International Symposium on Cluster, Cloud and Grid Computing (CCGRID)* (IEEE, New York, 2017), pp. 181–185
6. I.A.T. Hashem, I. Yaqoob, N.B. Anuar, S. Mokhtar, A. Gani, S.U. Khan, The rise of 'big data' on cloud computing: review and open research issues. Inf. Syst. **47**, 98–115 (2015)
7. C.J. Hsu, V. Nair, V.W. Freeh, T. Menzies, Arrow: low-level augmented bayesian optimization for finding the best cloud VM, in *Proceedings of the 2018 IEEE 38th International Conference on Distributed Computing Systems (ICDCS)* (IEEE, New York, 2018), pp. 660–670
8. W. Kuang, L.E. Brown, Z. Wang, Selective switching mechanism in virtual machines via support vector machines and transfer learning. Mach. Learn. **101**(1–3), 137–161 (2015)
9. Z. Li, X. Ma, J. Li, Q. Qiu, Y. Wang, Efficient cloud resource management using neuromorphic modeling and prediction for virtual machine resource utilization, in *Proceedings of the 2019 IEEE International Conference on Embedded Software and Systems (ICESS)* (IEEE, New York, 2019), pp. 1–8
10. W. Liu, K. Yue, M. Yue, Z. Yin, B. Zhang, A bayesian network-based approach for incremental learning of uncertain knowledge. Int. J. Uncertainty Fuzziness Knowledge Based Syst. **26**, 87–108 (2018)
11. A. Manekar, P. Gera, Studying cloud as IAAS for big data analytics: opportunity, challenges. Int. J. Eng. Technol. **7**(2.7), 909–912 (2018)
12. C. Meek, Causal inference and causal explanation with background knowledge (2013). arXiv preprint arXiv:1302.4972

13. F. Moradi, R. Stadler, A. Johnsson, Performance prediction in dynamic clouds using transfer learning, in *Proceedings of the 2019 IFIP/IEEE Symposium on Integrated Network and Service Management (IM)* (IEEE, New York,2019), pp. 242–250
14. A. Noraziah, M.A.I. Fakherldin, K. Adam, M.A. Majid, Big data processing in cloud computing environments. Adv. Sci. Lett. **23**(11), 11092–11095 (2017)
15. S.J. Pan, Q. Yang, A survey on transfer learning. IEEE Trans. Knowl. Data Eng. **22**(10), 1345–1359 (2009)
16. J. Pearl, Probabilistic reasoning in intelligent systems—networks of plausible inference, in *Morgan Kaufmann Series in Representation and Reasoning* (1988)
17. G.K. Shyam, S.S. Manvi, Virtual resource prediction in cloud environment: a bayesian approach. J. Netw. Comput. Appl. **65**, 144–154 (2016)
18. C. Stergiou, K.E. Psannis, B.G. Kim, B. Gupta, Secure integration of IoT and cloud computing. Future Gener. Comput. Syst. **78**, 964–975 (2018)
19. T. Verma, J. Pearl, Equivalence and synthesis of causal models, in *UAI* (1990)
20. Z. Xiao, W. Song, Q. Chen, Dynamic resource allocation using virtual machines for cloud computing environment. IEEE Trans. Parallel Distrib. Syst. **24**(6), 1107–1117 (2012)
21. H. Xiong, C. Wang, Cloud application classification and fine-grained resource provision based on prediction. Jisuanji Yingyong/J. Comput. Appl. **33**(6), 1534–1539 (2013)
22. M. Yannuzzi, R. Milito, R. Serral-Gracià, D. Montero, M. Nemirovsky, Key ingredients in an iot recipe: Fog computing, cloud computing, and more fog computing, in *Proceedings of the 2014 IEEE 19th International Workshop on Computer Aided Modeling and Design of Communication Links and Networks (CAMAD)* (IEEE, New York, 2014), pp. 325–329
23. K. Yue, Q. Fang, X. Wang, J. Li, W. Liu, A parallel and incremental approach for data-intensive learning of bayesian networks. IEEE Trans. Cybern. **45**(12), 2890–2904 (2015)

A Personalized Recommender System Using Real-Time Search Data Integrated with Historical Data

Hemanya Tyagi, Mohinder Pal Goyal, Robin Jindal, Matthew A. Lanham, and Dibyamshu Shrestha

1 Introduction

Businesses integrate recommendation systems on their websites to ease the search for the users. According to Salesforce, personalized product recommendations drive just 7% of visits but 26% of revenues on an e-commerce website [1]. Recommendations play an important role in other industries as well. For instance, two out of three movies watched on Netflix are recommended [2]. In fact, a research by Engagement Labs claims that personal recommendations are the #1 driver of consumer purchase decisions at every stage of the purchase cycle, across multiple product categories [3]. Furthermore, one can argue that the benefits of a recommendation system outweigh the cost for creating and maintaining recommendation systems [4]. However, deploying a recommendation system in the production environment comes with a variety of challenges. For instance, a recommendation system needs to be dynamic for being accurate and efficient. Moreover, by being dynamic it should incorporate the context of the user search at every instance. Other common issues are optimizing response times, frequently updating models, and predicting based on unseen data, also commonly known as the cold-start problem [5].

In this chapter, we present the idea to incorporate real-time search history in considering recommendations. This would allow the website to learn the context of what user wants and recommend resorts for them based on their past booking preferences and filtered by the current real-time search. This would allow website to recommend appropriate products/services for the client which would increase client satisfaction.

H. Tyagi (✉) · M. P. Goyal · R. Jindal · M. A. Lanham · D. Shrestha
Krannert School of Management, Purdue University, West Lafayette, IN, USA
e-mail: tyagih@purdue.edu; goyal62@purdue.edu; rjindal@purdue.edu; lanhamm@purdue.edu; dshresth@purdue.edu

© Springer Nature Switzerland AG 2021
R. Stahlbock et al. (eds.), *Advances in Data Science and Information Engineering*, Transactions on Computational Science and Computational Intelligence, https://doi.org/10.1007/978-3-030-71704-9_16

We specifically work on recommending resorts for a timeshare exchange company. Currently, a recommendation model is used by the company to recommend relevant resorts to the user based on the member data, resort amenities, and other factors. The model trained thus is deployed to the website, once a day. Hence, if a user searches for a resort in Denver today, the model can recommend resorts in Chicago, which is based on the search history of the user up until yesterday. If the user further filters the search to Aurora, Denver, the recommendation system would still recommend the resorts recommended in the previous search. This shows that the recommendation system is greatly affected by this inaccuracy. The context of the current search is a significant factor in determining which resort the user will go to next. Also, since the recommendations happen in real time, the recommendation model should return results in a matter of seconds. Therefore, speed also becomes a crucial factor in recommending relevant resorts. We work toward designing a model which incorporates the real-time search history and recommends eight relevant resorts to the user within 5 s.

Traditionally recommendation systems have used methods like collaborative filtering and RankBoost algorithms as their go-to methods. However, collaborative filtering suffers from a few problems, the most prominent of which is the cold-start problem. To overcome this, hybrid models have been used in the past. Still, there is little research on how to use real-time search data of users to recommend a relevant resort in a time-share industry. The problems are not just limited to which algorithms to choose, but how to measure the performance of the offline model and how to record and accommodate the implicit feedback of the user.

To develop the recommendation system, we develop models that diversify the recommendations and overcomes the problems of redundancy and irrelevancy. This is done by architecting models in a way that selectively values the recent searches and penalizes the older searches.

2   Data

In this study, we used the search and confirmation data provided by the client. The search data consists of searches conducted by around 1.3 million members in the past year, i.e., from January 2019 to January 2020. The dataset has a record of about 159 million searches in the past year. The search data has information on search details like date and time of search as well as more information about the region that was searched like region, destination, city, or resort ID of the search. It also has the information on the month and specific date the member is searching for.

The other primary table for this project is the confirmation data table, which has details about the bookings by around 1 million members for around 6000 resorts in the past 5 years. There are approximately 5 million recorded confirmations from January 2015 to January 2020. The dataset has information about the resort ID and resort amenities that a member has booked. It has details about when the booking was confirmed and the actual start and end date of the booking. The dataset also

Table 1 Data description

Table no.	Table name	Description
1	Confirmation data	Booking data for 1 M members in the past 5 years
2	Search data	Search data for 1.3 M members in the last year
3	Resort amenities	350+ amenities information of 20,000 resorts
4	Check in month table	Covert check in month ID in different languages to year-month of check-in
5	Member data	Information of 2.46 million members
6	Region ID table	Convert region ID in different languages to region ID independent of language and get details about the region
7	Destination ID table	Convert destination ID in different languages to destination ID independent of language and get details about the destination
8	City ID table	Convert City ID in different languages to City ID independent of language and get details about the city
9	Resort to XML	Information about all the resorts mostly its locations and all the associated IDs for that resort
10	Available data	Availability and details of 4061 resorts

contains a cancellation index, which indicates if a person made a booking and later canceled it.

The resort amenities table has information about 350+ features of all the resorts. Since the online portal has information in 17 different languages, we used tables that help us to convert language centric IDs to language-independent IDs so that the same resort in different languages has the identical IDs after using these tables. Availability information of resorts is also provided in the dataset but hasn't been used.

Table 1 provides a brief description of the different tables used in this study.

3 Methodology

The prime objective of the experiments is to find an algorithm which captures user preferences and recommends relevant resorts based on that. To develop the models, we need to investigate the following questions: What factors are users looking for while booking a resort? How to recommend resorts to first-time users who have no confirmation or search history?

The methodology used to answer the above questions is extensively described in Fig. 1.

Fig. 1 Methodology

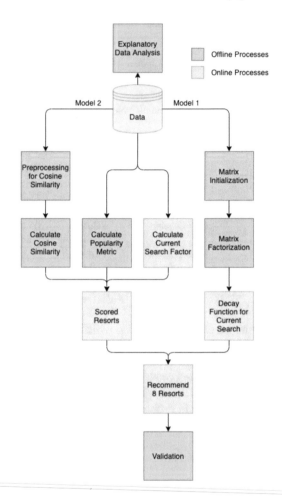

3.1 Explanatory Data Analysis (EDA)

The first step toward solving any data problem is visualizing the data and finding insights which might be useful in the modeling process. Some interesting insights that we got from the EDA are:

- The distribution of the confirmation data is right skewed in which more than 25% of the users have booked resorts less than three times till date.
- Two resorts in Kissimmee and Las Vegas are the most booked resorts with more than 26,000 bookings combined.
- There is seasonality in the data with most bookings being done in quarter 3. Furthermore, most users book resorts for 7 days.

3.2 Data Preprocessing

The following steps were carried out to preprocess the resort amenities table:

- Column selection: There are 165 columns in resort amenities table, out of which several columns have a majority of data missing. We selected 17 columns from this table in which a majority of the data was not missing.
- Missing data handling: For categorical columns, a new dummy encoded column was created to handle missing values. For the numerical column, Golf Distance, we substitute the missing values with 0, if golf is onsite or nearby, 200 otherwise.
- Min-max normalization: Since Golf Distance can cause a significant bias while calculating cosine if not normalized, Golf Distance was normalized using a min-max normalizer.

3.3 Modeling and Validation

We experimented upon two modeling approaches. The first model (*Model 1*) assumes that there are similarities among users and similarities among resorts and recommends resorts to user based on these similarities. The second model (*Model 2*) assumes that users book resorts on the basis of resort popularity, amenities, and search behavior.

To validate our models, we had data spanning across a year. To account for chronology and to accurately capture user behavior, we took the following steps:

- Data partitioning: We run two models: model 1 which is based on matrix factorization, and model 2 which is based on cosine similarity. Since matrix factorization is a computationally intensive process, the train-test split for both the models must be different.
- For the first model, we selected 3 sets of 1600 members who had 2 or more bookings. Then for each member, we split the data until the point of their second last booking and ran matrix factorization algorithm. For test, we selected members having 2 or more bookings in 2019 and recommended resorts to them based on each search.
- For the second model, we take a sample of 100 users who have more than 2 confirmed bookings between October 1, 2019 and December 31, 2019. We take a test window from October 1, 2019 to December 31, 2019. For each day, t in the test window, we run the model with confirmation data up to t-1 days and all search data available till day t. We then predict the resorts for each search on day t.
- Success criteria: To evaluate and compare the two models, we first need to define success. If a user books a resort that was recommended to him/her within 17 days prior to the confirmation date, we define it as a success.

- Evaluation metric (accuracy): The evaluation metric then can be defined simply as the ratio of the number of successes and the number of bookings done by the user.

4 Models

4.1 Model 1

Model 1 incorporates the technique of matrix factorization to recommend resorts.

Matrix factorization: Matrix factorization is the technique to predict the missing values in a sparse matrix. As the name suggest, it decomposes a matrix into product of two matrices.

In Fig. 2, Matrix X is the product of Matrix A and B. Matrix decomposition is nothing but the process of factorizing X into A and B.

Figure 3 shows the need of matrix factorization. We use matrix factorization when we need to fill in missing values in a matrix. This is specifically needed in recommendation systems, when we do not have the data about how much a user will value a resort, if s/he has not yet searched for or booked that resort.

Initially each search or booking adds points to a user-resort cell in the matrix. For example, if a member searches for a resort in Colorado, Denver, then for that member, all the resorts in that area will be given some points, that is, search factor. The formula used to allocate points for resorts based on member's searches and bookings was created for this purpose and tested:

$$\text{Search} \quad \text{factor} = (\text{surge} \quad \text{factor}) * \left(\frac{a^x - a^{-x}}{a^x + a^{-x}} \right) * \left(\frac{x}{y} \right)$$

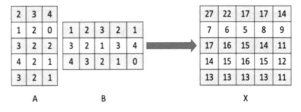

Fig. 2 Matrix multiplication

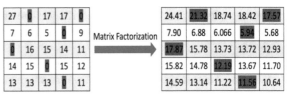

Fig. 3 Matrix factorization

Table 2 Surge table

Transaction type	Definition	Surge factor
Bookings	All historical bookings	5
Today's search	Searches done today	4
Recent search	Searches done after last booking	2
Past search	All searches done before last booking	1

Table 3 Weights for search type in model 1

Search type	a
Region search	1.05
Destination search	1.1
City search	1.5
Resort search/booking	2

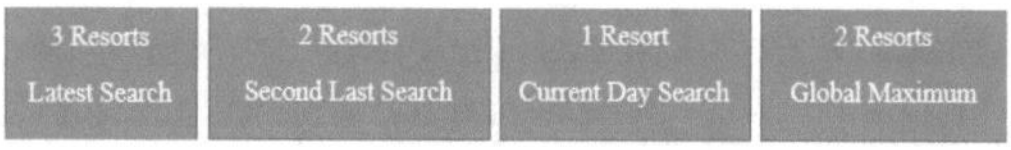

Fig. 4 Recommendation distribution

3 Resorts	2 Resorts	1 Resort	2 Resorts
Latest Search	Second Last Search	Current Day Search	Global Maximum

Here,

y = Average number of searches per booking of the user per search type
x = Number of times the transaction type is done
Surge factor can be computed from Table 2.
"a" can be computed from Table 3.

Matrix initialization and factorization: Resorts that do not show up in the member's searches or bookings are populated as 0 in the matrix initialization process. Using all transaction types except Today's Search, the matrix is populated using matrix factorization with all available resorts as column and members as rows. The matrix factorization model is run once per day.

Recommending using matrix factorization. Using the output of matrix factorization, when a member searches a region in the present day, search factor will be added to the resorts in that region and recommendation will be given as per the criteria in Fig. 4.

4.2 Model 2

Model 2 can be decomposed into four parts, each explained below:

(a) Popularity factor (P_i): For each resort i, P_i is nothing but the ratio of the number of times resort i was booked to the total number of confirmations.
(b) Cosine similarity (CS_{ij}): Cosine similarity, CS_{ij}, tells us how likely a customer j is to book resort i based on resort amenities. CS_{ij} is calculated using the following steps:

- Step 1: Calculate user vector u_j, which is the average of the resort amenities vector for the resort of each booking that the user has made.
- Step 2: Calculate cosine similarity between each resort vector (r_i) and user vector (u_j), CS_{ij} with the following formula:

$$CS_{ij} = \text{Cossim}\left(r_i, u_j\right) = \frac{A.B}{|A| \, |B|}$$

(c) Search factor (S_{ij}): If the user i has searched for a resort j before today, then the search factor of the resort for the user increases as follows:

$$Sij = Sij + \frac{\lambda}{T+1}$$

Where λ is the tuning parameter. In our model, we take it to be 1, and T is the difference in days between the search and today.

(d) Current search factor (C_{ij}): If the user searches for a region, destination, city or resort, the current search factor for all resorts in that region, destination, city and for the resort increases by 0.01, 0.02, 0.03, and 0.04, respectively.

After calculating all these four features, we finally calculate the resort score for each resort i and user j as:

$$\text{Score} = P_i + \beta_1 * CS_{ij} + \beta_2 * S_{ij} + \beta_3 * C_{ij}$$

where

$\beta_1 = $ Number of confirmations of user j.
$\beta_2 = $ Number of searches of user j till the previous day.
$\beta_3 = $ Number of searches of user j today.

After the resort score is calculated for each resort, the resorts are sorted by the resort score and the top eight resorts in the region searched are recommended to the user.

5   Results

Figure 5 compares the model accuracy of the two models implemented and the existing model. It is to be noted that the current context plays a crucial role in determining the booking behavior of a user.

Model 1, which uses matrix factorization, gets an accuracy of 50% on the test dataset. Model 2, which uses cosine similarity, is 65% accurate. This means that users on an average booked resort recommended to them in the last 17 days, 50% of the times when we use Model 1 and 65% when we use Model 2.

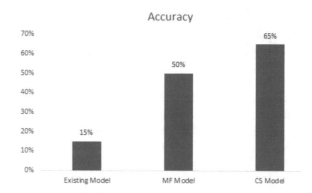

Fig. 5 Model accuracy comparison

Table 4 Surge table

	Model 1	Model 2
Average recommendation time	1.5 s	1.13 s
Accuracy	50%	65%
Diversification	Recommendations diversified by search history, search frequency, and historical bookings.	Recommendations confined to the current region searched for.
Factors considered for modeling	User-user similarity, resort-resort similarity	User search history, resort amenities, resort popularity

Table 4 compares the two models implemented. Due to the diversification involved in Model 1, Model 1 is a better model if one wants to promote under-booked hotels. Model 2, however, captures the user search behavior well and can be used to provide relevant recommendations to the user.

6 Conclusion

A personalized recommendation system based on the user's historic data and current searches influences the customer's decision. It leads to a better customer experience, boosts sales, and increase loyalty; and today customers expect it. Most of the existing models are time-consuming; therefore, there was a need for a more dynamic modeling approach to incorporate recent searches and come up with enhanced recommendations. Two models are developed to serve this purpose. Model 1 is based on matrix factorization, incorporating previous bookings, search weights added with decay function of time valuing recent searches and penalizing older searches, followed by machine learning ranking and filtering. This model gave significantly improved results and can be used for limited users. Its benefit is that it improves as more data inflows. Even though it takes time to run based on the size of the matrix, that can be reduced based on the clustering of the users and the items. Matrix factorization also works even if the products are not similar.

The second model is based on the features of the resorts and the historical booking for users and their respective searches. It is based on the concept that a user tends to book similar resorts every time depending on the amenities. It can be further improved if the price and availability can be incorporated. This model gave better accuracy in comparison to the previous models. This model takes more time for individual recommendations but takes much lesser time to calculate user vector at the end of the day and can cater to numerous customers. It assumes that all the products that we are recommending have a similar feature. In addition to giving personalized recommendations to individual users, this model can also be used to suggest the change in amenities of the resorts that can increase their sales, eventually leading to higher revenue.

Both models can be used for the recommendation engine based on the requirement of the client and the products. A hybrid model of the two algorithms can be developed and may lead to better results. All the testing has been done based on historical data, and for their true accuracy, it can be done in the live environment using A/B testing or click to conversion ratio. Overall a novel idea has been recommended in this chapter which recommends quickly as per the current search. Further research can be done based on the problem statement and data available by incorporating factors such as price, availability, etc.

References

1. Personalized Product Recommendations Drive Just 7% of Visits but 26% of Revenue (n.d.). Retrieved from https://www.salesforce.com/blog/2017/11/personalized-product-recommendations-drive-just-7-visits-26-revenue.html
2. How Netflix's Recommendations System Works (n.d.). Retrieved from https://help.netflix.com/en/node/100639
3. E. Keller, Recommendations are what drives your business. Remember to ask for them. (2012, July 27). Retrieved from https://www.forbes.com/sites/kellerfaygroup/2012/07/25/recommendations-are-what-drives-your-business-remember-to-ask-for-them/#62fa5f9539c6
4. H. Deng, Recommender systems in practice. (2019, December 5). Retrieved from https://towardsdatascience.com/recommender-systems-in-practice-cef9033bb23a
5. Challenges & Solutions for Production Recommendation Systems – Data Revenue Blog (n.d.). Retrieved from https://www.datarevenue.com/en-blog/building-a-production-ready-recommendation-system

Automated Prediction of Voter's Party Affiliation Using AI

Sabiha Mahmud Sumi

1 Introduction

The primary goal of any PEC is to increase the probability of victory. To reach this goal, all aspects of a PEC must be evaluated based on the efficiency and cost of a campaign or activity; i.e., will it motivate desired voters enough to cast their votes? This cost-to-benefit analysis can be an efficient process, especially when driven by high-performance ML models that predict specifically aligned sets of voters, with whom a political candidate may share their promises and messages with, and engage them further in joining their socio-political missions. Therefore, such predictions are crucial for promoting better communication between voters and their politically aligned candidates.

With each passing year, PECs in the United States are rapidly seen adopting various data-driven targeted-voter recommendation applications that are tailored to reduce the cost and increase the benefit in maximizing the likelihood of their victory [5]. The spread and access of big data has significantly improved the quality of empirical analysis in finding, aggregating, and identifying patterns and similarities across several domains. For example, recent successes of data-driven PECs have heavily relied on using knowledge from several domains (political, economic, medical-sciences, marketing etc.) to find novel patterns that suggest more efficient communication strategies with [20]. The underlying concept here is what the no-free-lunch (NFL) theorem suggests, that seeking a universal algorithm that solves all optimization problems is impossible, indicating that an optimization algorithm will perform better for some optimization problems than others [32]. To that end, NFL may also be considered to suggest that the success of any data-driven target-voter

S. M. Sumi (✉)
Department of Computing & Information Science, Mercyhurst University, Erie, PA, USA

© Springer Nature Switzerland AG 2021 257
R. Stahlbock et al. (eds.), *Advances in Data Science and Information Engineering*,
Transactions on Computational Science and Computational Intelligence,
https://doi.org/10.1007/978-3-030-71704-9_17

recommendation app requires unique integrations and applications of searching and optimizing that are developed from novel treatments of learned strategies across multiple domains.

The goal of this project is to develop a prediction model that accurately identifies any given voter's party affiliation based on their voting history, basic demographics, and their associated congressional district representative's party affiliation and CPVI score. The larger conceptual framework of this model is geared toward becoming a website-and-mobile-friendly platform, called LITICS360, that ultimately promotes the collection, archival, and retrieval of PEC and voter data through app-use by the PEC staff as well as voters. LITICS360 as a cross-platform and responsive app aims to facilitate transparent communication between PEC candidates and their voters to promote fair competition between all PECs alike. The principle of LITICS360 is motivated by the First Amendment of the US Constitution, "where the democratic right of every U.S. citizen guarantees the ability to exercise freedom of speech … and the right of the people peaceably to assemble and consult for their common good, and to apply the Government for redress of grievances" [31]. The ideals behind successful democratic governments are historically marked by transparent communication of voter grievances on socio-political issues and a PEC candidate's promises to address those issues through promises of political change.

The underlying requirements for achieving the idealistic and conceptual framework for LITICS360 can be divided into four general stages for prototyping:

 (i) Primary predictive model that will accurately identify a voter's party affiliation based on turnout history from previous elections, basic demographics, and their congressional district representative's party affiliation and CPVI score.
(ii) Secondary predictive model that will assign a CPVI score indicating a voter's likelihood to support their affiliated party, based on answers to survey questions and polls from marketing campaigns conducted by PECs.
(iii) Tertiary prediction model will evaluate their calculated CPVI score and predicted party affiliation, to further indicate a voter's likelihood to turnout at an upcoming election, based on their positive or negative interactions with PEC activities, i.e., page views and visits for a candidate's landing site, PEC event/rally attendance, donations, and direct communication with PEC and more.
(iv) Finally, these models will be deployed into a cross-platform app, which is developed with user experiences (UX) engineered from voter data and feature inputs by PECs, and delivered on an interactive user interface (UI) that visualizes the predictive analyses and competitive intelligence in real time, for PEC candidates, their staff, and their voters.

This chapter will be focusing on building the first stage of the larger conceptual framework of LITICS360, i.e., Pv1.0. While several dominating proprietary apps, as well as research papers, boast novel prediction models and share similar conceptual foundations, there will always be room for accommodating classification and clustering techniques that help uncover newer and more efficient patterns and similarities. Litics360 Pv1.0 aims to do just that, i.e., to contribute a model that accurately predicts a voter's party affiliation.

2 Paper Organization

This chapter aims to develop the groundwork for Pv1.0, which includes the following stages:

(i) *Research*: Historical background, evaluation of related works and apps.
(ii) *Data*: Collection, variable measures, feature variable selection and engineering, and wrangling methodologies.
(iii) *Model*: Architecture, performance measures, optimization tuning of hyperparameters.
(iv) *Evaluation*: Model and performance results.
(v) *Conclusion*: Comparison of models and concluding remarks.
(vi) *Future work*: Signification of this research and future work for building Litics360 app.

3 Background and Related Works

The culture of US-based PECs has dramatically evolved over the past decade. Historically (i.e., 1800s, 1900s and early 2000s), PECs and affiliated parties were known to use rudimentary standards for predicting voter turnout and support tendencies, in comparison to data-driven and media-powered communication standards used over the past decade. Turnout and party support predictions were previously based on the historical performance of precincts over the past four general elections, primarily measured by percentage of votes for any given party. Consequently, PEC-to-voter communication strategies would consist of re-connecting with previous donors and volunteer captains. Campaign strategies are more reliant on numbers-driven campaigns, implying poll numbers and policies in response to manually recorded surveys [11].

In the recent decade, previously used numbers-driven campaigns have evolved into data-driven campaigns. Scientific research of efficient data processing and improved computational power has introduced the notion of big data. Dynamic databases, improved analytic methods and development of prediction models using ML, are increasingly becoming specialized to compete with current PEC market standards. The 2012 and 2016 presidential elections saw the proliferation of data-driven PECs (for candidates Barak Obama vs. Mitt Romney and Donald Trump vs. Hillary Clinton). This big data revolution has not radically transformed PECs as much as television did in the 1960s, but in a close political contest, data-driven strategies can have enough impact to make the difference between winning and losing [22].

Powering existing PEC strategies with smart data-driven technologies has not been a popular choice, when state or county level elections are concerned. Inspiring grass-root and local PECs to adopt these data-driven prediction models require the app's front-end interfaces to be a brand-centric user-friendly experience that is appealing to the user. The bottom line for the mass-adoption of any novel PEC prediction algorithm is to keep in mind that the first customer of these apps will be the campaigners themselves. Therefore, the app must have an integrated approach to promote user-friendly experiences and interfaces to market the power of PEC prediction models using ML.

3.1 Evaluation of Reviewed Published Works and Apps

3.1.1 Big Data-Driven Classification Approaches

The primary concern with any given data-driven approach for PECs is the accuracy of prediction scores in forecasting the behaviors, preferences, and responses of voters. The secondary concern is a measure of the practicality of the app and its user-friendliness. A simplistic guide to successful predictive scoring models is to focus on creatively and critically thought-out variables that are sensibly linked to interesting predictions with the empirical validity [22].

There are various data-driven approaches to classifying voters that several research papers have identified. Table 1 categorizes several relevant works into five different approaches: Voter demographics; voter behaviors and preferences through means of surveys and polling; voter responsiveness through voting history, events, donations, marketing and advertising; and sentiment analysis through social media.

The first classification approach discusses voter demographic information available in registered voter files such as name, gender, age, geolocation, county, congressional district, registered party affiliation, among other basic demographic information. Several research papers, as shown in Table 1, have highlighted that these models simply predict or classify based on focal traits of interest, rather than why they voted/donated or showed support for a candidate. As such, figuring out causation is not the biggest concern, rather prediction accuracy is the main goal [2, 17, 23, 24].

Table 1 Reviewed published works categorized by approaches for classification

#	Classification approaches	Reviewed published works, citations
1	Classifying voter demographics	[2, 17, 23, 24]
2	Classifying voter behaviors and preferences (surveys, polling)	[1, 2, 5, 8]
3	Classifying voter responsiveness (voting history, events, donations, marketing/advertising)	[2, 3, 15, 19]
4	Sentiment analysis (social media)	[4, 5, 12, 13, 17, 18]

The second type is based on the behaviors, attitudes, and preferences of voters to reveal predictive scores for behavior or support scores based on surveys and polls. Several research papers, as shown in Table 1, have highlighted that these models too do not make causal claims about why the people turned up to vote or why they supported a particular candidate. Again, the intention is similar to the first category, i.e., to avoid overfitting the data [1, 2, 5, 8].

The third type is based on voter responsiveness to marketing or advertising campaigns, participation in events, rallies, demonstrations, giving donations, all in coordination with their turnout history. Research works, as highlighted in Table 1, have discussed these responsiveness scores, most of which are heterogeneous reactions to PECs in a randomized voter base as well as media-based voter base. The results derived from the effect of such marketing strategies are used as important variables that influence further communication strategies of PECs. Strong/weak positive/negative responsiveness is not causal to how the campaign was moderated or streamlined. Rather, it is about the observed differences and search for correlations across many subjects and variables used in the forming of messages that generate interesting results [2, 3, 15, 19].

The fourth type is based on text-based classification and analysis of publicly expressed sentiments on social media. Research works, as highlighted in Table 1, have discussed such sentiments to provide probable causal links of voter's candidate support score based on standings (i.e., for/against) on political issues. Opinion mining and sentiment analysis have rapidly become a popular data-driven approach for machine learning. However, more often than not, baseless causal links are grounded on the theoretic rationale of the model's architecture [4, 5, 12, 13, 17, 18].

3.1.2 ML Algorithms and Prediction Models

Currently, the vast majority of the predictive scores used by PECs are created by a PEC data analyst (or a team of them) using simple regression techniques: ordinary least squares for continuous outcomes; logistic regression for binary outcomes; and, rarely, to bid for truncated data like dollars donated or hours volunteered [8]. A wide variety of skills are needed for developing such models customizing them to specific political environments.

PEC data analysts have been searching for more systematic methods for selecting a preferred regression. The commercial marketing industry often uses k-means clustering or k-nearest neighbor classifier to divide consumers into categorical types such as blue collar, grilling, and SUV owners. However, such methods of clustering data based on voters or families are not as useful anymore for campaign data analysts, as strategic decisions in campaign planning are reliant on cost-to-benefit analysis and person-specific probabilities for particular outcomes. Thus, knowing that a set of citizens are similar in many dimensions does not assist with PEC-to-voter communication, if those dimensions are not highly correlated with voting behaviors, ideology, and propensity to donate [5, 7, 8, 26].

Table 2 Reviewed published works categorized by learning algorithms and models

ML algorithms/models	Reviewed published works, citations
Logistic regression (LR)	[8]
Natural language processing (NLP)	[5, 12, 13, 17]
K-nearest neighbor classifiers (kNN)	[5, 7, 26]
Random forest classifiers	[26]
K-means clustering (K-means)	[8]
Support vector machine (SVM)	[5, 8, 13, 26]
Naïve Bayes	[13, 26]

Supervised machine learning includes methods such as classification and regression trees. In a regression tree approach, "the algorithm grows a forest by drawing a series of samples from existing data; it divides the sample based on where the parameters best discriminate on the outcome of interest; it then looks at how regressions based on those divisions would predict the rest of the sample and iterates to a preferred fit" [6]. The payoff for this approach is that it generates estimates of what parameters are most important: that is, what parameters add the most predictive power when the group of other parameters is unchanged [26].

Other methods such as support vector machine (used to find maximal geometric margins that separate positive from negative predictions) and naïve bayes (predicting the likelihood of seeing feature vector when conditionally independent) are often used in supervised ensemble learning models to find correlations between census demographics and voter behavior features for support prediction [5, 8, 13, 26].

In Table 2, it can be seen that natural language processing (NLP) algorithms have become increasingly popular. A great many of the published works investigate the potential benefits that NLP brings to finding etymological trends and harnessing the power of such trends to motivate the public toward specific PEC strategies.

Compared to previously published work on PEC support and turnout probability prediction, the novel contribution of this paper is to combine voter demographic, behavior, and responsiveness for predicting scores for high and low probabilities of voter turnout. In addition, I will investigate the use of ensemble learning methods that will allow for more finetuning and organization of the predictive path rather than using a random search predictive model.

3.1.3 Reviewed Existing Proprietary Apps

Several existing proprietary apps, as briefly highlighted in Table 3, are available in the PEC app market that feature different capabilities including:

(i) Outreach capabilities: Canvassing, survey building, casework management, landing-website templates, targeted emails, and fundraising.
(ii) Analytic/statistic capabilities: User-activity and app-use, statistics, a breakdown of community information, databases with map views and filters,

Table 3 Reviewed published works categorized by existing proprietary applications

App	Plans/platforms/sub-apps	Citation
Ecanvasser	Essentials, Walk app, Go app, Leader	[10]
i-360	Walk, Call, Portal, Field Portal, Action, Text, Vote	[14]
L2 Political	Voter mapping, Constituent mapping, Voter outreach lists, Email/texting deployment, Digital advertising, Consumer Mapping, Automapping, V-count, Printable reports	[16]
Nation Builder	Software, Non-profit network, Enterprise organization, Advocacy, Politicians	[21]
The Optimizer	Native, Mobile, Display	[29]

email statistics (opens, clickthroughs, bounces, unsubscribes and spam reports), social media data capture, demographic data analytics, twitter-inferred political score, and data-driven similarity scores between comparing individuals/donors/volunteers/event-attendees using ML models.

(iii) Organization capabilities: Integrity of local or national campaign data with sub-campaigns for staff with a unified system for teamwork, database synced to supporter profiles and interactions, custom reporting, dashboards, goal setting/tracking/measuring.

Ecanvasser, a political canvassing campaign app, is primarily designed to help PEC organize their efforts by syncing electoral registers between its dashboard and canvassing mobile app, in addition to the management of issues and advocacy groups for community engagement through grassroots mobilization tools. Ecanvasser is used mainly by PECs that are in election mode or for managing constituency work [10].

i-360 is a solution for political campaigning, non-profits, and organization with grassroots technology that integrates management system and database for predictive models, digital/TV communication, and real-time analytics [14].

L2 Political is a PEC app that boasts a national voter file and selection platform at its core. This data-driven app provides access to their comprehensive voter-data to varied customers including local/state/federal campaigns, general consultants or direct response or media pollsters, and organizations such as (PACs/Super PACs/Associations/Unions). Their database is powered by five types of information records including voters with predictive attributes, non-registered voters, consumers, and constituents. The database enables voter mapping and filtration through varied demographic/psychographic/behavioral attributes that boast 600 behavioral, 400 demographic, and 91 predictive data fields. Outreach capabilities include mailing lists, canvassing/walk lists, phone/text lists, email lists, and digital communication [16].

Nation Builder (NP) is an app that is developed to facilitate organizations, movements, and campaigns alike. NP promotes several plans (primary, non-profit network, enterprise organization, advocacy organization, and political campaign) with capabilities for outreaching, measuring analytics and campaign organization [21].

The Optimizer is another performance-based automated optimization app for native advertising, push/pop/redirect traffic campaign optimization process, and banner campaigns [29].

4   Data and Methodology

4.1   Pv1.0 Problem Scope Definition

Pv1.0 aims to contribute a voter party affiliation identification model using learning algorithms. Independent variables include the voter's id number, residential markers, basic demographics, and election turnout history between years 2000 and 2020. It is important to also note that a voter's support for their congressional district incumbent is critical to understand whether a voter's stance on political issues (for or against) is aligned with that of their district representative. This is why it is important to also include voters' congressional district representative's affiliated party and CPVI score as another independent variable. The aim is to accurately predict a voter's party affiliation, as it is the target variable.

4.2   Data Collection, Feature Selection and Engineering

4.2.1   Data Selection Categories and Rationale

(i)   Voter identity and election turnout history:

- To develop a highly accurate party affiliation identification prediction model, a large database of voter information is required. The Ohio Secretary of State's statewide voter demographic database hosts publicly available datasets containing demographic and historical election turnout information of over seven million voters in the state [28].
- The database is a public record collection of registered voters in the state of Ohio, as submitted by each county Board of Elections. These records are submitted and maintained in accordance with the Ohio Revised Code for access to and use of voter registration lists which are open to public inspection and use for non-commercial purposes [27].
- Election voting history of the voters, for primary and general elections from year 2000 to 2020 as provided by the counties [28]. Voter turnout history is neither complete for each election nor for their party of choice.

(ii)   Ohio's congressional district representative's party affiliation and CPVI score:

- Using a simple website scrapping method, the congressional district representative's party affiliation and CPVI score were mined from the official US House of Representatives directory [9]. The scrapped data was then merged with each voter data using their associated congressional district number.

4.2.2 Feature Selection and Engineering

Preprocessing of all datasets has been done using Python and related libraries using Jupyter Notebooks. Data wrangling and cleaning applied to the Ohio voters' dataset includes the steps described below:

- Voters' missing data for voter ID and congressional district and county number were dropped as such information is vital for creating the model.
- Demographic features with over 90% missing values were also dropped as these were not important features.
- Voter's date of birth and registration date were converted to date time and then converted to a numeric age value.
- Features that were extracted from this dataset included each voter's voter ID, county number, DOB, party affiliation registration date, voter status, party affiliation, voter residential zip code, and congressional district number.
- Features containing voters' turnout between years 2000 and 2020 for general and primary elections were also extracted from the dataset. The values for these features represent their voting history and election turnout over the years.
- All values related to political parties, i.e., election turnout features, congressional party affiliation, and voter party affiliation, were converted to numeric values using a conversion key demonstrated in Table 4.

Table 4 Conversion key for parties to numeric values

Converted numeric value	Party affiliation marker	Party affiliation definition	Number of voters in total dataset
0	D	Democratic party	1,376,248
1	R	Republican party	1,930,234
2	C	Constitution party	0
3	E	Reform party	0
4	G	Green party	6222
5	L	Libertarian party	1009
6	N	Natural Law party	0
7	S	Socialist party	0
8	X	Voted without declaring party affiliation	4,108,403
9	NA	No voting record	–

4.2.3   Target Variable for Prediction

Pv1.0 model is a party affiliation identification prediction model, so the target class variable here is the registered party affiliation of the voter. For the purposes of training and testing the model, voters missing the records of their registered party affiliation were dropped before training the model, as the aim here is to develop and train a model that predicts the party affiliation. As the US presidential political system is more of a bipartisanship shared by the republican and democratic parties, the dataset was further stripped of any voters whose registered party affiliation was not republican or democratic. As mentioned in Table 4, out of the total datasets, there are 1.3 million registered democrats, and 1.9 registered republicans.

4.3   Data Preprocessing and Splitting

After cleaning and wrangling of the data, the data was first divided for modeling, training, and validation. The dataset included the information of a total of 7,772,371 registered voters from the state of Ohio. Among these, the number of voters who declared their registered party affiliation is a total of 3,207,039. The total dataset was split into two sets, a model set for building, training, and validation of the ML models (containing 90% of total or 6,994,611 voters), and a blind set for testing of the final model (containing 10% of total data or 777,179 voters). These were then processed further to retain only republican and democratic registered party voters.

5   PV1.0 Model

This section provides brief backgrounds on the three learning algorithms (i.e., decision tree classifier, random forest classifier, and gradient boosting XGBoost classifier with hyperparameter grid search) utilized in this chapter to accurately predict the voter's registered party affiliation.

5.1   Decision Tree Classifier

Decision tree classifier is one of the most powerful and popular algorithms, falling under the supervised learning algorithm umbrella [25]. The algorithm works great with categorical target variables as they are classified and sorted down from the root to the leaf node, providing a classification at each level.

In a binary tree, like the one developed for this case, classifiers are constructed by repeatedly splitting subsets of the learning sample into two descendent subsets, beginning with the learning sample itself. To split the learning sample into smaller

Fig. 1 Criterions – gini
index and entropy

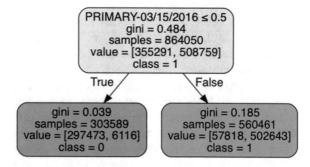

Gini Index

$$I_G = 1 - \sum_{j=1}^{c} p_j^2$$

p$_j$: proportion of the samples that
belongs to class c for a particular
node

Entropy

$$I_H = - \sum_{j=1}^{c} p_j log_2(p_j)$$

p$_j$: proportion of the samples that belongs
to class c for a particular node.

*This is the the definition of entropy for all
non-empty classes (p ≠ 0). The entropy is
0 if all samples at a node belong to the
same class.

Fig. 2 Decision tree result
using gini criterion and
max_depth of 1 for test
samples

PRIMARY-03/15/2016 ≤ 0.5
gini = 0.484
samples = 864050
value = [355291, 508759]
class = 1

True   False

gini = 0.039
samples = 303589
value = [297473, 6116]
class = 0

gini = 0.185
samples = 560461
value = [57818, 502643]
class = 1

subsets, the splits have to be selected in such a way that the descendent subsets are always purer than the parents [25]. The impurity function is based on the gini index criterion, which selects a test sample that maximizes the purity of the split. The information gain function is based on the entropy criterion, which selects a test that maximizes information gain [25]. These functions take the form of the equations in Fig. 1:

Figure 2 illustrates the binary tree developed using the decision tree classifier. This particular tree uses the gini index criterion to show the purity of the classification from the root node, which is the feature of the voter's turnout for primary election on March 15, 2016. In 864,050 voter test samples' predicted classes at the max_depth of 1.

5.2 Random Forest Classifier

Random forest classifier is an ensemble algorithm that combines more than one algorithm, in this case the decision tree algorithm, to classify objects. Furthermore, ensemble learning creates a set of decision trees (each of which is trained on a random subset of the training data), used to create aggregated predictions providing a single prediction from a series of predictions. The advantages include: Highly flexible and very accurate, naturally assigns feature importance scores, so can handle redundant feature columns, has the ability to address class imbalance by using the balanced class weight flag, scales to large datasets, generally robust to overfitting, data does not need to be scaled, and can learn non-linear hypothesis functions. The disadvantages include: Results may be difficult to interpret; the importance each feature has may not be robust to the variations in the training dataset [6].

5.3   Gradient Boosting Classifier Using XGBoost

Gradient boosting classification is a sequential technique, which works on the principle of an ensemble. It combines a set of weak learners and delivers improved prediction accuracy. A weak learner is one that is slightly better than random guessing. At any instant t, the model's outcomes are weighed based on the outcomes of previous instant t-1. The outcomes predicted correctly are given a lower weight and the ones miss-classified are weighted higher [30].

The effect of this is that the model can quickly fit, then overfit the training dataset. A technique to slow down the learning in gradient boosting is to apply a weighting factor for the corrections by new trees when added to the model, i.e., with XGBoost. When creating the gradient boosting model, XGBoost is a great tool for tuning the learning rate hyperparameters to control the weighting of the new trees added to the model. Grid search capability using scikit-learn can be used to evaluate the effect on the logarithmic loss of training a gradient boosting model with different learning rate values [30].

5.4   Performance Measures

The three learning algorithms, i.e., decision tree, random forest, and gradient boosting classifiers, help in developing the Pv1.0 model.

The decision tree classification model varies gini and entropy criterion as well as max_depth ranging from 1 to 9. The model's accuracy is defined as the fraction of correct predictions out of total number of data points. Finding the optimal value for max_depth is the tuning method used to find the best accuracy score and receiver operating characteristic (ROC) area under the curve (AUC) accuracy score.

The random forest model uses n_estimators to set a number of trees and uses ROC curves and precision, and recall scores show the probabilistic forecast for this binary classification model. These performance measures use values from both columns of the confusion matrix to evaluate the fraction of true positives among positive predictions.

Gradient boosting with XGBoost model varies hyperparameters with a range of learning rates from 0.01 to 1.0 and n estimators of 10 and 100 to find the best-tuned learning algorithm using the grid search method. The model's accuracy is defined using the ROC AUC score plotted to show the rate of true positives, which is the fraction of the elements of 1 that are classified as 1 correctly, as a function of the false positive rate, which is the fraction of the elements of 0 that are classified as 0 incorrectly. The sensitivity is given by the rate of true positives and anti-specificity by the rate of false positives. The anti-specificity, or false positives, correlates with the x-axis, and the sensitivity, or true positives, correlates to the y-axis, which forms the ROC AUC figure displayed. A subset of the data is also chosen through the confusion matrix for measuring the quality of the classification system.

6 PV1.0 Model Evaluation

The model developed for Pv1.0 uses the decision tree classifier at first in order for an easier demonstration of the model. Random forest model helps to create better evaluation on principle as it combines a number of weak estimators to form a strong estimator. Finally, the gradient boosting model with XGBoost is the solution model using grid search to tune hyperparameters for higher performance with higher learning rates and a larger number of trees. All the models have used a ten-fold cross validation to obtain the performance results that are shown. In this section, the performance results of these three models will be discussed.

6.1 Performance Results

6.1.1 Decision Tree Classification Model

The decision tree model had a range of max_depth inputs, in combination with gini and entropy criterion. The performance accuracy as well as the ROC AUC accuracy was calculated for each hyperparameter variation, to find the best combination that resulted in high accuracy and ROC AUC accuracy scores.

Figure 3 shows the performance results for gini criterion for the range of max_depths, and Fig. 4 shows the same for entropy criterion. The difference between gini and entropy is not far apart, as it can be seen in the Figs. 3 and 4. The performance accuracy and the ROC AUC accuracy scores are at their highest when max_depth ranges from 5 and above. Based on the performance results laid out in both Figs. 3 and 4, it can be seen that with higher max_depth the accuracy increases.

6.1.2 Random Forest Classification Model

The random forest model was set to run with n_estimators set to 100 trees. Table 5 contains the performance measures of the random forest learning algorithm, including recall, precision, and ROC scores, which show the probabilistic forecast for this binary classification model.

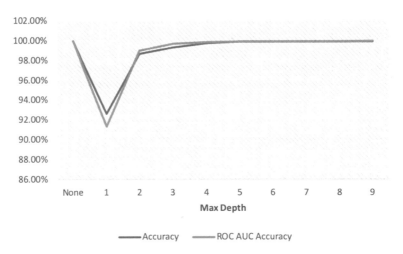

Fig. 3 Decision tree model performance results for test samples of model validation dataset using gini criterion

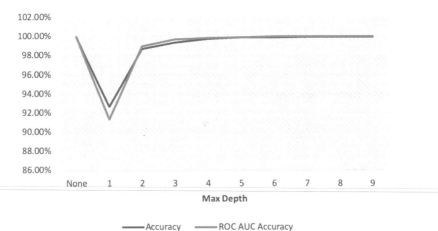

Fig. 4 Decision tree model performance results for test samples of model validation dataset using entropy criterion

Table 5 Random forest model performance results for model validation dataset	Score type	Baseline	Train samples	Test samples
	Recall	1.0	0.9936	0.9941
	Precision	0.5896	0.9978	0.9979
	ROC	0.5	0.9999	0.9999

The results listed in Table 5 show the recall, or sensitivity, score where the ratio of correctly predicted positive observations to the total predicted positive observations is at its highest level for both training and testing samples. These results show that all the registered voters' party affiliation was accurately predicted the in the random

Fig. 5 Random forest model
ROC performance results for
model validation dataset

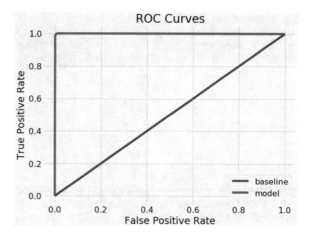

forest model. Table 5 also shows the precision score where the ratio of correctly predicted positive observations to the total predicted positive observations is also at its highest level for both training and testing samples. This high and near-perfect precision score communicates the low false-positive rate.

Figure 5 plots the false-positive rate on the x-axis versus the true positive rate on the y-axis for a number of different candidates, where the threshold values fall between 0.0 and 1.0. In Fig. 5, the false alarm rate is compared with the hit rate, which is demonstrated by the difference in the rates of true positives and false positives. The true positive, or the sensitivity, is calculated as the number of true positives divided by the sum of the number of true positives and the number of false negatives. The main takeaway from Fig. 5 is how well the model that is predicting the positive target class of the voter's party affiliation when the actual outcome is also positive. The false alarm, or the inverted specificity, rate summarizes how often a positive class is predicted when the actual outcome is negative. This inverted specificity is the total number of true negatives divided by the sum of the number of true negatives and false positives.

Represented at a point (0,1), Fig. 5 shows a line traveling from the bottom of the left of the plot to the top left and then across the top to the right. This representation of line shows a nearly perfect skillful model, where the probability of randomly chosen real positive occurrences versus negative occurrences, is at its highest.

The ROC curve in Fig. 5 plots a near-perfect skillful model, which is also demonstrated in Fig. 6's confusion matrix. Figure 6 displays the confusion matrix table describing the performance of the random forest classification model on the test data samples for which the true values are known. True positives and true negatives are the observations that are correctly predicted, shown in reddish-brown color representing a close to 1.0 accuracy.

Fig. 6 Normalized confusion matrix for random forest model with model validation dataset

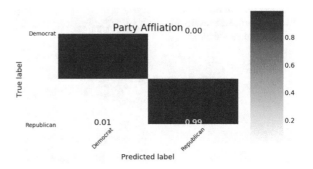

Fig. 7 Negative log loss results using hyperparameter grid search for XGBoost model

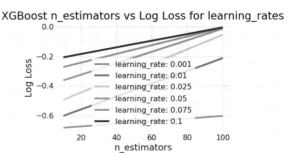

6.1.3 Gradient Boosting Classification Model with XGBoost Using Grid Search for Tuning

The gradient boosted classification model with XGBoost was set to run with n_estimators set to 10 trees and 100 trees. This model also had a range of learning rates including 0.001, 0.001, 0.025, 0.05, 0.075, 0.1. These hyperparameters were fitted using grid search to find the best XGBoost classifier.

Figure 7 shows negative logarithmic loss for hyperparameter grid search with n_estimators from 10 to 100 trees and varied learning rates, where the loss function is used to quantify the price paid for inaccuracy of predictions within this classification. Figure 7 shows that the best mean negative log loss score was -0.004291 with a standard deviation of 0.000002 for the hyperparameters of 0.1 learning rate and 100. Negative log loss performance results were calculated by doing a randomized grid search for the XGBoost classifier with varying hyperparameters of n_estimators and learning rates states above.

Table 6 contains the performance measures of the XGBoost model, including recall, precision, and ROC scores, which show the probabilistic forecast for this binary classification model. These scores were calculated after applying the best estimated hyperparameters for XGB classifier, using the grid search mentioned above.

The results listed in Table 6 are very similar to the random forest model as it shows the recall, or sensitivity, score, where the ratio of correctly predicted positive observations to the total predicted positive observations, is at its highest with both

Table 6 XGBoost performance results for model validation dataset

Score type	Baseline	Train samples	Test samples
Recall	1.0	0.9998	0.9998
Precision	0.5896	0.999	0.9989
ROC	0.5	0.9999	0.9999

Fig. 8 XGBoost ROC results for model validation dataset

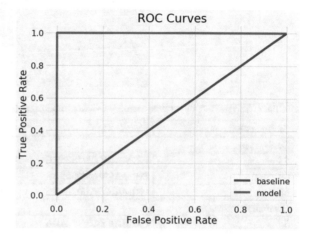

train and test samples. This high recall score indicates that all the registered voters' party affiliation were accurately predicted the in the XGBoost model. The precision performance measure shows the ratio of correctly predicted positive observations to the total predicted positive observations. As stated in Table 6, the high and near-perfect precision communicates the low false-positive rate.

Figure 8 shows the ROC curves for the XGBoost model which is very similar to Fig. 5 as the model results are near perfect in both the random forest and the XGBoost models. Again, represented at a point (0,1), Fig. 8 shows a line traveling from the bottom of the left of the plot to the top left and then across the top to the right. This representation of line similarly shows another perfect skillful model, where the probability of randomly chosen real positive occurrences versus negative occurrences, is at its highest.

Figure 9 displays the confusion matrix table describing the performance of the XGBoost classification model, which is similar to the confusion matrix of the random forest model shown in Fig. 6. As the ROC curve in Fig. 7 plots a perfect skillful model where the true positives and true negatives are predicted accurately, Fig. 8 reconfirms that the true positives and true negatives are correctly predicted, shown in reddish-brown color representing a close to 1.0 accuracy.

Fig. 9 Normalized confusion matrix for XGBoost model for model validation dataset

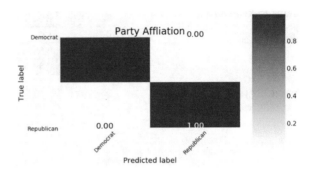

Table 7 Top three features influencing prediction models, in their order of importance

Feature column number	Feature name	Feature importance score
Decision tree model		
53	PRIMARY-03/15/2016	0.724
60	PRIMARY-05/08/2018	0.246
63	PRIMARY-05/07/2019	0.020
Random forest model		
53	PRIMARY-03/15/2016	0.509
60	PRIMARY-05/08/2018	0.237
26	PRIMARY-03/04/2008	0.048
Gradient boosting with XGBoost using grid search model		
53	PRIMARY-03/15/2016	0.706
60	PRIMARY-05/08/2018	0.166
63	PRIMARY-05/08/2018	0.189

6.2 Feature Importance

In order to increase the quality of predictive power of the learning models against the binary target of party affiliation, finding the variables that are strongly correlated with the target class is important. Table 7 shows the top three features, or the most predictive variables, that influence the highest accuracy performances of the binary classification predictions in the decision tree, random forest, and XGBoost models. It can be seen here that the feature that has the strongest correlation to the target class is the voter turnout at the primary election on March 15, 2016, with an importance score of over 70% for both decision tree and XGBoost models and over 50% for the random forest model.

Table 8 XGBoost performance results for blind test dataset

Score type	Baseline	Train samples	Test samples
Recall	1.0	0.9274	0.9257
Precision	0.5895	0.8793	0.8769
ROC	0.5	0.9616	0.9612

6.3 Testing XGBoost Model with Blind Test Dataset

With performance results of over 99% accuracy in all three models for the model validation dataset, the next step is to utilize the best ML model, while increasing the quality of its predictive power as well. This can be done by removing the more influential feature within the dataset, i.e., the feature with the maximum importance score for all three models, from the test dataset. After which, the best estimated hyperparameters (i.e., learning rate of 0.25 and 100 n_estimators) that were found using the randomized grid search were applied to calculate the XGBoost model's performance scores. Table 8 below contains the performance measures of the XGBoost model, including recall, precision, and ROC scores, for the blind test dataset.

The results listed in Table 8 shows the recall, or sensitivity, score, where the ratio of correctly predicted positive observations to the total predicted positive observations, is at over 92% with both train and test samples. This high recall score confirms that most of the registered voters' party affiliations were accurately predicted in the XGBoost model. The precision performance measure shows the ratio of correctly predicted positive observations to the total predicted positive observations at over 87%. These scores confirm that even without the most predictive feature, the model is performing with a highly successful accuracy score.

Figure 10 shows the ROC curves for the XGBoost model, which interprets the 96% accuracy scores for both train and test samples of the XGBoost model. Again, represented at a point (0,1), Fig. 10 shows a line traveling from the bottom of the left of the plot and curving to the top right well past the 87% mark. This representation of line similarly shows another high-performing skillful model, where the probability of randomly chosen real positive occurrences versus negative occurrences is high.

Figure 11 displays the normalized confusion matrix table describing the performance of the XGBoost classification model, where the 82% of predicted samples were true positives while 18% were false positives; and 92% of the predicted samples were true negatives while 7% were false negatives.

7 Conclusion

The objective of this research was to create a base model that predicts a voter's party affiliation. Learning algorithms, including decision tree, random forest, and gradient boosting classifiers were utilized toward this objective. Moreover,

Fig. 10 XGBoost ROC
results for blind test dataset

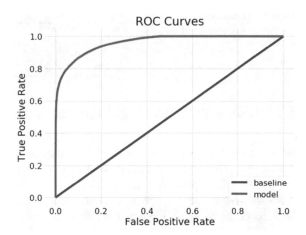

Fig. 11 Normalized
confusion matrix for
XGBoost model for blind test
dataset

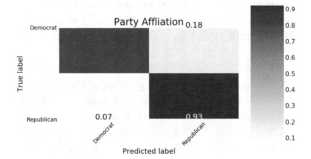

independent variables such as the voter's turnouts and party support in previous elections, residential demographic features, associated congressional district
representative's party affiliation, and CPVI scores, all had contributing factors in
accurately predicting a voter's party affiliation. By applying preprocessing methods,
data transformation and reduction led to high-performing and reliable models.
Furthermore, a comparison between the learning models (decision tree, random
forest and XGBoost) was carried out to identify that the XGBoost is the best model.
In conclusion, the comparison demonstrated that the random forest and the XGBoost
learning algorithms are very similar, as they have high and near-perfect accuracy
scores.

However, XGBoost is theoretically a better model overall for a number of
reasons: (a) It implements regularized boosting to reduce overfitting; (b) implements
parallel processing even though it is a sequential process resulting in greater speed;
(c) allows for higher optimization through a range of hyperparameter tuning through
grid search for evaluation criteria; (d) implements its in-built routine to handle
missing values which are important for this type of big data; (e) splits up until the
specified max_depth and then starts tree pruning process backward to remove splits
beyond which there is no positive gain; and finally (f) runs a cross-validation at

each iteration of the boosting process allowing for easily obtaining exact optimum number of boosting iterations in a single run.

The quality of the XGBoost model was also tested by removing the most predictive feature, proving that it continued to perform well with over 90% accuracy. With that in mind, the next step for improving this model would be to detect churners, voters who are likely to change party affiliation over the election years. Furthermore, applying the model on a set of churners would indicate another set of accuracy results, likely in promising an improved model.

8 Future Work

Imagine if a PEC appropriated their valuable resources for a marketing campaign that ended up mobilizing voters to cast votes for their opponent. Not only would this counterproductive marketing campaign be an inefficient use of a PEC's funds, but it would also be detrimental to potential voters who may have missed the chance to align themselves with candidates sharing their favored political interests, or even their chance to voice social grievances to PEC candidates in urging positive changes to their society. To curb these potentially flawed campaigns from happening, research as such must continue to find data-driven prediction models with higher accuracies and performance. Utilizing data about voters' socio-political preferences, expected behaviors, responses, past interactions with PECs, and historical record of turnouts at previous elections, is crucial in aligning voters with PECs that share similar socio-political agendas and goals.

The learning model contributed through this research provides a highly accurate model for identifying a voter's registered party affiliation based on a non-exhaustive election turnout history. This contribution has created the foundation for the first version of the prototype for Litics360.

The second step is to develop a secondary predictive model based on the first predictive identification of voter's party affiliation, which assign a specific CPVI number to the voter based on survey answers provided through PEC marketing strategies and polling questions, which is predictive of the voter's likelihood to support their registered party in an election. This secondary model will help identify voters who fall in the middle spectrum of the CPV index allowing for better utilization of PEC marketing and resources for garnering more efficient PEC-to-voter communication strategies.

The third step is to develop the tertiary prediction model to evaluate a voter's turnout probability based on the secondary predictive model's CPVI score, from polled answers, voters' interactions, i.e., page views and visits, for the candidate's landing site, PEC's event attendances, donations, and direct communication with PECs, from the secondary predictive model. This tertiary prediction model will help PECs further, to communicate with the right voters for greater turnout in upcoming elections.

Finally, developing a brand-centric and user-friendly front-end UI/UX, based on the features engineered for prediction models, will integrate interactive methods of visualizing this data in real-time through dashboard UI/UX for both PEC staff and voter users.

Acknowledgments I would like to give a special thanks to Dr. Afzal Upal, Chair of the Department of Computing and Information Science at Mercyhurst University, for his valuable guidance during my research. I would also like to acknowledge and give thanks to the data scientist Shraddha Dubey for always being a wonderful sounding board during the critical brainstorming sessions of this research.

References

1. J.H. Aldrich, R.D. McKelvey, A method of scaling with applications to the 1968 and 1972 presidential elections. Am. Polit. Sci. Rev. **71**(1), 111 (1977, March). https://doi.org/10.2307/1956957
2. B. Alexander et al., A Bayesian model for the prediction of United States presidential elections. Am. Politics Res. **37**(4), 700–724 (2009, June)
3. S. Ansolabehere et al., Candidate positioning in U.S. house elections. Am. J. Polit. Sci. **45**(1), 136 (2001, January). https://doi.org/10.2307/2669364
4. K. Benoit & M. Laver, Estimating Irish party policy positions using computer wordscoring: the election - a research note, Irish Political Studies, Routledge (Taylor & Francis Group), UK, **18**(1), 97–107 (2002), https://www.tandfonline.com/loi/fips20
5. A. Bonica, A data-driven voter guide for U.S. elections: Adapting quantitative measures of the preferences and priorities of political elites to help voters learn about candidates. Rsf **2**(7), 11–32 (2016). https://doi.org/10.7758/rsf.2016.2.7.02
6. L. Breiman, Random forests. Mach. Learn. **45**(1), 5–32 (2001). https://doi.org/10.1023/A:1010933404324
7. Catalist, Y.G. and Gelman, A. 2018. Voter Registration Databases and MRP Toward the Use of Large Scale Databases in Public Opinion Research
8. Challenor, T. 2017. Predicting Votes from Census Data
9. Directory of Representatives | House.gov: https://www.house.gov/representatives#state-ohio. Accessed 02 Feb 2020
10. Ecanvasser: 2012, http://www.ecanvasser.com/. Accessed 13 Dec 2019
11. D.M. Farrell, P. Webb, Political parties as campaign organizations, in *Parties Without Partisans: Political Change in Advanced Industrial Democracies*, (2002). https://doi.org/10.1093/0199253099.001.0001
12. J. Grimmer, B.M. Stewart, Text as data: The promise and pitfalls of automatic content analysis methods for political texts. Polit. Anal. **21**(3), 267–297 (2013, January). https://doi.org/10.1093/pan/mps028
13. A. Hasan et al., Machine learning-based sentiment analysis for twitter accounts. Math. Comput. Appl. **23**(1), 11 (2018, February). https://doi.org/10.3390/mca23010011
14. i360: 2009, http://www.i-360.com/. Accessed 13Dec 2019
15. P. Klimeka et al., Statistical detection of systematic election irregularities. Proc. Natl. Acad. Sci. U. S. A. **109**(41), 16469–16473 (2012, October). https://doi.org/10.1073/pnas.1210722109
16. L2 Political: 1990, https://l2political.com/. Accessed 13 Dec 2019
17. T. Louwerse, M. Rosema, The design effects of voting advice applications: Comparing methods of calculating matches. Acta Polit. **49**(3), 286–312 (2014, July). https://doi.org/10.1057/ap.2013.30

18. W. Lowe, Understanding Wordscores. Polit. Anal. **4**, 356–371 (2008). https://doi.org/10.1093/pan/mpn004
19. J.G. Matsusaka, F. Palda, Voter turnout: How much can we explain? Public Choice (Springer) **98**, 431–446 (1999)
20. A. Mian, H. Rosenthal, *Introduction: Big Data in Political Economy* (Rsf. Russell Sage Foundation, 2016)
21. NationBuilder: 2009, https://nationbuilder.com/. Accessed 13 Dec 2019
22. D.W. Nickerson, T. Rogers, Political campaigns and big data. J. Econ. Perspect. **28**(2), 51–74 (2014). https://doi.org/10.1257/jep.28.2.51
23. M. Peress, Estimating proposal and status Quo locations using voting and cosponsorship data. J. Polit. **75**(3), 613–631 (2013, July). https://doi.org/10.1017/S0022381613000571
24. K.T. Poole, H. Rosenthal, A spatial model for legislative roll call analysis. Am. J. Polit. Sci. **29**(2), 357 (1985, May). https://doi.org/10.2307/2111172
25. L.E. Raileanu, K. Stoffel, *Theoretical Comparison Between the Gini Index and Information Gain Criteria* (Kluwer Academic Publishers, 2004)
26. S. Smith et al., *Predicting Congressional Votes Based on Campaign Finance Data* (2012). https://doi.org/10.1109/ICMLA.2012.119
27. State Laws on Access to and Use of Voter Registration Lists: 2019, http://www.ncsl.org/research/elections-and-campaigns/access-to-and-use-of-voter-registration-lists.aspx. Accessed 1 Dec 2019
28. Statewide Voter Files Download Page, https://www6.ohiosos.gov/ords/f?p=VOTERFTP:STWD:::#stwdVtrFiles. Accessed 18 Apr 2020
29. TheOptimizer: 2016, https://theoptimizer.io/. Accessed 13 Dec 2019
30. Tune Learning Rate for Gradient Boosting with XGBoost in Python, https://machinelearningmastery.com/tune-learning-rate-for-gradient-boosting-with-xgboost-in-python/. Accessed 23 Apr 2020
31. US Constitution, *First Amendment Religion and Expression* (United States Senate, Office of the Secretary of the Senate, Library of Congress, 1791)
32. D.H. Wolpert, W.G. Macready, No free lunch theorems for optimization. IEEE Trans. Evol. Comput. **1**(1), 67–82 (1997). https://doi.org/10.1109/4235.585893

Part IV
Data Science, Deep Learning, and CNN

Deep Ensemble Learning for Early-Stage Churn Management in Subscription-Based Business

Sijia Zhang, Peng Jiang, Azadeh Moghtaderi, and Alexander Liss

1 Introduction

Customer churn, or loss of customers, is one of the most critical problems that subscription-based businesses must address in order to grow their customer base and remain competitive. Previous research [19] suggests that customers generate increasing revenue as they stay longer with the company. Retained customers tend to increase purchases over time, generate customer referrals, and help reduce operating costs. Therefore, improving retention is critically important for subscription-based businesses. Identifying users with high propensity to churn allows businesses to target the most vulnerable customers more efficiently.

In this paper, we apply machine learning to predict churn in subscription-based businesses. Specifically, we predict churn behavior in free trial periods that typically last for several days to a few weeks. This is referred to as *early-stage churn*. In this phase, customers test the product and determine if the service meets their expectations. Many businesses offer free trials to attract new customers. A large percentage of customers that sign up for free trials do not purchase the paid subscription, or cancel it soon after the start of their first term, leading to a much higher churn rate as compared to later periods [7]. Therefore, it is critical for businesses to identify potential early-stage churners and conduct necessary interventions to help increase bill-through and retention.

It is challenging to predict early-stage churn, because the observation window is short and limited information is available about the behavior and preferences of new users. Machine learning researchers have explored various models to predict mid- and late-stage churn [2, 3, 18, 20, 24]. Commonly used methodologies can

S. Zhang (✉) · P. Jiang · A. Moghtaderi · A. Liss
Ancestry.com Operations Inc., San Francisco, CA, USA
e-mail: sizhang@ancestry.com; amoghtaderi@ancestry.com; aliss@ancestry.com

© Springer Nature Switzerland AG 2021
R. Stahlbock et al. (eds.), *Advances in Data Science and Information Engineering*,
Transactions on Computational Science and Computational Intelligence,
https://doi.org/10.1007/978-3-030-71704-9_18

be categorized into two main classes: classification and survival analysis. While addressing the same problem, these two classes of methods are different. In classification, the target variable is binary. The trained classifiers output propensity scores to predict whether a customer churns or not within a specified time frame. In contrast, the outcome of survival analysis is a continuous variable that represents the time to event. Statistical models are typically developed to estimate the remaining time before the customer churns. In the case where churn does not occur within the observation window, the outcome variable is marked censored. Censorship allows survival analysis to make estimation for customers who have not churned in the observation window. Many different algorithms from the two categories of methods have been used in churn prediction, such as logistic regression, tree-based methods, deep learning, and semi-parametric and parametric survival models. However, the comparison between different methodologies and how they can be combined to improve predictive performance have not been well explored.

Inspired by ensemble learning that combines predictive capabilities of various models with different feature representations and utilization [4, 8, 23], we propose using stacked ensembles to integrate various classifiers and survival analysis methods for early-stage churn prediction. We first evaluate a simple, standard stacking model and test for performance boost over individual learners. To further advance the ensemble framework, we propose a novel deep ensemble model, Deep Stacking, to create deep learning ensemble of neural networks, boosted classifier, survival analysis, and potentially many other predictive models.

To test the effectiveness of our proposed ensemble models on early-churn prediction, we conduct a case study with data from Ancestry, the global leader in family history and consumer genomics. Customers rely on historical records and other resources to understand their heritage and family history. They can experience the genealogy service by signing up at Ancestry.com for a 14-day free trial and decide whether to purchase the subscription-based service by the end of the trial. In the experiments, we compare the performance of individual models and ensemble models for churn prediction of free trialers at Ancestry. We use two metrics to evaluate model performance, namely, AUC on all customers and hit rate (precision) on customers with high churn propensity. The reason to use hit rate is that companies typically offer retention incentives to a subset of users that have high propensity to churn instead of the entire customer base in order to control campaign cost [12, 14]. When we use tactics, such as offering discounts, to reduce churn, we need to consider the trade-off between campaign cost and revenue gained by retaining customers. Therefore, it is necessary to evaluate model performance on high-propensity customers and ensure that the false-positive rate in the target user segment is minimized.

We summarize our learnings and major contributions as follows:

– We compared performance of different classifiers and survival analysis for churn prediction and found that gradient boosting trees and deep learning models outperformed survival analysis algorithms in our application.

- We constructed a standard stacking model with a logistic regression meta-classifier and developed a new deep ensemble model to integrate the predictive capabilities of hybrid neural networks, gradient boosting trees, and survival analysis.
- Experiments on genealogy subscription data showed that both stacking models outperformed individual learners in terms of AUC and hit rate. Improvement in hit rate will increase the efficiency and business value of promotional campaigns for subscription-based businesses. We also observed that the deep ensemble model had overall better performance than standard stacking.
- We developed a churn prediction deployment pipeline that could be applied to other subscription-based businesses.

2 Related Work

Machine learning researchers have explored various models to predict churn. Commonly used methodologies can be categorized into two main classes: classification and survival analysis.

Classification models, including logistic regression, decision trees, neural networks, and boosting, are commonly used to predict future customer behaviors [16, 17]. Boosting has been shown to significantly improve accuracy in predicting customer churn in the wireless telecommunications industry [13]; combining it with neural networks or decision trees increased the model discriminability [16]. Extreme gradient boosting (XGBoost), a modern tree-based boosting algorithm [3], has been applied to address a wide range of machine learning challenges and proved to achieve state-of-the-art results in many cases [3, 10].

Deep learning methods are another type of widely used classifiers in customer churn prediction. Deep learning offers automatic generation of feature representations as well as flexible architecture [2, 20, 25]. Castanedo et al. [2] applied a four-layer feedforward model to predict churn in a mobile telecommunication network and improved AUC over random forests. Zhu [25] proposed a hybrid classifier which consists of a deep neural network (DNN) taking static features as input and a recurrent neural network (RNN) taking dynamic features as input. This model was applied to predict Microsoft Azure churn and increased the AUC by 0.9 percentage points as compared to DNN alone. In 2017, Chen et al. [4] used convolutional neural networks (CNN) as the meta-classifier to stack multiple predictions and achieved a 0.04 percentage point increase in accuracy over CNN alone for churn prediction in retail banking. Studies in [4, 25] inspired us to develop deep ensemble models to improve churn prediction.

Other than classification models, survival analysis is also widely used for churn prediction. Instead of predicting a binary output indicating whether customers churn or not like in classification models, statistical models are typically developed in survival analysis to estimate the remaining time before the customer churns. There are two key functions in survival analysis: survival function and hazard function.

Survival function, $S(t)$, indicates the probability that a user will remain a subscriber after time t. Hazard function, $f(t)$, is the instantaneous event rate over an infinitely small time period conditioned on that a user has stayed as a subscriber up to time t. It measures the potential risk of experiencing an event at time t. Depending on whether there is assumption about the data distribution, survival analysis models can be grouped into three types: parametric, nonparametric, and semi-parametric [5]. Parametric models assume certain distributions in the data, such as log-logistic or Weibull distribution. WTTE-RNN [15] is a recently developed model which estimates the distribution of time to event as having a discrete or continuous Weibull distribution. It uses RNN to estimate the two parameters of Weibull distribution. Nonparametric models like Kaplan-Meier method assume no distribution in data. Semi-parametric models like Cox regression make very few assumptions about the data.

3 Models

In this section, we describe two ensemble approaches to combine XGBoost, deep learning, and survival analysis methods to create better-performing predictive models. We first briefly introduce the standard two-layer stacking model with a meta-level classifier. Then we describe our proposed deep ensemble model. Base models considered for ensemble learning include XGBoost [3], hybrid RNN-DNN [25] and two survival analysis algorithms, namely, WTTE-RNN [15] and Cox regression [6].

3.1 Standard Meta Stacking Model

The standard Meta Stacking framework consists of two layers, as illustrated in Fig. 1 [11, 21]. Layer 1 is composed of individually trained base learners, whose predictions are then used as features in Layer 2, a meta-classifier, to generate the final prediction. Separate training sets are required to train models in each layer. The first dataset is used to train and validate individual base learners. Next, the trained models are applied on the second training set to generate churn propensity scores; these intermediate churn scores are used to train a meta-level logistic regression classifier. The advantage of using a meta-classifier instead of simply averaging Layer 1 predictions is that the logistic regression classier automatically determines the optimal weights for combining predictions from different Layer 1 models.

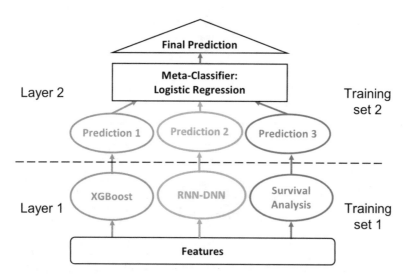

Fig. 1 Standard Meta Stacking model. Predictions of pre-trained XGBoost, RNN-DNN, and survival models are taken as input by a meta-level logistic regression classifier to generate the final prediction

3.2 Deep Stacking Model

As compared to tree-based models, deep learning classifiers, such as DNN and RNN, can generate different feature representations [2]. Inspired by the architecture of the hybrid deep learning model reported in [25], we propose a new deep ensemble method, which stacks XGBoost and survival models with a hybrid RNN-DNN model. Unlike [25], our framework is a two-step stacking model and can take advantage of different predictive models, such as gradient tree boosting and survival analysis, in addition to deep learning. As shown in Fig. 2, Deep Stacking merges the output of a RNN branch, output of a DNN branch, and predictions from other pre-trained models to generate the final churn propensity score.

The RNN branch of the Deep Stacking framework models sequences of user activities. We used long short-term memory (LSTM) units as building blocks of the RNN, because LSTMs are capable of learning long-term dependencies and allow the network to capture the entire context of the input sequence when making predictions. In our application, the RNN branch is composed of two LSTM layers and each layer contains 32 units.

The DNN branch of the ensemble model takes static features as inputs. It is composed of four identical hidden layers, each of which contains a dense layer. The number of neurons in each layer equals the size of the feature vector. This fully connected feed-forward network performs the following computation:

$$a^{(l+1)} = f(w^{(l)} * a^{(l)} + b^{(l)})$$

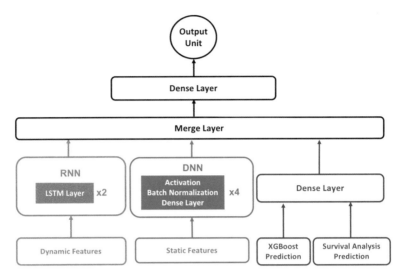

Fig. 2 Deep Stacking model. RNN and DNN branches are merged with scores from pre-trained XGBoost and survival models to make the final prediction

where $w^{(l)}$, $a^{(l)}$, and $b^{(l)}$ are model weight, activation, and bias of the l-th layer, respectively. f can be any activation function. In our application, we choose the popular ReLU activation to achieve faster learning and better performance. We employed batch normalization to standardize input to each layer for each mini-batch in order to reduce covariance shift.

The third branch of the Deep Stacking model contains a dense layer, which takes predictions from other pre-trained models, such as churn propensity scores from XGBoost and survival analysis, as input features. Predictions from additional base learners can be incorporated in this branch by increasing the number of units and the number of layers.

To integrate the three branches into one model, we use a merge layer to concatenate their outputs. We then add another dense layer and an output unit with a sigmoid activation function to generate the final prediction. We use cross-entropy loss as the loss function.

Training of the Deep Stacking model is carried out in two steps using separate training sets. Base learners, whose predictions are used as inputs by the third branch of the deep ensemble, are first trained and validated using the first dataset. Then using the second training set, all parts of the Deep Stacking model are jointly trained by back-propagating the gradients from the output to all three branches simultaneously with mini-batch stochastic optimization.

The final output of the Deep Stacking model is produced by jointly training all parts of the network, which is different from ensemble using a meta-level classifier. In the Meta Stacking model described above, the hybrid RNN-DNN model is one of the base learners that are individually trained. During training, RNN-DNN is not aware of the predictions of other models, and the outputs from different models are combined by the meta-level classifier as a separate step. However, in the deep ensemble framework, the parameters of RNN and DNN as well as the weights for summing other pre-trained models are optimized simultaneously. As compared to the Meta Stacking model, the deep ensemble framework is more efficient in terms of training, because joint training eliminates the need of a meta-level classifier. In addition, it provides the opportunity for RNN and DNN to complement the weakness of other models by knowing the predictions of pre-trained base learners at training time.

4   Case Study with Ancestry Data

4.1   Problem Formulation

Ancestry offers subscription-based service for customers to conduct genealogy research using rich online family history resources. The resources include a variety of records, such as census, birth, and marriage records, among others, and public family trees. Customers can experience the service by signing up for a 14-day free trial. By the end of the free trial, they decide whether to purchase the paid subscription based on the initial experience with the product. Customers can purchase or cancel the service during the 14-day free trial or they will be charged automatically at the end of the trial. Customers who are not willing to purchase a subscription by the end of the free trial are considered churned. In this paper, we focus on predicting churn of free trialers.

We use customer engagement data from the first half (7 days) of the free trial to predict churn and leave the rest of the trial as the intervention period. We build a model to predict whether a customer will churn between 7 and 21 days after they sign up for the free trial. The prediction window is extended by 7 days after the free trial ends, i.e., the period between day 14 and day 21. This is because a number of customers request to cancel within a few days after the free trial ends, which indicates their lack of interest in using our service in the long run, and thus we consider them as early-stage churners as well. In summary, we define churners as customers who cancel Ancestry service between day 7 and day 21 after the start of the free trial.

4.2   Features Generation

To develop churn prediction models, we extracted 96 features that capture diverse customer activities and preferences during the free trial period. The data collected includes both time-dependent and static variables. We categorize the features into the following three groups.

Genealogy Research Customers rely on a variety of resources provided at Ancestry to conduct genealogy research. They could build family trees by adding family members as tree nodes, perform a search within a large collection of records/public tree nodes and attach relevant search results to ancestors in their trees, get recommendations (called "hints") by the company about potentially relevant public tree nodes or records, and then decide whether to accept, reject, or pend these recommendations. They can also upload their own content, such as photos and stories of their ancestors, to their trees. They could exchange information with others via messages. We therefore extract features to illustrate these activities including page visits, tree building, searches, hints, uploads of content, and collaboration between customers. We engineer time-dependent features, such as number of record search and hints received on day 1, day 2, and up to day 7, to capture dynamic product engagements.

User Preferences In addition to features related to genealogy product engagements, we also extract user preference features that capture their choices of device and subscription package. For example, customers can use a mobile, tablet, or desktop to register. Subscription packages can differ in duration (monthly, half-year, and annual) and content coverage.

Other Ancestry Products Customers' prior relationships with other Ancestry products may affect their decisions about purchasing the genealogy subscription. For example, separate from the genealogy subscription, Ancestry also offers DNA kits as scientific and conclusive evidence to help customers trace their lineage. Customers who are familiar with Ancestry DNA product may have higher affinity for our genealogy service. Therefore, we construct features to reflect if the customer has already purchased or activated any DNA kit and their engagement with our DNA product.

After the three groups of features are collected, we encode and preprocess features before feeding them into machine learning models. For example, one-hot encoding is applied to all categorical features. We drop the first encoded column to avoid collinearity. Imputation is applied to handle missing values in numerical and categorical features.

4.3 Data

The training and test strategies are summarized in Fig. 3. We selected about 1M users from those who started free trial in 2018 as training data. The training set is randomly divided into two equally sized subsets. We use the first subset to train single models including classification models (such as logistic regression, XGBoost, hybrid RNN-DNN) and survival analysis models (such as WTTE-RNN and Cox). Training set 2 is used to train the second layer of the Meta Stacking model and the Deep Stacking model. We test all models on a random sample of approximately 75K customers that started their free trials in 2019.

Fig. 3 Training and test strategies for different models

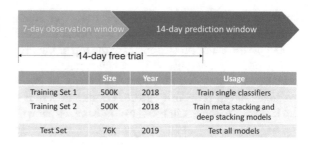

	Size	Year	Usage
Training Set 1	500K	2018	Train single classifiers
Training Set 2	500K	2018	Train meta stacking and deep stacking models
Test Set	76K	2019	Test all models

Classification models and survival analysis models define target labels differently. Binary labels are defined for classification models. Customers are labeled as churners or non-churners depending on whether they purchase a subscription and stay as a subscriber beyond day 21. Classification models are then trained to predict whether new customers will churn or not in the prediction window. In survival analysis, the target variable is no longer binary. In the training phase, we examine the cancellation behaviors of customers within the first 21 days since their initial signup date. Customers are then labeled by their survival time, i.e., the number of days between the signup date and the churn date if they churn during that time period. If they do not churn, their survival time is marked as censored and will be handled by survival analysis models. In the testing phase, we apply the trained models to predict how likely customers will churn in the next 2 weeks given that they have survived the first 7 days.

4.4 Evaluation Metrics

Overall model performance is evaluated using AUC, which is a standard metric that can evaluate the performance of predictive models using raw, unbinarized output scores. AUC is calculated on all customers, but retention campaigns typically only target customers with higher propensity scores who are predicted to be more likely to churn. In order to develop churn management strategies using estimated propensity scores, there are usually three steps involved [12, 14]:

1. Rank customers based on the predicted churn propensity.
2. Select a subset of customers at the top of the propensity ranking and offer them retention incentives.
3. Evaluate the effects of retention incentives on churn and profit via A/B tests.

A main reason to offer retention incentives only to customers with high propensity to churn is to reduce cost associated with retention campaigns. Churn management can become expensive depending on the strategies. For example, offering discount to a large group of customers or to users that are false positives in churn prediction model is suboptimal. To avoid high campaign costs, we must consider the trade-off between the number of churners captured and the number of customers to target and increase precision on the subset of customer with high churn propensity. Therefore, we employ an additional application-specific metric, hit rate, to assess model performance on customers at the top of the propensity ranking.

Hit rate is commonly used in churn management applications [9, 12, 22]. To calculate hit rate, we first rank customers in descending order based on their predicted churn propensity score. Hit rate at X% is computed as the precision of the top X% customers in the churn ranking, as shown below:

$$Hit\ rate(@X\%) = \frac{number\ of\ churners\ in\ top\ X\ percentile}{number\ of\ customers\ in\ top\ X\ percentile}$$

For instance, if the top 10% customers in the churn ranking contains 100 users and 60 of them are churners, then hit rate ($@10\%$) = 60/100 = 60%.

5 Experiment Results

In this section, we compare the performance of single and ensemble models using AUC and hit rate. We built logistic regression, XGBoost, and Cox models using Scikit-learn, XGBoost, and Lifelines Python packages, respectively. For WTTE-RNN, we speed up an existing Keras implementation [1] by transforming all features to tensors in parallel to handle millions of data in our application. We implement RNN-DNN and Deep Stacking model in Keras with a TensorFlow backend. Deep Stacking model uses stochastic gradient descent with learning rate of 0.05 and momentum of 0.9 in our application. All models except Cox use both static and dynamic features described in Sect. 4.2. The time-invariant Cox model takes only static features as covariates.

This section is structured as follows. First, we present the AUC and hit rate of single models. Second, we choose single models that have strong performances as base learners to build stacking models and evaluate the performance of those models using AUC and hit rate. Next, we add more weak learners to the stacking model and test how they affect the model performance. Last, we summarize insights gained by comparing Meta and Deep Stacking models.

5.1 Performance of Single Models

Table 1 summarizes the performance of different single models measured by AUC and hit rate @5%. Higher metric values indicate better performance. For survival analysis models, the AUC and hit rate of Cox regression model are higher than that of WTTE-RNN, although WTTE-RNN is more complex and uses additional time-dependent features. Comparing survival models and classification models, both metrics indicate that the two survival models perform better than logistic regression,

Table 1 AUC and hit rate @5% of various single models

Type	Model	AUC	Hit rate @5%
Survival analysis models	WTTE-RNN	0.6365	55.7%
	Cox	0.6470	61.8%
Classification models	Logistic regression	0.5678	54.1%
	RNN-DNN	0.6846	**78.9%**
	XGBoost	**0.6921**	78.2%

but worse than XGBoost and RNN-DNN. The difference in performance between survival analysis models and RNN-DNN and XGBoost classifiers is more evident when hit rate is used as the evaluation metric. The hit rate @5% of RNN-DNN and XGBoost is over 15% higher than that of the two survival analysis models. We also find that XGBoost outperforms RNN-DNN in terms of AUC, while RNN-DNN has higher hit rate @5% than XGBoost. Taken together, XGBoost and RNN-DNN classifiers are the strongest learners among all single models.

5.2 Performance of Stacking Models

In this section, we evaluate the performance of Meta Stacking and Deep Stacking models in comparison to single models. Logistic regression classifier has inferior performance as compared to the other four algorithms in Table 1, so it is no longer considered in our experiments with stacking models. Since XGBoost and RNN-DNN are the most competitive single models in our application, we first combine XGBoost and RNN-DNN using stacking models. Later, we add WTTE-RNN and Cox as additional base models to test the effects of adding additional learners on model performance.

Table 2 shows the performance of stacking models using XGBoost and RNN-DNN classifiers. For clear comparison between stacking and single models, we also include the performance of single classifiers in the same table. As shown in Table 2, the AUC of Meta Stacking is slightly lower than that of XGBoost but higher than that of RNN-DNN. In contrast, Deep Stacking outperforms both XGBoost and RNN-DNN in terms of AUC. Both Meta Stacking and Deep Stacking have increased hit rate @5% as compared to single models. The difference in hit rate @5% between Deep Stacking and its base learners is greater than 1%. 1–2% improvement in hit rate will result in significant increase in the efficiency and business value of promotional campaigns for subscription-based businesses. For example, if a business with 10M subscribers hopes to offer discounts to the top 5% customers that are most likely churn, a 1–2% increase in hit rate @5% means replacing 5K–10K false positives with true positives. Assuming the subscription price is $30 per month, this increase in hit rate may result in a maximum of 300K increase in revenue for one promotional campaign.

Table 2 Comparison of single models and stacking models with classifiers

Type	Model	AUC	Hit rate @5%
Single models	RNN-DNN	0.6846	**78.9%**
	XGBoost	**0.6921**	78.2%
Stacking models (XGBoost+RNN-DNN)	Meta Stacking	0.6918	79.9%
	Deep Stacking	**0.6925**	**80.0%**

Fig. 4 Hit rate at different percentiles within the top decile of churn ranking

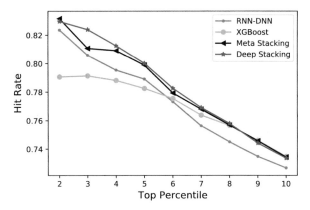

In addition to comparing hit rate @5%, we also show the hit rate at different percentiles in the top decile in Fig. 4. We observe that the hit rates of stacking models are consistently higher than that of a single RNN-DNN model within the top decile. The difference in hit rate between stacking models and XGBoost is more evident at smaller percentiles and does not drop to zero until 8%. Taken together, better performance is achieved by combining the predictive capabilities of RNN-DNN and XGBoost using either the standard Meta Stacking or the proposed Deep Stacking framework.

Both stacking models are flexible and are capable of taking predictions from base learners besides XGBoost and RNN-DNN as input. To test the effects of adding other learners, we incorporate survival analysis methods, WTTE-RNN, and Cox, into Meta Stacking and Deep Stacking models in addition to XGBoost and RNN-DNN.

Table 3 shows the performance of stacking models with additional survival analysis learners. For better illustration, we include the performance of stacking models with XGBoost and RNN-DNN only in the same table. We find that adding WTTE-RNN and Cox as additional base learners to Meta Stacking only slightly increases AUC, but including survival analysis models in Deep Stacking improves both AUC and hit rate @5% as compared to Deep Stacking with XGBoost and RNN-DNN alone.

Table 4 lists the changes in hit rate after including WTTE-RNN and Cox as additional pre-trained learners to the two stacking models at different percentiles. The baseline used for this comparison is stacking models with RNN-DNN and

Table 3 Performance of stacking models with additional survival models

Type	Model	AUC	Hit rate @5%
Stacking models (XGBoost+RNN-DNN)	Meta Stacking	0.6918	79.9%
	Deep Stacking	**0.6925**	**80.0%**
Stacking models (XGBoost +RNN-DNN +WTTE-RNN+Cox)	Meta Stacking	0.6920	79.9%
	Deep Stacking	**0.6938**	**80.3%**

Table 4 Change in hit rate after adding survival analysis

Top percentile	Meta Stacking (%)	Deep Stacking (%)
@1%	−0.53	0.40
@2%	0.00	0.33
@3%	0.57	0.17
@4%	0.46	0.18
@5%	−0.03	0.27

XGBoost only. A positive value indicates hit rate is increasing by using additional survival analysis models. Hit rate at top 1–5% is consistently increased by inputting predictions from survival models to the Deep Stacking model. In contrast, adding survival models as base learners in the Meta Stacking model does not consistently improve hit rate at top percentiles.

5.3 Insights on Model Performance

Comparing the two ensemble frameworks, the proposed Deep Stacking model has overall better performance than the standard Meta Stacking as measured by AUC of all customers and hit rate at top percentiles of churn ranking in our application. There are two possible reasons why the Deep Stacking model has the potential to achieve better performance than the standard Meta Stacking. First, deep learning offers automatic generation of feature representations with multiple levels of abstraction, which is more sophisticated than a logistic regression meta-classifier. Second, joint training in Deep Stacking provides the opportunity for RNN and DNN models to be aware of predictions of other models during network training and thus more likely to complement weaknesses of other models.

Our experiments suggest that the Deep Stacking method can be used as a flexible ensemble framework to combine a variety of different base learners and is able to utilize the predictive capabilities of deep learning, boosted trees, and survival analyses to improve overall performance. Although XGBoost and deep learning models have been proved to achieve state-of-the-art performance in customer churn prediction, the predictive performance can be further improved by combining those strong classifiers with weaker survival models in the Deep Stacking framework. In summary, the new Deep Stacking method is the best candidate model that should be served to predict churn in production.

6 System Implementation

In order to use churn prediction models in production, we developed a pipeline that automatically produces churn propensity scores daily based on the updated information of customers. This system utilizes generic building blocks and may be extended to predict churn in other subscription-based businesses. An overview of the system is provided in Fig. 5. We built this pipeline utilizing publicly available and popular AWS cloud service. We created AWS CloudWatch events to schedule automated actions that are triggered daily after the update of our online database. Feature extraction is performed by querying Ancestry's database. Monitoring procedures were set up to check if data collection is successfully completed. If the extraction step fails due to high load on the database, the system will log the error, send notifications, and retry feature collection at a different time. The generated features are fed into our preselected machine learning model to produce churn propensity scores. Both churn scores and features used to produce the predictions are stored in the cloud. These data can be used as features to retrain models and develop new models in the future. Churn propensity scores are also uploaded to our customer data platform.

7 Conclusions

Early-stage churn is an important business problem that relates to a customer's perceived value of the service. Accurate prediction of early-stage churn can provide a unique view of the underlying business issues and present an opportunity to

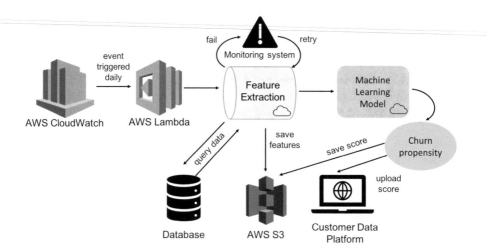

Fig. 5 Overview of churn prediction pipeline in production

improve service and to increase customer satisfaction. Despite these benefits, early-stage churn prediction is associated with great challenges due to short observation windows and limited customer information.

In this paper, we propose a new deep ensemble model that combines the predictive capabilities of popular churn prediction models, including gradient boosting trees, deep learning algorithms, and survival analysis methods, to make early-churn prediction with both static and dynamic features. XGBoost and RNN-DNN models have been proved to achieve state-of-the-art performance in customer churn prediction. They can outperform each other depending on the evaluation metric as shown in the experiments. Integrating predictions from XGBoost and RNN-DNN in our novel Deep Stacking model leads to higher overall AUC and hit rate at top percentiles of the churn ranking. The proposed deep ensemble framework is highly flexible and can incorporate additional base learners, such as WTTE-RNN and Cox. Although those survival analysis methods are inferior models in our application, adding them to the Deep Stacking model improved both generic and application-specific evaluation metrics. With this deep ensemble approach, we can utilize the strengths of a variety of models to create a better classifier. In the future, we will research resampling techniques and statistical tests that are applicable to stacking models in order to prove the statistical significance of the Deep Stacking model.

Acknowledgments The authors would like to thank Dipankar Ghosh for his help with the design and implementation of the churn prediction pipeline.

References

1. D. Batten, *Demo Weibull Time-to-Event Recurrent Neural Network in Keras* (2019). https://github.com/daynebatten/keras-wtte-rnn
2. F. Castanedo, G. Valverde, J. Zaratiegui, A. Vazquez, *Using Deep Learning to Predict Customer Churn in a Mobile Telecommunication Network* (2014). https://www.wiseathena.com/pdf/wa_dl.pdf
3. T. Chen, C. Guestrin, Xgboost: a scalable tree boosting system, in *Proceedings of the 22nd ACM SIGKDD International Conference on Knowledge Discovery and Data Mining, KDD '16* (ACM, New York, 2016), pp. 785–794
4. Y. Chen, Y.R. Gel, V. Lyubchich, T. Winship, Deep ensemble classifiers and peer effects analysis for churn forecasting in retail banking, in *Advances in Knowledge Discovery and Data Mining*, ed. by D. Phung, V.S. Tseng, G.I. Webb, B. Ho, M. Ganji, L. Rashidi (Springer, Cham, 2018), pp. 373–385
5. T.G. Clark, M.J. Bradburn, S.B. Love, D.G. Altman, Survival analysis Part I: basic concepts and first analyses. Br. J. Cancer **89**, 232–238 (2003)
6. D.R. Cox, Regression models and life-tables. J. R. Stat. Soc. Series B (Methodological) **34**(2), 187–220 (1972)
7. H. Datta, B. Foubert, H. van Heerde, The challenge of retaining customers acquired with free trials. J. Marketing Res. **52**(2), 217–234 (2015)
8. T.G. Dietterich, Ensemble methods in machine learning, in *Multiple Classifier Systems* (Springer, Berlin, 2000), pp. 1–15

9. A. Ghorbani, F. Taghiyareh, C. Lucas, The application of the locally linear model tree on customer churn prediction, in *Proceedings of the 2009 International Conference of Soft Computing and Pattern Recognition* (2009), pp. 472–477
10. B. Gregory, *Predicting Customer Churn: Extreme Gradient Boosting with Temporal Data* (2018). https://arxiv.org/pdf/1802.03396.pdf
11. F. Güneş, R. Wolfinger, P. Tan, *Stacked Ensemble Models for Improved Prediction Accuracy* (2017). https://support.sas.com/resources/papers/proceedings17/SAS0437-2017.pdf
12. S.Y. Hung, D.C. Yen, H.Y. Wang, Applying data mining to telecom churn management. Expert Syst. Appl. **31**, 515–524 (2006)
13. A. Lemmens, C. Croux, Bagging and boosting classification trees to predict churn. J. Marketing Res. **43**(2), 276–286 (2006)
14. A. Lemmens, S. Gupta, Managing churn to maximize profits. Marketing Sci. **39**(5), 956–973 (2017). http://dx.doi.org/10.2139/ssrn.2964906
15. E. Martinsson, *WTTE-RNN: Weibull Time to Event Recurrent Neural Network*. Master's thesis (Chalmers University Of Technology, Göteborg, 2016)
16. M.C. Mozer, R. Wolniewicz, D.B. Grimes, E. Johnson, H. Kaushansky, Predicting subscriber dissatisfaction and improving retention in the wireless telecommunications industry. IEEE Trans. Neural Networks **11**(3), 690–696 (2000)
17. E. Ngai, L. Xiu, D. Chau, Application of data mining techniques in customer relationship management: a literature review and classification. Expert Syst. Appl. **36**(2, Part 2), 2592–2602 (2009)
18. A. Periáñez, A. Saas, A. Guitart, C. Magne, Churn prediction in mobile social games: towards a complete assessment using survival ensembles, in *Proceedings of the 2016 IEEE International Conference on Data Science and Advanced Analytics (DSAA)* (2016), pp. 564–573
19. F. Reichheld, *Prescription for Cutting Costs* (Bain and Company, New York, 2001). https://www.bain.com/insights/prescription-for-cutting-costs-bain-brief/
20. P. Spanoudes, T. Nguyen, *Deep Learning in Customer Churn Prediction: Unsupervised Feature Learning on Abstract Company Independent Feature Vectors* (2017). https://arxiv.org/pdf/1703.03869.pdf
21. K.M. Ting, I.H. Witten, Issues in stacked generalization. J. Artif. Intell. Res. **10**, 271–289 (1999)
22. A. Tiwari, J. Hadden, C. Turner, A new neural network based customer profiling methodology for churn prediction, in *Computational Science and Its Applications, ICCSA 2010: International Conference* (Springer, Berlin, 2010), pp. 358–3694
23. A. Tsymbal, M. Pechenizkiy, P. Cunningham, *Diversity in Ensemble Feature Selection*. Technical report (Trinity College, Dublin, 2003). http://www.cs.tcd.ie/publications/tech-reports/reports.03/TCD-CS-2003-44.pdf
24. K.K.K. Wong, Using cox regression to model customer time to churn in the wireless telecommunications industry. J. Targeting Meas. Anal. Marketing **19**(1), 37–43 (2011)
25. F. Zhu, *Predicting Azure Churn with Deep Learning and Explaining Predictions with Lime* (2017). https://www.slideshare.net/FengZhu18/predicting-azure-churn-with-deep-learning-and-explaining-predictions-with-lime

Extending Micromobility Deployments: A Concept and Local Case Study

Zhila Dehdari Ebrahimi (iD)**, Raj Bridgelall** (iD)**, and Mohsen Momenitabar** (iD)

1 Introduction

The number of vehicles and trips is ever-increasing with population growth and technology evolution. Consequently, new modes of transportation are emerging to fill gaps in mobility and accessibility. Micromobility has emerged as a new form of transportation—a category of vehicles that weigh less than 500 kilograms and covers distances of less than 5 miles. Figure 1 shows how micromobility solutions fit within the spectrum of travel parameters.

Other modes that are emerging include app-hailed Robo-cab, air taxis, and hyper loops [1]. This chapter focuses on the history, impacts, and implications of the micromobility category where accessibility, flexibility, and affordability are its main advantages. Vehicles in the micromobility category are human- or electric-powered and include docked bikes, dockless bikes, e-scooters, and other emerging modes [2]. The main characteristic of docked bike sharing is that users must unlock the vehicle from a designated station and return it to any other designated station that has space. In contrast, users of dockless bike sharing use an app to enable the vehicle for usage, and then stow the vehicle anywhere after using it. The app tracks the position of available vehicles using the standard positioning service (SPS) of the global positioning system (GPS). The e-scooter is a more recent form of micromobility within the dockless category. Recent deployments of e-scooters brought both opportunities and challenges for cities around the globe. In some cases, companies are struggling to sustain their deployments as cities demand more attention to safety, clutter, and equity. As micromobility evolved, it has had a large

Z. Dehdari Ebrahimi (✉) · R. Bridgelall · M. Momenitabar
Department of Transportation, Logistics, & Finance, North Dakota State University, Fargo, ND, USA
e-mail: zhila.dehdari@ndsu.edu; raj@bridgelall.com; mohsen.momenitabar@ndsu.edu

© Springer Nature Switzerland AG 2021
R. Stahlbock et al. (eds.), *Advances in Data Science and Information Engineering*,
Transactions on Computational Science and Computational Intelligence,
https://doi.org/10.1007/978-3-030-71704-9_19

Fig. 1 Evolving modes of transportation

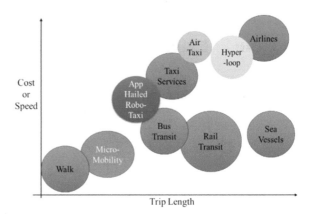

impact on the environment, society, and livability. The next three subsections review the history of bike sharing and its evolution toward e-scooter sharing, the benefits and deficiencies of the new mode, and factors in their adoption.

2 History

The first bike-sharing system emerged in the 1960s and has since completed three generations of evolution. The first generation began in 1965 with 50 unlocked bikes in Amsterdam for free use by the public [3]. The second generation began in the early 1990s as coin-deposit systems where users can unlock bikes with a small cash deposit [4]. A third generation replaced those systems by using membership cards to unlock bikes from docking stations. Transformations in service models took place over time because of differences in benefits, deficiencies, capital costs, and operating costs [4].

As of May 2011, there were estimated 136 bicycle-sharing programs in 165 cities around the world with 237,000 bikes on the roads [5]. By 2013, bike-sharing programs had been launched in more than 500 cities in 49 nations with a combined fleet of more than 500,000 bikes [6]. Bike sharing in the Americas emerged in Canada, Mexico, the United States, Argentina, Brazil, and Chile. Asia currently has the fastest-growing bike-sharing market with services deployed in China, South Korea, and Taiwan. The main pull for bike-sharing deployments are traffic management and air pollution control, which are major challenges in China. The network of urban roads in China is complex and people have many different choices of travel modes. The introduction of bike sharing in China spurred a mode shift from single occupancy motor vehicles. The availability of bike sharing, and the increasing accessibility of mobile technology influenced an integrated shared mobility app that combines cars and bikes [7]. Riders now mainly use public bicycles to commute, travel home during lunch breaks, and for after-work shopping

[56]. These developments reduced traffic congestion, environmental noise, the frequency of crashes, the severity of injuries, and the level of physical inactivity.

Bike share in the United States ramped up very quickly. There were 35 million trips taken in 2017, which was 25% more than trips in 2016. At the end of 2017, five major dockless bike share companies operated in 25 cities and suburbs. The inventory of shared bikes increased from 42,500 in 2016 to nearly 100,000 in 2017, with the majority of those being dockless. During the later half of 2017, companies added 44,000 new dockless bikes in cities around the United States, accounting for 44% of all shared bikes in the nation [9]. At the end of 2018, the popularity of e-scooters increased relative to shared bikes, and the use of dockless bike sharing decreased from 2017 to 2018 [10]. Reference [11] analyzed the challenge of using the new kind of public transportation mode in ten cities and found that among the urban transport modes, the e-scooter is most effective in reducing the use of private cars. However, e-scooters could create conflicts over space, speed, and safety when introduced without consideration of policies. The use of e-scooters started in Santa Monica and Austin in 2017. The number of e-scooters in Washington, D.C. increased from 5235 to at least 10,000. In San Francisco the increase was from 2500 to at least 4000. Urban communities like D.C. grant operational licenses through open application forms [12].

Reference [13] studied bike sharing in Seattle, USA, which is the first city to adopt dockless bike share. The case study found that inadequate system scale, station density, geographic coverage area, ease of use, and pricing structure contributed to the struggle for adoption. The study demonstrated that explicit options made by system designers and policymakers instead of local market or environmental factors lead to failure. Regardless of membership type, weather is less of a disutility for dockless scooter share (DSS) users than for Station Based Bike share (SBBS) users [14]. Gas prices have a positive impact on micromobility use. Reference [15] reviewed the barriers of dockless bike-sharing systems for travel behavior of user, user experience, and relevant social impacts of dockless bike-sharing systems. Reference [15] proposed the advantage of the dockless design of bike-sharing systems which could remarkably improve users' experiences at the end of their bike trips. They could conclude that people have not a responsibility to return their bikes to a designated dock. Furthermore, the high flexibility and efficiency of dockless bike sharing makes the integration with public transit an efficient option for first−/last-mile solutions. The GPS tracking device embedded in each dockless shared bike allows for the collection of large-scale route data, which allow researchers to evaluate travel behavior in new ways.

The first e-bike was introduced in 2017 and it was a new challenge for the government of Japan. They introduced a new "sandbox" program to support the innovative technologies and business in Japan. This program covers the different area like healthcare service, mobility, transportation, and financial systems. Reference [16] found that 129 companies were working under the program.

Micromobility is so new that public agencies have not yet included them as a transport mode in travel surveys, hospital admission records, and police crash databases [17]. Even so, the various forms of micromobility are proliferating based on differences in utilitarian needs [18]. An emerging mode of micromobility are hover boards which have two wheels and a cross-board that the user stands on. Another mode is VEEMO—a three-wheeled enclosed motorcycle-style vehicle that uses solar energy [19]. One advantage is that it shelters the user from bad weather. Evolving concepts could converge with the trends of electrification, sharing, connectivity, and autonomy [20]. Safety and well-being will likely improve as governments put in place guidelines for shared micromobility [21]. Reference [22] found that the COVID-19 caused a shift from public transit in New York City toward micromobility modes.

3   Benefits and Deficiencies

Recent analysis determined that bikes and e-scooters are best suited for short-distance trips or for connection to longer-distance public transit modes for journey completion [22]. Using micromobility for last-mile connectivity to transit will help cities to decrease traffic congestion and environmental pollution. The e-scooter mode of micromobility has benefits and deficiencies similar to shared bikes. The benefits are that they can spur a mode shift to public transportation. A report by the National League of Cities (NLC) found that micromobility helps communities save money on travel and leads to a more sustainable environment globally by reducing green-house-gas emissions [23]. Dockless modes of micromobility have the advantages of lower cost for portions of the trip and flexibility to transfer to other modes for longer legs of a trip. However, the vehicle must be safe, and the road infrastructure must be suitable for safe travel. The NLC study found that trip productivity can increase if various micromobility vehicles are close to metro stations, taxi stations, and ride-hailing stations. In general, micromobility modes promote physical activity, which provides health benefits [24].

The National Association of City Transportation Officials (NACTO) in the United States analyzed data for different modes of micromobility in six cities, including statistics such as the number of trips, the purpose of using a mode, the duration of each trip, and the average cost per trip. They found that more people used bike sharing than e-scooters for work trips and for connecting to other types of public transportation modes. The study found that most people used e-scooters for weekend shopping trips in the peak hours from 11 a.m. to 12 noon. The usage of e-scooters in the afternoon and into the evening is higher than that of shared-bikes. Most people use bike-share for week-day trips. Using trip distance and duration, NACTO calculated the average cost per trip for e-scooter and docked bike as $3.50 and $1.25, respectively. Consequently, companies have been offering discounts for low-income users of e-scooters.

The deficiencies of micromobility modes of transport are that individual or environmental factors contribute to the rate of accidents and injuries. Users of e-scooters tend to pay less attention to their surroundings while using their cellphones. Riders use e-scooters at uncontrolled speeds and drivers ignore them [23]. This increases the risk of accidents and injuries. This concern led Washington, D.C., to establish speed-limit rules for e-scooters. The use of dockless scooters also creates clutter in cities because users stow them anywhere and even discard some in lakes and oceans [23]. The lack of infrastructure such as special routes to accommodate e-scooters can interfere with pedestrian traffic and create safety issues. These shortcomings point to the growing need for micromobility companies to revise their usage and stowage policies to reduce interference with motor vehicle and pedestrian traffic. Consequently, cities need to have clearly defined paths to separate micromobility traffic flows from pedestrians, to separate faster electrically assisted vehicles from foot pedal bikes, and to designated stowage areas where clutter can be minimized without reducing their accessibility [25, 26].

4 Factors in Adoption

A study in Norway chose 66 persons at random to utilize an e-bike for a restricted period and compared the results with a control group of 160 persons [27]. The study found that e-bike trips expanded from 0.9 to 1.4 per day, distance from 4.8 to 10.3 km, and transport mode share from 28% to 48%, with the control group showing no change in bike usage. Their results also showed that the effect of e-bike increased with time and had more impact on women than on the men. Another study showed that in countries with a cycling culture, such as Denmark, e-bikes resulted in a mode shift from cars [28]. The study found that these impacts are lower in regions with no provisions for cycling, such as in north America and Australia. A study in Sweden found that an increased use of e-bikes for private journeys will result in energy efficiency gains [29].

Studies around the globe found that environmental conditions such as weather and terrain topography can influence adoption [30–32]. A Singapore study modeled the impact of fleet size, environment infrastructure, and weather condition on the preference for dockless bike sharing and found that weather conditions and infrastructure had the main impact on usage rate [33]. An Australian study showed that strong winds and rainfall deter the use of bike sharing and consequently decrease the number of trips [34]. A Canadian study found that weather conditions such as humidity, precipitation, and snow reduced the demand for bike sharing in Toronto, Canada [35].

Many studies sought to determine non-weather factors that influence the choice of bike sharing over other modes such as public transit and taxis. The main findings are that factors affecting micromobility usage rates include sociodemographic, population, density of buildings, proximity to the central business district, street geometry, proximity to water, accessibility to trails, length of bike lanes, distance to other micromobility stowage sites, transportation law and policies involving helmet and license requirements [8, 36–39]. A few studies examined gender attitudes in the use of bike sharing. Several studies revealed that more men used bikes than women [40, 41]. In contrast, other studies found that women were more interested in using bike sharing than men, especially on weekdays [42, 43]. However, women who care for children are more likely to rely on private vehicles [44].

The remainder of this chapter is organized into four sections. Section 2 presents best practices and various considerations for data analysis to inform deployment decisions. Section 3 describes relevant data from Washington, D.C., to provide some insights into the data analytics. Section 4 discusses the results from the case study in terms of recommendations for deployment and expansion. Section 5 summarizes the significance to the work and provides concluding remarks.

5   Method

Cities throughout the world have different designs, infrastructure, land use, budget, socio-demographics, climate, travel behavior, and levels of technology deployments that influence their approaches to micromobility adoption. Even within a locality, the urban landscape can differ across areas and influence land rent [45]. In general, the transportation network of a city center is complex. The density of transportation services tends to be heavier in city centers than in the outskirts. This geospatial variation in transport services impacts travel behavior and mode choice.

This research suggests a research strategy to determine opportunities for micromobility deployment and scaling in a selected city. The strategy starts by gathering available data about existing micromobility solutions deployed, their spatial-temporal distribution relative to public transit services, and areas of high trip generation and trip attraction. Figure 2 illustrates that such an analysis would begin with the synthesis of information from different social, behavioral, and geospatial data sources. Subsequently, decisions about where to deploy must be filtered with best practices that involve working closely with the city.

A synthesis of knowledge from the literature points to the following strategies as best practices for service providers to sustain and scale deployments:

1. Understand the needs of the city, the intentions behind them, and their compatibility with the business model of the service provider.
2. Work with cities upfront to implement solutions that:

    Complement public transportation services
    Minimize interference with motor vehicle, human-powered vehicles, and pedestrian traffi.
    Avoid stowage clutter in the middle of sidewalks or foot path.
    Enforce the avoidance of restricted access areas
    Make the service equitable and affordable for low-income riders
    Maintain a privacy policy that is compliant with state and federal laws
    Implement liability mechanisms that are fair

3. Implement usage policies that promote the safe operation of all types of vehicles provided.

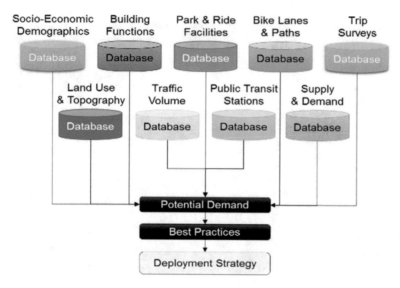

Fig. 2 Analysis of potential demand using available GIS data

4. Balance supply and demand, both spatially and temporally, to meet the needs of the city.
5. Remove disabled vehicles and maintain vehicles regularly to ensure operational safety and reliability.
6. Share usage data with cities to forge a partnership in both near-term and long-term urban planning.
7. Provide a multilingual website and all-hour customer support.
8. Collaboratively promote the service with the city's public engagement resources.

Common data sources are available from open-source data services that most cities currently host. Database services include information on the following:

1. *Land use*—derived from zoning maps that show the distribution of residences, businesses, neighborhood boundaries, street layout, bicycle lanes, pedestrian walkways, and public transit stations such as light rail, bus stops, subways, and taxi hubs.
2. *Trip generators*—derived from Geographical Information Systems (GIS) containing data on building usage, their location, and their relative distances to public transportation services. GIS also show a distribution of parks, community centers, and tourist attractions.
3. *Trip behavior*—derived from census data on population, sociodemographic, and trip surveys.

A central tenant of the finding from the literature review is that sharing micromobility usage data will help governments understand evolving travel behavior to inform long-term infrastructure planning that can support and sustain the adoption

Fig. 3 Distribution of Capital Bike Share stations and bike availability. (Capital Bike share, 2018)

of micromobility services. Currently, micromobility companies allocate stations for docked vehicles or rebalance dockless vehicles based on the population within each area. For example, Fig. 3 shows the current distribution of Capital Bike share docking stations in Washington, D.C [46].

Analyzing data about the type of micromobility usage for different sections of a city can reveal points where benefits sustained or stopped [47]. Demand analysis begins by identifying high-demand public transportation routes. Assessing demand at different times and stations will inform strategies for allocating micromobility services to minimize overall travel delay and to avoid overstocking. This tactic is based on the finding that many travelers use micromobility services to access public transportation for trip continuation. Databases on existing micromobility supply and demand includes pickup and drop-off times, travel trajectories, speeds, positions relative to transit locations, bike lanes, and bike trails. Knowledge about the spatial and temporal ebb and flow of supply and demand is an essential part of informing vehicle redistribution strategies.

6   Data for the Case Study

Dockless bike shares in Washington, D.C., began in September 2017 and reached 155,000 trips by 2018 [48]. This was in comparison to 1,220,000 docked bike trips from the company for the same period. The District Department of Transportation (DDOT) evaluates the success of shared micromobility deployments based on equity, data sharing, and company viability. The DDOT found that there is significant inequity of the current deployments. Analysis of a publicly available database indicates that the highest concentration of those aged 20–29 is in Noma, Dupont Circle, and downtown whereas the concentration for age 30–39 are around the Capitol Hill, National Mall, downtown, and southwest Washington [49]. These age groups currently tend to be the largest users of micromobility services.

Micromobility modes have become some of the most affordable forms of public transportation in major cities of the world [10]. Hence, analyzing the socio-demographic factors in each zone of a city can inform decisions about the redistribution of micromobility services. Areas that have a high to median income are Palisades, Chevy Chase, Friendship Heights, and Tenley town [50]. Most low- or moderate-income Washington, D.C. residents are in the south, east, middle, and north parts of the city. Hence, more bike lanes are needed in those areas. Micromobility companies can focus more on the east and south parts where there are fewer bike sharing or transit stations.

Micromobility usage has the potential to relieve traffic in various parts of Washington, D.C. Figure 4 shows relative bicycle traffic volumes in Washington, D.C [51]. The dataset associated with the chart contains the locations of permanent vehicle count stations and weigh-in-motion stations. The dataset includes the street segment ID, traffic sensor type, and a count of the number of bicycles passing those sensors. The highest bicycle traffic volume is around downtown, Capitol Hall, Georgetown, Fox hall Crescent, and Southwest Washington. Many public recreation sites like the National Park and Downtown exhibit high demand for micromobility services.

Accessibility to public transportation can help with trip completion where micromobility infrastructure is lacking, or the travel time is too long. The park-and-ride points shown in Fig. 5 contain lots where commuters can park their private vehicles and commute to the city center or other areas. The figure shows that park-and-ride points are only in the downtown and northeast areas. The dataset used to create the chart is from the Washington, D.C. Open-GIS Database [52].

Figure 6 shows that bike lanes are in downtown, Bloomingdale, 16th Street Heights, Capitol Hill, and toward the south-east [53]. Based on topographical databases, it is surmised that fewer bike lanes are in the north because of high elevations in that area. Figure 7 shows that most bicycle trips in Washington, D.C., are between 2 and 5.5 miles [54]. Figure 8 shows the current land use and zoning restrictions of Washington D.C [55].

The pattern indicates that once they have demonstrated viability in the city centers, micromobility companies can expand service in the residential areas of the

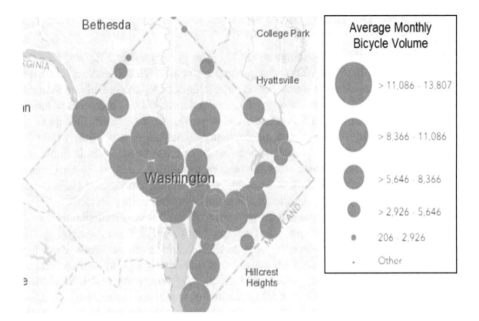

Fig. 4 Traffic monitor stations and average monthly bicycle volume. (DDOT, 2019)

north, east, and south zones where there is a high rate of apartment dwellings. In those areas, the residential zones are flat and have space to deploy micromobility services.

7 Results

A synthesis of the land use, demographic, traffic, ride path capacity, park-and-ride facility locations, public transit access, and demand factors for Washington, D.C., suggests that micromobility companies can focus, in the short term, on deploying most of their services in the areas around downtown, West, Fox hall Crescent, and Southwest Washington. This distribution will allow the companies to reach the highest number of potential customers and grow the market. A higher service rate in those areas would increase customer satisfaction and spur adoption. Based on the geospatial analysis, the areas that currently have a low level of biking facilities are Fort Totten, Bright wood, Chevy Chase, Tenley town, Hill East, and Carver/Langston. So, micromobility companies should work with the city to develop long-term strategies for service expansion in those areas. A phased deployment plan will minimize initial costs at the time when companies need to demonstrate viability.

Currently, the main cost drivers of dockless micromobility deployments are the man hours needed to charge and rebalance vehicles, modulate the fleet size, and

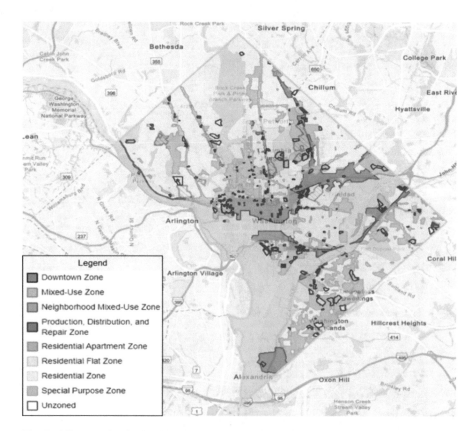

Fig. 5 Official zoning for Washington D.C. (DCOZ, 2019)

maintain support at all hours. Agencies can first focus longer term investments to build more bike lanes in the southern portion of Washington, D.C., where there are low elevations, high traffic volumes, and low incomes. Similarly, investing and creating more opportunities for deployments in the north zone where traffic is moderate, bike trails are few, and terrain is flat will help to enhance both mobility and accessibility. Adding micromobility services in the east where the rate of park-and-ride usage is high will help longer-distance commuters transfer to public transportation, decrease pollution, and promote greater equity for the area's low-income residents.

8   Conclusions

This study reviewed the different dimensions of micromobility services, factors that could influence demand, and best practices for extending and sustaining

Fig. 6 Park-and-ride locations. (DDOT, 2019)

Fig. 7 Trip Distance. (Newebcreations, 2019)

deployments. The authors used publicly available data from Washington, D.C., as a case study to examine the potential opportunities for growing micromobility services. The behavioral and geospatial datasets mined included the distributions of park-and-ride facilities, socioeconomic data, topography, land use, and current bicycle traffic. The best practices and analytical strategy of where and when to invest will help micromobility companies increase their profitability while helping to improve transportation sustainability. Increasing the rate of micromobility services by addressing their potential drawbacks can have positive consequences such as

Fig. 8 Bike lanes in Washington, D.C. (DDOT, 2019)

the reduction of traffic congestion, the reduction of pollution, and the enhancement of transportation equity. This work is the first exploratory stage of research in preparation for a more rigorous geospatial analysis of micromobility deployments as a function of land-use, socio-economic demographics, road networks, and traffic.

References

1. NGA, *Governors Staying Ahead of the Transportation Innovation Curve: A Policy Roadmap for States* (National Governors Association (NGA), Washington, DC, 2018)
2. NACTO, *Guidelines for Regulating Shared Mobility: Version 2* (National Association of City Transportation Officials (NACTO), Washington, DC, 2019)

3. F. Soriguera, V. Casado, E. Jiménez, A simulation model for public bike-sharing systems. Transp. Res. Procedia **33**, 139–146 (2018)
4. P. DeMaio, Bike-sharing: History, impacts, models of provision and future. J. Public Transp. **12**(4), 41–56 (2009)
5. S. Shaheen, S. Guzman, Worldwide bike sharing. Access Mag. **Fall 2011**(39), 22–27 (2011)
6. J. Larsen, Bike-sharing programs hit the streets in over 500 cities worldwide [Online] (2013). Available: http://www.earth-policy.org/mobile/releases/update112. Accessed 28 Oct 2019
7. B. Jiang, S. Liang, Z.-R. Peng, H. Cong, M. Levy, Q. Cheng, T. Wang, J.V. Remais, Transport and public health in China: The road to a healthy future. Lancet **390**(10104), 1781–1791 (2017)
8. Y. Zhang, T. Thomas, M. Brussel, M. Maarseveen, Exploring the impact of built environment factors on the use of public bikes at bike stations: Case study in Zhongshan, China. J. Transp. Geogr. **58**, 59–70 (2017)
9. NATCO, Bike share in the U.S.: 2017 (National Association of City Transportation Officials (NACTO), 2017)
10. NACTO, Shared micromobility in the U.S.: 2018 (National Association of City Transportation Officials (NACTO), 2018)
11. S. Gössling, Integrating e-scooters in urban transportation: Problems, policies, and the prospect of system change. Transp. Res. Part D: Transp. Environ. **79**, 102230 (2020)
12. D. Zipper, The Frenzied era of E-scooters is over [Online] (2020, February 27). Available: https://slate.com/business/2020/02/e-scooters-regulations-bird-lyft-lime-cities.html
13. D.M. Luke Peters, The death and rebirth of bikesharing in Seattle: Implications for policy and system design. Transp. Res. A Policy Pract. **130**, 208–226 (2019)
14. Z.Z.J.W.G.B. Hannah Younes, Comparing the temporal determinants of dockless scooter-share and station-based bike-share in Washington, D.C. Transp. Res. A Policy Pract. **134**, 308–320 (2020)
15. D. van Lierop, D. Ettema, Z. Chen, Dockless bike-sharing systems: What are the implications? Transp. Rev. **40**(3), 333–353 (2020)
16. G. O. JAPAN, How the Japanese government's new "sandbox" program is testing innovations in mobility and technology [Online] (2020, February 11). Available: https://hbr.org/sponsored/2020/02/how-the-japanese-governments-new-sandbox-program-is-testing-innovations-in-mobility-and-technology
17. E. Fishman, C. Cherry, E-bikes in the mainstream: Reviewing a decade of research. Transp. Rev. **36**(1), 72–91 (2015)
18. K. Lovejoy, S. Handy, Developments in bicycle equipment and its role in promoting cycling as a travel mode, in *City Cycling*, ed. by J. Pucher, R. Buehler, (The MIT Press, Cambridge, MA, 2012), pp. 75–104
19. MOVMI, The micromobility conferences, California 2019 [Online] (Movmi Shared Transportation Services Inc, 2019, October). Available: http://movmi.net/Micromobility-conference-2019/. Accessed 27 Oct 2019
20. E. Hannon, S. Knupfer, S. Stern, B. Sumers, J.T. Nijssen, *An Integrated Perspective on the Future of Mobility, Part 3: Setting the Direction Towards Seamless Mobility* (McKinsey & Company, 2019)
21. I.T. Forum, *Safe Micromobility* (International Transport Forum, Paris, 2020)
22. R. Zarif, D.M. Pankratz, B. Kelman, Small is beautiful: Making Micromobility work for citizens, cities, and service providers [Online] (2019). Available: https://www2.deloitte.com/us/en/insights/focus/future-of-mobility/Micromobility-is-the-future-of-urban-transportation.html. Accessed 27 Oct 2019
23. N. DuPuis, J. Griess, C. Klein, *Micromobility in Cities: A History and Policy Overview* (National League of Cities, Washington, DC, 2019)
24. N.A. Gallagher, *The Influence of Neighborhood Environment, Mobility Limitations, and Psychosocial Factors on Neighborhood Walking in Older Adults* (The University of Michigan, Michigan, 2010)
25. C. Nuworsoo, E. Cooper, K. Cushing, *Integration of Bicycling and Walking Facilities into the Infrastructure of Urban Communities* (Mineta Transportation Institute, San Jose, 2012)

26. A. Ajao, Electric scooters and micromobility: Here's everything you need to know [Online] (2019). Available: https://www.forbes.com/sites/adeyemiajao/2019/02/01/everything-you-want-to-know-about-scooters-and-Micromobility/#796971415de6
27. A. Fyhri, N. Fearnley, Effects of e-bikes on bicycle use and mode share. Transp. Res. Part D: Transp. Environ. **36**, 45–52 (2015)
28. S. Haustein, M. Møller, Age and attitude: Changes in cycling patterns of different e-bike user segments. Int. J. Sustain. Transp. **10**(9), 836–846 (2016)
29. L. Winslott Hiselius, Å. Svensson, Could the increased use of e-bikes (pedelecs) in Sweden contribute to a more sustainable transport system? in *9th International Conference on Environmental Engineering (ICEE)*, (2014)
30. B.E. Saelens, J.F. Sallis, L.D. Frank, Environmental correlates of walking and cycling: Findings from the transportation, urban design, and planning literatures. Ann. Behav. Med. **25**(2), 80–91 (2003)
31. R. Cervero, M. Duncan, Walking, bicycling, and urban landscapes: Evidence from the San Francisco Bay Area. Am. J. Public Health **93**(9), 1478–1483 (2003)
32. S.D. Lawson, B. Morris, Out of cars and onto bikes: What chance? Traffic Eng. Control **40**(5), 272–276 (1999)
33. Y. Shen, X. Zhang, J. Zhao, Understanding the usage of dockless bike sharing. Int. J. Sustain. Transp. **12**(9), 686–700 (2018)
34. J. Corcoran, T. Li, D. Rohde, E. Charles-Edwards, D. Mateo-Babiano, Spatio-temporal patterns of a public bicycle sharing program: The effect of weather and calendar events. J. Transp. Geogr. **41**, 292–305 (2014)
35. W. El-Assi, M.S. Mahmoud, K.N. Habib, Effects of built environment and weather on bike sharing demand: A station level analysis of commercial bike sharing in Toronto. Transportation **44**(3), 589–613 (2017)
36. C.M.D. Chardon, G. Caruso, I. Thomas, Bicycle sharing system 'success' determinants. Transp. Res. A Policy Pract. **100**, 202–214 (2017)
37. X. Wang, G. Lindsey, J.E. Schoner, A. Harrison, Modeling bike share station activity: Effects of nearby businesses and jobs on trips to and from stations. J. Urban Plann. Dev. **142**(1), 04015001 (2015)
38. A. Faghih-Imani, N. Eluru, Examining the impact of sample size in the analysis of bicycle-sharing systems. Transportmetrica A Transp. Sci. **13**(2), 139–161 (2017)
39. D.G. Chatman, Residential choice, the built environment, and nonwork travel: Evidence using new data and methods. Environ. Plann. A Econ. Space **41**(5), 1072–1089 (2009)
40. B. Caulfield, M. O'Mahony, W. Brazil, P. Weldon, Examining usage patterns of a bike-sharing scheme in a medium sized city. Transp. Res. A Policy Pract. **100**, 152–161 (2017)
41. E. Fishman, Bikeshare: A review of recent literature. Transp. Rev. **36**(1), 92–113 (2016)
42. Y. Zhao, L. Chen, C. Teng, S. Li, G. Pan, Green bicycling: A smartphone-based public bicycle sharing system for healthy life, in *Green Computing and Communications (GreenCom), 2013 IEEE and Internet of Things (iThings/CPSCom), IEEE International Conference on and IEEE Cyber, Physical and Social Computing*, (Atlanta, 2013)
43. D. Buck, R. Buehler, P. Happ, B. Rawls, P. Chung, N. Borecki, Are bikeshare users different from regular cyclists? A first look at short-term users, annual members, and area cyclists in the Washington, DC, region. Transp. Res. Rec. **2387**(1), 112–119 (2013)
44. S. Rosenbloom, E. Burns, Gender differences in commuter travel in Tucson: Implications for travel demand management programs. Transp. Res. Rec. **1404**, 82–90 (1993)
45. J. Wu, S. Wang, Y. Zhang, A. Zhang, C. Xia, Urban landscape as a spatial representation of land rent: A quantitative analysis. Comput. Environ. Urban. Syst. **74**, 62–73 (2019)
46. Capital Bikeshare, Find a station [Online] (2019). Available: https://secure.capitalbikeshare.com/map/. Accessed 26 Oct 2019
47. S. Bouton, E. Hannon, L. Haydamous, S. Ramanathan, B. Heid, S. Knupfer, T. Nauclér, F. Neuhaus, J.T. Nijssen, *Urban Commercial Transport and the Future of Mobility* (McKinsey & Company, 2017)

48. G. Chaffin, Dockless bikeshare helped grow the total shared bicycle trips in DC [Online] (2018). Available: https://ggwash.org/view/67638/dockless-bikeshare-helped-grow-the-total-shared-bicycle-trips-in-dc

49. Statistical Atlas, Age and sex in Washington, District of Columbia [Online] (2019). Available: https://statisticalatlas.com/place/District-of-Columbia/Washington/Age-and-Sex. Accessed 25 Oct 2019

50. Statistical Atlas, Household income in Washington, District of Columbia [Online] (2018). Available: https://statisticalatlas.com/place/District-of-Columbia/Washington/Household-Income. Accessed 25 Oct 2018

51. DDOT, Traffic monitoring stations [Online] (Department of Transportation (DDOT), Washington, DC, 2019). Available: https://opendata.dc.gov/datasets/a87c1b9a71e143a4914e3c384bda2d3a_92. Accessed 26 Oct 2019

52. DDOT, Park and ride points [Online] (Department of Transportation (DDOT), Washington, DC, 2019). Available: https://opendata.dc.gov/datasets/park-and-ride-points. Accessed 26 Oct 2019

53. DDOT, Open data DC [Online] (District of Columbia Department of Transportation (DDOT), 2019). Available: https://opendata.dc.gov/datasets/bicycle-lanes. Accessed 26 Oct 2019

54. Newebcreations, District of Columbia bicycle map [Online] (2019). Available: https://maps-washington-dc.com/dc-bike-map#&gid=1&pid=1. Accessed 26 Oct 2019

55. DCOZ, Official zoning map [Online] (2019). Available: http://maps.dcoz.dc.gov/zr16/. Accessed 27 Oct 2019

56. Y. Zhang, M. Brussel, T. Thomas, M.V. Maarseveen, Mining bike-sharing travel behavior data: An investigation into trip chains and transition activities. Comput. Environ. Urban. Syst. **69**, 39–50 (2018)

Real-Time Spatiotemporal Air Pollution Prediction with Deep Convolutional LSTM Through Satellite Image Analysis

Pratyush Muthukumar, Emmanuel Cocom, Jeanne Holm, Dawn Comer, Anthony Lyons, Irene Burga, Christa Hasenkopf, and Mohammad Pourhomayoun

1 Introduction

Air pollution is a silent killer. It is responsible for the early deaths of seven million people every year, around 600,000 of whom are children [1]. It means that every 5 s, somebody around the world dies prematurely from the effects of air pollution [1]. With the percentage of the global population living in urban areas projected to increase from 54% in 2015 to 68% in 2050 and in the United States up to 89%, the prevention of a significant increase in air pollution–related loss of life requires comprehensive mitigation strategies, as well as forecast systems, to limit and reduce the exposure to harmful urban air [2, 3].

In this chapter, we developed predictive models based on advanced machine learning algorithms to discover and classify patterns in urban air quality and predict air pollution in different areas. In designing the predictive models, we considered both temporal and spatial patterns in the data. The air quality data at a specific location is highly correlated to the past data at that location (temporal correlation) [4]. Also, it is highly correlated to air quality of adjacent areas because the air

P. Muthukumar (✉) · E. Cocom · M. Pourhomayoun
Department of Computer Science, California State University, Los Angeles, CA, USA
e-mail: pmuthuk2@calstatela.edu; ecocom@calstatela.edu; mpourho@calstatela.edu

J. Holm · D. Comer · A. Lyons · I. Burga
City of Los Angeles, Los Angeles, CA, USA
e-mail: jeanne.holm@lacity.org; dawn.comer@lacity.org; anthony.lyons@lacity.org; irene.burga@lacity.org

C. Hasenkopf
OpenAQ, Washington, DC, USA
e-mail: christa@openaq.org

© Springer Nature Switzerland AG 2021
R. Stahlbock et al. (eds.), *Advances in Data Science and Information Engineering*,
Transactions on Computational Science and Computational Intelligence,
https://doi.org/10.1007/978-3-030-71704-9_20

pollutants can transmit through the atmosphere from one area to other areas around it (spatial correlation).

The spatiotemporal machine learning problem has been studied recently in various contexts including weather-related applications [5, 6, 7]. Much of the previous research focuses on increasing either spatial or temporal correlations, but it is considerably more difficult to introduce an accurate measurement of both spatial correlation and temporal correlation in a highly complex model.

The convolutional long short-term memory (ConvLSTM) model is a complex machine learning model used for inputs that comprise sets of frames of data, which allows for unaltered video inputs. The ConvLSTM model is a variation of the traditional long short-term memory network, a time-series recurrent neural network. Recall the fully connected LSTM (FC-LSTM) structure with characteristic input, forget, and memory gates:

$$
\begin{aligned}
i_t &= \sigma \left(W_{xi} x_t + W_{hi} h_{t-1} + W_{ci} \circ c_{t-1} + b_i \right) \\
i_f &= \sigma \left(W_{xf} x_t + W_{hf} h_{t-1} + W_{cf} \circ c_{t-1} + b_f \right) \\
c_t &= f_t \circ c_{t-1} + i_1 \circ \tanh \left(W_{xc} x_t + W_{hc} h_{t-1} + b_c \right) \\
o_t &= \sigma \left(W_{xo} x_t + W_{ho} x_{h-1} + W_{co} \circ c_t + b_o \right) \\
h_t &= o_t \circ \tanh \left(c_t \right),
\end{aligned}
$$

where $\circ$ denotes the Hadamard product [8].

Notice that in traditional LSTM models, the input vector must be a one-dimensional vector, where each element of the vector is fed into different states of the model. Accordingly, the output of the LSTM is also a one-dimensional vector parameterized by time. Traditional LSTMs do not account for the spatial correlations in an environment, and in order to introduce these spatial correlations, we induced convolution throughout each of the gates and cell/hidden states.

There are two options to induce convolution in a traditional ConvLSTM, depending on the sequence of convolution. Either the convolution can interme-diately occur on the input tensor before being fed into a traditional FC-LSTM, or the convolution can accept input tensors as input and perform convolution operations when calculating the values of the cell and hidden states throughout the model. The first option refers to performing the convolution operation on a multidimensional input, transforming it into a single dimensional input, prior to its input in the FC-LSTM structure. Research in this field denotes the first option as a convolutional neural network – long short-term memory (CNN-LSTM), as the individual structures are not changed, and the second option as a convolutional LSTM (ConvLSTM), as the traditional LSTM's cell operations must be changed, referring to the addition of the convolution operation in each of the states of the FC-LSTM [9, 10, 11, 12]. If we allow convolution to replace the Hadamard products in the traditional LSTM structure, we can now input multidimensional input tensors to the structure and receive a multidimensional output tensor as output. The key equations for the ConvLSTM then are defined as

$$i_t = \sigma \left(W_{xi} x_t + W_{hi} h_{t-1} + W_{ci} * c_{t-1} + b_i \right)$$
$$i_f = \sigma \left(W_{xf} x_t + W_{hf} h_{t-1} + W_{cf} * c_{t-1} + b_f \right)$$
$$c_t = f_t * c_{t-1} + i_1 * \tanh \left(W_{xc} x_t + W_{hc} h_{t-1} + b_c \right)$$
$$o_t = \sigma \left(W_{xo} x_t + W_{ho} x_{h-1} + W_{co} * c_t + b_o \right)$$
$$h_t = o_t * \tanh \left(c_t \right),$$

where $*$ denotes the convolution operation [13].

2 Methods

2.1 Dataset

Our input data was sourced from the U.S. Geological Survey's EarthExplorer database records of the Sentinel 2 satellite, launched on June 23, 2015 [14]. The Sentinel 2 satellite launched by the European Space Agency in March 2015 images and records atmospheric and terrain data through 13 spectral bands based on the wavelength of the emitted light. Sentinel 2 operates along a 290-km orbital swath [15]. Of these 13 bands, we selected two spectral bands that provided data on the air pollution in the greater Los Angeles area. The first band with a 442.7 nm central wavelength measured the coastal aerosol levels allowing us to view fine particulate matters including dust, smoke, and general particulate matter. We also chose a finer spectral band at 945.1 nm central wavelength to measure specifically the Nitrogen Dioxide levels in the atmosphere. A sample input is shown in Fig. 1.

Fig. 1 Sample raw data. (Source: USGS EarthExplorer database of satellite imagery of Los Angeles taken on April 29, 2019 by ESA's Sentinel 2 satellite)

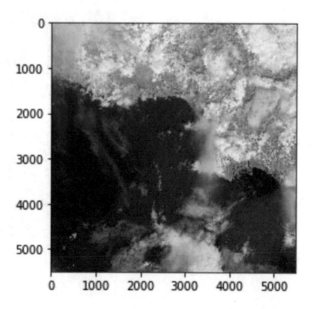

The blue structures correspond to strictly nitrogen dioxide air pollution, while the white, cloud-like structures correspond to general particulate matter.

2.2 Data Preprocessing

In order to feed our data into the ConvLSTM model, we transformed 225 GeoTIFF high-resolution images into a 5D tensor. The 225 GeoTIFF images corresponded to 1642 days of data, with each image being the imaging of the T11SLT 100 km 100 km2 tile, which is roughly the western 75% of Los Angeles County. Each of the 225 GeoTIFF images was taken 2 days apart, the orbit time of the Sentinel 2 satellite.

For our ConvLSTM input, we decided to focus only on the blue cloud-like structures that imaged nitrogen dioxide. To do so, we first imported our dataset into a JPEG format through the OpenCV python package. Since the size of the data is very large, we resampled the data into two batches, each with different lower resolutions. One resampling was a 400px by 400px JPEG image dataset of all 225 images, and the other was a 40px by 40px JPEG image dataset of all 225 images. For both of the lower resolution resamplings, we applied a mask that only showed the (0,60,60) to (225,255,255) RGB color scheme, roughly corresponding to only light blue shades. We set all other non-light blue shades to the (0,0,0) RGB color scheme or black. An example masked image is shown in Fig. 2.

Fig. 2 Masked image

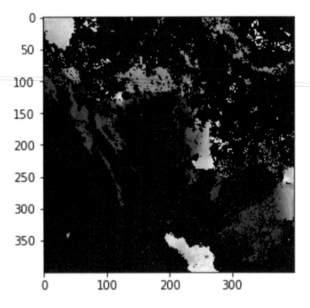

In this way, we had created four resamplings of the original image dataset: two masked 40px by 40px datasets and two masked 400px by 400px datasets. For two of the datasets with a mask, one 40px by 40px and one 400px by 400px, we binarized the data, so that all light blue pixels became 1, and all black pixels became 0. Thus, two of our datasets were binary arrays. For the remaining two image datasets, we kept the masked light blue and black color scheme intact. In this way, we had two datasets of masked color images (RGB) and two datasets of bit arrays. The color image datasets allowed us to use a ReLU activation function in the output layer, while the binary array datasets allowed us to use a sigmoid activation function.

After resampling the data into various groups based on binary image, masked image, or image resolution criterion, we began parsing our 225 image datasets into readable formats for our ConvLSTM models. In the current form, our image dataset was in the form of (225,400,400,3) for 400px by 400px masked images, (225,40,40,3) for 40px by 40px masked images, (225,400,400,1) for 400px by 400px binary images, and (225,40,40,1) for 40px by 40px binary images. We then took all of our datasets and batched every five image frame as one sample.

An overview of the input data preprocessing is described in Fig. 3.

2.3 Input Data Labeling

An important determinant of the success and practicality of a time series model is based on the selection of the training and testing sets from a continuous stream of time-indexed data. In our model, the label for a training set input is the frame that directly follows the current frame. In this way, the model is picking up on the pattern to base a current input image on the goal of accurately predicting a future image, which in this case means predicting the next 10 days in the future.

In our model, we use five input frames as a data sample to predict five future frames. In fact, since each frame was separated by 2 days, we use 10 days of previous satellite imagery data in order to predict 10 days in the future of satellite imagery data. For the input labels, we first shifted over all data by five frames, with the first 5 frames corresponding to the sixth through tenth frames and the last 5 frames corresponding to the 226th-230th frame. The 226th through 230th frames were filled with an averaging of the 30 days prior. Thus, we encoded sequential samples in training/testing inputs indexed by time and training/testing labels indexed by time, but did not overlap the training/testing inputs and the training/testing labels. We were able to perform a continuous prediction from the model's learned patterns. A visual representation of our labeling is shown in Fig. 4:

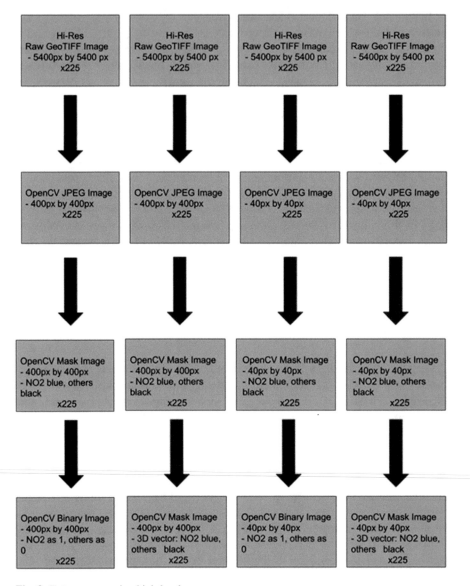

Fig. 3 Data preprocessing high-level process

3 Results

3.1 Aerial Satellite Image Model

We developed ConvLSTM models that learned correlations through spatial and temporal bounds. We predicted 5 frames in the future from 5 frames in the past, with

Fig. 4 Training/validation input and label creation process for sequential training and testing data without overlapping inputs and labels

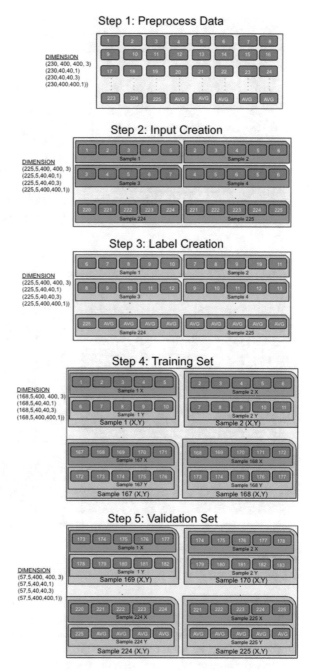

each frame being 2 days apart. For Figs. 5, 6 and 7, we visualized the prediction of 10 days worth of nitrogen dioxide levels through data from the previous 10 days.

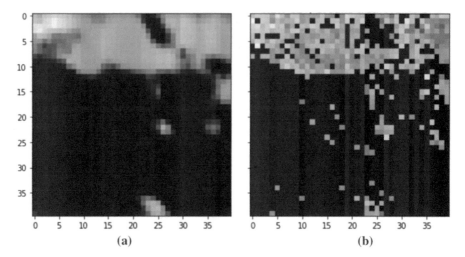

Fig. 5 Frame 1 prediction: second day in the future prediction of nitrogen dioxide air pollution in Los Angeles County from previous 10 days of data: (**a**) Prediction and (**b**) ground truth

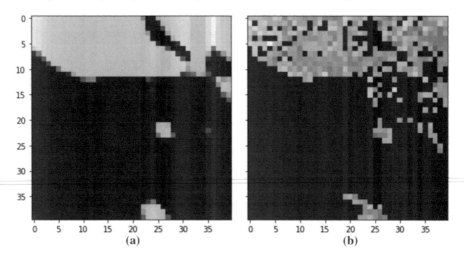

Fig. 6 Frame 2 prediction: tenth day in the future prediction of nitrogen dioxide air pollution in Los Angeles County from previous 10 days of data: (**a**) Prediction and (**b**) ground truth

Figures 5, 6, and 7 display the predictions on the masked 40px by 40px resampled dataset. We have recolored the figures to accentuate the nature of the prediction; however, its original form remains an RGB image with a light blue and black color scheme. We used our data to essentially predict early March 2020's nitrogen dioxide air pollution levels from late February 2020's nitrogen dioxide pollution data. For Figs. 5, 6 and 7, the left column displays our predictions, while the right column displays the ground truth.

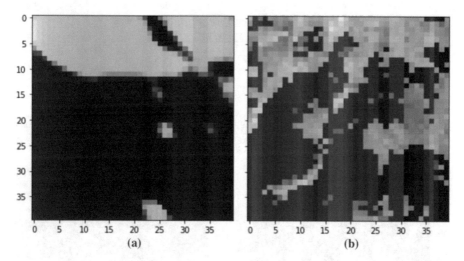

Fig. 7 Last frame prediction: tenth day in the future prediction of nitrogen dioxide air pollution in Los Angeles County from previous 10 days of data: (**a**) Prediction and (**b**) ground truth

As expected, the prediction of the near future is more accurate than farther in the future. In the context of the problem, this is because the nitrogen dioxide levels of tomorrow are more correlated to the past week's nitrogen dioxide levels than the future week's nitrogen dioxide levels. As we move to 10+ days in the future, the data from the previous 10 days are no longer strongly correlated to accurately model nitrogen dioxide particulate matter in the greater Los Angeles area.

3.2 Error Analysis

We used the structural similarity index measurement (SSIM) as an error measure to quantify the model's accuracy. SSIM is a very popular tool used for accurate assessment of weather prediction algorithms due to its ability to judge models on the similarities in the structure of a prediction [16]. Possible outputs of the SSIM metric range from 0 to 1, with 0 being completely dissimilar and 1 being exactly identical. With true pixel value p and predicted value $p^{\wedge}$, the SSIM of a single pixel is

$$\text{SSIM}\left(p, \hat{p}\right) = \frac{2\mu_p \mu_{\hat{p}} + c_1}{\left(\mu_p^2 + \mu_{\hat{p}}^2 + c_1\right)\sigma_p^2 + \sigma_{\hat{p}}^2 + c_2},$$

where $c1$ and $c2$ are constants relating to the relative noise of an image [17]. For a complete image I, the SSIM is [17].

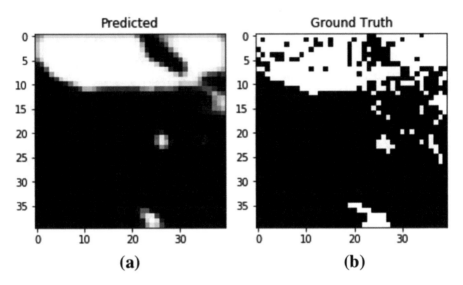

Fig. 8 First frame SSIM measurement: 77% structurally similar from SSIM error metric of second day in the future prediction of nitrogen dioxide air pollution in Los Angeles County from previous 10 days of binary input data: (**a**) Prediction and (**b**) ground truth

Table 1 SSIM values for first sample (first 5 frames): structural similarity percentages of 10 days in the future nitrogen dioxide predictions in LA County from previous 10 days of data

Sample 1	
	SSIM
Frame 1	0.77
Frame 2	0.70
Frame 3	0.63
Frame 4	0.56
Frame 5	0.51

$$SSIM(I) = \sum_{p \in I} \frac{2\mu_p \mu_{\hat{p}} + c_1}{\left(\mu_p^2 + \mu_{\hat{p}}^2 + c_1\right)\sigma_p^2 + \sigma_{\hat{p}}^2 + c_2}$$

We then ran our SSIM analysis metrics on our binary input and masked images with our first frame having the most accurate SSIM value in the two samples or ten frames of data we predicted. The future frames fell in accuracy by around 10–15% due to the ground truth being less correlated to the previous 10 days of nitrogen dioxide levels as compared to the first few frames of the ground truth (Fig. 8, Table 1).

4 Conclusion

In this chapter, we developed predictive models based on advanced machine learning algorithms to discover and classify patterns in urban air quality, and specifically predict nitrogen dioxide in greater Los Angeles area. In designing the predictive models, we took into account both temporal and spatial patterns in the data (i.e., the air quality correlation to the past and future data as well as the correlation to the adjacent locations).

To discover and learn both temporal and spatial patterns, we developed a convolutional long short-term memory (ConvLSTM) model, which is a complex machine learning approach used for inputs that comprise of sets of frames of multi-dimensional data. Our model was able to use the nitrogen ioxide air pollution data in the greater Los Angeles area of 10 days in order to predict nitrogen dioxide pollution anywhere in Los Angeles 10 days in the future.

This work can be used to alert researchers on the patterns of movement for nitrogen dioxide at any given time period in the next 5 years within the greater Los Angeles area.

5 Future Work

In the future, we would also like to analyze ground-based sensors with a wide variety of pollutant and atmospheric sensors including PM2.5, carbon monoxide, ozone, temperature, and wind speed.

This study could expand to larger areas past Los Angeles county and include satellite image data and ground-based sensor data in to order include more information when predicting air pollution.

Bibliography

1. With a premature death every five seconds, air pollution is violation of human rights, says un expert – united nations sustainable development. Website. https://www.un.org/sustainabledevelopment/
2. Earl Swigert. Unicef: An urban world. Website. https://www.unicef.org/sowc2012/urbanmap
3. 2018 revision of world urbanization prospects — multimedia library – united nations department of economic and social affairs. Website. https://www.un.org/development/desa/publications/2018-revision-of-world-urbanization-prospects.html.
4. N. Künzli, M. Jerrett, W.J. Mack, B. Beckerman, L. LaBree, F. Gilliland, D. Thomas, J. Peters, H.N. Hodis, Ambient air pollution and atherosclerosis in los Angeles. Environ. Health Perspect. **113**(2), 201–206 (2005)
5. Y. Lin, N. Mago, Y. Gao, Y. Li, Y.-Y. Chiang, C. Shahabi, J. L. Ambite. Exploiting spatiotemporal patterns for accurate air quality forecasting using deep learning. in *Proceedings of the 26th ACM SIGSPATIAL International Conference on Advances in Geographic Information Systems*, (2018), pp. 359–368.

6. C.-J. Huang, P.-H. Kuo, A deep cnn-lstm model for particulate matter (pm2. 5) forecasting in smart cities. Sensors **18**(7), 2220 (2018)
7. S. Roy, Y. Wan, C. Taylor, C. Wanke. A stochastic net- work model for uncertain spatiotemporal weather impact at the strategic time horizon. in *10th AIAA Aviation Technology, Integration, and Operations (ATIO) Conference*, (2010), p. 9348
8. S. Hochreiter, J. Schmidhuber, Long short-term memory. Neural Comput. **9**(8), 1735–1780 (1997)
9. S. Kim, J.-S. Kang, M. Lee, S.-K. Song. Deeptc: Con- vlstm network for trajectory prediction of tropical cyclone using spatiotemporal atmospheric simulation data, (2018)
10. R. C. Nascimento, Y. M. Souto, E. Ogasawara, F. Porto, E. Bezerra. Stconvs2s: Spatiotemporal convolutional sequence to sequence network for weather forecasting. *arXiv preprint arXiv:1912.00134*, (2019)
11. X. Shi, Z. Chen, H. Wang, D.-Y. Yeung, W.-K. Wong, W.-C. Woo. Convolutional lstm network: A machine learning approach for precipitation nowcasting. in *Advances in Neural Information Processing Systems*, (2015), pp. 802–810
12. X. Shi, Z. Gao, L. Lausen, H. Wang, D.-Y. Yeung, W.-K. Wong, W.-C. Woo. Deep learning for precipitation nowcasting: A bench- mark and a new model. in *Advances in Neural Information Processing Systems*, (2017), pp. 5617–5627
13. Y. Liu, H. Zheng, X. Feng, Z. Chen. Short-term traffic flow prediction with conv-lstm. in *2017 9th International Conference on Wireless Communications and Signal Processing (WCSP)*, (IEEE, 2017), pp. 1–6
14. USGS. Usgs earthexplorer satellite imagery database. Website. https://earthexplorer.usgs.gov/
15. M. Drusch, U. Del Bello, Ś. Carlier, O. Colin, V.-i. Fernandez, F. Gascon, B. Hoersch, C. Isola, P. Laberinti, P. Martimort, et al., Sentinel-2: Esa's optical high-resolution mission for gmes operational services. Remote Sens. Environ. **120**, 25–36 (2012)
16. B. Klein, L. Wolf, Y. Afek. A dynamic convolutional layer for short range weather prediction. in *Proceedings of the IEEE Conference on Computer Vision and Pattern Recognition*, (2015), pp. 4840–4848.
17. M.P. Sampat, Z. Wang, S. Gupta, A.C. Bovik, M.K. Markey, Complex wavelet structural similarity: A new image similarity index. IEEE Trans. Image Process. **18**(11), 2385–2401 (2009)

Performance Analysis of Deep Neural Maps

Boren Zheng and Lutz Hamel

1 Introduction

Deep neural maps (DNMs) [20] are unsupervised learning and visualization methods that combine autoencoders with self-organizing maps. An autoencoder (AE) is a deep artificial neural network that is widely used for dimensionality reduction and feature extraction in machine learning tasks [14]. The self-organizing map (SOM) is an artificial neural network designed for unsupervised learning [16]. It is often used for clustering and the representation of high-dimensional data on a 2D grid. Deep neural maps have shown improvements in performance compared to standalone self-organizing maps when considering clustering tasks [20, 22].

In diverse fields such as genomic data clustering [21] and cluster analysis of massive astronomical data [15], self-organizing maps are a good clustering approach since they not only accomplish the clustering task but also provide an accessible, visual clustering representation. However, because both genomic data and astronomical data are extremely high dimensional, convergence behaviors of SOMs tend to be very slow and erratic. In deep neural maps, an autoencoder is used to reduce the dimensionality of the original data as a preprocessing step before the training of the underlying self-organizing map begins thus improving the convergence behavior and speed of the map.

In all studies we are aware of, e.g. [4, 5, 20, 22], only consider a single autoencoder architecture as part of their deep neural maps. Here, we investigate five types of autoencoders as part of our deep neural maps using three different data sets illuminating the performance characteristics of the various autoencoders. The five autoencoder architectures we investigate are (1) basic, (2) sparse, (3) contractive,

B. Zheng · L. Hamel (✉)
Department of Computer Science and Statistics, University of Rhode Island, Kingston, RI, USA
e-mail: boren_zheng@uri.edu; lutzhamel@uri.edu

© Springer Nature Switzerland AG 2021
R. Stahlbock et al. (eds.), *Advances in Data Science and Information Engineering*,
Transactions on Computational Science and Computational Intelligence,
https://doi.org/10.1007/978-3-030-71704-9_21

(4) denoising, and (5) convolutional. Here autoencoder architectures (2), (3), and (4) can be considered regularized versions of the basic autoencoder architecture (1). Architecture (5), the convolutional autoencoder, is based on convolutional deep neural networks commonly used for image processing [18]. The data sets we used in our analysis are the digits data set derived from the MNIST database [19], the landsat data set [24], and a synthetic data set that contains 16 well-defined clusters in 64-dimensional space.

We show that DNMs improve the performance of standalone SOMs along two dimensions: (1) DNMs improve convergence behavior by removing noisy/superfluous dimensions from the input data. The improved convergence behavior can be observed by superior cluster homogeneity and convergence accuracy scores. (2) DNMs train faster due to the fact that the cluster detection part of the DNM deals with a lower-dimensional latent space.

It is perhaps a surprise that for nonimage data a DNM with a basic autoencoder outperforms all others. For image data we found that a DNM with a contractive autoencoder performs best and not a DNM with a convolutional autoencoder as one would expect.

The remaining sections of this paper are organized as follows. We give brief overviews of self-organizing maps, autoencoders, and the design of our deep neural maps in Sects. 2, 3, and 4, respectively. In Sect. 5, we detail our experiments, describe our data sets, and explain the evaluation methods we used. We discuss our results in Sect. 6. Finally, We summarize and propose future work in Sect. 7.

2   Self-Organizing Maps

Self-organizing maps were introduced by Kohonen in the 1980s as a way to visualize high-dimensional data on a 2D grid [16]. The 2D grid consists of high-dimensional neurons where the dimensionality of each neuron matches the dimensionality of the training data. Figure 1 illustrates the training process of a SOM. In the initial map, the neurons are initialized with small random values, and as the training data is repeatedly applied to the map, the neurons are starting to take on the form of the training data. What is particularly interesting is that certain regions of the map become sensitized to certain traits in the training data. Figure 4 shows typical 2D starburst visualizations of the final neuron map. The starbursts represent clusters in the high-dimensional training data space.

The basic SOM training algorithm can be summarized as follows [11]:

1. *Initialization*: initialize each neuron weight vector $\mathbf{m}_i$ with small random values.
2. *Selection step*: select a training data vector $\mathbf{x}_k$ from the training data.
3. *Competitive step*: find the best matching neuron $\mathbf{m}_c$ based on the Euclidean distance between the training data vector $\mathbf{x}_k$ and the neurons $\mathbf{m}_i$ on the map:

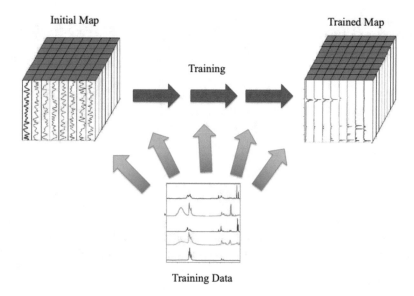

Fig. 1 Training a SOM

$$c = \arg\min_{i}(\|\mathbf{m}_i - \mathbf{x}_k\|). \tag{1}$$

4. *Update step*: update the winning neuron's $\mathbf{m}_c$ neighborhood using the following rule:

$$\mathbf{m}_i \leftarrow \mathbf{m}_i - \eta(\mathbf{m}_i - \mathbf{x}_k)h(c, i) \tag{2}$$

where $\eta(\mathbf{m}_i - \mathbf{x}_k)$ denotes the difference between a neuron and the training instance scaled by the learning rate $0 < \eta < 1$, $h(c, i)$ denotes the following loss function:

$$h(c, i) = \begin{cases} 1 & \text{if } i \in \Gamma(c), \\ 0 & \text{otherwise}, \end{cases} \tag{3}$$

where $\Gamma(c)$ is the neighborhood of the best matching neuron m_c.

Repeat steps 2, 3, and 4 until the map has converged.

3 Autoencoders

Autoencoders are deep neural networks that conceptually consist of three parts: (a) the encoder, (b) the neurons representing the latent space encoding the com-

Fig. 2 The basic
autoencoder architecture

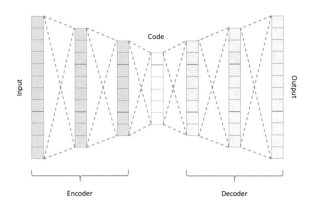

Code

Input

Output

Encoder

Decoder

pressed/encoded information, and (c) the decoder. Figure 2 shows the architecture of a basic autoencoder. Each colored column represents a layer of neurons in this illustration. Observe that both the encoder and the decoder are deep multilayer neural networks. In this basic architecture, the encoder maps the input into the hidden layer representing the latent space, and the decoder reconstructs the input from this hidden layer representation. An autoencoder where the latent space is of lower dimensionality than the input space is called undercomplete and we call an autoencoder for which the converse is true overcomplete. Regularization can be used to prevent autoencoders from simply copying information from the input to the latent space without learning anything useful [8].

The Basic Autoencoder Let $\phi : X \rightarrow F$ represents the encoder part of a basic autoencoder (AE) as shown in Fig. 2 that maps the input space X into a latent space F. Also, let $\psi : F \rightarrow X$ be the decoder part of a basic autoencoder that maps the latent space F into the input space X. Training a basic autoencoder can now be understood as the optimization problem:

$$\arg\min_{\phi,\psi} L\big(\mathbf{x}, (\psi \circ \phi)\mathbf{x}\big), \forall \mathbf{x} \in X. \tag{4}$$

That is, we want to find neural networks ϕ and ψ for the en- and decoders, respectively, that minimize the loss L between the original input and the reconstructed input available as output from the decoder. Individual layers ϕ_k and ψ_l in the en- and decoder networks, respectively, can be written as the following equations:

$$\phi_k(\mathbf{h}_{k-1}) = \sigma(\mathbf{h}_{k-1} \bullet \mathbf{W}_k) = \mathbf{h}_k$$
$$\psi_l(\mathbf{h}_{l-1}) = \sigma(\mathbf{h}_{l-1} \bullet \mathbf{U}_l) = \mathbf{h}_l \tag{5}$$

where $\bullet$ is the dot product extended to matrices, σ is the activation function, and the matrices $\mathbf{W}_k$ and $\mathbf{U}_l$ are the weight matrices of the corresponding layers. For the input layer of the encoder network we then have

$$\phi_0(\mathbf{x}) = \sigma(\mathbf{x} \bullet \mathbf{W}_0) = \mathbf{h}_0 \tag{6}$$

with $\mathbf{x} \in X$ and $\mathbf{h}_0$ an intermediate network representation of the input. For the output layer p of the encoder network we have

$$\phi_p(\mathbf{h}_{p-1}) = \sigma(\mathbf{h}_{p-1} \bullet \mathbf{W}_p) = \mathbf{h}_p \tag{7}$$

where $\mathbf{h}_p \in F$ is the representation of some point $\mathbf{x} \in X$ in latent space F. That is,

$$\phi(\mathbf{x}) = (\phi_p \circ \ldots \circ \phi_1 \circ \phi_0)\mathbf{x} = \mathbf{h}_p. \tag{8}$$

Similarly for the decoder network we have

$$\psi(\mathbf{h}_p) = (\psi_q \circ \ldots \circ \psi_{p+2} \circ \psi_{p+1})\mathbf{h}_p \approx \mathbf{x}. \tag{9}$$

with $q > p$. This optimization problem can be solved using a deep neural network library such as Keras [1].

The Sparse Autoencoder A sparse autoencoder (SAE) only has a small number of nodes that are activated in the hidden code layer at any particular time [2]. The objective function of an SAE is the objective function of the basic AE plus a sparsity penalty term:

$$L\big(\mathbf{x}, (\psi \circ \phi)\mathbf{x}\big) + \Omega(\mathbf{h}) \tag{10}$$

where ϕ and ψ denote the encoder and decoder networks, respectively, and $\mathbf{h} = (h_1, h_2, \ldots, h_n)$ with n the dimensionality of the latent space is the output vector of output layer of the encoder network defined as $\phi(\mathbf{x}) = \mathbf{h}$. The sparsity penalty term is defined as

$$\Omega(\mathbf{h}) = \lambda \sum_i^n |h_i| \tag{11}$$

where λ is a hyperparameter of the autoencoder model [8].

The Denoising Autoencoder The denoising autoencoder (DAE) does not directly add a regularization term to the objective function in order to avoid overfitting but instead adds noise to the input signal before the input is applied to the encoder network. The trick is that the output of the decoder is compared to the original signal rather than the noisy input, and therefore the network has to learn to ignore noise thereby preventing it from overfitting.

Let $\hat{\mathbf{x}} = \omega(\mathbf{x})$ for all $\hat{\mathbf{x}}, \mathbf{x} \in X$. Here $\mathbf{x}$ denotes an original training instance, and $\hat{\mathbf{x}}$ represents the original instance with noise added by process ω. We can now rewrite our objective function for the denoising autoencoder:

$$L\big(\mathbf{x}, (\psi_{\theta'} \circ \phi_\theta)\omega(\mathbf{x})\big) \tag{12}$$

where θ and θ' are additional parameters on the encoder and decoder networks, respectively, that are trained to minimize the average reconstruction error over the training set [8, 25].

The Contractive Autoencoder The contractive autoencoder (CAE) adds a regularization term to the loss function of the basic autoencoder based on the Frobenius norm of the Jacobian matrix of the features of the latent space [23]. This regularization term will make the autoencoder more robust to perturbations of the input and is encouraged to contract the input neighborhood to a smaller output neighborhood [8]. The objective function for the CAE is

$$L\big(\mathbf{x}, (\psi \circ \phi)\mathbf{x}\big) + \lambda \|J_f(\mathbf{x})\|_F^2 \tag{13}$$

where the regularization term is defined as

$$\|J_f(\mathbf{x})\|_F^2 = \sum_{ij}^{n} \left(\frac{\partial h_j}{\partial x_i}\right)^2. \tag{14}$$

Additionally, as before $\mathbf{h} = (h_1, h_2, \ldots, h_n)$ with n the dimensionality of the latent space is the output vector of the hidden code layer defined as $\phi(\mathbf{x}) = \mathbf{h}$.

The Convolutional Autoencoder A convolutional autoencoder (ConvAE) is built with convolutional layers rather than fully connected layers and hence tend to be well suited for image data sets. Here, single encoder and decoder layers are defined, respectively, as follows [9]:

$$\begin{aligned} \phi_k(\mathbf{h}_{k-1}) &= \sigma(\mathbf{h}_{k-1} * \mathbf{W}_k) = \mathbf{h}_k \\ \psi_l(\mathbf{h}_{l-1}) &= \sigma(\mathbf{h}_{l-1} * \mathbf{U}_l) = \mathbf{h}_l \end{aligned} \tag{15}$$

With respect to the basic autoencoder in Sect. 3 above, the only thing that has changed is that the dot product operation has been replaced with a convolution operator $*$. Our remarks on network composition from above also apply here with

$$\begin{aligned} \phi(\mathbf{x}) &= (\phi_p \circ \ldots \circ \phi_1 \circ \phi_0)\mathbf{x} = \mathbf{h}_p, \\ \psi(\mathbf{h}_p) &= (\psi_q \circ \ldots \circ \psi_{p+2} \circ \psi_{p+1})\mathbf{h}_p \approx \mathbf{x} \end{aligned} \tag{16}$$

where $\mathbf{x} \in X$, $\mathbf{h}_p \in F$ and q is the number of layers in the autoencoder with $p < q$.

The underlying optimization problem is to minimize the mean squared error between the input and output over all samples [9]:

$$\underset{\phi, \psi}{\arg\min} \frac{1}{|X|} \sum_{\mathbf{x} \in X} \|\mathbf{x} - (\psi \circ \phi)\mathbf{x}\|^2. \tag{17}$$

where $\mathbf{x} \in X$ represents the set of training instances in input space and $|X|$ the number of training instances.

4   Deep Neural Maps

The general architecture of our deep neural maps (DNMs) is shown in Fig. 3.
It consists of an autoencoder and a self-organizing map. The idea is that the
autoencoder maps the input data into a latent space and the SOM is trained using this
latent space. Pesteie, Abolmaesumi, and Rohling have shown that this architecture
performs well in their experiments [20]. Similarly, Rajashekar [22] proposed an
autoencoder-based self-organizing map framework that uses a basic autoencoder
with two hidden layers.

Training of DNMs consists of two phases shown with the red arrows in Fig. 3.
Phase I consists of training the autoencoder which in turn consists of solving the
optimization problems discussed in the previous section. Training autoencoders is
fairly fast. It takes about 200 epochs of batch size 128 to fully train the autoencoders
discussed here. This is very fast compared to training a fully converged SOM which
typically takes in the order of tens of thousands of iterations. Furthermore, training
time of the autoencoder is amortized over the SOM model evaluation steps. Phase
II is the training of the SOM using the latent space mapping of the input data.

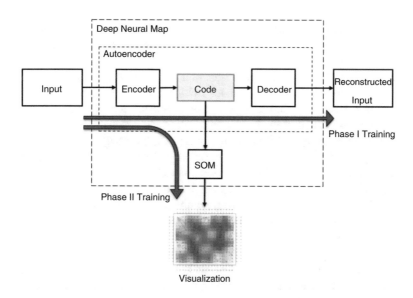

Fig. 3 Our deep neural map architecture

5   Experiments

We evaluate the performance of our deep neural maps with the various different encoders on three different real-world and synthetic data sets.

Data Sets For our experiments, we used the following data sets:

(1) The **dim064** [6, 7] is a 64-dimensional synthetic data set with 1024 observations that are well separated into 16 Gaussian clusters. We split the data set as follows: 60% data for training (614 instances) and 40% data for testing (410 instances). (2) The **landsat** satellite data set from the UCI machine learning repository [3] is a real-world data set with 6435 instances and 36 attributes and 6 classes. The data set consists of the multi-spectral values of pixels in a satellite image together with their classifications. The training set contains 4435 instances, and the test set contains 2000 instances, 500 of which were used for phase II training of the DNMs. (3) The **digits** database from the UCI machine learning repository [3] is derived from the MNIST database and represents hand-written digits as 8×8 pixel images [26]. The 1797 images are stored as vectors with 64 features. For our purposes we split this data set into a training set (1437 instances) and a test set (360 instances).

Model Evaluation and Selection As mentioned before, training a deep neural map consists of training two different models: one for the autoencoder and one for the self-organizing map.

For the autoencoders, our model selection criterion was the test loss error (or reconstruction error). We chose a model architecture and the number of epochs to train a model based on minimizing that loss. In this research we found that all our autoencoder models had properly converged after 200 epochs.

Once an autoencoder has been properly trained, we used its output to train the self-organizing map part of a deep neural map. Here we used the convergence accuracy [10] as a model selection criterion. For this research, we selected the model with the highest convergence accuracy. For the current work, we trained 40 models for each deep neural map with varying numbers of training iterations in order to compute the learning curve and select an appropriate model.

A fully trained deep neural network can then be used to produce a cluster representation of the input data on a 2D map. A typical DNM cluster presentation is shown in Fig. 4. It is a heat map where deep red colors represent cluster borders and yellow/white areas represent cluster centers. The starburst graphic overlay emphasizes the cluster structure of the data [12].

Remarks on Self-Organizing Map Architectures One of the most important hyper-parameters for SOMs is the size of the map. Here we use the following rule of thumb: *There should be as least as many neurons on the map as there are observations in the training data.* Since phase II training data consists of roughly 500 observations for all three experiments, we chose a map size of 25×20 for all experiments. Another hyper-parameters is the learning rate. Here we set the learning rate fairly aggressively at 0.6. Finally, the POPSOM package [13] uses a constant neighborhood which contracts over the duration of the training phase.

Remarks on Autoencoder Architectures Our basic autoencoder was imple-mented using a single fully connected layer as encoder and as decoder. We added an L1 regularizer to the basic AE in order to obtain the SAE. The CAE used the same architecture as the SAE except that we used a different penalty term. We implemented the penalty term of Eq. (13) as

$$\|J_f(\mathbf{x})\|_F^2 = \sum_{ij} \left(\frac{\partial h_j}{\partial x_i}\right)^2 = \sum_j [h_j(1 - h_j)]^2 \sum_i \left(\mathbf{W}_{ji}^T\right)^2 \tag{18}$$

where as before $\mathbf{x} \in X, \mathbf{h} = (h_1, \ldots, h_n) \in F$ is the input representation in latent space and $\mathbf{W}$ is the weight matrix of the encoder layer [17].

We set the noise factor to be 0.5 to create noisy input for the DAE. For the models of dim064 and landsat data sets, both the encoded layer and the decoded layer of the DAE are single fully connected layers. For the digits data set, we implemented the model as a Denoising Convolutional Autoencoder (DCAE). The encoder consists of three 2D convolutional layers followed by down-sampling layers (pooling size 2×2) and a flatten layer (encoded layer). The decoder consists of four 2D convolutional layers followed by three up-sampling layers (size 2×2); the last convolutional layer is the decoded layer.

We utilized 1D convolutional layers, 1D max-pooling layers, and 1D up-sampling layers to build the ConvAE models for the dim064 and the landsat data sets. For the model of the digits data set, the architecture of the ConvAE is the same as DCAE. However, it uses the original data as input rather than noisy data.

One more note, further dimension reduction was achieved by dropping columns in the latent space representation that were all zeros.

6   Results

We look at the performance of DNMs with different autoencoder architectures compared to standalone SOMs for each of our data sets. Our performance analysis uses five dimensions in order to describe the performances for both standalone SOMs and DNMs:

– **homog.**—Average homogeneity of the detected clusters. This is defined as

$$homog = \frac{1}{n} \sum_c l_c \tag{19}$$

Table 1 DNM performance with different AEs on the dim064 data set

	SOM	DNM(AE)	DNM(SAE)	DNM(CAE)	DNM(DAE)	DNM(ConvAE)
homog.	0.92	1.00	0.83	1.00	0.98	0.93
nclust	15	16	13	16	16	15
time (sec)	20.75	1.18	0.52	0.68	1.82	1.17
conv.	0.78	0.87	0.5	0.53	0.66	0.99
dim	64	11	7	9	12	8

where l_c is the number of majority label instances in cluster c and n is the total
number of observations in the training data set.

- **nclust**—the number of clusters detected
- **time**—phase II training time in CPU seconds (total training time for standalone
  SOM)
- **conv.**—phase II convergence accuracy (convergence accuracy for standalone
  SOM)
- **dim**—dimensions of the latent space (or in the case of the SOM it is the
  dimensionality of the original space)

Using these criteria, we validated the key advantages of deep neural maps over self-
organizing maps:

1. A better clustering behavior due to the removal of noisy and/or superfluous
   dimensions in the input data
2. Faster training times due to the lower dimensionality of the latent space

Performance Analysis for Dim064 Table 1 shows the typical performance values
of the standalone SOM and the phase II performances of our various DNMs when
trained with the dim064 data set. We can see that the SOM has an average homogeneity of 0.92, detected 15 clusters, took 20.75 s to converge with a convergence
accuracy of 0.78, and was trained with 64-dimensional data. Now, in order to find
the best DNM model we have to find a DNM model that:

1. Maximizes homogeneity
2. Minimizes the number of clusters
3. Minimizes training time
4. Maximizes convergence accuracy
5. Uses the smallest number of dimensions

Keep a couple of constraints in mind when looking for the best model: Maximizing
homogeneity is more important than minimizing the number of clusters (to a certain
degree—we can always achieve a homogeneity of 1 if we treat each point as a
cluster), and producing better values in all the other dimensions is more important
than reducing the number of dimensions.

Applying these criteria, we find that the deep neural map with the basic
autoencoder, DNM(AE), performs best with DNM(CAE) coming in as a close

Fig. 4 The graphical output of the DNM(AE) for the dim064 data set

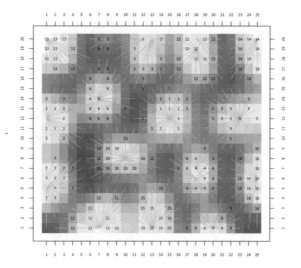

Table 2 DNM performance with different AEs on the landsat data set

	SOM	DNM(AE)	DNM(SAE)	DNM(CAE)	DNM(DAE)	DNM(ConvAE)
homog.	0.78	0.78	0.79	0.77	0.52	0.74
nclust	10	10	9	10	8	9
time (sec)	4.83	0.67	0.54	1.1	1.2	0.93
conv.	0.95	0.98	0.5	0.99	0.54	0.99
dim	36	3	2	3	2	5

second. The DNM(AE) has a homogeneity of 1 and finds all 16 clusters in the data set using an 11-dimensional latent space with a convergence accuracy of 0.87. Phase II training is about 17 times faster than training the standalone SOM.

This validates our first point above. The DNM(AE) has a better clustering behavior than the standalone SOM: a homogeneity value of 1 compared to 0.92 in the standalone SOM and a convergence accuracy of 0.87 compared to 0.78 in the standalone SOM. Furthermore DNM(AE) phase II training is about 17 times faster than training the standalone SOM validating our second point above.

Figure 4 shows the graphic output of the DNM(AE). The 16 homogeneous clusters are clearly visible under the starbursts, and we have mapped the labels of the training instances on top of the map.

Performance Analysis for Landsat Table 2 displays typical performance data for the landsat data set. Here we find that the standalone SOM detected 10 clusters with an average homogeneity of 0.78 and a convergence accuracy of 0.95. It took the standalone SOM 4.83 s to train on the 36 dimensional input data.

Applying the same analysis from the previous section to the performance of the phase II training of our various DNM architectures, we again find the DNM(AE) is the front runner. We chose it over the DNM(CAE) based on the fact that it trains about twice as fast and has a slightly higher average homogeneity. When comparing the phase II training of DNM(AE) to the standalone SOM, we find that it also detects 10 clusters with an average homogeneity of 0.78. However, its convergence accuracy of 0.98 is higher than the 0.95 of the standalone SOM and that it trains on 3D data compared to the 36 dimensions in the standalone. Most notably is that the phase II training of DNM(AE) runs about 7 times faster than training the standalone SOM.

We can observe again that our two evaluation criteria for DNMs are fulfilled. The DNM(AE) trains faster and has a higher convergence accuracy than the standalone SOM.

Figure 5 shows the map produced by our DNM(AE) for this data set. The clusters are clearly visible under the starbursts. What is interesting here is that the cluster with label 2 is set apart from all the other clusters.

Performance Analysis for Digits Table 3 shows the performance numbers for our digits data set. Here the deep neural map with the contractive autoencoder, DNM(CAE), is clearly the winner. It has the highest average cluster homogeneity of 0.78, detects a reasonable number of clusters, trains about 11 times faster than

Fig. 5 The graphical output of the DNM(AE) for the landsat data set

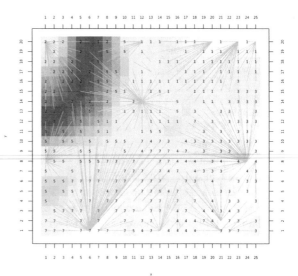

Table 3 DNM performance with different AEs on the digits data set

	SOM	DNM(AE)	DNM(SAE)	DNM(CAE)	DNM(DCAE)	DNM(ConvAE)
homog.	0.65	0.64	0.68	0.78	0.54	0.62
nclust	14	13	14	14	12	10
time (sec)	8.36	0.74	4.35	0.75	0.58	1.44
conv.	0.48	0.94	0.93	0.95	0.95	0.96
dim	64	12	12	12	5	11

Fig. 6 The graphical output of the DNM(AE) for the digits data set

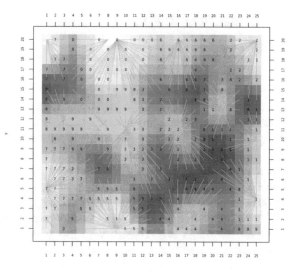

the standalone SOM, and has a convergence accuracy that is about twice that of the standalone SOM. The latent space for this DNM has 12 dimensions.

Therefore, we can observe once more that our two evaluation criteria are fulfilled: The DNM(CAE) trains faster than the standalone SOM and has a better convergence behavior in terms of higher homogeneity and convergence accuracy scores than the standalone SOM.

It is a bit disappointing that none of the models detected the ten clusters due to the ten digits in the data set. However, the data set seems very noisy, and the clusters that were found seem to be a reasonable approximation to the ideal clusters. Figure 6 shows the map produced by the DNM(CAE). What is perhaps noteworthy is that the digit 0 has the strongest cluster toward the upper left corner of the map. This is perhaps due to the fact that it is the most recognizable and unique digit compared to all the other digits.

Observations In all three of our data sets, we found that DNMs outperform standalone SOMs by training faster and exhibiting a better convergence behavior as defined by the average cluster homogeneity and convergence accuracy scores. To our surprise we found that deep neural maps with a convolutional autoencoder did not perform well on the digit image data. One way of interpreting this might be that the latent space here is geared toward reconstructability rather than preserving features for clustering. In order to remedy this, one would have to extend the objective function to include a clustering term which would penalize latent spaces that do not take preserving features for clustering into account.

7 Conclusions and Further Work

Deep neural maps are unsupervised learning and visualization methods that combine autoencoders with self-organizing maps. Recent work has shown that deep neural maps outperform standalone self-organizing maps when considering clustering tasks. The key idea is that a deep neural map outperforms a standalone self-organizing map in two dimensions: (1) better convergence behavior by removing noisy/superfluous dimensions from the input data and (2) faster training due to the fact that the cluster detection part of the DNM deals with a lower dimensional latent space. However, many different kinds of autoencoders exist such as the convolutional and the denoising autoencoder, and here we examined the effects of various autoencoders on the performance of the resulting deep neural maps. We investigated five types of autoencoders as part of our deep neural maps using three different data sets. Overall we show that deep neural maps perform better than standalone self-organizing maps both in terms of improved convergence behavior and faster training. Additionally we show that deep neural maps using the basic autoencoder outperform deep neural maps based on other autoencoders on nonimage data. To our surprise we found that deep neural maps based on contractive autoencoders outperformed deep neural maps based on convolutional autoencoders on image data.

Given the results from this limited study, we would choose the basic autoencoder as our autoencoder of choice in order to construct deep neural maps geared toward nonimage data corroborating earlier work done by Rajashekar [22]. We would like to develop a DNM R package for use by the non-deep learning specialists.

We need to further investigate why our DNM based on a convolutional autoencoder failed to deliver good results on image data. As we mentioned above, the first step is to investigate extending the relevant objective function with a clustering term.

References

1. F. Chollet, *Keras, GitHub* (2015). https://github.com/fchollet/keras
2. P. Domingos, *The master algorithm: How the quest for the ultimate learning machine will remake our world*. The master algorithm: How the quest for the ultimate learning machine will remake our world. (Basic Books, New York, 2015)
3. D. Dua, E. Karra Taniskidou, UCI machine learning repository, in *School of Information and Computer Science* (University of California, Irvine, 2019)
4. C. Ferles, Y. Papanikolaou, K.J. Naidoo, Denoising autoencoder self-organizing map (DASOM). Neural Netw. **105**, 112–131 (2018)
5. V. Fortuin, M. Hüser, F. Locatello, H. Strathmann, G. Rätsch, Som-vae: Interpretable discrete representation learning on time series (2018). arXiv preprint arXiv:1806.02199
6. P. Fränti, S. Sieranoja, K-means properties on six clustering benchmark datasets. Appl. Intell. **48**(12), 4743–4759 (2018)
7. P. Franti, O. Virmajoki, V. Hautamaki, Fast Agglomerative clustering using a k-nearest neighbor graph. IEEE Trans. Pattern Anal. Mach. Intell. **28**(11), 1875–1881 (2006)
8. I. Goodfellow, Y. Bengio, A. Courville, *Deep Learning* (MIT Press, New York, 2016)

9. X. Guo, X. Liu, E. Zhu, J. Yin, Deep Clustering with Convolutional Autoencoders, in *Neural Information Processing*, ed. by D. Liu, S. Xie, Y. Li, D. Zhao, E.-S.M. El-Alfy. Lecture Notes in Computer Science (Springer, Berlin, 2017), pp. 373–382

10. L. Hamel, SOM Quality Measures: An Efficient Statistical Approach, in *Advances in Self-Organizing Maps and Learning Vector Quantization*, ed. by E. Merényi, M.J. Mendenhall, P. O'Driscoll. Advances in Intelligent Systems and Computing (Springer, Cham, 2016), pp. 49–59

11. L. Hamel, VSOM: efficient, stochastic self-organizing map training, in *Intelligent Systems and Applications*, vol. 869, ed. by K. Arai, S. Kapoor, R. Bhatia (Springer, Cham, 2019), pp. 805–821

12. L. Hamel, C.W. Brown, Improved interpretability of the unified distance matrix with connected components, in *Proceedings of the International Conference on Data Mining (DMIN)*. The Steering Committee of The World Congress in Computer Science, Computer...(2011), pp. 1

13. L. Hamel, B. Ott, G. Breard, R. Tatoian, V. Gopu, *Popsom: functions for constructing and evaluating self-organizing maps* (2019)

14. G.E. Hinton, R.R. Salakhutdinov, Reducing the dimensionality of data with neural networks. Science **313**(5786), 504–507 (2006)

15. W. Jang, M. Hendry, Cluster analysis of massive datasets in astronomy. Stat. Comput. **17**(3), 253–262 (2007)

16. T. Kohonen, Self-organizing maps, in *Springer Series in Information Sciences*, 3 edn. (Springer, Berlin, 2001)

17. A. Kristiadi, *Deriving contractive autoencoder and implementing it in Keras—Agustinus Kristiadi's Blog*

18. A. Krizhevsky, I. Sutskever, G.E. Hinton, Imagenet classification with deep convolutional neural networks, in *Advances in Neural Information Processing Systems* (2012), pp. 1097–1105

19. Y. LeCun, C. Cortes, *MNIST Handwritten Digit Database* (2010)

20. M. Pesteie, P. Abolmaesumi, R. Rohling, Deep Neural Maps (2018). arXiv:1810.07291 [cs, stat]

21. K.S. Pollard, M.J. van der Laan, Cluster analysis of genomic data, in *Bioinformatics and Computational Biology Solutions Using R and Bioconductor*, ed. by W. Wong, M. Gail, K. Krickeberg, A. Tsiatis, J. Samet, R. Gentleman, V.J. Carey, W. Huber, R.A. Irizarry, S. Dudoit (Springer, New York, 2005), pp. 209–228

22. D. Rajashekar, *One-class learning with an Autoencoder Based Self Organizing Map* (2017)

23. S. Rifai, P. Vincent, X. Muller, X. Glorot, Y. Bengio, Contractive auto-encoders: explicit invariance during feature extraction. In *ICML* (2011)

24. C.J. Tucker, D.M. Grant, J.D. Dykstra, Nasa's global orthorectified landsat data set. Photogramm. Eng. Remote Sens. **70**(3), 313–322 (2004)

25. P. Vincent, H. Larochelle, I. Lajoie, Y. Bengio, P.-A. Manzagol, Stacked denoising autoencoders: learning useful representations in a deep network with a local denoising criterion. J. Mach. Learn. Res. **11**, 3371–3408 (2010)

26. L. Xu, A. Krzyzak, C.Y. Suen, Methods of combining multiple classifiers and their applications to handwriting recognition. IEEE Trans. Syst. Man Cybern **22**(3), 418–435 (1992)

Implicit Dedupe Learning Method on Contextual Data Quality Problems

Alladoumbaye Ngueilbaye, Hongzhi Wang, Daouda Ahmat Mahamat, and Roland Madadjim

1 Introduction

The definition of big data concept is usually based on the following five dimensions: volume, velocity, variety, veracity, and complexity. In this regard, volume refers to the management of large volumes of data, while the time needed to collect and process data is known as velocity. The types of data such as structured, semi-structured, and unstructured data are referred to as variety in big data concept. The reliability and quality of data are guaranteed by veracity. Finally, the complexity dimension deals with multiple sources of data processing, connecting, matching, cleansing, and transforming data across systems. However, to interconnect and correlate data relationships, hierarchies and multiple data connections can be uncontrolled. Therefore the combination of five dimensions can result in chaos and challenges. Hence, the focus of the present study is to address the veracity dimension of big data. There are many approaches for detecting, quantifying, and resolving data quality issues; however, data integration is often faced by numerous issues and do not inspire confidence in decision-making. Thus, to ensure data quality and beneficial use, the following factors should be considered [3, 7]:

A. Ngueilbaye · H. Wang (✉)
School of Computer Science and Technology, Harbin Institute of Technology, Harbin, China
e-mail: angueilbaye@hit.edu.cn; wangzh@hit.edu.cn

D. A. Mahamat
Departement d'Informatique, Universite de N'Djamena (Tchad), N'Djamena, Chad
e-mail: daouda.ahmat@uvt.td

R. Madadjim
School of Computer Science and Engineering, University of Nebraska-Lincoln, Lincoln, NE, USA
e-mail: madadjim@cse.uni.edu

© Springer Nature Switzerland AG 2021
R. Stahlbock et al. (eds.), *Advances in Data Science and Information Engineering*,
Transactions on Computational Science and Computational Intelligence,
https://doi.org/10.1007/978-3-030-71704-9_22

(a) Data views: the outlook of real world over the data, such as relevancy and granularity.
(b) Data values or dimensions: accuracy, duplication (redundancy), consistency, currency, and completeness.
(c) Data presentation: adequate format of data and simple interpretation.
(d) Other data problems: data privacy, security, and ownership.

The abovementioned data quality issues remain the fundamental characteristics of data to meet internal and external standards of organizations [31, 37]. Poor data quality negatively impacts the performance of an organization. According to Data Warehousing Institute research, $600 billion is lost annually as a result of poor data quality [5]. However, a related study observed that cost related to poor data quality is about 8–12% revenue of a typical organization or about 40–60% of service expenses [7]. Also, it was discovered by Meta Group that, due to the inadequate data quality, 41% of the data warehouse researches do not succeed and lead to wrong decision-making [26, 36] ("the principle of garbage in, garbage out"). Data quality issues usually come from errors, uncleaned data, and also including missing attributes values, incorrect attribute, or data inconsistency. It is often a case for operational databases to encounter 60%–90% of poor data problems [10]. Such problems hinder effective and efficient data usage which lead to an adverse effect on the obtained results and conclusions. Hence, data quality requirement must be checked before an analysis-oriented tool. Otherwise, data quality must be optimized in order to eliminate or repair any issues that may occur in the dataset as stated below [2, 6, 13, 17, 24]:

(a) Making data anomalies adequate in a single data source such as files and databases (e.g., elimination of redundant values in a file).
(b) Migrating poor data or unstructured data into the structured data.
(c) Integrating multiple sources of data into a new single data source (e.g., construction of the data warehouse).

The purpose of this study is to design an approach that can ensure a better quality of data sources, quality cost, and dimensions. Thus, questions regarding the true meaning of data quality arise significantly in seeking further clarification and formal definition to specify each data quality problem in an efficient manner [15, 27]. In this context, the field of data quality and machine learning remains broad, and many strides have been made in order to come up with the initial research viability [23]. Besides rigorous analysis of data quality problems, such kind of definition is significant because it has more supplementary information than the textual definition which explicitly shows that:

(a) It only concerns with a given data type (e.g., string data type).
(b) The knowledge of metadata is needed to identify the problem (e.g., the domain of the attribute).
(c) A mathematical model detects the data quality problem automatically (for illustration, we defined the domain violation as follows: $\exists\, r \in q$: v $(r, a) \notin Dom$ (a)).

(d) The result requires the function of detecting data quality problems (e.g., to detect the misspelling error, the function of a spelling checker must be available).

In this regard, a framework is established for a proposed automated tool to detect and correct data quality problems. This study intends to use it as a complement to the capabilities of today's data quality problem tools. To also support and enhance taxonomy of data quality problems as capital, (a) it is necessary to understand how much a given data quality tool can capture and correct data quality problems, i.e., it is used to estimate the coverage accuracy of a data quality tool, and (b) it makes it possible to apply research efforts that emphasize data quality problems that need further attention, i.e., in the case where data quality has no detection or correction support, this means that the research attention should be addressed as such:

Contributions In summary, the main contributions of this research paper are stated as follows:

- We propose dedupe learning as an effective and efficient method to eliminate the requirement for manual labeling, the challenging classification tasks, and also propose the Loan for Dummy Bank dataset for the implementation.
- We use the taxonomy representation of data quality problems (Fig. 1) to understand the semantics of data in order to elucidate the detection and correction of various anomalies in the data source.
- The Match and the Merge algorithms are presented for the similarity elimination of the data.

This research paper is structured as follows. Section 2 audits the related work. Section 3 describes the background of the study. Section 4 gives the detail description of the methodology. Section 5 explained the experimental setup. Section 6 addresses the results and discussions, and finally, Sect. 7 draws the conclusions.

2   Related Works

There are several research carried out on data quality problems; nonetheless, a minimum number exhaustively detects the set of anomalies that affect data quality and propose solutions to its diminishing value [14, 15, 21, 27]. Rahm and Do [26] underlined the significant difference between one-source (single) and many-source (multi) issues as we also stated in this paper. They separated the two categories into schema-related and instance-related problems. They determined schema-related problems as those sources exhibited by improving the schema design, schema translation, and schema integration while they determined instance-related as errors and inconsistencies in the current data contents that cannot be taken away at the schema level. In this aspect, we consider only the related problems with the instances of the data without making the separation among them. In single-source problems, for both schema-related and instance-related, they made a difference between the following scopes: attribute, record type, and source. In some observation points, this is very similar to the separation that we implement in our taxonomy (Fig. 1). Oliveira et al. [21] structured their taxonomy of data quality problems by the granularity levels of occurrences [36] in the sense that information is stored in multiple sources. Each of them interconnected to many relations. Kim et al. [15] structured a full taxonomy of data quality problems, explaining the logic behind its structure. They emphasized a successive hierarchical treatment approach in which the taxonomy is based on the hope that data quality problems manifest in three different ways: missing data, not missing but wrong data, and not missing and not wrong but unusable data. The perspective approach view of detected data quality problems stipulates that this taxonomy is closely related to our work. Müller and Freytag [14] liberally described a rougher classification of data quality problems into three categories: syntactical, semantic, and coverage anomalies. Coverage anomalies speculate in the decrements of the number of entities and their properties from where they are represented in the collected dataset (e.g., missing values). This work is limited at the level of data quality problems that occur in a single relation of a single source. Notwithstanding this research, the classification is too generic and raises some barriers to compare it with our work. However, it is evident that some critical data quality problems are addressed.

3   Background

3.1   Data Quality Fundamentals

The quality of the data is a general term used to outline certain characteristics of data such as complete, reliable, relevant and up-to-date, and coherent as well as the complete process of making it possible to assure its characteristics. The goal is to obtain data without duplication, misspellings, omission, overinflated variation and by the defined structure. The data are to be of good quality if it meets the requirements of its users which is dependent on its use as well as structure [1, 16, 19, 20, 29, 32]. The impact and the cost of poor data quality are not the same according to the type of entities.

3.2   Data Quality Problems Detection Techniques

This section involved the introduction of the different techniques used in identification (detection and classification) of the data quality issues highlighted. The methods are represented in the form of binary tree classification to allow the identification and classification of specific group of problems or levels and respect the hierarchical representation in the taxonomy. The methods emphasized the logical way that needs

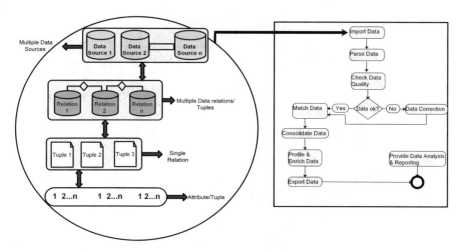

Fig. 1 Dedupe learning for detecting and correcting data quality problems

to be completed as the first way to detect data quality problems automatically or at least semi-automatically (human intervention is needed). It should be considered as future research directions to be explored, and there are no quality issues in a given group only if in the representation. The crossed tree and the arrived leaf at the destination where it is mentioned "in existence of quality problems (marked by Yes)" and the other situation where data quality problem exists (marked by No). Figure 2 shows how the method iteratively crosschecks whether other data quality problems exist or not [28, 34].

Data profiling is a weighty way of detecting anomalies in column data. Profiling is a collection of statistics. It consists of an exploratory analysis of the data on three levels: (i) analysis of the frequency indicator columns, the null values for detecting dubious data such as outliers, (ii) dependency analysis (e.g., functional dependencies), and (iii) analysis of duplicates and similar [11, 30]. Profiling is widely used in the industrial world. It is clear that this is a very manual task. Users do not receive any assistance or guidance. The user is supposed to know the semantics of the columns and to have an idea about the content of the manipulated source. The elimination of similar is the process of comparing lines of a data source in order to determine those which present the same object of the real world, and it consists of two stages: the comparison (Match) and the fusion (Merge). The match function allows us to define if two tuples in a table are similar, and the merge function allows us to generate a new one by merging the two supposed similar ones [8]. In account of the reconciliation of various inner data wellsprings of the foundation and external sources in research data frameworks, issues, as expressed in Sects. 3.1 and 3.2, need to be overcome. Presently it is imperative to contradict the causes in this step and to enhance the data quality. The techniques of distinguishing and remedying errors and irregularities with the point of expanding the nature of given data sources for the data quality management is alluded to as data purifying (or "data cleaning") [18].

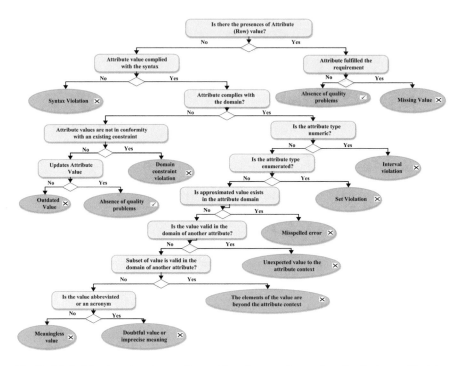

Fig. 2 Method for detecting data quality problems in an attribute value of a single row (Tuple)

Data cleaning incorporates every vital action to clean messy data (fragmented, off-base, not modern, irregularity, excess, and so forth). The data cleaning procedure can be generally organized as pursued [4, 33]:

1. Characterizing and deciding the genuine issue.
2. Finding and distinguishing defective cases.
3. Adjusting the discovered mistakes.

Data purging (cleansing) uses an assortment of particular techniques and innovations inside the data purifying procedure [26] and subdivides them into the accompanying stages: **Parsing** is the main fundamental part of information purifying, helping the client to comprehend and change the characteristics more precisely. This procedure finds, distinguishes, and segregates singular information components, e.g., names, addresses, postal division, and city. The significant issues are organized fields that must be recognized. **Correction and Standardization** are further essential to check the parsed information accuracy and afterward right it a short time later. Institutionalization is essential for effectively coordinating, and there is no chance to get around utilizing a second solid information source. For address information, postal approval is suggested. **Data Improvement or Enhancement** is the procedure that extends existing information with information from different sources. Herein extra information is added to close existing data holes. Run of the mill improvement esteems is a statistic, geographic, or address data. **Matching**: There are various sorts of coordinating—for reduplicating, coordinating to various datasets, merging, or gathering. The adjustment empowers the acknowledgment of the equivalent information. For instance, redundancies can be identified and dense for additional data. **Merge (Consolidation)**: Matching data attributes with contexts is achieved by uniting them. **Data Profiling**: Data profiling is a weighty way of detecting anomalies in column data. Profiling is a collection of statistics. It consists of an exploratory analysis of the data on three levels:

(i) Analysis of the frequency indicator columns, the null values for detecting dubious data such as outliers.
(ii) Dependency analysis (e.g., functional dependencies).
(iii) Analysis of duplicates and similar.

Profilers do not portray business runs and do not roll out any improvements. They are just for breaking down the information. It is frequently utilized toward the start of a data investigation; however, it can likewise better outline the consequences of the investigation. These means are fundamental for accomplishing and keeping up the highest data quality in research data frameworks. Errors in the integration of various data sources in a big data are dispensed with by the clearing up. The clearing up procedure includes missing observations, and complete observations are naturally changed by a particular arrangement concurring to set principles.

4   Materials and Methods

4.1   Dedupe Learning for Detecting and Correcting Contextual Data Quality Problems

Let introduce some notes now. Let T be a table and C all of its attributes; we denote T (C). We consider that $C = A \cup B$. A and B are the set of attributes that serve to eliminate similar tuples in the table $T \cdot B = C - A$ and $A \cap B = \emptyset$ (B can be empty). Let $[[D_k]]$ be the set of concrete values of a tuple of data such as strings Alphanumeric (String), Numeric (Number), Date, Boolean (Boolean), or still lists and ranges of values. A table T is interpreted by the set $[[T]]$ of all $\{A_1 \cdots A_n, B_1 \cdots B_m\}$-tuples defined on $[[D_k]]_{K=1,n+m}$. $[[T]]$ is the set of functions with $t : \{A_1 \cdots A_n, B_1 \cdots B_m\} \rightarrow \bigcup[[D_k]]_{K=1,n+m}$ such that $t(A_i)$ noted $t \cdot A_i$ (respect $t \cdot B_j$) which is an element of $[[D_i]]$ (respect $[[D_j]]$). Each element $t[[T]]$ is a tuple of the table T, and each $t \cdot A_i$ is a value of the attribute in t, which will also be noted v.

Definition 1 Let denote $(\approx)$ as the similarity between values. Suppose two values v_1 and v_2 are similar, we note $(v_1 \approx v_2)$, if the similarity distance d, calculated between these two values, satisfies a condition k for two tuples t_1 and t_2, for an attribute $A_{i(i=1,n)}, t \cdot A_i$ if the condition k_i is satisfied. The condition k_i is based on the calculation of the similarity distance d_i depending on the type of data. Several cases of figures can be envisaged. The user can thus give (provide) one or more several thresholds $k_i : d_i$ which is less than a maximum threshold: $(d_i < s_i)$ with $s_i \in [0, 1]$. The similarity ($\approx$) verifies the properties of reflexivity, commutativity, and associativity. That is, $[v \approx v]$, $[(v \approx v_1) \Leftrightarrow v_1 \approx v]$, and $[v \approx (v_1 \approx v_2) \Leftrightarrow (v \approx v_1) \approx v_2]$.

Example 1 Addr1 $\leftarrow t1 \cdot Addr$ and Addr2 $\leftarrow t2 \cdot Addr$; Name1 $\leftarrow t1 \cdot Name$ and Name2 $\leftarrow t2 \cdot Name$; Mail1 $\leftarrow t1 \cdot Mail$ and Mail2 $\leftarrow t2 \cdot Mail$; and Phone$\cdot t1 \leftarrow t1 \cdot$ Phone and Phone$\cdot t2 \leftarrow t2 \cdot$ Phone. Let us the two following addresses addr1 = "2021 FH DANIEL HENRY/BRUCE" and Addr2 ="2021 FH DANIEL 39750 HENRY-BRUCE" are similar, according to the Jaro-Winkler method since the calculated distance is 5.7% (less than the threshold t =0,2).

Definition 2 (Similarity Rule) let a r be a similarity rule defined as a conjunction of similarities which relate to attributes $A_{i(i=1,n)}$ of the set A of the table $T : r = (t \cdot A_1 \approx t_1 \cdot A_1) \wedge (t \cdot A_2 \approx t_1 \cdot A_2) \wedge (t \cdot A_i \approx t_1 \cdot A_i) \dots \wedge (t \cdot A_n \approx t_1 \cdot A_n)$.

Example 2 Two similarity rules r_1 and r_2:
$$r_1 = (Name_1 \approx Name_2) \wedge (Mail_1 \approx Mail_2) \wedge (Address_1 \approx Address_2).$$
$$r_2 = (Mail_1 \approx Mail_2) \wedge (Phone_1 \approx Phone_2).$$

Definition 3 (Similarity between Tuples) if two tuples t and t_1 are similar, then the relationship (disjunction) between all the similarity rules defined in table T is true. We can say that $t_1 \approx t_2$ if and only if $(r_1 \wedge r_2 \wedge r_k \dots \wedge r_q)$ is true with q being the number of similarity rules.

Example 3 t_1 and t_2 are similar if and only if $r_1 \vee r_2 : t_1 \approx t_2$ if and only if $((Name_1 \approx Name_2) \wedge (Mail_1 \approx Mail_2) \wedge (Addr_1 \approx Addr_2)) \vee ((Mail_1 \approx Mail_2) \wedge (Phone_1 \approx Phone_2))$. Note, however, that these rules may contain inconsistencies or the contradictions that will be checked.

4.2 Dedupe Learning Method

Dedupe learning is an integrated machine learning approach based on the traditional machine learning techniques proposed by this study, to handle entity resolution challenges, to find an adequate solution, and to understand entity resolution better. The concept is to perform deduplication known as record linkage. It works by joining records through the integration of fuzzy way, indexing, and blocking using data such as names, addresses, phones, dates, etc. This is because real-world data are often imputed by people without considering the following factors:

1. Not reviewed.
2. Not linked with related data.
3. Not properly normalized by the input system.

4. Not correctly inputted because people make mistakes: typos, mishearing, miscalculation, misinterpretation, etc.

Thus, the above factors usually trigger the following problems on data:

1. Lack of unique identifiers (making it difficult to detect duplicates in a dataset or to link with other datasets).
2. Duplications (e.g., multiple records refer to a single person).
3. Inconsistencies (e.g., a person appears with multiple addresses).
4. Bad formatting (e.g., birth dates appear with multiple formats like DD/MM/YY and YYYY-MM-DD).

All factors mentioned above affect the ability to properly extract knowledge from one or more datasets. The solution is to perform the record linkage. It works by joining records in a fuzzy way using data like names, addresses, phone numbers, dates, etc. The term record linkage is mostly used when the linkage is applied to multiple datasets, like joining a Restaurant Food Inspection dataset with an Employee Wage dataset. What we will discuss in this section is a specific application of record linkage, called deduplication, which is applying record linkage on the datasets against itself to find which records are duplicated. Geocoding street addresses, i.e., converting them to latitude/longitude, is very useful for matching as geocoders usually clean irrelevant address variations. Also, having lat/lng enables the calculation of geometric distances between addresses. During the process of dedupe learning, three (3) algorithms have been adopted: record linkage (Algorithm 1), matching (Algorithm 2), and the consolidation (Algorithm 3) as shown below.

4.3 Dedupe Learning (DDL) Setup

Our experimental setup was a specific application of record linkage, called deduplication, which is applying record linkage on a single dataset against itself to find which records are duplicates. Good deduplication on the datasets would find that:

(i) (0,1) are duplicates.
(ii) (2, 3) are duplicates.
(iii) But (0,1) and (2,3) are different, despite being similar.

5 Experimental Evaluation

This section illustrates the results of the experimental evaluation performed on two real world datasets. The results show that Dedupe Learning performed the best on the entire datasets with negligible proficient effort compared to the traditional methods with approximate accuracy values.

Algorithm 1 Deduplication (duplication) algorithm

Result: Cleaned Target Data SC
Input: a set I of tuples
Output: a set I' of tuples
1   Choose key attributes (deduplication attributes)
    I' $\rightarrow \phi$
    while $I \neq \phi$ **do**
2 | t $\rightarrow$ a tuple from I
     | remove t from I
     | **if** $I' \neq \phi$ **then**
3 | | **for** *all tuples t' in I' and t $\neq$ t'* **do**
4 | | | **if** *Match (t, t')* **then**
5 | | | | -t" $\rightarrow$ Merge (t, t')
     | | | | **if** $t'' \in /I \cup I' \cup t$ **then**
6 | | | | | add t" to I
7 | | | | **end**
8 | | | remove t' from I' endif endfor
9 | | **end**
10 | **end**
11 | add t to I'
12 | **end**
13 **end**
14 return I'

5.1   Experiment Setup

Intruments and Platform   Our experiments were performed on a laptop computer with a Processor Intel(R)Core (TM) i7-8750H CPU @ 2.2 GHz, 2208 Mhz, 6 cores(s), 12 Logical Processor(s), 8 GB of memory, a 512 GB SSD, and NVIDIA GTX 1060 with Windows 10 Operating System. We used TensorFlow [25], and Keras [9], a deep learning framework, to train and test our models.

Data Source   To investigate the efficiency and effectiveness of our approaches, we made use of Loan Data for Dummy Bank dataset with 887379 instances with 30 tuples and Restaurant dataset with 881 instances and 6 tuples. Talend open studio was used for data profiling analysis as well as the generation of some dataset in our experiment [12, 35].

Implementation   To illustrate the constraint and subsequent from such dependence, we propose an innovative learning model, called dedupe learning. It means machine learning techniques to handle data quality problems, and dedupe learning is derived from dedupe which is the python library that uses machine learning to execute deduplication or data linkage and entity resolution rapidly on structured data. Dedupe learning aims to enable adequate machine labeling deprived of the necessity for physical labeling exertion. Stimulated by the steady of human learning nature, which is an expert at solving the issues with growing inflexibility, it initiates with some accessible instances in a task, which can be repeatedly labeled by the

---

Algorithm 2 Match algorithm

Result: Cleaned Target Data SC
Input: Two tuples, t_1, $t_2 \in$ T and S(thresholds)
Output: = True if t1 $\approx$ t2
15 Result←False
   for *all Rule r_j j from 1 to q* **do**
16 | Rule j ←True I← 1
    | **while** *Rulesj and i<n* **do**
17 | | v1 ← t1.Ai   v2 ← t2.Ai
    | | **if** *v1 or v2 = NULL* **then**
18 | | | Result ← False
19 | | **else**
20 | | | **switch** dtype(A_i)
    | | | **Case** Date:
    | | | Result ← matchValues(v1,v2)
    | | | **Case** Numeric:
    | | | Result ← handleNumericValues(v1,v2)
    | | | **Case** String:
    | | | Result ← handelStringValues(v1,v2)
    | | | Result ← matchValues(v1,v2)
    | | | **end switch**
21 | **end**
22 | $Rules_j=Rules_j$ and Result; i++
23 | **end**
24 | Result ← Results or $Rules_j$
25 **end**

---

machine with adequate accuracy, and then steadily reasons about the labels of the most challenging instances constructed on the observations provided by the labeled structured instances. The results of machine labeling were bolstered to traditional model training. However, the following two properties of dedupe learning make it in general sense unique in relation to the current learning models:

- Distribution misalignment among structured and unstructured instances in a task. Dedupe learning processes the instances in increasing order of inflexibility. The distribution misalignment among the labeled and unlabeled instances reduces the current learning models unfit for dedupe learning.
- Dedupe learning in its ensemble task. The procedure of iterative labeling can be achieved in an unsupervised manner, deprived of needing any human intervention.

Algorithm 3 Merge algorithm

Input: Data Source S $\{i_1, \ldots, i_n\}$, Parameter P
Output: Cleaned Targe Data Source S
26 *identityMerges* ← *emptySet()*
    while *isNotEmpty(S)* **do**
27 | r ← firstElementFrom(S)
    | iMerge ← emptySet()
    | insert(iMerge, r)
    | remove(S,r)

28 **end**
29 **for** $i \in S$ **do**
30 | **if** *shouldInclud(iMerge,i,p)* **then**
31 | | insert(iMerge,i)
    | | remove(S,i)
    | | insert(identityMergess, iMerge);

32 | **end**
33 | **return** identityMerges;

34 **end**

5.2   Experiment for Matching Algorithm on the Datasets

The objective of this section is to measure the quality of dedupe learning with matching algorithm to approximate matching information. We create a set of random sample from Loan for Dummy Bank dataset and Restaurant dataset under different capacities: (i) minimum 10% of tuples for each dataset, (ii) medium 20% of tuples in the dataset, and (iii) maximum 30% of tuples for each dataset. The evaluation is executed by defining the thresholds from 0.5 to 0.65 with the gap of 0.005. The time complexity of the matching algorithm is $O(n^2)$. The results are shown in Table 1 and Fig. 3.

Table 1 Evaluation of dedupe learning for matching algorithm

Datasets	Threshold	TP	FP	FN	Precision	Recall	F-Score
Restaurant	0.50	91	17	9	84.26	91	87.50
	0.55	91	13	9	87.50	91	89.21
	0.60	86	3	15	96.62	85.14	90.52
	0.65	89	3	12	96.73	88.11	92.22
Loan for Dummy Bank	0.50	264	93	90	73.94	74.57	74.26
	0.55	482	70	60	87.31	88.92	88.11
	0.60	252	65	91	79.49	73.46	76.36
	0.65	689	74	92	90.30	88.22	89.24

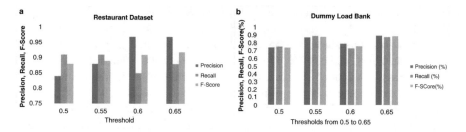

Fig. 3 Evaluation of Dedupe Learning on (**a**) Restaurant dataset and (**b**) Loan Dummy Bank Dataset

As indicated above, it is not trivial to that, so we had used different techniques to figure it out. Usually, processing a dataset needs the following techniques to be applied:

1. Preprocessing (input: dataset; output, cleaned dataset) First and foremost, we used string functions and regexes to normalize names and remove undesired variations. There are other classifiers from record linkage library we could try, but the truth is that:

    (i) It is challenging to build a good training set that takes into account all the important cases of matches/non-matches.
    (ii) It is possible to tune classifier parameters to get better results, but it is very difficult to decide the right parameters that will generalize well for future predictions.
    (iii) It is unsure if the indexing rules used are really sensible: we dropped "true positives" that are not being blocked together or even introducing "false negatives" that are being blocked together, but our classifier is not being able to classify them as non-matching.

2. Indexing (output: pairs to compare) Indexing is applied here to extract which pairs we want to run a comparison on.
3. We ran a comparison to get the comparison vectors and then classified them as matching or non-matching.
4. We fed a support vector machine classifier with our training data [22]. SVMs are resilient to noise, can handle correlated features, and are robust to imbalanced training sets. That last attribute is relevant for deduplication because we normally find more negative pairs than positive pairs to be added to a training set.

6 Discussions

This manuscript introduced dedupe learning method as traditional machine learning techniques to democratize the detection and correction of data quality problems in large dataset. The groundwork of this contribution is the identification of the data linkage as a key aspect block for developing effective entity classifiers. We also amended some algorithms to handle the issues at three stages such as record linkage, matching, and consolidation developing mindful classifiers and an efficient method based on dedupe learning. Our experimental results in Table 1 and Fig. 3 show that our approach is promising and already achieves or fit within the related work on various benchmarks datasets. We do believe that dedupe learning method is powerful and an effective tool that has supremacy beyond the entity resolution, and it is our hope to extend this idea to build practical and effective entity resolution systems. Dedupe learning method would be improved in the following ways: First, recognize and automatically propose an appropriate dedupe learning architecture for an intended dataset whether simple or complex. Secondly, we design a system that leverages both automatic features such as detection and correction dirty data cases need to be addressed.

7 Conclusions

This research paper investigated the method for detecting and correcting contextual data quality problems by means of the taxonomy of data quality problems represented by granularity levels and proposed dedupe learning method to deal with data quality problems for structured data in the context of big data. As a result, an extended formal complete taxonomy was designed and represented along with some proposed algorithms. The taxonomies of the existing data quality problems and dedupe learning method were formally evaluated on structured data. We have attempted throughout this research to list out all the different anomalies related to the schema of a data source or from the data themselves. List again the causes of these anomalies and some detection approaches that remain too manual and require some user expertise. The latter requires a minimum of semantics in the source. We presented the study/experiment cited in the literature for the treatment and correction of these data. Similarly, the presence of semantics will allow for better quality in data cleaning such as homogenization, unification, and recommendation of key attributes and the similarity elimination which its presence generates significant concern about data quality. Lastly, the objective in this study was to participate in the construction of techniques that can help users during the different phases of target data of the sizes that are generally very important (databases, data warehouse, and big data).

Acknowledgments This paper was partially funded by the National Key R&D Program of China under Grant No.2018YFB1004700 and NSFC Grant Nos. U1866602, 61602129, and 61772157.

References

1. N. Abdullah, S.A. Ismail, S. Sophiayati, S.M. Sam, Data quality in big data: a review. Int. J. Advance Soft Comput. Appl. **7**(3), 17–27 (2015)
2. M. Ahmed, S. Choudhury, F. Al-Turjman, Big data analytics for intelligent internet of things, in *Artificial Intelligence in IoT* (Springer, Berlin, 2019), pp. 107–127
3. D. Ardagna, C. Cappiello, W. Samá, M. Vitali, Context-aware data quality assessment for big data. Future Gener. Comput. Syst. **89**, 548–562 (2018)
4. O. Azeroual, M. Abuosba, Improving the data quality in the research information systems (2019). arXiv preprint arXiv:1901.07388
5. C. Batini, A. Rula, M. Scannapieco, G. Viscusi, From data quality to big data quality, in *Big Data: Concepts, Methodologies, Tools, and Applications* (IGI Global, New York, 2016), pp. 1934–1956
6. R.J.C. Bose, R.S. Mans, W.M. van der Aalst, Wanna improve process mining results?, in *Proceedings of the 2013 IEEE symposium on computational intelligence and data mining (CIDM)* (IEEE, New York, 2013), pp. 127–134
7. L. Cai, Y. Zhu, The challenges of data quality and data quality assessment in the big data era. Data Sci. J. **14**(2), 1–10, (2015). http://dx.doi.org/10.5334/dsj-2015-002
8. F. Chiang, R.J. Miller, Discovering data quality rules. Proc. VLDB Endowment **1**(1), 1166–1177 (2008)
9. F. Chollet, *Deep Learning MIT Python und Keras: Das Praxis-Handbuch vom Entwickler der Keras-Bibliothek* (MITP-Verlags GmbH and Co. KG, New York, 2018)
10. C. Cichy, S. Rass, An overview of data quality frameworks. IEEE Access **7**, 24634–24648 (2019)
11. T. Dasu, T. Johnson, S. Muthukrishnan, V. Shkapenyuk, Mining database structure; or, how to build a data quality browser, in *Proceedings of the 2002 ACM SIGMOD International Conference on Management of Data* (2002), pp. 240–251
12. M.N. Ferozi, *Loan Data for Dummy Bank* (2018). https://www.kaggle.com/mrferozi/loan-data-for-dummy-bank
13. T. Gschwandtner, J. Gärtner, W. Aigner, S. Miksch, A taxonomy of dirty time-oriented data, in *International Conference on Availability, Reliability, and Security* (Springer, Berlin, 2012), pp. 58–72
14. V.N. Gudivada, Data analytics: fundamentals, in *Data Analytics for Intelligent Transportation Systems* (Elsevier, Berlin, 2017), pp. 31–67
15. W. Kim, B.-J. Choi, E.-K. Hong, S.-K. Kim, D. Lee, A taxonomy of dirty data. Data Min. Knowl. Discovery **7**(1), 81–99 (2003)
16. R. Krishnan, A. Hussain, P. Sherimon, Conceptual clustering of documents for automatic ontology generation, in *International Conference on Brain Inspired Cognitive Systems* (Springer, Berlin, 2013), pp. 235–244
17. L. Li, T. Peng, & J. Kennedy, A rule based taxonomy of dirty data. GSTF Journal on Computing (JoC), **1**(2), 140–148 (2014)
18. S. Matook, M. Indulska, Improving the quality of process reference models: a quality function deployment-based approach. Decis. Support Syst. **47**(1), 60–71 (2009)
19. J. Merino, I. Caballero, B. Rivas, M. Serrano, M. Piattini, A data quality in use model for big data. Future Gener. Comput. Syst. **63**, 123–130 (2016)
20. M. Mezzanzanica, R. Boselli, M. Cesarini, F. Mercorio, A model-based evaluation of data quality activities in KDD. Inf. Process. Manage. **51**(2), 144–166 (2015)
21. H. Müller, J.-C. Freytag, U. Leser, Improving data quality by source analysis. J. Data Inf. Qual. (JDIQ) **2**(4), 1–38 (2012)
22. A. Ngueilbaye, L. Lei, H. Wang, Comparative study of data mining techniques on heart disease prediction system: a case study for the "republic of chad". Int. J. Sci. Res. **5**(5), 1564–1571 (2016)

23. A. Ngueilbaye, H. Wang, M. Khan, D.A. Mahamat, Adoption of human metabolic processes as data quality based models. J. Supercomputing 77, 1779–1817 (2021). https://doi.org/10.1007/s11227-020-03300-3

24. P. Oliveira, F. Rodrigues, P. Henriques, H. Galhardas, A taxonomy of data quality problems, in *Proceedings of the 2nd International Workshop on Data and Information Quality* (2005), pp. 219–233

25. S. Pattanayak, S. Pattanayak John, *Pro Deep Learning with TensorFlow* (Springer, Berlin, 2017)

26. E. Rahm, H.H. Do, Data cleaning: problems and current approaches. IEEE Data Eng. Bull. **23**(4), 3–13 (2000)

27. E. Rahm, E. Peukert, *Large Scale Entity Resolution* (2019)

28. S. Ram, J. Park, Semantic conflict resolution ontology (SCROL): an ontology for detecting and resolving data and schema-level semantic conflicts. IEEE Trans. Knowl. Data Eng. **16**(2), 189–202 (2004)

29. H.N. Roa, E. Loza-Aguirre, P. Flores, A survey on the problems affecting the development of open government data initiatives, in *Proceedings of the 2019 Sixth International Conference on eDemocracy and eGovernment (ICEDEG)* (IEEE, New York, 2019), pp. 157–163

30. A.B. Salem et al., Semantic recognition of a data structure in big-data. J. Comput. Commun. **2**(09), 93 (2014)

31. C. Samitsch, *Data Quality and Its Impacts on Decision-making: How Managers can Benefit from Good Data* (Springer, Berlin, 2014)

32. T. Schäffer, & D. Stelzer, Towards a taxonomy for coordinating quality of master data in product information sharing, *In Proceeding of MIT International Conference on Information Quality*, UA Little Rock, October 6-7, pp. 1–9.(2017)

33. M. Shiloach, S.K. Frencher Jr, J.E. Steeger, K.S. Rowell, K. Bartzokis, M.G. Tomeh, K.E. Richards, C.Y. Ko, B.L. Hall, Toward robust information: data quality and inter-rater reliability in the american college of surgeons national surgical quality improvement program. J. Am. Coll. Surgeons **210**(1), 6–16 (2010)

34. S. Soares, Big data quality, in *Big Data Governance: An Emerging Imperative* (2012), pp. 101–112

35. S. Tejada, C.A. Knoblock, S. Minton, Learning domain-independent string transformation weights for high accuracy object identification, in *Proceedings of the Eighth ACM SIGKDD International Conference on Knowledge Discovery and Data Mining* (2002), pp. 350–359

36. Y. Xiao, L.Y. Lu, J.S. Liu, Z. Zhou, Knowledge diffusion path analysis of data quality literature: a main path analysis. J. Inform. **8**(3), 594–605 (2014)

37. A. Zaveri, A. Rula, A. Maurino, R. Pietrobon, J. Lehmann, S. Auer, Quality assessment for linked data: a survey. Semantic Web **7**(1), 63–93 (2016)

Deep Learning Approach to Extract Geometric Features of Bacterial Cells in Biofilms

Md Hafizur Rahman, Jamison Duckworth, Shankarachary Ragi,
Parvathi Chundi, Venkata R. Gadhamshetty, and Govinda Chilkoor

1 Introduction

Microscopy image analysis tools including BiofilmQ [1], ImageJ [2], BioFilm Analyzer [3], and Imaris [4] have been successfully used in the past for feature extraction from microscopy images. However, these tools perform well when the microscopy images are characterized by homogeneous features and are not necessarily optimized for feature extraction from images with heterogeneous features. Biofilm images often display heterogeneities related to shape and size of bacterial cells, cell clusters, pores, and sometimes microbial debris. In case of the microbiologically induced corrosion (MIC), the heterogeneities multiply due to the existence of corrosion products as can be seen in Fig. 2a. The existing image analysis tools are not necessarily designed to automate the extraction of these heterogeneous features. For instance, BiofilmQ performs well in segmentation and data visualization but lacks the capability of segmenting individual bacterial cells

Md. H. Rahman (✉) · J. Duckworth · S. Ragi
Department of Electrical Engineering, South Dakota School of Mines and Technology, Rapid City, SD, USA
e-mail: mdhafizur.rahman@mines.sdsmt.edu; jamison.duckworth@mines.sdsmt.edu; shankarachary.ragi@sdsmt.edu

P. Chundi
College of Information Science & Technology, University of Nebraska Omaha, Nebraska, NE, USA
e-mail: pchundi@unomaha.edu

V. R. Gadhamshetty · G. Chilkoor
Department of Civil and Environmental Engineering, South Dakota School of Mines and Technology, Rapid City, SD, USA
e-mail: venkata.gadhamshetty@sdsmt.edu; govind.chilkoor@sdsmt.edu

© Springer Nature Switzerland AG 2021
R. Stahlbock et al. (eds.), *Advances in Data Science and Information Engineering*,
Transactions on Computational Science and Computational Intelligence,
https://doi.org/10.1007/978-3-030-71704-9_23

when present in a cluster [1]. ImageJ, a tool widely used in microscopy image analysis, also underperforms in separating individual cells from bacterial clusters or overlapping cells. These observations are discussed in more detail in the results section (Sect. 2.4).

1.1 Key Contributions

The main goal of this study is to develop an automated artificial intelligence (AI)-based model for extracting the geometric characteristics of bacterial cells present in an MIC biofilm image. To this end, we apply a novel combination of a deep convolutional neural network (DCNN) model [5] with an image processing technique called modified watershed algorithm [6] to perform image segmentation on the SEM images of the biofilm. The DCNN model is used to perform a primary segmentation of an input image followed by a modified watershed segmentation algorithm that is used to further improve the segmentation performance. We then apply a pixel counting approach to automatically extract the size properties of the individual bacterial cells in the images even when they overlap or exist in a cluster. Finally, we perform a numerical study to compare the performance of our DCNN-based biofilm feature extraction methods with ImageJ. We discuss the benefits of the proposed method using a case study of sulfate-reducing bacterial (SRB) biofilm grown on mild-steel surface.

The rest part of this paper is organized as follows. In Sect. 2, we discuss the DCNN training process using DeepLabv3+, separation process of the bacterial cells, and cells cluster segmentation process. At the end of Sect. 2, we perform qualitative comparison of our proposed model against ImageJ. Finally, we present conclusions in Sect. 3.

2   Methods and Results

In this study, we focus on extracting the size characteristics (also referred to as biofilm features henceforth) of bacterial cells in biofilm images. The feature extraction approach is outlined in Fig. 1. The input to the model is a raw unprocessed SEM image of a biofilm, and the outputs are segmented objects in the image and size characteristics of the bacterial cells present in the image. First, we train a deep convolutional neural network with labeled or annotated images as shown in Fig. 2. We hand-label the image regions into four classes: bacterial cells, pores, biofilm surface, and other corrosion products (henceforth called CPs) using a standard pixel annotation tool. Since we focus only on size characteristics of bacterial cells, we suppress the other object classes in the image by simply masking the pixels that belong to the other classes. Next, we determine if the image contains cell clusters (method explained later). If the clusters are present, we implement a modified

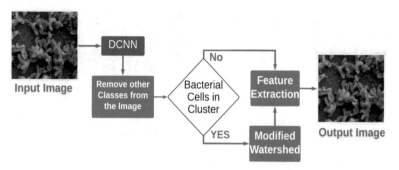

Fig. 1 Feature extraction from biofilm SEM images via DCNN and watershed methods

watershed algorithm to separate individual bacterial cells in the cluster. Next, we implement the feature extraction procedures to estimate the size properties of the bacterial cells using Matlab's regionprops tool [7].

2.1 Data Collection

The biofilm microscopy image datasets (SEM images) are obtained from the 2D-BEST center [8]. SEM was used to characterize the biofilm and the corrosion products of SRB-G20 on mild steel surfaces. Mild steel samples coated with the biofilm were briefly immersed in 3% glutaraldehyde in cacodylate buffer (0.1 M, pH 7.2) for 2 h. The treated samples were rinsed with sodium cacodylate buffer and distilled water. The samples were then dried using ultra-pure nitrogen gas. We used 70 SEM images of the biofilm to train the DCNN model as explained below.

2.2 DCNN Training

The size of each image is 229x256 pixels. First, using a pixel annotation tool [9], we label the bacterial cells, pores, and other corrosion products in the training images based on the guidance from domain experts in the 2D-BEST center. We train a deep convolutional neural network (DCNN) via DeepLabv3+ [5] which was developed particularly for semantic image segmentation. DeepLabv3+ uses atrous convolution [10], which is derived from a wavelet transform method a'trous (French for "hole algorithm"). One of the key components of this DCNN model is the encoder, which encodes the multiscale contextual information of the input image for the segmentation task by penetrating the incoming feature with atrous spatial pyramid pooling operations at different rates and feasible fields of view. The atrous convolution allows us to increase the field of view of the filters by

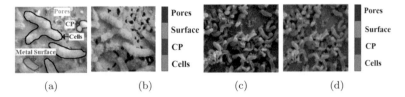

(a)   (b)   (c)   (d)

Fig. 2 Image segmentation via DCNN. (**a**) Raw SEM image of a biofilm where the objects in yellow, red, black, and blue boundaries indicate that bacterial cells, pores, corrosion products, and metal surface, respectively. (**b**) Annotated image sample used for training. (**c**) Input image for DCNN model. (**d**) Output of DCNN model with segmented objects

Fig. 3 Frequency of object classes appearing in the training images

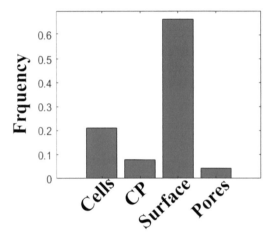

introducing holes into the filters and captures features at multiple scales. The second component decoder captures sharper object boundaries by gradually recovering the spatial information from the encoding phase and creates an output with the size of the original input image.

We use Matlab's pixelLabelDatastore [11] object to read the labeled image data and store pixel label data for semantic segmentation. This function encloses the pixel label data and the label ID to a class named colormap. We create a supporting function which provides RGB color values for user-defined specific values for cells, CP, pores, and surface. For example, the value of RGB 70 130 180 (blue) is used for the bacterial cells. The colormap of pixel-labeled images is then overlaid on the input image for the segmentation. If there are any missing areas with no overlap on the image, then those regions are not classified into any of the four objects and are not used for DCNN training.

In principle, all classes of the training dataset should have an equal number of observations. However, this is not the case in our dataset as can be seen in Fig. 3. This is, of course, an expected issue in biofilm image datasets as area (pixels) covered by biofilm surface is significantly higher than the other object classes. This inequality in the number of observations may have the learning process favor

Table 1 Class weight for balancing classes

Class	Cells	CP	Surface	Pores
Weight	0.6737	1.9395	0.2555	3.2782

Fig. 4 (**a**) Output after removing the objects of the other three classes, (**b**) the binary image of (**a**)

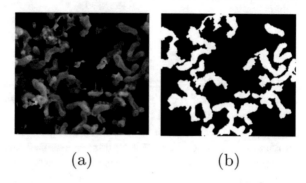

(a)   (b)

dominant classes. We need to balance these classes to improve the training model. To address this, we use the class weighting as shown in Table 1 which multiplies with the class weight of appearing in the training images.

The training parameters used for DeepLabv3+ are shown in the table below:

Parameter	Value
Initial learning rate	0.001
Maximum epoch	200
Learning rate schedule	Piecewise
Learning rate reduced by a factor	0.3 every 10 epochs
Min-batch size	8
Maximum iterations	1000

Adapting the learning rate of the model can increase performance and reduce training time. We used a piecewise learn rate schedule which multiplies the learning rate by a factor of 0.3 every 10 epochs from the initial learning rate of 0.001. This adaptive learning rate allows the model to learn quickly with a higher initial learning rate. A "ValidationData" parameter is used for testing the validation date for every epoch. To avoid the network from overfitting on the training dataset, the "ValidationPatience" parameter is set to 4 to stop training early if the validation accuracy converges. The "min-batch size" parameter helps reduce the usage of memory while training. We can either increase or decrease this value depending on the availability of GPU memory.

To increase the data volume for training, we implement an approach called data augmentation [12]. This method augments the images with a random combination of resizing, rotation, reflection, shear, and translation transformations. We implemented the reflection transformations in this study. DeepLabv3+ is developed with a pre-trained network to ease the training effort. ResNet-18 [13], ResNet-50 [13],

Table 2 Different model
accuracy at different
iterations by using biofilm
dataset

	Model accuracy at different iterations		
Name of network	150	500	1000
ResNet-18	70.65%	70.37%	71.12%
ResNet-50	74.78%	74.67%	74.24%
MobileNet-V2	61.09%	60.12%	61.17%

and MobileNet-V2 [14] are used to train the DeepLabv3+ model with different iterations to check the model efficiency, which is shown in Table 2.

Our original dataset has 44 images for training, 11 images for validation, and 11 images for testing. The validation images are used to minimize the overfitting problem of the model, and testing images are used only for testing the final result to check the model accuracy. We choose the ResNet-50 network for our application since it achieves the highest model accuracy, as is evident in Table 2. The trained DCNN performance in image segmentation is shown in Fig. 2, where the model is able to identify the four classes: cells, pores, surface, and CP.

2.3   Cell Cluster Segmentation

A key challenge with detecting bacterial cells in biofilm images is separating the cells from clusters. To address this, we implement a modified watershed transform method [6]. First, we compute the distance transform of the masked clusters in Fig. 4b. The distance transform computes the Euclidean distance transform of the binary image by allocating a number for each pixel that represents the distance between a pixel and the nearest nonzero pixel of the given image. Figure 5a shows the inverse of the distance transform of the masked image [15]. A maxima edge finding algorithm is then used to find dividing sections between all logged minima [16]. We observe that this standard watershed algorithm leads to undesirable over-segmentation, as seen in Fig. 5b. The reason is each small local minima becomes a catchment basin. To address this issue, we implement a modified distance transform as described in [17]. A Matlab function *imextendedmin* [18] is applied on the inverse of the distance transform image (Fig. 5a) to find local minima. This function helps the watershed segmentation to filter out the small local minima, which addresses the over-segmentation issue discussed earlier. The resulting image with local minima marked is shown in Fig. 6a. The figure shows the minima only at the desired locations. Next, with this new distance transformation method, we apply the watershed transform for cell segmentation. The resulting cell segmentation is shown in Fig. 6b, which shows that the modified watershed method overcomes the over-segmentation issue.

Fig. 5 (**a**) Inverse of the distance transform of image in Fig. 4 (**b**). (**b**) Watershed segmentation of the image

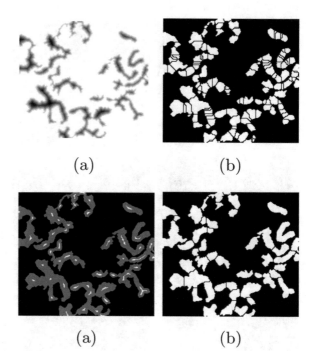

(a)　　　　　　　　(b)

Fig. 6 (**a**) Modified watershed algorithm with gray areas as the bacteria group and lighter gray areas as the centers of clusters found from the modification. (**b**) New segmentation applied from the modified watershed algorithm

(a)　　　　　　　　(b)

Table 3 Estimated size characteristics of bacterial cells

Image of Fig. 7	Model	Number of bacterial cells	Average length (μm)	Average width (μm)	Average perimeter (μm)
1st row	Proposed model	63	1.38	0.84	4.01
	ImageJ	96	1.07	0.65	3.32
2nd row	Proposed model	34	1.69	0.91	4.58
	ImageJ	70	1.20	0.90	6.03
3rd row	Proposed model	49	1.50	0.97	4.60
	ImageJ	95	1.05	0.61	3.12

2.4 Qualitative Comparison of the Proposed Model with ImageJ

We apply our segmentation approach on images not used for training with results shown in Fig. 7. However, not all of the cells are recovered which may be attributed to either cells that are cut off at the edges of the image or because of the presence of very large clusters. Even though this model is not perfect, it holds some improvement over that of the ImageJ analysis, as seen qualitatively in Table 3 and Fig. 7. We used three images to verify the proposed model in Fig. 7 and then compared the results of the proposed model with the results of the ImageJ analysis

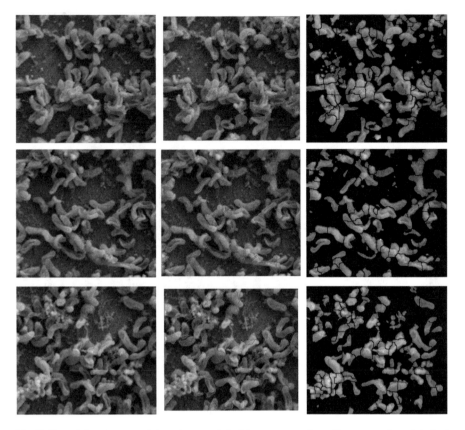

Fig. 7 From left column to right column: original images to evaluate the proposed model, final results of automatic segmentation by proposed method, and results of ImageJ segmentation

tool. The red contours in the middle column of Fig. 7 are the borders of bacterial cells. These results show that our approach outperforms ImageJ, which generated over-segmented images. Due to the over-segmentation in clustering cells, as seen in Table 3, ImageJ counted an excessive number of bacteria that affected the size characteristics of the cells.

3 Conclusions

In this paper, we developed a combination of deep learning and image processing algorithms to estimate the size properties of bacterial cells in microbially induced corrosion of metals. Particularly, we focused on sulfate-reducing bacteria (SRB) grown on mild steel. We trained a deep convolutional neural network with SEM images of the biofilm to perform image segmentation and detect classes of objects including bacterial cells, pores, surface, and other corrosion products. After segmentation, we retained the bacterial cells in the image while masking out the other object classes. Next, we applied a modified watershed algorithm to detect individual bacterial cells when they overlap or exist in clusters. Our numerical study showed that our methods were able to detect individual bacterial cells in the presence of other objects and even when the cells exist in clusters. Finally, we extracted the number of cells and their average width and length and compared the results with that of ImageJ. We provided qualitative evidence showing our methods performed relatively better in counting the number of cells and in segmentation of clusters. While the proposed models show promise, our models are unable to extract every individual cell in the clusters. In our future studies, we will further tune the parameters of the watershed approach for better segmentation of cells in a cluster and perform quantitative comparison of size characteristics against the ground-truth measurements.

Acknowledgments The authors would like to acknowledge the funding support from NSF RII T-2 FEC award #1920954 and NSF RII T-1 FEC award #1849206.

V. Gadhamshetty would like to acknowledge partial support from NSF CAREER award #1454102.

References

1. R. Hartmann, H. Jeckel, E. Jelli, P.K. Singh, S. Vaidya, M. Bayer, L. Vídakovic, F. Diaz-Pascual, J.C. Fong, A. Dragos, et. al., Biofilmq, a software tool forquantitative image analysis of microbial biofilm communities. BioRxiv (2019), p. 735423
2. D. Prodanov, K. Verstreken, Automated segmentation andmorphometry of cell and tissue structures. selected algorithms in imagej. Mol. Imaging, 183–208 (2012)
3. M.I. Bogachev, V.Y. Volkov, O.A. Markelov, E.Y. Trizna, D.R. Baydamshina, V. Melnikov, R.R. Murtazina, P.V. Zelenikhin, I.S. Sharafutdinov, A.R. Kayumov, Fast and simple tool for the quantification of biofilm-embedded cells sub-populations from fluorescent microscopic images. PloS One **13**(5) (2018)
4. Oxford Instruments plc: Imaris v9.5 (2019). https://imaris.oxinst.com/downloads
5. L.-C. Chen, Y. Zhu, G. Papandreou, F. Schroff, H. Adam, Encoder-decoder with atrous separable convolution for semantic image segmentation, in *Proceedings of the European Conference on Computer Vision (ECCV)*, Munich (2018)
6. R.C. Gonzalez, *Digital Image Processing Using MATLAB* (2004), MATLAB, USA.
7. Math Works, regionprops, The MathWorks Inc. (2006). https://www.mathworks.com/help/images/ref/regionprops.html
8. G. Chilkoor, S.P. Karanam, S. Star, N. Shrestha, R.K. Sani, V.K. Upadhyayula, D. Ghoshal, N.A. Koratkar, M. Meyyappan, V. Gadhamshetty, Hexagonal Boron nitride: the thinnest insulating barrier to microbial corrosion. ACS Nano **12**(3), 2242–2252 (2018)
9. A. Bréhéret, Pixel annotation tool (2017). https://github.com/abreheret/PixelAnnotationTool
10. L.-C. Chen, G. Papandreou, I. Kokkinos, K. Murphy, A.L. Yuille, Deeplab: semantic image segmentation with deep convolutional nets, atrous convolution, and fully connected CRFs. IEEE Trans. Pattern Anal. Mach. Intell. **40**(4), 834–848 (2017)
11. pixelLabelDatastore, The Mathworks, Inc. (2017). https://www.mathworks.com/help/vision/ref/pixellabeldatastore.html

12. C. Shorten, T.M. Khoshgoftaar, A survey on image data augmentation for deep learning. J. Big Data **6**(1), 60 (2019)
13. C. Szegedy, S. Ioffe, V. Vanhoucke, A.A. Alemi, Inception-v4, inception-resnet and the impact of residual connections on learning, in *Thirty-First AAAI Conference on Artificial Intelligence*, San Francisco (2017)
14. M. Sandler, A. Howard, M. Zhu, A. Zhmoginov, L.-C. Chen, Mobilenetv2: inverted residuals and linear bottlenecks, in *Proceedings of the IEEE Conference on Computer Vision and Pattern Recognition*, Salt Lake City (2018)
15. P. Soille, *Morphological Image Analysis: Principles and Applications* (Springer Science & Business Media, Berlin, 2013)
16. L. Najman, M. Schmitt, Watershed of a continuous function. Signal Proces. **38**(1), 99–112 (1994)
17. J. Serra, *Image Analysis and Mathematical Morphology* (Academic Press, New York, 1982)
18. imextendedmin, The MathWorks, Inc. (2006). https://www.mathworks.com/help/images/ref/imextendedmin.html

GFDLECG: PAC Classification for ECG Signals Using Gradient Features and Deep Learning

Hashim Abu-gellban, Long Nguyen, and Fang Jin

1 Introduction

Electrocardiogram (ECG) presents the electrical mobility of the heart. The ECG signal is essential for the arrhythmia disease identification. The disease is an irregular heartbeat that is too fast or too slow. The premature atrial complex (PAC) is a kind of heart arrhythmia which is also called the supraventricular premature beat (S type) [1]. PAC is the activation of atria over a pathway other than the sinus node, which is fairly common pervasive in adults who have or do not have heart disease [2, 3]. The common symptoms of PAC are palpitations and missing beats. PAC may be triggered by caffeine, alcohol, abnormal levels of magnesium in the blood, and stress. Furthermore, PAC may be a sign of the underlying heart conditions. In severe conditions, it may change to atrial fibrillation or supraventricular tachycardia that can cause sudden cardiac death.

Using adhesive ECG electrodes is common in care and patient monitoring [4–8]. Recently, considerable research focused on providing anomaly detection techniques to assist cardiologists with diagnosing ECG signals. Therefore, the objectives of PAC classification algorithms are developing an automated detection approach with high performance to increase the productivity of healthcare professionals and to save more lives by early PAC detection, especially for lonely elderly people who stay at home without help. Whereas, the automatic abnormal heart classification from ECG sequences is a challenging task because of imbalance data and other reasons related to the ECG signals (e.g., biomedical contamination, external noise, time-varying dynamics, morphological characteristics) [9].

H. Abu-gellban (✉) · L. Nguyen · F. Jin
Department of Computer Science, Texas Tech University, Lubbock, TX, USA
e-mail: hashim.gellban@ttu.edu; long.nguyen@ttu.edu; fang.jin@ttu.edu

© Springer Nature Switzerland AG 2021
R. Stahlbock et al. (eds.), *Advances in Data Science and Information Engineering*,
Transactions on Computational Science and Computational Intelligence,
https://doi.org/10.1007/978-3-030-71704-9_24

The performance of PAC classification using the ECG dataset [6] has been a critical issue in earlier research as a result of these challenges in the ECG signals. Wang et al. [7] extracted Shapelet features from multivariate time series (MTS) and using the boosting algorithm to combine weak classifiers into a single classifier, which gave 81% accuracy. Karim et al. [10] developed multivariate long short-term memory (LSTM)/attention LSTM (ALSTM) and fully convolutional networks (FCN) to learn long short-term behavior from the raw ECG data without extracting any new feature, resulting in 86% F1-score. Furthermore, Schäfer et al. [11] created a new algorithm called (WEASEL + MUSE) to extract features from MTS. After that, the logistic regression (LR) classifier was executed, which produced 88% accuracy. However, some important features may not be produced by this algorithm because of the complex multi-phase of filtering and selections flows. The previous works did not take advantage of combining good feature extraction techniques and proper neural network architectures for the given problem to mitigate the impact of the ECG signals' issues to foster the performance of their approaches. In general, deep learning approaches are more appropriate than the traditional machine learning classifiers (e.g., LR) in solving high-dimensional time series detection problems.

To address the issues related to the nature of ECG signals and imbalanced datasets, we proposed GFDLECG (gradient feature and deep learning approach for ECG classification). GFDLECG employs a gradient filter and a deep-learning-based model to detect abnormal (PAC) heartbeats of a human subject from their ECG signal behaviors with high performance. It applies an efficient feature generation method called the gradient feature generation (GFG) algorithm, to extract further features from ECG. Our developed approach also applies a novel neural network architecture called multivariate gated recurrent unit and residual fully convolutional GRU networks (MGRU-ResFCNGRU) to identify PAC ECG heartbeat from the original ECG multivariate sequences as well as from the gradient ECG signals. More specifically, our main contributions in this paper are:

- We are one of the first in employing the gated recurrent unit (GRU) and the residual fully convolutional networks with GRU (ResFCNGRU) in a multivariate time series. First, we utilize fast training in GRU while keeping the ability to learn temporal behaviors as in LSTM (long short-term memory). Then, the FCN component is employed as a latent feature extractor for our proposed classification model. We have also added GRU and residual components to FCN, to enhance the overall model performance.
- We proposed the gradient feature generation (GFG) algorithm for given ECG signals in our framework to generate additional important features based on the gradient algorithm. Extracting the features by computing the slopes of electrodes is effective to increase the performance of the model, as the gradient of signals provides the neural networks with the amounts of the signals' changes within a fixed time interval.
- Extensive experiments were conducted to present the capabilities of our approach in discovering PAC from ECG sequences with high performance. The results show that the ECG pattern based on the normal/abnormal identification

with GFDLECG is promising and outperforms the previous approaches. Several other neural networks and feature generation methods were applied, to show the effectiveness of our proposed approach.

2   Problem Formulation

Here, we present our multivariate time series classification problem as follows:

2.1   Multivariate Time Series (MTS)

MTS of M attributes are sequences of simultaneous observations. Let $X = [X^1, X^2, \ldots, X^M]$ be the set of the variables representing M time series and $X^m \in \mathbb{R}^T$. Each time series is called univariate time series (UTS) denoted by $X^m = [x_1, x_2, \ldots, x_T]$, where $x_t \in \mathbb{R}$ is the tth element in the time series.

2.2   The Input

The input MTS^i of the model consists of M variables performed by C class labels, where $i \in \mathbb{Z}_n$ and n is the number of examples. We can denote this as pairs: (X, y_i), where $y_i \in \{0, 1\}$ represents a class label. Zero is the normal class label and one is the abnormal class label.

2.3   Problem Definition

Given the input MTS^i which consists of M features $[X^1, X^2, \ldots, X^M]$, we have to find a function f that classifies the input as normal or PAC heartbeat, as follows:

$$y = f(X^1, X^2, \ldots, X^M) \tag{1}$$

3   Related Work

In this section, we focus on the previous work of the PAC feature generation/selection and classification.

3.1 Feature Generation and Selection from Multivariate Time Series of ECG

Werth et al. [12] employed Lomb-Scargle algorithm [13] to generate the frequency spectrum. The WEASEL + MUSE algorithm was employed [11] to extract features from ECG. [14] applied the Daubechies wavelet 6 filters algorithm [15] to remove noise in the ECG time series and detected the R-peak features using Pan-Tompkins algorithm [16]. Moreover, Li et al. [17] proposed using different techniques to extract features, such as generating statistical features, independent component analysis, discrete wavelet transform (DWT), discrete cosine transform (DCT), and continuous wavelet transform (CWT). Gutiérrez-Gnecchi et al. [18] preprocessed the raw data and extracted both the P and T ECG waves by the Mallat filter bank, the wavelet transformer, and wave detection algorithms. However, discovering P-wave is a complex task [19]. Xia et al. [20] preprocessed ECG using the wavelet method to decrease the noise and extracted features of median beat and 8 characteristics points using R-peaks and RandomWalk algorithms. Finally, Walinjkar et al. [21] used and extracted the human subject's age and instantaneous heartbeat, the ECG's amplitude/WABP, and RR interval from the ECG dataset.

3.2 Traditional Classification Algorithms Using Logistic Regression, Decision Tree, and KNN

Schäfer et al. [11] used the logistic regression classifier to identify the binary class label. Li et al. [17] applied a random forest algorithm to classify heartbeats. Walinjkar et al. [21] applied the bagged tree and the weighted KNN algorithms to detect four class labels including PAC, after dropping normal ECG examples from the training data.

3.3 Deep Learning for Time Series Classification

Kachuee et al. [5] applied five residual convolutional blocks followed by two dense layers to classify the ECG lead II dataset [8]. Werth et al. [12] used ResNet and ResNext architectures employing GRU and BiGRU to learn complex nonlinear patterns in ECG sequences. Acharya et al. [14] applied augmentation and CNN to identify five classes (N, S, V, F, Q). Karim et al. [10] employed MLSTM-FCN and MALSTM-FCN to classify PAC from ECG. Probabilistic backpropagation Neural Network (PNN) was employed to detect eight different ECG class labels, where PAC had a low performance with 76.82% accuracy [18]. Xia et al. [20] concatenated three deep learning network architectures which consist of different layers (e.g., CNN, BiRNN, and Dense) and attention with context blocks, resulting in F1-measure 89.7% for the PAC class label.

4 ECG Binary Classification Framework

4.1 Dataset Preprocessing

We normalize each example to be treated fairly during building the classification model and the evaluation process. The values of the raw data are normalized between [0, 1] by scaling each time series, using the following formula:

$$x_t' = \frac{x_t - \min_m}{\max_m - \min_m} \tag{2}$$

where the maximum and minimum of the mth feature are $\max_m$ and $\min_m$ for all time series in the dataset. After that, we shuffle the examples in the dataset since shuffling is an essential process for the training and testing datasets to represent the entire dataset.

4.2 Gradient Feature Generation (GFG)

We calculate the gradient for each time series (Lead0, Lead1) using the second-order finite central differences of the interior elements of the time series (X^m) [22–24]. We also use the second-order forward and backward differences for the first and last points, respectively. The length of the calculated gradient is the same as the length of the original time series. To compute the gradient, we assume that X^m has three continuous derivatives or more ($X^m \in C^3$). Let h_s and h_d be a heterogeneous step size. We need to minimize the consistency error (η_t). η_t is the differences between the gradient's actual value and estimate value from the adjacent points.

$$\eta_t = X_t^{m(1)} - [\alpha x_t + \beta x_{t+h_d} + \gamma x_{t-h_s}] \tag{3}$$

We compute the following linear system by using Taylor series expansion instead of x_{t+h_d} and x_{t-h_s}:

$$\begin{cases} \alpha + \beta + \gamma = 0 \\ \beta h_d - \gamma h_s = 1 \\ \beta h_d^2 + \gamma h_s^2 = 0 \end{cases} \tag{4}$$

Therefore, the estimation of $X_t^{m(1)}$ is calculated using the following formula:

$$\hat{X}_t^{m(1)} = \frac{h_s^2 x_{t+h_d} + (h_d^2 - h_s^2)x_t - h_d^2 x_{t-h_s}}{h_d h_s (h_d + h_s)}$$

$$+ \mathcal{O}\left(\frac{h_d h_s^2 + h_d^2 h_s}{h_d + h_s}\right) \tag{5}$$

In our experiments, we have used a homogeneous step size ($h_d = h_s$). Therefore, the second order of $\hat{X}_t^{m(1)}$ is computed as the following:

$$\hat{X}_t^{m(1)} = \frac{x_{t+h} - x_{t-h}}{2h} + \mathcal{O}(h^2) \tag{6}$$

We perform a local normalization for every generated gradient lead feature, which has enhanced the performance of our classification model better than the Z-normalization and the Min–Max normalization. We scale each time series individually where the values can be at most 1 and at least -1.

Algorithm 1 describes the steps in more detail. $\mid TS' \mid$ is the absolute value of each element in TS'. $\oplus$ is the operator to concatenate the original MTS which contains lead0 and lead1 signals with their gradients. GFG_MTS contains four sequences that are sent to NN to identify the class label. Figure 1 illustrates the steps of preprocessing from raw data to the input data of NN to detect the PAC class label.

4.3   The Proposed Neural Network

Figure 2 shows our proposed classification architecture using neural networks. We first explain the GRU layer and then the residual FCN with GRU. Finally, we combine all neural networks.

4.3.1   Gated Recurrent Units Network (GRU)

LSTM suffers from the vanishing gradient problem [25, 26]. GRU solves the problem by using the update gate which learns from the previous long short-term pattern, to anticipate the future sequence. We also employ GRU since it is faster to learn than LSTM and it discovers reliances of numerous time scales to enhance the performance of the model [27, 28]. The update formulas of GRU are:

Algorithm 1 The gradient feature generation algorithm

---

function gradient_feature_generation (MTS)

Input : $MTS = X \in \mathbb{R}^{M \times T}$.
Output: $GFG\_MTS = X' \in \mathbb{R}^{M' \times T}$, where $M' = 2 * M$.

$GFG\_MTS \leftarrow copy(MTS)$

foreach $TS \in MTS$ **do**
    /* Compute TS' (the slope of TS)   */
    $TS' \leftarrow copy(TS)$
    foreach $x'_t \in TS'$ **do**
        /* $x_{t+h}, x_t, x_{t-h} \in TS$ */
        if $t = 1$ **then**
          | $x'_t \leftarrow \frac{x_{t+h} - x_t}{h}$
        else
          if $t = T$ **then**
            | $x'_t \leftarrow \frac{x_t - x_{t-h}}{h}$
          else
            | $x'_t \leftarrow \frac{x_{t+h} - x_{t-h}}{2h}$
          end
        end
    end
    /* Scale TS'   */
    $max_{TS'} \leftarrow max(| TS' |)$
    foreach $x'_t \in TS'$ **do**
    | $x'_t \leftarrow \frac{x'_t}{max_{TS'}}$
    end
    $GFG\_MTS \leftarrow GFG\_MTS \oplus TS'$
end
return GFG_MTS

Fig. 1 Data processing pipeline for MTS classification

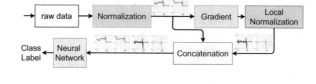

$$z_t^j = \sigma \left(W_z x_t + U_z h_{t-1}^j \right)$$

$$r_t^j = \sigma \left(W_r x_t + U_r h_{t-1}^j \right)$$

$$\tilde{h}_t^j = \tanh \left(W x_t + U \left(r_t^j \odot h_{t-1}^j \right) \right)$$

$$h_t^j = \left(1 - z_t^j \right) h_{t-1}^j + z_t^j \tilde{h}_t^j$$

(7)

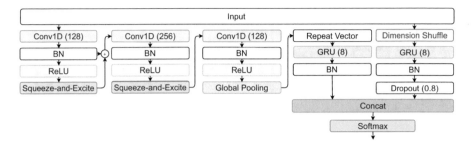

Fig. 2 The architecture of GFDLECG. The width of layer (number of nodes) and percentages of dropout are shown between parentheses

where W and U are the weight matrices. z_t^j, r_t^j, $\tilde{h}_t^j$, and h_t^j present the update gate, the reset gate, the candidate activation, and the activation units at time t, respectively.

4.3.2 Residual Fully Convolutional Networks (ResFCN) with GRU

We use three convolution layers to build the fully convolutional networks (FCN), since FCN is famous NN in finding the semantic segmentation of images [29]. The basic block of convolution:

$$y = W \otimes x + b$$
$$s = BN(y) \tag{8}$$
$$h = ReLU(s)$$

where $\otimes$ is the convolution operator. BN is important to decrease the building time and provide the generalization to NN. ReLU is the rectified linear unit. Squeeze-and-excite is used to extract interdependencies between feature channels [30]. The residual operation (h^{res}) is a shortcut connection to combine the outputs (h^s and h^{se}) before and after the first squeeze-and-excite block, where "+" is the residual operator. In our experiments, we have found that this shortcut improves the accuracy of the PAC classification model.

$$h^{res} = h^s + h^{se} \tag{9}$$

The repeat vector layer is applied to convert the dimensionality of the data from 2D to 3D, to be adequate for GRU layer. GRU is also employed after the global pooling (h^{gp}) to learn the long short-term pattern, which increases the efficiency of our proposed model.

$$h^r = \text{repeat\_vector}(h^{gp}) \tag{10}$$

4.3.3   Combining GRU and ResFCNGRU

Figure 2 shows the concatenation between ResFCNGRU and GRU (after the dropout) neural networks, as follows:

$$h^c = h^b \oplus h^d$$
$$y = \text{softmax}(h^c)$$

(11)

where $\oplus$ is the concatenation operator and h^c is the output of concatenating the left ResFCNGRU neural network (h^b) and the rightmost GRU with the dropout layer (h^d). The class label is predicted by the softmax activation function.

5   Experiments and Results

We start by describing the dataset as in Sect. 5.1. In all experiments, we set $h = 1$ for the GFG algorithm. Section 5.2 shows that the performance of our methodology outperforms the results of other previous approaches. Sections 5.3 and 5.4 brief the advantages of using GRU layer in MGRU-resFCNGRU architecture. At the end, we show the importance of adding the gradient features in the preprocessing phase of ECG signals in enhancing the performance.

5.1   Dataset Description

The ECG dataset has 200 examples (133 normal and 67 abnormal) [6]. Each example is a heartbeat and contains two heart electrodes (Lead0, Lead1) sequences. The abnormal class label represents cardiac pathology (supraventricular premature beat) known as PAC or S type. Time series is a sequence of values ordered by time (t) (equally gaps ECG records). ECG signals have variable lengths between 39 and 152. We unify the length for all time series to be the maximum length ($T = 152$), where empty values are replaced by zeroes. We split the dataset approximately into training (50%) and testing (50%) per a class label.

Figure 3 shows an example for each class label with different original time series lengths. The gradient sequences represent the new features extracted from Lead0 and Lead1 using GFG. We can see that the two class labels have clear differences for the gradient leads. The gradient lead0 of the normal heartbeat has at the end of the original time series (i.e., before adding zeros to have a fixed length) a short crest and a large trough, while the PAC heartbeat contains one high crest at the start of time series and a large trough at the end. Additionally, the gradient lead1 of the normal heartbeat consists of a small crest and a large trough at the end of the original

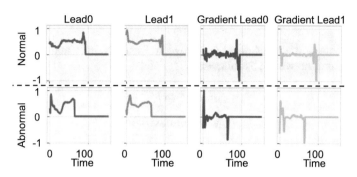

Fig. 3 Examples of normal and abnormal heartbeat sequences including the new gradient lead features

Table 1 Performance summary of our proposed approach compared with other algorithms from previous research using the same dataset. The boldface represents the best performance

Algorithm	Accuracy
Boosting [7]	0.81
WEASEL + MUSE [11]	0.88
MLSTM-FCN [10]	0.86
MALSTM-FCN [10]	0.86
GFDLECG	**0.97**

signal. Whereas, the gradient lead1 has a deep trough without any crest at the end of the original abnormal heartbeat.

5.2 Overall Performance Summary

Table 1 shows the comparative results of our proposed algorithm against the other algorithms [7, 10, 11] from the previous research using the same dataset [6]. Our approach using the gradient feature generation method and the MGRU-ResFCNGRU architecture outperforms the previous methods with 97% accuracy and 97% F1-score. Furthermore, the F1-score of the MDDNN method [31] is 88% while our approach performs better with 9% more (Fig. 4).

5.3 The Effect of GRU in GFDLECG

We ran the preprocessed dataset including the gradient leads features on the same neural network architecture MGRU-ResFCNGRU (GFDLECG), and we also replaced GRU with different layers (BiLSTM [32], LSTM [33], ALSTM [10]). The purpose of this experiment was to manifest the effectiveness of GRU in our proposed architecture. GRU was the best layer with F1-score (97%) as shown in Fig. 5. We

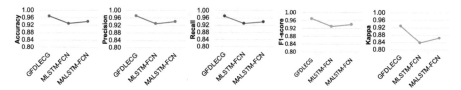

Fig. 4 Performance metrics for GFDLECG (i.e., using the MGRU-ResFCNGRU architecture) and other neural networks using the same gradient method employed in GFG. The y axes are the metrics and x axes are the neural networks for building the models

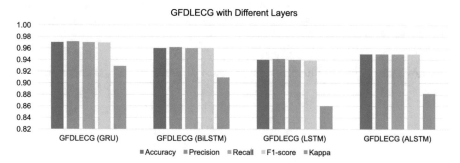

Fig. 5 Performance metrics by replacing GRU in GFDLECG with other layers (BiLSTM, LSTM, MALSTM). The best layer for GFDLECG is GRU which is employed in our neural network architecture for GFDLECG

can see that BiLSTM (96% F1-score) performed better than LSTM (94% F1-score) and ALSTM (95% F1-score).

5.4 GFDLECG vs Other Neural Networks Using Gradient Filter

To show the effectiveness of the new gradient features in our neural network architecture (MGRU-ResFCNGRU) and some other neural networks (MLSTM-FCN and MALSTM-FCN [10]), using GFG, the accuracy of MLSTM-FCN and MALSTM-FCN has been increased from 86% to 93% and 94%, respectively. The results are shown in Fig. 4.

5.5 The Effect of Gradient Feature

We generated different signal filtering methods (Gradient, Lomb-Scargle, Median) to compare the gradient feature generation with other features. Next, we built our classification model using MGRU-ResFCNGRU for all different filtering signals

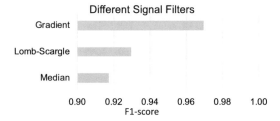

Fig. 6 F1-score of the classification models using different signal filter algorithms. We have also replaced the gradient algorithm in GFG with the Lomb-Scargle and the median algorithms. All preprocessed data have been sent to the MGRU-ResFCNGRU neural network to build the classification model

and raw data. Figure 6 illustrates the increase of performance using the gradient features.

6 Conclusion

In this research, we present a new method for ECG classification to discover PAC heartbeats based on the gradient signal processing and the neural networks. We combine the original signals with the new gradient features using our proposed algorithm called the gradient feature generation (GFG). We have trained the preprocessed ECG signals using a new neural network architecture (MGRU-ResFCNGRU) which takes advantage of the GRU layers, the CNN layers, and the residual operation to foster the performance. GFDLECG outperforms other approaches from previous research using the same dataset. Moreover, the results show that the gradient feature and the neural network architecture are more effective than the other methodologies conducted in the experiments. For future work, we will apply this approach to other multivariate time series problems, to study its effectiveness to enhance the performance.

References

1. A.S. Manolis, H. Calkins, B.C. Downey, Supraventricular premature beats. (2015). http://www.uptodate.com/contents/supraventricular-premature-beats
2. Cardiac health – providing information, risk assessment and advice about your heart health, 2011. http://www.cardiachealth.org/. Accessed 24 February 2020
3. Y. Wang, M.M. Scheinman, W.W. Chien, T.J. Cohen, M.D. Lesh, J.C. Griffin, Patients with supraventricular tachycardia presenting with aborted sudden death: incidence, mechanism and long-term follow-up. J. Am. Coll. Cardiol. **18**(7), 1711–1719 (1991)
4. J. Werth, X. Long, E. Zwartkruis-Pelgrim, H. Niemarkt, W. Chen, R.M. Aarts, P. Andriessen, Unobtrusive assessment of neonatal sleep state based on heart rate variability retrieved from electrocardiography used for regular patient monitoring. Early Hum. Dev. **113**, 104–113 (2017)

5. M. Kachuee, S. Fazeli, M. Sarrafzadeh, ECG heartbeat classification: a deep transferable representation, in *IEEE International Conference on Healthcare Informatics (ICHI)* (IEEE, Piscataway, 2018), pp. 443–444
6. Bobski's World. http://www.cs.cmu.edu/~bobski/. Accessed 8 January 2020
7. H. Wang, J. Wu, Boosting for real-time multivariate time series classification, in *Thirty-First AAAI Conference on Artificial Intelligence* (AAAI, Menlo Park, 2017)
8. A.L. Goldberger, L.A. Amaral, L. Glass, J.M. Hausdorff, P.C. Ivanov, R.G. Mark, J.E. Mietus, G.B. Moody, C.-K. Peng, H.E. Stanley, Physiobank, physiotoolkit, and physionet: components of a new research resource for complex physiologic signals. Circulation **101**(23), e215–e220 (2000)
9. R. Hoekema, G.J. Uijen, A.V. Oosterom, Geometrical aspects of the interindividual variability of multilead ECG recordings, in *IEEE Transactions on Biomedical Engineering*, vol. 48, no. 5 (IEEE, Piscataway, 2001), pp. 551–559
10. F. Karim, S. Majumdar, H. Darabi, S. Harford, Multivariate LSTM-FCNs for time series classification, in *Neural Networks 116* (Elsevier, Amsterdam, 2019)
11. P. Schäfer, U. Leser, Fast and accurate time series classification with weasel. *In Proceedings of the 2017 ACM on Conference on Information and Knowledge Management*, vol. 1711, no. 11343, pp. 637–646 (2017)
12. J. Werth, M. Radha, P. Andriessen, R.M. Aarts, X. Long, Deep learning approach for ECG-based automatic sleep state classification in preterm infants. Biomed. Signal Process. Control **56**, 101663 (2020)
13. T. Ruf, The lomb-scargle periodogram in biological rhythm research: analysis of incomplete and unequally spaced time-series. Biol. Rhythm. Res. **30**(2), 178–201 (1999)
14. U.R. Acharya, S.L. Oh, Y. Hagiwara, J.H. Tan, M. Adam, A. Gertych, R.S. Tan, A deep convolutional neural network model to classify heartbeats. Comput. Biol. Med. **89**, 389–396 (2017)
15. B.N. Singh, A.K. Tiwari, Optimal selection of wavelet basis function applied to ECG signal denoising. Digital Signal Process. **16**(3), 275–287 (2006)
16. J. Pan, W.J. Tompkins, A real-time qrs detection algorithm. IEEE Trans. Biomed. Eng. (3), 230–236 (1985)
17. T. Li, M. Zhou, ECG classification using wavelet packet entropy and random forests. Entropy **18**(8), 285 (2016)
18. J.A. Gutiérrez-Gnecchi, R. Morfin-Magana, D. Lorias-Espinoza, A. del Carmen Tellez-Anguiano, E. Reyes-Archundia, A. Méndez-Patiño, R. Castañeda-Miranda, DSP-based arrhythmia classification using wavelet transform and probabilistic neural network. Biomed. Signal Process. Control **32**, 44–56 (2017)
19. M. Potse, T.A. Lankveld, S. Zeemering, P.C. Dagnelie, C.D. Stehouwer, R.M. Henry, A.C. Linnenbank, N.H. Kuijpers, U. Schotten, P-wave complexity in normal subjects and computer models. J. Electrocardiol. **49**(4), 545–553 (2016)
20. Z. Xia, Z. Sang, Y. Guo, W. Ji, C. Han, Y. Chen, S. Yang, L. Men, Automatic multi-label classification in 12-lead ECGs using neural networks and characteristic points, in *Machine Learning and Medical Engineering for Cardiovascular Health and Intravascular Imaging and Computer Assisted Stenting* (Springer, Cham, 2019), pp. 80–87
21. A. Walinjkar, J. Woods, ECG classification and prognostic approach towards personalized healthcare, in *2017 International Conference on Social Media, Wearable and Web Analytics (Social Media)* (IEEE, Piscataway, 2017), pp. 1–8
22. B. Fornberg, Generation of finite difference formulas on arbitrarily spaced grids. Math. Comput. **41**(184) (1988)
23. D.R. Durran, *Numerical Methods for Wave Equations in Geophysical Fluid Dynamics*, vol. 32 (Springer Science & Business Media, Berlin, 2013)
24. A. Quarteroni, R. Sacco, F. Saleri, Foundations of matrix analysis, in *Numerical Mathematics* (Springer, Berlin, 2007), pp. 1–32

25. K. Cho, B. Van Merriënboer, C. Gulcehre, D. Bahdanau, F. Bougares, H. Schwenk, Y. Bengio, Learning phrase representations using RNN encoder-decoder for statistical machine translation. Preprint. arXiv:1406.1078 (2014)
26. H. Abu-gellban, L.H. Nguyen, F. Jin, LiveDI: an anti-theft model based on driving behavior, in *2020 ACM Workshop on Information Hiding and Multimedia Security (IH&MMSec'20)*. https://doi.org/10.1145/3369412.3395069 (In press, 2020)
27. J. Chung, C. Gulcehre, K. Cho, Y. Bengio, Empirical evaluation of gated recurrent neural networks on sequence modeling. Preprint. arXiv:1412.3555 (2014)
28. L.H. Nguyen, Z. Pan, O. Openiyi, H. Abu-gellban, M. Moghadasi, F. Jin, Self-boosted time-series forecasting with multi-task and multi-view learning. Preprint. arXiv:1909.08181 (2019)
29. J. Long, E. Shelhamer, T. Darrell, Fully convolutional networks for semantic segmentation, in *Proceedings of the IEEE Conference on Computer Vision and Pattern Recognition* (IEEE, Piscataway, 2015), pp. 3431–3440
30. J. Hu, L. Shen, G. Sun, Squeeze-and-excitation networks, in *Proceedings of the IEEE Conference on Computer Vision and Pattern Recognition* (IEEE, Piscataway, 2015), pp. 7132–7141
31. H.-S. Huang, C.-L. Liu, V.S. Tseng, Multivariate time series early classification using multi-domain deep neural network, in *2018 IEEE 5th International Conference on Data Science and Advanced Analytics (DSAA)* (IEEE, Piscataway, 2018), pp. 90–98
32. Y. Zhang, T. Xu, Y. Dai, Research on chatbots for open domain: using BiLSTM and sequence to sequence, in *The 2019 International Conference on Artificial Intelligence and Computer Science* (ACM, New York, 2019), pp. 145–149
33. F. Xiong, K. Zou, Z. Liu, H. Wang, Predicting learning status in MOOCs using LSTM, in *The ACM Turing Celebration Conference-China* (ACM, New York, 2019), pp. 1–5

Tornado Storm Data Synthesization Using Deep Convolutional Generative Adversarial Network

Carlos A. Barajas, Matthias K. Gobbert, and Jianwu Wang

1 Introduction

Forecasting storm conditions using traditional physics-based weather models can pose difficulties in simulating particularly complicated phenomena. These models can be inaccurate due to necessary simplifications in physics or the presence of some uncertainty. These physically based models can also be computationally demanding and time-consuming. In the cases where the use of accurate physics may be too slow or incomplete using machine learning to categorize atmospheric conditions can be beneficial [1]. For related works, see Appendix A of [2].

A forecaster must use care when using binary classifications of severe weather such as those which are provided in this paper. The case of a false alarm warning can be harmful to public perception of severe weather threats and has unnecessary costs. On the one hand, an increased false alarm rate will reduce the public's trust in the warning system [3]. On the other hand, a lack of warning in a severe weather situation can cause severe injury or death to members of the public. Minimizing both false alarms and missed alarms is key in weather forecasting and public warning systems.

With advances in deep learning technologies, it is possible to accurately and quickly determine whether or not application data is of a possibly severe weather condition like a tornado. Specifically one can use a supervised neural network such as a convolutional neural network (CNN) for these binary classification scenarios.

C. A. Barajas · M. K. Gobbert (✉)
Dept. of Mathematics and Statistics, University of Maryland, Baltimore, MD, USA
e-mail: barajasc@umbc.edu; gobbert@umbc.edu

J. Wang
Dept. of Information Systems, University of Maryland, Baltimore, MD, USA
e-mail: jianwu@umbc.edu

© Springer Nature Switzerland AG 2021
R. Stahlbock et al. (eds.), *Advances in Data Science and Information Engineering*,
Transactions on Computational Science and Computational Intelligence,
https://doi.org/10.1007/978-3-030-71704-9_25

These CNNs require large amounts, hundreds of thousands and even millions, of data samples to learn from. Without an ample amount of data to learn from, a CNN has no hope of achieving accurate predictions on anything except the original training data provided. Of the 183,723 storms in the data set used in this work, only around 9000 entries have conditions which lead to tornadic behavior in the future [4]. This imbalance of tornado versus no tornado results in a situation where a machine is very good at predicting no potential tornado but is very bad at predicting when there is a tornado imminent hence false negatives.

It is for these reasons there is a real motivation to acquire more data that would result in tornadic conditions. This heralds the need of synthetic data to bolster the amount of data used for training a neural network. Synthetic data must be generated such that it is indistinguishable from real data and can be used in conjunction with the natural data to train a neural network on a more balanced data set which produces less if any false negatives.

2   GAN-Based Data Augmentation

Each generative adversarial network (GAN) has not just one neural network but rather two networks which compete against each other to generate the best synthetic data possible. There are two parts of a GAN that make it an effective producer of synthetic data. The generator takes in random data and uses this to generate fake data. The discriminator understands what real data looks like, and because of this, it is capable of determining whether or not any given data is fake or real. Together these two pieces make a GAN capable of producing synthetic data similar to given data that is indistinguishable from the original data qualitatively. The process of training the GAN means providing the discriminator enough data that it can judge whether or not provided data is fake or real. Given this feedback, the generator must adapt by generating more realistic data such that the discriminator cannot tell the difference between the falsified data and the real natural data. If the generator can fool the discriminator, which is designed to be an expert on the data, then the synthetic data is considered just as good as naturally collected data. Data obtained from this properly tuned GAN is typically considered more robust and a much more effective method for training than the primitive method of duplication [5].

This allows for the generation of a plethora of new data which is distinct from all previously generated data and also unique in that it is not an augmented version of any of the original data. The more interesting prospect is that, given a data which is tornadic in nature, a GAN can be used to generate synthetic data that is also promised to be tornadic [6]. With a GAN, new original data can be generated for the training of the CNN that would be used for prediction. This CNN when given real natural data from an upcoming storm would be able to accurately and instantly deliver a verdict of tornadic or not, rather than waiting for a simulation to finish days later.

3 Data

The data set used in this analysis was obtained from the Machine Learning in Python for Environmental Science Problems AMS Short Course, provided by David John Gagne from the National Center for Atmospheric Research [7]. Each file contains the reflectivity, 10 meter U and V components of the wind field, 2 meter temperature, and the maximum relative vorticity for a storm patch, as well as several other variables. These files are in the form of $32 \times 32 \times 3$ images describing the storm. We treat the underlying data as an image and push it through the CNN as if it were a normal RGB image. This allows our findings to generalize to other non-specialized CNNs. Figure 1 shows two example images from one of these files. Storms are defined as having simulated radar reflectivity of 40 dBZ or greater as seen in Fig. 1b. Reflectivity, in combination with the wind field, can be used to estimate the probability of specific low-level vorticity speeds. In the case of Fig. 1a, the reflectivity and wind field were not sufficient enough to cause future low-level vorticity speeds. The data set contains nearly 80,000 convective storm centroids across the central United States.

We preprocessed the original NCAR storm data containing 183,723 distinct storms, each of which consists of $32 \times 32 \times 3$ grid points, and extracted composite reflectivity, 10 m west-east wind component in meters per second, and 10 m south-north wind component in meters per second at each grid point giving approximately 2 GB worth of data. We use the future vertical velocity as the output of the network. This gives us three layers of data per storm entry producing a total data size of $183,723 \times 32 \times 32 \times 3$ floats to feed into the neural network. We use 138,963 storms for training the model and 44,760 storms for testing the accuracy of the model. We track the total wall time for training and testing over both image sets.

With only a handful of tornadic cases present in the base data set, we used primitive augmentation techniques to bolster the number of tornadic cases to be fed into the DCGAN. The only primitive augmentation technique used was rotation. Reflection and translation could generate a storm that might not be physically possible.

Fig. 1 Sample images of radar reflectivity and wind field for a storm which (**a**) does not and (**b**) does produce future tornadic conditions

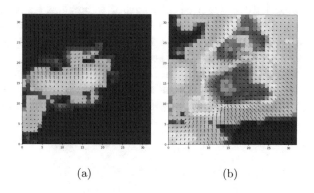

(a)   (b)

4 Results: DCGAN-Based Weather Data Synthesization

The input data was the original input data inflated by primitive augmentation techniques covered in Sect. 3. The constants for training were learning rate of 0.001, batch size of 128, data multiplier of 1, and 1 GPU. The images produced by the generator are logged every 25 epochs and evaluated qualitatively as they improve in realism. For implementation details, see Appendix B of [2].

Consider Fig. 2 which contains several rows of images. Each row represents three images generated at the listed epoch number. The first row of the figure contains three images generated before any training was done. whereas the last row represents three images generated after all the training had been done. The images in the first row are more like noise than real weather which is to be expected as the generator takes in raw noise from a Gaussian distribution. The second row of images are from the 25 epoch markers. Each image has some of the hallmarks of tornadic storm. There are clear attempts at nesting reflectivity levels such as putting higher reflectivity in centers or groups. Yet the generator has not been able to really gauge how sensitive the ranges should be and is mixing high and low reflectivity where one might expect the centers to be areas with the most reflectivity. Additionally the wind velocities appear to be very random and non-sensical. At 50 epochs, the generator has learned how to more properly gauge relative reflectivity levels. The sizes of the high reflectivity clusters seem very small and uneventful. Additionally there is not enough variation in reflectivity. The images are just mostly high levels of reflectivity rather than a concentrated area of high reflectivity which transitions to very low reflectivity over time. The wind patterns are being created in ways that they are moving in ways relative to the clusters of reflectivity which is a positive sign.

75 and 100 epochs are where the generator has a solid grasp of what it should be doing, and the differences between these stages are subtler than previous ones. The storms for 75 and 100 epochs have a rich variation of high and low reflectivity especially when compared to previous epoch counts. Each has a clear set of centers that smoothly transition into lower reflectivity as you move away them. The wind patterns have a clear dependence on the center's position and intensity. This is especially apparent when compared to previous epoch markers. At 75 epochs, the storms were either very mundane or had a single high activity center or two. The majority of the storms looked well-formed but still not as distinct as would be expected. The 100 epoch benchmark produced the most varied and interesting data set. All of the storms have different shapes, sizes, intensities, and centers. There is a clear and smooth transition from high reflectivity to low reflectivity. The wind patterns seem to interact with the centers and have purpose.

Fig. 2 Sample images of
GAN generator output at 0,
25, 50, 75, and 100 epochs

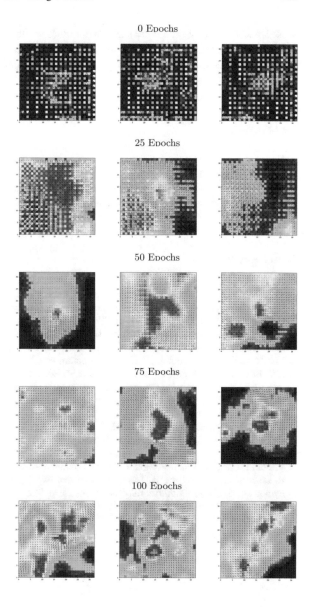

5 Conclusions

The DCGAN is the first complex network trained using the new framework from
[4]. The images started out as nothing more than a plottable noise. As the epochs
progress, every new set of generated images slowly gained additional qualities that
put them closer to the realm of realism. At the 75 epoch mark, most of the storms
were mostly indistinguishable from the real storms in the data set; however there

were small features which were still not accurate to the careful eye of a layperson. At 100 epochs, the images were completely indistinguishable from the real storm data set by a layperson. The transitions, the realistic concentrations of reflectivity, the obvious correlation of wind velocity relative to the concentrated activity, and the rich variety of storms produced were all quality. The data set might be able to be used to train the predictive network to predict real storms as the data is indistinguishable by the casual observer.

Acknowledgments This work is supported in part by the US National Science Foundation under the CyberTraining (OAC–1730250) and MRI (OAC–1726023) programs. The hardware used in the computational studies is part of the UMBC High Performance Computing Facility (HPCF). Co-author Carlos A. Barajas was supported as HPCF RA as well as CyberTraining RA.

References

1. V. Nourani, S. Uzelaltinbulat, F. Sadikoglu, N. Behfar, Artificial intelligence based ensemble modeling for multi-station prediction of precipitation. Atmosphere **10**(2), 80–27 (2019)
2. C.A. Barajas, M.K. Gobbert, J. Wang, Tornado storm data synthesization using deep convolutional generative adversarial network (DCGAN): related works and implementation details. Tech. Rep. HPCF–2020–19, UMBC High Performance Computing Facility, University of Maryland, Baltimore County (2020). http://hpcf.umbc.edu
3. L.R. Barnes, E.C. Gruntfest, M.H. Hayden, D.M. Schultz, C. Benight, False alarms and close calls: a conceptual model of warning accuracy. Weather Forecast. **22**(5), 1140–1147 (2007)
4. C.A. Barajas, M.K. Gobbert, J. Wang, Performance benchmarking of data augmentation and deep learning for tornado prediction, in *2019 IEEE International Conference on Big Data (Big Data)* (IEEE, Piscataway, 2019), pp. 3607–3615
5. F.H.K. dos Santos Tanaka, C. Aranha, Data augmentation using GANs. ArXiv abs/1904.09135 (2019)
6. V. Bok, J. Langr, GANs In Action. Manning Publications, USA, (2019)
7. R. Lagerquist, D.J. Gagne II, Basic machine learning for predicting thunderstorm rotation: Python tutorial. https://github.com/djgagne/ams-ml-python-course/blob/master/module_2/ML_Short_Course_Module_2_Basic.ipynb (2019)

Integrated Plant Growth and Disease Monitoring with IoT and Deep Learning Technology

Jonathan Fowler and Soheyla Amirian

1 Introduction

Artificial intelligence (AI) has permeated human life in all aspects, and utilizing it is beneficial in health, science, agriculture, economics, and finance. In agriculture, there are many applications for AI. For example, monitoring of health and disease on plants play an important role in farm profitability, as AI has the ability to reduce costs and time to market. AI alone does not achieve this goal; rather, an integrated solution employing a variety of technologies would best serve the market.

Sensors that detect environmental variables such as air temperature, light level, soil moisture, and CO_2 levels are available to consumers from a variety of vendors. This competition and innovation allow the consumer to build a solution at a nominal cost and expect further improvements in both capability and affordability. *Time-series databases* are built specifically for collecting metrics that change over time, such as CPU load or website traffic. In our integrated solution proposal, the database plays a central role, as we are measuring the condition of plants over time. Intervention data, sensor data, and image processing data all come together in this database and are synchronized by time stamp in order to show the effects of input and output across all variables.

Deep learning convolutional neural networks play a key role in many applications such as image recognition. Image recognition is used to perform a large number of visual tasks, such as understanding the content of images. Deep learning

J. Fowler (✉)
Colorado Technical University, Department of Computer Science, Colorado Springs, CO, USA
e-mail: jon@fowlercs.com

S. Amirian
The University of Georgia, Department of Computer Science, Athens, GA, USA
e-mail: amirian@uga.edu

© Springer Nature Switzerland AG 2021
R. Stahlbock et al. (eds.), *Advances in Data Science and Information Engineering*,
Transactions on Computational Science and Computational Intelligence,
https://doi.org/10.1007/978-3-030-71704-9_26

tasks for image recognition are categorized into classification, object detection, and segmentation. *Image classification* is a topic of image recognition to classify the contextual information in images. In our proposed system, the classification task is formally defined as the process of labeling plant images if healthy or sick, to detect diseases using the leaves' images. The classification is a principal component of our solution, one part of a whole system working in unison.

That comprehensive system utilizes a time-series database to collect all inputs regarding the plant state. It includes manual interventions, drone imagery, and plant sensors. We propose this system to protect our plants, save expenses, reduce labor, and mitigate risk. In Sect. 2, we review previous research in the various technology areas. Then, we define the problem in Sect. 3. In Sect. 4, the architecture and the methodology of our proposed system are explained. At last, we discuss the future of our research.

2 Literature Review

In the field of combining agriculture and technology, there are many works that have been done. In 2008, Vellidis et al. [1] developed a prototype temperature and moisture sensor for cotton plants. This solution involved Watermark© sensors and thermocouples, radio frequency identification (RFID), and a custom circuit board. Kassim et al. [2] designed an environmental Intelligent Greenhouse Monitoring System (IGMS) based on Wireless Sensor Network (WSN) technology which is used to monitor temperature, humidity, and soil moisture. They found that soil moisture is the key parameter to determine when to irrigate and the amount of water to supply. In addition, the authors determined that automated delivery of water fertilizer based on the soil moisture readings was more effective and efficient than scheduled delivery. TongKe [3] examined different components of a smart agriculture solution, bringing together cloud architecture, RFID technology, IoT plant sensors, and data visualization. The authors proposed this solution in response to issues facing rural farmers in China. Badhe et al. [4] introduced a soil monitoring system using a MCP3204 A/D converter, a DHT11 board, and a Raspberry Pi. This system collected temperature, moisture, light, humidity, and pH value. The authors proved out the architecture and design for collecting the information and alerting based on optimal thresholds.

3 Problem Statement

Care of plants in any large-scale agricultural setting is a heavily manual process that is time-consuming and dependent upon both the human eye and inaccurate estimates of soil conditions. The effect of water and nutrients are not known real-time, and any adverse effects show up as latent indicators. At present, the time, labor, and inaccuracies in plant care make scalability a major concern in large-scale agricultural operations. Research in the various component areas of this integrated solution has yielded a better understanding of their applications, which we explored in Sect. 2, but the research on an integrated solution bringing them all together is scant.

4 Architecture and Methodology

The system includes three primary points of data collection: (1) sensors assigned to the plant itself, (2) drones collecting high-resolution images, and (3) a log of manual activities such as watering or pruning. The data are compiled in a cloud VM and synchronized in a time-series database. The images are processed by a deep learning model to detect signs of disease, and an alerting protocol notifies the user via email or SMS text if the detection threshold is met. An overview of the solution is shown in Fig. 1.

4.1 Analytics Server

The analytics server itself is an Ubuntu virtual machine (VM) running on the Linode cloud service. It is a 4-core instance with 8 GB RAM and 160 GB of storage. A cloud VM was chosen over a physical on-premise server due to cost and scalability. Key components of the VM include the time-series database (InfluxDB), data collector (Telegraf), Python scripts, dashboard service (Grafana), and Apache web server.

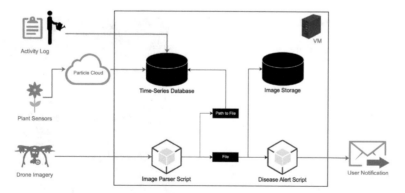

Fig. 1 The proposed architecture. The framework collects the data from: (1) a log of manual activities, (2) sensors, and (3) drones. The data are compiled in a cloud server and synchronized in a time-series database. The images are processed by a convolutional neural network model to detect signs of disease, and an alerting protocol notifies the user via email/SMS text if the detection threshold is met

4.2 Activity Log

Manual interventions, such as watering the plants, feeding them, pruning them, relocating them, etc., are recorded in an activity log and stored in the time-series database. It is necessary to maintain such a log of manual interventions in order to have as complete a picture as possible of cause and effect on the plants. For example, if soil moisture spiked at a particular time, we may expect to see a corresponding entry in the event log that indicates the plant was watered.

4.3 Sensors

The data collection from the individual plant is handled by a modified Particle Photon sensor [5]. An example of this assembly is shown in Fig. 2. The sensor collects ambient temperature, humidity, light level, soil moisture level, CO_2 level, and TVOC level. These readings are pushed to the Particle Cloud and then delivered to the analytics server, powered by InfluxDB and its related Telegraf data collector plugin [6]. The readings are made and sent every 10 s and can be downsampled at the database level to decrease complexity. A simple dashboard, shown in Fig. 3, illustrates the real-time delivery of information from the sensor package.

4.4 Image Intake and Analysis

High-resolution images of individual plants are taken at regular intervals by drones flying over the specified area. These images are fed to the analytics server, where

Fig. 2 The sensor package deployed on a test plant. It collects ambient temperature, humidity, light level, and soil moisture level

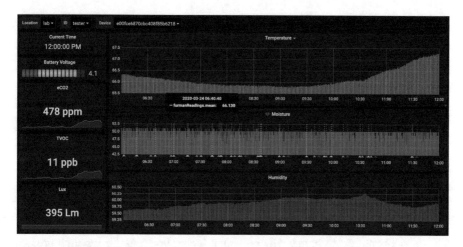

Fig. 3 A prototype dashboard showing readings from the sensor package shown in Fig. 2. The readings are made and sent every 10 s

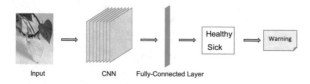

Fig. 4 Overview of the deep learning methodology for classifying images as healthy or sick

a Python script stores the file in block storage and pushes the file location string to the time-series database. A second Python script, known as the classifier, analyzes the image utilizing a deep learning model trained on photos of healthy and diseased plants like those present in the sample. The classifier script produces a likelihood of disease, and a score above a specific threshold triggers an alert via email or SMS message.

The classifier script itself is powered by Keras framework. The images taken by drones feed in to the classifier. The convolutional neural network system extracts the information from the image, and based on the trained process, the system classifies it as healthy or sick. The process of the classifier is shown in Fig. 4.

4.5 Alerting System

When an image is identified as a sick plant, an alert is sent via email or SMS to the designated recipients. This alert includes the image, the likelihood of disease score, and the time stamp. The intention is for the recipient(s) to query the database for activities and sensor readings from a specified time window, searching for variations in the sensor readings and manual activity logs to understand what conditions led to the sick plant.

In its current proposal form, the alerting system is responsive, not proactive. The system alerts a recipient, and the next steps are on the recipients, manually intervening in order to identify and change the conditions that led to the sick plant. Our intention is to catalog those manual interventions in order to build a model for various responses to the conditions that led to the sick plant. The ultimate goal is to have an automated response with minimal manual intervention. More on this future research is explained in Sect. 5.

5 Discussion and Further Research

The concepts presented in this paper are an ongoing investigative research project. Here we discussed what the proposed system hopes to achieve and implications for future research, possibly setting the stage for a follow-up paper in collaboration with an agricultural subject matter expert. An important next step is further developing the alerting system to transition to a semi-automated or fully-automated response system, based on the preferences of the grower. A fully automated response system would utilize the machine learning models produced from the process described in Sect. 4 to apply the correct interventions (e.g., more water, plant food, pest control measures, etc.). A semi-automated response system would leverage the same process but allow the grower to choose the intervention from a list recommended by the model. Further work will require a much more powerful VM or on-premise server with a large amount of RAM and at least one GPU for deep learning. Deployments of this solution will involve different plant species, which will require model training for these species and the necessary computing power to learn and compare.

References

1. G. Vellidis, M. Tucker, C. Perry, C. Kvien, C. Bednarz, A real-time wireless smart sensor array for scheduling irrigation. Comput. Elect. Agric. **61**(1), 44–50 (2008)
2. M.R.M. Kassim, I. Mat, A.N. Harun, Wireless sensor network in precision agriculture application, in *2014 International Conference on Computer, Information and Telecommunication Systems (CITS)* (IEEE, Piscataway, 2014), pp. 1–5
3. F. TongKe, Smart agriculture based on cloud computing and IoT. J. Converg. Inf. Technol. **8**(2) (2013)
4. A. Badhe, S. Kharadkar, R. Ware, P. Kamble, S. Chavan, IoT based smart agriculture and soil nutrient detection system. Int. J. Fut. Revol. Comput. Sci. Commun. Eng. **4**, 774–777 (2018)
5. Particle.io, Particle Photon. Accessed 30 December 2019
6. InfluxData, Telegraf 1.13 documentation. Accessed 30 December 2019

Part V
Data Analytics, Mining, Machine Learning, Information Retrieval, and Applications

Meta-Learning for Industrial System Monitoring via Multi-Objective Optimization

Parastoo Kamranfar, Jeff Bynum, David Lattanzi, and Amarda Shehu

1 Introduction

While industrial operators have utilized automated monitoring for several decades, advances in machine learning (ML) have increased both the capabilities and the complexity of monitoring systems for physical, industrial processes. In particular, automated systems for long-term industrial health monitoring are becoming increasingly important in the workplace across various industries [7]. Industrial health monitoring is a challenging application domain, and research on how to put together reliable ML-based frameworks that are capable of diagnosing degradations of mechanical systems is growing.

To illustrate the challenges involved, we point to a Selective Compliance Articulated Robot Arm (SCARA) that is a popular option for small robotic assembly applications. Selective compliance refers to the fact that the SCARA robot is compliant in the X-Y axis and rigid in the Z axis. Figure 1 shows a SCARA robot.

Obtaining data on such robots is an expensive process. Often, the only data we have available are passively collected sensor data. A fundamental problem we consider in this paper is the following: based on acoustic monitoring data, distinguish the movements of a SCARA robot. While this is a foundational step toward building systems for recognizing mechanical degradations, the problem is surprisingly challenging. The datasets available are small. The one provided to us by our industrial collaborators only contains 842 samples. Moreover, we do not have labels for all the data, as labeling requires manpower; out of the 842 samples, only 79 are labelled. Moreover, the SCARA's repetitive motions are broken down into 7 basic ones, that is, the labeled samples can have one of 7 available labels.

P. Kamranfar · J. Bynum · D. Lattanzi (✉) · A. Shehu
George Mason University, Fairfax, VA, USA
e-mail: pkamranf@gmu.edu; jbynum@gmu.edu; dlattanzi@gmu.edu; ashehu@gmu.edu

© Springer Nature Switzerland AG 2021   397
R. Stahlbock et al. (eds.), *Advances in Data Science and Information Engineering*,
Transactions on Computational Science and Computational Intelligence,
https://doi.org/10.1007/978-3-030-71704-9_27

Fig. 1 A SCARA robot in the workplace [16]. SCARA stands for Selective Compliance Articulated Robot Arm; the robot is compliant in the X-Y axis and rigid in the Z axis. Y-motions are base movements. X-axis motions are arm extensions/retractions

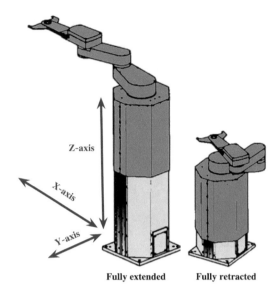

Z-axis

X-axis

Y-axis

Fully extended   **Fully retracted**

The small- and scarcely labeled data regime posed by this setting is typical of industrial settings. This regime forces attention on shallow models that utilize features, parameters, and hyper-parameters sparingly so as to avoid overfitting. Our focus in such a setting is not a superficial one. We are not just interested in putting together a framework that is shown, based on selected metrics, effective. Instead, we take a more deliberative approach here and consider this particular real-world problem important to our industrial collaborators as a motivating example for what is important to consider in the broader setting of industrial monitoring.

In the broader setting, systems designed for industrial monitoring utilize various algorithms in a step-wise fashion. At a minimum, a reduction/compression algorithm is put together (and possibly optimized over its parameter space) to reduce a large number of engineered, domain-specific, and/or domain-agnostic features. The resulting features are then utilized by a feature-based learning algorithm. In the context of scarcely labeled data, our attention is narrowed to unsupervised learning algorithms via clustering. There are many feature reduction algorithms, as there are many clustering algorithms operating over the reduced feature space. For a given algorithm, various models can be put together by considering different values for their parameters and hyper-parameters. The consideration of different algorithms, their interplay, and different models for an algorithm gives rise to a vast system configuration space whose exploration for optimal system design poses challenges well beyond basic model selection.

This paper considers the following problem: how to explore a vast system configuration space in search of optimal system design(s). This problem is known as meta-learning in the broader ML community. In industrial monitoring, the problem is not investigated due to the complexity of the system space. In this paper, we propose a Pareto-based, multi-objective optimization approach to explore the configuration space of an industrial system and select an optimal design (or many near-optimal designs) along various performance metrics.

To ground our investigation in a real-world setting that represents the challenges inherent in industrial system monitoring, we evaluate our approach on noninvasive acoustic data collected from a SCARA robot operating in the assembly floor. A detailed evaluation compares solutions obtained via this approach and those obtained via other existing approaches and shows them superior in distinguishing movements of the SCARA robot. More importantly, our contribution generalizes beyond the specific application motivating the work presented in this paper. Our multi-objective optimization approach for meta-learning can be easily generalized for monitoring systems in a variety of industrial processes.

2 Prior Work

Significant efforts have been paid to extracting features in industrial monitoring. These efforts can be categorized into domain-specific and domain-agnostic. There is rich literature in extracting features via domain-specific insight. Work in [12] reviews the benefits and limitations of common statistical, temporal, and spectral-based features for diagnosing machinery faults. Signal energy metrics and statistical features such as kurtosis, skewness, and crest factor have also been utilized in industrial settings [15]. Though often with little concern of overfitting in the presence of limited data, deep neural network-based approaches have also been employed to obtain domain-agnostic features in an end-to-end manner [25]. Dimensionality reduction via principal component analysis (PCA) has been useful in obtaining a lower number of collective features combining numerous domain-specific or domain-agnostic ones [15].

The intended purpose for the featurization of a dataset at hand is to employ the features to build a model over the data. The process of finding the best model out of generated/considered models is known as model selection [11, 19]. Some attention has been paid to model selection in industrial health monitoring. We highlight a Bayesian framework for auto-regressive model selection introduced in [22] and the approximate Bayesian computation algorithm in [1]. Some attention is also paid to understanding and optimizing the various hyper-parameters that determine performance or to evaluating the quality of model predictions in an objective manner [8].

Model selection is an active area in ML. It is worth noting that model selection, hyper-parameter optimization, and tuning are terms used interchangeably to explain the process of setting up an ML algorithm by tuning its hyper-parameters to obtain a well-performing model among available ones [10].

Extensive work has been carried out in model selection [4]. Work in [11] tackles the problem of hyper-parameter tuning by considering two strategies: better optimization and reduction of the number of hyper-parameters. Work in [17] suggests a model selection framework to automatically select the model type and kernel function and optimize the hyper-parameter of the selected function by making use of genetic algorithms as the selection mechanism.

Data analysis systems for physical, industrial processes are increasingly complex and often utilize several algorithms in a multistep process. Due to the scarcity of labeled data, the focus is on systems that utilize unsupervised learning algorithms (via clustering) to extract knowledge from data. Due to the small-data regime that rules much of the industrial and physical settings, such algorithms operate on features that combine engineered (domain-specific) features via some dimensionality reduction/compression algorithm. Therefore, for the purpose of demonstrating the complexity of a real-world application of ML technologies, the data analysis systems most suitable at the moment for industrial processes consist, at a minimum, of two algorithms, one that reduces engineered features and one that utilizes such features under the umbrella of unsupervised learning.

Limiting attention to model selection presupposes that one has decided the feature reduction (similarly, the unsupervised learning) algorithm and is seeking the best model for a specific algorithm. The complexity of data analysis systems for industrial processes necessitates going beyond model selection and carrying out some form of learning over the system space. This consists of learning over the algorithm involved, their interplays, and the model space for each algorithm. This problem is known as meta-learning [14]. To the best of our knowledge, no attention has been paid to meta-learning in the industrial health monitoring domain. However, meta-learning (as well as model selection) is an active research area in the ML community, though there is ambiguity in ML literature on what exactly a meta-learning problem is. Work in [14] has thoroughly reviewed the notion of meta-learning and proposes a consensus definition.

The idea of algorithm selection or algorithm recommendation was firstly formalized in Rice's work [21] to illustrate the selection/suggestion of a suitable algorithm depending on characteristics of the target problem, including calculation of appropriate features (known as meta-features) [23]. The focus of algorithm selection/recommendation is to systematically learn the relationship between data characterization and algorithm performance in terms of rules [3] or by using visual analysis [5]. Work in [19] reviews model selection, algorithm selection, and model evaluation under the umbrella of supervised learning from a statistical point of view.

The efforts we propose in this paper fall under the umbrella of meta-learning, as we consider a vast system design configuration space, with configurations consisting of design choices over algorithms for feature reduction, algorithms for unsupervised learning via clustering, and (hyper-)parameter optimization. It is not a surprise that meta-learning for industrial health monitoring has not been attempted; the size of the configuration space (as we illustrate in Sect. 4) is overwhelming. Moreover, it is unclear what an optimal configuration is, as complex design choices can be evaluated via various metrics. How does one select over a vast configuration space where each configuration can be evaluated according to different criteria?

This question is not limited to industrial monitoring but is of general utility for meta-learning. We address it here by leveraging multi-objective optimization. In particular, we have drawn upon Pareto optimality and utilize it to debut a novel, Pareto-based approach for meta-learning.

Pareto-based optimization has been employed to solve different optimization problems [2, 10]. Early work in [13] advocates posing an ML problem as a multi-objective one and utilizing Pareto-based optimality. However, due to challenges in recasting ML algorithms driven by minimization of a loss function, such work has not made a measurable impact.

As laid out in Sect. 1, we build over a framework of multi-objective optimization and debut a novel, Pareto-based meta-learning approach. The proposed approach balances multiple performance metrics to select Pareto-optimal design configurations over a vast combined algorithm and model space. The approach is grounded in a specific application in industrial monitoring to distinguish the movements of an industrial robot from its recorded acoustic signals. We now describe the approach in greater methodological detail.

3   Methodology

For the purpose of grounding the proposed Pareto-based, multi-objective meta-learning approach in a concrete setting, we consider the following monitoring system. The system's goal is to distinguish the motions of the SCARA robot from scarcely labeled acoustic data. Because of the small- and scarcely labeled data, the system consists of the following three steps: feature design, feature reduction, and clustering. We first describe in some detail each step and the various choices each step entails. We then relate the complexity of the configuration space resulting from these various choices and detail the Pareto-based optimization approach proposed to handle such a space and select an optimal configuration (or many optimal configurations).

3.1   Feature Design

Feature design is not the focus of this paper. Instead, we leverage existing work and expertise and capture complex harmonics present in vibration signals as follows: we first use six-level wavelet decomposition to obtain 51 features, which include wavelet skewness, kurtosis, crest factor, k-factor, peak amplitude, mean amplitude, and more [9]. These 51 features give rise to a 51-dimensional space.

3.2 Feature Reduction

It is well-understood that in high-dimensional space data becomes sparse, and distance-based clustering performs poorly. Dimensionality reduction algorithms allow reducing the number of features while capturing linear and nonlinear correlations among the original features. There are now many such algorithms. To illustrate the various choices involved, we focus here on three representative ones: PCA, as a representative of linear algorithms, and IsoMap and spectral embedding, as representatives of nonlinear algorithms. In each of them, an important decision concerns how many of the obtained, reduced features to retain for unsupervised learning.

In PCA, cumulative variance is commonly used to make this decision. Figure 2 shows the cumulative variance on the SCARA dataset employed in Sect. 4, highlighting the cumulative variance for different numbers of top principal components (PCs). Typically, the compromise is between desiring a low number of features while retaining sufficient variance. A cumulative variance of 80% is a threshold that is often employed.

Unlike PCA, IsoMap is a nonlinear algorithm that measures the distance between two data points via geodesic distance on a nearest-neighbor graph. An important decision in IsoMap is not only the number of components/features but also the number of nearest neighbors [24], thus adding an additional system design parameter. IsoMap is not a variance-preserving model; hence, variance analysis cannot be applied to select components. One needs to consider a varying number of components, as well as a varying number of neighbors in the context of some other metric that needs to be optimized.

Spectral embedding is another nonlinear algorithm that aims to find the underlying structure of data by learning the manifold where the data sits. The model computes the Graph Laplacian by means of an affinity matrix calculated using nearest neighbors [18]. Similar to IsoMap, one needs to consider a varying number of components/features, as well as a varying number of neighbors.

Fig. 2 Cumulative variance of PCs over SCARA dataset

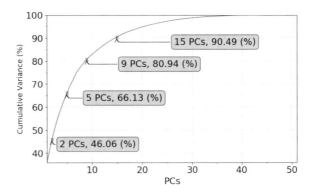

3.3 Unsupervised Learning via Clustering

Once features are extracted, the featurized data is then subjected to clustering. Among many available clustering algorithms, we consider here two widely adopted ones, agglomerative hierarchical clustering and k-means.

K-means is the most popular algorithm that generates clusters by finding cluster centroids. The algorithm starts with k centroids chosen at random from the data, with k being a parameter, and assigns data to clusters in a way that minimize the within-cluster scatter. Note that given a number of clusters, the algorithm can be run over many times to possibly obtain different cluster assignments of the data, since the results depend on the (at-random) initialization of the cluster centroids. Section 4 that evaluates k-means considers different variants resulting from different runs/initializations. One can also vary the number of clusters. The elbow finding approach is a popular one for identifying the number of clusters in an automated manner. Briefly, one varies k and measures the sum of squared errors (SSE) for each k. SSE sums the squared distances of each data point from its assigned cluster centroid. More clusters result in low SSE, but the goal is to find the elbow of the SSE curve, which is the region where further increasing k yields diminishing returns.

Agglomerative hierarchical clustering does not need prior specification of the number of clusters. Agglomerative clustering starts by considering each point as a cluster and continues by merging clusters until reaching the root of tree or dendrogram [18]. By cutting the dendrogram at different places, the desired number of clusters can be achieved.

3.4 Metrics for Evaluating Quality of Clusters

Evaluating the quality of discovered clusters is an active area of research. Clustering quality/validity metrics can be categorized into two main groups. When there is prior information about cluster labels in a dataset, external metrics can be measured. Otherwise, internal indices can be utilized to measure the quality of clusters by means of their structures and with no help from external information [20]. We consider both external and internal metrics here, since it may be the case that labels are available for a small portion of an industrial dataset or not available at all. For the sake of completeness, we explain silhouette score and cluster entropy as follows. We modify the cluster entropy definition to design a new index, label entropy, to capture additional information regarding the quality of clustering in the presence of some labeled data.

Silhouette Score (SC) This is an internal metric for evaluating the separability of clusters. $SC(i) = \frac{b(i)-a(i)}{max(b(i),a(i))}$ measures the inherent similarity between observation i and the cluster to which it is assigned; $a(i)$ is the mean Euclidean distance between observation i and others within its cluster, and $b(i)$ measures the minimum average Euclidean distance between observation i and all observations on the other clusters. The denominator $max(a_i, b_i)$ provides normalization. A higher score suggests high similarity, approaching 1, between observations and cluster assignments [20].

Cluster Entropy (CE) Cluster entropy is typically used to estimate the purity of clusters by using their labels [20]. This metric measures the homogeneity of a cluster by considering all participating data point labels within a cluster and finding the fraction of points with the same class label for each of the labels represented in a cluster: $CE(c) = -\sum_k p(k|c) \log p(k|c)$, where k represents class labels, c stands for a cluster, and $p(k|c)$ shows the frequency of class label k in cluster c. A lower cluster entropy presents a purer cluster; 0 is the minimum value, which is reached when all points in a cluster have the same class label.

Label Distribution Entropy (LDE) We propose this metric here as a modification of cluster entropy to take class label information into consideration. This leverages the existence of labels for a small portion of the dataset; its purpose is to track the distribution of labels among the clusters. We modify cluster entropy as follows: instead of finding the frequency of each class label among all available ones inside a cluster, we utilize the portion of a specific label captured in a cluster. This metric helps one to understand whether data points with the same class label are clustered together or spread among different clusters. We formulate it as $LDE(c) = -\sum_k q(k|c) \log q(k|c)$, where k represents class labels, c stands for a cluster, and $q(k|c)$ shows the distribution of class label k in cluster c. For label distribution entropy, to which we interchangeably refer as label entropy, as for cluster entropy, zero is the indicator of best value.

3.5   Approaches for Meta-Learning over System Space

As the above lays out, a monitoring system can be instantiated in various ways. For instance, one can be utilizing PCA as the algorithm for feature reduction, extracting the n top (variance-preserving) features from the PCA to featurize the data, and selecting k-means as the clustering algorithm to operate over the featurized data with the number of clusters set to a specific parameter value. Different choices for each of these steps (changing the algorithm or changing values of hyper-parameters and parameters in these algorithms) results in different configurations, giving rise to a vast (design) configuration space.

   Meta-learning here involves being able to determine an optimal configuration in a vast configuration space. Two fundamental questions need answering: (i) How does one quantitatively evaluate the quality of a design configuration? (ii) How does one use this information to automatically select from a possibly large configuration space an optimal configuration on which insight and interpretations can be drawn about the data at hand in an unbiased manner?

To answer the above questions, we suggest that one approach is to use evaluation metrics that characterize the quality of discovered clusters. This is similar to how feature selection techniques use the learner's performance as the objective function to assist in the evaluation of feature subsets [6]. Specifically, we consider the three evaluation metrics described above as the objectives/criteria for optimization over the system configuration space. More metrics can be considered, but our focus here is to illustrate how one can carry out meta-learning for complex, ML-based industrial monitoring systems. The novel approach we propose to do so leverages multi-objective optimization, which considers multiple metrics/criteria. To place this approach in the context of other approaches one can naively construct, we describe below several additional approaches based on one metric at a time or several metrics in an aggregate manner.

3.5.1 Single-Objective Optimization

If one considers one metric of interest as a single-objective function, then a simple meta-learning approach can rank configurations by the metric of interest and select the best configuration according to the single objective. Considering that we present above three potential metrics of interest, one can put together three selection mechanisms:

- SO-SC: Selects the best configuration with the highest silhouette score.
- SO-CE: Selects the best configuration with the lowest cluster entropy.
- SO-LDE: Selects the best configuration with the lowest label distribution entropy.

3.5.2 Single-Objective Optimization by Aggregating Metrics of Interest

A common approach to considering multiple criteria is to aggregate them in one objective function. We refer to this function as SO-Aggregate and define it for our specific setting here as

$$SO\text{-}Aggregate = w_1 \times SC + w_2 \times (1 - CE) + w_3 \times (1 - LDE)$$

In the above, $w_1, w_2,$ and w_3 are weights that determine the importance/contribution of each of the criteria. This readily illustrates the issue with aggregate functions. It is not clear how to weight the importance of one criterion relative to the other. When no prior information is available, equal weights can be set (e.g., $w_1, w_2, w_3 = 1/3$ here). Also note that while we desire a high silhouette coefficient score, we also desire a low cluster entropy score and a low label distribution score. The equation for SO-Aggregate addresses this and associates a higher aggregate score with top configurations.

3.5.3 Multi-Objective Optimization

The three metrics essentially associate three characteristics with each of the many possible design configurations. A configuration with the highest silhouette score may have a lower cluster entropy than another configuration with a lower silhouette score. In other words, if one uses these three metrics as optimization objectives, they may be conflicting objectives. For this reason, we leverage Pareto-based optimality via the concept of configuration dominance. A configuration i is said to dominate a configuration j if it is better than j along all metrics of interest. This is also known as strong dominance. If equality is allowed on one or more of the objectives, then this is known as soft dominance. In this paper, we employ strong dominance to discriminate among system configurations.

The Pareto rank of a configuration i is the number of configurations that dominate i. Configurations with Pareto rank 0 are said to belong to the Pareto front of the configuration space. Many configurations can belong to the Pareto front. In such a case, one can use Pareto count to further compare such configurations. The Pareto count of a configuration i is the number of configurations that i dominates. Therefore, configurations with high Pareto count are desired.

Our automated multi-objective selection strategy proceeds as follows: First, all configurations are ordered by Pareto rank (from low to high). Then, configurations with the same Pareto rank are further ordered by Pareto count (from high to low). One or more top configurations can then be selected in the obtained ordering. The top one reveals the best configuration that considers all metrics/objectives of interest. This configuration can be further analyzed. The clusters it produces can be further interpreted, and insight about a dataset can thus be obtained in an unbiased manner. This completes meta-learning via Pareto-based, multi-objective optimization, and we evaluate this approach to the other single-objective and aggregate-objectives explained above in Sect. 4.

4 Results

Considering the SCARA dataset that motivates this work, transitional motions between stations in a realistic manufacturing assembly process are used as ground-truth actuation states. The acoustic data are provided to us only for the Y-axis motions of the SCARA robot. These are divided into seven different sub-motions within four different stations with ranking distances of A, C, D, and B; A and B are farthest, while C and D are closest. Noninvasive acoustic data, sampled at 48 kHz, are recorded. By considering around 1 s data segmentation, 842 data samples are obtained. The ground-truth from these segments is known only for few data points (79); this is captured by a human operator via visual-based labeling.

The quality measurement of a signal reported as signal-to-noise ratio (SNR) indicates the amount of noise included in a signal. The SNR value between actuation data and background noise is unknown, but we estimate it to be 1.7dB based on an evaluation of labeled signals. Figure 3a shows a recorded SCARA actuation event. The distribution, potentially describing the variability in noise during a physical manufacturing environment, is displayed in Fig. 3b.

As described in Sect. 3, we further reduce these features via PCA, IsoMap, or spectral embedding. As is commonly practiced, we employ cumulative variance for PCA-based feature extraction. Figure 2 shows that around 66% of the data variance is covered by the top 5 PCs and more than 80% and 90% of the variance is covered by the top 9 and 15 PCs, respectively. Therefore, we consider projections of the data over the top 2, 5, 9, and 15 PCs as possible features. We consider this list of possible values for the number of features extracted via IsoMap and spectral embedding, as well. For IsoMap and spectral embedding, the number of neighbors is another important parameter, which we vary in a broad range, from 2 to 15 (11 figures).

Several clustering algorithms can be applied onto the featurized dataset. As described in Sect. 3, we utilize the popular k-means and hierarchical clustering algorithms (Ward's method). We do not utilize DBSCAN, as the SCARA dataset is both noisy and unbalanced; the latter information is derived from the distribution of labels on the 79 labeled data points. In addition, instead of relying on the popular sum-of-squared errors (SSE) approach to obtain the number of clusters for k-means, for the purpose of automated evaluation, the number of clusters is varied from 2 to 7, where 7 is the number of motion labels/classes available for 79 data points.

If one considers the setting where IsoMap is employed for featurization and k-means for clustering, the number of design configurations is 264 (considering the variation of parameters, such as the number of features and neighbors in IsoMap and the number of clusters in k-means). As explained in Sect. 3, each configuration is evaluated along three metrics, average silhouette score, cluster entropy, and label distribution entropy; we recall that these evaluate the quality of the obtained clusters. These metrics become objectives, and the Pareto-based method described in Sect. 3 can be utilized to compare and select Pareto-optimal design configurations. Figure 4a visualizes all these 264 configurations, highlighting the Pareto front (design configurations with Pareto rank 0) in red.

As Fig. 4a clearly shows, the determination of good design configuration(s) is nontrivial in a large configuration space. Consider further the nondeterminism of clustering algorithms, such as k-means, where different initializations for the cluster seeds can lead to different clusters. A design configuration may be in the Pareto front (have Pareto rank 0), but if we run its k-means component again, it may fall out of the front and have lower Pareto rank. It is indeed an interesting question how stable a design configuration is in the presence of nondeterminism.

We visualize this in Fig. 4b, where we show the average rank of design configurations; consider a configuration (x, y, z), where x and y denote the number of features and neighbors selected in IsoMap, respectively, and z denotes the number of clusters specified for k-means. Every configuration with Pareto rank zero is recorded in each run of k-means. Since this might result in more than one configuration with

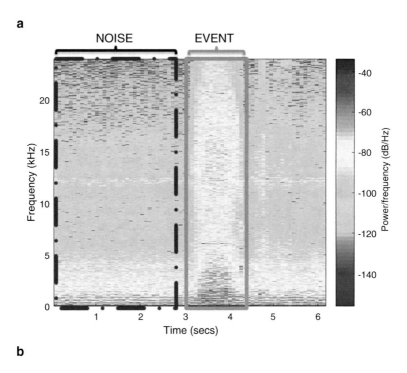

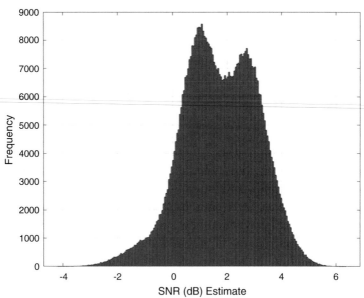

Fig. 3 (**a**) A SCARA spectrogram example, with noise and event regions annotated. (**b**) SNR event and noise distributions in SCARA dataset

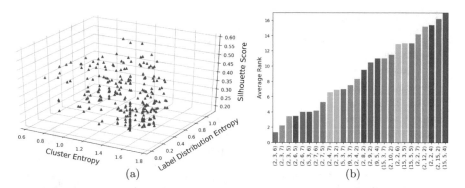

Fig. 4 (a) IsoMap and k-means design configuration space. Pareto front configurations are in red. (b) Average Pareto rank shown over ten variants of a design configuration. Configurations are denoted on the X axis by the number of features and neighbors in IsoMap and the number of k-means clusters

Pareto rank zero, Pareto count of these configurations is considered. We further sort these values in descending order. Thus, the configuration with the highest Pareto count is placed in the first ranking position. We continue this rank assignment for the rest of the configurations and report the average rank. As Fig. 4b shows, many design configurations are quite stable (low average rank). This is reassuring, as it increases confidence in the non-arbitrariness of evaluation-based conclusions. In particular, IsoMap with 2 selected features and 3 neighbors along with k-means with 6 clusters is the top configuration according to this analysis over the design space.

We note that Fig. 4a shows only a subset of the design configuration space. When considering the additional featurization models (PCA, IsoMap, spectral embeddings) with their various hyper-parameters and the different clustering algorithms, the design space contains 1104 configurations. This number swells further due to variants obtained by re-runs of k-means to account for their nondeterminism. To unravel the quality of these configurations, a Pareto-based selection is first conducted within each subspace, that is, configurations are grouped based on the featurizer and clustering algorithm employed.

Table 1 lists in column 1 the six subspaces resulting from considering three featurizers (PCA, IsoMap, spectral embedding) and two clustering algorithms (k-means, hierarchical clustering). Within each subspace, the configurations are ordered first by Pareto rank (from low to high), and those within each rank are further ordered by Pareto count (from high to low), as described in Sect. 3. The top configuration in such subspace is shown in Table 1, with its evaluation along the three performance metrics. To obtain one overall best design configuration, all the subspaces are merged, and the same ordering is applied. This best configuration is obtained with spectral embedding with two features and seven neighbors for spectral embedding and k-means with 6 clusters. Figure 5 shows the distribution of labels (motion information) in each cluster. Table 1 also confirms that this design

Table 1 Best design configurations in each subspace

Configuration	Features, neighbors, clusters	Silhouette coefficient	Cluster entropy	Label distribution entropy
PCA + k-means	5,-,5	0.25	0.96	0.45
PCA + hierarchical	9,-,4	0.21	1.1	0.66
IsoMap + k-means	2,3,6	0.45	0.82	0.14
IsoMap + hierarchical	5,3,7	0.38	0.83	0.14
Spectral + k-means	2,7,6	0.67	0.88	0
Spectral + hierarchical	5,7,6	0.58	0.88	0

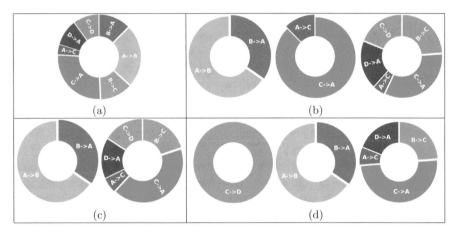

Fig. 5 Distribution of motions of known labels over clusters containing labeled data for the best overall configuration obtained with (**a**) SO-SC, (**b**) SO-CE, (**c**) SO-LDE, and (**d**) MO-Pareto

configuration is the one with the highest silhouette score, lowest label distribution entropy, and low cluster entropy.

We now compare the Pareto-based approach we propose in this paper with the other single-objective (including the aggregate) approaches SO-SC, SO-CE, and SO-LDE described in Sect. 3. Specifically, we compare the best configuration obtained by each approach.

Distribution of known labels (79) over the identified clusters for three single-objective optimizations and the Pareto-based optimization is visualized in Fig. 5. The motions are indicated by showing the direction from the source station to the destination station as source → destination. Each colored portion shows the percentage of participating labels.

Here is how one can visually interpret the results shown in Fig. 5. Each pie chart visually relates a clusters, and the colors track subsets of data points in a cluster with the same label. For consistency, a label is related with the same color across the different approaches. The presence of many colors in a pie chart indicates that the cluster entropy is high; as Sect. 3 relates, a configuration with high cluster entropy is of poor quality. The existence of color overlap among clusters illustrates that the

Table 2 Best design configurations of multi-objective Pareto-based and three single-objective optimizations

Method	Configuration	Features, neighbors, clusters	Silhouette Coefficient	Cluster entropy	Label distribution entropy
SO-SC	Spectral + k-means	2,4,7	0.992	2.60	0
SO-CE	IsoMap + hierarchical	15,7,7	0.22	0.56	0.62
SO-LDE	IsoMap + k-means	2,3,2	0.49	1.50	0
MO-Pareto	Spectral + k-means	2,7,6	0.67	0.88	0

label entropy is high. If one compares the various selection approaches in Fig. 5a–d, the quality of selected configurations improves as one goes from single-objective selection approaches to the multi-objective approach we propose for meta-learning over the system space.

Moreover, the careful reader will notice that the number of pie charts varies. The higher number of clusters (3) found to contain labeled data is associated with configurations selected by SO-CE (in Fig. 5b) and the multi-objective MO_{pareto} approach (Fig. 5d). Given that there are 7 labels and ideally one wants to obtain one cluster per label, a higher over a lower number of clusters containing labeled data is desired.

Table 2 compares the top configurations quantitatively. The top configuration selected by SO-SC is a configuration with two features, four neighbors for spectral embedding, and seven clusters. Distribution of known labels for this configuration is visualized in Fig. 5a, which shows that all labeled samples are clustered together (no label information is available regarding the other 6 clusters); this makes silhouette coefficient reach a high value of 0.99, but consider how non-informative this coefficient is by itself in this setting.

Table 2 shows that SO-CE selects as top a configuration with 15 features, 7 neighbors for IsoMap, along with 7 clusters by hierarchical clustering. This configuration is shown in Fig. 5b, where 79 samples are placed into 3 clusters; none of the clusters are pure in terms of labels. However, this configuration is more discriminating than the one obtained by SO-SC. Nevertheless, this configuration is not able to distinguish between the forward and backward motions, as A→B and B→A are clustered together; A→C and C→A are also grouped together (although a small portion of C→A motions are wrongly placed in another cluster highlighted in pink).

SO-LDE is not able to discriminate among 207 configurations, as they all evaluate to 0 according to label distribution entropy. One of them is drawn at random and visualized in Fig. 5c. Both detected clusters contain known labels and are obtained with IsoMap and k-means.

We analyze the top configuration obtained by MO-Pareto and shown in Fig. 5d in greater detail. Since the labels are located just in 3 clusters out of the 6 detected ones, 3 pie charts are illustrated. This Pareto-optimal design configuration distinguishes the farthest (A to B or B to A) and closest motions of the arm (from C to D) from other types of motions. It does not distinguish forward and backward motions in two stations; corresponding labels (A→B and B→A) are always grouped together in the same cluster. The other four remaining motions (D→A, B→C, C→A, and A→C) are also indistinguishable, as they are grouped in the same cluster. Altogether, these results show that MO-Pareto outperforms the single-objective approaches.

Now we compare MO-Pareto to the aggregate approach SO-Aggregate. The top configuration selected by SO-Aggregate is also the one selected by MO-Pareto. So, we compare the behavior of these approaches in greater detail but evaluate how close or far the top *m* configurations drawn by SO-Aggregate are to Pareto-optimal configurations (those in the Pareto front). Out of 1104 possible configurations, 258 are Pareto-optimal. Figure 6a shows the average Pareto rank of the top 1, 10, 100, 150, and 258 configurations, respectively, drawn by SO-Aggregate. Figure 6a indeed shows that the average Pareto rank of configurations drawn by SO-Aggregate goes up rapidly as *m* increases.

For completeness, the same analysis is carried out over configurations selected by SO-SC, SO-CE, and SO-LDE in Fig. 6b–d. These results show that SO-LDE performs better than the other single-objective approaches. All 207 configurations selected by this approach are Pareto-optimal, and the average rank of all 258 top configurations is 9.7; for comparison, the average rank of all 258 configurations

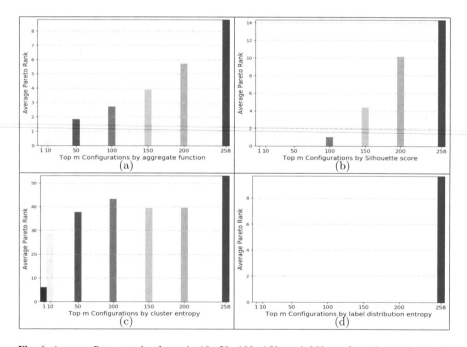

Fig. 6 Average Pareto rank of top 1, 10, 50, 100, 150, and 258 configurations selected by SO-Aggregate in (**a**), SO-SC in (**b**), SO-CE in (**c**), and SO-LDE in (**d**)

selected by SO-CE is 53.2. SO-Aggregate and SO-LDE average rank over all 258 top configurations are close to each other: 8.79 and 9.7, respectively. SO-Aggregate performs well in terms of top configuration, but as more configurations are considered, SO-LDE works better. Thus, both SO-Aggregate and SO-LDE are able to capture some aspects of configuration space, but none of them outperforms MO-Pareto.

5 Conclusion

We have proposed a Pareto-based, multi-objective optimization approach for meta-learning in industrial monitoring system design. Our study is motivated by a real-world problem of quantifying actuations of the SCARA robot, a ubiquitous industrial robot for the purpose of long-term industrial monitoring. A detailed evaluation shows that solutions obtained via this approach are superior over those obtained via other approaches in distinguishing movements from recorded acoustic signals. Our contribution generalizes beyond the specific application that motivates this work. The Pareto-based, multi-objective optimization approach can be easily generalized to diagnose degradations of mechanical systems in a variety of industrial processes and to carry out meta-learning in other domains.

It is worth noting that the proposed multi-objective approach did not find an ideal configuration where data were partitioned in 7 clusters, each associated with one of the 7 labels/motions. We speculate here on the various reasons. Noise may result in loss of inter-class separability. More crucially, we emphasize that the data and the labels are obtained via different modalities. The data for automated monitoring are acoustic data, whereas the labels in the small (labeled) subset are obtained by a human operator analyzing a video feed. This discrepancy between the process employed for labeling a portion of the data and the actual characteristics of the data further brings into focus that it is rather interesting that the Pareto-based method we propose is able to distinguish many of the motions from one another.

Motions that are not separable from acoustic signals may be distinguishable with further data sampling. Another direction of future work concerns additionally obtaining and considering motions along the X axis, which introduced a new modality and will invariably add to the complexity of the prediction task and the overall meta-learning one. Other future works focused on improving prediction can consider ML methods more robust to noise and semi-supervised methods. Taken together, the results show that while industrial monitoring is a challenging setting for ML, there is much room for learning and meta-learning methods and much promise toward an ambitious objective of automated diagnosis of mechanical degradation.

References

1. A.B. Abdessalem, N. Dervilis, D. Wagg, K. Worden, Model selection and parameter estimation in structural dynamics using approximate bayesian computation. Mech. Syst. Signal Process. **99**, 306–325 (2018)
2. N. Akhter, A. Shehu, From extraction of local structures of protein energy landscapes to improved decoy selection in template-free protein structure prediction. Molecules **23**(1), 216 (2018)
3. S. Ali, K.A. Smith, On learning algorithm selection for classification. Appl. Soft Comput. **6**(2), 119–138 (2006)
4. J. Bergstra, Y. Bengio, Random search for hyper-parameter optimization. J. Mach. Learn. Res. **13**(Feb), 281–305 (2012)
5. P.B. Brazdil, C. Soares, J.P. Da Costa, Ranking learning algorithms: using IBL and meta-learning on accuracy and time results. Mach. Learn. **50**(3), 251–277 (2003)
6. G. Chandrashekar, F. Sahin, A survey on feature selection methods. Comput. Elect. Eng. **40**(1), 16–28 (2014)
7. G.V. Demarie, D. Sabia, A machine learning approach for the automatic long-term structural health monitoring. Struct. Health Monit. **18**(3), 819–837 (2019)
8. G. Gui, H. Pan, Z. Lin, Y. Li, Z. Yuan, Data-driven support vector machine with optimization techniques for structural health monitoring and damage detection. KSCE J. Civil Eng. **21**(2), 523–534 (2017)
9. F. Hemmati, W. Orfali, M.S. Gadala, Roller bearing acoustic signature extraction by wavelet packet transform, applications in fault detection and size estimation. Appl. Acoust. **104**, 101–118 (2016)
10. D. Horn, B. Bischl, Multi-objective parameter configuration of machine learning algorithms using model-based optimization, in *2016 IEEE Symposium Series on Computational Intelligence (SSCI)* (IEEE, Piscataway, 2016), pp. 1–8
11. F. Hutter, J. Lücke, L. Schmidt-Thieme, Beyond manual tuning of hyperparameters. Künstl. Intell. **29**(4), 329–337 (2015)
12. A.K.S. Jardine, D. Lin, D. Banjevic, A review on machinery diagnostics and prognostics implementing condition-based maintenance. Mech. Syst. Signal Process. **20**(7), 1483–1510 (2006)
13. Y. Jin, B. Sendhoff, Pareto-based multiobjective machine learning: an overview and case studies. IEEE Trans. Syst. Man Cybern. Part C Appl. Rev. **38**(3), 397–415 (2008)
14. C. Lemke, M. Budka, B. Gabrys, Metalearning: a survey of trends and technologies. Artif. Intell. Rev. **44**(1), 117–130 (2015)
15. R. Liu, B. Yang, E. Zio, X. Chen, Artificial intelligence for fault diagnosis of rotating machinery: a review. Mech. Syst. Signal Process. **108**, 33–57 (2018)
16. K. Mathia, *Robotics for Electronics Manufacturing: Principles and Applications in Cleanroom Automation* (Cambridge University Press, Cambridge, 2010)
17. A. Mehmani, S. Chowdhury, C. Meinrenken, A. Messac, Concurrent surrogate model selection (cosmos): optimizing model type, kernel function, and hyper-parameters. Struct. Multidiscip. Optim. **57**(3), 1093–1114 (2018)
18. F. Pedregosa, G. Varoquaux, A. Gramfort, V. Michel, B. Thirion, et al., Scikit-learn: machine learning in Python. J. Mach. Learn. Res. **12**, 2825–2830 (2011)
19. S. Raschka, Model evaluation, model selection, and algorithm selection in machine learning. Preprint. arXiv:1811.12808 (2018)
20. E. Rendón, I. Abundez, A. Arizmendi, E.M. Quiroz, Internal versus external cluster validation indexes. Int. J. Comput. Commun. **5**(1), 27–34 (2011)
21. J.R. Rice, et al., The algorithm selection problem. Adv. Comput. **15**(65–118), 5 (1976)
22. T. Saito, J.L. Beck, Bayesian model selection for arx models and its application to structural health monitoring. Earthq. Eng. Struct. Dyn. **39**(15), 1737–1759 (2010)

23. K.A. Smith-Miles, Towards insightful algorithm selection for optimisation using meta-learning concepts, in *2008 IEEE International Joint Conference on Neural Networks (IEEE World Congress on Computational Intelligence)* (IEEE, Piscataway, 2008), pp. 4118–4124
24. L. Van Der Maaten, E. Postma, J. Van den Herik: Dimensionality reduction: a comparative. J. Mach. Learn. Res. **10**(66–71), 13 (2009)
25. Y. Yang, P. Fu, Y. He, Bearing fault automatic classification based on deep learning. IEEE Access **6**, 71,540–71,554 (2018)

Leveraging Insights from "Buy-Online Pickup-in-Store" Data to Improve On-Shelf Availability

Sushree S. Patra, Pranav Saboo, Sachin U. Arakeri, Shantam D. Mogali, Zaid Ahmed, and Matthew A. Lanham

1 Introduction

Customers demand a channel-agnostic, seamless shopping experience across physical stores, mobile, online, and other platforms. One of the strong omnichannel trends is the buy-online pickup-in-store model, which integrates online and offline operations by allowing customers to place orders online and collect them in their chosen stores. The retail store is the last mile of supply chain management, and error-free store execution ensures that the collective efforts of the whole supply chain yield desired results. Store execution primarily involves moving goods from the backdoor to the shelves to make the products available to end consumers. However, in-store logistics are highly labor-intensive. Store managers oftentimes ask store employees to perform shelf audits and enter them to the database for future reference. According to a study conducted by the University of Colorado, the implications of stock-out suggested that retailers on average lose 4% of their annual sales due to OOS items. The study also lost highlighted that, on an average, OOS items cost the manufacturers $23 million for every $1 billion in sales [6].

The data used in our study provided new opportunities for the use of analytics in improving the ease of business. To manage the scale of the problem, a machine learning model will predict the probability of having a product in stock when a customer order arrives. The retail giant Walmart recently admitted to a shelf-OOS problem and predicted a $3 billion opportunity in filling up the empty shelves created due to ineffective auditing and re-shelving operations [1]. It is one of the key performance components of customer service in retail. The complement to on-

S. S. Patra (✉) · P. Saboo · S. U. Arakeri · S. D. Mogali · Z. Ahmed · M. A. Lanham
Krannert School of Management, Purdue University, West Lafayette, IN, USA
e-mail: patras@purdue.edu; psaboo@purdue.edu; sarakeri@purdue.edu; smogali@purdue.edu; ahmed152@purdue.edu; lanhamm@purdue.edu

© Springer Nature Switzerland AG 2021
R. Stahlbock et al. (eds.), *Advances in Data Science and Information Engineering*,
Transactions on Computational Science and Computational Intelligence,
https://doi.org/10.1007/978-3-030-71704-9_28

shelf availability (OSA) is out of stock (OOS) which can be defined as: "a product not found in the desired form, flavor or size, not found in salable condition, or not shelved in the expected location" [7]. Simply put, an OOS occurs if a product is not available when a customer order arrives.

In recent years many machine learning techniques have been implemented to forecast sales and predict out-of-stock products, but the variety of approaches makes it challenging and time-consuming to pick the optimal methodology due to unpredictable consumer behavior. The objective of our study is to use multiple classification algorithms to predict OOS and improve on-shelf availability for retail stores. The dataset was obtained from a national grocer, and it contains transactional information for online orders at a store-product level. There are a lot of factors that can drive the OOS rate some of which are promotions, balance on hand, seasonality, etc. We developed a model which can incorporate some of these features to accurately predict OOS. We also created new features using the existing variables to capture the cyclic and seasonality in the dataset. We tried to present important issues like data cleaning, feature engineering, feature selection, and model evaluation criteria for the classification model.

The paper is divided into the following sections: Literature Review summarizes the prior research conducted in the area of OOS prediction and how our study is an extension to the prior research. The Data section summarizes the data used for our study. Data dictionary and the entity-relationship diagram provide an understanding of the variables used and how they are related to each other. The Methodology section elaborates on the steps taken to reach the end goal. This section summarizes the steps for data preprocessing, upsampling, feature engineering, target transformation, parallel computation, and deployment. The Model section summarizes the model evaluation criteria used for this study and various models that were used and implemented. Results section summarizes the result and compares with the baseline model, and the Conclusion section summarizes the application and some of the recommendations from our study. Finally, the Reference section contains all the journals and websites that we referred to during this research study.

2 Literature Review

A significant number of works of academic literature have been published regarding predicting out of stocks, demand forecasting, on-shelf availability, the causes, and the impact of stockouts and lost opportunity for retailers when a stockout occurs. Product availability is a measure of the service level a firm's supply chain offers to the end customer. High product availability means the consumers find and buy the products they want. The out-of-shelf measure is used to determine items that are not available on the store shelves. Out-of-stock cases are driving away revenue of multiple major retailers. Walmart recently admitted to a shelf-OOS problem and predicted a $3 billion opportunity in filling in empty shelves caused by ineffective auditing and re-shelving operations [3]. Not only that, Walmart even issued an urgent memo that demands store managers to improve grocery performance, which was seriously compromised by non-negligible shelf-OOS ratios [5]. Predicting demand is a challenging task for FMCG products as there can be multiple contributors for a sudden surge or dip in demand in the market. However, it becomes all the more difficult when a retailer is present in various channels. Our study is limited to predicting out of stocks for online transactions only. Nevertheless, we needed to understand the ecosystem of multi-channel retailers as this can have a confounding effect on demand and thus affect the OOS rate.

In the paper "Towards a predictive Approach for Omni-Channel Supply Chains" [9], the researcher aimed to find a predictive approach to deal with the complexities in demand forecasting of an omnipresent retailer by combining clustering with artificial neural network. In this paper, the products were clustered depending on the channels using the K-means method. Inaccurate demand forecasting and thus erroneous inventory levels can have a cascading effect on the bottom line of a retailer. Inaccurate store inventories hinder cross-channel fulfillment and also increase stockout possibilities for walk-in customers, while overstaffing for replenishment by overestimating demands will burden the retailer with higher costs [10].

To accurately understand the client's expectations and model accordingly, it was essential for us to understand the definition of OOS for our study correctly. As in the retail industry, OOS cases also refer to the events wherein an item is in-store (e.g., misplaced or stored in the backroom), but it is unavailable to customers [12]. Furthermore, we learned that it would be better to explore different models for each product category. The causes of the stockouts examined in various research studies indicate that the causes of retail stockouts are specific to the retailer, store, category, and product. No single solution works everywhere [11]. Coming to the approach, a classification model was our first choice to model the problem. The study "Predict on-shelf product availability in grocery retailing with classification methods" [8] gave us a high-level understanding of algorithms which will potentially perform well to identify "out-of-shelf" products. We also used an ensemble learning method to increase the performance of the base classifiers. Another challenge we tackled while modeling was to handle imbalanced data. Our random forest classifier, which, generally speaking, is robust to noise, also suffers from the curse of learning from an extremely imbalanced dataset. Because random forest tends to focus more on the prediction accuracy of the majority class to minimize the overall error rate, thus resulting in poor accuracy for the minority class. The paper [1] proposed two solutions, balanced random forest (BRF) and weighted random forest (WRF), to solve the issue of imbalance data in the random forest model. We got great insights from the paper "Forty years of Out-of-Stock research fi?! and shelves are still empty" [1] for our feature selections. A good portion of this paper dealt with supply-side issues and analyzed the extent and root causes of situations resulting in out-of-stock cases. Another valuable insight from this paper which we experienced in our data as well was that the degree to which the OOS rate depends on the characteristics of the category, i.e., OOS rates vary between categories, with the worst performing at about 15 to 16% and the best performing at OOS rates as low as

1%. One alternative approach for this project would have been time series analysis, but we passed on that considering the granularity of the data provided. In time series, we would have required to aggregate the data either weekly and monthly, similar to (9 M.W.T. Gemmink 2017) paper, but this would have resulted in a naive model with weak accuracy. Lastly, when modeling for the OOS rate, another important factor to incorporate is the timing of shelf audits for replenishment as it has a substantial influence on product availability, customer satisfaction, and sales performance [1]. Unfortunately, in our research, the data regarding an audit for replenishment was not available, and this can have some effect on our model's prediction accuracy.

3   Data

The data was collected from a national retailer which consists of online transactions from January 2019 to December 2019 for 3547 products across 246 stores through-out the country amounting to 32.4 million records for this period. The transactional data is at the store-product level which contains information on ProductID, Store ID, Number of units sold, Number of digital transactions per day, Out-of-stock (OOS) transaction, and Sales forecasted for the next day (Table 1).

The other data table used is the product hierarchy table which has information for a product with individual product level to the aggregated business segment level. The other levels of hierarchy are business area, product category, product subcategory, and product class. The entity-relationship diagram (ERD) shows the relationships between various attributes that explain the logical structure of all tables. The combination of Product ID, Day date, and Store ID makes the primary key for the OOS prediction table.

Table 1 Data dictionary

Variable	Type	Description
P_ID	Categorical	Product ID
Day_dt	Date	Day date
ut_id	Categorical	Store ID
promo_flg	Categorical	Item on promotion
DP_fcst	Numeric	Day prior forecast
dgt_txn_occ	Numeric	Digital transaction count
DP_Units	Numeric	Day prior units sold
oos_tn	Numeric	Out-of-stock transaction count
DP_OOS	Numeric	Day prior OOS
DP_dgt_tn	Numeric	Day prior digital transaction count
dp_oos_chain	Categorical	Day prior total OOS for all stores for a specific product
sell_thru	Numeric	dp_units dp_fcst
Boh	Numeric	Balance on hand
Mdq	Numeric	Minimum display quantity

4 Methodology

The methodology is executed in four consecutive steps. The first step describes the process of garnering raw data by physical store audit and retailers database encompassing chronological data for 3547 products dating back to January 2019. The second step illustrates the process of KDD [4], EDA, and closely assesses the next steps for data preparation, data selection, data cleaning, predefined business rules, the correct interpretation of results of information discovery, and its business translation. The third step deals with feature engineering, the performance of various classification models, parallel computation using multi-processing capabilities of all server cores, and narrowing down on potential models. The third step shortlists the model to be proposed for deployment, evaluates the findings with the national retailer, and involves multiple improvisations during the whole process. After evaluation and iterations, the last phase explains the steps to deploy the model and create a prototype to integrate with the retail chains existing information system and be ready for consumption.

4.1 Data Preprocessing

Multiple ad hoc data cleaning methods were applied to fix incorrect data types, remove extreme outliers due to typo entries, and inconsistencies from the data.

$$\text{OOS Rate} = \frac{\text{OOS}_{\text{txn}}}{\text{OOS}_{\text{txn}} + \text{Dgt\_Occ}_{\text{txn}}}$$

The below table provides an overview of the issues found during the data cleaning process. Post ad hoc data cleaning, to prioritize focus on business impact and prepare computationally feasible dataset, below business segments were removed based on an insignificant amount of sales and out-of-stock transactions. See Table 2.

4.2 Feature Engineering

To predict if a stock will go out of stock or not, certain exogenous variables could not be used such as unit, sales, etc. for the day to be predicted due to their unavailability on the day itself. The features were created to incorporate the cyclic nature of demand and on-shelf availability. For example, sales for the day to be predicted are not possible; therefore, to inculcate the information of sales for that day, we take a standard deviation and mean by aggregating day, week, month, and quarterly basis. The engineered features provided a significant increase in overall accuracy. Further, dimensionality reduction was performed to remove features that

Table 2 Issues discovered in dataset

Issue discovered	Root cause	Cleaning rule	Records affected
Products containing minimum display quantities (MDQ) over 100,000 units	Manual entry for the value	Removed records having MDQ > physical possible space on the shelf	0.07%
The balance on hand (BOH) containing negative values	Correction for products returned or replaced	Replaced with MDQ	0.01%
Sell-through containing extremely high values	NA	Removed all values beyond four standard deviations	0.04%
True duplicate	Data extraction and loading repetition	Removed true duplicates	0.01%

Table 3 Feature importance

Feature	Importance
week_mean_dgt_txn_occ	0.099476
month_mean_dgt_txn_occ	0.076283
quarter_mean_dgt_txn_occ	0.056487
day_mean_dgt_txn_occ	0.050176
month_sd_dgt_txn_occ	0.044761
week_num_sd_dgt_txn_occ	0.038701
week_num_sd_chain_oos	0.030058
day_mean_chain_oos	0.029998
day_sd_dgt_txn_occ	0.025305
quarter_sd_dgt_txn_occ	0.020243
day_mean_sales	0.006902
week_num_sd_units	0.005063
DAY_DT_Ordinal	0.004835
month_sd_units	0.004508
day_sd_units	0.004335

were highly correlated to avoid redundancy and risk of overfitting and optimize computational performance. Business segments that were extremely skewed such as alcoholic beverages, grocery, dairy, dry grocery, etc. observed increment in F1 score on average, by 60%. However, business segments that had OOS rate greater than 20% observed a 5–7% increment in F1 score.

Table 3 provides the information to identify features contributing to the model accuracy. We experimented with multiple features and found features listed in Table 4 as the most significant ones.

Table 4 Features created and selected for research study

Type	Feature name	Definition
Product-Time	DAY_DT_Ordinal	Converts the given date to proleptic Gregorian ordinal of the date
	day_mean_sales	Aggregate sales of product into mean for each day
	day_mean_chain_oos	Aggregate chain OOS of product into mean for each day
	day_mean_dgt_txn_occ	Aggregate digital transaction of product into mean for each day
	day_sd_units	Standard deviation of units for each day
	day_sd_dgt_txn_occ	Standard deviation of digital transaction for each day
	week_mean_dgt_txn_occ	Aggregate digital transaction of product into mean for each week
	week_num_sd_units	Standard deviation of units for each week
	week_num_sd_chain_oos	Standard deviation of chain OOS for each week
	week_num_sd_dgt_txn_occ	Standard deviation of digital transaction for each week
	month_mean_dgt_txn_occ	Aggregate digital transaction of product into mean for each month
	month_sd_units	Standard deviation of units for every month
	month_sd_dgt_txn_occ	Standard deviation of digital transaction for every month
	quarter_mean_dgt_txn_occ	Aggregate digital transaction of product into mean for each quarter
	quarter_sd_dgt_txn_occ	Standard deviation of digital transaction for every quarter
Product-OOS	not_dp_oos_since	Number of days since a product is not out of stock
Context features	quarter	The quarter of the year
	month	The month of the year
	week_of_year	The week of the year

4.3   Upsampling

OOS data skewness is not uncommon in the retail industry [2] with most figures ranging between 3.3% and 12%. The skewness intrinsically makes the data less informational. Our approach was to segment the data granular enough to obtain good accuracy and simultaneously aggregate the data to avoid computational challenges. Table 5 provides a brief overview of the OOS rate for different segments. Classification using skewed data is biased in favor of the majority class. The situation deteriorates when combined with high dimensional data. To tackle the imbalance

Table 5 Business segment considered

Rules	Business segment	OOS rate %	% of records	% of sales
Included	PRODUCE	6.10%	19.14%	38.71%
	DAIRY	3.70%	20.91%	19.66%
	GROCERY DSD	2.90%	18.62%	16.79%
	DRY GROCERY	3.30%	20.60%	8.53%
	DELI	37.40%	2.76%	3.68%
	PACKAGED MEAT	4.90%	4.38%	3.40%
	FROZEN FOODS	4.50%	6.92%	2.66%
	MEAT	36.00%	1.32%	2.43%
	CONSUMABLES	3.60%	3.16%	2.17%
	ALCOHOLIC BEVERAGES	1.60%	0.24%	0.62%
	BAKERY	7.90%	1.01%	0.60%
	SEAFOOD	22.60%	0.37%	0.58%
Excluded	PETS	4.20%	0.31%	0.08%
	OTC HEALTH CARE	2.00%	0.14%	0.05%
	BABY CONSUMABLES	1.50%	0.11%	0.05%
	BULK WATER AND COFFEE	21.70%	0.02%	0.01%
	BEAUTY CARE	25.00%	0.00%	0.00%

of classes, we used an oversampling technique which creates new minority class examples by extrapolating between existing examples.

4.4 Target Transformation

The dependent variable fioos_tnfi ranged between 0 and 35, which represented how many times the product was reported not on-shelf. Although this provides the intensity of demand for a product and could translate to sales, it becomes cumbersome and unintuitive to classify if a product is out of stock or not. Therefore, we transformed the dependent variable which can be expressed as follows:

$$\text{Out Of Stock}_{(t,s)} = 0, \text{OOS} = 0$$

$$\text{Out Of Stock}_{(t,s)} > 0, \text{OOS} = 1$$

4.5 Parallel Computation

With roughly 11 billion data points (32 million rows $\times$ 343 columns) to model, the computational system is bound to break at some point. During experimentation with various models, the kernel broke frequently despite segmenting and reducing data points substantially. The first solution to address this was to downsize the dataset until the kernel can parse, process, and return the results successfully. Although the kernel completed the task, the time taken to model was as high as 20,000 s (6 h). An alternative to run the models was to perform out-of-core learning method process data in chunks. However, only a few data models support partial learning (out-of-core), and models such as AdaBoost, logistic regression, etc. require the entire dataset to be fed to the model. To overcome the computational challenge, we leveraged all the cores of the server by exploiting multiprocessing capabilities. Each dataset was mapped to one core (total of 10) and executed in parallel. With this, we received results in 1/10th the time when we ran 10 datasets in parallel. Table 6 shows the comparison for two models executed with and without features and the time saved.

5 Model

5.1 Model Evaluation Criteria

The dataset is highly imbalanced with 94% majority class (not OOS) and 6% minority class (OOS). As our area of interest is to predict OOS accurately which is the minority class, we focus on precision and recall. Recall summarizes the fraction of examples assigned as OOS that belong to the OOS class and precision summarizes how well the OOS class was predicted. Precision and recall can be combined into a single score that seeks to balance both concerns, called the F-score or the F-measure. The F-measure is a popular metric for imbalanced classification. (See Table 7.)

Table 6 Processing time comparisons for series and parallel for deli and packaged meat

Dataset	Time taken in series	Time taken in parallel
Deli	2,468.63 s	1,455.28 s
Packaged meat	7,619.28 s	3,709.63 s

Table 7 Confusion matrix definition

		Prediction	
		n	p
Actual	n'	True Negative	False Positive
	p'	False Negative	True Positive

$$Precision = \frac{True\ Positive}{True\ Positive + False\ Positive}$$

$$Recall = \frac{True\ Positive}{True\ Positive + False\ Negative}$$

$$F\ Measure = \frac{2 * Precision * Recall}{Precision + Recall}$$

5.2 Model Experiments

As there are very few research studies which classify out of stock to improve shelf availability, we tried to examine linear models, neural networks, and decision tree-based models for classification on seafood and alcoholic beverages data to narrow down on promising models. Utilizing each classification algorithm, we evaluated the F1 score, AUC score, precision, recall, and time taken for each model. Based on these parameters, we found that random forest, logistic regression, and AdaBoost performed relatively well over others. Because of the ease of its interpretability, logistic regression was one of the potential candidate models. Tables 8 and 9 below show the results for multiple models run on the seafood and alcoholic beverages dataset (Figs. 1 and 2).

Because of data skewness, stratified fivefold cross-validation was performed to observe the consistency of model performance to avoid overfitting of the trained dataset. Cross-validation also results in less biased models over models without cross-validation. RF consistently performed better than logistic regression and AdaBoost regardless of the skewness in the data.

5.3 Logistic Regression

Logistic regression was our first choice for this project because of the elegance, simplicity, and interpretability of the model. Since in our data cleaning process, we categorized all the not OOS cases as 0 and all the OOS cases as 1. This makes our logistic regression model as a binary logistic regression. Similar to linear regression, logistic regression uses an equation as the representation. But the difference here is it uses a logistic function to fit the output of the equation between 0 and 1 instead of fitting a straight line or hyperplane. Below is the logistic function equation:

$$H(x) = \frac{1}{1 + exp(-x)}$$

Table 8 Model results on seafood dataset (with and without features)

SEAFOOD	F score		AUC Score		Precision		Recall		Time taken (s)	
	W/O Feat	W Feat	W/O Feat	W Feat	W/O Feat	W Feat	W/O Feat	W Feat	W/O Feat	W Feat
Random forest classifier	86%	90%	92%	95%	83%	85%	90%	95%	49	54
AdaBoost classifier	85%	88%	92%	94%	81%	84%	90%	92%	39	53
Logistic regression	84%	89%	92%	95%	78%	83%	92%	95%	15	13
Complement NB	84%	85%	92%	91%	78%	82%	91%	87%	0	0
Multinomial NB	84%	85%	92%	91%	78%	82%	91%	87%	0	0
Bernoulli NB	84%	78%	90%	89%	82%	69%	86%	90%	1	0
Passive aggressive classifier	84%	89%	92%	95%	77%	83%	92%	96%	1	2
Liner SVC	84%	89%	92%	95%	77%	83%	92%	95%	9	18
Linear discriminant analysis	84%	88%	92%	94%	77%	83%	92%	94%	11	11
Ridge classifier	84%	88%	92%	94%	77%	83%	92%	94%	2	2
Nearest centroid	82%	83%	88%	88%	84%	85%	80%	81%	0	0
Decision tree classifier	77%	83%	85%	90%	77%	83%	77%	84%	8	14
MLP classifier	77%	82%	85%	89%	77%	82%	77%	82%	713	692
Gaussian NB	67%	72%	82%	86%	57%	61%	82%	88%	1	1
Quadratic discriminant analysis	58%	71%	77%	86%	44%	59%	86%	90%	9	6

Table 9 Model results on alcoholic beverages dataset (with and without features)

ALCOHOLIC BEVERAGES	F score		AUC score		Precision		Recall		Time taken (s)	
	W/O Feat	W Feat	W/O Feat	W Feat	W/O Feat	W Feat	W/O Feat	W Feat	W/O Feat	W Feat
Random forest classifier	1%	67%	50%	77%	67%	85%	1%	55%	51	71
AdaBoost classifier	19%	61%	62%	84%	14%	56%	27%	69%	59	49
Decision tree classifier	12%	53%	55%	78%	11%	50%	12%	56%	10	49
MLP classifier	6%	51%	52%	72%	9%	60%	4%	44%	771	355
Extra trees classifier	3%	48%	51%	67%	33%	81%	1%	34%	109	110
Linear discriminant analysis	9%	31%	73%	87%	5%	19%	67%	79%	29	12
Ridge classifier	9%	31%	73%	87%	5%	19%	67%	79%	2	3
Passive aggressive classifier	10%	30%	77%	87%	6%	18%	75%	80%	4	3
Liner SVC	9%	29%	73%	87%	5%	18%	66%	81%	12	13
Logistic regression	10%	28%	74%	87%	5%	17%	67%	81%	14	16
Bernoulli NB	13%	18%	76%	86%	7%	10%	66%	85%	1	1
Complement NB	7%	8%	69%	71%	4%	4%	66%	66%	0	0
Multinomial NB	7%	8%	69%	71%	4%	4%	66%	66%	1	0
Nearest centroid	6%	7%	68%	68%	3%	4%	70%	63%	1	0
Gaussian NB	3%	3%	51%	52%	2%	2%	92%	94%	1	1
Quadratic discriminant analysis	3%	3%	50%	51%	2%	2%	92%	97%	26	7

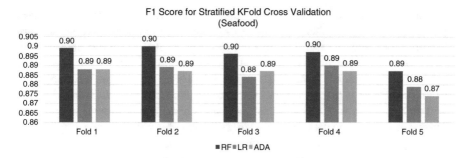

Fig. 1 F1 score for stratified K-fold cross-validation (OOS rate high)

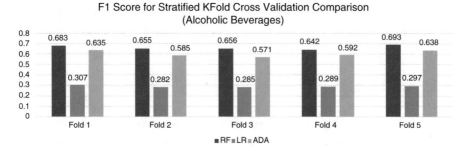

Fig. 2 F1 score for stratified K-fold cross-validation (OOS rate low)

The advantage of using logistic regression is it is very efficient to train, is easy to implement, and uses less computational resources. Moreover, it does not require input features to be scaled, does not require much tuning, and is easy to regularize. Also, the logistic regression model gives us probabilities instead of just classifying; this makes decision-making easy and straightforward. Few drawbacks of logistic regression are that we can't solve nonlinear problems. Also, it doesn't perform well when feature space is too large, doesn't handle a large number of categorical features/variables well, and is vulnerable to overfitting. Since feature engineering plays an essential role concerning the performance of logistic regression, the model does work better when we remove attributes that are unrelated to the output variable. The choice we made while selecting the features and synthesizing new ones resulted in good outcomes from the model. Also, being conservative in selecting features with multicollinearity helped us in getting good results.

5.4 Random Forest

Random forest is a method similar to the decision trees. In decision trees used for classification, we used recursive binary splitting to divide the predictor space into various regions. The classification error rate is used as the criterion for each binary split. The better approaches that can be used individually for the binary splits are the Gini index and entropy. This process of decision trees is further improved by using bootstrap aggregation, also called bagging. The intuition behind bagging is that averaging a set of observations reduces variance. One of the advantages of the random forest is that it is one of the most accurate decision models and works great on large datasets. Random forest also assists in extracting variable importance. Coming to the drawbacks of random forests, it is vulnerable to overfitting in case of noisy data, and its results are difficult to interpret. The reason why random forest gave us fantastic results is its ability to handle the high dimensionality in our data. Since we have (insert number of variables) variables and thus potentially high multicollinearity, random forest is an apt method for classification of out of stock (OOS) and non-OOS cases. Another reason is possibly a large quantity of dataset availability for training the model as random forest works great for large datasets.

5.5 *AdaBoost*

Another classifier that gave us a great result is the AdaBoost classifier. AdaBoost or adaptive boosting combines multiple classifiers to increase the accuracy of classifiers and is an iterative ensemble method. AdaBoost classifier builds a robust classifier by combining multiple poorly performing classifiers, which are generally a stump (a tree with just two leaves) to ultimately get a strong classifier. The intuition behind AdaBoost is to set the weights of classifiers and to train the sample data sample in each iteration in such a way that it makes accurate predictions of unusual observations conducive. Now any statistical learning algorithm can be used as a base classifier if it accepts weights on the training set. In our model, the base classifier used was a default option that is a decision tree classifier.

6 Results

By applying the classification algorithm on the dataset, the research concludes that the data without any enhanced features shows low F1 score for business segments with high OOS rate. Features engineered during the research establish the possibility of developing a novel method for capturing information on cyclic and seasonality of product's OOS behavior and utilize them to predict significantly better. The empirical results derived from the research show an improvement of 50% to 68% (Fig. 3) and Table 10 in F1 score for products with high OOS rate. The study also provides marginal increment (1%–3%) in F1 score for products that are already being predicted with accuracy as high as 90%. The results found from the study provide an innovative solution to the low on-shelf availability issues faced across retail industry especially for fast-moving goods. By alleviating the key hurdle of

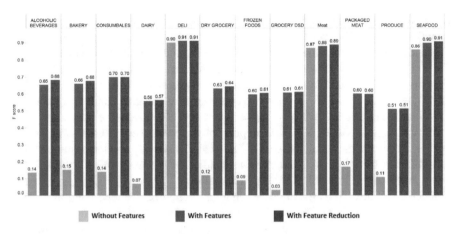

Fig. 3 Results comparing F1 score obtained using without features, with feature, and with reduced features

Table 10 Improvement on base model using features

Business segment	Base F1	Improvement using features
Deli	0.90	1%
Meat	0.87	1.5%
Packaged meat	0.04	56%
Dry groceries	0.04	61%
Grocery DSD	0.04	58%
Seafood	0.86	5.2%
Frozen food	0.04	57%
Alcoholic beverages	0.17	66%

low OSA, retailers can reduce significant source of loss of revenue, provide better customer experience, and maintain tighter supply chain management.

The figure below shows the business categories and the sub-categories within them where the model was able to perform relatively better than compared to other sub-categories. The figure summarizes the categories with F scores greater than 80%. Therefore, drilling down on each segment, the below sub-categories can be utilized to predict OOS with high confidence.

7 Conclusions

Engineered features showed significant improvement on model performance. Features like weekly mean, monthly mean, quarterly mean, monthly SD, and weekly SD for digital transactions proved to have the highest importance in identifying OOS rate at product-store level. We incorporated the costs of incorrect classifications to compare models based on business measures and found random forest to be the best performer. Based on the cost matrix, the estimated potential savings from our predictions was $50M per year Although the study provides significant benefits from both retailers' and customers' perspective, it still holds certain drawbacks.

The study could possibly perform better if provided with more data per store per product. Empirical results show that if dataset contains information of product that goes out of stock in at least 6 stores everyday (roughly 2000 OOS counts), the predicting capabilities give higher accuracy. Moreover, the data also lacks the information about how often a product is replenished at each store which generally affects the whole process of knowledge discovery and modeling training. Lastly, even though class imbalance was handled by creating synthetic records, the imbalance still affects the overall performance and possibly hinders the learning process. The next steps for the research entail garnering granular data, increasing data points containing OOS, and information about replenishment rate. Such value addition will help engineer better features and might provide more accurate results, ultimately tackling industry-wide problem and possibly extending into other sectors as well.

References

1. J. Aastrup, H. Kotzab, Forty years of out-of-stock research–and shelves are still empty. Int. Rev. Retail Distrib. Consum. Res. **20**(1), 147–164 (2010)
2. D. Corsten, T. Gruen, On shelf availability: An examination of the extent, the causes, and the efforts to address retail out-of-stocks, in *Consumer Driven Electronic Transformation* (Springer, Berlin, 2005), pp. 131–149
3. R. Dudley, Wal-Mart Sees $3 Billion Opportunity Refilling Empty Shelves. https://www.bloomberg.com/news/articles/2014-03-28/wal-mart-says-refilling-empty-shelves-is-3-billion-opportunity
4. U. Fayyad, G. Piatetsky-Shapiro, P. Smyth, From data mining to knowledge discovery in databases. AI Maga. **17**(3), 37–37 (1996)
5. S. Greenhouse, H. Tabuchi, Walmart Memo Orders Stores to Improve Grocery Performance. https://www.nytimes.com/2014/11/12/business/walmart-memo-orders-stores-to-improve-grocery-performance.html
6. T.W. Gruen, D. Corsten, *A comprehensive guide to retail out-of-stock reduction in the fast-moving consumer goods industry* (Grocery Manufacturers Association, Food Marketing Institute, and National Association of Chain Drug Stores, Washington, DC, 2007)
7. I. Moussaoui, B.D. Williams, C. Hofer, J.A. Aloysius, M.A. Waller, Drivers of retail on-shelf availability: Systematic review, critical assessment, and reflections on the road ahead. Int. J. Phys. Distrib **46**(5), 516–535 (2016). https://doi.org/10.1108/IJPDLM-11-2014-0284
8. D. Papakiriakopoulos, Predict on-shelf product availability in grocery retailing with classification methods. Expert Syst. Appl. **39**(4), 4473–4482 (2012)
9. M.M. Pereira, E.M. Frazzon, Towards a predictive approach for omni-channel retailing supply chains. IFAC-PapersOnLine **52**(13), 844–850 (2019)

10. Forrester, "Customer desires vs. retailer capabilities: Minding the Omni-Channel commerce gap," A Forrester Consulting Thought Leadership Paper Commissioned by Accenture and Hybris, an SAP company. (January, 2014). http://www.accenture.com/us-en/landing-pages/Documents/Seamless/Accenture-hybris-Forrester-new_2014.pdf. Accessed 23 Feb 2015
11. L. Sanchez-Ruiz, B. Blanco, A. Kyguolienė, A theoretical overview of the stockout problem in retail: from causes to consequences. Manag. Organ. Syst. Res. **79**(1), 103–116 (2018)
12. Z. Ton, A. Raman, The effect of product variety and inventory levels on retail store sales: a longitudinal study. Prod. Oper. Manag. **19**(5), 546–560 (2010)

Analyzing the Impact of Foursquare and Streetlight Data with Human Demographics on Future Crime Prediction

Fateha Khanam Bappee, Lucas May Petry, Amilcar Soares, and Stan Matwin

1 Introduction

Crime is one of the well-known social problems that affect the quality of life and slows down the country's economy. Nowadays, with the advancement of big data analytics, exploring diverse sources of data has gained increasing interest and attention that offers better crime analysis and prediction for crime researchers. Besides, identifying crime patterns and trends is of great importance for police and law enforcement agencies. Crime patterns tell us the story about the environment, demography, temporality, and how criminals interact with those factors.

In this paper, we address the problem of predicting future crime incidents for small geographic areas (i.e., dissemination areas as defined by Statistics Canada) in Halifax, Canada. Traditionally, criminology researchers study and analyze the historical crime data by focusing on sociological and psychological theories to obtain crime and criminal behavioral patterns. However, such strategies may intro-

F. K. Bappee (✉)
Dalhousie University, Halifax, NS, Canada
e-mail: ft487931@dal.ca

L. M. Petry
Universidade Federal de Santa Catarina, Florianópolis, Brazil
e-mail: lucas.petry@posgrad.ufsc.br

A. Soares
Memorial University of Newfoundland, St. John's, NL, Canada
e-mail: amilcar.soares@dal.ca; amilcarsj@mun.ca

S. Matwin
Dalhousie University, Halifax, NS, Canada

Polish Academy of Sciences, Warsaw, Poland
e-mail: stan@cs.dal.ca

© Springer Nature Switzerland AG 2021
R. Stahlbock et al. (eds.), *Advances in Data Science and Information Engineering*,
Transactions on Computational Science and Computational Intelligence,
https://doi.org/10.1007/978-3-030-71704-9_29

duce bias from the theory-ladenness of observation [4]. The literature states that, in the real world, crime has a mutual relationship with time, space, and population that complicates the researcher's study more [8]. Several works have explored the relationships between criminal activities and socioeconomic factors, for instance, educational facilities, ethnicity, income level, unemployment, etc., as well as human behavioral factors [5, 13, 14]. Crime rate or crime occurrence prediction has received considerable attention in many studies, including [25, 28, 35]. Several studies tried to predict specific types of crime for a specific region or time by detecting the patterns of that crime [34]. Spatiotemporal pattern plays a vital role in advanced research in crime analysis and prediction [18]. Nowadays, advanced techniques are applied to detect different crime patterns such as spatiotemporal, demographic, meteorological, and human behavioral patterns for crime prediction. However, it is challenging to make accurate estimations from diverse data sources due to nonlinear relationships and data dependencies.

Most of the studies that presented data-driven crime pattern detection and prediction approaches have focused on mega-cities like Chicago, New York, Greater London, etc. However, the physical characteristics, human impact characteristics, and their interactions are totally different for different regions and cities. Therefore, applying those models for predicting crime in a smaller city is challenging and may lead to different outcomes. Our study aims to build data-driven models for future crime incident prediction for smaller cities. The main hypothesis is that the relative scarcity of data, compared to mega-cities, can be compensated by using nontraditional datasets that can be derived from social media and the Internet-of-Things (IoT) infrastructure of a modern city. We extract five different categories of features from six different data sources. We propose to explore traditional demographics data with commuting features (e.g., commuting mode and time), IoT-like streetlight poles position data, as well as human mobility data with dynamic features from location-based social networks. To the best of our knowledge, employing demographics data with human mobility features for future crime prediction is the first attempt for a small city such as Halifax, Canada. For model building, we use ensemble learning methods such as random forest and gradient boosting. We conduct a performance comparison for all five categories of features. We also compare the prediction results generated from ensemble learning methods with a baseline method called DNN-based feature level data fusion [16].

In summary, the contributions of this paper are: (i) we propose the use of streetlight infrastructure data with demographic characteristics for improving future crime prediction. Its effectiveness is demonstrated in our experimental evaluation results; (ii) we propose data-driven models to predict future crime occurrences in smaller cities. This implies that fewer data points are applicable for training the models; and (iii) we experimentally show the effect of each feature group proposed in previous works and this paper on crime prediction, evaluating the classification performance of different feature combinations.

The rest of the paper is organized as follows. Section 2 reviews the related work. Section 3 provides the details of feature engineering approaches to improve the prediction accuracy in Halifax. After, in Sect. 4, the data source, data preparation activities, and experimental results are presented. Finally, Sect. 5 presents some concluding remarks with future research directions.

2 Related Work

The relationship between crime and various factors has been studied in many scientific and criminology works. Nowadays, researchers can use spatial information from the real world using Geographic Information Systems (GIS). Likewise, demographic information is easily accessible from different statistical sources. The use of historical facts and temporal dynamics between neighborhoods and crimes has also been broadly noted in criminology. Researchers have emphasized the feasible computation solutions for the urban crime after analyzing the factors related to different categories of crimes and their consequences. We have categorized the existing work of crime prediction from four aspects: temporal and historical, geographic, demographic and streetlight, and human behavioral aspects.

Temporal patterns of crime are learned from sequential crime data by analyzing the structure (various intervals) of temporal resources. Crime rates can be examined for hours of the day, different days of the week, months, seasons, years, and others. Many researchers have studied how to identify temporal patterns among criminal incidents [10, 16]. Several works also focus on historical information to predict future crime incidents [33]. In [29, 30], the authors presented a periodic temporal pattern with hourly crime intensity and holiday information for crime forecasting. A study [34] shows that drunk-driving incidents and other criminal incidents occur during Saturday nights, bar game nights close to the bar, and sports season close to the stadium. This implies that the temporal influence of crime may change over geographic regions.

Existing works also examined the geographic influence for future crime prediction or crime rate estimation [28]. Wang et al. [28] presented a crime rate inference problem for Chicago community areas by utilizing point-of-interest (POI) data as well as geographical influence features. Geospatial Discriminative Patterns (GDPatterns) was introduced in [26] to capture the spatial properties of crime. Furthermore, spatial autocorrelation is considered in [33] where the average number of neighbors is calculated for each grid. Besides, the authors in [1, 2] found spatial patterns (hotspots) for crime prediction using the Apriori algorithm and localized kernel density estimation (LKDE), respectively. Recently, another study [3] focused on the creation of spatial features to predict crime using geocoding and crime hotspot techniques. As the distributions of crime vary in time and space, several studies have identified spatiotemporal patterns for crime prediction [11, 35]. In [11], the authors investigated a spatiotemporal dynamic for Break and Entries (BNEs) crime incidents. However, considering the geographic influence may add a little help for crime prediction as the neighboring community shares similar demographics.

Traditional demographic features have been extensively used in many research for crime prediction [4, 5, 16]. In [8], the author applied population density, mean people per household, people in the urban area, people under the poverty level, and people in dense housing with some other features to detect community crime patterns. A study discovered the association of construction permits, foreclosures, etc. with crime tendency [19]. Researchers also explored residential stability, number of vacant houses, number of people who are married or separated, and education [4, 28]. However, using only traditional demographic feature is insufficient to understand the implicit characteristics of crime and criminals. Few works reported the impact of streetlight distributions on the criminal behavioral pattern and crime prediction. The researchers in 2018 [31] have found an inverse relationship between streetlight density and crime rates based on the census block groups in Detroit. In our study, we also consider extracting streetlight features but for crime incidents prediction. However, due to human mobility, a region's demographic characteristics may change for a short or long period of time.

Human behavioral pattern aims to obtain understanding from human behavior, mobility, and networks. In [5], the authors investigated the predictive power of aggregated and anonymized human behavioral data derived from a multimodal combination of mobile network activity and demographic information. Specifically, footfall or the estimated number of people within each cell is derived from the mobile network by aggregating every hour the total number of unique phone calls in each cell tower and mapping the cell tower coverage areas to the Smartsteps cells. Similar works have been done by Andrey et al. [6] and Traunmueller et al. [27] for crime hotspot classification and to find a correlation between crime and metrics derived from population diversity. In [28], the authors profile the crime rate by applying taxi flow data to understand the reflection of city dynamics. The authors considered taxi flows as "hyperlinks" in the city to connect the locations. Each taxi trip recorded pickup/dropoff time and location, operation time, and the total amount paid. The taxi flow features indicate how neighboring areas contribute much crime in the target area through social interaction. A data-driven approach is presented in [4] for crime rate prediction that also considers road network, transportation nodes, and human mobility. Recently, crime event prediction for Brisbane and New York is studied in [15, 23] using dynamic features extracted from Foursquare data. The authors measure the region's popularity by determining the total number of observed check-ins in that region for a specific time interval. Also, the number of unique users that checked in to a specific venue and the number of tips users have ever written about that venue are counted to measure the popularity, heterogeneity, and quality of the region.

In our study, we proposed a data-driven approach for a smaller city, Halifax, by investigating an extensive set of features from all different aspects. We mainly focus on human behavioral aspects, streetlight features, and the traditional demographic features for future crime prediction.

3   Feature Extraction

Aiming at predicting future crime incidents, we extract features for each dissemination area (DA). According to Statistics Canada, a DA is the smallest standard geographic area in their data, which consists of one or more adjacent dissemination blocks [24]. This section is organized as follows. Section 3.1 details the temporal and historical features used in this work. The demographics and streetlight features are explained in Sect. 3.2, while Sect. 3.3 shows the POI features used in this work. Finally, Sect. 3.4 shows some human mobility dynamic features extracted from social networks.

3.1   Temporal and Historical Features

According to criminology research, crime may change over a long period of time (e.g., season) as well as in a short period of time (e.g., day or week) [22]. Thus, the temporal features we extracted are month, day of the week, time interval in a day, and season. We arrange crime records in eight 3 h time intervals and four seasons (winter, fall, summer, and spring) for each DA. On the other hand, some research analyzed the relation of future crime incidents with the past crime history [33]. Therefore, we calculate crime frequency and crime density for each region based on historical crime data. As the area and population sizes are different for different regions, we normalize the crime frequency using the area and population size to obtain the crime density (D_{cr}).

$$D_{cr}(r) = \frac{CR(r)}{P(r)}, \tag{1}$$

$$D_{cr}(r) = \frac{CR(r)}{A(r)}, \tag{2}$$

where $CR(r)$ addresses the total number of crimes in DA r, $P(r)$ is the total number of population in region r, and $A(r)$ is the area of that region. We also compute the crime distribution based on each season.

3.2   Demographic and Streetlight Features

Researchers have widely used demographic and socioeconomic features for crime rate estimation [28] and crime occurrence prediction [5]. The main demographic features we consider for our study are population density, dwelling characteristics, income, mobility, the journey to work, aboriginals and visible minorities, age, and sex. The journey to work features measure two main things: (i) the time people leave for work and (ii) the primary mode of commute for residents aged more than 15 years. We consider six different measures for the time people leave for work, such as between 5 a.m. and 5:59 a.m, 6 a.m. and 6:59 a.m., 7 a.m. and 7:59 a.m., 8 a.m. and 8:59 a.m., 9 a.m. and 11:59 a.m., and 12 p.m. and 4:59 a.m. For the mode of commute, public transit, walk, bicycle, and other methods are considered. Mobility indicates the geographic movement of a population over a period of time; for instance, it shows the information if a person moved to the current place of residence or is living at the same place as 1 year or 5 years ago.

Besides demographic features, we observe the effect and graveness of streetlight distribution on future crime incidents prediction motivated by [31]. Given a dataset of streetlight locations, for each DA, we propose the use of three streetlight features: (i) the total number of streetlights, (ii) the streetlight density, and (iii) the average minimum distance between crime data points and streetlight poles. The streetlight density of region r is computed as follows:

$$D_{st}(r) = \frac{St(r)}{A(r)}, \tag{3}$$

where $St(r)$ denotes the total number of streetlights in DA r. To calculate the average minimum distance from crime location to streetlight poles, we use the Haversine distance metric with scikit-learn [20].

Figure 1 visualizes the crime (year 2013), population, and streetlight densities by most observable DAs in Halifax. Dark red color indicates high density, and light red indicates low density. The bin sizes for population and streetlight densities are the same; on the other hand, we choose smaller bin sizes to get a clear picture of crime density. As shown in Fig. 1, most of the criminal incidents happen in downtown Halifax.

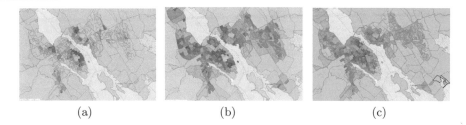

(a) (b) (c)

Fig. 1 Crime, population density, and streetlight density by most observable DAs in Halifax. (**a**) Crime density. (**b**) Population density. (**c**) Streetlight density

3.3 POI Features

In this work, we propose the use of POI features that can be obtained from location-based social networks (e.g., Foursquare). Our extracted POI features include (i) the total number of POIs, (ii) the POI frequency, and the density for different POI categories. Foursquare identifies ten major POI categories, such as food, arts and entertainment, college and university, nightlife spots, outdoors, and recreation, professional and other places, residence, shop, and service event, and travel and transport. The density of each POI category is defined as follows:

$$D_c(r) = \frac{N_c(r)}{N(r)}, \tag{4}$$

$$D_{1c}(r) = \frac{N_c(r)}{A(r)}, \tag{5}$$

where $N_c(r)$ is the total number of POIs of category c in a DA r, $N(r)$ is the total number of POIs in region r, and $A(r)$ is the area of that region. Figure 2 shows the POI and check-in count distributions of most observable dissemination areas (DAs) in Halifax.

3.4 Human Mobility Dynamic Features

Our study also explores dynamic human mobility data from location-based social networks to find if there is any relation with crime context. Social networks often have their users' location data, including their visits to different POIs in a city. We extract ten features for each DA based on the total number of user check-ins and check-in frequency for each POI category. Moreover, the check-in count for each DA at a time interval, the check-in density, region popularity, and visitor count

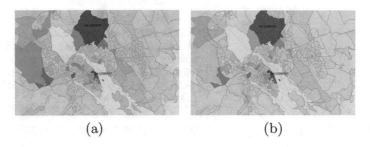

(a)   (b)

Fig. 2 The total POI and check-in count distributions by most observable DAs in Halifax. (a) POI count distribution. (b) Check-in count distribution

are also computed. For DA r at time interval t, the check-in density is defined as follows:

$$D_{ck}(r, t) = \frac{Ck(r, t)}{Ck(r)}, \tag{6}$$

$$D_{ck}(r, t) = \frac{Ck(r, t)}{A(r)}, \tag{7}$$

where $Ck(r, t)$ is the number of check-ins in DA r at time interval t, and $Ck(r)$ is the total number of check-ins in that region. Visitor count refers to the number of unique users that visited a DA at time interval t (i.e., region popularity).

$$R_{rp}(r, t) = \frac{Ck(r, t)}{Ck(t)}, \tag{8}$$

where $Ck(t)$ is the total number of check-in at time interval t for all regions.

We extract a total of 153 features for each dissemination area. Among them, we select 65 features that are more relevant to the crime prediction problem. The details of the total features chosen for each category appear in Table 1.

Table 1 Details of the selected features

Feature category	Extracted features	Selected features	Selected feature names
Temporal and historical	12	8	Month, weekday, time interval, season, crime frequency, crime density based on population, crime density based on area, crime density for season
Demographic	101	32	Population, population density, dwelling characteristics (11) mobility movers, mobility non movers, mobility migrants, mobility non migrants, aboriginals and visible minorities, primary mode of commute for residents (5), journey to work: the time people leave for work (5), low income (3), age and sex
Streetlight	3	2	Streetlight frequency, streetlight density
Foursquare POI	21	19	Total POI, food count, residence count, nightlife count, arts & entertainment count, college & university count, outdoors & recreation count, professional & other places count, shop & service count, travel & transport count, and the densities of all POI categories (9)
Foursquare dynamic	16	4	Total check-in for each time interval, check-in density, visitor count, region popularity

4 Experiments

We conducted experiments to evaluate the effectiveness of the different groups of features that can be aggregated to crime data for the task of crime prediction. In the following sections, we describe the datasets used (Sect. 4.1), the experimental setup (Sect. 4.2), and the achieved results (Sects. 4.3 and 4.4).

4.1 Datasets

We use crime data provided by the Halifax Regional Police (HRP) department, which includes records for all Dissemination Areas (DAs) in the Halifax Regional Municipality (HRM) in Nova Scotia, Canada. Our dataset was extracted from the Uniform Crime Reporting (UCR) survey, which was designed to measure the incidence of crime and its characteristics in Canadian society. For our experiments, we explore all crime occurrences from 2012 to 2014. The crime attributes extracted from the dataset include the geographic location, incident start time, month, weekday, and UCR descriptions (incident type). We have a total of 201,086 crime observations (excluding invalid and null information), where 69,340 data points happened in 2012, 65,785 in 2013, and 65,961 in 2014. We map all crime records to one of the 599 DAs collected for Halifax from Statistics Canada 2016 census, based on their geographic location. We group and index crime occurrences based on the DA where they happened, the year, month, day of the week, and the time interval of the day (we partition a day into eight 3 h time intervals).

In addition to the raw crime data, we collected demographic data for each DA from the Canadian Census Analyser [21]. We also extracted POI and dynamic features for Halifax from a dataset of Foursquare check-ins around the world, collected between April 2012 and January 2014 [32]. Lastly, streetlight information was obtained from the Streetlight Vision (SLV) API of HRM, which contains the location of 42,653 streetlight poles after removing null values and invalid data. We then computed the streetlight features proposed in Sect. 3 and mapped them to each DA.

Given that there are only records of crime occurrences in the dataset, we augment it to include "no crime" records. Thus, if there was no crime for a specific time interval, we labeled that observation as "no crime." The final size of the dataset, including crime and no crime records, is 1,207,584 (3*12*7*8*599).

As the occurrence of crime events is not frequent, most of the data (around 87%) are labeled with "no crime." To address this issue, we apply the under-sampling technique for "no crime" records to obtain a more balanced dataset [17].

We use the random under-sampling technique, which randomly selects a subset of observations from the major class (no crime) of the dataset. Applying random under-sampling might lead to a biased dataset; also, the deleted data points could have a beneficial or adverse impact to fit the model. However, this under-sampling

Table 2 Details of the datasets

Dataset	Source	Total data
Historical crime data	Halifax Regional Police	201,086
Dissemination area data	Statistics Canada	599
Demographic data	Canadian Census Analyser	599
Streetlight data	Halifax Regional Municipality	42,653
Foursquare POI data	Foursquare	2301
Foursquare checkin data	Foursquare	12,171

approach is compatible for our study as we are employing it for artificially creating "no crime" records only. The number of records for "crime" occurrences is sufficient despite having class imbalance. Table 2 shows the details of the dataset.

4.2 Experimental Setup

We run experiments with well-known ensemble learning classifiers, random forest (RF) [7] and gradient boosting (GB) [12], with scikit-learn [20] in Python.

We used randomized grid-search in preliminary experiments for the hyper-parameter optimization of each classifier evaluated. Besides evaluating the effect of each group of features, we compare our results to a DNN-based feature level data fusion baseline method [16]. Since the environmental context feature group used in [16] is unavailable for Halifax, we implement the DNN without those features. We use the same parameter settings reported in the corresponding paper for the baseline model, except for the activations of the DNN, which were replaced by sigmoid functions as they resulted in a better performance. We train the DNN for 300 epochs and select the best test scores. For evaluating the effectiveness of each feature group, we analyze the accuracy and F-score of the classifiers. At the same time, for the comparison with the baseline method, we also report precision, recall, and area under the ROC curve (AUC) scores.

We run a tenfold time-constrained cross-validation, similar to what was proposed in [9]. This is more appropriate since we guarantee that the records in the training set happened before the ones used for testing, and so we are effectively using data from the past to predict the future. We consider a sliding time window of 2 years, where the first 12 months are taken for training the models and the subsequent 12 months are used for testing. Thus, the models are still capable of capturing seasonality patterns as the training split always contains a full year of data. As our dataset includes 3 consecutive years of crime records from 2012 to 2014, for the first fold, we take all records from January 2012 to December 2012 for training, and the test split goes from January 2013 to December 2013. Next, for the second fold, we slide the window 1 month forward so that the training set spans from February 2012 to

Table 3 Results for average accuracy and F-score

No.	Model	Features					Random forest		Gradient boosting	
		R	D	S	F	P	Accuracy (%)	F-score (%)	Accuracy (%)	F-score (%)
1	MR	✓					59.60	63.93	59.85	64.65
2	MD	✓	✓				69.07	68.30	**69.94**	**69.45**
3	MS	✓		✓			68.51	67.46	68.52	68.25
4	MF	✓			✓		64.08	61.25	64.70	61.16
5	MP	✓				✓	66.75	64.19	67.61	64.19
6	MDS	✓	✓	✓			69.08	68.32	69.97	**69.51**
7	MDF	✓	✓		✓		69.15	68.29	69.95	69.33
8	MDP	✓	✓			✓	68.98	68.20	**70.00**	69.42
9	MSF	✓		✓	✓		68.06	66.48	69.04	67.50
10	MSP	✓		✓		✓	68.66	66.89	69.50	68.21
11	MFP	✓			✓	✓	66.84	64.02	67.60	64.07
12	MA	✓	✓	✓	✓	✓	69.09	68.16	**69.96**	**69.31**

January 2013, and the test spans from February 2013 to January 2014. We repeat this process for ten different folds.

4.3 Results for Our Proposed Features

In Table 3, we show the classification results with various feature combinations. We tested the addition of four different groups of features to the raw dataset (temporal + historic crime) (R): Demographic (D), Streetlight (S), Foursquare dynamic (F), and Foursquare POI (P) features. We compare 12 different models by adding all feature categories one by one with the raw features. Our first model is implemented based on the raw features only, named as model MR. We built models MD, MS, MF, and MP by adding demographic, streetlight, Foursquare check-in, and Foursquare POI data, respectively, with the raw data. Similarly, by combining two consecutive feature groups with the raw data, we built the models MDS, MDF, MDP, MSF, MSP, and MFP. Finally, model MA is implemented based on all of the feature combinations. Both RF and GB classifiers share a similar trend for all models based on classification accuracy and F-score. As GB performs better than RF for all combinations, in our discussions, we only consider the GB method. Model MR, trained only with raw features, is resulting in low accuracy of 59.85% and 64.65% F-score. Such behavior is expected since criminal behavior is affected by many different variables other than simple spatial and temporal factors [36].

By analyzing the addition of each group of features individually (top part of Table 3), the inclusion of demographic features (model MD) exhibits the best results, for which GB shows an improvement of almost 10% in accuracy (69.94%) and about 5% in F-score (69.45%) compared to only raw features. Similarly, streetlight features in model MS show an approximate 9% and 4% improvement for accuracy and F-score, respectively. Demographic variables reveal most of the characteristics of different regions, including social and economic factors, which are commonly correlated with criminality. Likewise, the installment of streetlight poles that reflects streetlight density feature also considers the same demographic profile for each corresponding area. Interestingly, Foursquare POI and dynamic features achieve less accuracy individually as compared to demographic and streetlight features. One of the reasons for this may be that there is missing information for places and check-in data for some dissemination areas.

Models 6 to 11 show the evaluation results for three feature categories combination. The accuracy and F-scores are better and almost consistent for all models except model MFP. The reason behind this is that all of them contain either demographic or streetlight features except MFP. In model MA, we combine all five categories of features. It gives us similar results as Model 6 (MDS), where we added both the demographic and streetlight categories. As Foursquare features do not lead to performance loss while combining others, in our study, we used all feature categories for building a model.

4.4   Comparison with a Baseline

Table 4 reports the accuracy, precision, recall, and AUC score for one of the best performing ensemble-based models, model MA with gradient boosting (GB-MA) and the baseline DNN model. Our proposed model performs significantly better than the baseline model based on precision, recall, and AUC scores. Though DNN can handle nonlinear relationships and data dependencies among different sources, it is challenging for the model to perform accurately for smaller domains or domains that suffer from data scarcity. This is the most likely reason for the baseline model to degrade performance. On the other hand, our model performed, on average, about 2% worse considering accuracy compared to the baseline model. This is due to the existence of a label imbalance in some of the testing folds.

Table 4 Performance evaluation

Model	Accuracy (%)	Precision (%)	Recall (%)	AUC
DNN (baseline) [16]	**71.82**	49.52	49.74	50.00
GB-MA	69.96	**70.13**	**68.53**	**69.95**

5 Conclusions and Future Work

In this paper, we study a fundamental problem of crime incident prediction for the future time interval. We have presented a data-driven approach to see how prediction accuracy can be improved by integrating multiple sources of data. Specifically, we focus on exploring population-centric features with streetlight and Foursquare-based features for each dissemination area in Halifax. Our problem also considers the temporal dimension of the crime profile in depth. We compare all five categories of feature combinations differently and unitedly. The results show that demographic and streetlight features have strong correlations with crime. Both of them show significant performance improvement for crime prediction individually and jointly. Though Foursquare data does not outperform demographic or streetlight data, it presents a satisfactory performance for crime prediction. Additionally, we compare our best ensemble model (i.e., model MA with gradient boosting in Table 3) with the DNN-based baseline model. Our results show that GB outperforms the DNN baseline for the same groups of features. Therefore, applying ensemble-based method leads to better performance in predicting future crime for smaller cities, such as Halifax.

In the future, we plan to extend this work in multiple directions. We want to integrate real-time streetlight data (e.g., light temperature, lux level, outages determined by power supply failures, etc.) with the current dataset. Moreover, identifying specific types of crime that might happen in the near future is our immediate concern. As it is very challenging to get accurate results for future crime prediction when sufficient data is unavailable, performing domain adaptation and some form of transfer learning using available data from a big city would be advantageous. Furthermore, the subject of data discrimination is another crucial concern for the study that focuses on real-world datasets. Investigating discrimination in socially sensitive decision records is state-of-the-art research to avoid biased classification learning.

References

1. M. Al Boni, M.S. Gerber, Automatic optimization of localized kernel density estimation for hotspot policing, in *Proceedings – 2016 15th IEEE International Conference on Machine Learning and Applications*, ICMLA 2016 (2017)
2. T. Almanie, R. Mirza, E. Lor, Crime prediction based on crime types and using spatial and temporal criminal hotspots. Int. J. Data. Min. Knowl. Discov. **5** (2015). https://doi.org/10.5121/ijdkp.2015.5401
3. F.K. Bappee, A. Soares Júnior, S. Matwin, Predicting crime using spatial features, in *Advances in Artificial Intelligence*, ed. by E. Bagheri, J.C. Cheung (Springer International Publishing, Cham, 2018), pp. 367–373
4. A. Belesiotis, G. Papadakis, D. Skoutas, Analyzing and predicting spatial crime distribution using crowdsourced and open data. ACM Trans. Spatial Algorithm. Syst. **3**(4), 12:1–12:31 (2018)

5. A. Bogomolov, B. Lepri, J. Staiano, N. Oliver, F. Pianesi, A. Pentland, Once upon a crime: towards crime prediction from demographics and mobile data. CoRR abs/1409.2983 (2014). http://arxiv.org/abs/1409.2983

6. A. Bogomolov, B. Lepri, J. Staiano, E. Letouzé, N. Oliver, F. Pianesi, A. Pentland, Moves on the street: Classifying crime hotspots using aggregated anonymized data on people dynamics. Big data **3**, 150814115029008 (2015). https://doi.org/10.1089/big.2014.0054

7. L. Breiman, Random forests. Mach. Learn. **45**(1), 5–32 (2001)

8. A.L. Buczak, C.M. Gifford, Fuzzy association rule mining for community crime pattern discovery. in *ACM SIGKDD Workshop on Intelligence and Security Informatics*, ISI-KDD '10 (ACM, New York, 2010), pp. 2:1–2:10

9. V. Cerqueira, L. Torgo, I. Mozetic, Evaluating time series forecasting models: an empirical study on performance estimation methods. CoRR abs/1905.11744 (2019). http://arxiv.org/abs/1905.11744

10. L. Duan, T. Hu, E. Cheng, J. Zhu, C. Gao, Deep convolutional neural networks for spatiotemporal crime prediction, in *Proceedings of the 2017 International Conference on Information and Knowledge Engineering*, IKE '17 (2017), pp. 61–67

11. J. Fitterer, T.A. Nelson, F. Nathoo, redictive crime mapping. Police Pract. Res. **16** (2015). https://doi.org/10.1080/15614263.2014.972618

12. J.H. Friedman, Greedy function approximation: a gradient boosting machine. Annal. Stat. 1189–1232 (2001)

13. C. Graif, R.J. Sampson, Spatial heterogeneity in the effects of immigration and diversity on neighborhood homicide rates. Homicide Stud. (2009). https://doi.org/10.1177/1088767909336728

14. D.E. Hojman, Inequality, unemployment and crime in Latin American cities. Crime Law and Soc. Change **41**, 33–51 (2004). https://doi.org/10.1023/B:CRIS.0000015327.30140.8d

15. C. Kadar, J. Iria, I. Pletikosa, Exploring Foursquare-derived features for crime prediction in New York City, in *KDD – Urban Computing WS '16* (2016). https://doi.org/10.1145/1235

16. H.W. Kang, H.B. Kang, Prediction of crime occurrence from multi-modal data using deep learning. PLOS One **12** (2017)

17. G. Lemaître, F. Nogueira, C.K. Aridas, Imbalanced-learn: a python toolbox to tackle the curse of imbalanced datasets in machine learning. J. M. Learn. Res. **18**(17), 1–5 (2017)

18. K. Leong, A. Sung, A review of spatio-temporal pattern analysis approaches on crime analysis. International E-Journal of Criminal Sciences **9** (2015) Isbn 1988-7949. Issn 1988–7949

19. Y. Mu, W. Ding, M. Morabito, D. Tao, Empirical discriminative tensor analysis for crime forecasting, in *Lecture Notes in Computer Science (including subseries Lecture Notes in Artificial Intelligence and Lecture Notes in Bioinformatics)* (2011)

20. F. Pedregosa, G. Varoquaux, A. Gramfort, V. Michel, B. Thirion, O. Grisel, M. Blondel, P. Prettenhofer, R. Weiss, V. Dubourg, J. Vanderplas, A. Passos, D. Cournapeau, M. Brucher, M. Perrot, E. Duchesnay, Scikit-learn: Machine learning in Python. J. Mach. Learn. Res. **12**, 2825–2830 (2011)

21. Profile of Census Dissemination Areas. http://datacentre.chass.utoronto.ca/cgi-bin/census/2016/displayCensus.cgi?year=2016&geo=da. Accessed 07 February 2019

22. J. Ratcliffe, The hotspot matrix: a framework for the spatio-temporal targeting of crime reduction. Police Pract. Res. **5**(1), 5–23 (2004)

23. S.K. Rumi, K. Deng, F.D. Salim, Crime event prediction with dynamic features. EPJ Data Sci. (2018). https://doi.org/10.1140/epjds/s13688-018-0171-7

24. Statistics Canada. 2016 census - boundary files. https://www12.statcan.gc.ca/census-recensement/2011/geo/bound-limit/bound-limit-2016-eng.cfm. Accessed 07 February 2019

25. P. Wang, R. Mathieu, J. Ke, H.J. Cai, Predicting criminal recidivism with support vector machine, in *International Conference on Management and Service Science, MASS 2010 International Conference* (2010)

26. D. Wang, W. Ding, H. Lo, M. Morabito, P. Chen, J. Salazar, T. Stepinski, Understanding the spatial distribution of crime based on its related variables using geospatial discriminative patterns. Comput. Environ. Urban Syst. (2013). https://doi.org/10.1016/j.compenvurbsys.2013.01.008

27. M. Traunmueller, G. Quattrone, L. Capra, Mining mobile phone data to investigate urban crime theories at scale, in *SocInfo, Lecture Notes in Computer Science*, vol. 8851 (Springer, Berlin, 2014), pp. 396–411

28. H. Wang, D. Kifer, C. Graif, Z. Li, Crime rate inference with big data, in *Proceedings of the 22nd ACM SIGKDD International Conference on Knowledge Discovery and Data Mining*, San Francisco, August 13–17 (2016), pp. 635–644

29. B. Wang, P. Yin, A.L. Bertozzi, P.J. Brantingham, S.J. Osher, J. Xin, Deep learning for real-time crime forecasting and its ternarization. CoRR abs/1711.08833 (2017). http://arxiv.org/abs/1711.08833

30. B. Wang, D. Zhang, D. Zhang, P.J. Brantingham, A.L. Bertozzi, Deep learning for real time crime forecasting. International Symposium on Nonlinear Theory and its Applications. 1707.03340, (2017), https://par.nsf.gov/biblio/10048857

31. Y. Xu, F. Cong, E. Kennedy, S. Jiang, S. Owusu-Agyemang, The impact of street lights on spatial-temporal patterns of crime in Detroit, Michigan. Cities **79** (2018). https://doi.org/10.1016/j.cities.2018.02.021

32. D. Yang, D. Zhang, B. Qu, Participatory cultural mapping based on collective behavior data in location-based social networks. ACM Trans. Intell. Syst. Technol. (TIST) **7**(3), 30 (2016)

33. C.H. Yu, M.W. Ward, M. Morabito, W. Ding, Crime forecasting using data mining techniques, in *2011 IEEE 11th International Conference on Data Mining Workshops* (2011)

34. J. Yu, W. Ding, M. Morabito, P. Chen, Hierarchical Spatio-temporal pattern discovery and predictive modeling. IEEE Trans. Knowl. Data. Eng. **28**, 1 (2015). https://doi.org/10.1109/TKDE.2015.2507570

35. X. Zhao, J. Tang, Modeling temporal-spatial correlations for crime prediction, in *Proceedings of the 2017 ACM on Conference on Information and Knowledge Management*, CIKM '17 (ACM, New York, 2017), pp. 497–506

36. X. Zhao, J. Tang, Crime in urban areas: a data mining perspective. CoRR abs/1804.08159 (2018). http://arxiv.org/abs/1804.08159

Nested Named Sets in Information Retrieval

Mark Burgin and H. Paul Zellweger

1 Introduction

Analysis of the database area shows that a focal problem of working with data in databases is amplifying efficiency of human-database interaction. This goal can be achieved on different levels, which form the following hierarchy:

queries&their algorithmic realization

↑

logic&its algorithmic realization

↑

algebras&their algorithmic realization

↑

data structures

The main efforts in this area have been directed at the development of logical systems for database management, which, as Benedikt explains in his Keynote Talk at PODS'18, became a leading application of logic within computer science [2].

While this explanation is undeniably true, the efficiency of human-database interaction depends not only on logic and corresponding query languages but to a great extent on the structuration of data and databases. Codd's work transforms

M. Burgin (✉)
Mathematics Department, UCLA, Los Angeles, CA, USA
e-mail: mburgin@math.ucla.edu

H. P. Zellweger
ArborWay Labs, Rochester, MN, USA
e-mail: pz@arborwaylabs.com

© Springer Nature Switzerland AG 2021
R. Stahlbock et al. (eds.), *Advances in Data Science and Information Engineering*,
Transactions on Computational Science and Computational Intelligence,
https://doi.org/10.1007/978-3-030-71704-9_30

data management in many ways because he suggested the relational table with an advanced relational algebra [14].

Three types of structures – structure of databases, structure of database procedures, and structures of data in databases – are critical for efficient and reliable functioning of databases. Mathematical modeling of these structures allows exploration of properties and functioning of databases.

The conventional approach has primarily been based on relational data structures represented by flat forms in relational databases. However, in many situations, practical information processing involves nesting and nested structures. As a result, researchers started exploring how to use nested structures in databases [13, 19, 20, 21, 22].

Here we use the mathematical structure called a nested named set to model structures determining database properties and functioning. Our approach is based on the evidence that named sets play an essential role in databases and knowledge bases. Relations are a special case of named sets. Consequently, all relational databases store named sets and work with them [17, 18]. Named set chains are key structures in temporal databases [6, 27]. Named sets were employed for building a mathematical model for efficient managing XML data [23]. All knowledge and information structures, from microstructures to megastructures, are formed from named sets [7, 9]. Named sets have been constructively utilized for data visualization and information retrieval in databases [12, 29, 30, 31], database management, and information acquisition on the Internet as it is possible to consider the Internet as huge database [11]. As a matter of fact, any network, especially the World Wide Web extensively, uses names and consequently name sets because names are related to what they name, which produces a named set. Utilization of names on the Internet has become explicit and as experts in network technology assert Named Data Networking (NDN) is a future internet architecture inspired by years of empirical research into network usage and a growing awareness of unsolved problems in contemporary internet architectures [26]. In addition, named sets provide a unified data metamodel that allows building suitable data models on all levels: from high-level or conceptual to representational or implementation to low-level or physical models. In turn, named sets and their chains form efficient high-level metadata for different purposes (cf., [28]).

At the same time, information on the Internet, as a rule, includes nested structures. For instance, when one web site contains a reference to another web site, it means that the second web site is nested in the first one. This feature also demonstrates the necessity of nested named sets for mathematical representation of information on the Internet.

Moreover, named sets provide a unified data metamodel that allows building suitable data models on all levels: from high-level or conceptual to representational or implementation to low-level or physical models. In turn, named sets and their chains form efficient high-level metadata for different purposes (cf., [28]).

In this paper, we develop and explore novel mathematical models aimed at an efficient unified representation for data structures, database schemas, database procedures, and database structures. Mathematical structures utilized for this purpose are called nested named sets. Specific characteristics of nested named sets are introduced and studied. Their application to uncovering database properties and describing database functioning is presented.

It is important to explain what novel technological advantages our approach has in comparison with other approaches to database technology. First, we use a more advanced model, which is called set-based nested named sets, for data structures and database organization. This model includes both relational and hierarchical databases as particular cases. At the same time, set-based named sets include not only relations but also multirelations, labeled graphs, and especially important algorithmic named sets, in which links between elements are specified and realized by algorithm (computational procedures).

Second, the theory of named sets provides a powerful collection of operations with data, data structures, and database structures.

Third, nested named sets give an adequate and efficient representation of nested structures in databases, such as nested lists.

Fourth, application of named sets makes available increasing transparency and efficiency of database construction, utilization maintenance, and management.

Fifth, named sets provide flexible unification of database architectures.

The chapter has the following structure. In Sect. 2, we describe basic constructions from named set theory used for the development and management of databases. In Sect. 3, we study basic relations between named sets in general and nested name sets in particular. In Sect. 4, we demonstrate applications of nested name sets to database development and management.

2 Basic Definitions and Constructions

A named set is the mathematical structure of entities, ties, and names [5]. The early literature on named sets goes back to the 1980s [3, 4]. Today this literature includes more advanced results, such as those of Burgin [8, 9, 10].

2.1 General Named Sets

Named sets, which are also called *fundamental triads*, have different types and categories [8]. The two names for the same structure reflect different features of this structure. Labeling this structure a named set, we direct attention to its inner structure. Dubbing this structure a fundamental triad, we reveal its unity. At first, three primary types – basic, bidirectional and cyclic named sets – are introduced.

Definition 2.1 A *basic named set* is a triad $\mathbf{X} = (X, f, N)$ with the following visual (graphic) representation:

$$X \xrightarrow{f} N \tag{1}$$

In this triad $\mathbf{X} = (X, f, N)$, components X and N are two objects and f is a correspondence (e.g., a binary relation) from X to N. With respect to $\mathbf{X}$, the object X is called the *support* of $\mathbf{X}$ and denoted $S(\mathbf{X})$, the object N is called the *component of names (reflector)* or *set of names* of $\mathbf{X}$ and denoted $N(\mathbf{X})$, and the object f is called the *naming correspondence (reflection)* of $\mathbf{X}$ and denoted $r(\mathbf{X})$. It means that $\mathbf{X} = (S(\mathbf{X}), r(\mathbf{X}), N(\mathbf{X}))$. Note that in $\mathbf{X}$, components X and N are not automatically sets, while f is not necessarily a mapping or a function even if X and N are sets. For instance, X and N are sets of words and f is an algorithm.

The standard example is a basic named set (basic fundamental triad), in which X consists of people, N consists of their names, and f is the correspondence between people and their names. Another example is a basic named set (fundamental triad), in which X consists of things, N consists of their names, and f is the correspondence between things and their names [16].

It is necessary to make a distinction between triples and triads. A *triple* is any set with three elements, while a *triad* is a system of three connected elements (components). It is worthy of note that mathematicians introduced the concept of a triple in an abstract category. In essence, such a triple is a triad that consists of three fundamental triads and thus is a triad of the second order [8]. Understanding of the complex nature of the *categorical triple* compelled mathematicians to change the name of this structure and now it is always called a *monad*. Interestingly, this shows connection between fundamental triads and Leibniz monads. However, in this case, a monad consists of triads and not the other way around.

Definition 2.2 A *bidirectional named set* is a triad $\mathbf{X} = (X, f, Z)$ with the following visual (graphic) representation:

$$X \xleftrightarrow{f} Z \tag{2}$$

It is also a triad $\mathbf{D} = (X, f, Y)$, in which the naming correspondence f goes in two directions.

In this triad $\mathbf{D} = (X, f, Z)$, components X and Z are two objects and f is a correspondence (e.g., a binary relation) between X and Z, which goes in two directions. With respect to $\mathbf{D}$, $X = S(\mathbf{D})$ is also called the *support* of $\mathbf{D}$, $Z = N(\mathbf{D})$ is called the *component of names (reflector)* or *set of names* of $\mathbf{D}$, and $f = r(\mathbf{D})$ is called the *naming correspondence (reflection)* of $\mathbf{D}$. It means that $\mathbf{D} = (S(\mathbf{D}), r(\mathbf{D}), N(\mathbf{D}))$. Note that in $\mathbf{D}, f$ is not necessarily a mapping or a function.

To understand the difference between basic and bidirectional named sets, let us consider set-theoretical named sets, in which the reflection is a binary relation [8]. A *directed binary relation* between sets X and Y is a set of pairs, in which the first element is for X and the second from Y, while an *undirected binary relation* between sets X and Y is a set of pairs, in which either the first element is for X and the second is from Y or the first element is for Y and the second is from X. Thus, a basic named set has a directed binary relation as its reflection. In contrast to this, a bidirectional named set has an undirected binary relation as its reflection.

Often binary relations are treated as subsets of a Cartesian product of two sets. However, in their super-formalized treatise on mathematics, Boubaki defines a binary relation as a set-theoretical named set. Namely, a binary relation R is defined as an ordered triple (X, G, Y) where X and Y are arbitrary sets (or classes), and G is a subset of the Cartesian product $X \times Y$. It means that a binary relation is a basic set-theoretical named set.

A bidirectional set-theoretical named set has a reflection, which consists of two binary relations. As one of these binary relations can be empty, any basic set-theoretical named set is a particular case of a bidirectional set-theoretical named set. At the same time, as we will see, any bidirectional set-theoretical named set can be composed of two basic set-theoretical named sets.

We have an example of a bidirectional named set when two people are exchanging messages, e.g., by e-mails, messaging, or talking to one to another. In this case, X and Z are people while f and g are messages that go from one person to another.

There is one more important class of named sets.

Definition 2.3 A *cyclic named set* is a fundamental triad or named set $\mathbf{X} = (X, f, X)$.

In essence, a named set $\mathbf{X}$ is cyclic when $N(\mathbf{X}) = S(\mathbf{XS})$.

A cyclic named set can be basic. Then it has the following form:

$$X \xrightarrow{f} X \tag{3}$$

The following graphic form can also describe it:

$$(4)$$

An example of a basic cyclic named set is a subatomic particle, such as an electron, which acts on itself.

A cyclic named set can also be bidirectional. Then it has the following form:

$$X \xleftrightarrow{f} X \tag{5}$$

The following graphic form can also describe it:

$$(6)$$

An example of a bidirectional cyclic named set is a computer network. In it, X consists of computers and f contains all connections between them.

2.2 Set-Based Named Sets

Set-based named sets are important in general because many mathematical constructions are set-based named sets. At the same time, set-based named sets are specifically significant for databases because data are conventionally presented as set-theoretical constructions, such as records, arrays, lists, or stacks, all of which are specific named sets.

Definition 2.4 A named set $\mathbf{X} = (X, f, N)$ is called *set-based* if $S(\mathbf{X})$ and $N(\mathbf{X})$ are sets.

Important classes of set-based named sets are:

- *Set-theoretical named sets*, in which $r(\mathbf{X})$ is a binary relation between sets X and N.
- *Algorithmic named sets*, in which $r(\mathbf{X})$ is an algorithm that transforms the set $S(\mathbf{X})$ into the set $N(\mathbf{X})$.
- *Automaton named sets*, in which $r(\mathbf{X})$ is an automaton that transforms the set $S(\mathbf{X})$ into the set $N(\mathbf{X})$.
- *Categorical named sets*, in which $S(\mathbf{X})$ and $N(\mathbf{X})$ are two objects from some category and $r(\mathbf{X})$ is a set of morphisms between these objects.

Let us consider some examples:

1. Graphs are cyclic set-theoretical named sets.
2. Labeled graphs are cyclic labeled set-theoretical named sets.

There are various relations between structures in general and named sets in particular.

Definition 2.5 A structure $\mathbf{A}$ is a *faithful representation* of a structure $\mathbf{B}$ if all elements and relations from $\mathbf{B}$ are represented by elements and relations from $\mathbf{A}$.

For instance, it is possible to faithfully represent an arbitrary graph by a binary relation.

Definition 2.6 Structures $\mathbf{A}$ and $\mathbf{B}$ are called *equivalent* if each of them is a faithful representation of another.

Let us establish equivalence of some data structures.

Proposition 2.1 Set-theoretical named sets are equivalent to binary relations as well as to bipartite graphs.

2.3   Nested Named Sets

There are three classes of nested named sets.

Definition 2.7 A set-based named set $\mathbf{X} = (X, f, N) = (S(\mathbf{X}), r(\mathbf{X}), N(\mathbf{X}))$ is called:

- *Nested from above* if elements of the set $N(\mathbf{X})$ are named sets.
- *Nested from below* if elements of the set $S(\mathbf{X})$ are named sets.
- *Amply nested* if elements of both sets $S(\mathbf{X})$ and $N(\mathbf{X})$ are named sets.

Example 2.1 Let us consider vectors from the three-dimensional vector space V over the field R of real numbers. When a basis B of V is specified, each vector v is represented by a triple of real numbers (a_1, a_2, a_3) from the space R^3. This representation gives us the first named set.

$$\mathbf{Repr} = \left(V, r_B, R^3\right)$$

Here r_B is the numerical representation of vectors from V determined by the basis B.

The named set **Repr** is amply nested. Indeed, the vector space V is not a pure set but has an elaborate structure including operations with its elements, relations in it and its properties, for example, those that are defined by identities in the vector space V. Thus, V is also a named set:

$$V = (V, p, \{O, R, P\})$$

Here V is the set of vectors from V; O is the set of operations in V; R is the set of relations in V; and P is the set of properties of V.

In turn, the named set V is also amply nested. Indeed, as it is demonstrated, for example, in [8], each operation, binary relation or property is a named set.

At the same time, R^3 is also a named set. Indeed, each triple (vector) of real numbers (a_1, a_2, a_3) is a named set of the form.

$$(X, f, \{1, 2, 3\})$$

Here X is the set of one, two, or three real numbers and connects each of these numbers with the digit denoting its position in the vector (a_1, a_2, a_3). Namely, a_1 is connected with 1, a_2 is connected with 2, and a_3 is connected with 3. For instance, in the named set of the vector $(2, 3)$, the number 2 is connected with the digit 1, while the number 3 is connected with the digits 2 and 3.

The named set $(X, f, \{1, 2, 3\})$ is also amply nested because each number is (is represented by) a named set and each digit is (is represented by) a named set as a symbol [8].

Example 2.2 Let us consider 3×3-matrices of real numbers. The same reasoning as before shows that each such a matrix is a nested named set.

3 Relations Between Nested Named Sets

There are two basic relations – membership and inclusion – between named sets in general and nested named sets, in particular.

Named set membership E is defined in the following way.

Let us consider two set-based named sets $\mathbf{X} = (X, f, N) = (S(\mathbf{X}), r(\mathbf{X}), N(\mathbf{X}))$ and $\mathbf{Z} = (Z, g, M) = (S(\mathbf{Z}), r(\mathbf{Z}), N(\mathbf{Z}))$. Then $\mathbf{Z}$ E $\mathbf{X}$ if $\mathbf{Z}$ belongs to $S(\mathbf{X})$ or to $N(\mathbf{X})$. In a formal description, we have.

$$\mathbf{Z} \, \mathrm{E} \, \mathbf{X} \iff \mathbf{Z} \in S(\mathbf{X}) \vee \mathbf{Z} \in N(\mathbf{X}))$$

For nested set-based named sets, there is a hierarchy of membership relations E_0, E_1, ..., which are defined recursively:

$$\mathbf{Z} \, \mathrm{E}_1 \, \mathbf{X} \iff \mathbf{Z} \in S(\mathbf{X}) \vee \mathbf{Z} \in N(\mathbf{X}))$$
$$\mathbf{Z} \, \mathrm{E}_n \, \mathbf{X} \iff \exists \mathbf{W} \, (\mathbf{W} \, \mathrm{E}_{n-1} \, \mathbf{X} \& \mathbf{Z} \, \mathrm{E}_1 \, \mathbf{W}))$$
$$\mathbf{Z} \, \mathrm{E} \, \mathbf{X} \iff \exists n \, (\mathbf{Z} \, \mathrm{E}_n \, \mathbf{X}))$$

There are two named set inclusion relations – direct inclusion $\sqsubseteq$ and nested inclusion $\sqsubseteq$.

To define direct inclusion for set-based named sets, we use the conventional inclusion of sets, i.e., $X \subseteq Y$ if all elements from X belong to Y. In addition, we utilize the *domination relation* $\preccurlyeq$ for naming correspondences of named sets. For different types of set-based named sets, there are different domination relations.

1. When $\mathbf{X}$ and $\mathbf{Z}$ are set-theoretical named sets, $r(\mathbf{Z})$ and $r(\mathbf{X})$ are sets and $r(\mathbf{Z}) \preccurlyeq r(\mathbf{X})$ if $r(\mathbf{Z}) \subseteq r(\mathbf{X})$.
2. When $\mathbf{X}$ and $\mathbf{Z}$ are labeled set-theoretical named sets, $r(\mathbf{Z})$ and $r(\mathbf{X})$ are labeled sets, i.e., set-theoretical named sets, and we define $r(\mathbf{Z}) \preccurlyeq r(\mathbf{X})$ if $r(\mathbf{Z}) \subseteq r(\mathbf{X})$.
3. When $\mathbf{X}$ and $\mathbf{Z}$ are algorithmic named sets, $r(\mathbf{Z})$ and $r(\mathbf{X})$ are algorithms and $r(\mathbf{Z}) \preccurlyeq r(\mathbf{X})$ if the restriction of $r(\mathbf{X})$ onto $S(\mathbf{Z})$ and $N(\mathbf{Z})$ coincides with $r(\mathbf{Z})$.
4. When $\mathbf{X}$ and $\mathbf{Z}$ are algorithmic named sets and K is a class of algorithms, $r(\mathbf{Z})$ and $r(\mathbf{X})$ are algorithms and $r(\mathbf{Z}) \preccurlyeq_K r(\mathbf{X})$ if $r(\mathbf{Z})$ is reducible in K to the restriction of $r(\mathbf{X})$ onto $S(\mathbf{Z})$ and $N(\mathbf{Z})$.

Now we define *direct named set inclusion* $\sqsubseteq$ in the following way.

Let us consider two set-based named sets $\mathbf{X} = (X, f, N) = (S(\mathbf{X}), r(\mathbf{X}), N(\mathbf{X}))$ and $\mathbf{Z} = (Z, g, M) = (S(\mathbf{Z}), r(\mathbf{Z}), N(\mathbf{Z}))$. Then $\mathbf{Z} \sqsubseteq \mathbf{X}$ if $S(\mathbf{Z})$ is a subset of $S(\mathbf{X})$, $N(\mathbf{Z})$ is a subset of $N(\mathbf{X})$, and $r(\mathbf{Z})$ is dominated by $r(\mathbf{X})$. In a formal description, we have.

$$\mathbf{Z} \sqsubseteq \mathbf{X} \iff S(\mathbf{Z}) \subseteq S(\mathbf{X}) \,\&\, N(\mathbf{Z}) \subseteq N(\mathbf{X})) \,\&\, r(\mathbf{Z}) \preccurlyeq r(\mathbf{X})$$

In nested named sets, their components contain other named sets. This induces the hierarchy of nested inclusions:

$$\mathbf{Z} \sqsubseteq_1 \mathbf{X} \iff S(\mathbf{Z}) \subseteq S(\mathbf{X}) \,\&\, N(\mathbf{Z}) \subseteq N(\mathbf{X})) \,\&\, r(\mathbf{Z}) \preccurlyeq r(\mathbf{X})$$
$$\mathbf{Z} \sqsubseteq_n \mathbf{X} \iff \exists \mathbf{W} \left(\mathbf{W} \sqsubseteq_{n-1} \mathbf{X} \,\&\, \mathbf{Z} \sqsubseteq_1 \mathbf{W} \right)$$
$$\mathbf{Z} \sqsubseteq \mathbf{X} \iff \exists n \, (\mathbf{Z} \sqsubseteq_n \mathbf{X}))$$

We see that relations $\sqsubseteq$ and $\sqsubseteq_1$ coincide.

Let us study properties of relations between named sets. We remind that a named set $\mathbf{X}$ is a *singlenamed set* if all elements in its support $S(\mathbf{X})$ have the same name [8]. Any set is a singlenamed set.

Proposition 3.1 When $\mathbf{Z}$ and $\mathbf{X}$ are singlenamed sets, then $\mathbf{Z} \sqsubseteq \mathbf{X}$ if and only if $S(\mathbf{Z}) \subseteq S(\mathbf{X})$ and all elements in $S(\mathbf{Z})$ and $S(\mathbf{X})$ have the same name.

Applying induction, we obtain the following results:

Proposition 3.2 If $\mathbf{Z} \, \mathbf{E} \, \mathbf{Y}$ and $\mathbf{Y} \, \mathbf{E} \, \mathbf{X}$, then $\mathbf{Z} \, \mathbf{E} \, \mathbf{X}$.

It means that relation $\mathbf{E}$ is transitive.

Proposition 3.3 If $\mathbf{Z} \sqsubseteq \mathbf{Y}$ and $\mathbf{Y} \sqsubseteq \mathbf{X}$, then $\mathbf{Z} \sqsubseteq \mathbf{X}$.

It means that relation $\sqsubseteq$ is transitive.

Proposition 3.4 If the dominance relation is transitive, $\mathbf{Z} \sqsubseteq \mathbf{Y}$ and $\mathbf{Y} \sqsubseteq \mathbf{X}$, then $\mathbf{Z} \sqsubseteq \mathbf{X}$.

Inclusion of sets $\subseteq$ is transitive. It gives us the following result.

Corollary 3.1 If $\mathbf{Z}$, $\mathbf{X}$ and $\mathbf{Y}$ are set-theoretical named sets, $\mathbf{Z} \sqsubseteq \mathbf{Y}$ and $\mathbf{Y} \sqsubseteq \mathbf{X}$, then $\mathbf{Z} \sqsubseteq \mathbf{X}$.

Applying Proposition 3.4 several times and using induction, we prove the following result.

Proposition 3.5 If the dominance relation is transitive, then $\mathbf{Z} \sqsubseteq \mathbf{X}$ if and only if $\mathbf{Z} \sqsubseteq \mathbf{X}$.

Often inclusion and membership relations are coordinated with operations with named sets [8].

Proposition 3.6 If $\mathbf{Z} \sqsubseteq \mathbf{X}$ and $\mathbf{Y} \sqsubseteq \mathbf{X}$, then $\mathbf{Z} \bigcup \mathbf{Y} \sqsubseteq \mathbf{X}$.

In a similar way, we have:

Proposition 3.7 If Z E $X \cap Y$, then Z E X and Z E Y.

Equally, we have:

Proposition 3.8 For any set-based named sets Z, X and Y, we have:

1. $Y \sqsubseteq Z \cup Y$
2. $X \cap Y \sqsubseteq Y$
3. Y E X implies Y E $Z \cup X$

These properties of nested named sets allow construction of efficient organization and transformation of data in databases as it is demonstrated in the next section.

4 Application to Databases

We employ nested named sets for the automated construction of the decision tree interface for the relational database [31, 33]. This interface extracts lists of raw data from the database and aggregates them into a single list, which itself consists of nested data lists. This novel approach is data-driven aiming at three goals: (1) setting a new standard of the database query tools for information consumers and providers alike; (2) providing efficient techniques for database developers; and (3) maintaining and further advancing the theoretical tradition, which has guided the development of the relational model over the past four decades. The first goal is achieved by displaying a point and click decision tree interface, which enables end-users to navigate over data relations support in the underlying database without their needing to know anything about its schema, SQL, or its data content. At the bottom of these navigation paths, the Decision Tree transfers control to a window that displays information related to the data topics selected for reaching this point. Figure 1 provides an overview of these details. For the information providers, the advantage of this technology is that it is generated entirely by the program logic.

The theory guiding the development of this interface, named set theory, is outlined in Sects. 2 and 3 of this chapter. As it is mentioned earlier, the modeling

Fig. 1 Window displaying information related to data topics

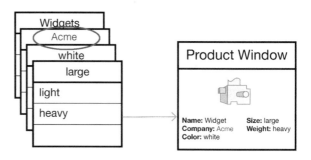

of structures in the relational model, including its schema, data content, and table operations, enables the explication and utilization of the Branching Data Model [32]. It is a uniform pattern of data relations, which exists throughout the relational model establishing a new level of abstraction over relational data. Describing the Decision Tree interface here, it will be discussed the Branching Data Model involving pairs of attributes within the same table. In addition, we will show how the Branching Data Model is expanded to the Nested Branching Data Model, which establishes menu data for the Decision Tree interface in the form of a list of nested data lists. Our future research is oriented at exploration of how the Branching Data Model establishes binding between pairs of key attributes in two different tables.

The automated Decision Tree interface has an authoring system on the server providing executive functions for this system. This system performs the following operations: (1) extracts lists of raw data from a remote database; (2) formats these lists of data on the server; and then (3) sends them to the client's interface for display. The authoring system has its own integrated services for managing the connections to the related databases and for generating a local data dictionary of their schema labels. This parallel data dictionary serves two purposes: 1) as a security filter which grants access to databases, tables, fields, and records; and 2) as an integrated service device which builds and manages end-user labels associated with schema labels coordinating their usage when it constructs SQL statements at runtime.

Other features in the authoring system include tools for generating bottom-up tabbing structures to facilitate the scalability of visualizing lists of data. With this tabbing tool, 1.5 M data items in an attribute can be accessed within 4 nested lists, when each menu list is set to display no more than 35 data topics. Lastly, the authoring system employs WYSIWYG editors to generate templates for information windows, thereby enabling a developer to design any number of combinations of interface displays.

In spite of the great abundance of theory in computer science, many of the ideas presented here on the Decision Tree interface represent what Charles Peirce would call a "logical leap of the mind" or abduction, and what students of innovation would view as the result of "design thinking." All the same, some of them are essentially consistent with the general trends in certain branches of the field. For instance, in the database theory, nested relations connect rows in one table with rows in a second table by pairing the former's primary key with its foreign key counterpart, in [17] and in database interfaces, like [1]. When these two values are equivalent, a one-to-many relationship occurs at the row level, and the tables connect. In visualization research, however, the same one-to-many data relationships exist between a pair of attributes in the same table, but the values are rarely equal. In systems like SNAP [24], built for the RDBMS, a visualization schema exists that treats these data relationships as visualization components [25]. In our application of named set theory, we establish a uniform way of modeling both of these one-to-many data relationships within the same expression. With this new modeling capability, the uniformity of the Branching Data Model in the relational data model becomes evident.

4.1 The Branching Data Model

In the relational database model, a uniform pattern of data relations that occurs between two table attributes exists. This pattern is called the Branching Data Model. It is based on the well-defined SQL SELECT statement, which simplifies the process of this retrieval operation. A single data condition and a single output attribute start the process of exposing the one-to-many data relationship. The algorithmic named set, which is defined in [8], establishes the mathematical framework for this operation by reducing its elements to their basic I/O patterns.

In the mathematical notation, an algorithmic named set has the form.

$$(A, \text{algorithm}, B)$$

where A is an input set, B is an output set, and the "algorithm" represents a series of non-contradictory rules for the transformation of the elements from the input set into the elements from the output set. The designation of the input and output sets by their schema labels, as the reader will see, presents an I/O perspective on the Branching Data Model.

The Branching Data Model was first discovered in the form of a relational table between a pair of attributes. One attribute in the pair corresponds to the input set. The other attribute is a channel that delivers the output data. The algorithm that binds the input data to its output set, like all mathematical operations, represents numerous implementations. That is, of course, true just as long as the input set always produces the same output set.

While the algorithm binds an input set to its output, we can turn to a high-level abstraction in relational model to encapsulate these details, namely, the ANSI standard SQL. Its elements, according to the SELECT operation, are combined in the following form.

$$\text{SELECT } B \text{ FROM table WHERE } A = a$$

This expression sets the stage for the discovery of the logical structure of the Branching Data Model. For instance, the database table consists of two sets of data, A and B, which correspond to the attributes A and B. Besides, the input attribute A has a single data condition consisting of the data element "a." It also declares a single output attribute B. In summary, the SELECT expression simplifies the three high-level components of the relational table: its permanent schema, its temporary data content, and the structure of the operations that unifies these two components.

In the database table, the SELECT statement casts a one-to-many (multivalued) mapping between the data in these two sets of data A and B with corresponding attributes A and B. To obtain this result, the data condition on the input attribute always refers to a known domain value. In effect, this known value self-references an element in the attribute's domain. The output attribute, as mentioned above, always refers to a single table attribute.

Fig. 2 Example – Branching
Data Model in the Widgets
Table

Widgets			
COLOR	SIZE	WEIGHT	ID
red	small	light	100
red	medium	light	101
red	medium	heavy	102
white	medium	heavy	103
blue	medium	heavy	104
blue	large	heavy	105

Below is an example of the Branching Data Model in the Widgets table with the SELECT SIZE From Widgets WHERE COLOR = "red" statement (Fig. 2).

The Branching Data Model is a primitive tree structure. It consists of a single data value, such as "red," which branches to output data "small" and "medium." In terms of its logical operations, the target data value "red" on the input attribute establishes a new dataset in the table, and its output data establishes a new subset. Therefore, the transition from input to output both reduces the original dataset and establishes the branching pattern.

In the table below, the well-formed query captures the uniformity of this one-to-many data pattern. For example, when we apply the "red," "white," and "blue" values to the "SELECT SIZE From Widgets WHERE COLOR = value" statement all of the results are logically consistent.

The well-defined SELECT statement initiates the formation of the Branching Data Model by extracting a list of raw data. These values are a subset of the output attribute. As indicated earlier, each output list is also a subset of the input dataset. However, it is the program logic managing the SELECT query that spatially arranges this output data into a single list of nested lists [30].

And finally, the query not only extracts the parts of these data patterns, but its input and output attribute labels model their details. For example, "(COLOR, SIZE)" models the uniformity of the one-to-many data patterns of the Branching Data Model in Widegts table of Fig. 3. Moreover, this modeling capability allows us to separate the I/O variables from the SELECT statement, thereby creating two pools of symbols for each query required. A template expression lays out the keywords, syntax, and tagged spaces for the I/O variables, and a separate arrangement of models of the Branching Data Model takes place in computer memory.

4.2 The Nested Branching Data Model

The theory of nested named sets opens up an opportunity for expanding the Branching Data Model to enhance its tree structure, in an organic way, which comprises each model. In turn, by creating the Nested Branching Data Model, the

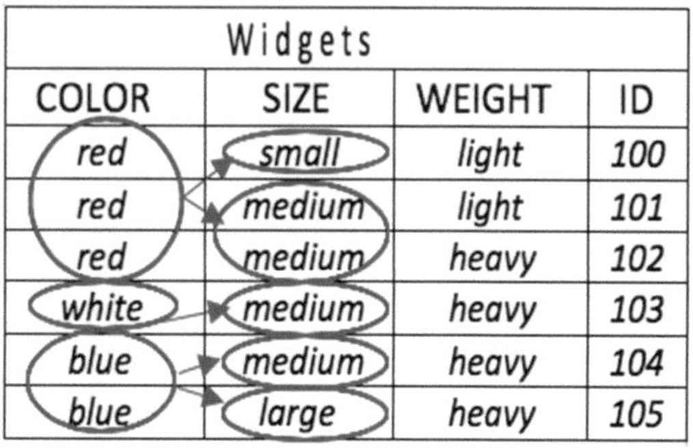

Fig. 3 Example –
Uniformity of the
one-to-many data patterns

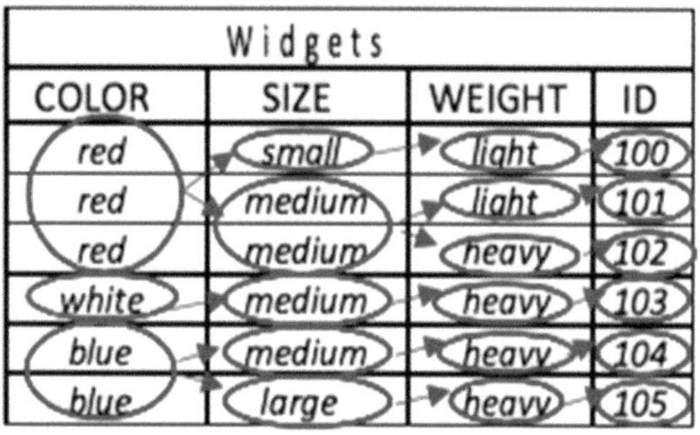

Fig. 4 Example – Data
network model presented in
the Widgets Table

database developers can generate and manage an abstract data network of relational data for the Decision Tree interface.

An example of a nested Branching Data Model expression is presented below:

"(COLOR, COLOR) (COLOR, SIZE) (SIZE, WEIGHT) (WEIGHT, ID)".

This expression consists of a chain of interrelated attribute pairs. It includes both cyclic and basic named set models, which merge with the SELECT template. Each pair of attributes represents an individual Branching Data Model. In one construction setting, recursion parses the expression, one model at a time, to generate menu data. Each iteration generates the SELECT statement that extracts output data from the database based on a new target value. After the initial call, all prior data conditions pass directly to the next iteration and become part of the WHERE clause. In terms of nested named sets, the new and prior conditions correspond to the comments made to "direct inclusion" and "nested inclusion" data relations that induce a hierarchy of nested inclusions. Subsequently, this expression models the data network in Fig. 4 presented in the Widgets table below.

For exposition purposes, we transform these individual tree structures into a single structure to highlight its hierarchical containment and nested nature. Each nested level of the Widgets table, in Fig. 5, represents an attribute domain and its subsets. It starts with the set of "red," "white," and "blue" labels. Each one of these labels, in turn, refers to an attribute domain subset. Operationally, this downward progression corresponds to the overlapping of one Branching Data Model with its

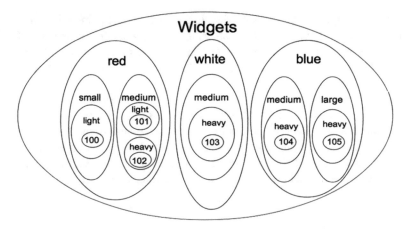

Fig. 5 Nested levels of the Widgets table: each represents an attribute domain and its subsets

successor model. The output data from one model flows to the new data condition in the next model. Fittingly, Fig. 5 is an Euler Syllogism diagram. This figure illustrates the nested structure of data sets in the database table predicted by the algorithmic named set.

In the Decision Tree interface, this list of nested data lists establishes a "data funnel" architecture. End-users select a data item from the first list in the interface to navigate down to more lists in the database table to locate the necessary information. Whenever the interface has less than two data values, it either generates a default response from the single value or none at all. At the other extreme, whenever the number of data values exceeds a maximum, which is set by the developer, the interface displays tabbing structures based on the data type. Each list of data shows a logical set of options that describe one property of this information. All of the data values in the menu list represent one table attribute or property at a time. With each end-user selection, control flows to a subset of the dataset. This process repeats itself until reaching the information window. This logical organization of data assures that each list of options connects to a unique primary key, and a different information window. Codd, the inventor of the relational model, describes its underlying mathematical structure of the table as a form of *applied* predicate calculus [15].

5 Conclusion

Thus, we can see that nesting patterns are a natural property of advanced data structures in the relational database. We demonstrate that the theory of nested named sets provides an ideal mathematical model for studying this data. This model simplifies the complexity of the query to its essential I/O components. It also reduces the complexity of relational data to a branching tree pattern. Outside of the query, program logic organizes this pattern into an expandable data structure for the decision tree interface. Subsequently, the paper demonstrates how the nested named set model can uncover uniform patterns of relational data that have remained, until now, hidden from direct view.

The research presented here opens new problems for exploration. For instance, the database system effectively uses chains of named sets. So, it would be interesting to study the relations between nested named sets and chains of named sets.

Another critical problem is the construction and investigation of operations with chains of named sets. This investigation would allow the further development of efficient database management systems as well as interface systems.

References

1. E. Bakke, D.R. Karger, Expressive query construction through direct manipulation of nested relational results. in *Proceedings of the 2016 Intl. Conference on Management of Data* (San Francisco, CA, USA, 2016), pp.1377–1392
2. M. Benedikt, How can reasoners simplify database querying (And Why Haven't They Done It Yet)?. in *PODS'18: 35th ACM SIGMODSIGACT-SIGAI Symposium on Principles of Database Systems* (Houston, TX, USA. ACM, New York, NY, USA, 2018), pp. 1–15
3. M. Burgin, Operations with named sets, in Ordered sets and lattices (Saratov, 1986), pp. 3–12 (in Russian)
4. M. Burgin, An axiomatic system for the theory of named sets, theory of semigroups and its applications, No. 11, (Saratov, 1993), pp. 8–17 (in Russian)
5. M. Burgin, *Theory of Named Sets as a Foundational Basis for Mathematics* (Structures in Mathematical Theories, San Sebastian, 1990), pp. 417–420
6. M. Burgin, Structural organization of temporal databases, in *Proceedings of the 17th International Conference on Software Engineering and Data Engineering (SEDE-2008)*, (ISCA, Los Angeles, California, 2008) pp. 68–73
7. M. Burgin, *Theory of Information: Fundamentality, Diversity and Unification* (World Scientific, New York, 2010)
8. M. Burgin, *Theory of Named Sets* (Nova Science, New York, 2011)
9. M. Burgin, *Theory of Knowledge: Structures and Processes* (World Scientific, New York, 2016)
10. M. Burgin, Triadic structures in interpersonal communication. Information **9**(11), 283 (2018)
11. M. Burgin, A. Tandon, Naming and its regularities in distributed environments. in *Proceedings of the 2006 International Conference on Foundations of Computer Science* (CSREA Press, 2006), pp. 10–16
12. M. Burgin, Zellweger, P. A unified approach to data representation. in *Proceedings of the 2005 International Conference on Foundations of Computer Science* (CSREA Press, Las Vegas, June 2005), pp. 3–9
13. G. Chu, P.J. Stuckey, Nested constraint programs, in *Principles and Practice of Constraint Programming (CP 2014), Lecture Notes in Computer Science*, v. 8656. (Springer, New York2014)
14. E.F. Codd, A relational model of data for large shared data banks. CACM **13**(6), 377–387 (1970)
15. E.F. Codd, Extending the database relational model to capture more meaning. ACM Trans. Database Syst. **4**, 397–434 (December 1979)

16. M. L. Dalla Chiara, G. Toraldo di Francia, Individuals, kinds and names in physics, in (Corsi, G. et al. eds.), *Bridging the gap: philosophy, mathematics, physics* (Kluwer Ac. Publ., 1993), pp. 261-283

17. C.J. Date, *An Introduction to Database Systems* (Addison Wesley, Boston/San Francisco/New York, 2004)

18. R. Elmasri, S.B. Navathe, *Fundamentals of Database Systems* (Addison-Wesley Publishing Company, Reading, Massachusetts, 2000)

19. A. B. Haddow, C. Grover, Recognising nested named entities in biomedical text. In *Proceedings of the Workshop on BioNLP 2007: Biological, Translational, and Clinical Language Processing*, (2007), pp. 65–72

20. J.R. Finkel, C.D. Manning, Nested named entity recognition, EMNLP '09 *Proceedings of the 2009 Conference on Empirical Methods in Natural Language Processing*, v. 1, pp. 141–150

21. M. Ju, M. Miwa, S. Ananiadou. A neural layered model for nested named entity recognition. in NAACL-HLT'18, (ACL, 2018), pp. 1446–1459

22. A. Katiyar and C. Cardie. Nested named entity recognition revisited. in NAACL-HLT'18, (ACL, 2018), pp. 861–871

23. A.S. Nocedal, J.K. Gerrikagoitia Arrien, M. Burgin, A mathematical model for managing XML data. International Journal of Metadata, Semantics and Ontologies (IJMSO) **6**(1), 56–73 (2011)

24. C. North, B. Shneiderman, Snap-together visualization: a user interface for coordinating visualizations via relational schema. AVI 2000, (Palermo, Italy), pp 128–135

25. C. North, N. Conklin, V. Saini, Visualization schemas for flexible Information Visualization. in *Proceedings of InfoVis* 2002, pp. 15–22

26. J. O'Toole, D.K. Gifford Names should mean what, not where. in *EW 5 Proceedings of the 5th workshop on ACM SIGOPS European workshop: Models and paradigms for distributed systems structuring*, 1992, pp. 1–5

27. R.T. Snodgrass, C.S. Jensen, *Developing Time-Oriented Database Applications in SQL* (Morgan Kaufmann, 1999)

28. A. Tannenbaum, *Metadata Solutions* (Addison-Wesley, Reading, 2002)

29. H. P. Zellweger. Unifying Data Relations to Enable End-Users to Navigate Over Relation Data, 21st International Conference on Software Engineering and Data Engineering 2012 (SEDE-2012), At Los Angeles, CA, pp. 87–92

30. H. P. Zellweger. Tree visualizations in structured data recursively defined by the Aleph Data Relation. in *Proceedings of the 20th International Conference of Information Visualization (IV2016)* (Lisbon, Portugal, July 19-22, 2016), pp. 21–26

31. H. P. Zellweger. A decision tree interface based on predicate calculus. in *Proceedings. of the 21st International Conference of Information Visualization (IV2017)*, (London, United Kingdom, July 11–14, 2017), pp. 188–193

32. H. P. Zellweger. The branching data model, the foundation for automated tree visualization. in *Proceedings of the 22nd International Conference of Information Visualization (IV2018)*. (Salerno, Italy July 2018), pp. 33–37

33. H. P. Zellweger. A demonstration of automated database application development. in *Proceedings. of the 24th International Conference on Intelligent User Interfaces (IUI'19)* (Marina del Rey, CA USA, March 17–20, 2019), pp. 65–66

Obstacle Detection via Air Disturbance in Autonomous Quadcopters

Jason Hughes and Damian Lyons

1 Introduction

Autonomous flying drones are currently available, meaning the drones can fly to a preprogrammed destination while avoiding objects with no input from a user. The current object avoidance technology comes in the form of a camera, laser, or ultrasonic sensor; this study looks into if the internal sensors can be used to detect such obstacles.

The most prevalent issue that is stopping drones from mass usage in industries is their battery life. Because the drones must be light to fly, they must have small batteries, and thus they have a short flight time. The previously mentioned sensors for object avoidance cut down on the drone's battery life substantially. From previous work in the Fordham University Robotics and Computer Vision lab, wind currents can be detected using a classifier [5, 9]. The wind makes the drone unstable, and this instability can be detected from the data gathered from the drone's internal sensors. The drone also flies unstably when it is close to large objects, like walls. This is because the wall interferes with the airflow created by the rotors of the quadcopter. This interference causes a similar instability as with a wind current.

The idea behind this paper is that the wind current created by a quadcopter interferes with the stability of the drone and that can be used to detect which side of the drone the interfering object is located. This eliminates the need for the camera, laser, or ultrasonic sensors which thus increases the battery life. The onboard computer would have to work no harder, since the data points would be gathered and sent to the home computer that is flying the drone via radio waves, and

J. Hughes (✉) · D. Lyons
Fordham University, Robotics and Computer Vision Lab, Bronx, NY, USA
e-mail: jhughes50@fordham.edu

R. Stahlbock et al. (eds.), *Advances in Data Science and Information Engineering*,
Transactions on Computational Science and Computational Intelligence,
https://doi.org/10.1007/978-3-030-71704-9_31

the home computer would do the calculation and prediction. The prediction can then be sent to the driver program, which can trigger the drone to avoid such an object.

This project looks at well-formed objects (walls perpendicular with the ground) because this provides a good surface for the air from the quadcopter's rotors to "bounce" off of. Walls were also used because it is likely what drones will encounter when they fly autonomously in buildings. The quadcopters were flown perpendicular to the walls to collect clear data. While this project is still in its infancy, it was taken on to ultimately eliminate the need for the aforementioned external sensors to increase flight time of autonomous drones.

The drone chosen to collect the data was the Crazyflie 2.0 because of its simple internal sensors and ease of programming[1]. The quadcopter has an inertial measurement unit sensor (IMU). This measures the gyroscope in the (x, y, z) planes and acceleration with an accelerometer again in the (x, y, z) planes. From the gyroscope and accelerometer, the drone can calculate its roll, pitch, and yaw angles for recording. There is also a barometer to measure air pressure. On the bottom of the quadcopter is a flow deck that measures the drone's (x, y, z) 3D Cartesian coordinates.

To gather the data, the drone was flown around a U-shaped wall. Data was collected from the drone every hundredth of a second. The dimensions of the U-shaped wall were 1.2 m by 1.2 m by 1.2 m. In total 12 s of data was collected from the left and right walls and 10 s from the front wall, giving a total of 34 s of wall data. This is added to one-third of the no wall data. After cleaning the data, there was a total of 30 s of wall data plus 10 s of no wall data chosen at random. This gives a total of 40 s of data, and with data collected at every hundredth of a second, there were a total of 4000 examples for training and testing.

2   Literature Review

There have been many studies done on ground effect in quadcopters. A ground effect occurs when the drone is flying near the ground and the air that is being pushed down from the rotors hits the ground and pushes back up on the drone making it unstable. This has been studied extensively to make quadcopter landing and takeoff more stable. Alternatively, wind effect in drones has not been studied very much. Drone-to-drone wind detection was studied in [9]. This project has one drone fly underneath the other and could successfully detect that there was a drone over top using only the internal sensors of the bottom quadcopter and a classifier. Additionally, the researchers in [5] were able to detect wind gusts using only the internal sensors of the drone. They had the drone fly in front of a fan in a different direction and built a classifier to determine if the drone was in a wind gust or not.

These two studies show that the internal sensors can be used to detect wind. Specifically, [9] shows that drones can detect wind currents created by rotors. These papers differ from this project in the respect that they do not predict which direction the wind is coming from. This is pertinent to building a truly autonomous drone that does not rely on outside sensors.

Yoon et al. and Diaz from NASA's Ames Research Center used computational fluid dynamics (CFD) to model the airflow of a drone's rotors in [3, 8]. They show that the velocity of the air underneath the rotor takes the form of a cylinder. The air moves faster within the cylinder in the shape of an hourglass. This is important because it is suspected when a drone is near a wall the air underneath its rotors is getting pushed outward and then interacting with the wall, causing an instability. Their work was done using the DJI Phantom 3 which is large and powerful. Thai et al. [7] also did CFD for the DJI Phantom 3. Unlike the others, their work shows one instance of the drone's propeller spinning and shows that the velocity has a helical flow pattern to it. They also show that the airflow from the rotors has the hourglass shape.

Other researchers at McGill University have worked on wall detection without using additional sensors in [2]. This study has drones with rotor guards fly directly at the wall. They are looking at what speed can the quadcopter travel at, hit the wall, and still recover and not crash from hitting the wall. The idea behind this study is to have an autonomously flying drone that hits walls, recovers, and moves in a different direction after the collision. This project wants to avoid the collision and just use the air disturbance from the wall to fly autonomously. Also, not all drones fly with rotor guards, and without them, the drone would just crash. Nonetheless, wall detection is still being worked on in the industry.

3 Modeling

In order to understand the air flow of the drone better, both mathematical and physical modeling was done to better understand the data was collected and lead to feature selection and feature generation.

3.1 *Mathematical Modeling*

Mathematical modeling was done to simulate the airflow under the rotors of the drone. [4] used three-dimensional Navier-Stokes equations to look at a rotor wake vortex. In forward flight, they calculated a spiral vortical wake geometry. The following three-dimensional Navier-Stokes equation for helical flow from [4] was solved.

$$\nabla \cdot \mathbf{u} = 0, \tag{1}$$

$$\frac{\partial \mathbf{u}}{\partial t} + (\mathbf{u} \cdot \nabla)\mathbf{u} = -\frac{\nabla p}{\rho} + \nu \cdot \nabla^2 \mathbf{u} + \mathbf{F} \tag{2}$$

The equation was transformed to include the curl expression represented by $\mathbf{w}$ in the following equations.

$$\nabla \cdot \mathbf{u} = 0, \tag{3}$$

$$\frac{\partial \mathbf{u}}{\partial t} = \mathbf{u} \times \mathbf{w} + \nu \cdot \nabla^2 \mathbf{u} - \left(\frac{1}{2} \nabla \left(\mathbf{u}^2 \right) + \frac{\nabla p}{\rho} + \nabla \phi \right) \tag{4}$$

By expanding the curl expression, a system of partial differential equations was created and then solved. The variables in the equations are represented as follows:

- $\mathbf{u}$: air velocity vector containing the radial, angular, and tangential velocities
- ∇p: change in pressure from above the rotor to underneath the rotor
- ρ: air density coefficient
- ν: kinematic air viscosity
- $\mathbf{F}$: force vector in x, y, z directions
- $-\nabla \phi$: force potential from $\mathbf{F}$

Ershkov explains that the curl field arises from the source of vorticity in the fluid field, which in this case would be the rotor of the drone. $-\frac{\nabla p}{\rho} + \nabla \phi$ is the x, y, z force vector. The system of equations gives the solution as:

$$\mathbf{u} = \exp \left(-\nu \cdot \alpha^2 \cdot t \right) \cdot \mathbf{u}(t_0). \tag{5}$$

where $\mathbf{u}(t_0)$ refers the velocities at the propeller which can be shown in as:

$$\mathbf{u}(t_0) = \begin{bmatrix} u_r(t_0) \\ u_t(t_0) \\ u_z(t_0) \end{bmatrix} \tag{6}$$

$u_r, u_t, and\, u_z$ refer to the radial, tangential, and downward velocities at the rotor of the quadcopter. The velocities can be calculated with the equations:

$$u_r(t_0) = b \cdot \pi n \cdot 2r, \quad u_t(t_0) = r \cdot \omega, \quad u_z(t_0) = \sqrt{\frac{T/A}{2\rho}} \tag{7}$$

where b is a blade swirling factor, n is the rpm, ω is the induced velocity which equals u_z, T is the force of thrust, and A is the area of the rotor disk. The answers in (5) are then used in (3) to solve the equation as time continues; the resulting vectors for u_r, u_t, and u_z were plotted as a vector field (Fig. 1).

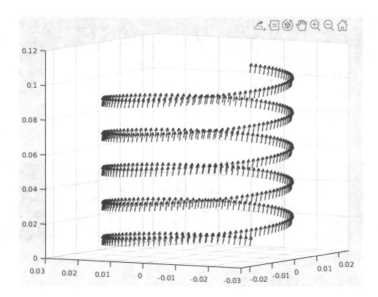

Fig. 1 Velocity vector field

Fig. 2 Experimental setup

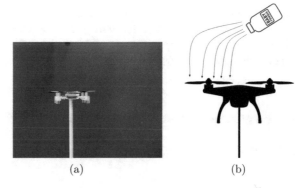

(a)   (b)

3.2 Airflow Testing

In order to validate the mathematics, airflow testing was done by looking at talcum powder in the drone's rotors. This was done to observe the helical airflow of the quadcopter's rotors and to see how the air could possibly interact with a nearby wall (Fig. 2).

First, a drone was mounted on a rod with a camera mounted 1 meter away to video and photograph the airflow. Initially, a fog machine was hung upside down with the nozzle pointing downward at the top of the drone's rotors. After some initial photos and videos, it was observed that the drone was not visible through the fog and thus the airflow could not be examined. Alternatively, talcum powder was dropped into the rotors from 0.5 meters above. The rotors were tested at varied

(a)   (b)   (c)

Fig. 3 Sequence of frames from video showing airflow

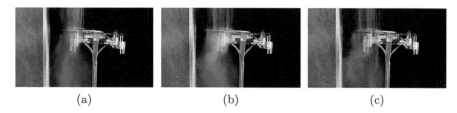

(a)   (b)   (c)

Fig. 4 Sequence of frames with wall

thrust starting at 25% and working up to 80%. The procedure was videoed for later analysis, and a snapshot from a video at 80% thrust is shown in Fig. 3.

The process was repeated but with a wall near the drone to examine how the airflow from the quadcopter interacts with the wall. The drone was placed two rotor diameter lengths away from the wall, which is about 0.1 m (Fig. 4).

4 Experimentation

4.1 Wall Data Collection

Data needed to be gathered from drones both near and away from walls. The drone used was a Bitcraze Crazyflie 2.0 with a flow deck and loco position system (LPS) deck (Fig. 2) [1]. The drone measures about 10 cm by 10 cm and has total weight, including the decks, of 42g. The flow deck uses a laser sensor to determine the height of the quadcopter, and the LPS deck gives off a signal to the LPS nodes in each corner of the flying area to determine its Cartesian position (x, y, z).

While the drone is flown by a base computer using a Python interface. In Python, a driver program was created to fly the drone in a U-shape. While the quadcopter is flying, it sends data to the computer from its internal sensors which is then stored for later analysis (Fig. 5).

The quadcopter was flown around a U-shaped wall within two rotor diameter lengths of the wall. The walls were constructed from particle board 1m tall by 1.2m long. It started with the wall to its left and flew toward the front-facing wall in the x-direction. When it reached the front-facing wall, it stopped and moved in the

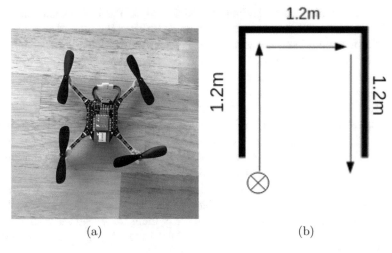

(a)   (b)

Fig. 5 (**a**) Crazyflie 2.0. (**b**) Flight path

y-direction toward the right-facing wall. Once the quadcopter reached the right-facing wall, it moved backward in the $-x$-direction until it reached the opening again where it landed.

While the quadcopter was flying, data was collected from the onboard sensors which include an MPU-9250 inertial measurement unit (IMU). The gyroscope sensor has digital output based on the x, y, z axes angular rates sensors (gyroscopes) with a user-programmable full-scale range of ±250, ±500, ±1000, and $\pm2000°$/s and integrated 16-bit ADCs. The accelerometer is a digital-output triple-axis component with a programmable full-scale range of $\pm2g$, $\pm4g$, $\pm8g$, and $\pm16g$ and integrated 16-bit analog-to-digital converters (ADCs). In addition to the gyroscope and accelerometer data, stabilizer (roll, pitch, yaw), Cartesian position (x, y, z), and barometer data were collected every hundredth of a second [1]. The data was stored in separate files based on the wall position (left, front, right). The test was repeated five times and was repeated another time with no walls for a control group.

4.2 Classifier Building

In order to predict whether a wall was to the left, right, or front of the drone, a classifier was built using Pandas and SKlearn packages in Python [6]. The data was collected based on which wall the drone was flying near, i.e., left, front, or right. This data was then divided into 1 s increments.

The data was first preprocessed manually by cutting out the seconds when the drone was near the corners of the U-shaped wall leading to 30 s of flight data. To test how accurately the wall side can be predicted, tests were conducted on each

1 s interval, meaning an individual second would be the test set and the model would train on the 29 remaining seconds and 10 s from the control flight with no walls. The ten control seconds were chosen by selecting the middle-most data from each direction the drone was moving. This was done so the classifier would not be affected by noise from the drone changing direction. The training and testing would be repeated for each individual second to get an accuracy score at each time interval. A K-NearestNeighbor, a RandomForest, and a GradientBoosting classifiers were then run on all the features with four distinct classes, left, front, right, and none. Next, the same three classifiers were run using the same format as above but with only two classes. The class that was the test class was kept the same, and all the data belonging to the three remaining classes was changed to "other." This gives the classifier an easier task of distinguishing between two classes rather than four. This leads to the problem of class imbalance. The class that is being tested will have 9 s left for training, and there will be 30 s for training for the non-testing class. To resolve this, the testing class was duplicated until there were an equal number of examples between the testing class and other classes.

Initially, a K-NearestNeighbor classifier was tested, but the results were less than optimal, so the GradientBoosting and RandomForest classifiers were also tested. These classifiers were chosen because of their ensemble capabilities, which is necessary because of the way the distinct features are affected by the walls' location relative to the drone. Based on the modeling, all features except for gyroscope x and y and position x and y were dropped. The gyroscope was most affected by being close to the wall, and keeping the other features only adds training and testing time for similar results.

5   Results

First, the wall and control data was run through the classifier from the previous study by [Gu] which was used to determine if a Crazyflie drone was in a wind current or not. The classifier performed very well achieving about 99% accuracy. This means that the drone is being affected by its airflow being near the wall.

The classifier tested on 1 s of data and used the other 29 s of data with the drone near the wall and an additional 9 s of no wall data to train on. This was repeated for the 30 s of flight time around the U-shaped wall. An accuracy score was taken at each second to give an overall average of the 30 s of testing. This was repeated ten times to get a more accurate average of the full 30 s.

When testing with all the features and using four classes, left, front, right, and none, the RandomForest classifier performed best with 98.21% accuracy followed closely by the GradientBoosting classifier at 97.47%. The K-NearestNeighbor classifier was significantly less accurate at 34.01%. The graphs in Fig. 6 show the accuracy score each of the 1 s intervals for the three classifiers. To complete a full test of the 30 s of flight time, the GradientBoosting was very slow, taking 72.14 s. RandomForest took 14.03 s and NearestNeighbor took just 2.67 s.

Fig. 6 One-second interval accuracy scores (four classes)

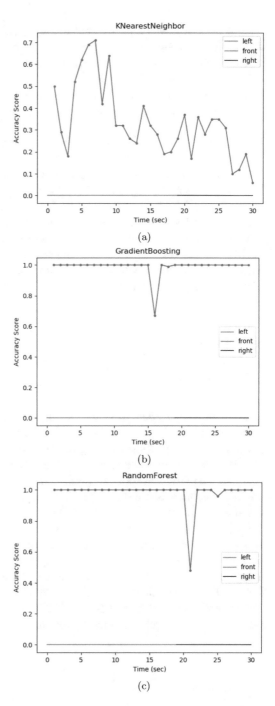

The same test was repeated, but this time only the x-gyroscope, y-gyroscope, x-direction, and y-direction were used for training and testing. Averaging ten runs of the classifier showed that the GradientBoosting was slightly more accurate than the RandomForest at 97.17% and 96.91%, respectively. NearestNeighbor improved to 51.85%. The testing time dropped significantly to 2.63, 33.20, and 8.86 s for NearestNeighbor, GradientBoosting, and RandomForest, respectively.

Next, the classifier was changed to test on a wall side, while all the data not belonging to the test class was labeled as other, meaning that the classifier was only testing on two classes. When testing on the four features listed previously, the GradientBoosting was most accurate at 96.65% followed by RandomForest at 94.45% and NearestNeighbor at 77.44%. An example of one test of the full 30 s is shown in Fig. 7. The times for testing of NearestNeighbor and RandomForest were similar at 2.62 and 9.35 s, while GradientBoosting was improved to 10.30 s. The test was repeated with all the features, and the results were 58.64%, 96.24%, and 96.18% for NearestNeighbor, GradientBoosting, and RandomForest, respectively. These accuracy scores were an average of ten tests of the classifier (i.e., if you were to average the points on the graphs for ten different runs).

6 Discussion

6.1 Classifier Accuracy

The classifier testing has 2 distinct results: first, a transition to 2 classes rather than 4 does not help the classifier, and second, 12 features are not necessary. The classifier performed very well when testing on four features using the RandomForest and GradientBoosting classifiers, showing that these are the only features that need to be used. The GradientBoosting performed almost as well on 4 features as it did on all 12 features, and it did it less than half the time. This was the most accurate classifier, but it was followed closely by RandomForest. This classifier was about 1.5% less accurate when testing on 4 classes rather than all 12 which is negligible. It did it in similar times. The only classifier that was affected by switching from four to two classes was the NearestNeighbor classifier, but it never performed accurately enough to be used practically. Overall, both the GradientBoosting and RandomForest classifiers are reasonable ones to use in an onboard classifier system. The speed of the test in real life will likely be a negligible difference, and they performed equally well when testing on four classes.

Fig. 7 One-second interval
accuracy scores (two classes)

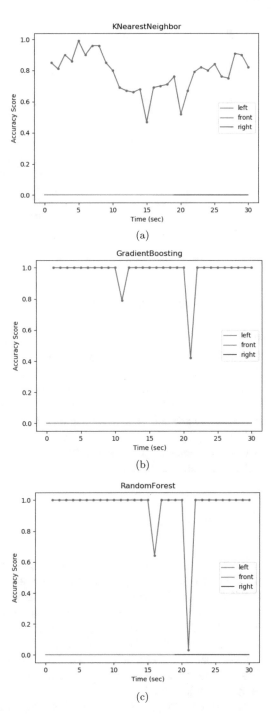

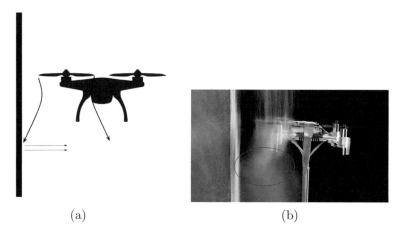

(a) (b)

Fig. 8 Artificial ground effect

6.2 *Artificial Ground Effect*

The unsteadiness of the drone occurs because of an artificial ground effect. It is well studied that when drones are near the ground, they become unsteady because the air from their rotors hits the ground and comes back at the quadcopter, thus making it unstable. In this case, the air is hitting the wall and then dispersing, as air does when it hits an object. This dispersion causes a mass of air underneath the drone acting as the "artificial ground." The proposition is shown in Fig. 8a and highlighted from the airflow imaging in 8b. Since the "ground" is only on one side of the drone, there is likely a difference in gyroscope, accelerometer and stabilizer reading between a left wall, right wall, and front wall.

6.3 *Lab Variances*

A limitation of this project was the amount of data. Only one test was completed with the drone flying around the U-shaped wall. The lab was shut down due to the COVID-19 pandemic before more tests could be completed. Ideally, there would be between five to ten tests of the drone flying near the walls. This would eliminate the need to duplicate the testing class when testing with two classes because the majority class could be undersampled if there was more data. It can also be noted that the very low accuracy can most likely be attributed to the limited amount of data. As the study stands right now, data is limited so all of the data needed to be used for training and thus the testing class had to be duplicated.

Upon further investigation, the way the data was collected meant the position was unique for each wall. When the wall was to the left, the drone was increasing in x-

position; when the wall was to the front, the y-position was increasing; and when the wall was to the right, the x-position was decreasing. When only the position data was fed to the classifier, it preformed equally as well as it did when the gyroscope data was also included. When testing on just the gyroscope data, the classifier preformed at about 27% accuracy. However, there is something to say about the direction the quadcopter is moving and its relation to objects. The ability to use pitch and roll angles to get a better sense of where the wall is is being worked on currently.

6.4 Future Work

Future work for this project includes rerunning these experiments with more data to get more accurate results. After that, data will be collected with drones at different angles relative to the wall and using a regression classifier to predict what that angle is or it could calculated from the raw roll and pitch data from the drone using Rodrigues angles. From this angle, the drone would be able to calculate and turn or do a feature transformation within a classifier to detect which side the wall is on. Future work will also include feature generation from the mathematical model in this paper for wall detection. Lastly, a driver can be built implementing a classifier to predict if the drone is near a wall while it is flying. This study highlights the early steps to reaching full autonomy in quadcopters without using camera or ultrasonic sensors to avoid objects.

References

1. Bitcraze Crazyflie 2.0, www.bitcraze.io/products/old-products/crazyflie-2-0/. Accessed 5 May 2020
2. F. Chui, G. Dicker, I. Sharf, Dynamics of a Quadrotor Undergoing Impact with a Wall Int. Conf. Unmanned Aircraft Sys, Arlington VA (2016)
3. P. Diaz, S. Yoon, "High-Fidelity Computational Aerodynamics of Multi-Rotor Unmanned Aerial Vehicles" AIAA SciTech Forum 2018, Kissimmee, Florida
4. S. Ershkov, Non-stationary helical flows for incompressible 3D Navier-Stokes equations. Appl. Math. Comput. **274**, 611–614 (2016)
5. S. Gu, M. Lin, T.-H. Nguyen, D. Lyons, Wind gust detection using physical sensors in quadcopters. arXiv:1906.09371 [cs.RO]
6. Scikit-Learn, https://scikit-learn.org/stable/. Accessed 20 April 2020
7. A. Thai, R. Jain, S. Grace, "CFD Validation of Small Quadrotor Performance using CREATE-AV Helios" Vertical Flight Society 75th Annual Forum & Technology Display, Philadelphia, Pennsylvania, May 13–16 (2019)
8. S. Yoon, H. Lee, T. Pulliam, "Computational Analysis of Multi-Rotor Flows" NASA Ames Research Center, Moffett Field, California 94035
9. Q. Zhou, J. Hughes, D. Lyons, Drone Proximity Detection Via Air Disturbance Analysis, SPIE Defense + Commercial Sensing 2020 Digital Forum

Comprehensive Performance Comparison Between Flink and Spark Streaming for Real-Time Health Score Service in Manufacturing

Seungchul Lee, Donghwan Kim, and Daeyoung Kim

1 Introduction

Predictive maintenance (PdM) refers to the intelligent monitoring system for a machine to avoid future failures. It makes it possible to avoid unnecessary equipment replacement and also to detect the root cause of the machine's failure, which results in saving costs and improving efficiency.

In the process of the PdM, data are collected from various sensors and then they are analyzed using the advanced data mining algorithms such as neural networks and regression models in order to calculate a health score.

In the PdM, the health score represents the equipment's condition containing a metric about how many days a machine lasts before a breakdown arises. This health scores are used as a standard measurement for judging the equipment's assessment so that it is important to compute accurate health scores for given datasets from machines.

Nowadays, more accurate health scores are computed because of the improvements of the sophisticated machine-learning algorithms, so they become used as a landmark for the measurement of equipment's condition. A representative learning algorithm used in the predictive maintenance is a neural network approach. Sampaio, Gustavo Scalabrini, et al. have applied an artificial neural network to the prediction of the motor failure to find out conditions of the industrial equipment. In this study, neural networks outperform other machine learning techniques showing comparative small RMSE performance index values [1]. A fuzzy neural network, a learning machine that finds the parameter's weight of a fuzzy system by exploiting approximation techniques from neural networks, also has been applied to machine

S. Lee · D. Kim · D. Kim (✉)
Research Inc, BISTel, Austin, TX, USA
e-mail: dhkim2@bistel.com; dykim3@bistel.com

© Springer Nature Switzerland AG 2021   483
R. Stahlbock et al. (eds.), *Advances in Data Science and Information Engineering*,
Transactions on Computational Science and Computational Intelligence,
https://doi.org/10.1007/978-3-030-71704-9_32

condition monitoring by aiding in the diagnosis of faults with high prediction accuracy in a manufacturing system [2].

Recently, this neural network approaches for predictive maintenance have been powered by the development of the deep-learning models. For example, the deep learning model is an advanced branch of neural network by making up more than three layers in neural network. The deep learning models is able to more accurately predict predictive maintenance lifetime outperforming the existing approaches which are based on simple statistics. Huuhtanen, Timo, and Alexander Jung have applied the convolutional neural network (CNN) to predict the power curve of the photovoltaic panel [3]. In addition, Long Short-Term Memory network, one of the recurrent neural networks, is applied to the sensor measurements, providing the probabilities of failure in different time horizons [4]. As mentioned in the studies above, machine-learning algorithms especially mainly focused on neural-networks are frequently applied to the predictive maintenance for helping the identification of the moment for preparing maintenance activities.

Although employing machine-learning algorithms has become a part of the predictive maintenance, computing a health score in real time is still a challenge in the industry. Since a huge number of large-scale data from sensors are produced continuously, processing these continuous streams of data without a moment's delay is an important task. To resolve this issue, many services have applied stream processing engines for real-time analysis. In stream system, data records are chunked into small static dataset based on hourly, daily, or monthly; and then they are processed according to the time-agnostic fashion. An example of real-time service is present in the Bansal's study [5]. This study shows a real-time-based predictive maintenance system using neural network, which is operated in continuous data pipeline. Bansal, Dheeraj et Al. also present the novel real-time predictive maintenance system for the prediction of the machine's parameters using the motion current signature. As predictive services are integrated with stream processing engines as shown in the studies above, it is essential to apply the stream engine to the PdM solutions to compute health score in real time.

Despite some efforts to applying data stream processing engines, there are still few descriptive literatures about implementing stream engines to predictive maintenance system. Many of the studies are still focused on learning algorithms, and some of them have tried to introduce the model and their APIs [6]. However, they are insufficient to identify in-depth techniques of learning models used in stream engines and to even harder to confirm detailed content of the computational performance of them in predictive maintenance system. Considering the increasing availability of data from sensors and lack of the detailed studies for stream engines for PdM, it is certain that a thorough research for stream processing engines is significantly required for an accurate health score service system.

Recognizing this problem, we extensively investigated two powerful stream engines, Apache Flink and Apache Spark Streaming, in calculating a health score, which can be obtained when machine-learning algorithms are integrated with stream processing. To measure the performance of them, we evaluated through several benchmarks using simulated records as follows. Firstly, we estimated the record throughput for two stream processing engines, Flink and Spark Streaming. By comparing two stream processing engines for processing the given 5000 data records with specific time intervals, we are able to identify each inherent system design for computing the health score in stream. Secondly, we measured the processing time of the computing health scores for given sensor data records by increasing the number of assets. Datasets containing 10,000 assets, which can be considered to be a large-scale size in an assembly line in PdM, are constructed and tested with two processing engines for computing the score. To achieve these experiments, we generated simulated data records and used them as an input to our health score service. In addition, we also increased the number of parameters with up to one hundred while preserving the number of assets and then tested our health score services with them. From these experiments, we identified the results of the processing time that two different stream engines output and also confirmed that the health score service is adequate to process large-scale sensor data for computing health scores in real time.

Other performance tests are also performed. We estimated two stream engines' processing time by varying the number of parallelisms and also estimated processing time for obtaining health scores on multi nodes. These experiments are also beneficial to understand the stream processing's characteristics in the parallel systems.

To compute the health score, we applied the two learning algorithms, the deep learning and the multivariate algorithm,to the stream processing engines. To achieve this connection where two different streams are to be joined, we used the broadcast stream to combine sensor data records with a model context. Because two stream engines have their own distinct APIs for combining streams, we investigated the method for combining multiple streams in a code level. In addition, considering that a model loaded in the stream might be updated with a new one when a user action occurs, we have implemented a separate web server that communicates with stream engines via the REST API call [7]. Specifically, to enhance the performance of the stream processing, we separated training a model logic from the stream, which enables low latency of processing data records and reduces the time of computing health scores as well. We integrated the above techniques into a single service including employing machine-learning algorithms so that we can serve a health score service in real time.

The contribution of our paper is as follows:

1. We introduce a health score service which serves a health score in real time using the stream processing engine.
2. We examine the stream code APIs that two stream engines implemented respectively.
3. We evaluate the two stream engines using various datasets as follows:

   (a) Dataset with different time intervals.
   (b) Maximum number of assets.
   (c) Varying the number of parallelisms.

This chapter begins with an overview of architecture of Spark and Flink and their difference in Sect. 2. We then discuss the method for obtaining health scores on stream in Sect. 3. Finally, we discuss the performance results of Spark and Flink using various sensor data records in Sect. 4, and conclude our chapter in Sects. 5 and 6.

2 Goals and Background

2.1 Goals

Our work targets estimating computational performance of the stream processing engines for obtaining the health scores. To evaluate the health score service, it should be estimated for the following goals:

1. Maximum throughput attainable under different incoming data interval.
2. Processing time for large-scale sensor data records.
3. Processing time when varying the number of the parallelisms.

We primarily tested two stream engines, Flink and Spark Streaming, with three goals as presented in the above. To our knowledge, previous studies did not extensively perform the system's computational performance for obtaining health scores. Thus, our study is more focused on evaluating computing performance while introducing stream APIs and models for achieving health scores in real time.

2.2 Characteristics of Streaming Process Engines

2.2.1 Flink

Flink is a framework and distributed streaming processing engine for stateful computations over batched and continuous data [8]. There are two main characteristics of Flink's architecture for computing data as follows. Firstly, the accurate time control and powerful stateful stream that the Flink employs make it possible to run any kind of application for continuously incoming dataset. Secondly, the stateful operations in Flink can be used for many kinds of applications including not only a classification tree or unsupervised learning model's context but also a simple counter or a sum in various situations. Especially, stateful Flink's applications are optimized for local state access, which is kept in memory when the state size does not exceed the available memory. This function yields very low processing latencies by accessing local and in-memory access.

2.2.2 Spark Streaming

Spark streaming is one of the earliest stream processing engines to provide a high-level of functional API [9]. It is based on writing directed acyclic graphs (DAG) of physical operators, which is also same architecture of the Flink streaming dataflow. The Spark Streaming operation is that it separates the input data streams into mini batches and store them in Spark's memory called RDD [10]. And then Spark job processes individual batches in a distributed system. Especially, the Spark Streaming is more focused on parallel recovery mechanism that is more efficient to the replication, second-scale recovery from faults. This advantage is powerful to operate more large commodity clusters in distributed system compared to the traditional stream services.

3 Methods

As we explained in the Sect. 1, it is difficult to see detailed descriptions for API instructions and benchmark performance of the stream engines when it comes to obtaining health scores even though there are many literatures on real-time health score applications. To address the challenge, we perform in-depth analysis of the two stream processing engines, Flink and Spark Streaming, for better understanding of underlying concept that they have applied for computing the health score. Firstly, we introduce the entire system of health score service in the Sect. 3.A. And then, we explain the APIs instructions for obtaining health scores in the following section. Especially, we identified the difference of the two streams methods for joining multiple data streams. Finally, we briefly explain the optimization techniques we accomplished for the performance of the health score service in the last section in Methods.

In addition, we introduce the model used in our service since the health score is computed using the machine-learning algorithms. For this reason, we also cover the detail of the two algorithms, the standard LSTM and multivariate algorithm, in Sect. 3.D.

3.1 A. System Architecture

We have implemented a health score service system which mainly consists of three different components as shown in a Fig. 1. Firstly, sensor data are collected from various sensors and then be sent to the Kafka topic. The sensor data records consist of timestamp, asset's name, and a number of parameters that contain their own value. To communicate between each component, the service uses a Kafka consumer to receive messages from the topic. This Kafka message system makes it easy to scale the system into multiple nodes. The received messages from Kafka

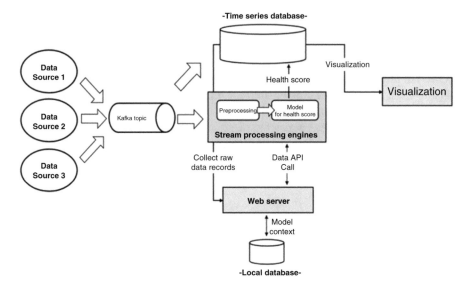

Fig. 1 An overview of the system architecture for the health score service

topic are sent to the stream processing engines and also sent to the time series database at the same time. The Kafka consumer of the stream engines takes the messages and then processes them for obtaining a health score in real-time while the time series database receives the messages for storing purpose. In our health score service, all the sensor data records are stored in the time series database, allowing us to search the raw sensor data with a short query according to the user's situation. Since the sensor data records are always produced with timestamp, which are used as a primary key in the database, the time-series database in our health score service is appropriate selection in this case. Once the messages arrive at the stream engine, it tries to pre-process them and to apply the models to the incoming messages for calculating the health scores. Considering the real-time performance of the streams, the logic for training a model is executed in a separate server. To exchange the model context between the server and stream, we used the REST API's call to transmit the serialized model context and other information over HTTP. This design is simple but an efficient way to transmit model context inside the stream logic with very low latency. For storing the state and data, we use two databases in the health score service for the different purposes. As mentioned above, the time series database is responsible to store raw data records as well as health score values that is obtained from model on stream. We also use the local database communicating with a server to keep meta data including model contexts and other specification values. All these data stored in the local database is very small size compared to the data stored in the time-series database. Finally, we connect the visualization tools so that we can easily monitor the series of health score values along with the raw sensor data.

3.2 Stream APIs

In the section, we explain the stream operators including the stateful operators by showing a short code example for how to process the sensor data records to compute the health score. Since our study is focused on conducting comparison between Flink and Spark Streaming, we will examine the difference of API instructions that two stream engines have implemented in the programming point of a view. Note that two main features that we should consider in our health score service are that we should join the raw sensor data records with the model context including the specification context. The specification context is the meta data consisting of a list of the parameters' names and assets' names. As shown in the left of the Fig. 2, there are mainly two streams, the sensor data stream that handles incoming sensor data records from various sensors and the specification broadcast stream that contains the asset's metadata that is the asset's name and its corresponding parameters' names. The challenge is that the service needs to combine these two streams into one single stream. To achieve this issue, we use the broadcast state to combine the sensor data stream with a broadcast stream containing the specification context. Note that the size of specification contexts is relatively small so that it is proper to broadcast all the elements on all tasks in nodes without degrading the performance. After connecting these two streams, we are able to manipulate any of the keyed state with broadcast variables. If a new specification context is added to the broadcast stream, we can discard the old specifications and apply the latest one by seeing the timestamps that each specification contains.

As shown in Fig. 3, there are two programming API instructions, Flink (left) and Spark Streaming (right), for processing two main streams in our health score service. In Flink operators shown in a Fig. 2 left, sensor data records are collected by the Kafka topic in the source and they are transformed using the stateful operators. In the transformation operators, we use the.keyBy() function to convert data stream to a keyed stream, where all the input data are partitioned by a key (the key is asset name in our health score service). We also connect the broadcast stream into the sensor data stream by calling. Connect() method followed by .process() function that applies any kind of the logic for the preprocessing. With the broadcast state in Flink, the contents of the broadcast state can be easily copied to all the task nodes over the network, which indicates that the two streams are joined together. Whereas the Flink offers us a simple broadcast method for continuous stream, the Spark Streaming has a limitation on converting stream into the broadcast variable. It is difficult to convert D-Stream (the name of the Spark Streaming system) into the local broadcast variable for combining two data streams. Therefore, we alternatively joined two D-Streams into one single D-Stream by applying one transformation before joining. As shown in the right of Fig. 2, there are two streams, the sensor data stream and the specification data stream. After the transformation operation where the input data is converted into paired data consisting of key-values, we are able to combine two streams by matching the same key. Considering the join operation needs to move data with executors requiring lots of shuffle-write, the Flink's broadcast system is

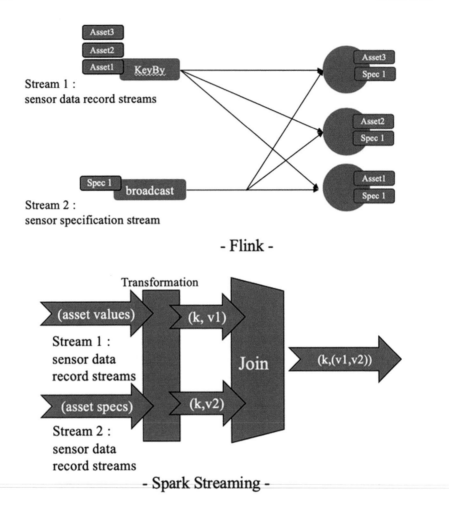

Fig. 2 A data stream flow of two stream processing engines for joining two data streams

more efficient than those of Spark Streaming system because the broadcast operation enables to join two tables without moving the sensor data records, which cause the large data shuffling over network. The Spark Streaming APIs for join can be simply done by calling. Join () method as shown in Fig. 3.

After combining these two data streams, we apply a machine-learning model into the incoming sensor data records from processing step. To enhance the performance of the stream, we operate training a model in a separate server from the stream engines. Thus, communicating between server and stream is necessary to transmit the model into the stream. To achieve this issue, the stream requests the server via REST API call to receive the model context. Using this web protocol, we can also combine the model context with incoming sensor data records. After receiving model contexts from the server over HTTP, the Flink converts them into broadcast

```
DataStream<String> sensorDataRecords = env.addSource(new FlinkKafkaConsumer<>(…));          ]- Source

DataStream<AssetParameter> dataStreamWithParameter = sensorDataRecords.
                                                    .keyBy("id")                            ]- Transformation
                                                    .connect(specBroadcastStream)
                                                    .process(new ParameterMapper();)

dataStreamWithParameter.addSink(new JsonSerilizationSchema());                               ]- Sink
                                        - Flink APIs -

JavaInputDStream<String> sensorDataDStream = KafkaUtils.createDirectStream(...)              ]- Source
JavaInputDStream<String> specificationDStream = KafkaUtils.createDirectStream(...)

JavaPairDStream<String,String> sensorDStreamPair = sensorDataStream.mapToPair(...)           ]- Transformation
JavaPairDStream<String,String> specificationDStreamPair = sensorDataStream.mapToPair(...)

JavaPairDStream<String,Tuple2<String,String>> joined = sensorDStreamPair.join(specificationDStreamPair)  ]- Join
                                        - Spark Streaming APIs -
```

Fig. 3 A two spark stream API's for processing sensor data

variables to copy them to all task nodes for computing health scores. The Spark Streaming also uses join method to combine the model contexts with sensor datasets in a similar way to the preprocessing step.

3.3 Optimizations for Stream Processing

3.3.1 Number of Parallelisms

A stream program consists of multiple tasks including transformations, operators, sources, and sinks. These tasks are subdivided into several parallel instances for execution and each parallel instance processes a subset of the task. Thus, setting the number of parallelisms is an important task in the stream application and it can be adjusted depending on the user's application. When you set the parallelism to one for the entire job, then it would prevent the system from scaling, which would affect the whole performance while the order of events is preserved. The number of parallelisms should be decided according to the application's needs of scalability. Clearly, it might be more considerable to specify large numbers of parallelisms for the performance. From this point of view, we identify the overall performance for obtaining health scores by adjusting the number of parallelisms in multi-nodes in the study.

3.3.2 Checkpointing

Checkpoint makes state in stream processing fault-tolerant at large-scale by allowing state to be recovered, resulting in identical stream's context for unprecedent exceptions in applications. Because checkpoint management varies significantly among stream processing solutions, we optimized the checkpointing configuration by estimating the recovery time for single-node backup for computing health score when the failure unexpectedly happens.

3.4 A. Models for Health Score

To compute a health score, various kinds of machine-learning algorithms can be used. The following sections cover two machine-learning algorithms for obtaining score for given sensor data records that contain multiple parameters.

3.4.1 Long Short-Term Memory (LSTM) Model

Long short-term memory (LSTM) is one of the artificial recurrent neural network architectures used in the deep learning applications [11]. The distinguishing feature of LSTM is that it introduces a new structure called a memory cell consisting of four main elements: an input gate, a neuron with a self-recurrent connection, a forget gate, and an output gate. This concept has been created to overcome the vanishing gradients problem where the gradient value becomes so small that learning the weights becomes slow and declined in the traditional recurrent neural network [12]. The memory cell which works as a conveyor belt makes gradient signal last longer, allowing the network to control gradient values in a more efficient way for every step. The model we used in the service is a standard LSTM model. Our model is composed of two LSTM layers with softmax on the top, yielding n probabilities proportional (the n is the number of parameters) to the exponential of the input data as shown in Fig. 4. For example, from an input sensor data record, the memory cells in the LSTM layer will produce a representation sequence. We specify the unit of final dense layer as the same number of input parameters, which produces the same number of predictions. Thus, we are able to compare the difference between input data records and the predicted output. By measuring the difference using the mean squared error (MSE), we can compute a single value used as a health score in the LSTM model. As we mentioned, training a model is performed in a local server, which enhances the overall performance of the health score service by separating training phases from the stream.

3.4.2 Multivariate Analysis Model

Since most of the sensor data records consist of multiple parameters, multivariate analysis is needed, which requires interpretation of relation among each parameter [13]. To compute health score for a dataset with parameters, we used the Mahalanobis distance, which is one of the most common distance metrics in multivariate statistics [14]. It measures the distance between a vector consisting of multiple

Fig. 4 The LSTM model layers for obtaining health score

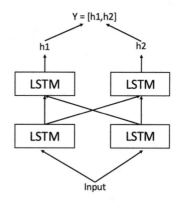

points and a distribution, accounting for the variance of each variables and the covariance between variables. The Mahalanobis distance can be defined as follows: $D^2 = (x - \overline{x}) \, S^{-1} \, (x - \overline{x})$ where x is an input data points, and $\overline{x}$ is a mean of input x, and S^{-1} is an inverse covariance matrix. Since the covariance matrix S^{-1} is obtained from the principal component analysis (PCA), applying PCA method is required prior to computing the Mahalanobis distance [15]. The Mahalanobis distance is computed based on both the mean and variance of the input variables, including covariance of all the variables, which would be more efficient to represent the distance as a health score. As training a LSTM model is performed in a local server, the multivariate model is also performed in a local server including the procedure of the PCA algorithm for obtaining the covariance matrix. After the model is built, the context is sent to the stream and then the equation of the Mahalanobis is applied for incoming data to obtain the health score.

4 Datasets

In our experiments, we used the tabular datasets for estimating the performance of two stream engines, Flink and Spark Streaming. Table 1 summarizes the datasets used in the following experiments. As shown in the table below, all the data records should contain their unique asset identification number. The massive data records are distributed across the multi-nodes using this id number, ensuring that the data records whose id number is identical are assigned into same node. Note that this behavior makes the algorithms do not consider the distant records which are stored on different nodes in order to merge them for applying algorithm. Along with the asset id, the dataset also includes the several parameters as shown in the table. Depending on the characteristics of the sensors in manufacturing, there are huge differences in the number of parameters ranging from 1 to 100. They are represented as numerical values and most of the system should apply normalization before

Table 1 Dataset description

Asset Id	Parameter1	Parameter2	Parameter3	Parameter 4	Others
#01	4.34	153	8	10,434	Vibration
#02	3.21	128	7	11,932	Current
#03	8.33	102	11	14,302	Current

applying some algorithms. Besides, other information such as data type are included in the datasets.

5   Results

Our evaluation has focused on addressing the following challenges:

1. Processing time for different time interval
2. Maximum number of assets
3. Varying the number of parallelisms

In doing above challenges, we compare Spark Streaming with Flink in the aspects of computational performance. We evaluated Flink and Spark Streaming using both several simulated sensor datasets. Our experiments were performed on both a single-node and multi-node (7 nodes) production clusters. The benchmark was performed on a YARN and Kafka cluster as well as a single node. We used a total number of 7 node YARN cluster, with four task node managers for Spark Streaming and Flink, one with zookeeper server, one with Spark Master, and one for Flink Master, respectively. Each node manager in the cluster was equipped with 16GB RAM, 4 core CPUs, a 200GB HDDs, and 1Gbps Ethernet network while the single node has specification with 16GB RAM, 2.5GHz Quad cores, and a 500GB SSDs. We tested our service for computing health score using the system.

The health score service receives continuous sensor data as a string format consisting of a timestamp, asset name, parameter name, and their values. For each input message, the asset *id* is extracted to be assigned as a key and all the sensor data records are distributed across multi-nodes based on the key we set. The data records with same asset id will be assigned into same partitions in a task, allowing machine-learning algorithm to be applied in a local state. Before submitting a streaming job, we specify the address of the separate web server so that the initial metadata including server host, specification context, and model context (if the model is already uploaded in the server) will be transmitted to the stream side, waiting for joining with incoming sensor data records.

5.1   Performance of the Health Score Service

5.1.1   Processing Time for Different Time Interval

We tested the performance of the health score service using simulated sensor data records. We first report the throughput of the Flink for different time intervals using the datasets and then compare it against the Spark Streaming. We set the total number of data records as 5000 for one asset and then calculates the throughput of the stream engines for computing health scores using 10 assets. We produce the input sensor data records with a different time interval, 10 ms, 20 ms, 50 ms, and 100 ms. Fig. 5 reports the throughput results that the Flink outperforms the Spark Streaming in processing data records for computing the health scores. The throughput of the Flink can be found that it can handle up to 10 ms time interval without delay, demonstrating that it is very effective in real-time processing data. However, the throughput of the Spark Streaming is relatively low compared to those of the Flink because of its stream design of adopting small batches. This requires a certain time period for mini batch size before running the Spark Streaming job, which means that the Spark Streaming is expected to be performed as near real-time stream engine. We can see that the Spark Streaming's processing time is about 1.62x and 1.21x slow compared to the those of the Flink for the interval of 10 ms and 20 ms, respectively. It is clear that the converting an input stream into a paired stream in the Spark Streaming results in degrading the overall performance of the service.

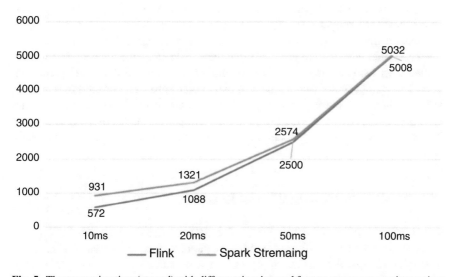

Fig. 5 The processing time (second) with different time interval for two stream processing engines

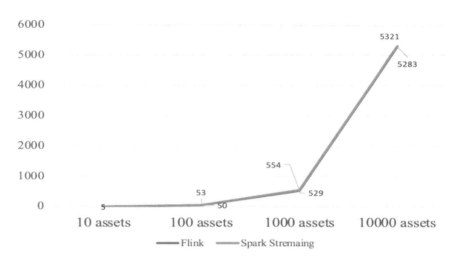

Fig. 6 The processing time (second) for two stream engines as increasing the number of assets

5.1.2 Maximum Number of Assets

We also evaluated how many assets they process in a limited computing resource. We performed the experiments on a single node and specified the number of parallelisms as one to make sure that there is only a single thread in a pipelined stream. The input datasets are constructed consisting of 10, 100, 1000, 10,000 assets with a small number of parameters. By using these five different input datasets, we estimated the running time for processing incoming input data records with 10 millisecond time intervals. Fig. 6 shows that the running time is increasing as a greater number of assets are given as input data. Especially, the running time greatly quadratically increases for all the experiments when we specified 50 records for each asset. The result seemed that a bunch of data records are well buffered for a certain period without timeout before being sent to the other subtask. There is no distinct difference between Spark Streaming and Flink for performance time, indicating that this experiment is more dependent on the computing resources (e.g., a buffer size in a memory).

5.1.3 Varying the Number of Parallelisms

To confirm the effect of parallelism, we tested the health score service on four nodes while specifying different number of parallelisms. As shown in Fig. 7, the processing time of the service on two nodes and four nodes outputs health score in double and nearly quadruple for both two stream engines for processing 5000 records with 100 millisecond intervals. We can see that the processing time of the Flink is slightly superior to those of the Spark Streaming when multiple nodes are used. It is because only the Flink implements converting data stream into broadcast

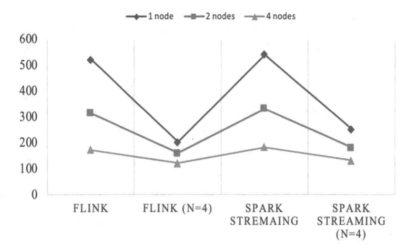

Fig. 7 The processing time (second) for different number of parallelisms

variable, allowing the data to be broadcasted to all downstream tasks. With the large number of parallelisms, we can see that the processing time is more improved for both Spark Streaming and the Flink. However, we cannot guarantee the order of the elements because task parallelism makes redistributing the entire data records across all nodes.

6   Discussion

We have studied two stream processing engines, Flink and Spark Streaming, for computing a health score in the perspective of the computational performance. By applying machine-learning algorithm into the real-time stream engine, our service enables to compute health score without a delay. Especially, it should be considered that the final results are obtained within in a time to successfully achieve real-time application service. Based on this perspective, we tried to estimate the computing performance of the two powerful streaming engines using various kinds of datasets.

Before benchmarking the stream engines, the entire system of architecture should be properly designed, reflecting the characteristics of the health score service. As explained in detail in Sect. 3.A, we used two databases for different purposes. The time-series database is responsible to store the data records that contains timestamp as well as health score that is computed from the stream engine. Since the database we used in the service is developed for a distributed system, it can accommodate large-scale data streams compared to the datasets we used in this study. We also use the local database storing metadata including specification and a model context in the server in order to make it possible to achieve low latency contexts by communicating with a separate server. This design makes a stream engine be more effective to calculate the health score by reducing the unnecessary workload to the stream. To communicate among various kinds of components used in the service, we adopted the Kafka service for effectively ingesting incoming messages between components. We believe that our design of the architecture in the health score service can be easily expanded to apply other machine-learning algorithms because it is easy to transmit the model context from the server to the stream. We evaluated two powerful stream engines to measure the processing time performance for a given dataset. We mainly tested them with three experiments and confirmed that the processing time of the Flink outperforms those of the Spark Streaming in computing health score with a very short interval. When the data with ten millisecond intervals are used as an input, the Flink achieved a comparable low latency while operating all the records, which can be considered to be real-time application. When varying the number of parallelisms, the Flink also shows slightly better performance results, indicating that processing time for computing 5 k records in four parallel instances is slightly short compared to the Spark Streaming. We think that the different results obtained from two engines are caused by their own designed architectures. The Spark Streaming adopts the micro-batching system which processes in a single mini batch with a delay of few seconds while the Flink selects the native streaming system. The Flink has the advantage of the processing data records because they process them as soon as it arrives, allowing the service to achieve minimum latency. This characteristic of the processing mechanisms is shown in our experiments, indicating that the overall processing time is slightly faster than those of the Spark Streaming. However, this point also means that the Flink is hard to achieve fault-tolerant requiring tracking and checkpointing a certain time when the data records processed. We did not experiment the performance of the processing time for fault-tolerant case to identify the recovery rate of two different stream engines in this chapter and this should be further studied.

7 Conclusions

We have investigated two powerful stream, Spark Streaming and Flink, for computing health score that represents a machine's lifetime. Even though there are lots of literature introducing real-time health score applications, they only focused on methodology for obtaining health score and do not explain sufficient description for real-time applications in computing health scores. Our study is powered for these reasons and we have studied two real-time stream engines and their APIs along with machine-learning algorithms. Moreover, we evaluated them using various simulated datasets for computing health scores, which can be helpful to select the stream engines considering your characteristics of applications.

Based on these experiments, we recommend the readers to use the Apache Flink in general case. As explained in the Methodology, the Flink provides a more variety of the stream APIs compared to the Spark Streaming and this functionality makes user be familiar with manipulating stream data records, allowing any results that user requires are obtained with minimal efforts. In addition, the native streaming characteristic of the Flink makes it possible to process every data record without waiting for others. Therefore, if the application needs to process data records with minimal delay and serves the results with very low latency, we carefully recommend the Flink because the Spark Streaming adopts the mini-batching stream service which requires a certain time for processing data. Although we did not experiment the case of the fault-tolerance and recovery the state for unexpected case, we anticipate that the Spark Streaming might show better performance of processing time over the Flink because of the powerful Spark Streaming's checkpointing mechanism and this should be further studied in our future work.

Competing Interests The authors declare that they have no competing interests.

Ethics Approval and Consent to Participate Not applicable.

Authors' Contributions SL, DK designed and coordinated this research. SL developed the main method and DHK studied the algorithm.

Acknowledgments This work was supported by the World Class 300 Project (R&D) (S2641209, "Improvement of manufacturing yield and productivity through the development of next generation intelligent Smart manufacturing solution based on AI & Big data") of the MOTIE, MSS(Korea).

References

1. G. S. Sampaio, et al. Prediction of Motor Failure Time Using An Artificial Neural Network. Sensors (Basel, Switzerland) 19.19 (2019)
2. H.V. Jagadish et al., Big data and its technical challenges. Commun. ACM **57**(7), 86–94 (2014)
3. T. Huuhtanen, A. Jung, Predictive maintenance of photovoltaic panels via deep learning. in *IEEE Data Science Workshop (DSW)*, (IEEE, 2018)
4. K.T.P. Nguyen, K. Medjaher, A new dynamic predictive maintenance framework using deep learning for failure prognostics. Reliab. Eng. Syst. Saf. **188**, 251–262 (2019)
5. D. Bansal, D.J. Evans, B. Jones, A real-time predictive maintenance system for machine systems. Int. J. Mach. Tools Manuf. **44**(7–8), 759–766 (2004)
6. S. Lee, D. Kim. A real-time based intelligent system for predicting equipment status. in *International Conference on Computational Science and Computational Intelligence (CSCI)*, (IEEE, 2019)
7. M. Masse. REST API design rulebook: designing consistent RESTful Web Service Interfaces. (O'Reilly Media, Inc., 2011)
8. P. Carbone et al., Apache flink: Stream and batch processing in a single engine. Bull. IEEE Comput. Soc. Tech Committee Data Eng **36.4** (2015)
9. M. Zaharia, et al. Discretized streams: Fault-tolerant streaming computation at scale. in *Proceedings of the twenty-fourth ACM symposium on operating systems principles*, (2013)
10. M. Zaharia, et al. Resilient distributed datasets: A fault-tolerant abstraction for in-memory cluster computing. Presented as part of the 9th
11. S. Hochreiter, J. Schmidhuber, Long short-term mem- ory. Neural Comput. **9**(8), 1735–1780 (1997)
12. S. Hochreiter, The vanishing gradient problem during learning recurrent neural nets and problem solutions. Int. J. Uncertainty, Fuzziness Knowledge-Based Syst **6**(02), 107–116 (1998)

13. J. F. Hair, et al. *Multivariate Data Analysis*. vol. 5 no. 3. (Prentice hall, Upper Saddle River, 1998)
14. R. De Maesschalck, D. Jouan-Rimbaud, D. L. Mas-sart. The Mahalanobis distance. Chemometrics and intelligent laboratory systems 50.1 (2000): 1–18. CARBONE, Paris, et al. Apache flink: Stream and batch processing in a single engine. Bull. IEEE Comput. Soc. Tech. Committee Data Eng., 36.4, (2015)
15. B. Chen, Z. Xian-yong, Z. Wen-jing. Eliminating outlier samples in near-infrared model by method of PCA-mahalanobis distance [J]. J. Jiangsu Univ. (Natural Science Edition) 4 (2008)

Discovery of Urban Mobility Patterns

**Iván Darío Peñaranda Arenas, Hugo Alatrista-Salas,
and Miguel Núñez-del-Prado Cortez**

1 Introduction

The ubiquity of information systems and the large number of mobile devices allow the capture of large amounts of heterogeneous data with a strong temporal repeatability (Big Data). This fact has allowed human activity to generate a fingerprint that is captured through different media, such as geo-social networks, call detail records (CDRs), and bank transactions, among others. These data that represent the behavior of people hide correlations and behavior patterns, which could be exploited by decision-makers to generate high-impact strategies. Indeed, the detection of urban mobility patterns is crucial for the development of urban planning policies and the design of business strategies, which helps to find answers to questions such as the following:

- Where do customers from certain businesses or commercial areas come from?
- What are the main businesses that compete to attract customers in each category?
- What factors influence the attractiveness of an establishment?

One of the approaches adopted in order to extract mobility patterns and answer the above questions is the analysis based on *trade areas*, originally proposed by Huff, who defines them as geographically delineated urban areas that contain the potential customers of certain businesses [8]. This paper intends to adapt this

I. D. Peñaranda Arenas (✉)
Pontificia Universidad Católica del Perú, Lima, Peru
e-mail: ivan.penaranda@pucp.edu.pe

H. Alatrista-Salas · M. Núñez-del-Prado Cortez
Universidad del Pacífico, Lima, Peru
e-mail: h.alatristas@up.edu.pe; m.nunezdelpradoc@up.edu.pe

© Springer Nature Switzerland AG 2021
R. Stahlbock et al. (eds.), *Advances in Data Science and Information Engineering*,
Transactions on Computational Science and Computational Intelligence,
https://doi.org/10.1007/978-3-030-71704-9_33

approach to the context of the city of Lima, using a database of bank transactions. The main contributions of our study are summarized below:

- Up to our knowledge, this is the first work in which urban mobility patterns are extracted from bank transaction data.
- The trade areas approach is adapted through the incorporation of bank transaction data, showing the importance of taking into account the business category and including individuals from all social classes.
- The importance of using some metrics used to determine attractiveness is analyzed according to the category to which the business belongs.
- The use of *influence diagrams* is proposed as a tool that simplifies the visualization of the place of origin of business customers and the preference for them according to socioeconomic classification.

The rest of the paper is organized as follows: Sect. 2 shows some of the most relevant works related to our problem. Section 3 presents a general description of the used dataset and the main challenges associated with our proposal, whose methodology is explained in Sect. 4. Then, in Sect. 5, the results obtained during the experiments are shown discussing their implications; finally, the conclusions and recommendations are presented.

2   Related Works

In recent years, the trend to take advantage of mobile phone data to carry out urban monitoring has gained strength, thanks to the ubiquity of mobile devices and the development of these networks. In the studies of Dobra et al. [3] and Young et al. [16], for example, systems that use aggregate traffic data in the cellular network are proposed for the spatiotemporal detection of behavioral anomalies, even being able to identify the location and extent of anomalies such as natural catastrophes, violence events, and popular celebrations.

Regarding the analysis of urban mobility patterns, particularly toward commercial areas, this task has been traditionally addressed through small-scale conditioned techniques or studies at the level of establishments in which information is collected manually through surveys [4]. Another method suggests the use of customer records in social networks as a low-cost alternative [14]. Among the main limitations of these approaches are the dispersion and bias presented on the data they use and the difficulty of inferring the place of origin of consumers. It is for these reasons that it is difficult to apply them at the level of a district or analyze patterns that require greater granularity over time.

A novel reference framework for the analysis of commercial areas that in turn allows the understanding of urban mobility patterns is CellTradeMap [17], which uses mobile flow records (MFRs) obtained from cellular phone data to extract reliable information on the location of consumers, thus adapting the conventional analysis of trade areas by offering an effective data collection methodology in terms of cost that is appropriate for analysis even at the level of a complete city. Other contributions that are worth highlighting from this work are its ability to describe commercial areas that can be explained by prior knowledge and the revelation of metrics that are important for determining commercial attractiveness as well as predicting commercial areas of new zones with a high degree of accuracy.

Although it is possible to infer through this last approach the place of residence of the clients, it is not possible, from the MFRs, to be sure if the client actually made a purchase at an establishment. On the contrary, the bank transaction data that we use in this work do have this last advantage, but with the inconvenience of uncertainty in determining the place of origin of the clients. However, this paper shows that it is indeed possible to carry out the detection of urban mobility patterns through these data by adapting the analysis of the trade areas approach.

3 General Aspects

In the used dataset, all the transactions made in Peru by the clients of a banking entity are recorded during the period between May 31, 2016, and October 31, 2017. Of the total of these transactions, more than 50 million were carried out in Lima, and it is in this subset that this work is centered. The most relevant attributes for our study can be grouped into the following categories:

- **Customer attributes** as the id (to preserve his/her privacy), age, gender, and social class.
- **Store attributes** that include the name, province, district, geographic coordinates, and commercial categorization. To establish the latter, three systems are used in the dataset: *mccg*, Visa's own classification; *coicop*, UN categorization; and a bank's own system called *merchant_type*.
- **Banking agency attributes**, among others, the province and the district where the agency in which the client opened his/her bank account is located.
- **Transaction attributes** such as the date and amount of the transaction in dollars.

Regarding the social class or income level of the clients, it is worth noting that this was previously calculated from the level of consumption, taking as a reference the method proposed by Leo et al. [10, 11] and defining an order consisting of nine categories in which social class 9 corresponds to the one with the highest purchasing power.

On the other hand, in Fig. 1, the ten categories of shops registering the largest number of transactions in Lima are shown. There is a marked predominance of the categories that are part of the *top 5*, which is why emphasis will be placed on these ones in addition to Technology for the extraction of mobility patterns.

The biggest challenge posed by the extraction of mobility patterns from the bank data we have is undoubtedly the uncertainty associated with the place of origin of customers, particularly their place of residence. In fact, this information is not available in the database for privacy reasons. While there are techniques that have

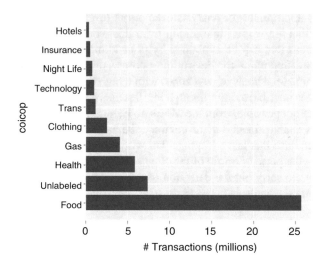

Fig. 1 Categories of businesses with the highest number of transactions

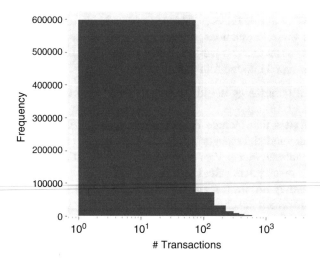

Fig. 2 Distribution of the number of transactions made by customers

been used for the extraction of permanence points either from GPS traces [18] or for mobile phone data using CDRs [15] or MFRs [17], these cannot be applied to bank transaction data given their peculiarities. In fact, our dataset has a very high dispersion, as shown in Fig. 2. Indeed, close to 600,000 customers, representing a proportion greater than 81% of users residing in Lima, made a maximum of 70 transactions during the entire period, which is equivalent to an approximate frequency of 1 transaction per week. Therefore, given the low transaction frequency observed for most of the clients, it is inadequate to use the same techniques applied to infer the place of residence in the case of mobile phone data. Consequently, we hypothesize that the district where the place of residence of the clients is located limits or coincides with the district where the banking agency in which they opened their accounts is located.

4 Methodology

For the development of this work, the methods proposed by Henretty et al. [6] and by Zhao et al. [17] were taken as references. The following is a description of the main components of our proposal, which are represented in Fig. 3.

1. **Feature engineering.** Once the data is cleaned and after choosing those features that will be examined for the discovery of mobility patterns, the needed transformations are carried out; these include normalization of the number of transactions for the calculation of probabilities, the conversion of latitude/longitude values to another coordinate system, and the creation of new attributes such as the accessibility of commerce for the consumers of each district, expressed in terms of distance or time. For the calculation of accessibility, the location of the closest point of the customer's district of origin to the respective store is first obtained, and the same is used to estimate the distance, while in the case of time, the duration of a car trip is calculated between both points. Regarding discretization,

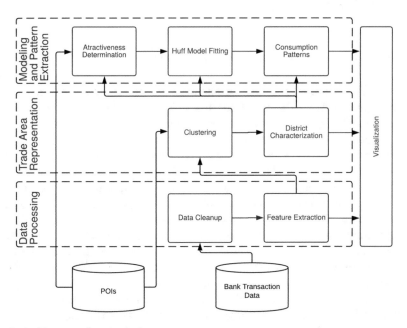

Fig. 3 Architecture of our method

this determines the level of spatial granularity; in fact, for the calculation of accessibility, it was decided to manage 13 decimal figures in the values of the coordinates of the shops, given that the coordinates of the closest points of each district returned by PostGIS use the aforementioned number of figures.

2. **Clustering and identification of districts.** In this phase, the geographic coordinates data of the commercial establishments where the bank transactions were made were used in order to segment the city into a set of regions or disjoint districts. To carry out this task, the following methods were tested: DBSCAN, HDBSCAN, and K-Means. Since the input data corresponds to geographical coordinates, it is necessary to use a suitable metric to calculate the distance between the shops; for this reason, the use of the Euclidean distance is discarded, and instead the large circle distance was employed which was calculated by means of the haversine formula [13]. In order to select the optimal number of clusters, a cross-validation process was carried out using the silhouette coefficient and the Davies-Bouldin index (DBI) as quality criteria. In fact, in a previous work [2], it was shown that the latter index correlates with the F-measure obtained in the extraction of points of interest (POIs) in human mobility through various clustering techniques such as DBSCAN.

3. **Huff model.** The Huff model has been widely used in spatial economic analysis in order to determine the likelihood of visit or purchase in a trade area. It is based on the assumption that this probability can be estimated based on the attractiveness of the stores, their accessibility to consumers, and the attractiveness as well as the accessibility of competing stores [9]. The original version of the model is expressed as follows:

$$P_{ij} = \frac{A_j D_{ij}^{\lambda}}{\sum_{i=1}^{N} A_i D_{ij}^{\lambda}} \tag{1}$$

where P_{ij} represents the visit or purchase probability of a customer from a geographic area i in a store or trade area j, A_j is the measure of attraction from the business j, D_{ij} measures the accessibility of the business to customers located in the area i, λ is a parameter that corresponds to the sensitivity of P_{ij} to D_{ij}, and N is the number of businesses or commercial areas that compete with each other to attract customers. It should be noted that in the original version of the Huff model, it is generally assumed that attractiveness is equivalent to the area of the commercial establishment and accessibility is expressed in terms of the distance or time required by consumers to reach the business.

4. **Determination of attractiveness.** For the business categories that register a greater number of transactions according to the *coicop* classification, Pareto diagrams were elaborated, in order to select a representative sample of shops of each class. Regarding the Food, Health, Gas Stations, and Clothing Stores categories, which registered the largest number of transactions, this sample corresponds to 83, 24, 47, and 22 businesses, respectively. Next, the attractiveness of the selected

businesses was determined, taking as a metric to estimate it the rating and the number of ratings that each store registers on Google Maps.

5. **Calibration of the Huff model.** The adjustment of the Huff model allows, among other things, to identify which are the factors of attraction of the businesses that are most influential. Since we are considering more than one characteristic to determine the attraction of each shop, it is necessary to use the advanced version of the Huff model whose mathematical expression is:

$$P_{ij} = \frac{\prod_{h=1}^{H} A_{hj}^{\gamma_h} D_{ij}^{\lambda}}{\sum_{j=1}^{N} \prod_{h=1}^{H} A_{hj}^{\gamma_h} D_{ij}^{\lambda}} \tag{2}$$

In the previous equation, H is the number of attraction factors of each store, and γ_h represents the sensitivity of P_{ij} to each attraction characteristic. The rest of the variables or parameters have the same meaning as the original version shown in Eq. (1). The model in its advanced version can be expressed in a linear version after applying the following logarithmic transformation [7]:

$$\log\left(\frac{P_{ij}}{\tilde{P}_i}\right) = \sum_{h=1}^{H} \gamma_h \log\left(\frac{A_{hj}}{\tilde{A}_j}\right) + \lambda \log\left(\frac{D_{ij}}{\tilde{D}_i}\right) = \hat{y}_{ij} \tag{3}$$

with $\tilde{P}_i$, $\tilde{A}_j$, and $\tilde{D}_i$ the geometric mean of P_{ij}, A_{hj}, and D_{ij}, respectively. After adjusting this last linear version using least squares, it is possible to obtain the parameters γ_h and λ that define the model. In order to obtain predictions $\hat{P}_{ij}$ of the purchase probabilities and calculate the degree of fit of the model, it is necessary to apply the softmax function to the values estimated by (3) using the following expression:

$$\hat{P}_{ij} = \frac{e^{\hat{y}_{ij}}}{\sum_{j=1}^{N} e^{\hat{y}_{ij}}} \tag{4}$$

6. **Consumption patterns.** To facilitate the extraction of mobility and/or consumption patterns, *influence diagrams* of those businesses located in the top 3 of the selected commercial categories were elaborated, by either the number of transactions, unique customers, or amount spent. In the *influence diagrams*, the distribution by income level of each of these variables is represented for each district, which appears ordered according to the geodetic distance to store, and the variable in question is normalized by dividing it by the total number of transactions, customers, or amount spent in the district. For the same businesses, and taking into account the distribution of social classes in each district according to the indicated variables, the Jensen–Shannon distance and Gini index diagrams were generated.

The Jensen–Shannon distance is a metric that allows, in our study, to estimate how far apart the distributions by social class of the commerce and the district are according to the number of transactions, customers, and amount spent. According to the Python SciPy library documentation (used to calculate this distance), for two probability vectors p and q, this is defined as follows:

$$JSD(p, q) = \sqrt{\frac{D(p||m) + D(q||m)}{2}} \tag{5}$$

where m is the mean in terms of the points of p and q and D is the Kullback-Leibler divergence.

On the other hand, the Gini coefficient allows us to quantify the level of inequality in the distribution by social class of transactions, clients, and amount spent, both for commerce and for the district. There are several ways to calculate it; in our case, a variant is used in which the n observations x are sorted in ascending order and have a ranking i, as proposed by David [1] and Guest [5].

$$G = \frac{\sum_{i=1}^{n} (2i - n - 1) x_i}{n \sum_{i=1}^{n} x_i} \tag{6}$$

5 Experimental Results and Discussion

5.1 Clustering

Table 1 shows the best results obtained during cross-validation for each of the clustering techniques tested. It should be noted that a value of the silhouette coefficient close to 1 shows that the observations have been very well classified; on the other hand, the Davies-Bouldin index is an indicator of the degree of similarity between clusters; therefore, the smaller its magnitude, the better the set of clusters that will be formed [2]. The results of Table 1 suggest that the performance achieved by the DBSCAN and K-Means methods is similar, although there is a notable difference in the number of clusters that are generated. Likewise, the poor performance achieved with HDBSCAN and the huge number of establishments that are classified as noise when using this technique are striking.

Table 1 Cross-validation results with the tested clustering methods

Method	Silh-Coef	DBI	#Clusters	Noise points
DBSCAN	0.535948	0.977980	17	35
HDBSCAN	0.058971	3.634427	16	5939
K-Means	0.492291	0.683749	29	NA

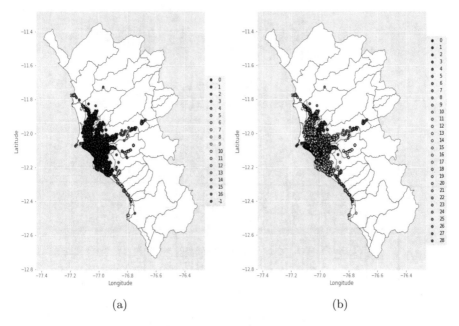

Fig. 4 Generated clusters. (**a**) DBSCAN. (**b**) K-Means

Although similar results are achieved with DBSCAN and K-Means in terms of the silhouette coefficient, the DBI value indicates that a better configuration is obtained when using K-Means. In fact, the maps in Fig. 4 show that almost all businesses in the province of Lima are assigned to a single cluster in the configuration generated by DBSCAN, that is the reason why this method is not suitable in our case to identify districts or commercial areas, unlike K-Means.

5.2 Huff Model

Table 2 contains the results of the calibration of the Huff model carried out with the selected businesses of each category. As performance metrics of the model developed for each social class, the coefficient of determination and the RMSE were chosen, and the symbol (*b*) refers to the model adopted as a baseline, which assumes that all businesses have the same attractiveness to customers in all districts.

In the first place, it is appreciated that for all the business categories considered, the model that includes all social classes in the analysis and assumes different attractiveness per business obtains better performance than the baseline, although there are categories such as Food and Gas Stations where the difference is not so considerable, which suggests that in these categories the shops have an attractiveness that is not far from each other. Nevertheless, in classes like Health,

Table 2 Huff model calibration per social class and business category

Business category	Metric	Model for social class									All
		1	2	3	4	5	6	7	8	9	
Food	$R^2(b)$	−0.433	−0.092	−0.164	−0.037	−0.118	−0.100	−0.132	−0.457	−0.801	0.156
	RMSE(b)	3.719	3.322	3.595	2.994	3.138	3.149	3.174	3.326	3.856	2.369
	R^2	−0.625	−0.185	−0.284	−0.099	−0.182	−0.139	−0.118	−0.440	−0.648	0.158
	RMSE	3.961	3.459	3.776	3.082	3.226	3.204	3.154	3.307	3.689	2.366
Health	$R^2(b)$	−0.264	−0.371	−0.289	−0.166	−0.111	−0.132	−0.158	−0.172	−0.340	−0.074
	RMSE(b)	13.777	12.355	11.547	9.973	9.101	9.211	8.722	8.657	8.535	7.607
	R^2	0.181	−0.208	−0.405	−0.905	−1.308	−0.993	−0.954	−0.539	−0.290	0.314
	RMSE	11.089	11.600	12.056	12.744	13.117	12.222	11.331	9.919	8.376	6.078
Gas	$R^2(b)$	0.064	−0.042	0.099	0.076	0.117	0.080	0.111	0.140	0.110	0.182
	RMSE(b)	4.374	5.007	4.093	4.739	4.468	5.030	4.576	3.911	4.570	3.281
	R^2	−0.063	−0.038	0.097	0.070	0.144	0.098	0.118	0.162	0.153	0.210
	RMSE	4.661	4.996	4.097	4.754	4.397	4.980	4.558	3.859	4.456	3.224
Clothing	$R^2(b)$	−0.173	−0.481	−0.089	−0.143	−0.035	−0.004	−0.051	−0.018	−0.053	0.001
	RMSE(b)	10.894	12.371	10.685	10.622	10.113	9.898	9.386	8.697	8.426	8.719
	R^2	−0.751	−0.918	−0.149	−0.270	−0.043	0.027	−0.049	0.020	−0.016	0.033
	RMSE	13.312	14.080	10.974	11.198	10.152	9.741	9.379	8.533	8.279	8.574
Technology	$R^2(b)$	−0.350	−0.429	−0.378	−0.255	−0.198	−0.151	−0.131	−0.225	−0.184	0.083
	RMSE(b)	10.239	10.604	10.405	10.110	9.128	9.643	10.319	10.565	11.637	7.814
	R^2	−0.258	−0.256	−0.171	−0.137	−0.037	−0.045	−0.067	−0.121	−0.082	0.188
	RMSE	9.883	9.943	9.591	9.624	8.491	9.189	10.023	10.105	11.126	7.351

Table 3 Significance of the employed variables in the Huff model

Business category	Significance for variable		
	Attr_1	Attr_2	Distance
Food	0.000	0.000	0.000
Health	0.000	0.644	0.000
Gas	0.169	0.089	0.000
Clothing	0.096	0.196	0.000
Technology	0.210	0.010	0.000

Clothing, and Technology, there is a dramatic difference in the performance of both models in terms of the magnitude of R^2, which is a sign that in these business classes there are more significant differences in the attractiveness when estimating the probability of purchase.

Secondly, if the model that includes all social classes and assumes different attractiveness is compared with the analogous models that take each social class separately, it is observed that it obtains a better performance in all categories of businesses, with some exceptions that they show up in the clothing one. This demonstrates that a better calibration of the Huff model is achieved when individuals of all social classes are included in each geographic area of study.

Finally, the low magnitude of the coefficient of determination achieved by the models evidences the need to include a greater number of attraction factors for the businesses considered such as the surface, the age, and the availability of parking places or the retail market clustering, local accessibility of the street network, and topographic slope of the terrain [12], among others, particularly in view of the fact that they are not always significant. Indeed, in Table 3, the p-value calculated in the hypothesis contrasts on the coefficients associated with the variables of the models that include all social classes in each category is observed. At a significance level of 5%, the distance variable is the only significant one in all models. On the other hand, the rating assigned to businesses (Attr_1) is only significant in the Food and Health categories, a sign of the importance that customers of these types of businesses assign to this factor, while the number of ratings (Attr_2) is only significant in the Food and Technology categories.

5.3 Consumption Patterns

People tend to buy more likely in those businesses located closer to their place of residence than in the distant ones. This consumption pattern is modeled by the known distance decay function [7]. In Fig. 5a, this trend is clearly observed for stores in the Food category. However, this pattern is not observed so markedly in all categories. Indeed, for establishments in the Health sector (Fig. 5b), there is a considerable proportion of cases that deviate from the pattern, which suggests that these consumers do not care about the distance they should travel in order to

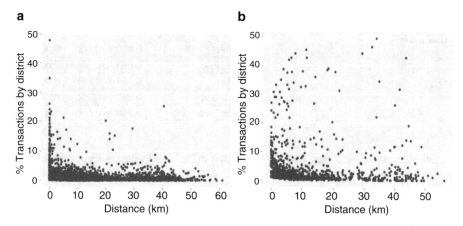

Fig. 5 Transactions percentage by district in terms of distance. (**a**) Food. (**b**) Health

accessing medical services. In addition, the *influence diagrams* of the businesses with the highest number of transactions in these categories, shown in Fig. 6a, b, confirm the aforementioned. It should be noted that in these diagrams, the first letter of the code assigned to each business represents the category, while the number refers to the place it occupies in the ranking by transactions; in this way, a trade identified by F-3 corresponds to the establishment that occupies the third place in the Food category by number of transactions. Also, as mentioned in Sect. 4, each bar represents in this case the distribution by social class of the number of normalized transactions, and the districts are arranged according to their distance to the respective business, so that the nearest districts are the ones at the left end of the diagram.

The *influence diagrams* allow to appreciate not only the districts where the clients of each business come from but also the preference of the social classes toward it. If, for example, they are compared, the diagrams in Fig. 6 show that there is a marked predominance of the upper income levels in most districts for the businesses of the Food category, while in the case of Health sector, the distribution is more balanced.

In the diagram shown in Fig. 7, there is an increasing pattern in Jensen–Shannon distances as the district departs from commerce. This graph compares the distribution of transactions by social class of commerce with that of the district when it acts either as origin (blue bars) or as destination (red bars). This consumption pattern indicates that the way in which transactions are distributed in the establishment as the distance increases becomes less and less similar to that of the district regardless of whether it acts as the issuer or receiver of transactions.

Similar to the calculation made for the Jensen–Shannon distances, in Fig. 8, the Gini coefficient of the distribution of transactions corresponding to the business (blue bars) versus the distribution of transactions of the district (red bars) is compared. This diagram shows an increasing pattern in the difference between the Gini coefficients of commerce and the district as the district departs from the

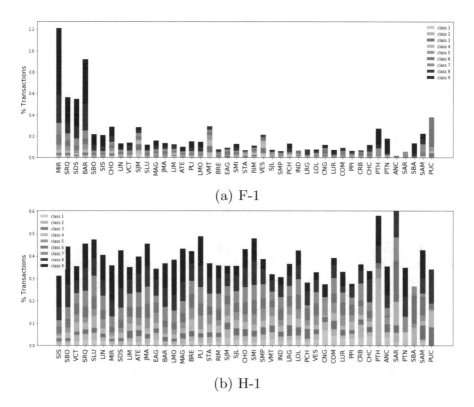

Fig. 6 *Influence diagrams* according to the distribution of number of transactions per social class. (**a**) F-1. (**b**) H-1

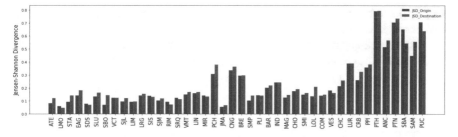

Fig. 7 The Jensen–Shannon distance diagram for G-1 according to the distribution of number of transactions

establishment. Likewise, the Gini coefficients of transactions in commerce tend to be higher for the most remote districts, which indicates that there is a greater inequality in the number of transactions according to social class when they come from more remote districts.

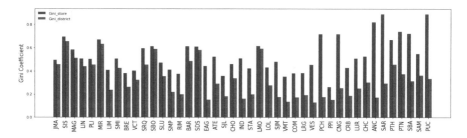

Fig. 8 Gini coefficient diagram for C-4 according to the distribution of number of transactions

6 Conclusions and Recommendations

This work showed that it is possible through bank transaction data the discovery of urban mobility patterns, which can be very useful tools to facilitate business decision-making such as the location of a new retail store or the design of marketing campaigns.

Although a satisfactory delimitation of commercial areas was not achieved using DBSCAN, the results obtained after using K-Means allow us to have an idea about the distribution of certain business categories in the city.

The extent to which the distance between the place of origin of the customers and the store affects the probability of making a transaction depends on the category of the business. Likewise, in order to achieve a better performance in the models that allow explaining this variable, it is necessary to include individuals of all social classes and a greater number of attraction factors.

The *influence diagrams* developed in this work allow us to visualize in a simple way for each business not only the origin of the clients but also the preference of each social class.

Future work suggests the inclusion of other business attraction factors in the model such as age and the area of commerce, as well as the analysis of urban mobility patterns combining data from bank transactions with mobile phone data.

References

1. H. David, Miscellanea: Gini's mean difference rediscovered. Biometrika **55**(3), 573–575 (1968)
2. M.N. Del Prado Cortez, H.A. Salas, Quality metrics for optimizing parameters tuning in clustering algorithms for extraction of points of interest in human mobility, in *SIMBig* (Citeseer, 2014), pp. 14–21
3. A. Dobra, N.E. Williams, N. Eagle, Spatiotemporal detection of unusual human population behavior using mobile phone data. PLoS One **10**(3), e0120449 (2015)
4. E. Dramowicz, Retail trade area analysis using the Huff model. Dir. Mag. **2** (2005)

5. O. Guest, O., B.C. Love, What the success of brain imaging implies about the neural code. Elife **6**, e21397 (2017)
6. T. Henretty, M. Baskaran, J. Ezick, D. Bruns-Smith, T.A. Simon, A quantitative and qualitative analysis of tensor decompositions on spatiotemporal data, in *2017 IEEE High Performance Extreme Computing Conference (HPEC)* (IEEE, Piscataway, 2017), pp. 1–7
7. D. Huff, B.M. McCallum, Calibrating the Huff model using ArcGIS business analyst. ESRI White Paper (2008)
8. D.L. Huff, A probabilistic analysis of shopping center trade areas. Land Econ. **39**(1), 81–90 (1963)
9. D.L. Huff, Parameter estimation in the Huff model. ESRI, ArcUser (2003), pp. 34–36
10. Y. Leo, E. Fleury, J.I. Alvarez-Hamelin, C. Sarraute, M. Karsai, Socioeconomic correlations and stratification in social-communication networks. J. R. Soc. Interface **13**(125), 20160598 (2016)
11. Y. Leo, M. Karsai, C. Sarraute, E. Fleury, Correlations of consumption patterns in social-economic networks, in *Proceedings of the 2016 IEEE/ACM International Conference on Advances in Social Networks Analysis and Mining* (IEEE Press, New York, 2016), pp. 493–500
12. E. Paroli, C. Maraschin, Locational Attractiveness Modelling of Retail in Santa Maria, Brazil. Urban Sci. **2**(4), 105 (2018)
13. F. Pedregosa, G. Varoquaux, A. Gramfort, V. Michel, B. Thirion, O. Grisel, M. Blondel, P. Prettenhofer, R. Weiss, V. Dubourg, J. Vanderplas, A. Passos, D. Cournapeau, M. Brucher, M. Perrot, E. Duchesnay, Scikit-learn: Machine learning in Python. J. Mach. Learn. Res. **12**, 2825–2830 (2011)
14. Y. Qu, J. Zhang, Trade area analysis using user generated mobile location data, in *Proceedings of the 22nd International Conference on World Wide Web* (ACM, New York, 2013), pp. 1053–1064
15. F. Wang, C. Chen, On data processing required to derive mobility patterns from passively-generated mobile phone data. Transp. Res. C: Emerg. Technol. **87**, 58–74 (2018)
16. W.C. Young, J.E. Blumenstock, E.B. Fox, T.H. McCormick, Detecting and classifying anomalous behavior in spatiotemporal network data, in *Proceedings of the 2014 KDD Workshop on Learning About Emergencies from Social Information (KDD-LESI 2014)* (2014), pp. 29–33
17. Y. Zhao, Z. Zhou, X. Wang, T. Liu, Y. Liu, Z. Yang, Celltrademap: delineating trade areas for urban commercial districts with cellular networks, in *IEEE INFOCOM 2019-IEEE Conference on Computer Communications* (IEEE, Piscataway, 2019), pp. 937–945
18. V.W. Zheng, Y. Zheng, X. Xie, Q. Yang, Collaborative location and activity recommendations with GPS history data, in *Proceedings of the 19th International Conference on World Wide Web* (ACM, New York, 2010), pp. 1029–1038

Improving Model Accuracy with Probability Scoring Machine Learning Models

Juily Vasandani, Saumya Bharti, Deepankar Singh, and Shreeansh Priyadarshi

1 Introduction

Binary classification is the simple task of making a prediction as to which of two groups a given observation belongs to on the basis of a specified classification rule. They also exist everywhere in society – from hospitals predicting high-risk vs low-risk patients to credit card companies categorizing applications as good vs bad credit, and even quality control in manufacturing processes (pass/fail). Its goal is simply to categorize data points into one of two different buckets, making it one of the simplest kinds of supervised learning problems, and machine-learning models are quickly becoming the immediate solution across different industries. These models extract information automatically using computational and statistical methods, giving software the ability to build knowledge from experience, derived from patterns and rules extracted from a large volume of data [1].

Typical classifiers commonly used for these problems include decision trees, logistic regression, support vector machines, and neural networks. Each method performs best under specific circumstances – depending on the number of observations, the dimensionality of the predictors, the noise in the data, and many other factors. To add to the complexity, there are also several known evaluation metrics that can be used to measure the performance of a classification model ranging from simple accuracy or misclassification measures to precision, recall, log loss, ROC curve, and the AUC score, which can show more insightful information about the classifier's performance [2]. Deciding on the eventual performance measure tends to be subjective since they have to align with the project objectives, and different

J. Vasandani (✉) · S. Bharti · D. Singh · S. Priyadarshi
Krannert School of Management, Purdue University, West Lafayette, IN, USA
e-mail: jvasanda@purdue.edu; bhartis@purdue.edu; singh681@purdue.edu; spriyad@purdue.edu

© Springer Nature Switzerland AG 2021
R. Stahlbock et al. (eds.), *Advances in Data Science and Information Engineering*,
Transactions on Computational Science and Computational Intelligence,
https://doi.org/10.1007/978-3-030-71704-9_34

measures may alter final predictions, making this choice a significant part of the overall classification problem.

While much of the research in this space has focused on the accuracies produced by different classifiers utilizing various probability scoring methods, no comparison of other accuracy improving methods (e.g., hyperparameter tuning, feature engineering, or ensemble techniques) has yet been performed. This chapter presents a comprehensive comparison of different machine learning models designed to solve a binary classification problem, using proprietary data from our partner company to understand the drivers that impact their conversion rate. Though traditionally defined within the context of digital marketing and e-commerce, a conversion generally occurs when users (or clients) take a desired action [3], not limited to a sale or a purchase. A conversion can be any key performance indicator (KPI) relevant to a firm, and for the purpose of this study, it is defined as the instance a sales representative completes a scheduled home visit with a lead generated by the company's marketing team. Our study not only evaluates each model, but various methods to improve each classifier's accuracy as well, particularly those that have not yet been addressed in the literature. Based on our experimental results, binary classification problems can be solved using different combination of models and accuracy tuning methods, which can be then evaluated using different criteria that works with the overall business problem. The motivation behind our study is to evaluate different ways to improve models already in production, since many organizations would likely already have basic models in place to solve their classification problems. Our aim is to answer the following research questions:

1. What industry-agnostic factors have a significant impact on conversion rates?
2. Which accuracy improvement techniques (e.g., sampling methods, hyperparameter tuning, feature engineering, etc.) can significantly increase a model's predictive power?

The following section reviews the past literature on various criteria and methods used to solve binary classification problems, including evaluating different models and performance measures traditionally used in this space. It is immediately followed by a deeper look into the context and background of the data used in our study, before a dissection of our methodology in the third section. We outline the different models we investigated, as well as the various methods tested against them to improve their accuracy. Finally, we present our results, discuss our conclusions, and lay out the framework for further research and potential future scope of this project.

2   Literature Review

2.1   Model Selection

Regardless of the industry, every person – whether they have made a single purchase or have been a loyal customer for many years – will eventually cease their relationship with a business. This loss, commonly known as customer attrition or churn, is one of the most popular business applications of the binary classification problem, with a wealth of resources dedicated to studying various solutions and their accuracy. Churn models aim to identify early churn signals and recognize customers with an increased likelihood to leave voluntarily, using various parameters and different evaluation methods to predict the attrition occurrence. A performance comparison between Monte Carlo simulations of five well-established techniques used for churn prediction showed that the best classifier was the SVM-POLY model tuned with AdaBoost [4]. However, later conclusions studying telecommunications churn found tree-based algorithms to be the best models to apply toward a real-time attrition problem [5, 6]. These contradicting results serve to show us that while algorithm selection is very important, there are several other factors that impact the effectiveness of a chosen machine learning solution.

Although the appropriate choice for a supervised learning algorithm depends on the task at hand, each of these algorithms has its own pros and cons, and there are cases when practitioners may select an inappropriate algorithm for their solution. No single model can uniformly outperform the others across all datasets, and the comparative summary of classification techniques included in Table 1 provides a truncated overview of model performance across several features [7]. When dealing with this type of machine learning problem, the key question is not whether a learning algorithm is superior to others, but under which conditions can a model outperform others on an applied problem. Although it may not give us a deterministic answer for model choice, it gives us a preliminary idea of which models to start fitting to our data based on the given problem.

An ensemble model combines classifiers as an extension of algorithm selection, improving the performance of individual classifiers. There are many ways to build an ensemble, and some of these mechanisms include: (i) using different subsets of training data with a single learning method, (ii) using different training parameters with a single training method, and (iii) using different learning methods [7]. Despite the many varied base models used to predict an outcome within an ensemble model, the ensemble acts and performs as a single model. During this process, the generalization error of the prediction is reduced, as long as the base models are diverse and independent of each other [8].

2.2 Probability Scoring

Model fit depends on the kind of data available and interpretation depends on the evaluation measure chosen, a model's performance can also be tuned toward better performance. Tuning a machine learning model is very much like turning the switches and knobs of an antique TV in order to get a clearer picture. The eventual goal is to improve model accuracy, which can be achieved by optimization techniques such as sampling methods, feature extraction and engineering, metric creation and selection, hyperparameter tuning, and ensemble methods, among

Table 1 Comparing algorithms (**** represents the best and * the worst performance)

	Algorithms		
	Decision trees	*Neural networks*	*Naïve bayes*
Accuracy in general	**	***	*
Speed of learning with respect to number of attributes and instances	***	*	****
Speed of classification	****	****	****
Tolerance to missing values	***	*	****
Tolerance to irrelevant attributes	***	*	**
Tolerance to redundant attributes	**	**	*
Tolerance to highly interdependent attributes (e.g., parity problems)	**	***	*
Dealing with discrete, binary, or continuous attributes	****	*** (not discrete)	***(not cont.)
Tolerance to noise	**	**	***
Dealing with danger of overfitting	**	*	***
Attempts for incremental learning	**	***	****
Explanation ability/transparency of knowledge/classifications	****	*	****
Model parameter handling	***	*	****

others. The ability to effectively gauge the impact of each optimization technique on overall model accuracy is a significant objective to address the research questions in this study, making our choice of evaluation metric very important.

A classification algorithm's prediction score indicates its certainty that the given observation belongs to the positive class. To make the decision about whether the observation should be classified as positive or negative, a classification threshold (cut-off) is selected and compared against its score. Any observations with scores higher than the threshold are then predicted as the positive class and scores lower than the threshold are predicted as the negative class. Once observations are classified, the prediction falls into four categories, depending on the actual observed answer and the predicted answer, where true positives and negatives imply that the observed and the predicted class are the same, while false positives and negatives indicate that misclassification has occurred for the given observations.

Likely the most widely accepted performance measurement metric for classification, the AUC – ROC curve, represents the ability of a model to distinguish between classes. The ROC curve is plotted with the true-positive rate (TPR) on the y-axis and the false-positive rate (FPR) on the x- axis. A higher AUC value indicates a model accurately predicting 0 s as 0 s, and 1 s as 1 s – effectively distinguishing between a customer that will churn vs one who will not.

While the goal of traditional binary classifiers revolves around predicting an observation's class label, certain problems require the additional nuance of measuring the likelihood that each example belongs to a given class. In the empirical work on probabilistic prediction in machine learning, the most standard loss functions are log loss, Brier loss, and spherical loss. In determining which loss function is likely to lead to better prediction algorithms, it was found that log loss is more selective and is likely to lead to better algorithms as a result – if a model is optimal under the log loss function, it will be optimal under the other two loss functions, but the opposite may not be true [9]. Log loss takes into account the uncertainty of a prediction based on how much it varies from the actual label by assigning a probability to each class rather than simply yielding the most likely class, quantifying accuracy of the prediction and then penalizing false classifications. By placing heavy penalties on classifiers that are confident about an incorrect classification, the model learns that it is better to be somewhat wrong rather than completely wrong [10].

For the purpose of our work, we have selected the log loss function as our evaluation metric because it incorporates probabilistic confidence as a measurement of an algorithm's prediction accuracy. In binary classification settings, as in our paper, the log loss for each observation equals:

$$H\left(p,q\right) = -\sum p_i \log q_i = -y \log \hat{y} - (1-y) \log\left(1-\hat{y}\right) \qquad (1)$$

Here, y refers to the actual class label, while $\hat{y}$ represents the predicted class label. By doing so, we will be able to see the incremental changes in accuracy and determine the impact of each individual optimization technique, clearly determining which technique can be applied to operationalized models already integrated into a company's production process.

3   Data

Utilizing proprietary data from our industry partner, the database consists of tables that capture business information regarding their customers and their geographic location, company projects, lead sources, employees, and their offered product selections. We used the entirety of the data for the initial analysis, excluding the tables irrelevant to our given problem. A master query allowed us to work only on the latest 5 years of data in order to capture the variation within the data and incorporate specific business constraints, leading to a final dataset with over 3,425,646 observations. We also maintained a strict coverage threshold of 50% for the model's attributes.

In Table 2, we have grouped a total of 52 features into 7 categories, outlined below:

Table 2 Data used in study

Variable categories	Description
Lead ID	Primary key for the master table, unique identifier for a lead
Homeowner information	Zip code, city, state
Product	A list of the product estimates for the lead
Source information	Description of the which category or group the lead sources comes under
Datetime information	Date and time of lead call, date and time of the impending appointment, date and time when the lead was sourced
User performance	Prior performance of the salespersons involved in the lead process, based on the zip code and time periods
Conversion rates	Prior conversion rates in the area, prior leads in the area

4 Methodology

4.1 Data Exploration and Pre-Processing

1. *Data Cleaning and Health*

The initial stage of any machine learning project involves ensuring the relevance of your data to the current problem, identifying and treating outliers as well as any missing values. Our raw data contained junk values across tables that needed to be removed before the analysis. We filtered and excluded data points older than 5 years and removed features with less than 50% coverage to ensure results applicable to our industry partners and their problems during feature importance experiments.

2. *Exploratory Data Analysis*

Once we filtered our data to meet our requirements, we explored the data to understand the following:

- Distribution and fill-rate of features
- Correlations between predictors
- Associations between predictors and the target variable

4.2 Feature Engineering and Data Partitioning

1. *Hypothesis Development*

Once we developed our master dataset and have gained an understanding of the business impact associated with the features, we developed hypotheses to understand the internal and external properties that directly impact the relationship between our predictors and the target.

- H_1: Impact of *intrinsic* predictors on the conversion rate

  - Past employee performance
  - Employee experience
  - Appointment details
  - Past neighborhood performance

- H_2: Impact of *extrinsic* predictors on the conversion rate

  - Lead source information
  - Customer details
  - Preferred service details

2. *Custom Metric Creation*

During this phase of the project, we generated several features that could be directly inputted into our model as a predictor. They were engineered to address our hypotheses regarding factors that impact our target variable. Some of the metrics we created are:

- **Lead gap:** Difference in time when the lead was created to when the actual appointment was set.
- **Previous status:** Identifying information on whether the lead was a prospect approached previously or a customer already once converted.
- 6- and 12-month conversion rates split by geography
- 6- and 12-month records of employee performance handling that lead in that location.

3. *Data Partition*

In order to perform tests on the model's predictive power, we split our data 80/20% into a training and testing set, respectively. To allow our model to learn better and given that the target class populated only 28% of the times in the original data, we knew we needed to downsample the data from our majority class to have the same number of observations of both the target and non-target class. Even after downsampling, we were still left with nearly 1.5 million observations for our training dataset.

4.3 Model Building, Evaluation, and Ensemble Creation

1. *Base Model Training*

Given the properties of our data, we chose 7 initial classifiers to fit to our data:

1. Logistic regression
2. Random forest
3. Multilayer perceptron (MLP) artificial neural network

4. 4x Boosting models (Gradient boosting machine, Light Gradient Boosting, CatBoost, and XGBoost)

2. *Probability Scoring*

We use the log loss model to estimate the performance of the classification model to gauge incremental improvements.

3. *Ensemble Creation*

Based on the algorithm's log loss score, we selected the four best-performing models to ensemble. In this case, the **XGBoost**, **Light Gradient Boosting**, **Cat-Boost**, and the **artificial neural network** had both the lowest log loss values as well as the highest AUC.

5   Models

As explored in the Literature Review section, there is no deterministic choice when it comes to selecting the right classifier. Rather, it is important to understand under which conditions an algorithm would outperform others, whether this is the quality and type of data or the type of application problem. The nature of a binary classification problem combined with the special attributes of our partner's proprietary data meant that we had several choices for technique, detailed below.

5.1   Logistic Regression

A statistical method borrowed for machine learning problems, logistic regression, is used when our target variable is categorical and is often considered the go-to method for binary classification problems. A logit model estimates the probability of the binary target using a linear combination of the predictor variables as arguments of the sigmoid function, outputting a number between 0 and 1 – given a threshold that establishes which values belong to class 1 and to class 0 [11]:

$$y = e^{b_0 + b_1 x} / 1 + e^{b_0 + b_1 s} \tag{2}$$

Similar to linear regressions, input values (x) are linearly combined using weights or coefficients (represented by b) to predict an output value (y). The main difference between them relates to their modeled output, with logistic regression being categorical rather than continuous [12].

5.2   Random Forest

Random forests are a type of learning technique for classification that operates by constructing a multitude of decision trees with low correlation that operate as a group, with each tree classifying an observation into a class. The class which ultimately has the greatest number of votes across the forest is used to classify the observation [13]. It uses bagging and feature randomness when building each individual tree to create an uncorrelated forest of trees whose collective prediction is more accurate than that of any individual tree. The basic premise behind their success is that as a collective group, the trees can protect each other from their individual errors [14].

There are many different evaluation metrics for tree-based models, but the most common and the one we have chosen to use for the purposes of our research is the Gini Impurity measure, which calculates the probability that a randomly chosen sample in a node would be incorrectly labeled if it was labeled by the distribution of the sample in that given node:

$$I_G(n) = 1 - \sum_{i=1}^{J} (p_i)^2 \tag{3}$$

At each node, the decision tree searches through the features for the value to split on that results in the greatest reduction in the Gini coefficient, repeating the splitting process in a greedy, recursive procedure until its pre-assigned maximum depth, or if each node contains only samples from a single class [15].

5.3   Artificial Neural Network

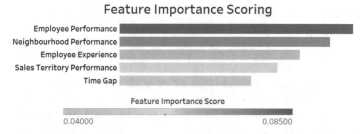

A multilayer perceptron (MLP) is a class of artificial neural network that can be viewed as a logistic regression classifier. The input is first transformed using a learnt non-linear transformation and are composed of at least three layers [8]: the input layer (receives the signal), the output layer (to make a prediction about the input), and an arbitrary number of hidden layers in between them (the computational engine of the neural network). They train on a set of input-output pairs to model the correlations between the pairs, tuning the parameters by adjusting the weights and biases of the model to minimize errors [16]:

$$y = \varphi \left(\sum_{i=1}^{n} w_i x_i + b \right) = \varphi \left(w^T x + b \right) \tag{4}$$

In the equation above, the perceptron produces a single output based on several real-valued inputs by forming a linear combination using its input weights (whose vectors are x and w, respectively) and sometimes passing the output through a nonlinear activation function – represented by φ. Our model in particular has four hidden layers with nodes sized 120, 60, 30, and 10, respectively. All nodes have the RELU (rectified linear unit) activation function.

5.4 Boosting Models

The term "boosting" in machine learning refers to a family of algorithms that focus on training a series of weak learners to convert them into strong learners [17], rather than traditional algorithms that focus on high-quality predictions done by a single model. For the purposes of our study, we focused particularly on these models: Categorical Boosting (CatBoost), Extreme Gradient Boosting (XGBoost), Light Gradient Boosting (LGBoost), and Gradient Boosting Machine (GBM). Despite their similarities, each algorithm boasts their own pros and their own cons, including speed, accuracy, or ability to handle outliers and missing data, further reiterating the importance of validation to determine the appropriate – and the most effective – model for the given problem.

5.5 Ensemble Model

Ensemble models – a collection of several models working together on a single dataset – tend to be reliable and accurate due to the diversity of the models that are contained within them. There are two main ways to determine the final prediction of the ensemble: hard vs. soft voting. Hard voting is when a model is selected by the ensemble to provide the final prediction by a simple majority vote for accuracy. On the other hand, soft voting can only be done when classifiers calculate probabilities for the outcomes, relevant to our probability scoring metric [18]. Soft voting arrives at the best result by averaging out the probabilities calculated by individual algorithms.

6 Results

6.1 Feature Importance

Despite our data source coming from within the home improvement space, we paid particular attention to converting our results into industry-agnostic insights. To do this, we group our features into buckets that were more applicable outside our partner company, rather than highlight individual variables.

In particular, we found both **employee performance** and **employee experience**, as well as **neighborhood performance** to be the biggest contributors to a lead conversion.

It is interesting to note that amongst the top 5 features that contribute ~33% toward the probability of the final predicted class are **all extrinsic features**. This is particularly significant as these are the factors that companies have a higher level of control over, giving our partner company and any other businesses hoping to maximize their conversion rates a large scope for optimization and overall improvement.

6.2 Model

Once we had our models (Light Gradient Boosting, XGBoost, CatBoost, the MLP, and the ensemble), we calculated log loss values to study the improvements across each tuning method. At each stage of the process, the values decrease, highlighting the value gained directly from each technique (Table 3).

Throughout our research, our ensemble model consistently performed the best, with a **final log loss value of 0.5784**. However, one very surprising finding was the performance of our artificial neural network, outperforming the ensemble after incorporating our engineered metrics into the model – with a log loss score of 0.5832 compared to the ensemble's score of 0.5897 – a relatively significant difference. However, after the final accuracy improvement technique of hyperparameter tuning, the neural network's final log loss was 0.5785, falling short of the best-performing log loss score by just one unit.

This further emphasizes the need to test various models before making a conclusive decision on which supervised learning algorithm should be used to

Table 3 Incremental log loss improvements across models

	LGB	XGB	CB	ANN	E
Base model	0.8452	0.8524	0.8298	0.8269	0.8183
Downsampling	0.6531	0.6672	0.6547	0.6535	0.6525
Feature engineering	0.5927	0.6045	0.5938	0.5832	0.5897
Parameter tuning	0.5781	0.5871	0.5789	0.5785	0.5784

address a given problem. In addition, it is clear that the marginal improvement from the base model toward the later stages of improving accuracy declines sharply, with an average decrease of just under 0.03 across the models. On the other hand, the decrease in log loss after performing feature engineering experts across models reached an overage of 0.3, over **10x more significant than the impact of hypertuning**. This is because the amount by which log loss can decrease is constrained, while increases in the log loss value are unbounded [10].

7   Conclusion

Throughout our research project, our objective was to build and improve upon supervised learning predictive models by comparing their probability score calculated through the log loss function. In doing so, we hope to provide insights to businesses who have already operationalized a base model and help to develop heuristics to improve those models already in production. Our focus for this research project was to gain industry-agnostic insights that can be applied to any binary classification problem that requires a conversion to occur.

Our exploratory analysis found that the top 5 most important features, accounting for approximately 33% of the probability of the prediction, are all **extrinsic factors** – which are the features that a company has more control over, including **employee performance and experience**, and **time gaps** (between touchpoints in the sales funnel). With clearer insight into the drivers of the prediction, our results illustrate that optimizing these features would likely lead to boosted conversion rates.

Further, by evaluating our classifiers with the log loss probability scoring method, we are able to see the incremental improvements in accuracy after each improvement technique is applied to the model. From our results, it is clear that **downsampling our data** and **feature engineering** both significantly improved the model's predictability, reducing log loss over **10x more** than through hyperparameter tuning. Across all the models, the ensemble came out as the clear winner, with a final log loss score of **0.5784**, followed very closely by our artificial neural network model – scoring **0.5785**.

These results tell us that while the ensemble has performed the best, there may be scenarios and cases where a neural network would make predictions quicker, and more accurately, further reiterating that there is no fixed model that will outperform others across all circumstances. Instead, finding the best solution depends on your data, as well as your model choice, what tuning methods have been applied, and what evaluation metric you are using to measure your success.

Although our research culminated in very relevant insights for both our partner company and other organizations looking to optimize their predictive classifiers, there is additional scope for improvement mainly relating to increasing the diversity of our data sources. Using more data on customer demographics, such as age or occupation, as well as data from competitors and other external sources, our partner company will be able to further improve the model as they operationalize it and integrate it into their current production process.

Acknowledgments We would like to thank both our industry partner and Professor Lanham for their continued support and guidance.

References

1. N. Prasasti, H. Ohwada, Applicability of machine-learning techniques in predicting customer defection, in *2014 International Symposium on Technology Management and Emerging Technologies* (Bandung, 2014), pp. 157–162
2. A. Keramati, R. Jafari-Marandi, M. Aliannejadi, I. Ahmadian, M. Mozaffari, U. Abbasi, Improved churn prediction in telecommunication industry using data mining techniques. Appl. Soft Comput. **24**, 994–1012., ISSN 1568-4946 (2014). https://doi.org/10.1016/j.asoc.2014.08.041
3. J. Nielsen, Conversion Rate: Definition as used in UX and web analytics, *Nielsen Norman Group*. [Online]. Available: https://www.nngroup.com/articles/conversion-rates/. Accessed 14 Mar 2020]
4. T. Vafeiadis, K.I. Diamantaras, G. Sarigiannidis, K.C. Chatzisavvas, A comparison of machine learning techniques for customer churn prediction. Simul. Model. Pract. Theory **55**, 1–9., ISSN 1569-190X (2015). https://doi.org/10.1016/j.simpat.2015.03.003
5. J. Pamina, J. B. Raja, S. S. Peter, S. Soundarya, S. S. Bama, M. S. Sruthi, Inferring Machine Learning Based Parameter Estimation for Telecom Churn Prediction, in *ICCVBIC*, eds. by S. Smys, J. Tavares, V. Balas, A. Iliyasu, 2019, Advances in Intelligent Systems and Computing, vol 1108. Springer, Cham
6. A.K. Ahmad, A. Jafar, K. Aljoumaa, Customer churn prediction in telecom using machine learning in big data platform. Big Data **6**, 28 (2019). https://doi.org/10.1186/s40537-019-0191-6
7. S.B. Kotsiantis, I. Zaharakis, P. Pintelas, Supervised machine learning: A review of classification techniques. Artif. Intell. Rev. **26**(3), 159–190 (2006)
8. V. Kotu, B. Deshpande, *Data Science: Concepts and Practice* (Morgan Kaufmann is an imprint of Elsevier, Cambridge, MA, 2019)
9. V. Vovk, The fundamental nature of the log loss function, in *Fields of Logic and Computation II Lecture Notes in Computer Science*, 2015, pp. 307–318.
10. A. B. Collier, Making Sense of Logarithmic Loss, *datawookie*, 14-Dec-2015. [Online]. Available: https://datawookie.netlify.com/blog/2015/12/making-sense-of-logarithmic-loss/. Accessed 18 Mar 2020
11. A. Urso, A. Fiannaca, M.L. Rosa, V. Ravì, R. Rizzo, Data mining: Prediction methods. Enc. Bioinforma. Comput. Biol., 413–430 (2019)
12. R. Vasudev, How are Logistic Regression & Ordinary Least Squares Regression (Linear Regression) Related?, *Medium*, 05-Jun-2018. [Online]. Available: https://towardsdatascience.com/how-are-logistic-regression-ordinary-least-squares-regression-related-1deab32d79f5. Accessed 25 Mar 2020
13. L. Breiman, Random forests. Mach. Learn. **45**(1), 5–32 (2001)
14. T. Yiu, Understanding Random Forest, *Medium*, 14-Aug-2019. [Online]. Available: https://towardsdatascience.com/understanding-random-forest-58381e0602d2. Accessed 22 Mar 2020
15. W. Koehrsen, An Implementation and Explanation of the Random Forest in Python, *Medium*, 31-Aug-2018. [Online]. Available: https://towardsdatascience.com/an-implementation-and-explanation-of-the-random-forest-in-python-77bf308a9b76. Accessed 18 Mar 2020

16. C. Nicholson, A Beginner's Guide to Multilayer Perceptrons (MLP), *Pathmind*. [Online]. Available: https://pathmind.com/wiki/multilayer-perceptron. Accessed: 14 Mar 2020
17. J. Brownlee A gentle introduction to the gradient boosting algorithm for machine learning. *Machine Learning Mastery*. Nov, 9. 2016
18. S. Howal, "Ensemble Learning in Machine Learning: Getting Started," *Medium*, 15-Dec-2018. [Online]. Available: https://towardsdatascience.com/ensemble-learning-in-machine-learning-getting-started-4ed85eb38e00. Accessed 19 Mar 2020

Ensemble Learning for Early Identification of Students at Risk from Online Learning Platforms

Li Yu and Tongan Cai

1 Introduction

Online learning platforms are changing the way people acquire knowledge. Instead of going to a classroom, students nowadays can get access to online study materials anytime and anywhere by taking massive open online courses (MOOCs). Take Coursera as an example; Coursera is now the world's largest online learning platform for higher education, with more than 200 of the world's top universities and industry educator partners offering courses, certificates, and degrees. It has had over 45 million learners around so far.[1] Another online learning platform, the Open University, is one of the largest universities in Europe with 174,898 students. More than 2 million people worldwide have been learning with the Open University.[2]

As [14] suggested decades ago, online learning is superior for distributing course materials efficiently and granting easy access to students without space or time constraints, yet the online learning platform can cause students lack of support from the instructor—the instructors may overlook students in need. Since the online courses often have hundreds or thousands of students in one session, it is of great importance for instructors to identify students who are at risk of failing and withdrawing timely and accurately.

We would like to leverage some novel statistical and machine learning ideas to identify students who are at risk of failing and withdrawing in an online study

---

[1] https://about.coursera.org/press.

[2] http://www.open.ac.uk/about/main/strategy-and-policies/facts-and-figures.

L. Yu · T. Cai (✉)
The Pennsylvania State University, University Park, PA, USA
e-mail: luy133@psu.edu; cta@psu.edu

© Springer Nature Switzerland AG 2021
R. Stahlbock et al. (eds.), *Advances in Data Science and Information Engineering*,
Transactions on Computational Science and Computational Intelligence,
https://doi.org/10.1007/978-3-030-71704-9_35

platform. With achieving accurate identification of such students, we would also like to identify them at an early stage of the course. Specifically, we validate our framework on the Open University Learning Analytics Dataset (OULAD) [6]. Comparative results show that the proposed method can accurately identify students at risk at an early stage.

The following paper is organized as follows: Sect. 2 reviews both early and recent related works in this domain, Sect. 3 introduces in detail about the proposed method, and Sect. 4 gives a comprehensive presentation and analysis of our experiment and result. A conclusion is then drawn and future possibilities are discussed.

2   Related Works

Although online learning platforms have been running since the 1970s, learning analytics have not shed much light on it until recent years. Traditional learning analytic works like [11] focus on cognitive aspect of learning, including learning patterns and decision approaches, but online learning analytics favors a data-driven way. In 2015, the Knowledge Discovery in Databases (KDD) Cup was held on a dataset collected on a massive open online course (MOOC) in order to predict the dropout (withdrawal) of students. The data contains students' daily activity logs. Li et al. [9] adopts a multi-view semi-supervised learning on behavior features, [10] utilizes a clustering method on learning events to stratify student with different levels of activity, and [15] introduces an optimization of a joint embedding function to represent both students and course elements into a single shared space. While some of them achieves nearly 90% of classification accuracy, further improvement was limited due to the size of the dataset.

In 2017, a complete version of learning analytics dataset was published by the Open University [6] with richer features. This dataset is soon studied by various works. Rizvi et al. [13] uses a decision tree method to study the relation of demographics information to student performance. Hassan et al. [3] considers the temporal continuity in the data, models it as a time series, and deploys a long short-term memory (LSTM) deep learning model for the task. Both of them reported around 95% classification accuracy. Kuzilek et al. [8] and Peach et al. [12] adopt clustering idea to stratify students into groups with different risk levels. Heuer and Breiter [4] considers a binary feature of student daily activities and first assessment result to classify the student of pass/fail, which achieves 93% precision rate. There are also works using Gaussian mixture model [1] and Markov chain [7]. However, these works use the full dataset. Considering that withdrawn students have clearly missing activity logs during a later time period, the results of them are arguable. Some works start to model an "early identification" process by slicing the features into weeks, intervals, or portions. Hlosta et al. [5] developed a self-learning framework for incremental training on the data to achieve early identification, which achieves similar result to those with legacy data. Haiyang et al. [2] adopts a time

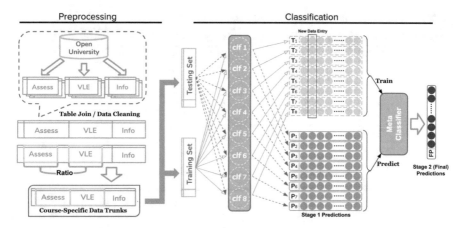

Fig. 1 Flow of the proposed framework

series forest on the activity data for student withdrawal prediction using different percentages of data.

We argue that previous works either missed the significance of course progress modelling or didn't use all data available (activity, assessment, demographics). Specifically, we would like to present a way to fully use the rich data available and also formulate the course progress to achieve early identification. We also propose a novel ensemble classification model for the task.

3 Method

The proposed framework can be stratified into two stages. As shown in Fig. 1, there are a processing stage for data engineering and a classification stage for decision-making. The following sections will elaborate on the details of data acquisition and analysis, information fusion and ratio formulation, choice of base classifiers, and the stacking ensemble method.

3.1 Data Acquisition and Analysis

As aforementioned, the dataset we use is the OULAD released by Open University in 2017. It is a comprehensive dataset that includes 22 courses and the learning analytics data of 32,593 students. The courses on OU are organized as modules and modules can be presented multiple times in a year. There are a total of 7 courses and 22 presentations of them. As can be seen in Table 1, each presentation is named as the year and the order of month it was presented. For example, 2013A means that

Table 1 Summary of presentations and students' performance

Module	Presentation	Distinction	Pass	Withdrawn	Fail	Total
AAA	2013J	20	258	60	45	383
	2014J	24	229	66	46	365
BBB	2013B	155	648	505	459	1767
	2013J	176	896	644	521	2237
	2014B	166	561	490	396	1613
	2014J	180	972	749	391	2292
CCC	2014B	192	471	898	375	1936
	2014J	306	709	1077	406	2498
DDD	2013B	54	456	432	361	1303
	2013J	98	731	681	428	1938
	2014B	119	360	490	259	1228
	2014J	112	680	647	364	1803
EEE	2013J	127	482	243	200	1052
	2014B	72	285	173	164	694
	2014J	157	527	306	198	1188
FFF	2013B	118	664	411	421	1614
	2013J	187	908	675	513	2283
	2014B	107	547	462	384	1500
	2014J	258	859	855	393	2365
GGG	2013J	141	451	66	294	952
	2014B	128	350	100	255	833
	2014J	127	317	126	179	749

the module was presented in January of 2013, whereas 2013J indicates the starting month to be October. The number of enrolled students for each module presentation ranges from several hundreds to a few thousands. Students in each module have final results as either "Distinction" and "Pass" or "Withdrawn" and "Fail," and the numbers of students ending with each of the four results are 3024, 12361, 7052, and 10156, respectively. It can be seen that the number of Distinction/Pass almost equals to that of Fail/Withdrawn.

The learning analytics data of students in OULAD can be categorized into three parts. The first part is the demographics of students which include the basic information such as gender, region, education levels, age, etc. The second part is their assessment results on tests, exams, and finals over the course time. Their assessment is evaluated as scores from 0 to 100. The third part is the clickstream recordings (10,655,280 entries) of students on each day of a given module presentation. It is the logs of interactions students made with the virtual learning environment (VLE). We average the number of clicks on each day of the module AAA 2013J for the four different final results and plot them in Fig. 2. It is clear that students who managed to pass (Distinction/Pass) the course tended to be more active than those who failed or withdrew, in terms of daily clicks on VLE.

Fig. 2 Average per-day clicks

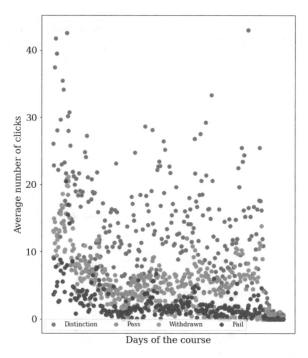

Those who finished with distinction also had more clicks than those who just passed. One interesting observation is that the number of clicks from the withdrawn category gradually diminished and reduced to zero toward the end of the course, which agrees with our expectation that students would not access the course materials after they dropped out.

3.2 Information Fusion and Ratio Formulation

OULAD is presented as a collection of tabular data linked with identifiers like student IDs and presentation codes. Figure 3 shows a typical structure of one module presentation. We can see that the VLE data of students can be retrieved from the tables studentVle and vle on site ID, which is the ID of sites belonging to 1 of the 20 activity types such as homepage, resource, and forum on VLE. Similarly, assessment results of students are the combination of table studentAssessment and assessments on assess ID, which is the ID of assessments such as tests, exams, or finals. We have to join the tables in order to get a full representation of student's learning statistics on one of the three categories: Info, VLE, and Assess. During the joining, we find that some assessment and VLE data are missing for students who dropped even before the class started, so their data are filled with zeros. For the demographics, we perform one-hot encoding on each columns of students' information since they are

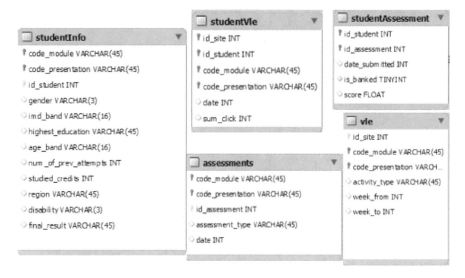

Fig. 3 Data schematic of a typical course

basically categorical data. We also exclude the info of IMD band due to the large portion of missing values.

As mentioned previously, our goal is to identify students at risk of failing or withdrawing a course at an early stage. To accommodate for this, we extract the VLE and Assess data of students based on a ratio of the module presentation length. The ratio starts at 5% and increments every 5% till it reaches 100%. Take AAA 2013J as an example; it spans a total of 268 days, and we choose the first 134 days of VLE and Assess data if the ratio is 50%. The demographics data is used as a whole because it won't change during the course presentation. This gives us a ratio-dependent student learning statistics so that we can perform classification at different stages of a course and identify students at risk before the course ends.

3.3 Base Classifiers

We choose eight base classifiers with different characteristics and decision boundaries and tune the hyperparameters individually on our extracted dataset. They are:

- K-nearest neighbor (KNN) classifier with 185 neighbors
- Random forest (RF) classifier with 170 estimators
- Support vector machine (SVM) with RBF kernel
- Multi-layer perceptron (MLP) classifier
- Logistic regression (LR) with L2 penalty

- Bernoulli naive Bayes (BNB) classifier
- Gradient boosting decision tree (GBDT) classifier with maximum depth of 4
- XGBoost (XGB) classifier with default hyperparameters

3.4 Stacking Ensemble

Stacking (meta) ensemble method was proposed by David H. Wolpert in 1992 [16] as "stacked generalization," but it is until recent years that stacking ensemble method is widely used in machine learning tasks. The idea of stacking ensemble method is intuitive: determine the final decision based on the results given by multiple base classifiers. It is expected that the stacking ensemble result should be better (at least no worse) than any one of the base classifiers. First, each base classifier is trained in a cross-validated fashion so that every data entry gets a preliminary classification result as probability ranging from 0 to 1. The results from multiple classifiers are then regarded as new features and stacked as a new data entry. The new data entries are then feed forward to a second stage "meta" classifier for a final classification decision.

In this proposed framework, the stacking ensemble model consists of the eight base classifiers mentioned and has one logistic regression classifier as the "meta" classifier. Each of the hyperparameters of the base classifiers is tuned on the full dataset using fivefold cross-validation. For the validation of the stacking ensemble model, we split the data into test and train sets.

4 Experiment and Result

4.1 Deployment and Experimental Settings

We deploy our proposed ensemble model with Python scikit-learn's implementation of nearest neighbor, random forest, SVM, neural network (MLP), logistic regression, and Bernoulli NB classifiers and XGBoost package. When performing hyperparameter tuning, we construct fivefold validation with StratifiedKFold cross-validation method to make each fold to resemble the data distribution of the whole dataset. During validation phase, the training and testing data is split randomly with ratio 80–20%. We run our proposed method using a quad-core computer.

Table 2 Confusion matrix

	Predicted	
Actual	Negative	Positive
Negative	True negative (TN)	False positive (FP)
Positive	False negative (FN)	True positive (TP)

4.2 Performance vs. Percentage of Data Used

The performance metrics we selected are accuracy, precision, and recall. In this binary classification problem, a confusion matrix can be formulated as Table 2.

Where we note

$$\text{Accuracy} = \frac{\text{TP} + \text{TN}}{\text{TP} + \text{FP} + \text{FN} + \text{TN}}$$

$$\text{Precision} = \frac{\text{TP}}{\text{TP} + \text{FP}} \qquad \text{Recall} = \frac{\text{TN}}{\text{FN} + \text{TN}}$$

We treat each module presentation separately because they have different length of days. The curves of prediction accuracy, precision, and recall with respect to the ratio of data being used are plotted as follows, with Fig. 4a showing the prediction accuracy, Fig. 4b the prediction precision, and Fig. 4c the prediction recall rate.

The proposed framework can achieve very high classification accuracy, since the accuracy rate for the courses can converge to about 90–95%. The framework is also able to identify the students at risk accurately in a relatively early stage. Around the middle of the course progress, the accuracy rate rises to higher than 85%, which means it can identify students who are at risk of failing or withdrawing (or have withdrawn) prematurely. For validation, we also included the precision and recall curves to support our observation.

We also plot the ROC curves with respect to different stages of the courses. Three ratios of data are chosen as milestones of typical course presentations. They are, respectively, 40% (Fig. 4d), 70% (Fig. 4e), and 100% (Fig. 4f). With 40% data incorporated, the AUC score for each of the course is already reasonable, with most curves showing higher than 0.8 scores. About half of the cases achieve 0.95 AUC score with 70% data used, indicating a convincing prediction is fulfilled at 70% time of course period. It is not surprising that perfect left-cornered ROC curves are spotted if all data are used, when the course has finished and results are released. From the analysis of accuracy and ROC over different timestamps of the course presentations, we validate that the proposed framework is effective to identify students at risk, at a relatively early stage.

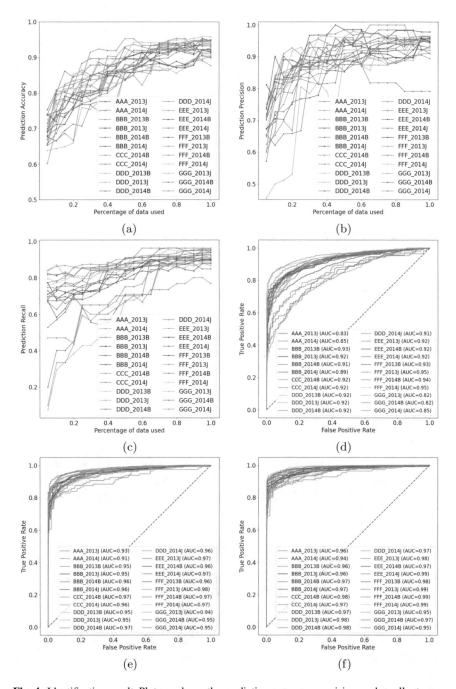

Fig. 4 Identification result. Plot a–c shows the prediction accuracy, precision, and recall rate over incremental amount of data. Plot d–f shows the ROC curve using 40, 70, and 100% data. (**a**) Accuracy vs. percentage of data used. (**b**) Precision vs. percentage of data used. (**c**) Recall vs. percentage of data used. (**d**) ROC curve using 40% of data. (**e**) ROC curve using 70% of data. (**f**) ROC curve using all of data

Table 3 Comparison of ensemble with base classifiers

Model	Accuracy	Precision	Recall
Nearest neighbor	0.8605	0.8006	0.8929
Gradient boosting	**0.9253**	0.9397	**0.9056**
XGBoost	0.9237	0.933	**0.9078***
Random forest	**0.9293***	**0.9653***	0.8935
SVM	0.9035	**0.9437**	0.8643
Neural network	0.908	0.9079	0.8982
Logistic regression	0.8991	0.8736	0.9092
Bernoulli NB	0.894	0.8928	0.8838
Stacking ensemble	**0.9293***	**0.9494**	**0.9056**

*denotes the most significant results.

4.3 Base Classifiers vs. Stacking Ensemble

We compare the stacking ensemble model with the base classifiers. The comparison is based on the full dataset with each base classifier tuned on separate classification task. Top values are marked, and top 3 for each metric are blackened and shown in Table 3. There are very strong classifiers like random forest, XGBoost, and gradient boosting that achieve very good result on one or more metrics. The stacking ensemble seems to be seeking a balance between accuracy, precision, and recall, achieving the "generally" best result for the classification.

4.4 Comparison with Related Works

A comparison of our proposed method with related works on the same dataset is shown in Table 4. For the first two works in the table [2, 3], they only considered the VLE files and excluded the fail cases. One thing worth noting here is the way we calculate accuracy, precision, and recall. We calculate and average them over all the 22 module presentations, whereas the accuracy reported in [2, 3] is the highest score among the 22 presentations. Considering the long sequences of inactivity in later period of a class for the withdrawn cases, their result can be trivial to get. The third work worked on only the demographic data, and the fourth one only worked on binarized daily VLE features. Both of them excluded the withdrawn cases. The fifth work validates the method only on a specific course and may not be suitable for all the courses. Compared to the related works, we can conclude that the proposed method shows improvements over the previous works in terms of (1) a better coverage of cases and features and (2) a globally good result on all the course data.

Table 4 Comparison with related works

Work	Method	Accuracy	Precision	Recall	Note
[2]	Time series forest	0.9399	–	–	VLE, excl. "fail"
[3]	LSTM	0.9725	0.9279	0.8592	VLE, excl. "fail"
[13]	Decision tree	<0.85	<0.87	<0.35	Demographics, excl. "withdrawn"
[4]	DT/RF/LR/SVM	0.8797	0.9333	0.8950	Binary VLE, excl. "withdrawn"
[1]	Mixture model	0.9200	–	0.9350	Specific course
Ours	Stacking	0.9293	0.9494	0.9056	–

Table 5 Comparison of ensemble with base classifiers

Model	Accuracy	Precision	Recall
Nearest neighbor	0.8605	0.8006	0.8929
SVM	0.9035	**0.9437***	0.8643
Neural network	0.908	0.9079	**0.8982**
Logistic regression	0.8991	0.8736	**0.9092***
Bernoulli NB	0.894	0.8928	0.8838
Reduced model	**0.9143***	**0.9249**	**0.8967**

*denotes the most significant results.

5 Conclusion and Discussion

In this project, we present an ensemble learning method for the early identification of students at risk of failing and withdrawal in an online learning environment. The proposed information fusion utilizes as much available information as possible by formulating the student activity data as daily numerical features and combining with the assessment results for the corresponding time period. The student demographics information is also considered as categorical features and included in training. The classification module utilizes various classifiers of different types in order to improve the robustness and adaptability of the model.

The proposed method is validated on the Open University Learning Analytics Data and achieves great accuracy in identifying the students at risk. It achieves 94.94% classification precision and maintained 90.56% recall for the identification of students at risk using full available data. The proposed method also achieves the goal of "early identification" that achieves higher than 85% accuracy with only half of the data incorporated, indicating that the proposed framework can correctly identify students who are at risk around the mid-term of the course.

We also recognize that there are some weaknesses in this project. First, the proposed stacking ensemble is not presenting a drastic improvement from the strong XGBoost and random forest classifiers. With limited data, the base classifiers can be good enough to do the job. A trial is made ruling out the XGBoost, gradient boosting, and random forest classifier. The overall result of classification decreased, but the ensemble model shows the improvement from the baselines (Table 5).

Another problem is that the proposed method, like other previous works, can't distinguish between the students who are likely to withdraw or who are likely to fail. Considering that the withdrawal cases have long sequences of inactive states, the identification of withdrawal is trivial if it is not required to also predict the actual withdraw timestamp of the student. Future works may try to work on early identification of the withdrawal and failing cases separately before it is too late and study the factors that result in the withdrawal/fail.

References

1. R. Alshabandar, A. Hussain, R. Keight, A. Laws, T. Baker, The application of gaussian mixture models for the identification of at-risk learners in massive open online courses, in *2018 IEEE Congress on Evolutionary Computation (CEC)* (IEEE, Piscataway, 2018), pp. 1–8
2. L. Haiyang, Z. Wang, P. Benachour, P. Tubman, A time series classification method for behaviour-based dropout prediction, in *2018 IEEE 18th International Conference on Advanced Learning Technologies (ICALT)* (IEEE, Piscataway, 2018), pp. 191–195
3. S.-U. Hassan, H. Waheed, N.R. Aljohani, M. Ali, S. Ventura, F. Herrera, Virtual learning environment to predict withdrawal by leveraging deep learning. Int. J. Intell. Syst. **34**(8), 1935–1952 (2019)
4. H. Heuer, A. Breiter, Student success prediction and the trade-off between big data and data minimization, in *DeLFI 2018-Die 16.E-Learning Fachtagung Informatik* (2018)
5. M. Hlosta, Z. Zdrahal, J. Zendulka, Ouroboros: early identification of at-risk students without models based on legacy data, in *Proceedings of the Seventh International Learning Analytics & Knowledge Conference* (ACM, New York, 2017), pp. 6–15
6. J. Kuzilek, M. Hlosta, Z. Zdrahal, Open university learning analytics dataset. Sci. Data **4**, 170171 (2017)
7. J. Kuzilek, J. Vaclavek, V. Fuglik, Z. Zdrahal, Student drop-out modelling using virtual learning environment behaviour data, in *European Conference on Technology Enhanced Learning* (Springer, Berlin, 2018), pp. 166–171
8. J. Kuzilek, J. Vaclavek, Z. Zdrahal, V. Fuglik, Analysing student vle behaviour intensity and performance, in *European Conference on Technology Enhanced Learning* (Springer, Berlin, 2019), pp. 587–590
9. W. Li, M. Gao, H. Li, Q. Xiong, J. Wen, Z. Wu, Dropout prediction in MOOCs using behavior features and multi-view semi-supervised learning, in *2016 International Joint Conference on Neural Networks (IJCNN)* (IEEE, Piscataway, 2016), pp. 3130–3137
10. Z. Liu, J. He, Y. Xue, Z. Huang, M. Li, Z. Du, Modeling the learning behaviors of massive open online courses, in *2015 IEEE International Conference on Big Data (Big Data)* (IEEE, Piscataway, 2015), pp. 2883–2885
11. T.J. Mock, T.L. Estrin, M.A. Vasarhelyi, Learning patterns, decision approach, and value of information. J. Account. Res. **10**(1), 129 (1972)
12. R.L. Peach, S.N. Yaliraki, D. Lefevre, M. Barahona, Data-driven unsupervised clustering of online learner behaviour (2019). arXiv preprint arXiv:1902.04047
13. S. Rizvi, B. Rienties, S.A. Khoja, The role of demographics in online learning; a decision tree based approach. Comput. Educ. **137**, 32–47 (2019)
14. R.W. Taylor, Pros and cons of online learning—a faculty perspective. J. Eur. Ind. Train. **26**(1), 24–37 (2002)
15. M. Teruel, L.A. Alemany, Co-embeddings for student modeling in virtual learning environments, in *Proceedings of the 26th Conference on User Modeling, Adaptation and Personalization* (ACM, New York, 2018), pp. 73–80
16. D.H. Wolpert, Stacked generalization. Neural Netw. **5**(2), 241–259 (1992)

An Improved Oversampling Method Based on Neighborhood Kernel Density Estimation for Imbalanced Emotion Dataset

Gague Kim, Seungeun Jung, Jiyoun Lim, Kyoung Ju Noh, and
Hyuntae Jeong

1 Introduction

One of the most common problems in data analysis or learning is imbalanced data
[1]. Specially, in the imbalanced dataset, recognition of minority class is often more
important than majority class. For example, in medical diagnosis, it may bring
serious consequences if a patient is diagnosed as a normal or a wrong disease.
In general, the classification performance of majority class is better than minority
class. Nevertheless, the classification of minority class is often more important as in
medical diagnosis. This problem occurs because traditional classification algorithms
have been developed to increase overall accuracy without considering imbalanced
data. For example, suppose there is a dataset with 3 defects out of 100 samples.
On this imbalanced dataset, even if all the defects are incorrectly recognized as
normal, the overall accuracy is still high at 97%. This is not a desirable approach in
imbalanced data in which the classification of minority class is more important.

Many studies have been conducted to solve imbalanced data problem [2, 3, 4,
5, 6, 7]. In recent years, approaches can be divided into two main categories. One
is an algorithmic level approach and the other is a data level approach [1]. The
algorithmic approach aims to learn so that the bias introduced by the class imbalance
be reduced and misclassification of minority class be avoided. On the other hand,
data level approach aims to quantitatively equalize minority and majority classes in
the training dataset by oversampling the minority class, undersampling the majority
class [2], or both. Which of the two approaches to choose usually depends on the
class distribution of imbalanced dataset. In general, the oversampling approach may

G. Kim (✉) · S. Jung · J. Lim · K. J. Noh · H. Jeong
Electronics and Telecommunications Research Institute, Daejeon, South Korea
e-mail: ggkim@etri.re.kr; schung@etri.re.kr; kusses@etri.re.kr; kjnoh@etri.re.kr;
htjeong@etri.re.kr

© Springer Nature Switzerland AG 2021   543
R. Stahlbock et al. (eds.), *Advances in Data Science and Information Engineering*,
Transactions on Computational Science and Computational Intelligence,
https://doi.org/10.1007/978-3-030-71704-9_36

be more effective on a dataset with a small size and a diverse distribution. The HRV (heart rate variability) dataset to be handled for emotion classification in this study is the case. Chawla [3] proposed SMOTE which interpolates between minority class samples by randomly resampling with original minority class. Hui Han [4] proposed the borderline SMOTE that generates synthetic minority samples only for a set of dangerous samples that may be classified as majority class, which is called Danger set. However, the SMOTE family may miss some valuable samples that are essential for classification. Haibo [5] proposed ADASYN (adaptive synthetic sampling) approach that synthesizes new samples in proportion to the density of the majority class samples around each minority class sample. However, critical minority samples selected by ADASYN may sometimes act as unnecessary noise that degrades classification performance. Chen [6] proposed a method for selecting boundary samples of support vector machine based on alien k-neighbors. However, this method generates a lot of redundancy, and may still misclassify samples that overlap on the boundary of the two classes. Wenhao [8] proposed the improved alien k-neighbors algorithm that applies oversampling only to cluster centers rather than to all minority class samples. This method can lead to distortion of the support vector because it considers the overlapping samples on the boundary of the two classes as noise samples and remove them. Lu [7] proposed an oversampling method based on the combination of probability density function estimation and Gibbs sampling to overcome the problems of the previous support vector machines. This method generates new samples based on the probability density function that follows the distribution of minority class. However, it may cause overfitting and lead to poor classification performance for minority class samples on the boundary of the two classes.

In this study, we overcome the existing oversampling problems by introducing kernel density estimation (KDE) based on statistics related to the density of majority class samples around each minority class sample. In other words, KDE allows oversampling technique to handle the probability that each minority class sample belongs to Danger set, not noise. Therefore, the proposed method is expected to show better performance than ADASYN on imbalanced dataset that contains some noise samples. In experiments, HRV (heart beat variability) dataset extracted from ECG (electrocardiogram) signal is used. This dataset is a typical imbalanced dataset that contains HRV samples for four emotion classes and has different amounts of samples in each class. To verify the efficiency of the proposed method, experiments are performed on this imbalanced HRV dataset.

2   Related Works

The oversampling technique aims to increase the number of samples so that the samples belonging to minority class can be well classified. The model trained on the resulting balanced dataset generally improves overall accuracy as well as recall rate of minority class.

2.1 Random Oversampling

Random oversampling is one of the earliest methods and is relatively easy to implement. This technique increases the number of minority class samples by randomly selecting them and then simply copying them. However, this method often results in overfitting, which degrades the classification performance for minority class samples around the border between classes. Therefore, a technique for controlling the distribution of synthetic samples around the border is required. SMOTE (synthetic minority oversampling technique) was developed to meet these realistic needs.

2.2 SMOTE (Synthetic Minority Oversampling Technique)

SMOTE [3] is one of the main oversampling techniques used to solve imbalanced data problems. This method generates synthetic samples by interpolating between each minority class sample and other minority class samples. The basic steps of SMOTE are as follows:

- *Step 1.* For each minority class sample x_i (i = 1, 2, $\cdots$, $|S_{min}|$) belonging to minority class set S_{min}, K-NN (k-nearest neighbors) is obtained by calculating the Euclidean distance between it and all other samples in S_{min}, where $|S|$ represents the size of the set S.
- *Step 2.* The number N of samples to be synthesized for each x_i is calculated depending on IR (imbalance ratio) between majority class set S_{maj} and minority class set S_{min}. From the K-NN obtained for each x_i, N samples $(x_{k1}, x_{k2}, \cdots, x_{kN})$ are randomly selected.
- *Step 3.* For each x_{kj} (j = 1, 2, $\cdots$, N), synthetic samples are generated as follows:

$$x_{new} = x_i + \delta \cdot \left\| x_i - x_{kj} \right\| \tag{1}$$

where δ represents a random number between 0 and 1.

2.3 Borderline-SMOTE

SMOTE leads to overgeneralization because synthetic samples are generated for random minority class samples without consideration the importance of samples on the boundary between two classes. To solve this problem, Borderline-SMOTE [4] controls the number of synthetic samples according to the following steps.

- *Step 1*. For each x_i belonging to S_{min}, m-NN which means m nearest neighbors around x_i is obtain, where the nearest neighbors are regardless of S_{min} and S_{maj}, while SMOTE deals only with S_{min}.
- *Step 2*. Among minority class samples, x_i satisfying $\frac{m}{2} \leq N_h < $ m is chosen, where N_h is the number of samples belonging to S_{maj} among m-NN samples. The resulting set of x_i is called Danger set.
- *Step 3*. For samples belonging to the Danger set, synthetic samples are generated according to the SMOTE algorithm. If all m-NN samples belong to S_{maj}, x_i is considered as noise, and a synthetic sample is no longer generated.

2.4 ADASYN (Adaptive Synthetic Sampling)

ADASYN [5] is also one of the adaptive techniques to solve the overgeneralization problem of SMOTE. ADASYN is a method to more formally adjust the number of synthetic samples to be generated depending on the density of majority class samples around each minority class sample. The basic steps of ADASYN are as follows:

- Step 1. The number N of synthetic samples to be generated for S_{min} and S_{maj} is calculated as follows:

$$N = \alpha \cdot \left(\left| S_{maj} \right| - \left| S_{min} \right| \right) \tag{2}$$

where α is between 0 and 1 to balance the amount of data between two classes after resampling.
- Step 2. K-NN is obtained for each x_i, and $P_s(x_i)$ indicating on the density of majority class samples around x_i is calculated as follows:

$$P_s(x_i) = \frac{N_h(x_i)}{Z}, i = 1, 2, \cdots, |S_{min}| \tag{3}$$

where $N_h(x_i)$ represents the number of samples belonging to majority class among K-NN samples of x_i and Z is a normalization constant to make $P_s(x_i)$ to be probability density function, that is, satisfying $\sum_{i=1}^{|S_{min}|} P_s(x_i) = 1$.
- Step 3. For each x_i, $N_{syn}(x_i)$, which is the number of synthetic samples to be generated, is determined as follows:

$$N_{syn}(x_i) = P_s(x_i) \cdot N \tag{4}$$

Then, synthetic samples are generated according to the SMOTE algorithm.

As shown in (4), the larger $P_s(x_i)$, the more synthetic samples are generated. In other words, the more samples belonging to S_{maj} around a minority class sample, the more synthetic samples are generated.

3 The Proposed Oversampling Method

In general, the balanced dataset obtained through oversampling does not have the same distribution as the original dataset. For this reason, some synthetic samples may actually act as noise in the learning process. In particular, minority class samples belonging to Danger set may be real samples or noise samples which degrade learning. So, the minority class samples in the existing Danger set need to be checked for real or noise. In this study, KDE (kernel density estimation) is introduced to deal with the probability of being the right sample belonging to Danger set. Then, depending on this probability, the number of synthetic samples to be generated is determined. The following Gaussian kernel function is used to achieve our goal:

$$P_d(x_i) = \frac{1}{Z} e^{-\frac{(x_i-\mu)^2}{2\sigma^2}}, i = 1, 2, \cdots, |S_{min}| \tag{5}$$

where μ and σ represent the mean and standard deviation of $N_h(x_i)$ with $i = 1, 2, \cdots, |S_{min}|$, respectively, and Z is a normalization constant such that $P_d(x_i)$ becomes probability density function similarly in ADASYN. In the proposed method, $N_{syn}(x_i)$ which is the number of synthetic samples to be generated for each x_i is determined as follows:

$$N_{syn}(x_i) = P_d(x_i) \cdot N \tag{6}$$

Now it's time to train the classifier on the resulting balanced dataset. Figure 1 shows the process of classifying samples in imbalanced dataset with the proposed oversampling approach and evaluating the performance of classification.

As shown in Fig. 1, after the oversampling process for the minority classes, a balanced training set is obtained. To train the balanced training set, various machine learning algorithms can be used. In this study, KNN (k-nearest neighbors), one of the major machine learning algorithms, is used. Finally, classification performance is evaluated on the validation set.

4 Experimental Setup

To evaluate the performance, the proposed method will be compared with the other existing oversampling techniques, including those obtained without oversampling process.

Fig. 1 Classification using proposed oversampling method

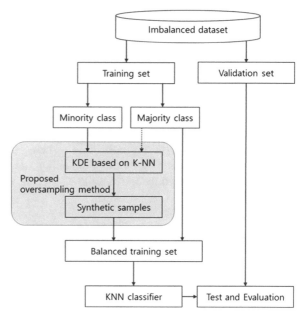

Table 1 The selected HRV features

Domain	Features	Definition	Unit
Time domain	meanNN	Mean of all NN intervals	ms
	SDNN	Standard deviation of all NN intervals	ms
	RMSSD	The square root of the mean of the sum of the squares of differences between adjacent NN intervals	ms
Frequency domain	VLF	Power in very low frequency range (0–0.04 Hz)	ms2
	LF	Power in low frequency range (0.04–0.15 Hz)	ms2
	HF	Power in high frequency range (0.15–0.4 Hz)	ms2
	LF/HF	Ratio of HF to HF	ms2

4.1 HRV Dataset for Emotion Classification

The imbalanced dataset to be used in this study is composed of HRV features extracted from ECG signals collected for emotion classification. The ECG signal was obtained using sensors attached to 30 voice actors' bodies after immersed in reading the sentences corresponding to the 4 emotions (Angry, Happy, Sad, Neutral). The collected ECG was extracted as HRV features in the time domain and frequency domain after cutting into 1-minute segments. Table 1 shows 7 HRV features used in emotion classification [9].

We obtained a total of 1735 segments from the collected ECG signal, and built HRV dataset by extracting HRV features from these segments. The HRV dataset was randomly split at a 7:3 ratio for training and validation, respectively. The amount of data per emotion class in the resulting HRV dataset is shown in Table 2.

Table 2 The amount of data per emotion class

Emotional class	Neutral	Happy	Angry	Sad	total
Training set	244	248	327	395	1214
Test set	105	118	139	159	521
Total	349	366	466	554	1735

Table 3 Confusion matrix

		Predicted	
		Positive	Negative
Actual	Positive	TP(true positive)	FN(false negative)
	Negative	FP(false positive)	TN(true negative)

As shown in Table 2, our HRV dataset is imbalanced dataset with a difference in the amount of data for each class. This difference comes from the fact that it takes a different amount of time to read the scripts for each emotion in immersive state. The IR (imbalance ratio), which means the proportion of the majority with respect to the minority, is about 1:1.05:1.34:1.59, which is not serious enough to be difficult to use in emotion classification. For this dataset, the proposed oversampling technique is applied to improve the performance of emotion classification.

4.2 Performance Measures for the Evaluation of Classification

In general, accuracy is used when dealing with classification performance. However, it is not reasonable to evaluate the imbalanced dataset by accuracy alone. This is because when the amount of minority class samples is too small, even if all the minority samples are classified as majority classes, the overall accuracy is still high. In addition, this kind of classifier cannot be used for applications where it is important to correctly classify minority class samples. Due this problem, many performance measures have been developed to deal with classification performance for imbalanced dataset. Among them, F-measure [10] is well known as an effective indicator for evaluating classification performance of imbalanced dataset [7, 8]. To illustrate performance measures, consider the confusion matrix shown in Table 3.

For the confusion matrix shown in Table 3, performance measures are defined as follows:

1.

$$precision = \frac{TP}{TP + FP}$$

2.

$$recall = \frac{TP}{TP + FN}$$

3.

$$\text{Accuracy} = \frac{(\text{TP} + \text{TN})}{(\text{TP} + \text{FP} + \text{TN} + \text{FN})}$$

4.

$$\text{F} - \text{measure} = \frac{\left(1 + \beta^2\right) \times \text{recall} \times \text{precision}}{\beta^2 \times \text{recall} + \text{precision}}$$

As shown in (4), F-measure is a measure that considers both precision and recall. It depends on β to decide whether to weight precision or recall. When β is 1, it means that recall is treated as importantly as precision. F-measure with $\beta = 1$ is called F1-measure which is commonly used for classification evaluation of imbalanced dataset.

5   Results and Discussion

The proposed oversampling method was applied to the imbalanced HRV dataset. For comparison, classification performance was also measured for 5 oversampling methods (random oversampling, regular SMOTE, borderline-SMOTE, borderline-SMOTE2, ADASYN). For performance evaluation, we calculated the four performance measures described above. Figure 2 compares recall rates to evaluate the performance for minority classes in the experimental results.

As shown in Fig. 2, the recall rates for "Happy" class and "Neutral" class, which are minority classes, have been roughly improved when the proposed method is applied. In particular, the recall rate for "Happy" class was greatly improved to 0.602 with the proposed method, while it was measured to be 0.492 with ADASYN. The improvement of the recall rates for these minority classes has led to an improvement in overall accuracy. In more detail, the overall accuracy by the proposed method was

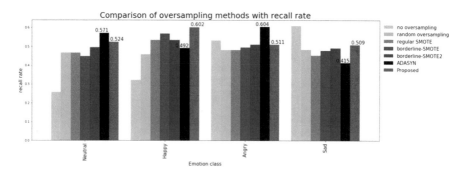

Fig. 2 Comparison of oversampling methods with recall rate

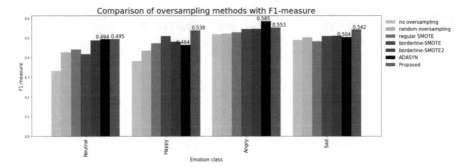

Fig. 3 Comparison of oversampling methods with F1-measure

0.534, showing improved performance compared to 0.514 of ADASYN. Figure 3 shows the performance compared with F1-measure.

As mentioned earlier, it is desirable to evaluate the classification performance for imbalanced dataset by F-measure. As shown in the Fig. 3, the proposed oversampling method brings a higher F1-measure than the other oversampling methods in "Happy" and "Neutral" which are minority classes. Also, considering F1-measures in "Angry" and "Sad," the performance is roughly improved over other oversampling methods. Compared with the average value for all classes, 0.532 of the proposed method is slightly higher than 0.512 of ADASYN. Therefore, the experimental results prove that the proposed method can improve the classification performance for minority class as well as overall accuracy.

6 Conclusion

In data analysis or recognition, imbalanced data is often encountered. Applying the existing classification algorithm to the imbalanced dataset, without modification, can reduce the recognition accuracy of minority class, which may cause serious problems in some applications where classification of minority class is important. In this study, we proposed a method to improve ADASYN, one of the existing adaptive oversampling techniques. It statistically deals with the probability that minority class samples belong to Danger set when deciding how many samples to generate. This prevents excessive resampling for minority class samples that may be noise samples. The experiments on the HRV dataset collected for emotion classification showed that the proposed method improved classification performance compared to the existing methods. In particular, the proposed method improved overall accuracy for all classes by increasing recall rate evenly for minority classes. In the future, in addition to the HRV dataset, we plan to demonstrate the effectiveness of the proposed method for imbalanced datasets that are frequently encountered in everyday life.

Acknowledgments This work was supported by Electronics and Telecommunications Research Institute (ETRI) grant funded by the Korean government. [20ZS1100], Core Technology Research for Self-Improving Artificial Intelligence System].

References

1. N. Rout, D. Mishra, M.K. Mallick, Handling Imbalanced Data: A Survey, in *International Proceedings on Advances in Soft Computing, Intelligent Systems and Applications* pp. 431–443, (2018)
2. J. Zhang, I. Mani, KNN approach to unbalance data distributions: A case study involving information extraction, in *Proceedings 12th Int. Conf. Machine Learning - Workshop on Learning from Imbalanced Datasets II* (Washington DC, USA, 2003), pp. 42–48
3. N.V. Chawla, K.W. Bowyer, L.O. Hall, W.P. Kegelmeyer, SMOTE: synthetic minority over-sampling 7 technique. J. Artif. Intell. Res. **16**(1), 321–357 (2002)
4. H. Han, W. Wang, B. Mao, Borderline-SMOTE: a new over-sampling method in imbalanced data sets learning, in *Proceedings of the International Conference on Intelligent Computing (ICIC'05)*, vol. 3644 of Lecture Notes in Computer Science, pp. 878–887, (2005)
5. H. He, Y. Bai, E.A. Garcia, S. Li. ADASYN: Adaptive synthetic sampling approach for imbalanced learning, in *IEEE International Joint Conference on Neural Networks (IEEE World Congress on Computational Intelligence)*, pp. 1322–1328, (2008)
6. C. Jingnian, H. Shunxiang, X. Li, Speeding up algorithm for support vector machine based on alien neighbor. Comput. Eng. **44**(5), 19–24 (2018)
7. L. Cao, Y.-K. Zhai, An over-sampling method based on probability density estimation for imbalanced datasets classification, in *Proceedings of the 2016 International Conference on Intelligent Information Processing*, (2016)
8. W. Xie, G. Liang, Z. Dong, B. Tan, B. Zhang, An improved oversampling algorithm based on the samples' selection strategy for classifying imbalanced data. Math. Probl. Eng. **3**, 1–13 (2019)
9. G. Kim, S. Jung, J. Lim, K. Noh, H. Jeong, ECG-based Emotion Classification using Optimized Machine Learning Techniques, in International Symposium on Advanced Intelligent Systems (ISIS) & International Conference on Biometrics and Kansei Engineering (ICBAKE), pp. 500–507, 2019
10. D. Powers, Evaluation: From precision, recall and F-measure to ROC, Informedness, Markedness and correlation. J. Mach. Learn. Technol. **2**(1), 37–63 (2011)

Time Series Modelling Strategies for Road Traffic Accident and Injury Data: A Case Study

Ghanim Al-Hasani, Md. Asaduzzaman, and Abdel-Hamid Soliman

1 Introduction

Road traffic accident (RTA) is one of the prime reasons for fatalities and disabilities globally. It has been considered as one of the significant health problems in terms of death and disability [1]. Over 50 million injuries occur, and more than 1.2 million people die in roadway-related accidents yearly. The RTA is going to be the fifth main cause of death in the world by 2030 [5]. Since over 50% of young adults, aged 15–44 years, die due to RTA, a significant economic impact is discernible through the loss of earning upon their families [8]. Moreover, road crashes including deaths and injuries cost from 1 to 2% ($100 bn) of the gross national product in low- and middle-income countries in addition to the total development aid received by these countries [7].

Although time series models for continuous variables have been well-studied, autoregressive moving average (ARMA) and autoregressive integrated moving average (ARIMA) models by Box and Jenkins (1970) have also been used to model count data recently [10]. However, finding an appropriate model for RTA and RTI time series data through suitable criteria and diagnostic checking are yet nontrivial tasks [3]. The major essential steps involved in time series modelling process are model specification, fitting and diagnostics of the model [2].

Due to the high concern of the government departments in major developed and mid-developed countries and the change of road safety policies, there has been a significant shift of trend of the number of accidents and injuries in recent years. Although there is a decline in the trend of the number of accidents in many Gulf

G. Al-Hasani · Md. Asaduzzaman (✉) · A.-H. Soliman
Staffordshire University, Stoke on Trent, UK
e-mail: ghanimalikhalfansalim.al-hasani@research.staffs.ac.uk; md.asaduzzaman@staffs.ac.uk; a.soliman@staffs.ac.uk

© Springer Nature Switzerland AG 2021
R. Stahlbock et al. (eds.), *Advances in Data Science and Information Engineering*,
Transactions on Computational Science and Computational Intelligence,
https://doi.org/10.1007/978-3-030-71704-9_37

countries including Oman, the severity and number of injuries have been found yet to be significant. Recently a substantial decrease in the number of injuries observed for Oman. A number of accidents and injuries are high volatility, there are also shifts of trends in accidents and injuries over time, and they possess high seasonality, which makes the analysis, model selection and forecasting complex tasks. In this paper, we study the times series model identification from a set of models, performed suitable diagnostic checks for the case study datasets on RTA and RTI in Oman and forecasted accidents and injuries for the next 2 years. The rest of the paper is organised as methodology in Sect. 2, data description in Sect. 3, results and discussion in Sect. 4 and some concluding remarks in Sect. 5.

2 Methodology

Time series is a sequence of values of a variable recorded over time, most often at a regular time interval. Time series are usually decomposed into four components: trend (T_t), cyclical pattern (C_t), seasonal variation (S_t) and random error (I_t). A time series can be expressed by an additive model defined as

$$X_t = T_t + C_t + S_t + I_t,\tag{1}$$

which can be used when the variation around the trend does not vary with the series. The multiplicative model defined as

$$X_t = T_t \cdot C_t \cdot S_t \cdot I_t,\tag{2}$$

is appropriate when the trend is proportional to the series. Often graphical approach (plot of a series) is used to identify whether a time series is additive or multiplicative.

There are three parts in the ARIMA model: AR is the autoregressive part, I is the differencing part, and MA is the moving average part.

An ARIMA (p, d, q) model for a time series sequence $\{X_t, t = 1, 2, \ldots, n\}$ can be written as

$$\phi(B)(1 - B)^d X_t = \theta(B)A_t,\tag{3}$$

where p is the order of the AR process, d is an order of differences, q is the order of the MA process, A_t is the white noise sequence, ϕ is a polynomial of degree p, B is a backshift operator and θ is a polynomial of degree q.

However, an ARIMA model could not analyse time series with seasonal characteristics; therefore, seasonal autoregressive integrated moving average (SARIMA) models have been developed [12]. SARIMA models perform better than the historical average, linear regression and simple ARIMA models for data with seasonal variations. In fact, SARIMA models are capable of taking into account the trend and seasonality. A SARIMA$(p, d, q)(P, D, Q)_s$ model can be written by the following equation:

$$\phi(B)\Phi(B^s)(1 - B^s)^d X_t = \theta(B)\Theta(B^s)A_t, \tag{4}$$

where Φ, Θ, P, D and Q are seasonal counterparts of ϕ, θ, p, d and q, respectively, and s is the seasonality.

There are several approaches to fit time series models such as the least square method, method of moment and maximum likelihood method. However, choosing the most suitable model can be cumbersome. Several criteria such as Akaike information criterion (AIC) and Bayesian information criteria (BIC) have been used as an essential tool to choose the best model from a set of models [11].

Once a suitable model is fitted, diagnostic checking of the model is performed, which concerns evaluating the quality of the model. This study assesses time series models through three residuals, root mean square error (RMSE), mean absolute percentage error (MAPE) and mean absolute scaled error (MASE), which are frequently used in time series analysis. RMSE is the standard deviation of the residuals, defined as

$$\text{RMSE} = \left[\frac{\sum_{i=1}^{n}(x_{f_i} - x_{o_i})^2}{n} \right]^{1/2}, \tag{5}$$

where n is a simple size, x_{f_i} are the forecasted values and x_{o_i} are the observed values. The mean absolute percent error (MAPE) measures the size of the error in percentage terms. It is calculated as the average of the unsigned percentage error, as defined in the equation below

$$\text{MAPE} = \frac{1}{n} \sum_{i=1}^{n} \frac{|x_{f_i} - x_{o_i}|}{|x_{f_i}|} \times 100. \tag{6}$$

The mean absolute scale error (MASE) is used to compare models of a time series through scale-free for assessing forecast accuracy across series [4]. MASE is defined as

$$\text{MASE} = \frac{1}{n} \sum_{i=1}^{n} \left(\left| \frac{e_t}{\frac{1}{n-1} \sum_{i=2}^{n} |x_{o_i} - x_{o_{i-1}}|} \right| \right), \tag{7}$$

where $e_t = x_{o_i} - x_{f_i}$ and the outcome values are independent of the data scale. However, if the outcome value less than one, then it indicates better forecasting. Alternatively, when the MASE value is greater than one, that means, the forecast is worse for the data.

3 Data

The data of road traffic accident and injuries (RTA and RTI) in Oman are maintained by the Royal Omani Police (ROP), and 'Statistical Summary Bulletins' are published annually by the Directorate of Road Traffic as part of the ROP. Summary data are published by the National Centre for Statistics and Information (NCSI) in monthly reports, called 'Monthly Statistical Bulletin' in Oman.

The RTA and RTI data for this study have been collected from two sources: 'Statistical Summary Bulletins' from the Directorate of Road Traffic and 'Monthly Statistical Bulletin' from the National Centre for Statistics and Information (NCSI). The data cover all road accidents and injuries for the period of 2000–2016 published in [9]. Additional data from 2016 to 2019 have been collected from the 'Monthly Statistical Bulletin' of the National Centre for Statistics and Information (NCSI) [6]. As a result, we consider monthly time series data of road traffic accidents and injuries from January 2000 to June 2019.

4 Results and Discussion

The time series data in this study represent the number of monthly road traffic accidents (RTA) and injuries (RTI) in Oman from January 2000 to June 2019. The resulting data consist of a total of 234 observations for both RTAs and RTIs.

Over the past two decades, the incidence of RTAs in Oman has fallen from a high of 1283 RTAs in October 2001 to a low 156 in February 2019. The mean is 660 for RTAs with a standard deviation of 247.5. Similarly, RTIs varied from a high of 1273 in March 2012 to a low of 125 in December 2014 with a mean of 620 RTI and standard deviation 265. The decomposition of trend, seasonality and random error components are shown in Fig. 1.

Different SARIMA models have been fitted in R and compared using the values of AIC, BIC, RMSE, MAPE and MASE. The values of AIC, BIC, RMSE, MAPE and MASE for different SARIMA models for RTA are given in Table 1. While developing different models for RTA data, models with a first-order difference ($d = 1$) are considered, as the data found to be stationary at lag 1. The analyses indicate that the best model for the RTA data in Oman is SARIMA $(3, 1, 1)(2, 0, 0)_{12}$ as the model has the lowest AIC (2744.69) and BIC (2768.84) values. For the RTI data, a number of models have been compared (Table 2), and the model SARIMA$(0, 1, 1)(1, 0, 2)_{12}$ is found to be the best. Although BIC value (2821.58) is not the lowest for the SARIMA$(0, 1, 1)(1, 0, 2)_{12}$ model due to more parameters than other models, the AIC value (2804.33) is the lowest. Additionally, this model has the lowest value in RMSE (96.73), MAPE (13.16) and MASE (0.83), which suggest that the model is better than the other models for the RTI time series in Oman.

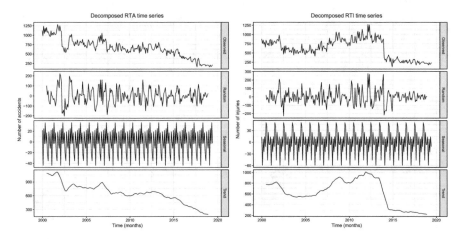

Fig. 1 Time series decomposition of (**a**) RTA and (**b**) RTI data

Table 1 Assessment of different models for RTA data in Oman

Model	AIC	BIC	RMSE	MAPE	MASE
$(4, 1, 1)(2, 0, 0)_{12}$	2746.56	2774.17	84.47	10.93	0.89
$(4, 1, 1)(1, 0, 0)_{12}$	2748.10	2772.25	85.19	11.12	0.91
$(3, 1, 1)(2, 0, 0)_{12}$	2744.69	2768.84	84.49	10.95	0.89
$(5, 1, 1)(2, 0, 0)_{12}$	2747.42	2778.21	84.17	10.91	0.89
$(4, 1, 0)(2, 0, 0)_{12}$	2754.42	2778.58	86.35	10.85	0.91

Table 2 Assessment of different models for RTI data in Oman

Model	AIC	BIC	RMSE	MAPE	MASE
$(1, 1, 2)(0, 0, 2)_{12}$	2810.48	2831.19	97.75	13.66	0.85
$(0, 1, 1)(0, 0, 2)_{12}$	2807.22	2821.03	97.92	13.65	0.85
$(0, 1, 1)(1, 0, 2)_{12}$	2804.33	2821.58	96.73	13.16	0.83
$(0, 1, 1)(0, 0, 1)_{12}$	2808.89	2819.24	98.76	13.53	0.85
$(1, 1, 1)(0, 0, 2)_{12}$	2808.72	2825.97	97.80	13.68	0.85
$(0, 1, 2)(0, 0, 2)_{12}$	2808.74	2825.99	97.80	13.68	0.85

Although the model $(3, 1, 1)(2, 0, 0)_{12}$ for RTAs have slightly higher RMSE and MAPE, the model is adequate and better than the other models considering AIC and BIC values. Moreover, the residual diagnostics, more specifically, the autocorrelation of the residuals, were checked by the Ljung–Box test ($Q = 22.5$, p-value $= 0.21$), which shows that the test is insignificant. Figure 2 shows that residuals for both models are white noise and ACF residuals fall near to the zero. It can be deduced from further goodness of fit analysis that the SARIMA$(3, 1, 1)(2, 0, 0)_{12}$ model fitted the data reasonably well.

Diagnostic checking and model validation were also performed as the procedures as mentioned earlier for the RTI data. The model SARIMA $(0, 1, 1)(1, 0, 2)_{12}$ has

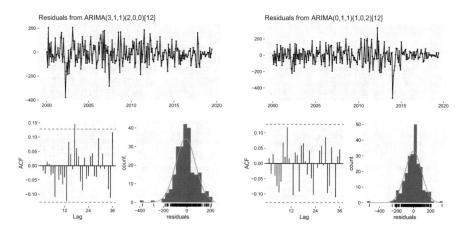

Fig. 2 Residuals of the fitted models. (**a**) Residuals of the fitted model in RTA. (**b**) Residuals of the fitted model in RTI

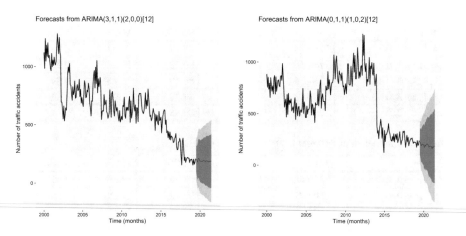

Fig. 3 Observed (black) and forecasted values (blue) of RTA and RTI in Oman. (**a**) Observed (black) and forecasted values (blue) of traffic accidents in Oman. (**b**) Observed (black) and forecasted values (blue) of traffic injuries in Oman

shown the highest adequacy than other models considering RMSE, MAPE and MASE as shown in Table 2. Results of the Ljung–Box test ($Q = 21.6$, p-value $= 0.4136$) suggest that autocorrelation coefficients are not significantly different from zero. The ACF residuals indicate that autocorrelation is near to zero and do not deviate from a zero mean, i.e. the residuals follow a white noise process. These suggest that the SARIMA$(0, 1, 1)(1, 0, 2)_{12}$ model fitted injuries data in Oman well. Based on the final models for RTA and RTI, we have forecasted the number of accidents and injuries for the next 24 months shown in Fig. 3.

5 Conclusion

This study has aimed to develop a time series model to forecast road traffic accidents and road traffic injuries in the Sultanate of Oman. A peak occurred with 1283 RTAs which then declined to the lowest point of 156 RTAs in February 2019. Based on the Box and Jenkins approach, SARIMA$(3, 1, 1)(2, 0, 0)_{12}$ model was selected to forecast RTAs for the next 24 months in Oman. This model forecasted the high occurrence of RTA in June, July and August in the following years. On the other hand, SARIMA$(0, 1, 1)(1, 0, 2)_{12}$ model was selected for predicting the number of traffic injuries, which shows a downward trend. The policymakers in Oman should keep under their consideration the results of this study. Both models in RTA and RTI show that there is a higher chance of accidents and injuries during the summer season. For future work, we would like to study the fatality-related crashes in Oman. Furthermore, investigation of the spatial factors and the socio-economic impact of the RTAs in Oman would be considered.

References

1. A. Boulieri, S. Liverani, K. de Hoogh, M. Blangiardo, A space–time multivariate Bayesian model to analyse road traffic accidents by severity. J. R. Stat. Soc. Ser. A **180**(1), 119–139 (2017)
2. G.E. Box, G.M. Jenkins, G.C. Reinsel, G.M. Ljung, *Time Series Analysis: Forecasting and Control* (Wiley, London, 2015)
3. J.D. Cryer, K.S. Chan, *Time Series Analysis with Application in R* (Springer, Berlin, 2008)
4. R.J. Hyndman et al., Another look at forecast-accuracy metrics for intermittent demand. Foresight Int. J. Appl. Forecast. **4**(4), 43–46 (2006)
5. F. Mannering, C. Bhat, Analytic methods in accident research: methodological frontier and future directions. Anal. Methods Accid. Res. **1**, 1–22 (2014)
6. NCSI, *Monthly Statistical Bulletin* (National Centre for Statistics & Information, Sultanate of Oman, 2000–2019)
7. M. Peden, A. Hyder, Road traffic injuries are a global public health problem. BMJ **324**(7346), 1153 (2002)
8. M. Peden, R. Scurfield, D. Sleet, D. Mohan, A.A. Hyder, E. Jarawan, C.D. Mathers, et al., World report on road traffic injury prevention (2004)
9. R.O. Police, *Traffic Statistic* (Director General of Traffic, 2013–2019)
10. M.A. Quddus, Time series count data models: an empirical application to traffic accidents. Accid. Anal. Prev. **40**(5), 1732–1741 (2008)
11. R. Raeside, D. White, Predicting casualty numbers in Great Britain. J. Transp. Res. Board (1897), 142–147 (2004)
12. X. Zhang, Y. Pang, M. Cui, L. Stallones, H. Xiang, Forecasting mortality of road traffic injuries in China using seasonal autoregressive integrated moving average model. Ann. Epidemiol. **25**(2), 101–106 (2015)

Toward a Reference Model for Artificial Intelligence Supporting Big Data Analysis

Thoralf Reis, Marco X. Bornschlegl, and Matthias L. Hemmje

1 Introduction and Motivation

Reference models are a powerful instrument of software engineering to master the complexity of developing comprehensive software systems [1]. They empower architects and developers to reuse established concepts, equip them with refined specifications and guidelines [1], and enable precise information exchange [2]. Successful reference models such as the ISO 7-Layer Reference Model manage to describe key artifacts (e.g., exchanged information) as specific as necessary yet as generalizable as possible [2] and sharpen scientific and industrial research focus and progress.

Artificial intelligence (AI) is a trending topic in various application domains with big industrial relevancy as 85% of CIOs will be piloting AI programs by 2020 according to Gartner Inc. [3]. Popular application areas reach from health care [4, 5], trading [6], driverless cars, humanoid robots [5] to people's everyday life where shopping websites recommend products, video streaming providers such as Netflix suitably recommend films, and social media platforms individually decide upon content relevancy for each user [6]. The term and application area of Big Data are closely connected to AI [7] since Big Data provides AI with the possibility of deriving, validating, applying, and enhancing AI models through processing (or learning from) huge amounts of data, whereas AI and its statistical approach of machine learning (ML) provide algorithms to exploit Big Data and its potential (Fig. 1). A definition for Big Data can be derived from Doug Laney [8] who describes data management challenges regarding three dimensions (the three v's):

T. Reis (✉) · M. X. Bornschlegl · M. L. Hemmje
University of Hagen, Faculty of Mathematics and Computer Science, Hagen, Germany
e-mail: thoralf.reis@fernuni-hagen.de; marco-xaver.bornschlegl@fernuni-hagen.de;
matthias.hemmje@fernuni-hagen.de

© Springer Nature Switzerland AG 2021
R. Stahlbock et al. (eds.), *Advances in Data Science and Information Engineering*,
Transactions on Computational Science and Computational Intelligence,
https://doi.org/10.1007/978-3-030-71704-9_38

Fig. 1 Use cases interconnecting Big Data analysis and artificial intelligence

volume (big amount of data), velocity (high frequent data inflow), and variety (ambiguous data manifestations regarding, e.g., data format, data structure, or data semantics) [9]. Visualization of Big Data in the context of AI applications offers the chance to meet the increasing demand for explainability and transparency of AI as in-transparent decision paths cause difficulties regarding human comprehensibility, debugging, fairness, and accountability [10] and raise concerns regarding data-driven discrimination [11]. According to a Gartner Inc. research, transparency is crucial for industrial application of AI as "enterprise AI projects with built-in transparency will be 100% more likely to get funding from CIOs" by 2022 [3].

Big Data user stereotypes vary regarding their organizational and technical knowledge level [12]: there exist domain experts such as data engineers, data analytics experts, data visualization specialists to management-level end user stereotypes [12]. According to Bornschlegl et al., visualization of information, human interaction, and perception by these different Big Data user stereotypes are pivotal elements for Big Data analysis [8]. To focus on this fact, the theoretical reference model Information visualization for Big Data (IVIS4BigData) that "close[s] the gap in research with regard to information visualization challenges of Big Data analysis as well as context awareness" was introduced [8]. IVIS4BigData covers the whole workflow of Big Data analysis from *Raw Data Sources* into *Views* that finally enable the different user stereotypes participating in the process to gain insight [12]. For this purpose, necessary advanced visual user interfaces as well as the underlying data storage, computation, and service infrastructure are considered to be decisive for the reference model [12]. The IVIS4BigData reference model summarizes AI models and algorithms supporting the analysis of Big Data besides other mechanisms as *Analytics* processing step that consists of transforming data and mapping it into an analyzed and structured form [12]. The further use case scenarios interconnecting AI, Big Data, and visualization (Fig. 1) of Big Data analysis-enabled (data-driven) model development and the support of human users within their Big Data exploration journey through AI (AI-supported data exploration) are not covered by this reference model.

Targeting the development of a Big Data analysis system that covers all aspects connecting Big Data, AI, and visualization raises the requirement for a reference model. In general, if there is no available reference model for this combined application domain, one can proceed as follows: The first option is to utilize a general AI reference model and gradually extend it on Big Data analysis and visualization aspects, e.g., identify necessary adaptions and extensions in the course of development or validation. Further approaches are to do it the other way around: design the system according to a Big Data analysis, visualization, or combined Big Data analysis and visualization reference model and extend it on the remaining application domains' specifics during design and development. There is no reference model covering all aspects of the process of AI supporting Big Data analysis. This publication targets to change that. Starting with the examination of AI's state of the art including standards, user stereotypes, and system lifecycles (Sect. 2), a systematic methodology based on the Big Data analysis and visualization reference model IVIS4BigData is followed (Sect. 3) in order to introduce a new reference model that covers all workflows, artifacts, and user stereotypes involved in AI-supported Big Data analysis (Sect. 4).

2 Artificial Intelligence

Although AI is a trending topic, it is a topic that has fascinated humans since ancient times [5]. Examples are the Greek Mythology where statues were brought to life or Mary Shelly's Frankenstein's monster, a man-made creature that acquires human behavior [5]. The modern computer-based AI is closely linked to the 1950s and Alan Turing's question regarding thinking capabilities of machines [10]. Important milestones were the introduction of artificial neural networks by Bernard Widrow in the 1950s [13] and the success of both IBM's Deep Blue chess AI [5] and Alphabet's AlphaGo Zero AI for the Asian game of Go [10] defeating the at the time world champions in the respective game. The latter AI learned its skills in a self-contained manner by playing against itself with neither human instructions nor historical data [10].

AI and AI-relevant terminology are standardized in ISO/IEC JTC 1/SC 42 standard [7] which declares AI to be a form of intelligence displayed by machines other than natural intelligence which is displayed by humans and animals [7]. In order to establish a clear terminology, the taxonomy of AI regarding existing approaches provided by the Internet Policy Research Initiative at MIT (IPRI) [10] divides the existing approaches into two categories: symbolic and statistical approaches [10]. The symbolic approach "requires that researchers build detailed and human-understandable decision structures" [10], while "machines induce a trend from a set of patterns" [10] within the statistical approach. ML is a manifestation of a statistical approach as ML "algorithms are characterized by the ability to learn over time without being explicitly programmed" [7].

In 2019, the AI Group of Experts at the OECD (AIGO) defined a lifecycle for AI systems consisting of four phases. The phases are visualized in Fig. 2 and are often conducted rather in an iterative way than strictly one after another [10]. The first phase *Data and Model Design* marks the starting point and consists of planning and designing of all, data collection, data processing, as well as model building and interpretation [10]. The thereby generated models are subject to *Verification and*

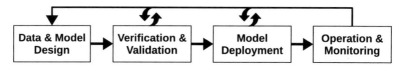

Fig. 2 AIGO's AI system lifecycle [10]

Validation in phase 2 as well as to *Model Deployment* in phase 3 in order to apply the model in live production [10]. Within the fourth phase *Operation and Monitoring*, the productive AI system's "recommendations and impacts" [10] are reviewed and assessed to maintain and if necessary adapt the system [10].

3 Methodology

In [12], Bornschlegl et al. derived the reference model IVIS4BigData through applying the information flow and key artifacts of Kaufmann's Big Data Management (BDM) reference model onto the information visualization (IVIS) reference model considering the user stereotypes involved in practical Big Data analysis use cases [12]. Orienting oneself on this methodology, this publication targets to systematically derive a new reference model for AI supporting Big Data analysis as follows: application of the information flow and key artifacts of AIGO's AI system lifecycle [10] onto Bornschlegl's IVIS4BigData reference model [12] considering the user stereotypes involved in practical AI use cases.

In order to reach the target reference model, the followed methodology is divided into four consecutive steps:

1. Mapping of AI lifecycle phases onto IVIS4BigData reference model elements: All of the four AI lifecycle phases are investigated regarding suiting IVIS4BigData processing steps or equivalent IVIS4BigData information artifacts. In case of a match, the existing element within the reference model is extended with information of the examined lifecycle phase. If the mapping was not successful, the reference model itself is complemented by a new element.
2. Identification of AI model types that facilitate AI to support Big Data analysis and incorporate them into reference model elements: In the first step, relevant categories of AI models are identified through examination of the use case scenarios in Fig. 1. Afterward, the reference model is investigated whether there are elements to which the identified AI models match, are content-related, or exchange information with. To conclude the second step, the result of the investigation is utilized to include information on the AI model in exactly the right spot of the reference model.
3. Extension of the reference model on AI input and output (IO) data: Examination of the AI lifecycle phases and the identified AI models (step 2) in order to identify the input data that is required by the respective phase or model as well as

the output data that is produced as the phase or AI model result. Eventually, the reference model is extended on the identified IO data as well as on a representation of its logical information flow.

4. Mapping of Big Data and AI user stereotypes to reference model elements: To enable software engineers and architects to utilize the reference model to tailor system implementations to the target audience, the reference model is required to contain the information which Big Data and AI user stereotypes are involved in the different processing steps or related to the information artifacts. To this end, all reference model elements are investigated whether they can be related to the respective user stereotypes and, if so, are extended with a link to it. Finally, all user stereotypes within the by-then derived reference model are compared to all Big Data and AI user stereotypes. If there exists a user stereotype that is not yet represented, the reference model is extended on further elements, links, or information artifacts which conclude the fourth step and ensure that all relevant user stereotypes are represented.

Each step results in an evolved version of the original reference model with the results of steps 1 to 3 being fed into the following step, while the result of the fourth and final step yields the proposed new reference model.

4 Conceptual Modeling

Following the proposed methodology, three preliminary steps need to be conducted before the reference model can be designed: The different categories of AI models for supporting Big Data analysis, AI-related IO data, as well as the AI user stereotypes need to be derived. The results of these preliminary steps as well as the proposed reference model are presented within the following sub-sections.

4.1 AI Models for Supporting Big Data Analysis

Starting the examination of the use case scenarios in Fig. 1 with the data transformation use case, the first AI model can be identified: Within this use case, a Big Data user stereotype utilizes an AI model to transform huge amounts of data to ease its exploration, e.g., utilize clustering to identify outliers. Since this data transformation processing step is part of the IVIS4BigData analytics phase, this first AI model is called *Analytics Model*. The data-driven model design use case differs from the first use case regarding the objective and the involved user stereotype: the AI model creation is conducted by an AI user stereotype, and the purpose of Big Data analysis is the creation of the AI model, not the other way around as for data transformation. Nevertheless, the representation within the reference model of both use cases is located within analytics phase. Hence, the introduced *Analytics Model* can be reused for the second use case. The last use case of AI-supported data exploration is relevant for both Big Data and AI user stereotypes which need to be supported in closing the knowledge gap between data science and the design of AI models. One way to ease comprehension is the intelligent user interface (e.g., through pre-selecting relevant tools or explaining difficult processing steps). The corresponding AI models are introduced as *UI models*. A second way to bridge the knowledge gap between the two user groups is to utilize automation where applicable: the introduction of *Automation Models* enables predicting repetitive or foreseeable user activities or system calculations to save time and to reduce infrastructure peak load probabilities. The demand for automation in Big Data analysis and AI application is backed by Gartner Inc. analysts Pieter den Hamer, Peter Krensky et al. according to whom "more than 40% of data science tasks will be automated" by 2020 [3].

4.2   AI Input and Output Data

Going through the different phases within AI system lifecycle (Fig. 2), the first IO data that is relevant to all phases obviously is a representation of the model itself. Since we derived already the relevant models for our reference model within the previous sub-section, we focus on the following types of IO data being exchanged between the different phases:

1. AI Result Data: Applying AI models generates meta information from processing the actual input data. This meta information is the primary result of AI models. Examples are class predictions, forecasting by regression, and confidence or cluster allocations.
2. AI Metrics: Validating or operating AI models requires quantitative control parameters that enable relevant AI user stereotypes to evaluate the performance of both, the AI model itself and the environment it is deployed to. Examples for this IO data are the AI model's accuracy or the time and resource consumption for model training.
3. Label Information: Supervised ML requires the presence of assessed data. This assessed data enables AI algorithms in training phase to adapt their calculation (e.g., adjust weights) and in testing phase to evaluate how good the adapted AI model performs on new data. In ML, the information accompanied with assessed data is called *label*. The label information is generated by human assessment (e.g., in Verification and Validation lifecycle phase) or through algorithms.

Naturally, AI models require the actual data to perform their calculations on, e.g., regress, categorize, or cluster. In the context of a Big Data reference model, this data can be anything from medical measurements [4] to sensor measurements of autonomous cars [5]. Yet this data needs to be distinguished from the following data that is produced within the Big Data analysis system itself and fed into the UI and automation in AI models:

4. User Interactions: To ease the system's usability, information on user activities such as the activity itself as well as the user who performed it are an important factor for the success of UI models. Based on this historic data, UI models are able to predict in a user-specific manner and adapt over time.
5. System Activities: Besides the user-triggered activities captured via user interaction data, further system activities are an important input to automation AI models. Examples for this data are time-stamped system logs.

4.3   AI User Stereotypes

Deriving the different user stereotypes involved in the application of AI, AIGO's AI system lifecycle [10] serves as a starting point. For every lifecycle phase, straightforward deduction of a user stereotype is possible: a *Model Designer* user stereotype for *Design, Implementation, and Training* of an AI model in *Data and Model Design* phase, a *Domain Expert* user stereotype for *Data Selection, Verification, and Validation* in *Verification and Validation* phase, a *Model Deployment Engineer* user stereotype for *Model Deployment* phase, as well as a *Model Operator* user stereotype for *Operation and Monitoring* phase.

In order to identify further AI user stereotypes, the model's maturing process is assessed more closely: as introduced in [10], it's not common to consecutively pass the system lifecycle phases from *Data and Model Design* to *Operation and Monitoring*. The different phases are rather passed iteratively with every phase being visited more than once. With increasing time and iterations, the AI model and its evaluation, deployment, or monitoring are enhanced and fine-tuned until the AI model reaches production readiness and can be released. From the target user group of the released AI model, we can derive the *Model End User* stereotype. This user stereotype can but does not have to be equivalent with one of the already derived user stereotypes: in case of the AI-supported data exploration use case from Fig. 1, the *Model End User*, e.g., the Big Data user stereotype data scientist, differs from the further involved AI user stereotypes. Growing concerns regarding AI's legal responsibility [14], ethical judgment, and non-transparent exploitation of personal data [15] instigate legislators to demand regulation and auditing processes for AI [14, 15]. This raises the necessity for a further AI user stereotype with focus on the AI system lifecycle phases before releasing the model to the *Model End User* stereotype. We propose the *Model Governance Officer* user stereotype for this role to ensure legal compliance, privacy, and data security as well as to maintain ethical standards through monitoring and documentation of the AI model design process. Examples are a documentation of the data an AI model was trained with or the reasoning of feature selection if the selected features include sensitive information such as religion or sexual orientation [15].

The resulting AI user stereotypes are visualized as an XY diagram in Fig. 3. The phases of the AI system lifecycle are arranged according to their publicness alongside the x-axis from the least end user-relevant Data and Model Design phase

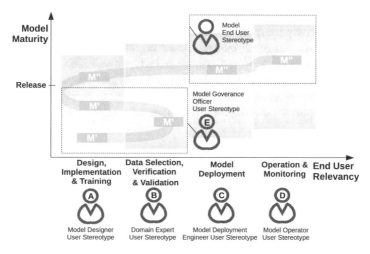

Fig. 3 AI user stereotypes alongside model lifecycle and maturing process

to the most end user-relevant Operation and Monitoring phase. The model maturity is drawn onto the non-quantified y-axis with a specific *Release* milestone. The four lifecycle-connected user stereotypes are drawn below the respective lifecycle phase (A to D). The gray color gradient within their column determines in what regions of model maturity the most activities of the respective user stereotype take place (from bottom left in early model design to top right in released model operation). The *Model End User* (F) stereotype's main area of activity is visualized through dashed lines in the top-right corner of the diagram. This user group focuses on released and deployed AI models that run in productive systems (with operation and monitoring), while the *Model Governance Officer* user stereotype (E) focuses on non-released models until they are production-ready and deployed for the end users. Hence, this user group's activity takes place from model design to model deployment of non-released models, visualized via a dashed rectangle in the bottom-left area of the diagram. In the background of the diagram, an s-shaped curve visualizes an exemplary AI model lifecycle and maturing process of a model M' that is adjusted as a model M'' after an initial data selection, verification, and validation process before it is released to the end users and reaches Operation and Monitoring phase.

4.4 Reference Model

Starting the reference model design with step 1 of the described approach, the Data and Model Design phase of the AI system lifecycle [10] is examined first in order to map it to IVIS4BigData reference model elements. This step consists of the selection of algorithms, data preparation, as well as the training and fine-tuning

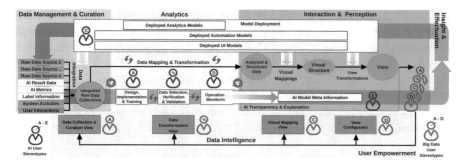

Fig. 4 AI2VIS4BigData: A reference model for artificial intelligence supporting Big Data analysis

of the model itself in case of ML and other statistical AI approaches, whereas it consists of the creation of a knowledge representation in symbolic AI approaches. Since these elements are part of the IVIS4BigData *Analytics* processing step, it is extended on a *Design, Implementation, and Training* reference to this lifecycle phase, while the user visualizes the intermediate and final results via *Interaction and Perception* processing step. Within *Validation and Verification* lifecycle phase, a user with the necessary domain knowledge selects suiting data and compares the actual behavior of the AI model with the model objectives or with the user's own subjective expectation. In order to do so, different evaluation data (e.g., KPIs) are calculated within *Analytics* processing step and visualized via *Interaction and Perception* processing step. As the actual calculation is located within *Analytics* processing step, it will be linked as *Data Selection, Verification, and Validation* to this lifecycle phase. The third lifecycle phase consists of the application of AI models in an execution environment on real data in an automatized form. No element of the so far derived reference model covers these aspects. Therefore, the extension of the reference model with a *Model Deployment Layer* is proposed. The last lifecycle phase of maintenance and monitoring contains activities to observe the correct AI model execution through evaluation of metrics and KPIs in numeric or visual forms. It is closely related to *Validation and Verification* lifecycle phase and consequently also located within the *Analytics* reference model processing step.

Within the second step of our proposed approach, the three derived AI models for supporting Big Data analysis have to be included to the reference model itself. All of the three AI models have to be deployed in order to be of use to the AI and Big Data user stereotypes. Therefore, an extension of the *Model Deployment Layer* is a good fit. Furthermore, the three AI models are closely connected to the *Analytics* processing step (*Analytics Models*), the user interfaces (*UI models*), as well as system activity information (*Automation Models*).

The third step of reference model design consists of the allocation of AI's IO data to reference model elements. The five different types of IO data from *AI Result Data* to *System Activities* have one thing in common: All of these data are produced within either the Big Data analysis system itself or third-party systems and fed (back) as input data for further analysis. As an example, *AI Result Data* is generated when executing AI models either within *Model Deployment Layer* of the reference model itself or by other systems whose activities shall be monitored within the Big Data analysis system. Therefore, the reference model's input processing step *Data Management and Curation* is extended on information of the different IO data types. Furthermore, there exists a logical connection of the IO data types to the different AI models. An example is the *System Activities* data's connection to the *Automation Models*.

To conclude the reference model design with step 4, the derived AI user stereotypes from A to F in Fig. 3 have to be mapped to reference model elements. Since the Data and Model Design, Verification and Validation, and Operation and Monitoring system lifecycle stages are located within the *Analytics* processing step, this step is extended on links to the respective user stereotypes A, B, and D. The *Model Deployment Engineer* user stereotype can easily be mapped to the *Model Deployment* layer of the reference model as well as to the AI models located within it. As there is no reference model element to which the *Model Governance Officer* user stereotype can be directly mapped to, the reference model needs to be extended to change that. Focus of this user stereotype is to document legal compliance and therefore create understanding and trust by the end user through transparency and explanation; the *Interaction and Perception* processing step is extended on an *AI Transparency and Explanation* layer that extends visual mapping and view transformations on AI model meta information. This information serves to explain decisions and document relevant environment and design conditions for the respective AI models. The last remaining AI user stereotype, the *End User*, does not need to be explicitly named within the reference model, since this user group is represented by the different Big Data user stereotypes in the context of the target application domain.

The resulting reference model is visualized in Fig. 4. It consists of the three processing steps *Data Management and Curation*, *Analytics*, and *Interaction and Perception* affiliated with a *Data Intelligence* layer for user interaction and user interfaces of the original IVIS4BigData reference model [12]. The newly introduced *Model Deployment* layer spreads over all of the three processing steps with only the *Analytics Models* being displayed exclusively within *Analytics* processing step. This layer is placed directly within the data and information transmission and feedback loop which ensures that deployed models get the input data they need for execution and produce output data that is directly fed back into the Big Data analysis system. Data integration is extended on this AI-relevant IO data. The deployed AI models impact all four information processing steps of the reference model from *data integration* to *view transformation*. The remaining three AI system lifecycle phases are displayed within *Analytics* layer and interconnected through bidirectional arrows emphasizing the iterative nature of AI model design.

5 Remaining Challenges and Outlook

The introduced theoretical reference model needs to proof its necessity, generalizability, and benefit to the state of the art in science and technology through extensive evaluation as well as practical applications. Consequently, the immediate next steps in AI2VIS4BigData reference model research are a case study and a survey. The case study's objective is to validate the motivation for the reference model through observing the presence and urgency of challenges for the different user stereotypes in AI, Big Data analysis, and visualization research. The conduct of a survey with participants from both, science and industry, targets to evaluate all reference model elements as well as the different assumptions and decisions that were made in the course of deriving the reference model. Further medium-term future works are a reference implementation and the experimental analysis of the same.

6 Conclusion

Interconnecting Big Data analysis, AI, and visualization offers the chance to enhance both, Big Data analysis itself and AI model design. Most important, it has significant potential to eliminate high barriers for users when getting started in both application domains as well as to maximize the exploitation of Big Data's assets. Yet it lacks of an important tool: a reference model for AI supporting Big Data analysis. Such a reference model eases overcoming software design complexity [1] and accelerates scientific, technological, and economic progress through establishing a common ground of precise system information [2].

This publication rectifies the situation of a missing reference model in the state of the art through introduction of the AI2VIS4BigData reference model for AI supporting Big Data analysis. Besides introducing the reference model, this paper comprehensively describes the different AI models for supporting Big Data analysis, AI IO data, AI user stereotypes, as well as the different steps of systematically deriving the reference model from AI system lifecycle combined with IVIS4BigData reference model under consideration of the user stereotypes involved in the application of AI. This reference model's objective is to architecture-independently specify the logical information flow, relevant artifacts, and involved user stereotypes. Consequently, it is independent of any choice of technology and system partitioning.

References

1. G. Engels, R. Heckel, G. Taentzer, H. Ehrig, A combined reference model- and view-based approach to system specification. Int. J. Softw. Eng. Knowl. Eng. **7**(4), 457–477 (1997)

2. C. Gunter, E. Gunter, M. Jackson, P. Zave, A reference model for requirements and specifications. IEEE Softw. **17**(3), 37–43 (2000)
3. P. Krensky, P. den Hamer, E. Brethenoux, J. Hare, C. Idoine, A. Linden, S. Sicular, F. Choudhary, *Critical Capabilities for Data Science and Machine Learning Platforms*, vol. G00391146 (Gartner, Stamford, 2020)
4. M. Healy, *A Machine Learning Emotion Detection Platform to Support Affective Well Being* (2018), pp. 2694–2700
5. R. Bond, F. Engel, M. Fuchs, M. Hemmje, P.M. Kevitt, M. McTear, M. Mulvenna, P. Walsh, H.J. Zheng, Digital empathy secures Frankenstein's monster, in *CEUR Workshop Proceedings*, vol. 2348 (2019), pp. 335–349
6. P. Roehrig, F. Malcolm, B. Pring, *What To Do When Machines Do Everything: How to Get Ahead in a World of AI, Algorithms, Bots, and Big Data* (Wiley, London, 2017)
7. ISO/IEC JTC 1/SC 42 Artificial Intelligence (2018). https://isotc.iso.org/livelink/livelink/open/jtc1sc42
8. M.X. Bornschlegl, Fakultät für Mathematik und Informatik Advanced Visual Interfaces Supporting Distributed Cloud-Based Big Data Analysis, DIssertation, University of Hagen, 2019
9. D. Laney, 3D Data Management: Controlling Data Volume, Velocity, and Variety, META Group, Tech. Rep., 2001
10. OECD, *Artificial Intelligence in Society* (2019)
11. P. Buxmann, Ein neuer Hype? Zur Ökonomie der künstlichen Intelligenz, in *Forschung & Lehre* (2020), pp. 22–23
12. M.X. Bornschlegl, K. Berwind, M. Kaufmann, F.C. Engel, P. Walsh, M.L. Hemmje, R. Riestra, IVIS4BigData: A reference model for advanced visual interfaces supporting big data analysis in virtual research environments. Lecture Notes in Computer Science (Including Subseries Lecture Notes in Artificial Intelligence and Lecture Notes in Bioinformatics), vol. 10084 (2016), pp. 1–18
13. Y. Wang, On abstract intelligence: toward a unifying theory of natural, artificial, machinable, and computational intelligence. Int. J. Softw. Sci. Comput. Intell. **1**(1), 1–17 (2009)
14. C. Wendehorst, Ist der Roboter haftbar? Künstliche Intelligenz und gültige Rechtsnormen, in *Forschung & Lehre* (2020), pp. 24–25
15. C. Woopen, I. Lohaus, Grundlegende Rechte und Freiheiten Schützen: Ethische Implikation der künstlichen Intelligenz, in *Forschung & Lehre* (2020), pp. 18–20

Improving Physician Decision-Making and Patient Outcomes Using Analytics: A Case Study with the World's Leading Knee Replacement Surgeon

Anish Pahwa, Shikhar Jamuar, Varun Kumar Singh, and Matthew A. Lanham

1 Introduction

Every year there are thousands of cases of chronic knee pain and disability due to different types of arthritis in people from different age groups. Athletes, accident victims, and aged people are among the most vulnerable. On top of this, the severity of an arthritis case tends to increase over time causing the patient progressively more pain. Knee replacement is the only effective and long-lasting course of treatment. Knee replacement has become so common in America that knees are replaced at a rate of more than 600,000 per year. In brief, the aim of the surgery is to minimize pain and restore mobility. However, there are many risks associated with the surgery; and a few complications like infection, blood clots, implant loosening, and continued pain can make matters even worse. These complications generally require revision surgery. Revision surgeries often have associated risks and might even add to complications, on top of the fact that there are extra associated costs. *Forbes* talks about how Medicare's bundling of fees hit knee replacement surgeries, affecting not only the patients but also physicians and insurance companies. Therefore, it becomes crucial to understand the drivers of these complications and control them to mitigate risk and minimize complications. Finding statistically significant but controllable decision factors that could lead to complications is an industry-wide challenge.

This study tries to address this problem by collaborating with a surgeon who has performed more than ten thousand knee replacement surgeries over their extensive career, using cutting-edge minimally invasive technologies. The study tries to

A. Pahwa (✉) · S. Jamuar · V. K. Singh · M. A. Lanham
Krannert School of Management, Purdue University, West Lafayette, IN, USA
e-mail: pahwaa@purdue.edu; sjamuar@purdue.edu; singh675@purdue.edu;
lanhamm@purdue.edu

© Springer Nature Switzerland AG 2021   573
R. Stahlbock et al. (eds.), *Advances in Data Science and Information Engineering*,
Transactions on Computational Science and Computational Intelligence,
https://doi.org/10.1007/978-3-030-71704-9_39

build a system that predicts the most important factors leading to post-operative complications by utilizing patient demographics, patient comorbidity data, surgery details, doctor details, and procedure specifics. A recent *Wall Street Journal* article found that exercise and increased muscle strength lead to better surgical outcomes. Building upon that we investigate factors such as age, height of a patient, distance from the clinic, and marital status affects the outcome of the surgical procedure. Surgery is classified as a failure depending on various parameters some of which are the number of post-operation visits and direct complication codes. The output of our model can provide insight regarding potential surgery complications and their expectations. This can help mitigate and minimize issues by warning doctors and patients in advance and even giving them recommendations to avoid realizing these complications later. The study also tries to map patient clusters to various doctors for improving the success rate and minimizing complications. Since the data we had was historical (ranging over 10 years), we had limited surgical specifics data. We believe that some of these surgical features could have also been significant in finding various controllable factors that affect the outcome, but some factors were ignored.

The remainder of the chapter discusses the following: The next section has a literature review of previous studies who have investigated factors measured and considered from successful and unsuccessful total knee replacement surgeries. Sections 3 and 4 present our data and methodology used in the study. We summarize our steps chronologically and provide reasonings for our assumptions in our analysis. In Sect. 5, we describe the different models that we built and tested. Since failure rate is less than 50%, we show how we defined this problem as an imbalanced binary classification problem and how we addressed this from a healthcare context. Section 6 summarizes the performance of all the models and the last section concludes the chapter with a discussion of the implications of this study, our recommendations from the clinical perspective, the future scope, and concluding remarks.

2 Literature Review

Many researchers, recent or otherwise, have focused on the problem of post-operative total knee replacement surgery complications. However, most of studies have only considered pain score as the surgery outcome indicator. While this measure has its merits, collecting and using this data require additional efforts and time such as subjective or perception-based surveys.

When predictors such as various pain or knee function scores are not present in the historical data, we must find a new approach to define surgery failures or complications. With an exponential increase in the number of knee replacement surgeries, the total number of surgery failures or complications is also increasing with the same magnitude. On an average, at least five out of one hundred total knee replacement surgeries develop complications. This not only leads to added costs to the health insurance providers and surgeons, but also decreases the quality of life of patient. Therefore, it becomes important to address this orthopedic industry wide challenge. A successful predictive model based on historical patient data could help us predict post-op complications before they occur and thus save costs and improve the success rate of these type of surgeries. The studies we based the foundation of our study on focused on some factors such as different success metrics, complications, use of post-op data, and building predictive models.

A 5-year prospective study by A.K. Nilsdotte of patient outcomes after total knee arthroplasty takes into consideration the knee injury and osteoarthritis outcome score (KOOS) pre-operatively once and then 6 months, 12 months, and at 5 years post-operatively [1]. The result showed significant improvement in all KOOS and scores 6 months post-op. The best post-operative result was reported at the 1-year follow-up. The 5-year follow-up again showed decline in KOOS scores. Age, comorbid conditions, and sex were some of the factors that affected post-operative KOOS pain scores.

Lingard et al. studied *"Predicting the outcome of knee arthroplasty"* relief of pain and the restoration of functional activities that are used as the outcome parameters of primary total knee arthroplasty. Pre-operative predictors of pain and functional outcome at 1 and 2 years following the surgery were used. The study recruited patients from three different countries. The authors employed hierarchical regression models and found that the most significant predictors of failures were low pre-operative pain scores, a higher number of comorbid conditions, and a low mental health score [2]. Country was also a significant factor for the functional status of patients.

Sanchez-Santos et al. studied the *"Development and validation of a clinical prediction model for patient-reported pain and function after primary total knee replacement surgery"* which focused on building a prediction model of patient-reported pain and function after undergoing total knee replacement (TKR). Pre-operative predictors such as patient characteristics and clinical factors were considered. The study employed bootstrap backward linear regression analysis. Low pre-operative knee score, living in poor areas, high BMI, and anxiety or depression were associated with worse outcome [3]. This is the first clinical prediction model for predicting self-reported pain and function 12 months after a total knee replacement surgery.

Brander et al. investigated predicting total knee replacement pain and used excessive post-operative pain, clinical and radiographic variables as predictors for total knee arthroplasty outcomes. Measures were VASP pain index, patient health, psychological state, and surgical component reliability [4]. They concluded that greater pre-operative pain, depression, and anxiety were associated with greater post-operative pain. Also, those factors corresponded to more home therapy and post-operative manipulations.

Another study by Tolk et al. created and validated predictive models that predicted residual symptoms on 10 specific outcome parameters at 12-month follow-up for patients undergoing primary TKA for knee osteoarthritis [5]. The predictive algorithms employed showed acceptable discriminative values (AUC 0.68–0.74) for predicting complications/residual symptoms.

Jørgensen et al. constructed a pre-operative risk score for patients in high risk of potentially preventable complications. The study concluded that pre-operative identification of patients at risk of preventable "medical" complications was statistically possible [6].

The study *"Predicting individual knee range of motion, knee pain, and walking limitation outcomes following total knee arthroplasty"* by Pua et al. concluded that statistically significant predictors for TKA outcomes were age, sex, race, education level, diabetes mellitus, pre-operative use of gait aids, contralateral knee pain, and psychological distress [7].

Our study aims to build a predictive model and a clinical recommender system using patient characteristics, patient health data, and doctor data to predict a failure of a total knee replacement surgery. We strive to identify the drivers to success of a surgery in order to minimize complications and revision surgeries, thereby reducing total costs associated. Our study is novel because we have developed a tailored outcome variable that is a combination of number of post-op visits, direct complications from ICD codes, and whether a patient underwent a knee surgery. If any of the required conditions is satisfied, then the surgery is flagged as a failure. Other studies used regular or hierarchical linear regression models in their investigations since their predictor variable (pain score or knee function score) was a continuous variable. This study however had a binary predictor variable. Thus, we framed our problem as a classification-type problem and utilized logistic regression, a decision tree (CART), and a random forest model. Table 1 shows some of the related studies found in the academic literature and how they compare to our study.

Table 1 Literature review and comparison

Author	Target variable used	Model type	Data used
Lingard, Elizabeth A. et al.	WOMAC Pain score	Hierarchical linear regression	Mental health, pain score, comorbid conditions
Brander, Victoria, A. et al.	Pain score	Multiple linear regression	Pain scale, health data, psychological state, device reliability
Sanchez-Santos, M.T. et al.	Pain score	General linear model	Pain score, previous history, weight data
Our study	Derived metric using 3 conditions	Logistic regression, CART, random forest	Patient demographics, comorbid conditions, doctor data

3 Data

The data that we used was extracted from a database of 400 tables from knee surgery patients over 20 years. We extracted patient-related demographics, health, and billing data from twelve different tables. We also used surgeon data in our analyses. We had patient age, sex, weight, and height measurements. We also had patient surgical financial data. Patient smoking history and their blood pressure data were also integrated into our analysis.

We derived variables such as BMI (body mass index) and surgery failure or success (the predictor variable) from the existing data. BMI was calculated using patient weight and height data. The outcome variable was created based on three conditions: the number of post-operative visits, direct complications (based on ICD medical codes corresponding to specific complications), and whether a patient had a revision surgery after the initial surgery. Any number of post-operative visits greater than six was considered as a failure in the surgery outcome based on feedback from the surgical domain expert. If any of those three conditions was met, the surgery was considered a failure. Table 2 depicts the variables used in our study.

4 Methodology

We followed the Cross Industry Standard Process for Data Mining (CRISP-DM) process in the development of our model as shown in Fig. 1. The most time was spent parsing the several hundred tables to gather relevant data. In collaboration with the surgeon, we filtered down the dataset to around 400 features with more than 6000 observation. We also derived variables like BMI and age which could make our analysis more comprehensive.

We defined surgical success as the combination of the following three rules:

1. Number of post-operation visits that a patient has pertaining to a surgery (> 6)

Table 2 Data used in the study

Variable	Type	Description
Sex	Categorical	The sex of the patient (either M or F or U)
age	Numeric	Age of patient at the time of surgery
BMI	Numeric	Body mass index of patient
Financial status	Categorical	The financial class for this employer
smoke	Binary	1 if patient smokes, 0 if patient does not smoke
height	Numeric	Height of patient
weight	Numeric	Weight of Patient
bp	Binary	If patient has high blood pressure
Surgery outcome	Binary	Derived variable. 1 if surgery is a failure, 0 if it is a success

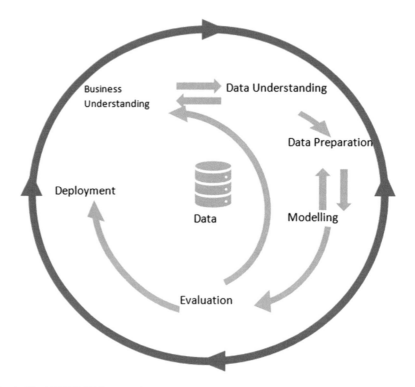

Fig. 1 Used CRISP-DM approach

2. If a patient has a revision surgery corresponding to a surgery
3. If a patient has direct complication recorded (according to ICD-9 and ICD-10 codes)

If any of these conditions was true, the surgery was deemed unsuccessful. Outliers in age and BMI variables were removed using the interquartile range (IQR) formula [Q1 − 1.5(IQR), Q3 + 1.5(IQR)]. In order to manage the missing data (missing variable measurements), we used decision tree imputation methods and checked for any abrupt changes in the pre- and post-imputation distributions of these variables to make sure our imputations conformed to the known distribution of each variable. The cleaned data was randomly partitioned into a 75/25% train and test sets.

As the data corresponds to surgery success and failure, it was obvious that the number of failures would be very less compared to surgery success. To encounter the target variable class imbalance, we rebalanced training sets target variable using the popular Synthetic Minority Oversampling Technique (SMOTE), which increased the minority class from 11% to 25% as shown in Fig. 2.

Framing the problem into an analytics problem, this is a binary classification problem trying to predict failure or no failure. While several potential classification algorithms could have been investigated, our goal was to provide interpretability

Fig. 2 Target variable
response distribution

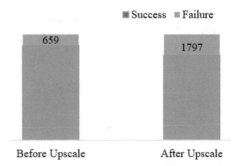

■ Success ■ Failure

659

1797

Before Upscale   After Upscale

Table 3 Pros and cons of each model

	Decision tree	Random forest	Logistic regression
Graphically interpretable	*****	***	**
Nominal predictor handling	*****	*****	***
Low variance	**	***	*****
Highly interpretable	**	*	*****
No hyper parameter tuning	**	**	*****
Easy to implement	***	***	****
Test error estimation	***	****	***

to improve decision-making rather than focus on identifying a model that had the highest predictive power. This is why we used a logistic regression and decision tree (CART) model. Ensembling methods are often used when the goal is to achieve a more accurate prediction, so we explored the random forest model as well just to show the relationship in accuracy of a more sophisticated model compared our interpretable models.

Due to a significant class imbalance in the dataset even after rebalancing, we did not evaluate the model using accuracy, but rather used F-1 score as our model performance evaluation meausre, as it accounts for class imbalance.

5   Model

For our case model interpretability was prioritized over model predictability to provide a solution that would be an early warning system that would support doctors and their patients in decision making. For a doctor to take some insights from the model it was crucial that the model be as simple and explainable as possible. In light of this we used the models described in Table 3.

Table 4 shows the tuning parameters for the decision tree and random forest model we found to be via cross-validation.

The logistic regression model by nature was highly interpretability in our case and proved to be very accurate as well. Therefore, we chose to use it:

Table 4 Tree models hyperparameters

Model	Hyper parameter
Decision tree	Maximum depth: 6 Interval Target Criterion: Reduction Gini index Leaf size: 5
Random forest	Number of trees: 200 Number of variables at each split: 4 Maximum depth: 10 Proportion of sample in each sample: 75%

Fig. 3 Model performances

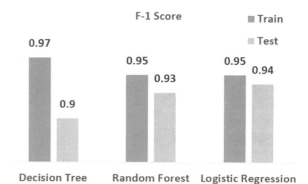

$$\ln\left(\frac{P}{1-P}\right) = \beta_0 + \beta_1 x => P = \frac{e^{\beta_0+\beta_1 x}}{1 + e^{\beta_0+\beta_1 x}}$$

6   Results

As shown in Fig. 3 the logistic regression was the least overfit, most accurate, and most generalizable model among the four models we investigated.

$$\ln\left(\frac{Failure}{Success}\right) = -0.92 + 0.99 * Doctorid16 + 0.8 * Doctorid496$$
$$+ 0.27 * Doctorid1478 + 0.34 * Doctorid1933$$
$$- 3.22 * Age + 0.29 * Smoke + 0.85 * BP + 0.9$$
$$* BMI$$

Figure 4 shows the receiver operating characteristics (ROC) curve for the logistic regression model. The ROC is a graph showing the false-positive rate versus the true-positive rate for all potential confusion matrix cut-off thresholds. The curve indicates models' predictive power across both the classes. The graph shows that the model in in consistent and can discriminate among a successful surgery and surgical failure fairly well (AUC = 87.8).

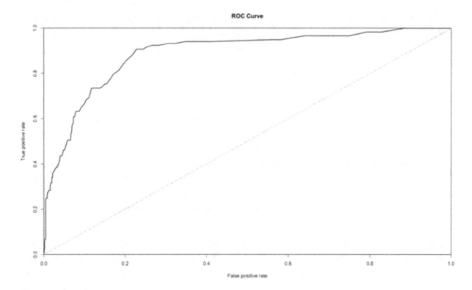

Fig. 4 ROC curve of final model

Table 5 Confusion matrix
for final model

		Actual	
		Success	**Failure**
Prediction	**Success**	1297	47
	Failure	15	152

Table 5 depicts the confusion matrix of our logistic regression model using a classification cut-off threshold of the typical 0.50 value. This shows the model was 95% accurate, with a precision of 82% and recall of 97%. A good precision and recall show that the model can correctly classify failures as failure and success as success.

7 Conclusions

There has been a constant rise in the number of total knee replacement surgeries all over the world and consequently the number of post-operative complications have added contributed to upward trending healthcare costs. It is therefore important to discover the statistically significant drivers of surgery outcomes to address this challenge. This study tried to do this by analyzing over six thousand similar knee replacement surgeries and develop an interpretable predictive model based on those significant drivers to support better decision-making. This could help decrease the associated avoidable industry-wide costs and bring down the failure rate of surgeries.

We found a patients' BMI is a significant factor contributing to the post-operative failure of surgeries. Higher BMI is associated with increased risks of developing complications after a surgery. Age of the patient significantly contributes toward failure of a surgery. Interestingly, as people increase in age, they are less likely to develop complications post-surgery. Having a smoking habit and/or having a higher blood pressure significantly contributes toward failure of a surgery. Smoking and high blood pressure increase the risk of infections in the artificial joints and increase the risk of developing blood clots or causing deep vein thrombosis (DVT), which could lead to a life-threatening situation. Additionally, surgeons and insurance carriers were also found to be significant factors impacting the outcome of a surgery.

The average cost of a total knee replacement surgery is $57,000 US dollars. There will be approximately 550,000 total knee arthroplasty procedures in the year 2021. Our model predicts up to two-thirds of complications accurately. When we factor in two-thirds of the complications requiring a revision, even if we ignore the extra costs associated with complications such as hospitalization, medications, imaging, and radiology costs, and just accounting for the raw cost of a revision surgery more than 2.1 billion USD could be saved in the United States alone in 1 year, if predicted complications/failures could be prevented. In addition to the economic costs, patient quality of life can improve drastically, and surgeon and patient time could be saved. By extrapolating we can conclude that our model can use patient demographics data, EHR and clinic data to predict complications before a surgery help clinics and insurance companies save up to $11B USD every year by the year 2030.

There are however some underlying assumptions based on the business context. The number of post-operative visits above which a surgery could be classified as a failure is a subjective number and a surgical failure might be defined differently surgeon to surgeon. We are assuming that this sample dataset has the same characteristics as would the entire population in the US. We also assume that the components used in the replacement surgeries are the same or have the same performance, since we did not have data on this.

Further, more investigation could be done on surgical factors, such as the technique used in the surgery (whether the ligaments were cut), the angle of cut, the gait of the knee, etc., to see if any of these factors significantly affects the outcome of the procedure. Collecting and using these data could improve our model further.

Acknowledgments We would like to thank Dr. Peter Bonutti and Justin Beyers from Bonutti Technologies for their collaboration of this work.

Bibliography

1. A.-K. Nilsdotter, S. Toksvig-Larsen, E. Roos, A 5-year prospective study of patient-relevant outcomes after total knee replacement. Osteoarthr. Cartil. **17**(5), 601–606 (2009). https://doi.org/10.1016/j.joca.2008.11.007

2. E.A. Lingard, J.N. Katz, E.A. Wright, C.B. Sledge, Kinemax Outcomes Group, Predicting the outcome of total knee arthroplasty (2004, October). Retrieved from https://www.ncbi.nlm.nih.gov/pubmed/15466726
3. M.T. Sanchez-Santos, C. Garriga, A. Judge, R.N. Batra, A.J. Price, A.D. Liddle, . . . , N.K. Arden, Development and validation of a clinical prediction model for patient-reported pain and function after primary total knee replacement surgery. Sci. Rep. **8**, 1–9 (2018, February 21). Retrieved from https://www.ncbi.nlm.nih.gov/pubmed/29467465
4. V.A. Brander, S.D. Stulberg, A.D. Adams, R.N. Harden, S. Bruehl, S.P. Stanos, T. Houle, Predicting total knee replacement pain: A prospective, observational study. Clin. Orthop. Related Res. **416**, 27–36 (2003, November). Retrieved from https://www.ncbi.nlm.nih.gov/pubmed/14646737
5. J. Tolk, J. Waarsing, R. Janssen, L.V. Steenbergen, S. Bierma-Zeinstra, M. Reijman, Outcome Prediction For treatment of knee osteoarthritis with a total knee arthroplasty. Development and validation of a prediction model for pain and functional outcome using the Dutch arthroplasty register (LROI) data. Poster Present. (2019). https://doi.org/10.1136/annrheumdis-2019-eular.1654
6. C.C. Jørgensen, M.A. Petersen, H. Kehlet, Preoperative prediction of potentially preventable morbidity after fast-track hip and knee arthroplasty: A detailed descriptive cohort study. BMJ Open **6**(1) (2016). https://doi.org/10.1136/bmjopen-2015-009813
7. Y.-H. Pua, C.L.-L. Poon, F.J.-T. Seah, J. Thumboo, R.A. Clark, M.-H. Tan, et al., Predicting individual knee range of motion, knee pain, and walking limitation outcomes following total knee arthroplasty. Acta Orthop. **90**(2), 179–186 (2019). https://doi.org/10.1080/17453674.2018.1560647

Optimizing Network Intrusion Detection Using Machine Learning

Sara Nayak, Anushka Atul Patil, Reethika Renganathan, and K. Lakshmisudha

1 Introduction

With the accelerated development in technology, there is a considerable increase in the number of security violations. To face this ever-growing issue, there is a need to keep all the data protected from network attacks. A system or software that supervises network traffic for dubious activities and raises alerts when any such activity is noticed is an IDS. In our work, we have used machine learning to increase the accuracy level of the IDS and to help improve the ability of risk assessment. With the usage of ML, large amounts of high-speed data can be processed in a lesser amount of time as compared to traditional systems.

2 Literature Review

The authors of [1] provide an insight into how the data set UNSW-NB15 was created and put forth the drawbacks of previously existing datasets used for network intrusion detection systems.

A two-staged anomaly-based network intrusion detection is used in [2]. Two algorithms that are used in the model are recursive feature elimination and random forests followed by binary classification out of which SVM has the highest accuracy.

S. Nayak (✉) · A. A. Patil · R. Renganathan · K. Lakshmisudha
Department of Information Technology, SIES Graduate School of Technology, Navi Mumbai, Maharashtra, India
e-mail: sara.nayak16@siesgst.ac.in; anushka.patil16@siesgst.ac.in; reethika.renganathan16@siesgst.ac.in; lakshmi.sudha@siesgst.ac.in

© Springer Nature Switzerland AG 2021
R. Stahlbock et al. (eds.), *Advances in Data Science and Information Engineering*, Transactions on Computational Science and Computational Intelligence, https://doi.org/10.1007/978-3-030-71704-9_40

In [3] a deep learning model is proposed where the number of hidden layers and neurons correspond to the number of features categories and the number of features respectively. The model is trained on the full UNSW-NB15 dataset.

In [4] they have combined feature learning and anomaly detection methods. Autoencoder and principal component analysis are used for feature learning. The paper concludes that OCSVM shows significant improvement in anomaly detection for multiple datasets including UNSW-NB15.

In [5], the dataset UNSW-NB15 is preprocessed by imputing missing values and by transforming categorical values to numerical values. It concludes that random forest classifier gave the most accurate results.

Artificial intelligence (AI) and computational intelligence (CI) are used in [6]. In AI, the supervised algorithms show better classification accuracy on the data with known attacks. In CI, techniques such as genetic algorithms, ANN, fuzzy logic, and artificial immune systems (AIS) were used.

[7] reviews supervised and unsupervised types of ANN as well as generative and discriminative deep network intrusion detection systems. It concludes that deep networks show better results than shallow networks.

[8] proposes a model, for OCSVM which is trained on the most appropriate features of the dataset and pre-processes the input data. For the sources that generate inflated traffic in the training model, the paper suggests the creation of a split OCSVM model, which is then followed by clustering the split OCSVM models.

3   Methodology

Data Preprocessing   We have used the UNSW-NB15 dataset to build our system. The dataset consists of 45 features and 257673 tuples split into two csv files called training and testing sets. As the training and testing files were separate in the dataset, they were concatenated into a single file for data preprocessing to maintain consistency. We preprocessed the data by handling missing values, performing label encoding to convert categorical attributes to numerical attributes, and performing random shuffling followed by min-max normalization.

Mathematically, it is shown as Eq. (1):

$$z = \frac{x - \min(x)}{\max(x) - \min(x)} \tag{1}$$

After normalization, we split the dataset into training and testing sets. We also separated the independent variables from the dependent variable as X and Y, respectively. The obtained data frames were converted to NumPy arrays for further implementation of the following algorithms:-

1. **Support Vector Machine Algorithm:** We have implemented SVM using the scikit-learn library. We wanted to gain linear separation in our data, which is

currently non-linear. Therefore, we have used RBF (radial basis function) kernel which behaves like a regularization parameter in SVM. The inner product kernel is defined by Eq. (2):

$$K_{RBF}\left(x, x^{'}\right) = \exp\left[-\gamma \left\| x-x^{'} \right\|^2\right] \tag{2}$$

We have generated a confusion matrix to understand the percentage of correctly and incorrectly classified examples. We have also generated a classification report which contains parameters like precision, recall, F1 score, accuracy, etc.

2. **Artificial Neural Network Algorithm**: Our ANN system architecture consists of 42 input neurons in the input layer and 2 hidden layers, each containing 12 neurons. The output layer contains one neuron (as it is a binary classification problem). The model has been developed using Keras. After preprocessing the data, it is sent to the regularized ANN model. Rectified linear unit or ReLU activation function has been used in the hidden layers and the output layer utilizes the sigmoid activation function. We use a binary cross-entropy/loss function to understand how accurate the probabilities predicted by the model are for a data point belonging to a specific class. Adam optimization algorithm has been used to update the weights in the network iteratively. A confusion matrix, as well as a classification report, was generated for the ANN model.

3. **One Class Support Vector Machine Algorithm**: In the present work, we trained the model on the data related to "normal" or "non-attack" data packets only. Thus, the model learns to flag-off any other packet that it encounters as an "attack" data packet. Using scikit-learn library's OCSVM classifier, we trained our model on the "normal" packets. In the OCSVM approach as well, we have used the RBF kernel, with gamma = 0.01 and nu = 0.95. The testing set consisted of a few "normal" packets and all the "attack" packets.

GUI Using Tkinter The user can give a CSV file as an input to the model from the GUI. After the CSV file is uploaded, the model tests this data using the trained model and generates the results after classification. For a better understanding of the results, we have used Matplotlib, which is a visualization library of Python (Fig. 1).

GUI Using Website GUI in the form of a website has been developed which can be used by all the users who have access to the internet. Since our work is Input/Output

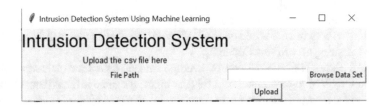

Fig. 1 Upload CSV using Tkinter

Fig. 2 Upload testing csv file
using website

Table 1 Analysis of proposed system

Algorithm	Accuracy (average)	Precision (average)	Recall (average)	F1-score (average)
Support vector machine	0.89	0.91	0.89	0.89
Artificial neural networks	0.91	0.91	0.91	0.91
One class support vector machine	0.93	0.93	0.93	0.90

Table 2 Comparative analysis

Reference no.	Algorithm used and accuracy
[2]	Stage 1 classification accuracy for SVM – 82.11%, Stage 2 multiclassification f1 score accuracy SVM – 75.5%
[4]	One-class support vector machine (OCSVM): PCA [Acc: 0.72, Precision: 0.99, Recall:0.51, F1: 0.67]; Encoder [Acc: 0.79, Precision: 0.96, Recall:0.64, F1: 0.77]

bound, using Node.js proved to be essential. The website asks the user to upload the test data and runs the backend Python file to generate the results. A windows toast notification system has been implemented in the model as an alert system to notify the user (Fig. 2).

4   Results

A comparative analysis of the three algorithms – SVM, ANN, and OCSVM – is done based on the following parameters: average accuracy, precision, recall, and F1-Score as shown in Table 1.

We also performed a comparative analysis of the accuracies of the papers reviewed by us which were based on the UNSW-NB15 dataset and utilizing different Machine Learning algorithms (Table 2).

Figure 3 shows the comparison of average rate of detection in case of SVM, ANN, and OCSVM algorithms. From this bar chart, it can be inferred that OCSVM has the highest average detection rate.

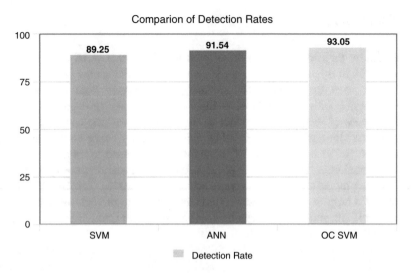

Fig. 3 Comparison of detection rates of SVM, ANN, and OCSVM algorithms

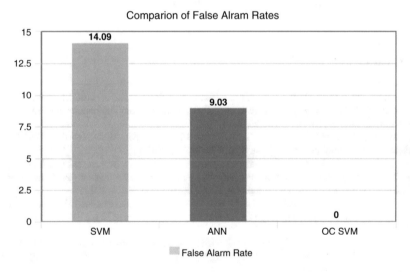

Fig. 4 Comparison of false alarm rates of SVM, ANN, and OCSVM algorithms

As shown in Fig. 4, which graphically compares the average false alarm rates of SVM, ANN, and OCSVM, we can infer that OCSVM has the lowest false alarm rate.

OCSVM outperforms both SVM and ANN in the two crucial aspects of the model – average detection rate and false alarm rate. Thus, this proposed method has enabled an optimized approach to detect intrusions in the network.

5 Conclusion

In this study, three different machine learning algorithms were used to predict intrusion within the network. The SVM model was implemented and a confusion matrix was used to interpret the percentage of examples that were classified correctly along with a classification report. This model generated an average accuracy score of 89.25%. The model based on ANN was implemented with two hidden layers along with the MLP-BP technique which produced a testing average accuracy rate of 91.54%. The OCSVM algorithm was used to train the model on data associated with "normal" data packets. Using the accuracy score parameter in OCSVM, the model reached an average accuracy of 93.04%. The OCSVM model can predict a large number of attacks and can be used to predict unknown attacks in the future. The entire system can be accessed by a user through online as well as offline GUI.

We intend to extend the scope of this work by learning and implementing the use of real-time data handling techniques. By handling live data, we can make our system more practical and robust. The alert system is currently Windows-based. It can be extended to other operating systems as well.

References

1. N. Moustafa, J. Slay, UNSW-NB15: a comprehensive data set for network intrusion detection systems (UNSW-NB15 network data set). 2015 military communications and information systems conference (MilCIS) (IEEE, 2015)
2. S. Meftah, T. Rachidi, N. Assem, Network based intrusion detection using the UNSW-NB15 dataset. Int. J. Comput. Digital Syst. 8(5), 478–487 (2019)
3. M. Al-Zewairi, S. Almajali, A. Awajan, Experimental evaluation of a multi-layer feed-forward artificial neural network classifier for network intrusion detection system. 2017 International Conference on New Trends in Computing Sciences (ICTCS) (IEEE, 2017)
4. D. Pérez, et al., Comparison of network intrusion detection performance using feature representation. International Conference on Engineering Applications of Neural Networks (Springer, Cham, 2019)
5. S. Kaiser, K. Ferens, Machine learning classifiers for network intrusion detection (2018)
6. M. Zamani, M. Movahedi, Machine learning techniques for intrusion detection. arXiv preprint arXiv:1312.2177 (2013)
7. E. Hodo, X. Bellekens, A. Hamilton, C. Tachtatzis, R. Atkinson, Shallow and deep networks intrusion detection system: a taxonomy and survey. arXiv preprint arXiv:1701.02145 (2017)
8. L.A. Maglaras, J. Jiang, T. Cruz, Integrated OCSVM mechanism for intrusion detection in SCADA systems. Electr. Lett. 50(25), 1935–1936 (2014)

Hyperparameter Optimization Algorithms for Gaussian Process Regression of Brain Tissue Compressive Stress

Folly Patterson, Osama Abuomar, and R. K. Prabhu

1 Introduction

Traumatic brain injury (TBI) is the most common cause of death and disability in the world [1]. The mechanical properties of brain tissue are a significant component of TBI yet are not fully understood. To this end, *in vitro* mechanical testing of post-mortem brain tissue is performed. Tension, compression, and shear tests are conducted at a wide range of strain rates. In the case of compression, a cylindrical sample of post-mortem brain tissue is compressed between two platens, usually at a constant strain rate, i.e., rate of deformation. However, the stress response of the brain can vary by several orders of magnitude because of the varying testing protocols. Brain samples can be tested from within a few hours of extraction to many days after [2] (post-mortem preservation time), and may be stored at ice cold, room, or physiological temperature in the meantime [3, 4] (storage temperature). The temperature at which specimens are tested may also vary from room to physiological temperature (testing temperature). The aspect ratio, or ratio of specimen diameter to thickness/height, varies due to variations in brain size between species. Understanding how these and other protocol factors affect the stress response of the brain is imperative to translating these results to the clinic.

The compressive response of brain tissue can be predicted from testing protocol parameters and brain specimen properties using regression methods [5]. Gaussian process regression provides a good estimation of the compressive stress from

F. Patterson (✉) · R. K. Prabhu
Center for Advanced Vehicular Systems, Mississippi State University, Starkville, MS, USA
e-mail: fdc33@cavs.msstate.edu; rprabhu@abe.msstate.edu

O. Abuomar
Department of Computer and Mathematical Sciences, Lewis University, Romeoville, IL, USA
e-mail: oabuomar@lewisu.edu

© Springer Nature Switzerland AG 2021   591
R. Stahlbock et al. (eds.), *Advances in Data Science and Information Engineering*,
Transactions on Computational Science and Computational Intelligence,
https://doi.org/10.1007/978-3-030-71704-9_41

the input parameters compared to multiple linear regression [unpublished data], but the performance can be improved by optimizing the hyperparameters using a search algorithm. Here, Bayesian optimization, grid search, and random search are compared for their ability to optimize Gaussian process regression hyperparameters.

2 Materials and Methods

2.1 Dataset

Data were collected from the literature on brain *in vitro* compression testing [3, 6–13] using a plot digitizer [14]. The strain, strain rate, brain matter composition (categorical variables), aspect ratio, storage temperature, testing temperature, and post-mortem preservation time were considered the input variables. Compressive stress was the output variable, and is the force divided by the current area of the brain specimen. Strain, strain rate, and stress were log-transformed to linearize the response. A total of 1559 observations were used.

2.2 Gaussian Process Regression

Gaussian process regression (GPR) was used to estimate the response, compressive stress. An output y that varies from a function $f(x)$ with Gaussian noise can be represented by

$$y = f(x) + \varepsilon \tag{1}$$

where the noise $\varepsilon \sim \mathcal{N}(0, \sigma_n^2)$ and the function $f(x)$ follow a Gaussian distribution with mean

$$m(x) = E[f(x)] \tag{2}$$

and covariance

$$k(x, x') = E[(f(x) - m(x))(f(x') - m(x'))]. \tag{3}$$

The function $f(x)$ can thus be written as [15]

$$f(x) \sim \mathcal{GP}(m(x), k(x, x')). \tag{4}$$

The joint distribution of the observed values y and the test outputs f_* under the prior probability is given as [15]

$$\begin{bmatrix} y \\ f_* \end{bmatrix} \sim \mathcal{N} \left(\mathbf{0}, \begin{bmatrix} K\,(X, X) + \sigma_n^2 I & K\,(X, X_*) \\ K\,(X_*, X) & K\,(X_*, X_*) \end{bmatrix} \right). \tag{5}$$

The predictive distribution of f_* given the training inputs X, observed outputs y, and testing outputs X_* is

$$f_* \mid X, y, X_* \sim \mathcal{N} \left(K\,(X_*, X) \left[K\,(X, X) + \sigma_n^2 I \right]^{-1} y, \right.$$
$$\left. K\,(X_*, X_*) - K\,(X_*, X) \left[K\,(X, X) + \sigma_n^2 I \right]^{-1} K\,(X, X_*) \right). \tag{6}$$

where the covariance function $K(X, X)$ is given by

$$K\,(X, X) = \begin{bmatrix} k\,(x_1, x_1) & \cdots & k\,(x_1, x_n) \\ \vdots & \ddots & \vdots \\ k\,(x_n, x_1) & \cdots & k\,(x_n, x_n) \end{bmatrix} \tag{7}$$

which can be parameterized by hyperparameters θ which are determined by maximizing the log likelihood function [15]:

$$\log P\,(y|X) = \frac{1}{2} y^T \left(K\,(X, X) + \sigma_n^2 I \right)^{-1} y - \frac{1}{2} \log \mid K\,(X, X) + \sigma_n^2 I \mid - \frac{n}{2} \log 2\pi \tag{8}$$

Here, the exponential kernel with a linear basis function was used:

$$k\,(x_i, x_j|\theta) = \sigma_f^2 \exp \left(-\frac{\|x_i - x_j\|}{\sigma_l} \right) \tag{9}$$

where the hyperparameters are σ_f, the signal standard deviation, and σ_l, the characteristic length scale.

2.3 Search Algorithms

2.3.1 Bayesian Optimization

Bayesian optimization fits a probabilistic model for the objective function $f(x)$. The predictive distribution contains evaluations of the objective function over the whole input space and is used by the algorithm to focus the search only on the most promising regions of the input space [16].

Bayesian optimization starts by evaluating $f(x)$ at random points. The Gaussian process model for $f(x)$ is updated to obtain a posterior distribution over the objective functions and a new input point is found that maximizes the acquisition function $a(x)$ [16]. This process continues until the maximum number of function evaluations has been reached.

The expected improvement-per second-plus acquisition function is used, where the expected improvement is

$$a_{EI}(x) = E\left[\max\left(0, \mu\left(x_{best}\right) - f(x)\right)\right] \tag{10}$$

where $\mu(x_{best})$ is the minimum of the posterior mean [16].

If the time taken to evaluate the objective function varies depending on the region of the input space, then the acquisition function is

$$a_{PS} = \frac{a_{EI}(x)}{\mu_S(x)} \tag{11}$$

And $\mu_S(x)$ is the posterior mean of the timing GP model [16].

2.3.2 Grid Search

For each hyperparameter, in this case σ_n (the noise standard deviation) and σ_l (the length scale), a set of values is chosen. Here, the range for the noise standard deviation is

$$\left[10^{-4}, \max\left(10^{-3}, 10\sigma_y\right)\right] \tag{12}$$

where σ_y is the standard deviation of the output variable. The range for the length scale is

$$\left[10^{-3}\max\left\{range\left(x_j\right)\right\}, \max\left\{range\left(x_j\right)\right\}\right] \tag{13}$$

Grid search divides each range into a set of m values, and searches among every possible combination of values so that the number of combinations is m^2 for two hyperparameters [17]. Here, $m = 10$.

2.3.3 Random Search

Random search searches the same space as grid search but uses a uniform distribution function to independently pick values at which to assess the objective function at each iteration. The resulting points chosen for evaluation are not uniformly distributed as they are for grid search [17].

2.4 Assessment

One hundred iterations of each algorithm were run with 5-fold cross-validation at each iteration [18], and each algorithm was assessed five times. The minimum objective function determined at each iteration and averaged over the five assessments. The total function evaluation time, RMSE, and log likelihood were found and averaged over the five assessments as well.

3 Results

The performance of the regression models can be found in Table 1. Grid search had the longest function evaluation time at almost two hours, but outperformed Bayesian optimization and random search in terms of RMSE (0.0429) and log likelihood (1420). Bayesian optimization had the fastest function evaluation time at about 90 min.

On average, Bayesian optimization reached the minimum objective function value after 63 evaluations (Fig. 1). Grid search found the minimum value after 29 evaluations and random search after 90 evaluations (Fig. 1).

The residuals plots for Bayesian optimization and grid search appear similar (Fig. 2). For random search, the residuals are more spread about the mean then for the other algorithms, corresponding with the algorithms' RMSE values (Table 1).

Table 1 Optimization results of each algorithm (S.D.)

Search algorithm	Function evaluation time (s)	RMSE	Log likelihood
Bayesian optimization	5436 (165)	0.0454 (0.0023)	1373 (41)
Grid search	6999 (109)	0.0429 (n/a)	1420 (n/a)
Random search	6981 (345)	0.0482 (0.0049)	1313 (95)

Fig. 1 Average minimum
objective function value (±
S.D.) at each iteration

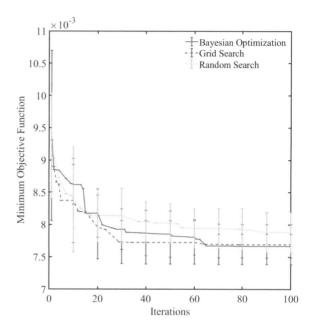

4 Discussion

This work focuses on the use of three algorithms to optimize the hyperparameters of
GPR: Bayesian optimization, grid search, and random search. Overall, grid search
optimizes the hyperparameters of Gaussian process regression in fewer iterations
than Bayesian optimization and random search, as well as results in a lower RMSE
and higher log likelihood. However, grid search has the longest function evaluation
time, while Bayesian optimization was the fastest. All algorithms completed 100
iterations in two hours or less.

Grid search can be further improved by searching among a finer grid mesh around
the optimal solution found in this work. Other algorithms such as genetic algorithms
or particle swarm optimization may provide better performance than the algorithms
used here. Further, the dataset used could benefit from additional data, as there is
a lack of variability in some of the input parameters. Here, only the compressive
stress of brain tissue was predicted, but there are other stress states and material
properties of interest for characterizing brain mechanical responses such as fatigue,
stress relaxation, and creep.

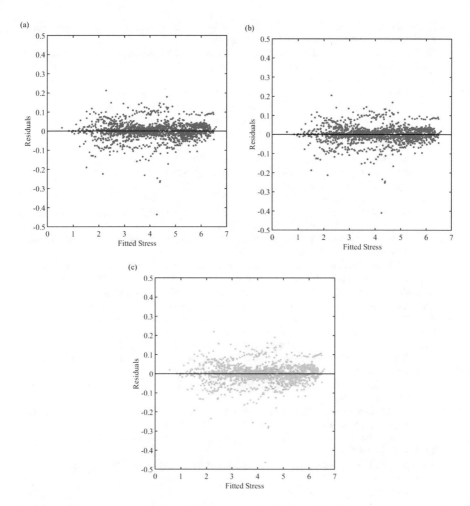

Fig. 2 Residuals vs. fitted compressive stress for (**a**) Bayesian optimization, (**b**) grid search, and (**c**) random search

Acknowledgment The authors are grateful to the Center for Advanced Vehicular Systems, Mississippi State University for their support.

References

1. M.C. Dewan et al., Estimating the global incidence of traumatic brain injury. J. Neurosurg. **130**(4), 1039–1048 (2018)
2. S. Budday et al., Mechanical properties of gray and white matter brain tissue by indentation. J. Mech. Behav. Biomed. Mater. **46**, 318–330 (2015)

3. B. Rashid, M. Destrade, M. Gilchrist, Temperature effects on brain tissue in compression. J. Mech. Behav. Biomed. Mater. **14**, 113–118 (2012)
4. M. Hrapko, J.A.W. Van Dommelen, G.W.M. Peters, J.S.H.M. Wismans, The influence of test conditions on characterization of the mechanical properties of brain tissue. J. Biomech. Eng. **130**(3), 03100301–03100310 (2008)
5. F. Crawford, J. Fisher, O. Abuomar, R. Prabhu, A multivariate linear regression analysis of in vitro testing conditions and brain biomechanical response under shear loads, in *International Conference on Data Science*, pp. 227–230 (2018)
6. K. Miller, K. Chinzei, Constitutive modelling of brain tissue: Experiment and theory. J. Biomech. **30**(11–12), 1115–1121 (1997)
7. F. Shen, T. Tay, J. Li, S. Nigen, P. Lee, H. Chan, Modified Bilston nonlinear viscoelastic model for finite element head injury studies. J. Biomech. Eng. **128**(5), 797–801 (2006)
8. F. Pervin, W.W. Chen, Dynamic mechanical response of bovine gray matter and white matter brain tissues under compression. J. Biomech. **42**(6), 731–735 (2009)
9. F. Pervin, W.W. Chen, Effect of inter-species, gender, and breeding on the mechanical behavior of brain tissue. NeuroImage **54**(1), 98–102 (2010)
10. J. Zhang et al., Effects of tissue preservation temperature on high strain-rate material properties of brain. J. Biomech. **44**(3), 391–396 (2011)
11. B. Rashid, M. Destrade, M.D. Gilchrist, Mechanical characterization of brain tissue in compression at dynamic strain rates. J. Mech. Behav. Biomed. Mater. **10**, 23–28 (2012)
12. B. Rashid, M. Destrade, M.D. Gilchrist, Determination of friction coefficient in unconfined compression of brain tissue. J. Mech. Behav. Biomed. Mater. **14**, 163–171 (2012)
13. Z. Li, H. Yang, G. Wang, X. Han, S. Zhang, Compressive properties and constitutive modeling of different regions of 8-week-old pediatric porcine brain under large strain and wide strain rates. J. Mech. Behav. Biomed. Mater. **89**(August 2018), 122–131 (2019)
14. A. Rohatgi, WebPlotDigitizer (2016). [Online]. Available: http://arohatgi.info/WebPlotDigitizer/app/. Accessed 12 Dec 2016
15. C.E. Rasmussen, C.K.I. Williams, *Gaussian Process for Machine Learning* (MIT Press, Cambridge, MA, 2006)
16. J. Snoek, H. Larochelle, R.P. Adams, Practical Bayesian optimization of machine learning algorithms. Adv. Neural Inf. Proces. Syst. **4**, 2951–2959 (2012)
17. J. Bergstra, Y. Bengio, Random search for hyper-parameter optimization. J. Mach. Learn. Res. **13**, 281–305 (2012)
18. T. Sergios, K. Koutroumbas, *Pattern Recognition*, 4th edn. (Academic, Burlington, 2009)

Competitive Pokémon Usage Tier Classification

Devin Navas and Dylan Donohue

1 Introduction

The gameplay of the Pokémon video game series focuses on the raising and battling of fictional creatures called Pokémon. Pokémon battles follow a simple format—two teams of Pokémon take turns attacking each other until all of one team's Pokémon have been knocked out. All Pokémon have a typing that determines what Pokémon types they are weak to and strong against, as well as unique numerical stats that help determine the level of damage they can give to and take from other Pokémon. *Pokémon Showdown* is a website used by competitive Pokémon players to build and battle with teams of Pokémon. Since it is much easier to build competitive Pokémon on *Pokémon Showdown* than it is in the Pokémon video games, many competitive Pokémon players use *Pokémon Showdown* to build and test teams that they plan to use in official Pokémon battle tournaments, whereas others use it as a method for competitive battling on its own. Pokémon on *Pokémon Showdown* are ranked into usage tiers based on how commonly they are used on players' teams. Generally, a Pokémon with a high usage is considered to be a good competitive Pokémon. A Pokémon's competitive viability takes into account many factors, most important being their stats, and their typing. It is typically the case that the Pokémon that are most commonly used in competitive battles have either high stats, good typing, or both.

Every competitive season for *Pokémon Showdown* determines the Pokémon available for use on teams based on what Pokémon are available in the most recently released Pokémon game. For this study, we used Weka to classify all the Pokémon that are currently available for use in competitive battle into the usage tiers from

D. Navas (✉) · D. Donohue
Department of Computer & Information Science, Fordham University, Bronx, NY, USA
e-mail: dnavas1@fordham.com; ddonohue7@fordham.edu

© Springer Nature Switzerland AG 2021

R. Stahlbock et al. (eds.), *Advances in Data Science and Information Engineering*,
Transactions on Computational Science and Computational Intelligence,
https://doi.org/10.1007/978-3-030-71704-9_42

Pokémon Showdown, listed here from most to least popular: Ubers, Overused (OU), Underused Borderline (UUBL), Underused (UU), Rarely Used Borderline (RUBL), Rarely Used (RU), Never Used Borderline (NUBL), Never Used (NU), and Partially Used (PU). The features evaluated for each Pokémon were each of their individual base stats, the sum of all their base stats—also known as their base stat total (BST)—how many type weaknesses the Pokémon has, and how many type resistances the Pokémon has.

Our motivation for pursuing this study was not to try and determine what Pokémon are considered the "best" to use, because the *Pokémon Showdown* usage tiers already give insight into this. Rather, since the metagame of competitive Pokémon is affected by more than just how high a Pokémon's stats are, we wanted to see how Pokémon might be classified based on their stats, and compare this to the reality of their usage by looking at how accurate the classifications were. Though we did expect to see many of the Pokémon from the Ubers and OU tier classified correctly, we were interested to see if the Pokémon from higher tiers might be misclassified into lower tiers, and what factors outside of stats and typing cause these Pokémon to be in a higher usage tier than predicted. In addition, in the case of Pokémon who were classified into tiers higher than their actual tier, we wanted to evaluate what aspects of the current metagame may cause the Pokémon's usage to be lower than expected. This information is useful, because it gives insight into how future trends may evolve in the competitive Pokémon scene, and what kinds of Pokémon may begin to rise or fall in usage. In addition, this model could be applied to new Pokémon that are added in future games, to assess how good they might perform on competitive teams, or how they might change the metagame.

There is no related work on the specific topic covered in this paper, but one study did use the random forest classifier to identify whether a Pokémon was a legendary based on its stats and type [1]. Legendary Pokémon are not categorized by any metrics, rather a Pokémon is just stated to be legendary according to the storyline of the game they appear in. However, many legendary Pokémon have very high stats and particular typings, so you can try to identify if a Pokémon is legendary by examining these features. In this way, Cardorelle's experiment is related to this study, since he also used stats and typing as features in his data set. Interestingly, our results tended to classify legendary Pokémon into the highest usage tier—Ubers—which is likely a result of many of them having high stats.

2 Background

2.1 Pokémon Typing

The most basic component to Pokémon battle strategies have to do with what "type" the Pokémon is, and consequently, that Pokémon's type weaknesses and type resists. A Pokémon can have up to two types, and can have any combination of types. A Pokémon's attacks also have a typing that corresponds to a single type—the move Water Gun is a water-type move, Flamethrower is a fire-type move, and so on. Type weaknesses are when a Pokémon type takes high damage from the moves of another type. For example, grass-type Pokémon are weak to fire-type moves, fire-type Pokémon are weak to water-type moves, and water-type Pokémon are weak to grass-type moves. The same goes for a Pokémon's type resists, which are the moves of a certain type that are not very effective against that Pokémon's type. One example of this is that a water-type move would do very little damage against a grass-type Pokémon. Therefore, continuing with this example, if a water-type Pokémon and a grass-type Pokémon face each other in battle, the grass-type Pokémon will have a natural advantage, since it has grass-type moves that are strong against water Pokémon, and is resistant to water-type moves. This is an abridged explanation, since there are many other factors that come into play regarding what types of moves a Pokémon could learn, but for simplicity's sake, we consider the typing of a Pokémon to encapsulate all of this information. Overall, if a Pokémon has many type resists, and few type weaknesses, this would be considered an ideal scenario.

2.1.1   Base Stats

The next most basic component to interpreting a Pokémon's viability in competitive play is the Pokémon's base stats. Base stats are numerical values that range from 1 to 255. Every Pokémon has six stats: HP, Attack, Defense, Special Attack, Special Defense, and Speed. The actual stats of a Pokémon in the games would be much higher than their base stats—oversimplifying the definition a bit, base stats are the essence of that Pokémon's potential for a specific stat. For example, since the Pokémon Diglett has a base HP stat of 10, it can be inferred that that Pokémon's in-game HP stat would be extremely low. Conversely, if a Pokémon's base Defense stat was rather high, something like 150, that would indicate that their in-game Defense would be high. Each stat contributes something different to a Pokémon's power, and while it is generally best to have the highest stats possible, any single base stat having a value greater than 100 is considered to be good in the competitive sphere. A Pokémon's base stat total (BST) is the sum of each of its base stats, and as is the case with base stats, generally the higher a Pokémon's BST, the better it is in competitive play.

3   Experiment Methodology

3.1   Data Sets

Since the number of type resists and type advantages, and the value of base stats are the most easily quantifiable values relating to a Pokémon's competitive viability and therefore their usage on *Pokémon Showdown*, these are the features we decided to use when compiling our data set. We were unable to find any readily available sets online with these specific features for the most up-to-date competitive Pokémon, so we compiled the data ourselves. Our data set consists of 297 Pokémon, each of their base stats, their BST, their number of type resists, and their number of type weaknesses. We gathered this data using information from *Smogon*, an online forum that helps run *Pokémon Showdown* and has statistics for every Pokémon currently usable on *Pokémon Showdown*. In total, there were 17 Pokémon in Ubers, 39 Pokémon in OU, 9 Pokémon in UUBL, 59 Pokémon in UU, 5 Pokémon in RUBL, 44 Pokémon in RU, 4 Pokémon in NUBL, 43 Pokémon in NU, and 77 Pokémon in PU, making PU the majority class for this data set.

3.2 Algorithms

The three algorithms we used in this experiment are Weka's J48 Decision tree, Weka's lazy IBk 1-nearest neighbor, and Weka's logistic regression algorithm. Logistic regression was advantageous to use because it provided us with numerical weights for our features and thus indicated which features were considered the most important when classifying Pokémon into the usage tiers. Seeing which features had the highest weights also allowed us to try and determine what was the main cause of misclassification errors for this algorithm.

3.3 Setup Methodology

We used two combinations of methods to prepare our data to be run through each of the algorithms. Overall, we did two different sets of runs with each of the algorithms, for a total of 6 runs overall. For both sets of runs, we used 10-fold cross validation. One set of runs was, additionally, put through some preprocessing. Since the majority class, the PU class, had a far greater amount of Pokémon in it than many of the other classes, we decided to use the Weka preprocessing filter "Resample" to resample without replacement. This reduced the size of some of the classes to balance the data by undersampling the OU, UU, RU, NU, and PU classes.

3.4 Evaluation Metrics

After running each of the different algorithms on the data sets, we looked at the confusion matrix, and other measures such as accuracy and precision for each of the runs to evaluate our results. The confusion matrix allowed us to see where misclassification errors occurred, and what the nature of these outliers was. Though we did aim to get the highest precision and accuracy possible, we were also interested in seeing what Pokémon were commonly misclassified into the wrong usage tier. If a Pokémon is often classified as a lower tier than its actual tier, it could imply that Pokémon has a value on competitive teams that is not discernible by looking only at its stats and typing. Conversely, if a Pokémon is usually put into a higher tier than its actual tier, it could imply that it might be used less for very specific reasons. An example of this would be if a Pokémon has good typing and stats, but just happens to be weak to many of the Pokémon in the higher usage tiers. Analyzing these outliers and trying to interpret why a Pokémon may have been misclassified in a certain way was one of the most engaging parts of our evaluation process.

4 Results

The runs without undersampling gave the overall highest accuracy results, and out of all the algorithms, logistic regression had the highest accuracy (Table 1). However, despite it performing the best in the metric of accuracy, the highest accuracy it achieved in each of its runs was still relatively low at 32.65%. We looked at what feature the logistic regression algorithm was giving the highest weight to try and figure out what might have been causing these numbers, and learned that it was weighing the type resist feature the most heavily with a weight .3, whereas the second highest weight was BST with a weight of .1. Having a high number of type resists is good for a Pokémon to have, but having high stats is certainly more important in the consideration of whether a Pokémon is competitively viable or not, so having the classification consider the wrong feature the most important is likely what caused the accuracy to remain low for logistic regression.

For the runs using cross validation without undersampling, the accuracies were a bit low, with the highest accuracy being logistic regression with an accuracy of 32.65% (Table 1). For each of the algorithms, the most precise predictions made were in Ubers and PU (Table 2). This makes sense, because Pokémon in Ubers are generally Pokémon with the highest BSTs and are easy to identify as strong, although there were some cases where Ubers Pokémon were placed in lower tiers despite their high stats. These types of misclassifications were the most notable since several of the higher tier Pokémon from Ubers and OU were classified into the lower tiers of PU or NU. One reason PU may have had such a high precision is

Table 1 Accuracy(%) using cross validation

Methods	Algorithm		
	J48	Lazy IBk	Logistic regression
Without undersampling	32.32	29.60	32.65
With undersampling	26.00	27.00	29.50

Table 2 Precision of classifications—cross validation without undersampling (vs with undersampling)

Usage tier	Algorithm		
	J48	Lazy IBk	Logistic regression
Ubers	.737 (.743)	.867 (.867)	.750 (.824)
OU	.333 (.333)	.205 (.300)	.237 (.200)
UUBL	0.00 (0.00)	0.00 (.429)	0.00 (0.00)
UU	.196 (.182)	.288 (.237)	.304 (.281)
RUBL	0.00 (0.00)	0.00 (0.00)	0.00 (0.00)
RU	.250 (.115)	.136 (.103)	.188 (.233)
NUBL	0.00 (.250)	0.00 (0.00)	0.00 (0.00)
NU	.116 (.194)	.116 (.194)	.148 (.233)
PU	.506 (.289)	.506 (.323)	.378 (.300)

that PU Pokémon often have very low stats, so any Pokémon with lower stats could have been correctly assumed to be in PU. In addition, since PU is the majority class, it is expected that many Pokémon might be misclassified into it.

Some Pokémon were misclassified in the same way by all of the algorithms in the set of runs without undersampling. The Pokémon Ditto and Mimikyu were consistently classified as PU and NU, respectively, but both of these Pokémon belong to the OU tier. In addition, the Ubers tier Pokémon Darmanitan-Galar was put into the PU tier by each algorithm, a drastic discrepancy in classification. The borderline tiers, UUBL, RUBL, and NUBL, were never predicted correctly for this set, but there are so few of these Pokémon, only 18 in total, that we expected these tiers to have low precision for most of the runs.

The runs using cross validation with undersampling had lower accuracies for each of the algorithms than the runs without undersampling, with the highest accuracy being logistic regression again, with an accuracy of 29.50% (Table 1). Although the overall accuracy was lower than the runs without undersampling, certain classes were predicted with higher precision, and the lazy IBk from this run had the best precision out of any of the algorithms in this and the other run. Ubers and PU were classified with the highest precision again, but some of the borderline tiers were actually predicted correctly this time, with J48 predicting NUBL with a precision of .250 and Nearest Neighbor predicting UUBL with a precision of .429 (Table 2).

Many of the outliers in this run were similar to the runs without undersampling, though Darmanitan-Galar was placed into the slightly higher NU tier rather than the PU tier. PU was predicted with a lower precision than it had in the other run, likely due to the undersampling. This is also likely what allowed for other classes, like Ubers and the borderline tiers, to be predicted correctly more often.

4.1 Interpreting Misclassifications

Since certain Ubers and OU Pokémon were misclassified into the lower tiers so consistently, we analyzed these cases to try and figure out if there were any common characteristics between these misclassified higher-tier Pokémon. The cases we found the most interesting were the Pokémon Ditto, Mimikyu, and Darmanitan-Galar, all of whom are high-tier Pokémon that are known to be very popular on competitive teams, but were never classified correctly throughout all of the sets of tests. Darmanitan-Galar is in Ubers, while Ditto and Mimikyu are in OU, but it makes sense for all of them to have been frequently misclassified because the strength of all of these Pokémon cannot be captured purely by their stats or typing. Each of these Pokémon have a unique skill called their "ability," which gives these Pokémon certain tactical advantages once in battle. Every Pokémon has an ability, though each Pokémon only has a certain pool of abilities that they can choose from. Most Pokémon share the same abilities among each other, but certain Pokémon have abilities unique to only that Pokémon. This is the case for Ditto, Mimikyu, and Darmanitan-Galar, all of whom have an ability that no other Pokémon has, with all of these abilities also being extremely useful in battle.

(a) *Ditto*: In Ditto's case, all of its stats have a value of 48, giving it a very low BST of 288, and while it only has one type weakness, it also has only one type resist. As such, it would make sense for Ditto to be classified into a lower tier based purely upon its stats and typing. However, Ditto's unique ability Imposter allows it to transform itself into the opponent's Pokémon once it enters a battle. This then grants Ditto the stats and typing of the opposing Pokémon, essentially giving Ditto the potential to match the strength level of any opponent. This versatility allows Ditto to be extremely flexible on any competitive team, and is thus why Ditto is in the OU tier.

(b) *Mimikyu*: Mimikyu's stats are generally average, with three of its stats being near or just above 100 (Attack: 90, Sp.Def: 105, Speed:96), and a BST of 476. It has 4 type resists and only 2 type weaknesses, which is a pretty good ratio, but overall Mimikyu's typing and stats are pretty average in comparison to most competitively viable Pokémon. What sets Mimikyu apart is its ability, Disguise, which protects it from taking damage from the first attack it receives from an enemy Pokémon. Getting one turn to take a free hit without any damage gives Mimikyu a huge advantage, since it can utilize this damage-free turn to use moves that increase its stats, or some other set-up based move that will give it the advantage over its opponent.

(c) *Darmanitan-Galar*: Darmanitan-Galar's stats are generally better than those of both Mimikyu and Ditto, but its BST of 480 is not much higher than Mimikyu's, and though its Attack of 140 is impressive, its ratio of 3 type weaknesses to 1 type resist make it understandable that it might have been classified into a lower tier. However, Darmanitan-Galar's ability Gorilla Tactics helps give it a major advantage in battle by multiplying its Attack by 1.5, which is a huge boost to this already high stat. Darmanitan-Galar is often given certain stat-boosting items in battle that increase its stats even higher in addition to the boost it gets from its ability. Pokémon can each hold one item while in battle, and these items are generally used to give Pokémon certain stat boosts or status effects. The items commonly given to Darmanitan-Galar in battle, the Choice Scarf and Choice Band, are both items that increase stats; Choice Scarf multiplies its holder's Speed by 1.5 while Choice Band multiplies its holder's Attack by 1.5. As a result, a Darmanitan-Galar could enter a battle with its stats already ridiculously boosted, having either its Attack and Speed both multiplied by 1.5 while holding the Choice Scarf, or, outrageously, its Attack multiplied by 3 while holding the Choice Band. These factors are what cause Darmanitan-Galar to be in the Ubers tier, since it can, given the right circumstances, sweep the entire enemy team by itself with the help of these stat boosts.

Other notable classification errors were those of Pokémon that do have objectively high stats and good typing, but do not fit into the current competitive metagame of its respective tier. An example of this is the Pokémon Escavelier, which has 9 type resists and only one type weakness, and three stats over 100. It would make sense that a Pokémon like this might belong to the OU tier, where Escavelier was commonly misclassified into. However, Escavelier's one type weakness is that it is 4x weak to fire type attacks, meaning any fire-type move used against it is multiplied by 4. This makes Excavelier much less viable despite its good stats and high number of type resists, since many commonly used Pokémon in the current metagame can easily defeat it as a result of this weakness, which is why it is currently placed in the RU tier.

5   Conclusion

Overall, logistic regression without undersampling yielded the highest accuracy of classifications, though this value was relatively low at 32.65%. As such, it is clear that there are factors that made it difficult to reliably predict the usage tiers, likely based on the parameters we set for this study, and there is certainly room for improvement in this area. In future studies, getting more accurate results would be beneficial, because this would allow us to use this model more efficiently if we wanted to try and predict the usage tiers of new Pokémon, or to apply this model of usage classification to other competitive esport games.

One of the more interesting aspects of analyzing the misclassification results was to see what kinds of Pokémon were misclassified most frequently. There was certainly a noticeable trend of Pokémon who are competitively viable thanks to the effects of their ability rather than just their stats or typing being misclassified into much lower tiers than their actual tier, as we saw with Ditto, Mimikyu, and Darmanitan-Galar. As such, it is clear that not incorporating abilities as a feature is one of the shortcomings of this study. Initially, we had decided not to use abilities as one of our features because they are not easily quantifiable, and we feared it might create excess noise as a result of us not being able to quantify the usefulness of a Pokémon's ability properly. However, after seeing how big of an impact not considering abilities had on the misclassification of many of the higher-tier Pokémon, if we were to redo this study in the future, it is clear that abilities should be taken into account as one of the features we evaluate on. This could possibly be done by trying to "rank" every ability by assigning it a numerical value on a small scale, like 1–5, that would try to assess how useful a Pokémon's ability is in battle. Given this example, Pokémon like Ditto, Mimikyu, and Darmanitan-Galar would all have a 5 on this scale, and perhaps this could help to increase the precision of the higher-tier classifications. Other qualitative aspects to a Pokémon's competitive viability, like the strength of what moves they can use, could also be taken into account using a scale like this if we were to expand upon this study in the future.

This study allowed us to create a method for analyzing the potential strength of Pokémon for use on competitive teams. As the *Pokémon Showdown* tiers are updated over time, this could give insight into what kinds of Pokémon have the potential to become more popularly used. For example, if the Pokémon that counter Escavelier began to fall out of popularity, we could predict Escavelier's rise in usage, based on the fact that it has the stats and typing to be ranked higher than it currently is. The greatest potential for applying this study, though, is when new Pokémon are added in future games, in which case, we could use the data from this study to try and predict how effective these Pokémon could be in battle based on their stats, even before they become used in official competitive matches. While there are no competitive games with mechanics particularly similar to Pokémon, our methods for this study could be applied to any competitive esports game with quantifiable character or usage stats, like League of Legends or Overwatch, to assess how changes to a character's stats might affect their usefulness, how one character's increased usage might affect the usage of another character, or to predict how newly introduced characters might fit into the metagame.

Reference

1. S. Cardorelle, Identifying legendary Pokémon using the random forest algorithm. Medium, Towards Data Science, 6 June 2019, towardsdatascience.com/identifying-legendary-pok%C3%A9mon-using-the-random-forest-algorithm-ed0904d07d64. Accessed 5 Apr 2020.

Mining Modern Music: The Classification of Popular Songs

Caitlin Genna

1 Introduction

Throughout history, what is considered "popular" music has changed periodically as new genres emerge and older genres fall out of fashion. Modern-day pop music is often criticized for being uninspired. Every song is seen as formulaic, using the same chord progressions, the same background beat track, etc. The goal of this project is to use data mining to create a model that can effectively represent what makes a song popular, or a "hit" in this modern era, and from that model be able to assess just how similar modern pop songs really are. Other interesting relationships could potentially be discovered through such a model, such as what musical elements (for example key, tempo) are most important in making music enjoyable, and how those elements combine and interact to create a successful song.

Spotify's API allows developers access to a list of audio feature values for every song on their platform, and this project is based on a compilation of 19,000 songs catalogued with various attributes as well as a popularity rating for each [3]. Spotify's audio features have been used in the past for data mining projects, such as a project done by Juan De Dios Santos of *Medium,* who used Spotify data from his own playlists and that of his friend to assess how "boring" his music taste is [4].

Since the goal of this project is to determine what makes a song popular or not, it was appropriate to make a clear distinction between the classes and treat the data as nominal rather than a numerical spectrum. This is more suited to decision tree algorithms like J48 and Random Forest, and other rule-based algorithms such as RIPPER are helpful in identifying feature relationships, e.g., any commonalities in what makes "hit" music popular.

C. Genna (✉)
Fordham University, New York City, NY, USA

© Springer Nature Switzerland AG 2021
R. Stahlbock et al. (eds.), *Advances in Data Science and Information Engineering,*
Transactions on Computational Science and Computational Intelligence,
https://doi.org/10.1007/978-3-030-71704-9_43

2   Attribute Definitions [2]

Song_name: title of the track as it appears on Spotify.

Song_popularity: measure of popularity based on the number of "listens" for each song and user ratings on a scale of 0–100.

Acousticness: measure of acousticness on a decimal scale of 0–1, 1 being the most acoustic.

Danceability: measure of danceability (based on a number of factors such as tempo, beat strength, etc.) on a decimal scale of 0–1, 1 being the most danceable.

Energy: measure of intensity, on a decimal scale of 0–1, in which 1 is the most energetic (fast, loud, etc.)

Instrumentalness: the amount of vocals vs. instruments in the track on a decimal scale of 0–1, 1 being the most instrumental.

Key: the overall key of the track, represented in accordance with standard pitch class notation. (0=C, 1=C#/D♭, 2=D, 3=D#/E♭, etc.)

Loudness: average loudness of track measured in decibels (dB), on a scale of -38.8 to 1.58.

Audio_mode: indicates the type of scale, 0 being minor and 1 being major.

Liveness: measure of probability that a track was recorded live, on a decimal scale of 0–1, 1 being the highest probability that the track is live.

Speechiness: measure of spoken word in a track on a decimal scale of 0–1, 1 being the most speech-like.

Tempo: average tempo measured in beats per minute (BPM).

Time_signature: overall time signature of the track, on a scale of 0–5.

Audio_valence: measure of the overall mood of the track on a decimal scale of 0–1, 1 being the most positive (happy, upbeat, etc.), and 0 being the most negative (sad, angry, etc.)

3   Pre-Processing

The dataset was originally in .csv format and had to be converted to .arff format. This involved the addition of headers and proper formatting for feature listings, as well as the removal of commas in elements of the song_name feature. All further processing and experimentation was done using Weka [1] software, and any filters and algorithms mentioned are referred to as they are named in the program.

Once the data was correctly formatted, the first pre-processing step was to convert the data to nominal type because most of the features were of numerical type, which is not ideal for decision tree algorithms such as J48. This was done by applying the Discretize filter to all but one of the features. The song_name feature was of string type and almost every element was unique, so discretizing the data was not productive. In addition, because of its uniqueness, song_name was not valuable in formulating generalized rules for classification (and in theory should not have great influence over whether a song is popular or not) and therefore the feature was removed and not included in the dataset. Features that had very few distinct values were merged into two groups (audio_mode, for example, with values 0 or 1) to eliminate unnecessary categories that would have no instances.

4   Experimentation

4.1   Testing Phase I

To begin testing, the values of the class attribute song_popularity were merged into four groups, to make the dataset more compatible with decision tree algorithms. The ranges of these groups were 0–24, 25–49, 50–74, and 75–100.

The first algorithm ran on the dataset was ZeroR, to establish a baseline for accuracy. The resulting accuracy was 47.29%, with 52.71% of instances classified incorrectly. This resulted from the simple rule of the majority class being chosen every time. The confusion matrix (Table 1) for ZeroR revealed that the majority class was c = medium, which was the 50–74 song_popularity range, making up 47.29% of the data.

With that baseline in mind, the J48 decision tree algorithm was run on the dataset. The model was built in 0.84 seconds and resulted in a 52.22% classification accuracy. In addition, the precision rates were extremely low for each class, the lowest being 0.197 (for the range of 0–24) and the highest being 0.607 (for the range of 75–100). As revealed by the confusion matrix, the vast majority of errors occurred in classifying the middle range, 50–74. An unpruned version of the J48 algorithm was tested as well, resulting in an even lower accuracy (48.73%).

In addition, the Naive Bayes classifier was run, resulting in the lowest accuracy of the three methods, 45.99%. In fact, the algorithm was unable to make any correct predictions for the lowest popularity range. Since Naive Bayes assumes that all attributes are independent of each other, such poor performance of the algorithm would suggest that the musical elements that make a song popular *do* have a relationship.

Table 1 Confusion matrix

A	B	C	D	←classified as
0	0	1854	0	A = veryLow
0	0	3959	0	B = low
0	0	8893	0	C = medium
0	0	4098	0	D = high

Classification results of the ZeroR algorithm, the target class being song_popularity

4.2   Testing Phase II

Since the first round of algorithms performed rather poorly – the best performance (J48 pruned) resulted in only a 4.93% accuracy increase from the baseline – it was clear the data needed to be processed differently in order to improve predictive performance. Instead of four classes for song_popularity, the feature was split into two: notPop (range of 0–69) and pop (range of 70–100). These groups were chosen to represent songs that were not popular and those that were popular, in other words "hit" songs. This division of the feature was much less even than the division during the first testing phase; the new division resulted in 14,706 instances (78% of data) in the notPop class and 4098 instances (22% of data) of the pop class. This division seemed reasonable, with the notPop range ending around the 75th percentile and considering approximately a quarter of the songs (the most highly rated ones) as "popular," as naturally it is rarer for a song to be a hit than to place average or less than average on the musical charts. Once this grouping change was made using the MergeManyValues filter, the algorithms were run a second time.

First, a new baseline for accuracy had to be established for the reconfigured data. The ZeroR algorithm was used once again and resulted in a 78.20% success rate (as expected in relation to the number of instances making up the majority class).

The J48 decision tree algorithm saw a drastic increase in accuracy, with 81.78% of instances classified correctly and an average precision rate of 0.80. In addition, the RandomForest algorithm was applied (using J48 trees) and resulted in an even higher classification accuracy of 92.50% with an average precision rate of 0.924. Interestingly, the majority of errors resulted from incorrectly classifying popular songs as non-popular.

The Naive Bayes algorithm was also given a second chance, resulting in a 76.77% accuracy. The algorithm had high precision (0.80) in predicting the notPop class, but poor precision (0.41) in predicting the pop class. While this performance is an improvement from the first round of testing, it does exceed or even meet the baseline for accuracy established by ZeroR.

K-means, a clustering algorithm, was also tested but resulted in poor accuracy (56.76%) even after the normalization of the data scales.

The JRIP algorithm resulted in an accuracy of 80.35% and was able to predict the pop class with a false-positive rate of only 0.019 by utilizing 31 rules for what makes a song popular (to be discussed later).

Further experimentation was done in an attempt to increase the accuracy of the above algorithms even more, but manipulation of the data in many cases resulted in a slight decrease in accuracy. For example, the merging of the loudness feature into three groups (low, medium, loud) decreased the accuracy of classification with a J48 decision tree by approximately 0.3%. It should also be noted that the above classification algorithms were tested with 10-fold cross validation, as a lower amount of folds were found to be less effective (5 folds resulted in a 1% decrease in accuracy).

5 Feature Relations

Through utilization of Weka's attribute evaluator and association functionalities, some interesting relationships between the features were revealed. The Apriori algorithm was used to generate common itemsets and produced several connections to the notPop class. In summary, the notPop class was frequently correlated with low instrumentalness (< 0.1) and low speechiness (< 0.1). These groups were generated through the Apriori algorithm using a minimum confidence of 0.8 and a minimum support of 0.55. Unrelated to the class feature, the Apriori algorithm also identified the group of song_duration_ms of 190,734.6–369,469.2, instrumentalness of -inf-0.0997 and time signature of 4/4 as a "best rule" or frequently occurring group of attributes, with a confidence value of 0.9. In simpler terms, those attribute values mean a song length of 3–6 min, low instrumentalness, and a 4/4 (one of the most commonly used) time signature. This makes sense, as it is what we would normally imagine as the modern "hit," or something you would hear on the radio.

The InfoGainAttributeEval algorithm was used in conjunction with the Ranker algorithm to produce an ordered list of attributes that offer the most information gain in the classification task. The list produced in order of most to least information gain is as follows: instrumentalness, loudness, acousticness, danceability, energy, audio_valence, key, speechiness, song_duration_ms, tempo, liveness, time_signature, audio_mode.

The JRIP algorithm discussed above in the second phase of testing also revealed interesting relationships between the features through the 31 rules generated for the pop class. Some attributes appeared very frequently in the rules. Figure 1 shows some of the most frequently appearing attributes and the number of rules they appeared in.

Many of the most frequently occurring features in the JRIP generated ruleset are also the features that resulted in the most information gain, and so they are reinforced as the most relevant features in determining whether a song will be popular or not.

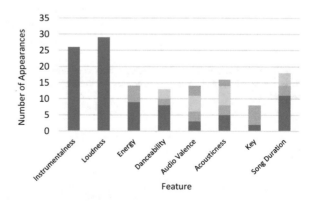

Fig. 1 Rule appearances Colors represent different values within a feature, amounting to each feature's total number of appearances in 31 rules generated by the JRIP algorithm

It is worth giving special attention to some of the rules produced by the JRIP algorithm, as they seem to be indicative of certain genres of music that are popular today. For example, one rule consisted of the following attributes: instrumentalness < 0.98, key of C, danceability > 0.89, and speechiness 0.58–0.47.

This rule stood out because of the high speechiness feature, which is a minority in the dataset as a whole. The other features of medium instrumentalness, high danceability, and the key of C, combined with the high speechiness, suggest this may be a rule describing popular rap songs. Another interesting rule pertains to longer songs (which are relatively rare in the dataset), which the rule dictates are popular if the song is live.

Other valuable relationships were revealed through the SimpleKMeans algorithm. Although the classification itself had low accuracy, the related visuals were useful in identifying relationships. For example (as seen in Fig. 2) songs with high acousticness are popular at lower loudnesses; then once a certain loudness is reached, all levels of acousticness have very high popularity rates, but any louder than that and the probability of the song being popular decreases dramatically across all acousticness levels. The relationship between acousticness and energy is similar (Fig. 3); high acousticness goes well with low energy, then as energy increases the relationship flips and favors low acousticness, but too much energy results in an unpopular song.

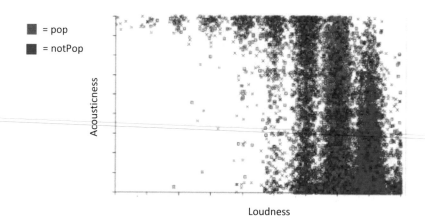

Fig. 2 Acousticness v. Loudness A comparison of songs classified as pop and notPop based on their levels of acousticness and loudness, increasing to the right and up on the x and y axes respectively

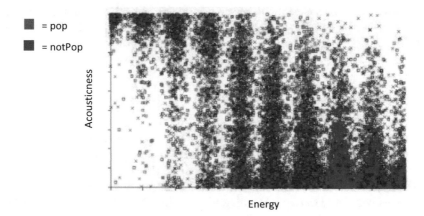

Fig. 3 Acousticness v. Energy A comparison of songs classified as pop and nonPop, based on their levels of acousticness and energy, increasing to the right and up on x and y axes respectively

Table 2 Accuracy results

	Accuracy	
Algorithm	(Phase 1)	(Phase 2)
ZeroR	47.29%	78.20%
J48	52.22%	81.78%
J48 Unpruned	48.73%	–
Random forest (J48)	–	92.50%
Naive Bayes	45.99%	76.77%
K-Means	–	56.76%
JRIP	–	80.35%

Summary of accuracy results for all algorithms tested in Phase 1 and Phase 2 of experimentation

6 Results

After two phases of testing and re-processing the data to make it more agreeable to decision tree algorithms and consolidating extraneous ranges for more accurate classifications, the random forest algorithm performed the best. As seen in Table 2, random forest was able to classify songs as popular or not popular with an accuracy of 92.50%.

Although other algorithms such as k-means and JRIP did not perform as well in terms of accuracy, they were useful in determining and visualizing relationships between the features and the class. Interesting correlations were revealed, such as popular songs being related to low instrumentalness and high danceability. Other features were correlated to the class through their relation to each other, such as levels acousticness and energy. Furthermore, some features had sweet spots that corresponded to popularity, such as being in the key of C or being 3–6 minutes long.

Of course, there were exceptions, specific instances where less common features come together to classify as popular, such as high liveness and a longer song duration.

7 Conclusion

Overall, the class of song popularity was able to be predicted to a high degree of accuracy, and several hard rules for what makes a song popular were identified. This all serves to support this project's original idea that modern day popular music is very similar in its composition, and perhaps creating a hit song is less about creativity and more about putting together the right combination of musical elements. Some aspects of technology and musical culture help explain this, such as the popularity of radio as a music outlet relating to a 3–6-min song duration as the acceptable length for a popular song (although streaming services are now taking over as the popular outlet). However, the fact that there were 31 different rules produced by the JRIP algorithm to cover what makes a popular song suggests that there is some variation in the class, and the musical community still enjoys songs from different genres, moods, etc.

One limitation of this project would be the use of nominal attributes and the necessity of delineating a point on the 0–100 scale as popular and unpopular. There may have been a song with a 69 popularity rating that was extremely similar to a song with a 70 popularity rating, but was considered unpopular because of the scale distinction and therefore influenced the algorithms in a less favorable way. Further research could be done on this dataset as a regression task by considering the popularity scale in its purely numerical state to see if more precise popularity ratings could be predicted. All in all, the study of popular trends in music has become easier through the use of data mining and availability of streaming data, and can potentially reveal relationships that inform us about our musical culture.

Acknowledgments I thank Dr. Gary Weiss (Fordham University) for sparking my interest in the powerful field of data mining.

References

1. F. Fibe, I.H. Witten, *Weka 3.8* (Hamilton, New Zealand, 1993)
2. "Get Audio Features for a Track," Spotify for Developers
3. E. Ramirez, *19,000 Spotify Songs* (Kaggle, 2019)
4. J.D.D. Santos, Is my Spotify music boring? An analysis involving music, data, and machine learning. Towards Data Sci (2017)

The Effectiveness of Pre-trained Code Embeddings

Ben Trevett, Donald Reay, and Nick K. Taylor

1 Introduction

Transfer learning is used to improve the performance on task T_i by using a model that has been first trained on task T_j, i.e., the model has been *pre-trained* on task T_j. For example, a model is trained to predict missing words in a sentence and then trained to predict the sentiment of a sentence. Pre-trained machine learning models are commonly used in natural language processing (NLP). Traditionally, transfer learning in NLP only pre-trained the *embedding layers*, which transform words into vectors using methods such as word2vec [11]. It is now common in NLP to pre-train all layers within a model and use a task-specific "head" that contains the only parameters which are not pre-trained [5, 7, 13].

In this paper, we explore transfer learning on programming languages. We test the transfer learning capabilities on two tasks, code retrieval and method name prediction, using models pre-trained on datasets with different characteristics. We show that transfer learning provides performance improvements on both tasks. Our results using models pre-trained on the different datasets suggest that semantic similarity between the variables and method names is more important than the source code syntax for these tasks.

Our contributions are as follows: (1) We propose a method for performing transfer learning in the domain of programming languages. (2) We show that transfer learning improves performance across the two tasks of code retrieval and method name prediction. (3) We show that the programming language of the pre-training dataset does not have to match that of the downstream task language. (4) We show

B. Trevett (✉) · D. Reay · N. K. Taylor
Heriot-Watt University, Edinburgh, UK
e-mail: bbt1@hw.ac.uk; d.s.reay@hw.ac.uk; n.k.taylor@hw.ac.uk

© Springer Nature Switzerland AG 2021
R. Stahlbock et al. (eds.), *Advances in Data Science and Information Engineering*,
Transactions on Computational Science and Computational Intelligence,
https://doi.org/10.1007/978-3-030-71704-9_44

that the pre-training dataset English language data provides comparable results to pre-trained on programming languages data.

As far as the authors are aware, this is the first study into the use of pre-training on code which investigates the use of datasets containing data that does not match that of the downstream task and also datasets which do not contain data in the downstream task language.

2   Related Work

For transfer learning, the traditional methods, such as word2vec [11] and GloVe [12], are only able to pre-train the embedding layers within a model. These methods have been succeeded by recent research on contextual embeddings using language models, such as ULMFiT [7], ELMO [13], and BERT [5], which have shown to provide state-of-the-art performance in many NLP tasks, but have not yet been widely applied to programming languages.

The use of machine learning on source code has received an increased amount of interest recently [1]. There has been work on learning embeddings from code [3, 16], yet research using applying pre-trained source code embeddings to downstream tasks is limited. Two examples are NL2Type [9] and DeepBugs [14], both of which only use word2vec for pre-training the embeddings but do not perform transfer learning. Recently Feng et al. [6] pre-trained a model for the code retrieval task, but did not test the ability to perform transfer learning when there is a mismatch between pre-training and downstream task languages.

3   Methodology

We perform two tasks: code retrieval and method name prediction. Both tasks use the CODESEARCHNET CORPUS [8] which contains 2 million methods and their associated documentation. The dataset contains six programming languages, but we only test our models on the Java examples.

The code retrieval task is to accurately pair each method, c_i, with query, d_i. We encode both the code and query tokens into a high-dimensional representation and then measure the distance between these representations. The goal is to have $f(c_i) \approx g(d_i)$, where f and g represent the code and query encoders. Performance is measured in MRR (Mean Reciprocal Rank) as in [8]. The method name prediction task is to predict the method name tokens, n_i, given the method body tokens, b_i. If the method name appears in the method body, it has been replaced by a <blank> token. Performance is measured by F1 score as in [2].

We perform our experiments using the *Transformer* [15] and neural bag-of-words (NBOW) models with task-specific heads. In the code retrieval task, the head performs a weighted sum over the outputs of the model [8]. For the method name prediction task, the head is a gated recurrent unit (GRU) [4] using a weighted sum over the outputs of the model as its initial hidden state.

We pre-train the models as masked language models, following [5], with an affine layer head used to predict the masked token. To perform transfer learning, we take the pre-trained model, replace its head with the task-specific head, and fine-tune it on the desired task. Each experiment is ran five times with different random seeds, the results of which are averaged together. We pre-train the model on four different datasets: Java, 6L, 5L, and English. Java is only the Java code within the CODESEARCHNET CORPUS and is the same data our model will be fine-tuned on for each task, i.e., first the Transformer is pre-trained as a masked language model and then trained on the same data for the desired task. 6L is comprised of all six languages in the CODESEARCHNET CORPUS. 5L is made up of five languages from the CODESEARCHNET CORPUS except Java. The English data is the WikiText-2 dataset [10].

4 Results

The test results for the code retrieval and method name prediction tasks are shown in Tables 1 and 2, respectively. The training curves for the code retrieval and method name prediction tasks are shown in Figs. 1 and 2.

For the code retrieval task, all four forms of pre-training achieve at least a relative 12% performance increase for the Transformer model and 7% for the NBOW model. The 6L data provides the best performance improvement for each model, 16% and 10% for the Transformer and NBOW models, respectively. For the method name prediction task, again, all four forms on pre-training provide an increase in

Table 1 Code retrieval MMR for Transformer (left) and NBOW (right)

Initialization	MRR	Initialization	MRR
Random	0.6069	Random	0.5191
Java	0.6849	Java	0.5598
6L	**0.7068**	**6L**	**0.5721**
5L	0.6967	5L	0.5643
English	0.6789	English	0.5548

The most significant results are shown in bold.

Table 2 Method name prediction F1 score for Transformer (left) and NBOW (right)

Initialization	F1	Initialization	F1
Random	0.4114	Random	0.2844
Java	0.4895	Java	0.3579
6L	**0.5106**	**6L**	**0.3815**
5L	0.5022	5L	0.3703
English	0.4796	English	0.3511

The most significant results are shown in bold.

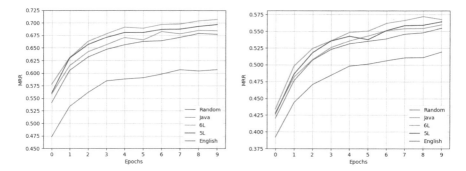

Fig. 1 Code retrieval MMR for the Transformer (left) and NBOW (right) models

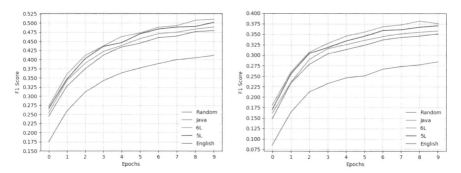

Fig. 2 Method name prediction F1 score for the Transformer (left) and NBOW (right) models

performance over the randomly initialized parameters, with at least 16% relative improvement for the Transformer and 23% for the NBOW model. Again, the 6L data provides the best performance increase, 24% and 34% for the Transformer and NBOW models.

5   Discussion

Intuitively, the datasets that contain Java code, Java and 6L, should give the best results. This is because they have been pre-trained on examples in the same language as the downstream task. However, pre-training on data without any Java code or on data that is not in a programming language give comparable results. One potential reason for this is that models learn to ignore programming language-related tokens. Consider the Java methods in Fig. 3. For the function getSurfaceArea, a fitting query would be "a method that calculates surface area"; similarly the function getAspectRatio would match with the query *"a method that calculates aspect ratio."* The semantic similarity between the tokens in the function and query gives a strong indication about how well they match. The programming language-related

```
float getSurfaceArea (int radius) {
      return 4 * Math.PI * radius * radius;
}
```

```
float getAspectRatio (int height, int width) {
      return height / width;
}
```

```
def get_surface_area (radius):
      return 4 * math.pi * radius * radius
```

```
def get_aspect_ratio (height, width):
      return height / width
```

Fig. 3 Two example functions in Java and their Python equivalents

tokens, such as the type definitions and semicolons, can virtually be ignored for this task. Similarly, for the method name prediction task, the model has to learn that *pi* and *radius* semantically relate to *surface area* and that *height* and *width* are semantically related to *aspect ratio*. Thus, the code-specific tokens are seemingly irrelevant, which makes the Python functions shown in Fig. 3 identical to the Java functions.

Thus, the reason why the 6L, 5L, and English datasets provide a performance increase is that they contain more examples of the context in which tokens appear. This increased number of contexts allows the model learn more examples of semantic similarity between tokens when pre-training and then transfer this knowledge when the model is being fine-tuned.

6 Conclusions

We have shown that transfer learning improves performance on code retrieval and method name prediction tasks. This is even true when the pre-training dataset does not contain the downstream task language and when the dataset is solely made up of languages from a different domain, namely, natural language.

Acknowledgments Ben Trevett is funded by an Engineering and Physical Sciences Research Council (EPSRC) grant and the ARM University Program.

References

1. M. Allamanis, E.T. Barr, P.T. Devanbu, C.A. Sutton, A survey of machine learning for big code and naturalness. ACM Comput. Surv. **51**(4), 81:1–81:37 (2018)

2. U. Alon, S. Brody, O. Levy, E. Yahav, code2seq: generating sequences from structured representations of code, in *Seventh International Conference on Learning Representations, ICLR 2019*, New Orleans, LA, USA, May 6–9, 2019

3. Z. Chen, M. Monperrus, A literature study of embeddings on source code (2019). CoRR abs/1904.03061. http://arxiv.org/abs/1904.03061

4. K. Cho, B. van Merrienboer, Ç. Gülçehre, D. Bahdanau, F. Bougares, H. Schwenk, Y. Bengio, Learning phrase representations using RNN encoder-decoder for statistical machine translation, in *Proceedings of the 2014 Conference on Empirical Methods in Natural Language Processing, EMNLP 2014*, October 25–29, 2014, Doha, Qatar, A meeting of SIGDAT, a Special Interest Group of the ACL, pp. 1724–1734

5. J. Devlin, M. Chang, K. Lee, K. Toutanova, BERT: pre-training of deep bidirectional transformers for language understanding, in *Proceedings of the 2019 Conference of the North American Chapter of the Association for Computational Linguistics: Human Language Technologies, NAACL-HLT 2019*, Minneapolis, MN, USA, June 2–7, 2019, vol. 1 (Long and Short Papers) (2019) pp. 4171–4186

6. Z. Feng, D. Guo, D. Tang, N. Duan, X. Feng, M. Gong, L. Shou, B. Qin, T. Liu, D. Jiang, M. Zhou, Codebert: a pre-trained model for programming and natural languages (2020). http://arxiv.org/2002.08155

7. J. Howard, S. Ruder, Universal language model fine-tuning for text classification, in *Proceedings of the 56th Annual Meeting of the Association for Computational Linguistics, ACL 2018*, Melbourne, Australia, July 15–20, 2018, vol. 1: Long Papers (2018), pp. 328–339

8. H. Husain, H. Wu, T. Gazit, M. Allamanis, M. Brockschmidt, Codesearchnet challenge: evaluating the state of semantic code search (2019). http://arxiv.org/abs/1909.09436

9. R.S. Malik, J. Patra, M. Pradel, *Nl2type: Inferring Javascript Function Types from Natural Language Information* (IEEE Press, New York, 2019)

10. S. Merity, C. Xiong, J. Bradbury, R. Socher, Pointer sentinel mixture models, in *Fifth International Conference on Learning Representations, ICLR 2017*, Toulon, France, April 24–26, 2017, Conference Track Proceedings

11. T. Mikolov, I. Sutskever, K. Chen, G.S. Corrado, J. Dean, Distributed representations of words and phrases and their compositionality, in Advances in *Neural Information Processing Systems 26: 27th Annual Conference on Neural Information Processing Systems 2013*. Proceedings of a Meeting Held December 5–8, 2013, Lake Tahoe, Nevada, USA (2013), pp 3111–3119

12. J. Pennington, R. Socher, C.D. Manning, Glove: global vectors for word representation, in *Proceedings of the 2014 Conference on Empirical Methods in Natural Language Processing, EMNLP 2014*, October 25–29, 2014, Doha, Qatar, A meeting of SIGDAT, a Special Interest Group of the ACL (2014), pp. 1532–1543

13. M.E. Peters, M. Neumann, M. Iyyer, M. Gardner, C. Clark, K. Lee, L. Zettlemoyer, Deep contextualized word representations, in *Proceedings of the 2018 Conference of the North American Chapter of the Association for Computational Linguistics: Human Language Technologies, NAACL-HLT 2018*, New Orleans, Louisiana, USA, June 1–6, 2018, vol. 1 (Long Papers) (2018), pp. 2227–2237

14. M. Pradel, K. Sen, Deepbugs: a learning approach to name-based bug detection. PACMPL 2(OOPSLA) (2018), pp. 147:1–147:25

15. A. Vaswani, N. Shazeer, N. Parmar, J. Uszkoreit, L. Jones, A.Z. Gomez, L. Kaiser, I. Polosukhin, Attention is all you need, in *Advances in Neural Information Processing Systems 30: Annual Conference on Neural Information Processing Systems 2017*, 4–9 December 2017, Long Beach, CA, USA (2017), pp. 5998–6008

16. Y. Wainakh, M.P. Moiz Rauf, Evaluating semantic representations of source code (2019). CoRR abs/1910.05177. http://arxiv.org/abs/1910.05177

An Analysis of Flight Delays at Taoyuan Airport

S. K. Hwang, S. M. Horng, and C. L. Chao

1 Introduction

In January 2016, Taoyuan Airports Company executives vowed to improve its on-time performance (OTP) [8]. The 2015 annual punctuality rankings released by OAG (Official Aviation Guide of Airways) show Taiwan's gateway airport ranked 170 out of 199 major airports in the world, down from 160th the previous year. The company announced that it was opening a second runway in 2016 and that it would improve overall performance at the airport. Ever since, 2016 has come and gone and the rankings for 2016 have also been published without even a mention of Taoyuan Airport in the overall performance category or the overall improvement category.

One of the most common problems experienced by passengers when traveling by air is a delayed flight [12, 13, 24]. Late arriving flights cause passengers to reschedule their meetings, miss important events, or even worse, miss connecting flights. The effects of flight delays also take a toll on airlines; in addition to dealing with passengers' unhappy emotions with customer service, flights have to be rescheduled, passengers need to be placed on different flights or hotels, and the entire network may experience a knock-on effect.

Airports are the resource providers of the aviation industry, with its main resources being runway and gate allocation (via the aircraft control tower and navigation systems). It also serves as a meeting point between passengers and

S. K. Hwang · S. M. Horng (✉)
Department of Business Administration, National Chengchi University, Taipei City, Taiwan
e-mail: 104363120@mail2.nccu.tw; shorng@nccu.edu.tw

C. L. Chao
Department of Accounting and Information Technology, National Chung Cheng University, Chiaji, Taiwan
e-mail: actact@ccu.edu.tw

© Springer Nature Switzerland AG 2021
R. Stahlbock et al. (eds.), *Advances in Data Science and Information Engineering*,
Transactions on Computational Science and Computational Intelligence,
https://doi.org/10.1007/978-3-030-71704-9_45

airlines. Airports also feel the impact of flight delays; the tarmac gets congested, gates need to be reassigned, and the airport needs to handle the excess passengers still waiting or stranded for flights. Resources at airports should be reallocated in case of a delay, an operation that is both time- and cost-consuming. Airports that experience frequently delays deter new airlines from operating to the airport and thus hinders market growth.

Another concern regarding flights delays is the growth of low-cost carriers (LCCs). Spurred by tourism and economic growth, a growing market has led many LCCs to enter the market at low prices, raising demand for air traffic and consequently leading to airport congestion [29, 41]. Since low-cost carriers (LCCs) entered the market in 1967, the airlines industry had some changes. LCCs operate under tight financial and time constraints to gain a competitive advantage. When carried out successfully, the LCC business model has enjoyed high growth in the market, often forcing full service carriers (FSCs) to rethink their business models.

This study focuses on Taiwan Taoyuan International Airport (Taoyuan or TPE for short), the largest operating airport in Taiwan managed by the Taoyuan International Airport Corporation and overseen by the Civil Aeronautics Administration (CAA). It serves passenger, cargo, and charter flights.

Taoyuan Airport has one of the lowest airport charges in the Asian region, less than half of airports such as Singapore and Tokyo Narita. This further attracts airlines, particularly LCCs, to operate to and from Taoyuan as it provides the ideal combination of low airport fees and taxes usually found in secondary airports and access to the primary airport passenger market. Drawing from the conclusions of Dziedzic and Warnock-Smith [16], this study tends to explore airport-specific factors that may contribute to airline delays. The study aims to answer the following questions:

- Can passengers avoid flight delays by trying different airline, destination, or timing options?
- Which factors can Taoyuan Airport influence to reduce delays?
- Which factors can airlines influence to reduce delays?
- Which factors show common trends leading to a flight delay?
- Does scheduling practices between FSCs and LCCs give rise to different performances?

This chapter is divided into the following sections: (1) a brief overview of past literature relating to flight delays, (2) an overview of the data obtained and used for this study, (3) an explanation on data-mining methods used in this study, and, finally, (4) a presentation and analysis of results.

2 Literature Review

In this section, we will first discuss the importance, cost effects, and models of flight delays, followed by a review regarding influential factors, and then end with the discussion between FSCs and LCCs.

2.1 Importance, Cost Effects, and Models of Flight Delays

Three forces are important in the aviation industry: the airline, the passenger, and the airport. Each of them has their specific set of demands to be met, so most research on delays were from these standpoints. Studies have been carried out across the globe from different regions such as the United States ([25, 30, 44]), Europe [14, 16], Brazil [34], and Taiwan [42]. Topics include analysis of flight delays ([4, 34]), delay estimation and simulation [23], delay optimization [10, 37, 45], and delay propagation [5, 22, 42].

In literature, many studies have shown that delay is an important factor for passengers to consider when purchasing a ticket, especially if the previous experience was negative [12, 36]. Similarly, passengers with transfers in their itinerary tend to avoid transferring at airports they associate delays with to minimize the risk of not making a connecting flight [20]. For airlines, it is important to curb flight delays as they also affect flight cancellation decisions, which lead to higher costs to the airline [43].

Methods used in the past are logarithmic models, Bayesian regression [30, 44], Cox regression [42] and the Hotelling model [9, 35]. However, only Sternberg et al. [34] have made use of data-mining techniques to find trends in flight delays. Nowadays, computers' calculating and restoring capabilities have enhanced dramatically to be able to process data that were considered infeasible in the past. Therefore, by applying data mining on the massive data of the airport, valuable information could be retrieved to help the operations and improve its efficiency.

2.2 Influential Factors of Flight Delays

Based on the literature, many factors contributed to the flight delay including seasonal effects, route characteristics, and turnaround and scheduling practices. The following section will introduce these factors in order.

Flight delays display a seasonal or cyclical behavior from past researches. External environment, weather and meteorological events influence delays at an airport ([2]; [42]). In a study of domestic flights in Brazil, Sternberg et al. [34] found that flights follow a one-day cycle of delays, noting that meteorological conditions also contribute to the "time of day" effect. Due to delay propagation, the spread or knock-on effect of delays across a flight network, flights in the evening or late night have a higher chance of being delayed [22].

On a larger scale, Perez-Rodriguez [30] and Rupp et al. [33] found that day-of-the-week and seasonal effects are also relevant in flight delay analysis. Although the former found certain days to be more prone to delays, no set rule or trend was discovered. In contrast, the latter discovered that winter months are more prone to flight delays in the United States as weather conditions are more extreme and happen more frequently than in summer months. Consequently, these results imply seasonality or cyclical behavior shown in airline delays [1].

Even though such cyclical information is publicly available, airlines however are faced with another dilemma when scheduling flights: should airlines choose preferred traveling times or flock to the quiet times and create another overcrowding situation? Sun [35] found that as competition increases, scheduling differentiation decreases. Since our study makes use of gate-to-gate data, crowding is more likely to have an effect on taxi time as opposed to push-back time.

Route characteristics also play an important role in flight delays. In the literature, it is not surprising that departures' delays are significantly correlated to incoming arrival delays [34, 42], thus an arrival delay could affect the flights later in the network [30, 42]. The results of Xu et al. [44] showed that airport of origin is a major factor in determining the arrival delay, which was further expanded by Perez-Rodriguez [30] that the distance from and the size of the airport of origin may be correlated to arrival delay. Interestingly, medium-sized airports have a high probability of contributing to airline delays compared to small and larger airports.

Furthermore, the amount of airlines operating the same route, or the competition that an airline faces on the route also has an impact on its on-time performance [32]. There is, however, no concrete evidence as to whether its impact is positive or negative, as Mazzeo [26], Rupp et al. [33], Mayer and Sinai [25], and Prince and Simon [31] found conflicting results.

Literature has shown that aircraft size and type also affect delays and that airlines also take this into consideration when scheduling flights ([42]). Smaller aircraft tend to have shorter flight distances, meaning airlines can operate more frequently at different times during the day. Although this increases the chance of delays as airports become more congested, the economic benefit outweighs the cost of delay as noted by Zou and Hansen. Perez-Rodriguez [30] also noted that aircraft size and delay probability are negatively correlated.

Studies on aircraft size and delays have often been linked with productivity or asset frontiers [32]. LCCs operating closer to their asset frontier due to their tighter business model constraints have worse on-time performance than their FSC peers that often have more flexibility with their aircrafts. Berster et al. [7] also found that airlines counter-congested airports by increasing aircraft size and reducing frequencies instead. This is viable from an economic standpoint as airlines offset the cost of flight delays by operating at lower unit costs.

Yimga [47] found that airlines cooperating under a codeshare or alliance network reduce the rate of delays as an increase in cooperation leads to more coordinated flight schedules in order to create a seamless travel experience for passengers. In contrast, Min and Joo [27] cautioned that the bigger the alliance is, the more time-consuming it becomes to operate. This may be of particular interest to this study as Taoyuan Airport's top three airlines belong to each of the top three airline alliances.

Flight delays can be affected by turnaround and scheduling practices. Past researches investigated the effect of turnaround times on delay departures. Wu and Caves [40] recommended that airlines can improve on-time performance through efficient ground operation and turnaround management. Mayer and Sinai [25] found that airlines schedule the least time possible for their flights as close to their productivity frontier as possible in order to maximize profitability as opposed to improving passenger satisfaction.

Arriving earlier than scheduled also means that airline face the possibility of gate unavailability, which puts strain on airport resources. A recent study investigated airlines' willingness to trade on-time performance of longer block times [21]. The results showed that FSCs value earliness more than on-time performance in order to improve passenger satisfaction and may hence have longer turn times scheduled into their networks.

2.3   FSC and LCC Behavior

Due to their efficient business models, the presence of LCCs have proven to improve overall on-time performance at airports [6, 11]. However, although LCCs decrease the occurrence of delays, the exact impact on the duration of the delay remains unknown [6].

Pioneered by Southwest, LCCs operate to secondary airports to enjoy lower fees, taxes, and less congested traffic, which contribute to quicker turnaround times and fewer delays [3, 39]. As the aviation environment evolved, their business models have also changed and LCCs have increasingly diversified their route network by operating to primary airports as well [15, 18].

FSC airlines also impose different scheduling rules depending on which airport they operating to and from. Airline operating a hub-and-spoke system, where airlines schedule all their flights from their "hub" around the same time to depart to their respective "spokes" destinations, may schedule more time in-between flights in order to allow sufficient time for passengers to connect between flights [21, 25]. During such peak times of travel, airport and runway congestion occurs contributing to increased probability of delays [5].

LCCs, on the other hand, display a more flexible business model by operating a point-to-point network as opposed to a hub-and-spoke network [25]. While FSCs still need to focus on customer service and transfer connectivity, LCCs mainly operate for passengers' first and final destination, thus eliminating the time added to delays by waiting for connecting passengers. This implies that LCCs can also succumb the congestion problem by scheduling their flights outside peak times, gaining market share without giving up passenger satisfaction in the form of flight delays.

3 Methodology

This study applied two data-mining techniques; regression and association rules to study the factors influencing flight delays. Variables used in this study will be defined first, followed by the introduction of the two techniques.

3.1 Variable Definitions

Variables used in this study include, Date of flight (further categorized by their weekdays and hours), Airline and alliance, Airline service type (LCC or FSC), Aircraft type, and Destination and Origin of flight (representing airports or cities). Although airlines have little control over the taxi time, they are able to control the turnaround time. During the turnaround, airlines need to disembark passengers, unload the cargo, clean the aircraft, prepare the aircraft for the next flight, refuel, carry out technical inspections, board passengers, and load the cargo for the next flight. In addition to the variables above, further variables were introduced or derived for the purposes of this study.

The on-time performance (OTP) variable, the dependent variable in this study, was created by subtracting the actual time of departure (arrival) by the scheduled time of departure (arrival).

$$OTP.DEP = ATD - STD \tag{1}$$

$$OTP.ARR = ATA - STA \tag{2}$$

where

OTP.DEP: on-time performance for departure flights
ATD: actual time of departure
STD: scheduled time of departure
OTP.ARR : on-time performance for arrival flights
ATA: actual time of arrival
STA: scheduled time of arrival

Load factor and passengers per flight was calculated from the statistics made available publicly and published annually by The Civil Aeronautics Administration of Taiwan (CAA). Annual averages per destination by airline were used as load factors for arriving and departing flights and thus does not take into account monthly differences in load factors. Data for Siem Riep (Angkor Wat) and Yangon were unavailable and thus annual airport averages were used as a replacement for these routes. Annual average load factors were obtained by destination for every airline serving the network and multiplied by the number of passengers each aircraft is able to carry to obtain the amount of passengers per flight.

$$\text{Passengers} = \text{load factor} \times \text{aircraft capacity} \qquad (3)$$

As this study deals primarily with gate-to-gate data, the turnaround time is calculated as the time an aircraft arrives at the gate and when it exits the gate again for a departure. The definition only accommodates aircraft leaving the gate for a departure and does not include time needed for taxiing, moving to a resting station or maintenance hangar. Although turnaround times are not publicly available, flights were matched by aircraft registrations to calculate turnaround times.

Turnaround times less than 8 h were considered scheduled. Turnaround times over 8 h were considered as long turnarounds; maintenance or overnight which could absorb flight delays [14] were assumed for these turnarounds. This provides enough time for A-checks, the shortest maintenance check that happens up to every 600 h for up to 10 h per check and also controls for unexpected maintenance that has a greater propagation effect as noted earlier by Kafle and Zou [22]. Even though there are certain airlines that schedule longer turnaround times, (e.g., Emirates has a schedule turnaround time of 7 h), this study assumes that these longer ground times are calculated to meet airlines' need elsewhere in their network.

$$\text{Turnaround}_{\text{scheduled}} = \text{STD} - \text{STA} \qquad (4)$$

$$\text{Turnaround}_{\text{actual}} = \text{ATD} - \text{ATA} \qquad (5)$$

3.2 ANOVA and Regression

The ANOVA results found that all categorical variables are significant and can be included in the regression model. Missing in these results are the LCC and Alliance factors. The effects of these factors are captured in the Airline variable. Similarly, each destination also has a competition factor, a measure of the amount of airlines serving the route, unique to that destination.

Combining the nonnumerical factors and numerical variables forms the multi-linear regression models for arrival and departure delays, respectively. These two models are essentially the same except for the departure model having an additional arrival delay variable. The models are as follows:

$$\text{Arrival Delay} = \text{Airline} + \text{Aircraft} + \text{Type} + \text{Origin} + \text{Year} + \text{Month} + \text{Weekday}$$
$$+ \text{Hour} + \text{Load factor} + \text{Passengers} + \text{Scheduled turnaround time}$$
$$(6)$$

$$\text{Departure Delay} = \text{Airline} + \text{Aircraft Type} + \text{Destination} + \text{Year} + \text{Month}$$
$$+ \text{Weekday} + \text{Hour} + \text{Load factor} + \text{Passengers}$$
$$+ \text{Arrival Delay} + \text{Scheduled turnaround time}$$

$$(7)$$

3.3 Data-Mining Process

A data-mining method using association rules was suggested by Sternberg et al. [34] to find hidden patterns and relationships within variables and will be used in this study. While regression shows us correlation, data-mining methods aim to show us the factors that are most likely to occur that will ultimately lead to a flight delay. Associations rules express a set of events, known as the antecedent or the left-hand side, that lead to another, also known as the consequent or right-hand side, in conditional probability form [38].

Association rules can be expressed as $X \rightarrow Y$, where X and Y are disjoint subsets (i.e., $X \cup Y = X + Y$), and X is an item set that leads to Y. Therefore, X is called the antecedent and Y the consequent term. Originally used as a technique to determine shopping basket behavior, association rules were used to determine which items were simultaneously purchased [19, 38]. Given an item set $I = \{i\_1, i\_2, ... i\_m\}$ and transaction set $T = \{t\_1, t\_2, ... t\_n\}$, every transaction t contains items from I. Furthermore, a transaction contains X if X is a subset of t [38].

Support count, denoted by $\sigma(X)$ for any subset X is defined as the number of times X appears in a transaction. This can be mathematically represented as follows:

$$\sigma(X) = |\{t\_i \mid X \subseteq t\_i, t\_i \in T\}|$$

$$(8)$$

Applying this definition to our original definition of an association rule, we can define support and confidence as for $X \rightarrow Y$ as

$$s(X \rightarrow Y) = \sigma(X \cup Y)/N, \quad c(X \rightarrow Y) = \sigma(X \cup Y)/\sigma(X)$$

$$(9)$$

Tan et al. [38] stressed that looking only at confidence can be misleading as it only takes into account support for the antecedent and not the consequent. Therefore, to address this problem, another metric is introduced known as lift that can be defined as

$$L(X \rightarrow Y) = (c(X \rightarrow Y))/(s(Y)).$$

$$(10)$$

Also known as the interest factor, the lift of an association rule takes into account the confidence of the rule in relation to the support of the consequent. The lift measure is also able to show whether there is a positive or negative correlation between X and Y. A lift of 1 implies that the probability of occurrence of X and Y is independent of each other. If the lift is larger than 1, it implies that the rule of $X \rightarrow Y$ is useful for predicting the consequent in future data sets. If a lift is less than one, it would imply that the two events are negatively correlated.

Using association rules, we are able to determine what factors are common and related to on-time performance by manipulating the consequent term to contain an on-time performance measure and letting the antecedent term contain the factors that could lead to such a performance measure.

$$L \ (\{\text{factors}\} \ \rightarrow \ \{\text{on} - \text{time performance measure}\}) . \qquad (11)$$

The data-mining process consists of (i) data indexing, (ii) rules generation, and (iii) rules analysis. Each process is discussed in detail below.

The first step is data indexing. When dealing with large amounts of data it is useful to index values, dividing values into manageable groups or bins. This process "standardizes" the data in two ways: continuous variables are discretized and categorical variables that take on too many values get grouped or clustered [19, 46].

Indexing techniques used in this study include conceptual hierarchy, binning, and categorization as outlined by Han et al. [19]. A summary of all data indexing is shown in Table 1.

The second step is rules generation. This study uses the R package called *arules*, and makes use of the *apriori* algorithm. This algorithm finds frequent item sets and limits those that do not meet the minimum confidence requirements. In order to avoid too many or too little rules, the support and confidence thresholds need to be maximized. For confidence threshold, Sternberg et al. [34] used the probability of a flight being delayed. If this study would apply the same principle then the weighted average of flights delayed would be 31.7%, as a confidence threshold. However, given that confidence level can be interpreted as a conditional probability [34], this would imply rules that would lead to an on-time performance of 68.3%, much lower than the regional top 10, which has been fluctuating around 80% in recent years [17]. As a result, the minimum support used for this study is 0.2 in order to generate rules that could aid Taoyuan Airport's overall performance. The maximum length for the algorithm has been left unconstrained and left up to algorithm to run until rules no longer meet the minimum support requirement.

We apply the variables outlined in Table 1 to the package *arules* with set minimum of minimum length, minimum support and confidence. Over the three years covered from 2014 to 2016, 292,221 departure flight tuples and 290,883 arrival flight tuples will be studied. Lift rules will first be obtained by setting a maximum rule length of 2 to find correlations between factors and departure delays. Finally, to determine if there are any interactions between factors, a top-down approach will be taken by including two or more classification groups in the antecedent. To find rules where factors are correlated with one another, rules will be filtered through where the factor concerned is set in the antecedent.

Table 1 Data indexing

Dimension	Indexed factor	Indexed values
Airline	Airline	Airline's ICAO 3-letter code
	Alliance	ONE: oneworld SKY: Skyteam STA: Star Alliance VAL: Value Alliance NON: None
	Service Type	FSC or LCC
Aircraft	Aircraft type	Airbus narrow (3 types), Airbus wide (5 types), Boeing narrow (5 types), Boeing wide (5 types), Regional (3 types)
Route	Airport	Airport IATA 3-letter code
	Airport size	Small (< 10 million passengers a year), Medium (11–30 million passengers a year), Large (31–50 million passengers a year), Mega (>50 million + passengers a year)
	Distance	Short: 0–999 km Regional: 1000–2999 km Medium: 3000–4999 km Long: 5000 km+
	Compe-tition	Monopoly: 0 Duopoly: 1–2 Multi: 3–4 Crowded: 5+
Scheduled time	Year	2014–2016
	Month	1–12
	Weekday	Monday–Sunday
	Hour group	Dawn (0–6), Early morning (6–9), Late morning (9–12), Midday (12–15), Afternoon (15–18), Early evening (18–21), Late evening (21–23).
Passenger loading	Load factor	Empty (0–50%), Medium (51–70%), High (71–85%), Full (>86%)
	Passenger	Low (0–100), Medium (101–200), High (201–300), Very High (>301).
Operation	Turn type	Very short (0–59 min), Short (60–119), Medium (120–239), Long (240–479), Very long (>480).
	On-time perf.	Early (before scheduled time), On time (scheduled time until 15 min thereafter), Delay (16 min–1 h after scheduled time), Overtime (more than 1 h after scheduled time).

Table 2 Numbers and frequencies of early, on time, and delayed flights

Flight	Definition	Arrival	Departure
Early	Before scheduled time	95,110	88,316
On time	Less than 15 min after the scheduled time	82,580	132,212
Delayed	15 min after scheduled time	113,192	71,693

The third step is to analyze the data based on the rules derived previously. The section below will illustrate the results of regression and data indexing in detail.

4 Results

By studying flights operating to and from Taoyuan airport alone, we control for airport size, as Perez-Rodriguez et al. [30] pointed that airport size is a significant factor in determining flight delays. Totals of 337,268 arrival flights and 337,254 departure flights were obtained for this period. After removing cargo, training, charter, and outlier flights, 290,882 arrival flights and 292,221 departure flights remained.

In order to restrict the study to significant airlines that could provide a large enough sample, only airlines that operated a minimum of 150 scheduled frequencies and flight numbers operating more than 30 flights in the three-year period, or one flights every month for three years within the three-year period were considered. This resulted on average, one flight a week for three years or three flights a week for a year.

4.1 Descriptive Analysis

Flights are evenly distributed across all weekdays, peaking somewhat between Saturdays to Mondays. The flight distribution throughout the day displays different behavior for arrival and departure flights. Departure flights display a morning rush hour between 6:00 and 8:00 whereas the busiest time for arrivals would be 21:00 at night. The two peaks may be indicative of the presence of hub airlines at TPE, where most flights depart in the mornings to their respective spoke cities and arrive back home in the evening for turnaround times, ideal for overnight maintenance and on-time departure the next day.

Although the CAA defines a delay time as any flight operated 30 min after scheduled time of operation, OAG and other literature defined a flight delay as a flight departing (arriving) 15 min after scheduled time of departure (arrival) [30, 34]. The latter definition will be used in this study (Table 2).

Both arrival and departure delay displayed skewness with arrivals more skewed but both are centered on zero. The arrivals delay distribution is flatter and displays a higher variance than the departure delay. Descriptive statistics would show that Taoyuan's negative on-time performance would come from arrivals as opposed to departures as the average arrival time is 16.55 min, outside the 15-min requirement for delays. For departures, most flights depart before 15 min after the scheduled time with under 25% leaving the gate 15 min after the scheduled time of departure.

Delays grouped by departing hour showed common time effects as in previous literature. Since 6:00 is the peak time for flight departures, the average departure delay also increases for this time frame. Throughout the day, the average flight delay remains around 10–25 min, with the average arrival delay increasing as the peak arrival evening rush hour approaches. Interestingly, flights operating around midnight have the highest delay, which may be the culmination of an entire day's delays. At 02:00 when the airport is least busy, flights arrive early on average. Furthermore, delays peak from Saturday to Monday every week. The above-mentioned observations may indicate that congestion at the airport may have a direct effect on flight delays.

Delays grouped by month showed high delays during the typhoon season between July and September in 2014 and 2015. For the years 2014 and 2015, arrival delays have been higher than departure delays. In 2016, however, arrival delays have been brought down even lower than departure delays. Although the opening of a new runway in 2016 has lowered the number of arrival delays, departure delays remain roughly the same.

As mentioned in previous sections, the LCC business model allows for greater flexibility when it comes to scheduling, turnaround time, and aircraft utilization. At Taoyuan, LCCs have a lower delay rates than FSCs (9.84 vs. 17.53 min for arrivals and 11.53 vs. 12.86 min for departures).

4.2   Regression Results

The results of regression analysis are shown in Table 3 in which Arr and Dep represent the results for arrival and departure regression models, respectively. Factors are categorized by city, airline, carrier type, month, weekday, hour, and others. Because of the large sample size, only variables with a p-value less than 0.001 are considered significant and will be presented in this Table where "+" sign indicates significantly positive and "-" sign shows significantly negative. Numbers in the parentheses at the column of Factor represent the variables. For example, of the 97 airports included in the model, 78 and 6 are significantly positively and negatively correlated, respectively, for arrival flights. All of the variables are categorical except "others." Both arrival and departure regression models have significant R-square values and the adjusted R-square is not significantly deviated from R-square indicating no overfitting problem for both models. Departure model shows better predictive capability than arrival model (0.295 vs. 0.110), because it is

Table 3 Results of regression models for arrival and departure delays

Factor	Arrival	Departure
Airport (City) (97)	78(+), 6(−)	6(+), 28(−)
Airline (44)	12(+), 18(−)	26(+), 5(−)
Carrier type		
Airbus (6)	6(+), 0(−)	0(+), 1(−)
Boeing (5)	4(+), 0(−)	0(+), 2(−)
McDonnel MD90	1(+), 0(−)	1(+), 0(−)
Month (12)	10(+), 0(−)	8(+), 0(−)
Weekday (7)	1(+), 2(−)	0(+), 2(−)
Hours (24)	14(+), 4(−)	1(+), 18(−)
Others		
Load factor	0(+), 1(−)	0(+), 0(−)
Passengers	1(+), 0(−)	1(+), 0(−)
Arrival delay	0(+), 0(−)	1(+), 0(−)
Turnaround Time	1(+), 0(−)	0(+), 1(−)

infeasible to retrieve few important factors of the latter model, such as the scheduled and actual departure times from the previous airport.

When analyzing the significant numerical factors in Table 3, the largest estimation coefficient for departure is the arrival delay. Given all other variables remained constant, every minute of arrival delay will have departure delay increased by 0.507 min, roughly 30 s. Another interesting observation is that contrary to expectations load factor is negatively correlated to arrival delay. One possible explanation is that high load factors may cause a departure delay from previous city and therefore pilots are motivated to fly faster so that they can make up the departure delay. This finding is similar to a study that customer satisfaction is higher when flights are delayed [28]. Investigations concluded that knowing the flight would be delayed, the cabin crew or airline company usually provided special treatment to the passengers to minimize the effect, and thus raised the customer satisfaction.

4.3 Data-Mining Results

The ten highest lift values overall for delayed arrival and departure flights are displayed in Table 4. For arrival flights, five of the six highest lift rules are related to origin airports. Airlines that depart to these airports also show higher lift values. In departure flights, a variety of dimensions represent the top ten highest lift values. Variables in both the top ten lift factors for both arrival and departure are airports of Wuxi, Nanjing, Shanghai, and Shenzhen; airlines of Shenzhen Airlines and China Eastern; and aircraft type of Boeing 747-400.

Next, we had to consider other factors that were correlated with departure delay. A delayed arrival had 120.4% chance of resulting in a delayed departure, with the next highest lift rule being the negatively correlated on-time arrival rule that was

Table 4 Top 10 lift variables
for single factor antecedents

Arrival delay		
Dimension	Variable	Lift
Origin	Wuxi	2.041
Origin	Nanjing	1.924
Airline	Shenzhen Airlines	1.744
Origin	Bali	1.583
Origin	Shanghai	1.506
Origin	Shenzhen	1.474
Origin	Jakarta	1.432
Airline	China Eastern	1.402
Aircraft type	Boeing 747-400	1.383
Airline	Asiana	1.354
Departure delay		
Destination	Wuxi	2.729
Destination	Nanjing	2.480
Operation	Incoming delay	2.204
Airline	China Eastern	1.819
Airline	Shenzhen Airlines	1.656
Destination	Shanghai	1.589
Aircraft type	Boeing 747-400	1.533
Destination	Shenzhen	1.503
Time of day	15:00-16:59	1.448
Destination	Los Angeles	1.437

obtained only by lowering the confidence level to 0.1. This implies that flights that arrive on time followed by a late departure do not occur frequently enough to be considered at the confidence threshold. On the other hand, on-time departures are positively correlated to early and on-time arrivals and, as expected, negatively correlated to delayed arrivals. It was also found that whatever on-time performance behavior arrivals show, there is a high probability that it would display the same behavior for the departing flight. Most notable, early and on-time arrivals show a positive correlation to their departure counterparts, with early arrivals more likely to depart early or on time.

Lift rules grouped by time of the day generally follow that of the amount of flights operated in that hour. However, toward the end of the day as the number of flights decrease lift values do not recover in the same way. Flights occurring after 15:00 seem to be positively correlated to delay whether arrival or departure.

Destinations that are more prone to delays are mainly from Asia and particularly China. The only destination on the list that is not from China is Los Angeles. Further analysis of airports by size and distance revealed that flights operating to and from big airports, (airports over 50 million passengers per annum) are usually delayed regardless of arrival or departure. On the other hand, short flights below 1000 km also have a higher probability of being delayed whether arriving or departing.

Six of the top ten airlines associated with delayed arrivals are also associated with delayed departures, further strengthening the observation that departure and arrival delays are correlated. Comparing the list of airlines associated to arrival delays with airlines associated with delayed departures, some differences in airlines occur. Of particular note is Asiana, which is associated with arrival delays but not departure delays. This implies that there is possibly some behavior with regard to turnaround times which can be investigated.

A clear pattern emerges when analyzing turnaround times. The probability of a departure delay increases as turnaround time decreases, yet the probability of an arrival delay decreases as turnaround times become shorter. This is possibly due to the fact that the same amount of work needs to be done in a shorter time. Airlines, or more specifically pilots, may be incentivized to arrive earlier to create a longer turnaround if turnaround times are shorter. Flight schedules with longer turnaround times allow airlines more flexibility to adjust their schedules should things go wrong during the turnaround.

No clear trend or pattern for passenger loads can be observed from lift rules. Lift values do, however, confirm the results found by regression that arrival delay decreases as load factor increases and that there is no significant relationship with departure delays. To investigate the effects of the number of passengers further, we look at aircraft size and aircraft classifications.

For aircraft size, narrow-body displayed lower lift values than their wide-body counterparts. The highest lift values came from flights operated by Boeing 747-400, followed by Airbus wide-body such as A330 and A340. This may be because these are the oldest aircrafts on the market and may frequently run into mechanical problems resulting in delays. Interestingly, the Boeing 777 which is a somewhat newer wide-body displayed similar lift values as narrow-body aircrafts.

5 Discussion and Conclusions

5.1 Discussion

This study has shown that one of the main factors resulting in delays is out of the airport's control: origin airports. Both regression and data mining show that flights departing to and from the eastern China region are likely to face high delays. The airports in this region include Wuxi, Shanghai, and Nanjing. A possible reason for this may be congested airspace, with many airlines operating to and from these airports within a restrictive airspace.

Despite some airlines having high delay rates, due to low frequencies they may not necessarily add to Taoyuan's overall performance. Taoyuan Airport should in fact turn its focus on airlines that impact their overall on-time performance. Besides hub airlines, bigger airlines such as China Eastern, China Southern, and Air China have an impact on overall on-time performance at Taoyuan.

A comparison between airlines and airports associated with flight delays shows a higher lift value and regression coefficients for airports than with airlines. Many of the airlines operate to closer destinations meaning that it is not necessarily just the origin airport but rather also the distance of the airport that is the reason for the delay. Most notably such airports are all within a 1,000 km radius from Taoyuan. Such delays can be explained by Taipei also being in the direct flight path of a busy air corridor for many airlines flying between South East Asia to Japan, Korea and North America, further congesting an already busy airspace.

The difference in results for the two models can be found with long-haul destinations. In regression, Amsterdam was a destination likely to be delayed whereas it was not significant in data mining. In regression, Los Angeles was a significant airport for arrival delays but in data mining it was a significant departure delay airport.

Airlines also attempt to schedule their flights accordingly to avoid rush hours and times of hubbing. At Taoyuan, hub airlines depart mostly during 7:00–10:00 in the morning with a second wave of arrival and departures between 14:00 and 17:00; finally, a third wave of arrivals occurs in the evening from 19:00 to 22:00. Airlines, especially LCC not based in Taoyuan try to avoid these hub times with many airlines opting to arrive or depart between 10:00 and 14:00 largely without much hindrance. However, many airlines not based in Taoyuan also start to arrive at 16:00 and a steady flow of flights that arrive at Taoyuan until the evening, colliding with hub airlines' rush hours. Airports can use this kind of information to control which airlines operate at which time at the airport to ease the traffic flow and reduce delays.

Both regression and data-mining models show that flights have a greater chance of being delayed in the afternoon, although in regression this is clearer for arrivals than for departures. While data-mining methods show that many flights in the late afternoon are delayed, the result is less clear than using regression methods. Regression results go one step further to support findings in previous studies that one of the main reasons why flight delays occur is due to flight delay propagation throughout the network.

Both regression and data-mining models found that a late incoming flight generally leads to a delayed departure for the next flight. This supports Kang and Hansen's [21] findings as to why airlines value earliness more than passenger satisfaction and therefore try avoid a delayed departure to minimize the probability of a delayed arrival. Regression results also show that for every minute of delayed arrival approximately adds 30 s to the departure delay.

Regression also shows that flight delays may be airline specific but does not state which factor makes it specific. Using lift rules, operations and turnaround dimensions seem to be the main reason why airlines are delayed or on time. Although airlines are not able to adjust their competitors' flight schedules, they are able to adjust their turnaround schedules. Regression results do not show that turnaround time affect delays. Lift rules, however, show that airlines with longer planned turnaround times are more likely to experience delayed incoming arrival flights as the longer ground time allows for more buffer to make up the time lost in the schedule.

Although many airlines try to utilize shorter turnaround times to decrease delays, only few have managed to succeed. A wide range of LCC delays were found, meaning that short turnarounds can either be very efficient or very detrimental to the airline. Short turnarounds do not work well with FSCs as this combination is generally associated with delays, further strengthening the case of the LCC business model.

Although arrival results are not clear in the regression model, arrival lifts for LCCs are lower than that of departure lifts, indicating that LCCs try to land as early as possible to extend or create a buffer to their short turnaround times. Since LCCs have shorter turn times they are able to schedule more flights in a day, yet with the risk of a single delay affecting the whole system.

Surprisingly, passenger load factors have not had a significant impact on delays as expected. Both regression and data-mining models could not indicate any trend or pattern. Airlines know how many passengers will arrive beforehand and can adjust their schedule accordingly such as start boarding earlier. This kind of flexibility is lost when airlines opt to operate under a short turnaround time, most notably LCCs. In order to counteract this loss in flexibility, LCCs usually operate narrow-bodies such as Airbus A320s and Boeing 737-800s, with little variation in passenger numbers. Looking at departure delays, especially when dropping confidence threshold, a semi-pattern of passenger capacity tends to develop with aircraft carrying more passengers being more likely to delay a flight.

In-between the three major airline alliances, there is a relationship between the size of the alliance and the lift rules. This confirms Min and Joo's [27] study that integration among alliance members may be beneficial but too much can be detrimental.

The findings for LCCs suggest that Taoyuan Airport is on the correct path for attracting more LCCs to operate to the airport, confirming results found by Dobruszkes et al. [15] and Graham [18] that LCCs are starting to operate to primary airports. Findings show that Taoyuan Airport provides the ideal advantage for LCCs to thrive, as all variables negatively correlated with flight delays have pointed toward the LCC business model.

The above analysis shows that although Taoyuan is wanting to reduce flight delays, it would have to do so by changing the way airlines operate at the airport. Airports have introduced slot restrictions in the past, restricting the amount of flights that can take off or land in a certain time period. However, since Taoyuan is still a developing airport and not a major airport in the region, this policy will deter airlines rather than attract and thus restrict growth.

With Taoyuan Airport also intending to raise airport fees, it may also use a model to incentivize airlines to operate on time. As many factors are airline specific, an airline specific fine model can be introduced. We propose a model where when an airline applies to operate to the airport, it applies for scheduled turnaround time as well. Any operation that happens after the scheduled turnaround time will result in a fine per minute to the airline. Airlines willing to schedule longer turnaround times to avoid costs, do so at the cost of losing aircraft utility as the aircraft will not be able to be scheduled on other flights. Although such a model would work for FSCs, the effects of such a model on LCCs would either be very positive in the sense that they have an edge over FSCs for better overall on-time performance, or it would have a very negative effect for LCCs due to short turnaround being exposed to high probability of delay risks. Such a model could deter poor-performing LCCs entrants to Taoyuan and thus improve its overall on-time performance.

5.2 Conclusions

This study used two models to address the factors affecting flight delays from two different managerial viewpoints. Firstly, linear regression was used that allows airlines to analyze delays in a detailed way, even to the minute, and allows airlines to draw comparisons to their peers. Secondly, a data-mining model used association rules to find probabilities of flight delays at Taoyuan Airport, which can be used from an airport's perspective to improve on-time rates.

This study found various factors that match previous studies such as the "time of day" effect and incoming delay affecting departure delays. As Taoyuan Airport is host to hub airlines, non-hub airlines try to schedule around the hub times. However, this is not completely avoidable and airlines are bound to overlap with hub times causing delays at the airport.

The models applied in this study show that many factors related to airline delays are actually related to the way airlines operate. At Taoyuan Airport, the LCC business model has successfully undercut their FSC peers to produce better on-time performance. Flight delays are not necessarily airline-destination-specific but rather airline-origin-specific. One of the main reasons an airline experiences a departure delay is due to an arrival delay that is related to the airport of origin. Although passengers can choose a specific airline–airport pairing, it may only reduce departure delay but not necessarily arrival delay.

Furthermore, this study has shown that Taoyuan Airport is a primary airport that LCCs are willing to operate to and from given Songshan Airport's operation limitations. The trade-off is that flight delays are more likely to occur due to congested airspace.

This study aimed to find trends and probabilities of factors resulting in better on-time performance. However, such measures would always just remain a point of reference and cannot replace the values of pure quantitative or qualitative studies. Although regression and data-mining models may be quite different in nature and target audience, future studies can investigate how these two models can be used together for a more generic target audience without the airport–airline separation.

This study kept the airport variable constant and did not consider the effects of data mining on a network as a whole. This, and a comparison of other airports using the same research methods, may be studied in the future.

References

1. M. Abdel-Aty, C. Lee, Y. Bai, X. Li, M. Michalak, Detecting periodic patterns of arrival delay. J. Air Transp. Manag. **13**(6), 355–361 (2007)
2. K.F. Abdelghany, S.S. Shah, S. Raina, A.F. Abdelghany, A model for projecting flight delays during irregular operation conditions. J. Air Transp. Manag. **10**(6), 385–394 (2004)
3. A.Z. Acar, S. Karabulak, Competition between full service network carriers and low cost carriers in Turkish Airline market. Procedia Soc. Behav. Sci. **207**, 642–651 (2015)
4. M. Ball, C. Barnhart, M. Dresner, M. Hansen, K. Neels, A. Odoni, et al., *Total Delay Impact Study: A Comprehensive Assessment of the Costs and Impacts of Flight Delay in the United States* (Institute of Transportation Studies, University of California, Berkeley, Berkeley, 2010)
5. P. Baumgarten, R. Malina, A. Lange, The impact of hubbing concentration on flight delays within airline networks: An empirical analysis of the US domestic market. Transp. Res. Part E Logistics Transp. Rev. **66**, 103–114 (2014)
6. W.E. Bendinelli, H.F. Bettini, A.V. Oliveira, Airline delays, congestion internalization and non-price spillover effects of low cost carrier entry. Transp. Res. A Policy Pract. **85**, 39–52 (2016)
7. P. Berster, M.C. Gelhausen, D. Wilken, Is increasing aircraft size common practice of airlines at congested airports? J. Air Transp. Manag. **46**, 40–48 (2015)
8. C.F. Bien, Y.F. Low, Taoyuan airport vows to improve on-time performance. Central News Agency (2016, January 14). Retrieved from http://www.taiwannews.com.tw/en/news/2868374
9. S. Borenstein, J. Netz, Why do all the flights leave at 8 am?: Competition and departure-time differentiation in airline markets. Int. J. Ind. Organ. **17**(5), 611–640 (1999)
10. J.O. Brunner, Rescheduling of flights during ground delay programs with consideration of passenger and crew connections. Transp. Res. Part E Logistics Transp. Rev. **72**, 236–252 (2014)
11. B. Bubalo, A.A. Gaggero, Low-cost carrier competition and airline service quality in Europe. Transp. Policy **43**, 23–31 (2015)
12. H.T. Chen, C.C. Chao, Airline choice by passengers from Taiwan and China: A case study of outgoing passengers from Kaohsiung International Airport. J. Air Transp. Manag. **49**, 53–63 (2015)
13. C.K.W. Chow, Customer satisfaction and service quality in the Chinese airline industry. J. Air Transp. Manag. **35**, 102–107 (2014)
14. A.J. Cook, G. Tanner, Innovative Cooperative Actions of R&D in EUROCONTROL Programme CARE INO III: Dynamic Cost Indexing: Aircraft maintenance–marginal delay costs (2008).
15. F. Dobruszkes, M. Givoni, T. Vowles, Hello major airports, goodbye regional airports? Recent changes in European and US low-cost airline airport choice. J. Air Transp. Manag. **59**, 50–62 (2017)
16. M. Dziedzic, D. Warnock-Smith, The role of secondary airports for today's low-cost carrier business models: The European case. Res. Transp. Bus. Manag. **21**, 19–32 (2016)
17. FlightStats, (2017, May 20). Retrieved from http://www.flightstats.com/company/monthly-performance-reports/airports/
18. A. Graham, Understanding the low cost carrier and airport relationship: A critical analysis of the salient issues. Tour. Manag. **36**, 66–76 (2013)
19. J. Han, J. Pei, M. Kamber, *Data Mining: Concepts and Techniques* (Elsevier, New York, 2011)
20. C.Y. Hsiao, M. Hansen, A passenger demand model for air transportation in a hub-and-spoke network. Transp. Res. Part E Logistics Transp. Rev. **47**(6), 1112–1125 (2011)
21. L. Kang, M. Hansen, Behavioral analysis of airline scheduled block time adjustment. Transp. Res. Part E Logistics Transp. Rev. **103**, 56–68 (2017)
22. N. Kafle, B. Zou, Modeling flight delay propagation: A new analytical-econometric approach. Transp. Res. B Methodol. **93**, 520–542 (2016)
23. S. Khanmohammadi, S. Tutun, Y. Kucuk, A new multilevel input layer artificial neural network for predicting flight delays at JFK airport. Proc. Comput. Sci. **95**, 237–244 (2016)

24. Y.K. Kim, H.R. Lee, Passenger complaints under irregular airline conditions–cross-cultural study. J. Air Transp. Manag. **15**(6), 350–353 (2009)
25. C. Mayer, T. Sinai, *Why do airlines systematically schedule their flights to arrive late?* (The Wharton School, University of Pennsylvania, Philadelphia, 2003)
26. M.J. Mazzeo, Competition and service quality in the US airline industry. Rev. Ind. Organ. **22**(4), 275–296 (2003)
27. H. Min, S.J. Joo, A comparative performance analysis of airline strategic alliances using data envelopment analysis. J. Air Transp. Manag. **52**, 99–110 (2016)
28. A. Neely, M. Al Najjar, Management learning not management control: The true role of performance measurement? Calif. Manag. Rev. **48**(3), 101–114 (2006)
29. J. Pearson, J.F. O'Connell, D. Pitfield, T. Ryley, The strategic capability of Asian network airlines to compete with low-cost carriers. J. Air Transp. Manag. **47**, 1–10 (2015)
30. J.V. Pérez-Rodríguez, J.M. Pérez-Sánchez, E. Gómez-Déniz, Modelling the asymmetric probabilistic delay of aircraft arrival. J. Air Transp. Manag. **62**, 90–98 (2017)
31. J.T. Prince, D.H. Simon, Multimarket contact and service quality: Evidence from on-time performance in the US airline industry. Acad. Manag. J. **52**(2), 336–354 (2009)
32. K. Ramdas, J. Williams, An empirical investigation into the tradeoffs that impact on-time performance in the airline industry. Wash. Post, 1–32 (2006)
33. N. Rupp, D. Owens, L. Plumly, Does competition influence airline on-time performance. Adv. Airline Econ. **1**, 251–272 (2006)
34. A. Sternberg, D. Carvalho, L. Murta, J. Soares, E. Ogasawara, An analysis of Brazilian flight delays based on frequent patterns. Transp. Res. Part E Logistics Transp. Rev. **95**, 282–298 (2016)
35. J.Y. Sun, Clustered airline flight scheduling: Evidence from airline deregulation in Korea. J. Air Transp. Manag. **42**, 85–94 (2015)
36. Y. Suzuki, The relationship between on-time performance and airline market share: a new approach. Transp. Res. Part E Logistics Transp. Rev. **36**(2), 139–154 (2000)
37. N. Takeichi, Nominal flight time optimization for arrival time scheduling through estimation/resolution of delay accumulation. Transp. Res. Part C Emerg. Technol. **77**, 433–443 (2017)
38. P.N. Tan, M. Steinbach, V. Kumar, *Introduction to Data Mining* (Person Education, New Delhi, 2006)
39. D. Warnock-Smith, A. Potter, An exploratory study into airport choice factors for European low-cost airlines. J. Air Transp. Manag. **11**(6), 388–392 (2005)
40. C.L. Wu, R.E. Caves, Flight schedule punctuality control and management: a stochastic approach. Transp. Plan. Technol. **26**(4), 313–330 (2003)
41. C. Wu, How aviation deregulation promotes international tourism in Northeast Asia: A case of the charter market in Japan. J. Air Transp. Manag. **57**, 260–271 (2016)
42. J.T. Wong, S.C. Tsai, A survival model for flight delay propagation. J. Air Transp. Manag. **23**, 5–11 (2012)
43. J. Xiong, M. Hansen, Modelling airline flight cancellation decisions. Transp. Res. Part E Logistics Transp. Rev. **56**, 64–80 (2013)
44. N. Xu, G. Donohue, K.B. Laskey, C.H. Chen, Estimation of delay propagation in the national aviation system using Bayesian networks. In *6th USA/Europe Air Traffic Management Research and Development Seminar* (2005, June)
45. Y. Xu, X. Prats, Effects of linear holding for reducing additional flight delays without extra fuel consumption. Transp. Res. Part D: Transp. Environ. **53**, 388–397 (2017)
46. M. Yaghini, *Data Discretization and Concept Hierarchy Generation* (2008). Retrieved May 26, 2017, from http://webpages.iust.ac.ir/yaghini/Courses/Application_IT_Fall2008/DM_02_07_Data%20Discretization%20and%20Concept%20Hierarchy%20Generation.pdf
47. J.O. Yimga, Airline code-sharing and its effects on on-time performance. J. Air Transp. Manag. **58**, 76–90 (2017)

Data Analysis for Supporting Cleaning Schedule of Photovoltaic Power Plants

Chung-Chian Hsu, Shi-Mai Fang, Yu-Sheng Chen, and Arthur Chang

1 Introduction

To reduce the extent of dependence on nuclear power and thermal power, Taiwan's government has aggressively promoted the use of green energy in recent years, including solar power and wind power. Especially, due to advancement in the technology and reduction in cost over the years, photovoltaic (PV) power plants have grown rapidly in the last decade in Taiwan and worldwide as well. Solar energy has in fact become an indispensable part of everyday human life [1], particularly in Taiwan. All over the world, corporations as well as governments are exploiting the solar energy market [2]. In order to respond to the ever-increasing demand on energy consumption, many countries have begun to address the problem of energy production. As a result, green energy has attracted attention from many governments. Green energy includes wind power, tidal power, solar energy, etc.

One critical issue of solar power plant operation is to determine when to clean dirty solar panels caused by dust or other pollutants. Overly frequent cleaning can lead to additional costs. On the other hand, insufficient cleaning will lead to reduced

C.-C. Hsu (✉) · S.-M. Fang
Department of Information Management, National Yunlin University of Science and Technology, Douliu, Yunlin, Taiwan
e-mail: hsucc@yuntech.edu.tw

Y.-S. Chen
Reforecast Technology Co. Ltd, Douliu, Yunlin, Taiwan
e-mail: anson@reforecast.com.tw

A. Chang
Bachelor Program in Interdisciplinary Study, National Yunlin University of Science and Technology, Douliu, Yunlin, Taiwan
e-mail: changart@yuntech.edu.tw

© Springer Nature Switzerland AG 2021
R. Stahlbock et al. (eds.), *Advances in Data Science and Information Engineering*,
Transactions on Computational Science and Computational Intelligence,
https://doi.org/10.1007/978-3-030-71704-9_46

production. Especially in a tropical island-type climate such as that in Taiwan, it rains frequently in some seasons, resulting in cleaning of dirty solar panels, referred to as *natural cleaning* in contrast to *manual cleaning* by maintenance personnel. It is necessary to investigate the cleaning issue of solar plants in Taiwan.

Pollution loss varies in different areas. In the UK, if there is no cleaning in a month, the sunshine intensity will decrease by 5–6% [3]. In Sudan, the reduced intensity of sunshine will be nine times that of the UK. In cooler and wet environments, the accumulated pollutants may include dust in the air, feces from birds and other animals, dust from burning materials, leaves, or pollen.

To the best of our knowledge, there has been no research on massive, operational solar power plants that have connected grid networks. Most of the research on cleaning frequency like those mentioned above was conducted under controlled environment with carefully maintained solar panels. In an uncontrolled environment, additional issues need to be taken into consideration, including module failure, the angle of the pyranometer, the orientation and tilt angle of the solar panels, etc.

In addition, for management companies of solar power plants that usually have hundreds of plants located over various regions, the environment surrounding each plant and the climate over the regions may be different, leading to different extent of soiling. For instance, the solar arrays set up on the rooftop of vegetable farmhouses are less polluted than those on the rooftop of animal farmhouses. In the winter, central Taiwan is more polluted than other regions due to the north-east monsoon from Mainland and the sandy environment of the region. Furthermore, in the summer of Taiwan, there are usually typhoons with heavy rains that naturally clean the solar panels. In summary, solar energy production highly depends on geographic locations due to different types of weather conditions. It is desirable to conduct research regarding this issue for PV systems installed under Taiwan's environmental conditions.

In this study, we investigate the panel cleaning issues of solar power plants in Taiwan under uncontrolled, operational configuration via the techniques of data analysis and data mining. We propose methods to measure power production improvement that resulted from panels cleaning via manual cleaning and natural cleaning. We develop an approach to real-time monitor output performance and soiling progress of power plants. The research results will be valuable and can help to determine the proper time for panels cleaning. We considered the additional issues mentioned above and avoided the need for parameters of the solar plants that are expensive to obtain. In essence, we monitor daily power production and compare with the expected power output estimated under the assumption of no dust on the panels. Cumulated revenue loss is calculated, which can support the manager on scheduling panel cleaning.

2 Methods

We propose a model for estimating the expected power generation under the condition of no dust. We compare the measured power output and the expected output. The difference between the two output values can be considered as the power loss due to dust on the panels. If the amount due to the loss is greater than the cleaning cost, the cleaning shall be scheduled.

For the estimation of the expected power generation under the condition of no dust, we resort to a machine-learning technique: linear regression and historical power output data. In particular, we use the three-day data of power output after manual panel cleaning as the training data to train a predication model P_{exp}, as shown in Eq. 1.

$$P_{exp} = R_0 + (R_1 \times T_{air}) + (R_2 \times W) + (R_3 \times G) \tag{1}$$

where P_{exp} is the expected power output (kW), T_{air} is temperature (°C), G is the irradiance (kW/m^2), W is wind speed (m/s), and R_0, R_1, R_2, and R_3 are regression coefficients.

Equation 2 calculates the difference between the expected and the measured power output during the period T and for N PV string.

$$P_{gap} = \frac{1}{12} \sum_{t \in T} \sum_{n \in N} \left(P_{t,n}^{exp} - P_{t,n}^{mea} \right) \tag{2}$$

where $P_{t,n}^{exp}$ and $P_{t,n}^{mea}$ denote the expected and the measured power output at time t for PV string n. Since the power output is measured at a time step of 5 min, the amount is divided by 12 to convert to the unit of kWh. In this study, T is set to the range of 6 AM to 6 PM, which is a proper setting considering the climate of Taiwan.

To arrange the cleaning schedule, cumulated revenue loss due to dust on panels since the last manual cleaning is a major concern. In addition, rain can certainly clean the PV plant, referred to as natural cleaning in this study, and shall be taken into account as well. However, the extent of cleaning depends on how heavy the rain is. We propose a method to model the effect of natural cleaning.

Table 1 presents the effect of natural cleaning from several studies. Daher et al. [4] found 5 mm rain can clean 6.8% when PV plants were heavily dust-laden while 0.87–4.33% when lightly dust-laden. Kimer et al. [5] found that 5.08–10.16 mm can clean up to 10% while 20 mm upto 40%. You et al. [6] found <5 mm can clean 30% of the dust while >10 mm 80%.

It can be observed that the cleaning effect resulting from rain was quite different depending on different areas and climate types. Consequently, we shall propose a model that can adapt to the local climate and consider the effect of both manual cleaning and natural cleaning. The proposed model is defined as follows:

$$K_{dirty}^d = (1 - C_{natural}(m)) \times (1 - C_{manual}) \tag{3}$$

Table 1 The cleaning effect on PV panels due to rain

References	Rain	Cleaning effect
Daher et al. [4]	5 mm	6.8% in heavy dust0.87–4.33% in light dust
Kimber et al. [5]	5.08~10.16 mm	0–10%
	= 20 mm	40%
You et al. [6]	<5 mm	30%
	>10 mm	80%

$$C_{natural}(m) = \begin{cases} C, & m > R \\ \frac{m}{R} \times C, & m \leq R \end{cases}$$

where K_{dirty}^d represents the dirt rate, $C_{manual} = 1$ after manual cleaning and otherwise 0, $C_{natural}(m)$ denotes the clean rate after rain, m denotes the amount of rain (mm). R and C are both predetermined constants. When the amount of rain exceeds R, the natural cleaning reaches rate C. R and C can be learned in advance based on historical, training data.

Cumulative revenue loss is calculated as follows: Eq. 4 computes the loss in day d; Eq. 5 is the cumulative loss up to day d since the last manual cleaning. Note that K_{dirty}^d takes manual cleaning and natural cleaning into account in Eq. 3.

$$E_{loss_{day}}^d = P_{gap}^d \times P_{sell} \tag{4}$$

$$E_{loss}^d = \left(E_{loss}^{d-1} + E_{loss_{day}}^d \right) \times K_{dirty}^d \tag{5}$$

where P_{sell} denotes sale revenue of solar power per kWh, $E_{loss_{day}}^d$ denotes revenue loss in day d, E_{loss}^d is the cumulative revenue loss for d days due to dust. The cumulative loss is discounted by K_{dirty}^d representing the cleaning effect.

3 Preliminary Result

A preliminary experiment was conducted on a PV power plant in central Taiwan. The period of the data was from January 2017 to March 2019. During the period, the plant was manually cleaned five times as indicated by the red vertical lines in Fig. 1.

Figure 1 shows the daily expected loss due to dust. It can be observed that the expected loss dropped after manual cleaning (indicated by the vertical red lines) or heavy rain (indicated by the vertical black lines). The left and the right portion of the figure depictsthe period around January, February, November, and December. During this period, it is the dry season in Taiwan. As a result, it can be seen that the

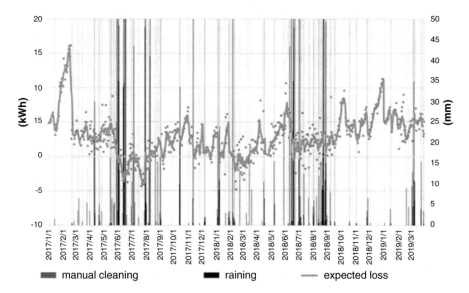

Fig. 1 The daily expected power loss on a solar power plant during the period of Jan. 2017 to Mar. 2019

loss increased gradually due to no or little rain. In the experiment, R and C of Eq. 3 were empirically set to 20 mm and 0.8, respectively.

Figure 2 shows the cumulative loss due to the dust on PV panels. The horizontal yellow line marks the cleaning cost. It can be seen that the loss rapidly increased during the dry season and exceeded the cleaning cost. In contrast, during the summer, the heavy rains reduced the revenue loss, which was far below the cleaning cost. According to the results, it is promising to develop a decision support system for scheduling manual cleaning based on the result of the data analysis.

4   Remarks and Future Tasks

To support managers on properly scheduling PV panel cleaning, we propose an approach to estimate the revenue loss due to cumulative dust on the panels. The expected power output according to the weather condition is first estimated. The difference between the expected and the measured output is calculated as the power loss due to the dust. The cumulative revenue loss is computed based on the daily differences. The preliminary result indicates that the proposed method is promising. In the future, we will conduct experiments on more power plants to test robustness of the method and also to fine-tune the model parameters based on more real-world historical data, such as the R and C values in Eq. 3.

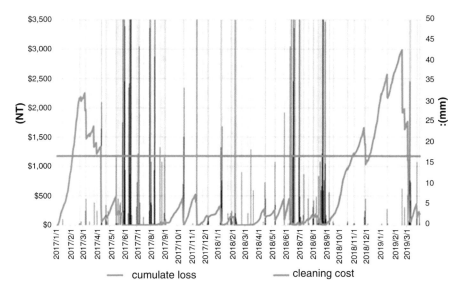

Fig. 2 The cumulative loss in revenue due to dust on PV panels

Furthermore, in the next steps, we will consider other models for estimating the expected power output from recent deep-learning techniques such as RNN, LSTM, and GRU, which are models concerning data sequences.

Acknowledgment This research is supported in part by the Ministry of Science and Technology, Taiwan under project MOST 108-2410-H-224-038 and Reforecast Technology Co., Ltd, Taiwan.

References

1. M.Q. Raza, M. Nadarajah, C. Ekanayake, On recent advances in PV output power forecast. Solar Energy **136**, 125–144 (2016)
2. M. Oliver, T. Jackson, The market for solar photovoltaics. Energy Policy **27**, 15 (1999)
3. S. Ghazi, A. Sayigh, K. Ip, Dust effect on flat surfaces – A review paper. Renew. Sustain. Energy Rev. **33**, 742–751 (2014)
4. D.H. Daher, L. Gaillard, M. Amara, C. Ménézo, Impact of tropical desert maritime climate on the performance of a PV grid-connected power plant. Renew. Energy **125**, 729–737 (2018). https://doi.org/10.1016/j.renene.2018.03.013
5. A. Kimber, L. Mitchell, S. Nogradi, H. Wenger, *The Effect of Soiling on Large Grid-Connected Photovoltaic Systems in California and the Southwest Region of the United States.* Paper presented at the 2006 IEEE 4th World Conference on Photovoltaic Energy Conference (2006, 7–12 May)
6. S. You, Y.J. Lim, Y. Dai, C.-H. Wang, On the temporal modelling of solar photovoltaic soiling: Energy and economic impacts in seven cities. Appl. Energy **228**, 1136–1146 (2018). https://doi.org/10.1016/j.apenergy.2018.07.020

Part VI
Information & Knowledge Engineering Methodologies, Frameworks, and Applications

Concept into Architecture: A Pragmatic Modeling Method for the Acquisition and Representation of Information

Sebastian Jahnen, Stefan Pickl, and Wolfgang Bein

1 Introduction

The strategic alignment of IT to the business processes and thus the targeted support of a company by IT is one of the highest prioritized goals of a company and is outlined by the term business IT alignment [1]. To achieve this goal, documentation is required that shows the relationship between IT and business processes; an or Enterprise Architecture (EA). Zachman [2] provided the first approaches and a first framework for creating such an EA. Here, EA takes on more than just the role of actual documentation. EA provides the basis for various analyses and related decisions in management area or when defining interfaces between areas, software, or technical solutions [3]. In addition, architectural models should also support the planning, development, and introduction of new systems and processes, as well as their strategic optimization [4]. Even if the requirements regarding the orientation of a company, or the company goals, are specified by the highest level of management, the list of stakeholders when creating an EA is significantly longer. This creates an interface between the modeling experts and the subject matter experts in the area to be modeled (SME) [5]. The gained information as part of the model must be obtained and presented in the required form of the EA.

An EA consists of many different views, which is why a starting point or a view to start with has to be selected when creating an EA. It is based on certain procedural patterns, which can also be found in different practices (e.g., goal-oriented requirements definition) [6] outside the creation of EA. After answering

S. Jahnen (✉) · S. Pickl · W. Bein
Fakultät für Informatik, Universität der Bundeswehr München, Neubiberg, Germany

Department of Computer Science, University of Nevada, Las Vegas, NV, USA
e-mail: Sebastian.Jahnen@unibw.de; Stefan.Pickl@unibw.de; Wolfgang.Bein@unlv.edu

© Springer Nature Switzerland AG 2021
R. Stahlbock et al. (eds.), *Advances in Data Science and Information Engineering*,
Transactions on Computational Science and Computational Intelligence,
https://doi.org/10.1007/978-3-030-71704-9_47

the question why something should be implemented, i.e., aligning with a vision or formulating a vision, in a next step procedural patterns recommend to show which steps are necessary to define an action. This procedure can also be found in the Architecture Development Method (ADM) of The Open Group Architecture Framework (TOGAF) [7]. In this process model the workflow of an action or activity should be defined after the vision has been created [8]. The NATO Operational View (NOV), which is divided into individual subviews [9], is used to represent these processes, where the NOV-5 represents the process model. All entities involved in the modeled process are represented in the NOV-2. These two subviews of the NAF often form the starting point for creating an EA.

In this contribution, a method is developed that should simplify the generation of information for the creation of a process model. This work also includes the development of a software that transforms the data into the form required by the named subviews of the NAF. For this purpose, the information is to be presented in a form that, in addition to the computer-aided transformation, also enables the SME to control and check the content of the information represented in the EA model. Also, this should enable the SME to be able to use this model.

The goal pursued was the development of artifacts that make it easier for SME to gain information for a process, enable them to start modeling an EA, and give them the opportunity to read and evaluate models and views of an EA and to be able to evaluate and control the content.

According to the principles of Action Design Research (ADR) [10, 11], common methods for generating information are described in a first step. Building on this, their strengths and weaknesses are highlighted and the problem is formulated. A concept is developed to address the problem, from which a first solution is implemented. This was evaluated through experiments in different cycles.

2 Research Methodology

This work tries to solve an important and relevant problem of an organization with artifacts and is therefore to be found in the field of design-oriented business informatics [12]. The solution to the relevant problems is given special weighting, since the primary design goal is measured on the basis of usefulness, since the construction of IT artifacts forms the starting point for scientific considerations in a practically motivated problem [13]. In order to be able to address the mentioned usefulness accordingly when evaluating the artifacts, the procedure model of the ADR shown in Fig. 1 was used.

The principles of the ADR focus on a classification of the problem in the corporate context [10]. Since this is crucial for the measurement of the usefulness and thus for the evaluation of the artifacts, the perceived usefulness and the perceived ease of use of the TAM [14] (see Fig. 2) were defined as measurement criteria.

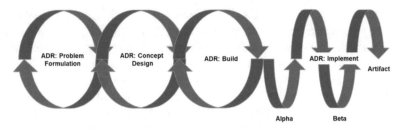

Fig. 1 Procedural Model for ADR process [11]

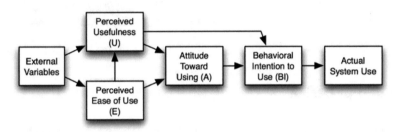

Fig. 2 TAM [14]

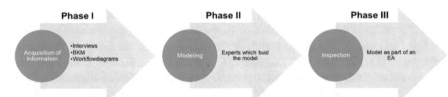

Fig. 3 Current Process for creating EA

3 Problem Formulation: Information Acquisition and Representation

The German armed forces (GAF), as part of the NATO forces use the NAF to create their EA. To model the required views, both modelers and SME are necessary. The start of modeling begins, according to common procedural models, with the creation of the operational view or the process model.

Through various interviews, both with modelers and with SME, a general process for the creation of these views was outlined (Fig. 3).

Object of the first phase is information acquisition and data collection. In this phase, different tools are used, such as methods that are based on given scenarios [15] (brainstorming [16] or group discussions [17]). Methods are also used that are initially based on process models. Mainly the image card method (BKM) [18] was mentioned as method/tool.

It was shown here that the GAF is coping very well with the challenge of obtaining information. However, there is a problem with the presentation of the information, since it does not exist in the form required by the NAF and must be transferred. The transfer is carried out by modelers who are not part of the area to be modeled [19]. This transfer happens without the presence of the SME.

In the last phase, the SME should check the transformed model about content accuracy, which is not possible due to the lack of knowledge in the EA area. This is related to the form of the model, since the presentation of information from Phase I is fundamentally different from the model to be checked after the transformation. It also follows from this that the further use of the model by the SME is nearly not possible, which creates a defensive stance against EA (lack of "stakeholder commitment" [20]).

4   Concept Design

The solution design must take up the current strengths of the existing tools and methods (best practice) and address the identified problems.

To develop a solution design, in addition to the practical orientation of the research, a literature evaluation has also to be carried out in order to be able to include possible solution approaches or to exclude falsified approaches [10]. This principle of Action Design Research is derived from the meaning of "Rigor & Relevance" [21, 22]. However, this does not mean that an evaluation of the literature alone satisfies a claim of scientific rigor and form; the applied methods must also be used correctly and conscientiously in research in the field of practice and the associated data collection [23].

4.1   Operational Context

As shown, the entry into EA is begun in an operational context. The focus is on the process and those involved in the process. A solution-oriented requirement analysis [24] in the modeling is necessary to implement this type. Since the focus of a process is on the question "What," the focus of the modeling has to be on the functional architecture and the "functional specification" [6], which also corresponds to the requirements of the TOGAF. The information required to address the functional specification can be found in the NAF in views NOV-5 (operational process) and in NOV-2 (information exchange relationships) [9].

4.2   User Requirements

In order to generate sufficient acceptance on the part of the end-user (addressing the perceived usefulness according to TAM) for a possible solution based on the solution concept, the interviews to determine the existing procedure also asked for ideas for improvement. User stories were recorded for a more precise definition of the suggestions for improvement and the desired functions or possibilities of a procedure [25]. The evaluation of this method led to various user stories, which were aggregated in a Kano model [26]. For this purpose, the prioritization was used to determine which function can be assigned to which characteristic type (threshold, performance, excitement). The fulfillment of all threshold features is essential for the acceptance of a solution. As a threshold feature, all functions that were perceived as positive when examining the existing procedure model could be identified. Performance features form functions that were outlined as possible improvements or missing functions in the interviews and that do not have the highest priority in the user stories.

Functions whose goal is to be summarized under the concept of facilitating work were recorded as a feature of excitement. The results are shown in Fig. 4.

To address the perceived ease of use, basic research was carried out using literature research and existing approaches examined. A look at the requirements for forms of representation in process modeling shows that more than seven different

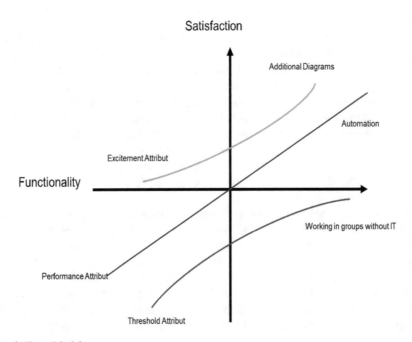

Fig. 4 Kano-Model

Table 1 Order and number of elements

	Order and number of elements
1	Max 3 per layer
2	Max 7 different types of elements
3	Max 30 elements per model
4	Readable in 30 s
5	Identifiable in 3 min
6	Understandable in 30 min

Table 2 Graphical form of elements

	Graphical form of elements
1	Elements have to be uniform
2	Not too much information
3	Arrangement from left to right
4	Unique identifier for elements in different layer
5	Connections between elements with arrows
6	No text outside

pieces of information or different elements of information are difficult to process for humans [27]. Furthermore, a diagram or a graphically represented architecture should be read in 30 s, the relationships recognized in 3 min and all details fully understood in 30 min. To achieve such timelines, it is necessary to limit the number of elements available to approximately 30 [28]. In relation to the usability with regard to a hierarchy in the graphical representation of the modeling method, the requirements are shown in Table 1.

Individual elements should be arranged in writing direction [29], from left to right, to enable the, already mentioned, 30 s to read a diagram/architecture. In addition to this structured procedure, the procedure should always be iterative [30], which is given by the aforementioned procedure of introducing and defining, as well as by the use of different levels of abstraction. The relationships between the individual elements should be depicted by arrows, which represent the reading and flow direction of the process. It must be noted here that the arrows should not overlap in order to ensure that they are legible and to comply with the required timelines [31]. In order to achieve or maintain this legibility, unnecessary text in the diagram should be avoided. Table 2 shows the collected requirements relating to usability with a focus on the illustration.

An easily understandable and analogously applicable method for information acquisition must be obtained from the solution design. The gained information which are presented in an easily understandable process model, has to be automatically converted into the form defined in NAF and contains additionally an information exchange model. It must enable the SME to understand the NAF-compliant model and to check that it is correct in terms of content. A holistic approach is derived from this, which is obtained from the interplay of humans, computers, and tasks and thus contributes to the goal of scientific findings [32].

5   Implementation of Requirements and Development of an Algorithm

A first step of developing the modeling method was to identify all necessary information for the creation of the required views [33]. From these, an element was developed that represents the node types in the modeling method. The exchange of information or the flow is realized by directed edges. It is based on the properties of the Petri nets [34, 35], but with the combination of positions and transitions in one single node type.

This results in the following representation as a 4-tuple (K, F, W, m_0) and where

$K = \{k\_1, k\_2, \cdots k\_n\}$ the finite set of nodes,
$F \subseteq (K \times K)$ the amount of edges (flux relation),
$W: F \rightarrow \{1,2,3, \cdots\}$ the edge weighting function
$m_0 \rightarrow \{0,1,2,3, \cdots\}$ the initial marking.

The use of this structure thereby achieves maximum readability and the associated usability through the planar representation of the graph [36]. Planar graphs show no overlap of edges, which contributes considerably to the required usability [37]. Figure 5 shows an example of the graphical implementation.

The results of the developed modeling method, the design of which has been optimized for manual application with maps, must be digitized for further use after successful modeling. For this purpose, templates (Fig. 6) were developed in the software yEd (a software developed by yWorks especially for graph visualization).

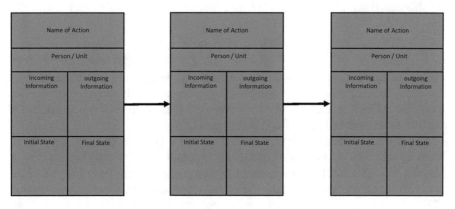

Fig. 5 Form of method with identified content

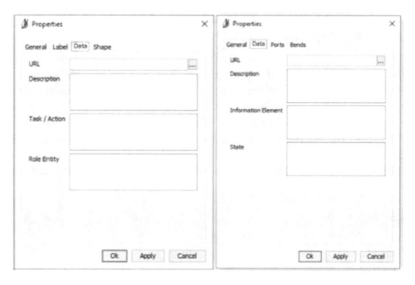

Fig. 6 Templates yEd

These templates allow the model to be converted 1:1 into an XML-based form of the Graph Markup Language (GraphML) [38].

In a software implemented in Java, an algorithm was developed that converts the GraphML file into an XML file [33], which can be read via the XML interface of the "Sparx Enterprise Architect" (Sparx EA).[1] The algorithm adds all information for the automated creation of the NOV-2 to this XML file.

6 Evaluation

To evaluate the developed artifacts, they were used in a project that aimed to implement a newly created process related to innovation management. The evaluation spanned several cycles and was always based on the fundamentals of Canonial Action Research according to (Diagnosis, Action Planning, Intervention, Evaluation, Learning) [39]. The use of this method accompanied the project for 6 months, from the first idea to the first EA model.

An evaluation framework was created for each cycle in order to be able to focus specifically on individual attributes. In addition to the TAM and UTAUT [14, 40, 41], ISO 9241 (ergonomics of human–system interaction) [42] was also taken into account. Based on this, a frame of reference for evaluation was created [43]. When creating these frameworks, additional concepts for measuring the

---

[1]https://www.sparxsystems.de/uml/neweditions/

Table 3 Evaluation cycles

	Evaluation cycles		
	Inspected part	Addressed attributes	Used methods
1	Modeling Artifact	Perceived Ease of Use	Experiment
2	Modeling Artifact	Perceived Ease of Use Perceived Usefulness Effort Expectancy	Experiment
3	IT-Artifact	Perceived Ease of Use	Cognitive Walkthrough
4	IT-Artifact	Perceived Ease of Use	Experience Research Co-Discovery-Method
5	Use of the Artifacts	Perceived Usefulness Performance Expectancy	Expert Discussion
6	Use of the Artifacts	Performance Expectancy Perceived Usefulness Effort Expectancy Perceived Ease of Use Self-Efficacy	Field Experiment Interviews Expert Discussion

Fig. 7 Parts of results during the cycles, using artifact

usability of modeling languages [44] were taken into account, which supplement the mentioned models and standards. Furthermore, the categories Syntactic, Semantic, and Pragmatic Quality [45, 46] were included in the design of the frame of reference with regard to the evaluation of the modeling language itself, its correctness, and the recognition value, which should enable the SME to control content accuracy. When evaluating the developed software, common methods of software evaluation, such as cognitive walkthrough [47] and user experience research (UER) [48], were used. Table 3 shows an overview of the evaluation cycles.

The outcomes of each cycle resulted in the redesign and adaptation of the artifacts. Figures 7 and 8 show the results of individual cycles.

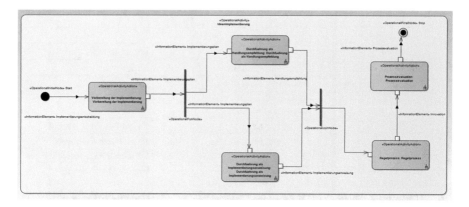

Fig. 8 Parts of results during the cycles, transformed model

7 Conclusion

The aim of this contribution was the development and provision of a method including a software concept to make a contribution to process modeling as part of an enterprise architecture in the area of operational architecture in accordance with NAF. The focus here was on adapting the previous process by empowering the individual users to find an entry into EA. The starting point was the previous procedure including best practice and the involvement of the SME in terms of information acquisition and final control.

The problem was derived from a practical problem that forms the basis for the development of the artifacts. Through the use of the artifacts on the part of SME across the entire research process, the development of these artifacts could be represented in a comprehensible manner. The validation of the artifacts also helps to demonstrate their usefulness. The artifacts were used as a method in a project. The associated application revealed the benefits of the artifacts, since the users created a first possible process of innovation management. It was also shown that the use of the artifacts made all users more accessible to EA, which is due to the fact that all users were able to find their delivered information in the views of the operational architecture. This also resulted in a general understanding of the creation of EA by the users, because by recognizing the process in one view, it was possible for them to recognize the link to other views.

Typical research results of design-oriented business informatics are available in the form of artifacts [49], which contribute to answer the question, "How well?"

Derived from the evaluation of the artifacts and their adaptation and integration into the context of the environment/organization, knowledge must be gained and formulated ("research through development" [50]), in accordance with the final principle of the ADR (cf. [10]).

The focus on process modeling to create certain views of an EA and the construction of an artifact in the form of a semi-formal language for modeling, geared toward usability, combined with an IT artifact for converting models into defined standards, provide an innovative conceptual contribution as a holistic view of task–human–computers in the field of information acquisition for the modeling of processes [51]. The implementation of the artifacts as instantiation already provides an innovative contribution [52, 53].

In conclusion, the contribution provides artifacts that make it possible to create a process model such as NOV-5 without knowledge about the required form and syntax related to the NAF. It is also possible to recognize the similarity between the original model, created with one of the artifacts and the NOV-5, transformed with the IT artifact. This enables SME to provide information for parts of an EA, facilitate them to control the provided content, and ensures that the SME can be an active part of the process in creating an EA.

References

1. J. Luftman, T. Brier, Achieving and sustaining business-IT alignment. *Calif. Manage. Rev.***42**(1), 109–122 (1999)
2. J.A. Zachman, A framework for information systems architecture. *IBM Syst. J.***26**(3), 276–292 (1987)
3. S. Aier, C. Riege, R. Winter, Unternehmensarchitektur – Literaturüberblick und Stand der Praxis. *Wirtschaftsinformatik***50**(4), 292–304 (2008)
4. M. Farwick, B. Agreiter, R. Breu, S. Ryll, K. Voges, and I. Hanschke, "Requirements for Automated Enterprise Architecture Model Maintenance - A Requirements Analysis based on a Literature Review and an Exploratory Survey.," in *ICEIS 2011 - Proceedings of the 13th International Conference on Enterprise Information Systems*, 2011, vol. 4, pp. 325–337.
5. E. Nowakowski, M. Farwick, T. Trojer, M. Häusler, J. Kessler, R. Breu, Enterprise Architecture Planning: Analyses of Requirements from Practice and Research. *Proc. 50th Hawaii Int. Conf. Syst. Sci.*, 4847–4856 (2017)
6. D.T. Ross, K.E. Schoman, Structured Analysis for Requirements Definition. *IEEE Trans. Softw. Eng.***SE-3**(1), 6–15 (1977)
7. V. Haren, "TOGAF Version 9.1," 2011.
8. M. Lankhorst, *S. J. B. A* (Robert Hoppenbrouwers, and J. Campschroer, "Enterprise Architecture at Work,", Slagter, 2017), pp. 141–170
9. NATO CONSULTATION COMMAND AND CONTROL BOARD, *NAF v3.1 Chapter 5.* 2010.
10. Sein, Henfridsson, Purao, Rossi, and Lindgren, Action Design Research. *MIS Q.***35**(1), 37 (2011)
11. M.T. Mullarkey, A.R. Hevner, *Entering action design research*, in International Conference on Design Science Research in Information Systems, pp. 121–134 (2015)
12. H. Österle, R. Winter, W. Brenner, *Gestaltungsorientierte Wirtschaftsinformatik: Ein Plädoyer für Rigor und Relevanz, vol. 62, no. 6* (Infowerk, Nürnberg, 2010)
13. J. Becker, D. Pfeiffer, Beziehungen zwischen behavioristischer und konstruktionsorientierter Forschung in der Wirtschaftsinformatik, in *Fortschritt in den Wirtschaftswissenschaften: Wissenschaftstheoretische Grundlagen und exemplarische Anwendungen*, ed. by S. Zelewski, N. Akca, (DUV, Wiesbaden, 2006), pp. 1–17
14. F.D. Davis, *A technology acceptance model for empirically testing new end-user information systems: theory and results* (Massachusetts Institute of Technology, Cambridge, MA, 1986)

15. M. Herzhoff, "Zum Zusammenspiel von Frühaufklärung und Szenariotechnik," in *Perspektiven des Strategischen Controllings*, no. Abschnitt 2, Wiesbaden Springer 2010, pp. 313–328.
16. M. Diehl, W. Stroebe, Productivity loss in brainstorming groups: Toward the solution of a riddle. *J. Pers. Soc. Psychol.***53**(3), 497 (1987)
17. P. Loos, B. Schäffer, *Das Gruppendiskussionsverfahren: Theoretische Grundlagen und empirische Anwendung*, vol 5 (Springer, Wiesbaden, 2013)
18. M. Gappmaier, C. Gappmaier, Alles Prozess?!: Einfach wirksame Prozessoptimierung in jeder Situation mit der Bildkartenmethode (BKM), vol. 3, in *Books on Demand*, (UMI, Ann Arbor, 2011)
19. G. Tapandjieva, A. Wegmann, *Ontology for SEAM Service Models*, in 20th International Conference on Enterprise Information Systems (ICEIS), no. EPFL-CONF-233594 (2018)
20. C. Lucke, M. Bürger, T. Diefenbach, J. Freter, U. Lechner, Categories of enterprise architecting issues - An empirical investigation based on expert interviews. *Multikonferenz Wirtschaftsinformatik 2012 - Tagungsband der MKWI2012*, 999–1010 (2012)
21. R. Gulati, Tent poles, tribalism, and boundary spanning: The rigor-relevance debate in management research. *Acad. Manag. J.***50**(4), 775–782 (2007)
22. A.R. Hevner, A Three Cycle View of Design Science Research. *Scand. J. Inf. Syst.***19**(2), 87–92 (2007)
23. H. Österle et al., Memorandum zur gestaltungsorientierten Wirtschaftsinformatik. *Schmalenbachs Zeitschrift für betriebswirtschaftliche Forsch.***62**(6), 664–672 (2010)
24. P. Harmon, *Business process change: A guide for business managers and BPM and Six Sigma professionals* (Elsevier, Cambridge, MA, 2010)
25. M. Cohn, *User stories applied: For agile software development* (Addison-Wesley Professional, Boston, 2004)
26. F. Bailom, H.H. Hinterhuber, K. Matzler, E. Sauerwein, Das Kano-Modell der Kundenzufriedenheit. *Mark. ZFp***18**(2), 117–126 (1996)
27. G.A. Miller, The magical number seven, plus or minus two: some limits on our capacity for processing information. *Psychol. Rev.***63**(2), 81 (1956)
28. W. Horton, *Illustrating computer documentation: The art of presenting information graphically on paper and online* (Wiley, New York, 1991)
29. H.-J. Haecker, Neue Überlegungen zu Schriftrichtung und Textstruktur des Diskos von Phaistos. *Kadmos***25**(2), 89–96 (1986)
30. R. Knackstedt, J. Pöppelbuß, J. Becker, Vorgehensmodell zur Entwicklung von Reifegradmodellen., in *Wirtschaftsinformatik (1)*, pp. 535–544 (2009).
31. H. Koning, C. Dormann, H. Van Vliet, Practical guidelines for the readability of IT-architecture diagrams, in *ACM SIGDOC Annu. Int. Conf. Comput. Doc. Proc.*, pp. 90–99, (2002)
32. E.J. Sinz, Konstruktionsforschung in der Wirtschaftsinformatik: Was sind die Erkenntnisziele gestaltungsorientierter Wirtschaftsinformatik-Forschung, *Wirtschaftsinformatik Ein Plädoyer für Rigor und Relev.*, p. 27 (2010)
33. S. Jahnen, S. Pickl, *Information Exchange Diagrams for Information Systems and Artificial Intelligence in the Context of Decision Support Systems*, in International Conference on Modelling and Simulation for Autonomous Systesm, pp. 393–401 (2018).
34. T. Murata, Petri Nets : Properties , Analysis and Applikations. Proceedings of the IEEE **77**(4), 541–580 (1989)
35. W. Reisig, *Petri nets: an introduction*, vol 4 (Springer, Berlin, 2012)
36. C. Gutwenger, M. Jünger, K. Klein, J. Kupke, S. Leipert, P. Mutzel, A new approach for visualizing UML class diagrams. Proc. 1st ACM Symp. Softw. Vis. **1**(May 2014), 179–188 (2003)
37. M. Jünger, P. Mutzel, *Graph drawing software* (Springer, Berlin, 2012)
38. M. Eiglsperger, U. Brandes, J. Lerner, C. Pich, Graph Markup Language (GraphML). *Handb. Graph Draw. Vis.*, 517–541 (2013)
39. R. Davison, M.G. Martinsons, N. Kock, Principles of canonical action research. *Inf. Syst. J.***14**(1), 65–86 (2004)

40. F.D. Davis, Perceived usefulness, perceived ease of use, and user acceptance of information technology. *MIS Q. Manag. Inf. Syst.* **13**(3), 319–339 (1989)
41. V. Venkatesh, M.G. Morris, G.B. Davis, F.D. Davis, User Acceptance of Information Technology: Toward a Unified View. *MIS Q.* **27**(3), 425–478 (2003)
42. T. Jokela, N. Iivari, J. Matero, M. Karukka, The standard of user-centered design and the standard definition of usability: analyzing ISO 13407 against ISO 9241-11, in Proceedings of the Latin American conference on Human-computer interaction, pp. 53–60 (2003)
43. E. Zur, V. Von, P. Zur, O. Von, "Ein Bezugsrahmen zur Evaluation von Sprachen zur Modellierung von Geschäftsprozessen," no. 36 (2003)
44. C. Schalles, J. Creagh, M. Rebstock, Ein generischer Ansatz zur Messung der Benutzerfreundlichkeit von Modellierungssprachen. *Modellierung* **161**, 15–30 (2010)
45. D.L. Moody, G. Sindre, T. Brasethvik, A. Sølvberg, Evaluating the quality of process models: Empirical testing of a quality framework. *Lect. Notes Comput. Sci. (including Subser. Lect. Notes Artif. Intell. Lect. Notes Bioinformatics)* **2503**, 380–396 (2002)
46. P. Rittgen, "Quality and perceived usefulness of process models," *Proceedings of the Symposium on Applied Computing*, pp. 65–72, 2010.
47. J. Rieman, M. Franzke, and D. Redmiles, "Usability evaluation with the cognitive walkthrough. *Conference Companion on Human Factors in Computing Systems – CHI '95*, pp. 387–388, 1995.
48. T. Hynek, User Experience Research—treibende Kraft der Designstrategie, in *Usability*, (Springer, Berlin, 2002), pp. 43–60
49. U. Frank, Die Konstruktion möglicher welten als chance und herausforderung der Wirtschaftsinformatik, in *Wissenschaftstheorie und gestaltungsorientierte Wirtschaftsinformatik*, (Springer, Cham, 2009), pp. 161–173
50. U. Frank, *Towards a Pluralistic Conception of Research Methods in Information Systems Research* (ICB, Essen, 2006)
51. M. Dumas, M. La Rosa, J. Mendling, H.A. Reijers, others, *Fundamentals of Business Process Management*, vol 1 (Springer, Berlin, 2013)
52. R. Winter, Design science research in Europe. *Eur. J. Inf. Syst.* **17**(5), 470–475 (2008)
53. S. Gregor, A.R. Hevner, Positioning and presenting design science types of knowledge in design science research. *MIS Q.* **37**(2), 337–355 (2013)

Improving Knowledge Engineering Through Inter-Organisational Architecture, Culture, Agility and Change in E-Learning Project Teams

Jonathan Bishop and Kamal Bechkoum

1 Introduction

E-Learning products are very different from other software products in that software compatibility is a big issue due to the regular use of multimedia content [26]. The Web initially solved the problem of having to produce different applications for different platforms, such as Mac OS and IBM PC CD-ROMs, but when smartphones became popular the problem resurfaced. E-Learning can be defined as the delivery of instructional content via electronic means. When developing any information system, it is essential to account for culture, especially if the system is used in an international context [28, 46–48]. There is, therefore, a need to develop better cognitive tools for use in learning environments [40].

1.1 Effective Project Management

Whether one is managing a single project or multiple projects, effective project management is the key to the success of any project [3]. Essential to efficacious project management is effective organisational architecture, especially in organisations that operate within set geographical boundaries [32], which includes

J. Bishop (✉)
Congress of Researchers and Organisations for Cybercommunity, E-Learning and Socialnomics, Swansea, Wales
e-mail: jonathan.bishop@crocels.ac.uk

K. Bechkoum
University of Gloucestershire, Cheltenham, UK
e-mail: kbechkoum@glos.ac.uk

© Springer Nature Switzerland AG 2021
R. Stahlbock et al. (eds.), *Advances in Data Science and Information Engineering*,
Transactions on Computational Science and Computational Intelligence,
https://doi.org/10.1007/978-3-030-71704-9_48

Crocels that is based in the UK, Republic of Ireland, and the United States. Adapting project management methodologies to accommodate such organisational architecture is thus essential (ibid). Understanding organisational theory [36] can assist in the process as can understanding organisational behaviour [50]. Indeed, if an organisation is to adopt a specific approach to project management, such as agile development, then it needs to be suited to that way of doing things [8]. Even in small organisations, taking a systematic approach to information systems design, development and evaluation can yield results [25].

It is argued that software projects are investments by an organisation as by committing to one it uses time, money and resources within the project with an expectation that there will be a return on investment in terms of a software product ([49], p.1). Software products are achieved through software engineering, which is a discipline concerned with all aspects of the software development process, from systems specification in the early stages to systems maintenance after the system has been deployed [70]. When a software project is going wrong it is called a "software crisis" and project managers need to adopt different approaches to overcome the situation [58]. One way of achieving this is through neuro-linguistic programming [68], which allows project managers to better influence their team and themselves through using language that directs behaviour towards a specific outcome.

1.2   Selection of Methodology for Project Management

Project management teams often find it problematic to choose from a range of standardised methodologies and this often results in project failure [38]. This chapter will, on the one hand, show how standardised methodologies can be adapted to the customs of project management teams and organisations and, on the other hand, provide an equation to monitor the human-element of project management to ensure projects are successful from the outset and do not fail, including by extending established equations in this area, such as by Meredith, Shafer, Mantel, and Sutton [55]. Implementing a project management methodology to accommodate organisational architecture might not mean that organisational architecture staying static if the organisation itself is not based around a culture of delivering projects, such as if it is volume-based or operations-orientated [3].

1.3   Project Management and Organisational Value Creation

The main purpose for any successful project is argued to be to deliver value for the stakeholders that will use what is delivered by that project. It is also argued that an important part of delivering value is for the project team itself to feel the benefits of a project, as their behaviour is influenced by what they gain from a project [19]. Furthermore, it is suggested that the value of a project should be considered in organisational terms as well as the financial return on investment. The project methodologies that will be produced by this research project will focus on the human and organisational side of project management as SSADM did [7, 34]. This will consider the culture and organisational architecture where the e-learning products will be developed and used.

1.4 Project Failure and Recovery in the Project Management Process

An important part of project management is being able to deal with crises – on the one hand, avoiding them where possible and dealing with them appropriately when it is not possible [9]. It is the project manager's responsibility to ensure these crises are managed, even if they are ones as common as project delay [44]. There are several early warning signs of a project crisis, some relating to a deterioration in the business case, difficulties among those involved in delivering the project, failure in following the project management methodology, and unplanned events [38]. An essential part of the toolkit that will be developed from this project will assist managers in dealing with project failure, including from learning from past projects carried out by the author based on Classroom 2.0 [10, 13, 14] and anything that arises from the workshops in this study based on School 3.0. Classroom 2.0 was based heavily on technology to enhance education [31], whereas School 3.0 is not just about the technology, but pedagogy and location of education also [11].

2 Background

Many organisations have dedicated, and sometimes tailored, project management methodologies for different operations within that organisation ([38], p. 39). It is increasingly common for organisations to use projects to structure how they deliver their operational objectives [75]. The impact of the organisation can be more effectively monitored by utilizing the structuring on a project delivery basis [71]. It is known that an organisation's structure and processes are an essential part of project management, particularly in e-learning projects, as the way in which a project team, platform and processes are constructed affect the success of a project [30].

2.1 Project, Portfolio and Programme Management

Project management is argued to be the means by which an organisation can make a significant impact on its competitiveness [63]. If one, therefore, sees a project as part of an organisation's processes then it becomes essential to have portfolio management, where those processes are managed to reduce risk to the organisation and properly and efficiently allocate resources [81]. When one's organisation is multi-dimensional, such as the organisations discussed in this chapter, which all have more than one location for their operations, then programme management becomes essential [53]. Programme management is about ensuring the benefits of project management are received by the whole organisation, beyond the individual objectives of each project [69].

2.2 Organisational Architecture

Project management is argued to be integral to an evolving organisational architecture [41]. In terms of project management, organisational architecture is often considered a fusion between organisational design and organisational development [74]. The way in which an organisation is designed is often integral to the success of the project management methodology chosen [78]. One key factor in organisational architecture when integrating it with project management is the effective optimisation of human and other resources [21]. Essential to organisational architecture, it is argued, is developing infrastructure, creating knowledge, changing minds and managing for value capture [42].

2.3 Organisational Agility

Understanding organisational agility in the case of project management often means understanding the technical skills required in the organisation and any gaps in those required skills in the workforce [76]. It is important to note that organisational agility has a different meaning to that typically understood in agile project management [39]. Organisational agility refers to the ability of an organisation to adapt to stay ahead of its competitors [79]. It is the ability of those within an organisation to adapt and change to maintain a competitive edge to deliver better value for stakeholders [20]. In terms of the development of e-learning systems, adopting new technologies and tools as they come along can enhance learner outcomes [51], but some have suggested waiting for technologies to mature before adopting them [59, 60].

2.4 Organisational Culture

Organisational culture goes hand-in-hand with project management, with one influencing the other [5]. In organisations relying on project management, it is known that if the chosen approach does not fit the organisation's culture then outcomes for systems and strategy can be ineffective [17]. In organisations that are based around project management even those who are not part of the project team can notice improvements to the organisational culture [56]. However, it is essential that the project management approach chosen fits with the organisation's existing culture as otherwise imposing an unsuitable approach can have adverse effects [22]. Indeed, if choice of project management approach is not anchored to an organisation's culture then it can be counter-productive [67]. Even beyond the initial choice of project management methodology, when the software is produced, especially e-learning products, the satisfaction of those using it is a core part of the organisation's culture [33].

2.5 Organisational Change

Organisational change can affect the success of projects for there are often greater training needs for project managers [37]. By the same token, projects can be effective at value creation and economic enhancers [52]. One of the biggest impacts affecting organisations based around project management is that the changing goals and objectives of the organisation can directly impact on the scope of the projects being undertaken [62]. It has been argued that involving the management of an organisation in the project management process can help them achieve the organisational change they want without compromising the projects that organisation undertakes [54]. This is especially the case if an operations management approach is taken to organisational change by managers [35]. Indeed, it is argued that library managers, and other information professionals, have a leading role to play in organisational change [65]. Even so it has been shown that information systems strategy has little impact on how effective organisational change is [80].

3 Problem Statement

There has long been conflict between project management and the way in which an organisation is designed and operates [16]. However, it has been argued that embedding project management into the way an organisation operates is essential for its success [17]. Even so, implementing standardised project management methodologies is no guarantee of success, as projects can fail due to management mistakes, planning mistakes as well as external influences ([38], p.291). Table 1 explores these in greater detail.

Figure 1 presents a model that shows how using a common project management methodology that is adapted based on organisational culture and architecture means that organisations can more effectively accommodate their common goals [12]. This chapter goes further, by also considering organisational agility and organisational change in the project management process. It is argued these are important factors

Table 1 The three root causes of project failure

Reason	Description
Management mistakes	These are due to a failure in stakeholder management, such as by allowing too many unnecessary scope changes, failing to provide proper governance, refusing to make decisions in a timely manner and ignoring the project manager's quest for help. This can also be the result of wanting to gold-plate the project. This is also the result of not performing project health checks.
Planning mistakes	These are the result of poor project management, perhaps not following the principles of the chosen project management approach, not having health checks and not selecting the proper tracking metrics.
External influences	These are normally the failures in assessing the environmental input factors correctly. This includes the timing for getting approvals and authorisation from third parties and poor understanding of the host country's culture and politics.

Fig. 1 An adaptation of Bishop's Project management for co-operative advantage model

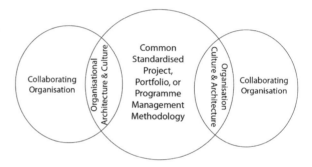

in project success and are inseparable from traditional measures, such as whether a project is delivered on time, within budget and at the required quality [2, 6, 18, 19].

4   A Revised Model for Integrating Organisational Architecture and Culture into Standardised Project, Portfolio and Programme Management Methodologies for E-Learning Product Development

The Crocels Community Media Group pioneered two projects in Wales, namely the Digital Classroom of Tomorrow (DCOT) Project [10, 14], which led to the Classroom 2.0 movement and the Clicks and Mortar Environment for Learning and Leisure Experiences (CAMELLE) Project [11, 13], which led to the School 3.0 movement.

Figure 2 presents a conceptual framework, called the 'Cornerstone Teaching as School 3.0 Model' (Cornerstone-TS 3.0), which is a working model for understanding how to manage projects in service-orientated organisations that rely on

Fig. 2 The Cornerstone
Teaching as School 3.0
Model (Cornerstone-TS 3.0)

e-learning product development. It was arrived at through the grounded theory literature review above which identified four aspects to effective project management, which were organisational culture, organisational change, organisational agility and organisational architecture. It is important for ensuring that even an organisation that does not have set project management methods in place is able to take a standardised approach to project management, which may be one used by an organisation they are collaborating with, and to tailor it to their organisation's way of doing things. That is what is intended by the Cornerstone-TS 3.0 Model in terms of organisations that produce e-learning products, whether a simple PowerPoint presentation through to a large-scale virtual learning environment (LSVLE).

4.1 Applying the Cornerstone-TS 3.0 Model

Applying the Cornerstone TS 3.0 Model in depth will require testing the model in a live project management environment in which organisational culture, change, agility and architecture can be captured, measured and analysed. This section discusses some of the factors that could form part of such measuring.

4.1.1 Organisational Culture

Organisational culture varies between organisations, but there are known commonalities between how organisational culture manifests [4]. It is known that management style and stakeholder performance are important factors in measuring organisational culture [15]. Indeed, leadership when it comes to performance can be a key factor when it comes to assessing organisational culture [72].

4.1.2 Organisational Change

Performance of staff is also linked to organisational change measurement [66]. Social and emotional intelligence are further factors in organisational change [73]. Important factors in organisational change are time, energy and resources [61]. Organisational change is sometimes measured against key performance indicators such as turnover [57] as well as commitment of key stakeholders [77].

4.1.3 Organisational Agility

The ability of different aspects of an organisation to inter-operate is a key factor in organisational agility [27]. Strategic management and research are known to be key to effective monitoring of organisational agility [43]. Effectiveness of learning among personnel is known to affect organisational agility [23]. Complexity is also important to measuring organisational agility [76]. In addition, evolutionary fitness and technical fitness are important factors when measuring organisational agility [24].

4.1.4 Organisational Architecture

The more staff that is involved in developing an aspect of a piece of software, the more complex the organisational architecture will need to be [64]. In an educational infrastructure, the quality of research and teaching are directly affected by organisational architecture [29]. Organisational architecture in many cases can be measured through management support, work discretion, reward and reinforcement, time availability and organisational boundaries [1]. Other factors key to measuring organisational architecture include structure, division of labour, resource allocation mechanisms and interdepartmental coordination [45].

5 Discussion

This study has investigated organisational culture, change, agility and architecture from the perspective of project management. These four factors were identified in response to the problem statement, namely that the embedding of project management into an organisation is essential yet very difficult to achieve, especially in organisations that do not naturally see themselves as carrying out projects. Equally important is identifying how to assist organisations in working with other organisations by identifying their commonalities and differences, such as in terms of the way the human resources are organised and deployed.

It has proposed a conceptual framework for managing projects based on the four factors identified from the literature, namely organisational culture, organisational change, organisational agility and organisational architecture. In order to test this proposed model, some of the variables that might contribute to these four factors have been explored. This study has shown that the four factors could form an important part in the development of customised project management methodologies in order to deliver e-learning products within a range of organisations. Future research will have to investigate the proposed conceptual framework in practices, including by designing questionnaires that may be based on some of the variables identified by this study to measure organisational culture, change, agility and architecture.

References

1. N.H. Ahmad, A.M. Nasurdin, S.R.M. Zainal, Nurturing intrapreneurship to enhance job performance: The role of pro-intrapreneurship organizational architecture. J. Innov. Manag. Small Medium Enterp. **2012**, 1 (2012)
2. J. Alvarez Dickinson, No title. *Pocho Humor: Contemporary Chicano Humor and the Critique of American Culture* (2009)
3. K. Artto, J. Kujala, P. Dietrich, M. Martinsuo, What is project strategy? Int. J. Proj. Manag. **26**(1), 4–12 (2008)
4. N.M. Ashkanasy, G.J. Nicholson, Climate of fear in organisational settings: construct definition, measurement and a test of theory. Aust. J. Psychol. **55**(1), 24–29 (2003)
5. R. Atkinson, L. Crawford, S. Ward, Fundamental uncertainties in projects and the scope of project management. Int. J. Proj. Manag. **24**(8), 687–698 (2006)
6. E.M. Bennatan, *On Time, within Budget: Software Project Management Practices and Techniques*, 1st edn. (Wiley, Chichester, 1996)
7. C. Bentley, *Introducing SSADM4+* (NCC Blacwell Ltd, Oxford, 1996)
8. H. Berger, Agile development in a bureaucratic arena—A case study experience. Int. J. Inf. Manag. **27**(6), 386–396 (2007). https://doi.org/10.1016/j.ijinfomgt.2007.08.009
9. J. Bernstein, *Manager's Guide to Crisis Management* (McGraw-Hill, New York, 2011)
10. J. Bishop, The potential of persuasive technology for educating heterogeneous user groups. MSc, University of Glamorgan. (Available online) (2004)
11. J. Bishop, Lessons from the emotivate project for increasing take-up of big society and responsible capitalism initiatives, in *Didactic Strategies and Technologies for Education: Incorporating Advancements*, ed. by P. M. Pumilia-Gnarini, E. Favaron, E. Pacetti, J. Bishop, L. Guerra, (IGI Global, Hershey, 2012), pp. 208–217
12. J. Bishop, Evaluating two successful and two failed multi-organisation e-learning software development projects using service-orientated approaches, in *Perspectives on the Information Society*, ed. by J. Bishop, 1st edn., (The Crocels Press Limited, London, 2019), pp. 147–213
13. J. Bishop, Evaluation-centred design of E-learning communities: A case study and review, in The 2nd International Conference on Internet Technologies and Applications (ITA'07), Wrexham, GB. pp. 1–9 (September 09 2007)
14. J. Bishop, R. Kingdon, M. Reddy, Cooperative e-learning in the multilingual and multicultural school: The role of 'classroom 2.0' for increasing participation in education, in *Didactic Strategies and Technologies for Education: Incorporating Advancements*, ed. by P. M. Pumilia-Gnarini, E. Favaron, E. Pacetti, J. Bishop, L. Guerra, (IGI Global, Hershey, 2012), pp. 137–150
15. U.S. Bititci, K. Mendibil, S. Nudurupati, P. Garengo, T. Turner, Dynamics of performance measurement and organisational culture. Int. J. Oper. Prod. Manag. **26**(12), 1325–1350 (2006)
16. M. Bresnen, Conflicting and conflated discourses? Project management, organisational change and learning, in *Making Projects Critical*, (Palgrave Macmillan, New York, 2006), pp. 68–89

17. C. Brown, A comprehensive organisational model for the effective management of project management. S. Afr. J. Bus. Manag. **39**(3), 1–10 (2008)
18. R. Burke, S. Barron, *Project Management Leadership* (Wiley Online Library, New York, 2007)
19. R. Burke, S. Barron, *Project Management Leadership* (Wiley, Chichester, 2014)
20. A.M. Carvalho, P. Sampaio, E. Rebentisch, J.Á. Carvalho, P. Saraiva, Operational excellence, organisational culture and agility: the missing link? Total Qual. Manag. Bus. Excell. **30**(13–14), 1495–1514 (2019)
21. D.M. Clayton, *The Presidential Campaign of Barack Obama: A Critical Analysis of a Racially Transcendent Strategy* (Taylor & Francis, Abingdon, 2010)
22. D. Comninos, E. Frigenti, Business focused project management. Br. J. Adm. Manag. **34**(December), 12–15 (2002)
23. K. Conboy, B. Fitzgerald, Toward a conceptual framework of agile methods: A study of agility in different disciplines. Paper presented at the Proceedings of the 2004 ACM Workshop on Interdisciplinary Software Engineering Research, pp. 37–44 (2004)
24. C. Crick, E. Chew, Understanding the role of business-IT alignment in organisational agility. Paper presented at the *Iceis (3)),* pp. 459–464 (2014)
25. D. Cunliffe, Developing usable web sites–a review and model. Internet Res. **10**(4), 295–308 (2000)
26. D. Cunliffe, G. Elliott, *Multimedia computing* (Lexden Publishing Ltd, Newcastle under Lyme, 2005)
27. A.H. Dekker, Measuring the agility of networked military forces. J. Battlefield Technol. **9**(1), 19 (2006)
28. E.M. Del Galdo, J. Nielsen, *International Users Interface* (Wiley, Chichester, 1996)
29. E. D'Souza, Contractual arrangements in academia: implications for performance. Econ. Polit. Wkly. **39**, 2165–2168 (2004)
30. G. Elliott, J. Cook, D. Monk, S. Burnett, M. Lynch, *Building an e-learning development team, platform and process from scratch,* in The 2003 World Conference on Educational Multimedia, Hypermedia & Telecommunications (EdMedia'03), Honolulu, HI. pp. 2813–2820 (2003)
31. F. Falcinelli, C. Laici, ICT in the classroom: New learning environment, in *Didactic Strategies and Technologies for Education: Incorporating Advancements*, ed. by P. M. Pumilia-Gnarini, E. Favaron, E. Pacetti, J. Bishop, L. Guerra, (IGI Global, Hershey, 2012), pp. 48–56
32. A. Gordon, J. Pollack, Managing healthcare integration: adapting project management to the needs of organizational change. Proj. Manag. J. **49**(5), 5–21 (2018)
33. M. Graff, *Evaluating Online Interaction in the Context of Instruction.* in Proceedings of the 5th European Conference on E-Learning (ECEL'06), Winchester, GB. pp. 131–136 (September 12 2006)
34. J. Hall, C. Slater, *Introducing SSADM4+* (Blackwell Publishing, Oxford, 1996)
35. A. Hammadi, T. Reiners, R. Taylor, J. Earnest, L. Wood, We have to Integrate to Engage in Change: Exploring sustainable project management. Paper presented at the 26th EurOMA Conference (2019)
36. M.J. Hatch, *Organization Theory: Modern, Symbolic, and Postmodern Perspectives* (Oxford University Press, Oxford, 2018)
37. H.A. Hornstein, The integration of project management and organizational change management is now a necessity. Int. J. Proj. Manag. **33**(2), 291–298 (2015)
38. H. Kerzner, *Project Recovery: Case Studies and Techiques for Overcoming Project Failure* (Wiley, Chichester, 2014)
39. M.R. Khan, W.D. Fernandez, J. J. Jiang, Is there Such a Thing as Agile IT Program Management. Paper presented at the International Research Workshop on IT Project Management (2016)
40. P.A.M. Kommers, D.H. Jonassen, J.T. Mayes, A. Ferreira, *Cognitive Tools for Learning* (Springer, New York, 1992)
41. E. Larson, C. Gray, *Project Management: The Managerial Process with MS Project* (McGraw-Hill Education, New York, 2013)

42. M. Laursen, Project networks as constellations for value creation. Proj. Manag. J. **49**(2), 56–70 (2018)
43. S. Lim, F. Mavondo, *The Structure of Strategic Capabilities, Implications for Organisational Agility and Superior Performance: A Conceptual Framework* (Conceptual Framework, Department of Marketing, Monash University, Monash University, Melbourne, 2000)
44. S. Madhavan, Delays and failures in projects: Using soft systems and action research methods to explore integration issues in research and practice. *Proceedings of the 22nd World Multi-Conference on Systemics, Cybernetics and Informatics (WMSCI 2018),* Orlando, FL (July 11 2018)
45. G.M. Magnan, M. Day, C. Hillenbrand, S.E. Fawcett, The role of relational architecture in developing relational capability: Organizational levers for strategic relationships (2013)
46. G. Mantovani, *New Communication Environments: From Everyday to Virtual* (Taylor & Francis, London, 1996a)
47. G. Mantovani, Social context in HCI: a new framework for mental models, cooperation, and communication. Cogn. Sci. **20**(2), 237–269 (1996b)
48. G. Mantovani, *Exploring Borders: Understanding Culture and Psychology* (Routledge, Abinbton, 2000)
49. J.T. Marchewka, *Information Technology Project Management: Providing Measurable Organizational Value*, 5th edn. (Wiley, Chichester, 2015)
50. J. Martin, *Organizational Behaviour and Management* (Cengage learning EMEA, Hampshire, 2005)
51. D. Martland, E-learning: What communication tools does it require? in *E-Learn: World Conference on E-Learning in Corporate, Government, Healthcare, and Higher Education*, (Association for the Advancement of Computing in Education, Waynesville, 2003), pp. 2313–2316
52. H. Maylor, *Project management*, 4th edn. (Pearson Education Limited, Harlow, 2010)
53. R. Mc Lean, K. Theodore, A. La Foucade, S. Lalta, C. Laptiste, R.B. St. Martin, et al., Austerity, and funding cuts: implications for sustainability of the response to the caribbean HIV/AIDS epidemic. Glob. Public Health **14**(11), 1612–1623 (2019)
54. W. McElroy, Implementing strategic change through projects. Int. J. Proj. Manag. **14**(6), 325–329 (1996)
55. J.R. Meredith, S.M. Shafer, S.J. Mantel, M.M. Sutton, *Project Management in Practice* (Wiley, Chichester, 2014)
56. A. Miklosik, Improving project management performance through capability maturity measurement. Procedia Econ. Financ. **30**, 522–530 (2015)
57. K.M. Morrell, J. Loan-Clarke, A.J. Wilkinson, Organisational change and employee turnover. Pers. Rev. **33**(2), 161–173 (2004)
58. K.E. Nidiffer, D. Dolan, Evolving distributed project management. IEEE Softw. **22**(5), 63–72 (2005)
59. J. Nielsen, *Usability engineering* (Morgan Kaufmann, Amsterdam, 1994)
60. J. Nielsen, *Designing Web Usability* (New Riders, New York, 2000)
61. J. Oakland, S. Tanner, A new framework for managing change. *The TQM Magazine* (2007)
62. S. Pellegrinelli, Programme management: organising project-based change. Int. J. Proj. Manag. **15**(3), 141–149 (1997)
63. A.R. Peña, F.A. Muñoz, Soft skills as a critical success factor in project management, in *Handbook of Research on Project Management Strategies and Tools for Organizational Success*, (IGI Global, Hershey, 2020), pp. 376–392
64. G. Peng, J. Mu, Do modular products lead to modular organisations? Evidence from open source software development. Int. J. Prod. Res. **56**(20), 6719–6733 (2018)
65. H. Preston, M. Allmand, Discovering the information professional: organisational culture in a digital world. Online Inf. Rev. **25**(6), 388–396 (2001)
66. I.D. Rajamanoharan, P. Collier, Six sigma implementation, organisational change and the impact on performance measurement systems. Int. J. Six Sigma Comp. Adv. **2**(1), 48–68 (2006)

67. P. Roberts, *Guide to Project Management: Getting it Right and Achieving Lasting Benefit* (Wiley, Chichester, 2013)
68. Y.S. Rudall, NPL for project managers: make things happen with neuro-linguistic programming. Kybernetes **41**, (2012)
69. S. Saadi, G. Bell, Exploring the use of soft systems methodology (SSM) in front-ending public-funded rural bridge construction projects in Bangladesh, in *Problem Structuring Approaches for the Management of Projects*, (Springer, Cham, 2019), pp. 161–214
70. I. Sommerville, *Software engineering*, 10th edn. (Pearson Education Limited, London, 2016)
71. P. Stephenson, The holy grail of biodiversity conservation management: monitoring impact in projects and project portfolios. Persp. Ecol. Conserv. **17**(4), 182–192 (2019)
72. L.T. Tuan, Organisational culture, leadership and performance measurement integratedness. Int. J. Manag. Enterp. Dev. **9**(3), 251–275 (2010)
73. M. Vakola, I. Tsaousis, I. Nikolaou, The role of emotional intelligence and personality variables on attitudes toward organisational change. J. Manag. Psychol. **19**(2), 88–110 (2004)
74. A. Van Der Merwe, Project management and business development: integrating strategy, structure, processes and projects. Int. J. Proj. Manag. **20**(5), 401–411 (2002)
75. S. Venkatachalam, A. Marshall, U. Ojiako, C.S. Chanshi, Organisational learning in small and medium sized south african energy project organisations. Manag. Res. Rev. **43**, 595–623 (2019)
76. S. Vinodh, S. Aravindraj, Agility evaluation using the IF–THEN approach. Int. J. Prod. Res. **50**(24), 7100–7109 (2012)
77. C.M. Visagie, The Relationship between Employee Attitudes towards Planned Organisational Change and Organisational Commitment: An Investigation of a Selected Case within the South African Telecommunications Industry. (MTech in Business Administration, Cape Peninsula University of Technology, 2010)
78. A. Walker, *Project Management in Construction* (Wiley, Chichester, 2015)
79. D.H. Wolf, G.R. Fink, Proteinase C (carboxypeptidase Y) mutant of yeast. J. Bacteriol. **123**(3), 1150–1156 (1975)
80. M.G. Wynn, Information systems strategy development and implementation in SMEs. Manag. Res. News **32**(10), 78–90 (2009)
81. H. Yun, M. Lee, Y.S. Kang, J. Seok, Portfolio management via two-stage deep learning with a joint cost. Expert Syst. Appl. **143**, 113041 (2020)

Do Sarcastic News and Online Comments Make Readers Happier?

Jih-Hsin Tang, Chih-Fen Wei, Ming-Chun Chen, and Chih-Shi Chang

1 Introduction

Since online newsvendors have encouraged readers' participation globally, online commenting has become an important phenomenon. Both civil and uncivil comments have appeared below online news articles; however, whether these comments affect readers' emotional experiences is a critical question.

Previous research focused on the reasons for online comments or on the emotional response to online comments; less attention has been paid to the interplay of online news and comments on a reader's emotion. We would like to fill this gap by examining the influence of news type (sarcastic or neutral) and online comments (civil, uncivil, or none) on a reader's emotion.

The concept of Web 2.0 has encouraged online users to share their photos, diaries, videos, and opinions online, and social media services such as blogs, YouTube, and Facebook have greatly impacted societies. To cultivate their own online communities and encourage discussion, many newsvendors such as CNN and the BBC have invited their users to comment on the news articles they post online. Online commenting on news sites has become a widespread phenomenon and has attracted scholars to examine what types of news articles is most likely to garner comments, and how anonymity in comments and news site commenting rules have shaped the types of comment that people post. Some scholars have focused on the readers' emotional responses to civil or uncivil comments.

J.-H. Tang (✉)
Department of Information Management, National Taipei University of Business, Taipei, Taiwan
e-mail: jefftang@ntub.edu.tw

C.-F. Wei · M.-C. Chen · C.-S. Chang
Department of Psychology and Counseling, University of Taipei, New Taipei City, Taiwan
e-mail: cfwei@utaipei.edu.tw; cmcchen@utaipei.edu.tw

© Springer Nature Switzerland AG 2021
R. Stahlbock et al. (eds.), *Advances in Data Science and Information Engineering*,
Transactions on Computational Science and Computational Intelligence,
https://doi.org/10.1007/978-3-030-71704-9_49

1.1 Incivility in News Comments

Although online newsvendors encourage their readers to exchange ideas freely, both civil and uncivil comments appear online. Incivility has become a central concern for citizens; for example, a survey found that >80% of Americans consider "the lack of civil or respectful discourse in our political system" as either a "somewhat serious" or "very serious" problem.

Incivility in news comments has attracted different domain experts to study the causes and effects of this phenomenon. Of course, mass communication scholars view this as an important research agenda and have attempted to define the "incivility" of news comments. For example, [3] defined incivility as features of discussion that convey an unnecessary disrespectful tone toward a discussion forum, its participants, or its topics. They further operationalized incivility into five forms: (1) name-calling: mean-spirited or disparaging words directed at a person or group of people; (2) aspersion: mean-spirited or disparaging words directed at an idea, plan, policy, or behavior; (3) lying: stating or implying that an idea, plan, or policy is disingenuous; (4) vulgarity: using profanity or language considered improper (e.g., "pissed" or "screw") in professional discourse; (5) pejorative for speech: disparaging remarks about the way a person communicates. They examined a three-week census of articles and comments posted to a local news website and analyzed >300 articles and 6400 comments and found that incivility occurs frequently (22%) and is associated with key contextual factors such as the article topic and the sources it quotes.

Communication researchers are more concerned about whether incivility in news comments affects readers' willingness to participate in discussions [8], open-mindedness, and attitude certainty [1].

Social psychologists are more concerned about the impact of incivility in news comments on an individual's cognition, emotion, and behavior. Rösner et al. [6] conducted an online experiment to study the effects of uncivil comments on readers' cognitive, emotional, and behavioral reactions. They designed a one-factorial between-subjects experiment that included four experimental conditions and a control group. Subjects were exposed to a news article and six user comments of which zero, one, three, or all six were uncivil. They found that exposure to uncivil comments can lead to an increase in readers' hostile cognitions. However, no significant effects were found for hostile emotions or the use of incivility in readers' own comments.

2 Research Hypothesis

Hypothesis 1: Sarcastic news articles impact readers' emotional state whereas neutral articles do not.
Hypothesis 2: Uncivil comments impact readers' emotional state whereas neutral or no comments do not.

Hypothesis 3: Types of news article (sarcastic vs. neutral) and types of comments (uncivil, neutral, and none) impact readers' emotion.

3 Research Method

3.1 Research Design

We took a 2*3 complete between-subject online experiment to investigate the type of news article (sarcastic vs. neutral) and corresponding comments (uncivil, neutral, and none) on subjects' emotion.

3.2 Experimental Materials

The researchers chose eight pieces of news and invited 95 voluntary online subjects (44 males and 51 females) to judge the sarcasm of the articles; the most sarcastic one was chosen as "sarcastic news" and the least as "neutral news." The sarcastic article was about the upcoming 2020 presidential election; the Taipei mayor considered himself almost marginalized. The neutral news was about "The opening ceremony of the Cheng-chen playground park; the Taipei mayor had invited 40 "little mayors" to have fun together. Once the target had been selected, they were posted on Taiwan's largest bulletin board system, PTT, to collect reader comments. Twenty reader comments from the sarcastic and neutral news were selected based upon their popularity and we then invited 149 voluntary online users (52 males and 94 females) to judge the extent of incivility in each comment using a Likert-type scale with the range 1–5, the higher the score, the less civil. The top four uncivil comments were chosen as uncivil comments and the four most civil were chosen as neutral comments. The uncivil comments included "you have been a marginal person since child, is there any difference?", "if you would like to participate in presidential election, declare it. How disgusting!" The neutral ones included "to be a mayor attentively, you can make it." and "I still don't think he would like to be a candidate of 2020 presidential election." The news (sarcastic vs. neutral) and comments (uncivil, neutral, and none) were set out (Fig. 1).

3.3 Emotion Scale

The Positive and Negative Affect Scale (PANAS) was originally developed by Watson, Clark, and Tellegen [7] and translated and adapted by [4]. The 20-item scale was divided into two parts: 10 for positive affect and 10 for negative. The scale

Fig. 1 Experimental stimuli (Sarcastic News with uncivil comments vs. Neutral News with civil comments)

was chosen for its high reliability at 0.79–0.89 and fair validity. The positive-affect items included "I am proud of myself" and "I am interested in many things." The negative-affect items included "I felt sad" and "I felt irritated." The overall score was computed as the positive affect minus the negative affect; the higher the score, the more positive the individual.

3.4 Control Variables

Since the experimental news was about politics, the subjects' preferences might have impacted the outcomes. Therefore, we also designed three items to measure subjects' preference for the Taipei mayor, their familiarity with the news, and the influence of the news as control variables.

3.5　Formal Experiment

The formal experiment was administered online and each participant was randomly assigned into one of six experimental conditions: news (sarcastic vs. neutral) and comments (civil, uncivil, and none). Participants were required to fill in their gender, age, and average time spent online. Afterward, they had to answer PANAS before reading the news and comments. After reading the assigned news and/or comments, they had to respond to PANAS again. Finally, all subjects had to answer three questions about their views about the Taipei Mayor, this piece of news, and its influence.

4　Research Results

4.1　Subjects

Five hundred and twenty-nine subjects (274 males and 255 females) participated in this online experiment; each was assigned to one of the six conditions. Their ages were 18–35 years with mean 27.6 years and SD 6.9. Sixty-two percent of subjects were on the Internet for >4 h per day, implying that most of them were heavy Internet users.

4.2　Control Variables

All three control variables (preference for the Taipei City Mayor, familiarity with the news, and influence of the news) were significantly correlated with the subjects' negative affect at $r = -0.104, -0.228, -0.284$ with $p < 0.01$, respectively. Therefore, we conducted a two-way ANCOVA to remove the influence of control variables.

As shown in Table 1, the main effects of sarcasm in the news and incivility of comments were significant at $p < 0.001$, confirming Hypotheses 1 and 2. This implies that sarcasm in news and incivility in comments might put subjects in a bad mood after reading them. Interestingly, the interaction between the sarcasm in news and the incivility of comments was also significant at $p < 0.01$.

4.3　Emotion Difference

Each subject's emotion was measured before and after each experimental condition with PANAS. The individual's emotional difference was the dependent variable of

Table 1 Two-way ANCOVA

Source	SS	df	MS	F	p
Likability	173.15	1	173.15	1.87	.172
Familiarity	34.78	1	34.782	0.38	.540
Influence	2795.9	1	2795.9	30.25	.000
Level of sarcasm	1873.88	1	1873.88	20.27	.000
Incivility of Comments	2973.16	2	1486.58	16.08	.000
Sarcasm * Incivility	1389.19	2	694.595	37.52	.001
Residual	66390	520	92.43		

Table 2 Emotional Difference for Sarcastic News and Uncivil Comments

	Comments					
	No		Uncivil		Neutral	
News	M	SD	M	SD	M	SD
Neutral	0.81	0.94	−6.46	1.07	1.70	1.12
Sarcastic	−2.81	1.19	−6.86	0.96	−6.66	1.00

this research, and the positive score implied that subject was happier than before, and negative score the contrary. As shown in Table 1, only the condition of neutral news without comment and with neutral comments made subjects happier. Subjects were not happier in any other conditions (sarcastic news and neutral news with uncivil comments). It is interesting to note that each of the news being sarcastic and users posting uncivil comments make subjects less happy.

As shown in Fig. 2, all slopes of the three lines (blue, gray, and green) were lower from left to right, meaning that subjects were unhappy in the sarcastic news condition regardless of user comments. However, the most striking difference was the neutral/sarcastic news with neutral comments. The subjects might explain the neutral comments (Table 2).

As shown in Fig. 2, on the conditions of reading neutral news, the subjects' emotions were highest with neutral comments (1.70), followed by no comments (0.81), and then uncivil comments (−6.46). The results demonstrated that the subjects' emotions were affected by the uncivil comments effectively even if the news they had read was neutral. When reading sarcastic news, the subjects' emotions were lower without comments (−2.81), followed by neutral comments (−6.66), and the lowest with uncivil comments (−6.86). These results demonstrate that reading sarcastic news puts subjects in a bad mood; however, the effect of corresponding comments was quite different. Interestingly, the corresponding neutral comments of sarcastic news made subjects as sad as if they had read uncivil comments. When subjects read neutral news articles with no comments or neutral comments, they were in high spirits; however, when they read neural news articles with uncivil comments, they were in low spirits. Likewise, when subjects read sarcastic news article with or without comments, they were in low spirits; however, the corresponding neutral comments were as effective as uncivil comments.

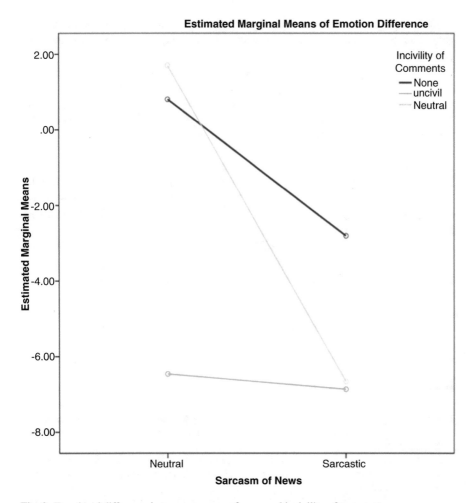

Fig. 2 Emotional difference between sarcasm of news and incivility of comments

5 Discussion

The purpose of this study is to investigate the interplay of sarcasm in news articles and incivility in online comments on readers' emotion. Sarcasm in news articles *per se* and the incivility of comments were confirmed to make readers unhappy, in line with previous studies. However, the most interesting finding in this study is the interaction between sarcasm in news and incivility in comments. While readers browse neutral news with neutral online comments, they can have a positive affect; however, when they browse the same news with uncivil comments, they feel unhappy. Similarly, when readers browse sarcastic news articles with or without uncivil comments, they feel less happy.

Only a set of online uncivil comments make the reading experience uncomfortable. The phenomenon could be explained by framing theory in communication literature [2].

Gregory Bateson posited the concept of framing in 1972. He defined psychological frames as a "spatial and temporary bounding of a set of interactive messages" that operates as a form of metacommunication [5]. Framing theory suggests that how something is presented to an audience (called "the frame") influences what choices people make about how to process information.

The findings of this study could be explained this way. When readers browse neutral news articles with uncivil comments, the bounding of these two pieces of information makes a "sarcastic news frame" that makes most readers uncomfortable and unhappy. Similarly, while readers browse sarcastic news articles with neutral comments, the bounding of these two pieces of information creates another "sarcastic news frame," which means that most readers perceive the "neutral comments" in terms of the "sarcastic news frame" that fosters a negative emotional state. If this is the case, the only way to avoid negative emotion while browsing news online is to avoid a "sarcastic news frame"; that is, stop browsing sarcastic news and uncivil comments.

The limitations of this study are that target news is politics and all subjects are Taiwanese; therefore, this study's external validity might be limited.

References

1. P. Borah, Does it matter where you read the news story? Interaction of incivility and news frames in the political blogosphere. Commun. Res. *41*(6), 809–827 (2014). https://doi.org/10.1177/0093650212449353
2. D. Chong, J.N. Druckman, A theory of framing and opinion formation in competitive elite environments. J. Commun. *57*(1), 99–118 (2007). https://doi.org/10.1111/j.1460-2466.2006.00331.x
3. K. Coe, K. Kenski, S.A. Rains, Online and uncivil? Patterns and determinants of incivility in newspaper website comments. J. Commun. *64*(4), 658–679 (2014). https://doi.org/10.1111/jcom.12104
4. T.-W. Fang, The relations among perfectionism, learning problem, and positive and negative affect: the mediating effect of rumination. Bull. Educ. Psychol. *43*(4), 735–762 (2012). https://doi.org/10.6251/BEP.20110302
5. K. Hallahan, Need for cognition as motivation to process publicity and advertising. J. Promot. Manag. *14*(3), 169–194 (2008)
6. L. Rösner, S. Winter, N.C. Krämer, Dangerous minds? Effects of uncivil online comments on aggressive cognitions, emotions, and behavior. Comput. Hum. Behav. *58*, 461–470 (2016). https://doi.org/10.1016/j.chb.2016.01.022
7. D. Watson, L.A. Clark, A. Tellegen, Development and validation of brief measures of positive and negative affect: the PANAS scales. J. Pers. Soc. Psychol. *54*(6), 1063–1070 (1988). https://doi.org/10.1037/0022-3514.54.6.1063
8. M. Ziegele, P.B. Jost, Not Funny? The Effects of Factual Versus Sarcastic Journalistic *Responses* to Uncivil User Comments. *Communication Research*, October, 009365021667185 (2016). https://doi.org/10.1177/0093650216671854

GeoDataLinks: A Suggestion for a Replacement for the ESRI Shapefile

Vitit Kantabutra

1 Introduction

Recently this author was reading through an ESRI Shapefile from a well-known geodatabase, GSHHG, using his own Ruby code. Trying to make sense of the data, this author suddenly realized that he was just encountering one polygon after another, none of which was identified in any way within the Shapefile itself. The IDs are, of course, in a DBF file and are associated with the polygons simply by the order in which they appear. This linking of the ID with the polygons by their ordering is a very insecure, error-prone link. By the way, the IDs are just running numbers (integers) with some gaps in the number sequence.

One thing is clear: the Shapefile is a legacy data structure that seems to violate modern data structuring principles. Different geometric entities of a single type are stored together in a file without any attributes, not even their identifiers. The attributes are stored in a separate database file and are very loosely linked to the associated entities. In the particular case of the GSHHG, the identifiers are simply running numbers (with gaps), and the other attributes don't have any clear meaning that can be understood simply by looking at the entries in the DBF file. In fact, this author still can't find any documentation that would help to clarify where each polygon is or to which river, lake, or another body of water each polygon belongs.

This author is not the only one frustrated with the Shapefile system of geodata storage, that is, storage using Shapefiles and other necessary or helpful accompanying files. There is a well-known Web site [1], where "members of the geospatial IT industry" encourage a move away from the Shapefile system of geodata storage and gave reasons for which the Shapefile system is "a bad format."

V. Kantabutra (✉)

Department of Electrical and Computer Engineering, Idaho State University, Pocatello, ID, USA

e-mail: vititkantabutra@isu.edu

© Springer Nature Switzerland AG 2021

R. Stahlbock et al. (eds.), *Advances in Data Science and Information Engineering*,

Transactions on Computational Science and Computational Intelligence,

https://doi.org/10.1007/978-3-030-71704-9_50

The [1] Web site offers several Shapefile alternatives, all of which were open formats. Of the ones suggested, the OGC GeoPackage appears to be the most promising.

In this paper, this author will critique the OGC GeoPackage and other formats suggested on that Web site and suggest a new replacement for the Shapefile based on the ILE (Intentionally-Linked Entities) database system [3–5].

2 About the Shapefile

The ESRI Shapefile [2] is a file capable of storing one or more geometric shapes, all of the same type.[1] No identifying attribute or any other attribute is stored with the shape in the Shapefile itself. With every Shapefile, one should also store a file of search indices, with an "shx" extension, and also a database file, with a "dbf" extension, for storing attributes pertaining to the Shapefile entries. Pay particular attention to the fact that even the identifiers of the geographic shapes are in the DBF file, not where the shapes themselves are. There is also a file with a "prj" extension that describes the coordinate system.

One problem with Shapefile just illustrated in the previous paragraph is a form of fragmentation. This type of fragmentation is said to occur when essential information about one entity can't be found in a single place. When even the identifier of the geometric entities can't be stored with the entity itself, then we have a serious problem with the integrity of the information stored in the database. In the Shapefile storage system, we have an array of entities of one type (e.g., polygons) in the Shapefile itself and the entities' identities in a DBF file. The entities are associated with their respective identities merely by matching the order of occurrence of the identities with the order of occurrence of the entities themselves. That is, the entities appear in the Shapefile in a linear sequence, and the identities likewise appear in the DBF file in a linear sequence, and the ith identifier in the sequence of identifier in the DBF file identifies the ith entity in the sequence of entities in the Shapefile.

This kind of association between the identifiers and the entity identified is very prone to errors. For example, if one identifier or one entity is removed or misplaced by mistake, a large number of entities are likely to be misidentified, leading to all sorts of problems.

A second problem with Shapefile is its reliance on a Relational database system for storing the attributes. The Relational system determines relationships by matching attributes, and that procedure of determining relationships is error-prone. For example, keys and other attributes can be mistyped easily. Also, blanks and other invisible characters can be lurking in an attribute string and not be detected.

---

[1]A Shapefile can also hold a NULL SHAPE, but we will ignore this because it is not useful to our discussion.

This weakness can be avoided by using the ILE database system, which indicates relationships by using pointer (or references) or data structures containing pointers (or references). More on ILE later, when the present result is introduced.

More details about the contents and formats of all the files in the Shapefile system are given in [2].

The Web site [1] lists things that the authors of the site found "good" and "bad" about the Shapefile storage system. On the "good" side, all they listed were that Shapefile is by far the most widely supported format in existing software packages, that the specification is open, and that the format is "good enough" for many use cases. It should also be mentioned that the Shapefile format appears likely to require less space than its competitors.

On the "bad" side of Shapefile, the Web site's authors listed many items, as follows:

- No coordinate reference system definition. This, however, can be done with a prj file, but is not mandatory even though it should be to make the coordinates meaningful. The Cepicky2019 site also claims there are other issues with projection, but their link to that information is broken.
- It's a multifile format. Having to zip and unzip several files for transportation is error-prone. My own complaint about this aspect of Shapefile is not that one needs several files, but that files are not well-organized because, as noted earlier in the Introduction, the attributes (including the identifiers) are not stored with the shapes themselves, making the storage system even more error-prone. In fact, if the data could have been divided into logical geographical units, such as having a single file containing all data for an administrative unit or a body of water, etc., then it would be appropriate to have multiple files in the database.
- Attribute names are limited to 10 characters.
- Only 255 attributes. The file does not allow you to store more than 255 attribute fields.
- Limited data types. Data types are limited to float, integer, date, and text with a maximum 254 characters.
- Unknown character set. There is no way to specify the character set used in the database.
- It's limited to 2 GB of file size. Although some tools are able to surpass this limit, they can never exceed 4 GB of data.
- No topology in the data. There is no way to describe topological relations in the format.
- Single geometry type per file. There is no way to save mixed geometry features.
- More complicated data structures are impossible to save. It's a "flat tablet" format.
- There is no way to store 3D data with textures or appearances such as material definitions. There is also no way to store solids or parametric objects.

- Projections definition. They are incompatible or missing.
- Line and polygon geometry type, single or multipart, cannot be reliably determined at the layer level; it must be determined at the individual feature level.

3   Previously Proposed Replacements for Shapefile

The Web site [1] discussed possible replacements for the Shapefile. Although there are more than 80 vector data formats in use according to the Web site, only a few can be considered as candidates for Shapefile replacement. Note that the site's authors only considered open formats. The alternatives important enough to list at the site are as follows:

- OGC GeoPackage. This is a container for SQLite. SQLite is a Relational system (may be not in the strict sense by Codd's definition) that also allows up to one data element of the type BLOB (Binary Large Object) per row. As one might correctly guess, this BLOB data type is used by GeoPackage for all sorts of geometric objects because such objects as polylines and polygons are of variable length. The idea then is to put each geographical element in each row of a Relational database, complete with all the element's attributes. This solves a major problem with Shapefile that we call "fragmentation" in this paper, namely, the problem that each geographic element has to be stored alone in a data structure without its attributes.

    However, GeoPackage suffers from the same problems that all Relational systems do. One main problem, as detailed in [5], is that all the relationship linkages are done by means of data value matching, which is very error-prone. The scheme to be presented in this paper is based on the ILE database system and doesn't have that problem.
- GeoJSON. To quote the Wikipedia article on GeoJSON directly, "GeoJSON is an open standard format designed for representing simple geographical features, along with their non-spatial attributes. It is based on the JavaScript Object Notation (JSON)." It is an easy-to-use format, but doesn't appear to allow representing complex relationships among entities.
- OGC GML. This XML-based format allows for sophisticated descriptions of geographic entities. However, being XML-based, the format allows for only a limited description of relationships. See [6].
- SpatiaLite. For the purposes of this paper, this is similar to GeoPackage.
- CSV. This is not a well-organized way to store spatial data and will not be discussed further.
- OGC KML. This, like GML, is another XML-based storage scheme and is subjected to the same restrictions on relationship representations.

In the next section, we will discuss the new storage scheme and how it is a better alternative to Shapefile than the others we have just discussed.

4 GeoDataLinks, the New Storage Scheme for Geographic Data

In coming up with a proposed replacement for Shapefile, the author proposes a solution to the two most troublesome properties of Shapefile that are difficult to improve on, namely:

1. Disorganized file structure, that is, the geographic entities can't be stored in the same file with their identifiers and other attributes,
2. Inability to express relationships among geographic entities.

Our proposed data storage scheme, GeoDataLinks (henceforth to be also called GDL), will solve these problems. GDL is based on ILE (Intentionally Linked Entities) [3–5], which can be thought of as a direct implementation of the Entity-Relationship (E/R) model, without support for weak entities, but with some features that are more general, more powerful features than E/R. ILE implements entities as objects and naturally supports binary relationships as well as relationships with higher arities. As for binary relationships, it can easily handle one-to-one, one-to-many, many-to-one, and many-to-many binary relationships. It is widely believed that the E/R model cannot be directly implemented, but that's what ILE is, namely, a direct implementation of the E/R model, for one user at a time at least.

An ILE database is represented by a database object, which has two different kinds of objects directly referenced inside it, namely:

1. entity set objects
2. relationship set objects

An entity set object then has entity objects references contained in it, and the meaning of this situation is that these entities belong to the entity set.

Likewise a relationship set object has relationship objects referenced in it.

ILE has been implemented in prototypical form in Ruby, an object-oriented programming language. Despite needing the object-oriented properties of Ruby, ILE is not an objected-oriented database system because the entities are not identified by their locations or object IDs. Instead, the entities are identified by their key attributes like in a Relational database system.

In an ILE database, each set of entities of like kind will have an entity set object defining them and bonding them together. In other words, an entity set object serves the same function as a table in an RDBMS. An entity will have two kinds of attributes, namely, key attributes and non-key attributes. All these attributes will be specified in the entity set object to which the entity belongs. Each type of entity can have an unlimited number of attributes of both keys and non-key types.

Now we consider what data types should be allowed for the attributes of an entity. This author would suggest that unsigned integers and strings of printable characters be the allowed types. This, of course, could be changed if it appears that other types of attributes are suitable for use as identifiers or parts of identifiers.

Non-key attributes can be of any type that the programming language used for implementing ILE supports, and more, relying on compositing more primitive types as needed. For example, all the useful geographic types must be supported, including the variable-length ones such as polylines and polygons.

A relationship is an object that relates two or more entities. Each relationship is represented by a relationship object, and all relationship objects of like kind are grouped together physically and logically with a relationship set, implemented with a relationship set object. This is similar to the (unimplemented) E/R model where relationships and relationship sets are distinct from entities and entity sets. This is to be contrasted with the Relational model, where both entity sets and relationship sets are modeled and implemented as tables.

Each relationship is defined by its arity, or number of roles. Each role in the relationship is played by an entity belonging to a specific entity set. ILE also permits each role to (optionally) be played by more than one entity, all belonging to the same entity set. Let's assume for simplicity that each role is played by one entity. Then the relationship object will have a pointer (reference) field for each role that points to the entity playing that role, and the entity will also have a pointer field that points back to the relationship object. This is illustrated in Fig. 1.

In case an entity is involved in more than one relationship, we can associate with the entity not one pointer but an array of pointers. Likewise, in a relationship where a role can be played by more than one entity, we can associate an array of pointers with the role in the relationship object, so that we can point from that role to many entities. Figure 2 illustrates how to implement a one-to-many binary relationship.

Fig. 1 An example of how ILE implements a relationship among entities using a relationship object

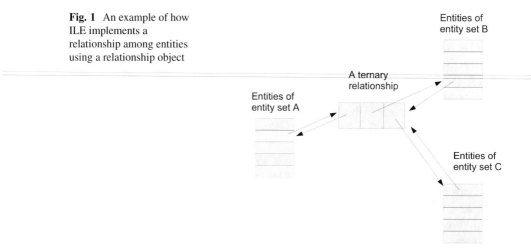

Fig. 2 An example of a one-to-many or many-to-one relationship in ILE

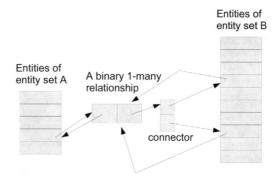

5   GeoDataLinks: Using ILE as a Replacement for Shapefile

Having introduced ILE, we are now ready to see how it can be used as a replacement for Shapefile. There are two cases to consider, the static map case and the temporal case. We will discuss the static map case in this paper, leaving the temporal case for future work.

Henceforth, we will consider entity sets of places such as the set of all states of the United States, all the provinces in Thailand, or all the secondary schools in Québec, etc. The information on all the entities in all the entity sets is presumed to be all valid at a certain point in time. If the name uniquely identifies the entity within the entity set, then the name alone would be enough as the identifier, which would be the sole key field needed for the entities in the entity set. if the names are not unique, then a second key field can be used to help distinguish the various places with duplicate names.

Now comes perhaps the most important part of this discussion, that is, how a variable-length geographic feature of an entity should be represented. There is more than one solution. One possibility is to make the entire border a non-key attribute of the entity. This is simple to do because in Ruby, the language in which ILE is currently implemented in, the dynamically allocated array is a common, well-managed data structure. Not only that, but it is also possible to implement the polygon as a linked list if necessary to achieve better memory management.

All other non-key attributes are just implemented in a routine fashion.

There is one flaw in the implementation just discussed—when two geographic entities border each other, the shared border are duplicated, and if we stick with Shapefile's directional convention, then the vertices on the shared border will occur in opposite orders. This duplication is not a good situation, because it can be difficult to make sure the entries are identical, especially if there's any correction involved.

Following this line of thinking, we introduce another solution that doesn't involve duplication. Instead of making the borders an attribute, make them entities and

Fig. 3 RA and RB are examples of reflexive ternary relationships that makes certain polylines borders between two administrative units. Every polyline B0, B1, B2, etc. has such a relationship associating the border with two administrative units. However, only two such relationships were illustrated in this figure

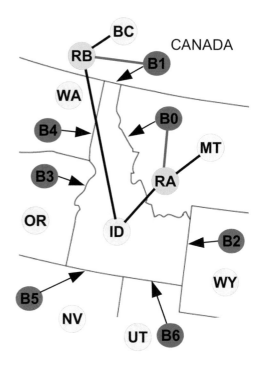

therefore "first-class citizens" of the database system.[2] Furthermore, instead of making the border of a geographic administrative unit an entity, make the border between two administrative units an entity. To say that a border entity makes the border between administrative unit A and administrative unit B, we need a ternary relation that relates A and B and the border entity. Figure 3 illustrates this situation. Note that this ternary relationship is, of course, reflexive with respect to the two roles A and B that are filled by administrative units.

Holes should also be modeled as entities and can be modeled with a binary relationship between the geographic unit with the hole and the hole itself. To model an administrative unit that borders the sea or ocean, we can treat the ocean as if it were an administrative unit and use the ternary relationship.

There will be an entity that represents all the pieces of the border put together. To make this possible, the author invents a new kind of one-to-many relationship not in existence in ILE or in RDBMS before this. This relationship is called the *ordered one-to-many relationship*, where unlike the usual one-to-many relationship, the entities on the "many" side are considered to be ordered. This ordering is implemented naturally by using the inherent ordering already in existence because

---

[2]Following [5], we say that an entity in an ILE database system is a first-class citizen, whereas an attribute is a second-class citizen.

Fig. 4 The ordered
one-to-many relationship
shown here says that the
boundary around Idaho is
composed of the seven
inter-administrative-unit
boundaries in clockwise order

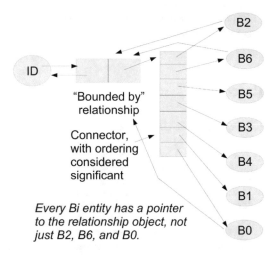

of the connector array that is used for pointing to the elements on the "many" side.
Figure 4 illustrates this relationship.

6 Conclusions

We have discussed GeoDataLinks, a possible replacement for the ESRI Shapefile, an
historically important but now obsolete system of files for describing geographical
entities for geographical information processing. In Shapefile, the stored information is fragmented in the sense that the geographic shapes are not stored with
their identifiers and other attributes. This kind of fragmentation can easily cause
a large number of geographical entities to be misidentified. Furthermore, Shapefile
doesn't permit the description of relationships among geographic entities, which
means the entities' full properties can't be represented. The existing candidates for
replacing Shapefile don't have the fragmentation problem, but are at best based
on the Relational database system or an extension thereof, which means that the
power to express relationships among geographic entities is limited. GeoDataLinks,
our new candidate for replacing Shapefile, is based on ILE (the Intentionally-
Linked Entities database system), which uses pointer-based data structures to
implement relationships between entities. In addition to not having the fragmentation problem, GeoDataLinks enables a natural expression of complex relationships
among geographic entities which is important for accurately and fully representing
geographical information. The work that culminated in GeoDataLinks includes the
advent of a new kind of one-to-many relationship, called the *ordered one-to-many
relationship,* where the entities on the "many" side of the relationship are linearly
ordered. In summary, GeoDataLinks makes use of modern, object-oriented pro-

gramming technology to provide a robust, improved way of representing complex geographic data.

References

1. J. Cepicky, Switch from shapefile. http://switchfromshapefile.org (Last Modified 2019)
2. Environmental Systems Research Institute, ESRI Shapefile technical descriptions (1998)
3. V. Kantabutra, A new type of database system: intentionally-linked entities—a detailed suggestion for a direct way to implement the entity-relationship data model, in *CSREA* (IEEE, Piscataway, 2007), pp. 258–263
4. V. Kantabutra, D.P. Ames, A suggestion for a data structure for temporal GIS, in *2009 Fifth IEEE International Conference on E-Science Workshops* (2009), pp. 158–161
5. V. Kantabutra, J.B. Owens, D.P. Ames, C.N. Burns, B. Stephenson, Using the newly-created ILE DBMS to better represent temporal and historical GIS data. Trans. GIS **14**, 39–58 (2010)
6. K. Williams, XML for data: modeling many-to-many relationships (2002). https://www.ibm.com/developerworks/library

Nutrition Intake and Emotions Logging System

Tony Anusic and Suhair Amer

1 Introduction

Obesity and overweight are significant health risks. It has become a widespread phenomenon and it is defined based on the Body Mass Index of an individual. A person is considered obese if his or her Body Mass Index is higher than or equal to 30 kg/m2. Obesity may lead to chronic diseases such as high cholesterol, type II diabetes, heart disease, and many others [1]. Some researchers say that in order to lose weight in a healthy way and maintain it, the individual needs to measure daily food intake [2]. This is because of the lack of balance between the amount of intake and the amount of energy spent [3]. In recent years, people are becoming more interested in watching their weight, watching their intake, and eating healthier. Health monitoring using mobile devices is becoming more accessible by individuals.

Some research is based on image-processing techniques and nutritional fact tables. It uses a special calibration technique and built-in camera of mobile devices and records a photo of the food before and after eating it in order to measure the consumption of calorie and nutrient components. The algorithm then extracts important features and then uses a support vector machine, as a classifier, to achieve accurate results [4].

Another approach combines speech processing, natural language processing, and text mining in a unified platform to extract nutrient information such as calorie intake from spoken data. After converting the voice data to text, food name and portion size information within the text are identified. A tiered matching algorithm

T. Anusic · S. Amer (✉)
Computer Science Department, Southeast Missouri State University, Cape Girardeau, MO, USA
e-mail: samer@semo.edu

© Springer Nature Switzerland AG 2021 695
R. Stahlbock et al. (eds.), *Advances in Data Science and Information Engineering*,
Transactions on Computational Science and Computational Intelligence,
https://doi.org/10.1007/978-3-030-71704-9_51

is used to search the food name in a nutrition database and to accurately compute calorie intake [5].

Machine-learning methods are used to measure calories of fruit and vegetables. The model helps patients and dieticians to compute daily intake of calories. The machine-learning models are used to predict classification accuracy. A camera and intelligent mat are used to capture the picture of the fruit/vegetable, in order to calculate the consumption of calorie [6].

Another approach uses a food diary that a user can use on his/her mobile phone. For example, Pattern-Oriented Nutrition Diary is a mobile-phone food diary designed using a theory-driven approach. It addresses a common challenge of food diaries on mobile phones, which is related to the amount of effort required to create food entries. The study showed that people preferred different approaches to creating entries, which reflected their self-reported nutrition concerns [7].

In general, effective communication can improve weight loss maintenance. Having a conversation is an emotional experience, and some research is using chatbots that are conversation-driven intelligent systems. Chatbots have been reported to increase compliance with health interventions. One study identified the needs of adults aged 18+ who were maintaining weight loss using semi-structured interviews. Their findings identified five key themes: *(1) Weight loss maintenance is challenging; (2) social contact is beneficial but may also reinforce unhealthy habits; (3) Apps should be convenient and support progress tracking; (4) personal messages should be specific and relevant; (5) chatbots have potential for weight loss maintenance* [8].

In addition, mobile telephone technology can aid weight management interventions by delivering support in real-time and real-world settings. There are technologies used in weight management such as text- message-based systems that have shown to be beneficial for self-reporting and motivation but these systems require a human operator to read and respond to messages from participants, thus 24/7 availability is not always possible [9].

In summary, a literature review was conducted that searched electronic databases to locate publications dated between 2006 and 2018 about using digital technologies for weight loss maintenance. The review highlighted that digital technologies have the potential to be effective communication tools for significantly aiding weight loss maintenance, especially in the short term [10].

2 Problem Statement

This chapter discusses establishing requirements, designing, and developing a Personal Nutrition Intake and Emotions logging simple interactive system. The logging system would be like a form where the user will enter the data, which will then be stored to a file. A graphical user interface should make the process of entering the information fun. The system will motivate its users to log their food intake over a long period of time and how it correlated with their emotion and mood.

The aim is to use this information in the future and correlate what they ate with their mood that day. For example, did they eat more when they were happier or sad? They eat more around holidays, etc. It is important to note that unbalanced diet is a major risk for chronic diseases such as cardiovascular disease, metabolic diseases, kidney diseases, cancer and neurodegenerative diseases. Current methods to capture dietary intakes may include food-frequency questionnaires, 7-day food records and 24 h recall. These methods are expensive to conduct, cumbersome for participants to use, and prone to reporting errors. These are also examples of ways doctors or centers are asking their patients to record their food intake. For example, patients with diabetes would be given a notebook and are asked to record what they ate throughout the day and it is divided as breakfast, snack , lunch, etc. and then other information. In their next visit, notes are checked and are given directions to what to do next. The problem with these recording techniques is that they are boring, time consuming, and many patients forget and don't fill it correctly.

The system needed to have a simple joyful design and interface that is easy to navigate and use and that will make them want to record/enter their data. It is necessary to keep in mind that people have patterns and routine. For example, a person will repeat eating the same thing over time. So, in total, let us say they have 20 food options. Users may be asked to enter the options on first use and let the app convert it to a list that they can choose from. An "add a new item" button always available. Application should allow users to print/retrieve the information.

The next sections will elaborate on the work of the student author of this chapter. The student needed to complete three phases of the project: (1) establish requirements and design a simple interactive system, (2) implement a simple interactive system, and (3) Data analysis and evaluation of the simple interactive system.

3 Establishing Requirements and Designing a Simple Interactive System

3.1 Usability and User Experience Goals

Usability Goals include:

- Effectiveness (is it good at what it is supposed to do)
- Efficiency (is it able to carry out the task)
- Utility (does it provide functionality to do what needs to be done)

User Experience Goals include:

- Fun
- Helpful
- Enjoyable to use

3.2   Questions Using Design Goals

Usability Goals as Questions are listed as follows:

- Effectiveness – Does the application allow for users to log their food intake over a long period of time and correlate it with emotions?
- Efficiency – Does the application allow for users to use the recorded data in the future to correlate what they ate with their mood?
- Utility – Does the application provide the functions that allow the user to log their food intake and emotions with simplicity?

User Experience Goals as Questions are listed next:

- Fun – Is the application fun to use or provide interactivity with the user?
- Helpful – Does the application accomplish the task in a simple way?
- Enjoyable – Is the application enjoyable to use or provide good feedback to the user?

3.3   Users' Needs, Requirements, and Main Tasks

The user needs to be able to record, over time, their food intake and emotions during the day by using a fun and easy to use application. That meant that the application must provide an easily accessible GUI to enter food intake and select emotion options easily. The application should also be fun to use and have a simple design for the user to use it daily. To do that the main tasks is to create an application to allow users to record their food intake and emotions by providing them with a fun and simple GUI that they can use over a span of time. It should include buttons and text boxes to create a more intuitive way to record information.

3.4   Scenarios and Use Cases

Several scenarios were developed. For example: Scenario: User1 wants to record and enter nutrition intake information into a logging system to track what they eat and how they feel every day. He/she opens the application and is greeted to input their data. The user will type into a box what they ate at the start of the week and select an emotion that they felt on that day. For the rest of the remaining time, the user will input or select previously typed information to record their intake and emotions. When they are finished, they can review what they ate and how they felt during the period they used the system. The user studies this information and may decide to change their intake.

A use case:

1. The app displays a GUI for inputting what was eaten and how they felt that day.
2. The user types into a text box what they ate that day.
3. The system records what they ate and prompts them to select an emotion.
4. The user selects the emotion for the day.
5. The system records the emotion and the user can add more logs and review saved logs.

3.5 Requirements Using Volere shell

Requirement #: 1
Requirement Type: 9
Event/use case #: 1-5
Description: User should be able to easily enter food intake and select emotion options.
Rationale: A simple yet fun design with easy to locate items allows the user to achieve their goal of recording nutrition intake and emotions quickly and effortlessly for future analysis.
Originator: XXXX
Fit Criterion: The app shall communicate to the user the common food they ate and how they feel everyday over a span of time.
Customer Satisfaction: 5
Customer Dissatisfaction: 5
Priority: High
Conflicts: None
Supporting Materials: None
History: Created April 12, 2018

Requirement #: 2
Requirement Type: 10
Event/use case #: 1
Description: The app should have a fun and simple design for the user to use.
Rationale: A simple design with fun interactions lets the user accomplish their task easily and will want to use the app more.
Originator: XXXX
Fit Criterion: The app shall provide a simple and working functional GUI that is fun to use.
Customer Satisfaction: 5
Customer Dissatisfaction: 5
Priority: High
Conflicts: None
Supporting Materials: None
History: Created April 12, 2018

3.6   Conceptual Model

An intake logging system is a nutrition record form in a simple virtual environment. The operations that can be performed are record keeping of food intake and emotions for future analysis. The application focuses on providing an enjoyable user experience using a virtual environment rather than plain paper forms and questionnaires.

3.7   Mental Model

The application represents a virtual version of a nutrition form someone may fill out for themselves or for a doctor but in a simpler and fun way. There are text boxes for users to enter what they ate on specific days, options to choose what day it is, and buttons to choose and lock in emotions that they felt. Rather than using only text boxes for entering data, the application provides more interactive methods to allow the user to record data.

3.8   Analyzed Findings and Enhanced Conceptual Model

The application needs to record users' data without being boring or time consuming. It must provide them a way to input data quickly and easily by focusing on what the user needs most and getting rid of any unnecessary additional clutter. The application is based on popular fitness apps that do a great job of recording and showing data in a simple way. It is interactive by using images, buttons, and text boxes for data input.

3.9   Interface Design Issues

When creating an interface, there are some issues that needs to be considered when attempting to figure out a suitable design. The application needs to be simple and fun and display/require what is most needed to the user. This is to avoid GUI clutter and managing how much to put on the interface without diminishing the user experience. Another issue is figuring out the best way to provide interactions to the user without making things too complicated for a simple application.

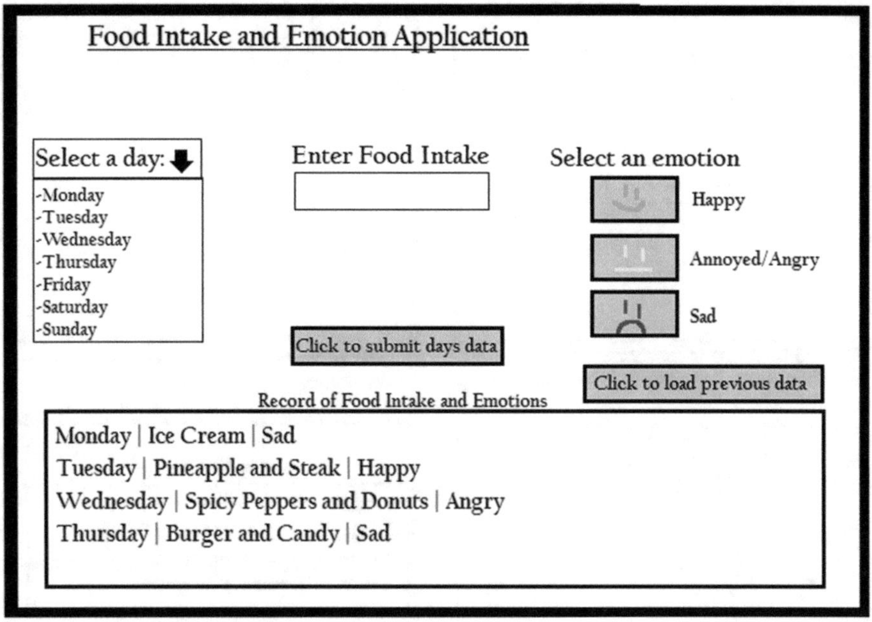

Fig. 1 Card-Based Prototype Design

3.10 Initial Designs and Evaluation of Designs

This section will show some of the initial design that was developed. Figure 1 is the Card-Based Prototype Design. Figure 2 is the Storyboard Design. The user is asked to select a day, enter food intake, select an emotion, submit data, and view previous data. Overall, the initial designs will help achieve the goal of recording food intake and emotions with an easy and fun to use interface. Two users were asked to evaluate the design and if it can achieve the task. For the most part, the users said that the initial design does achieve what the task is asking for and it seems fun to use. They also mentioned that it can look better by spacing GUI components better and maybe adding an emotion counter in the display box of data for better analysis capability.

4 Implementing a Simple Interactive System

The system was created using Visual Basic C#, which provides easy to use GUI options and a clean design. When running the system, the main page, as shown in Figs. 3 and 4, consists of a text box to enter week number, a dropdown menu to select the week day, text box to enter food intake, dropdown menu to select mood/emotion, buttons to submit or retrieve stored information. When the user clicks the submit

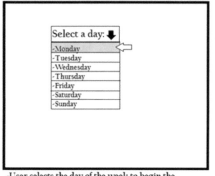

-User selects the day of the week to begin the recording process using a drop down menu

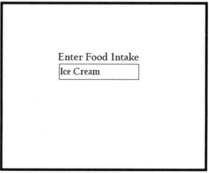

-User enters the food they ate for that day in a textbox

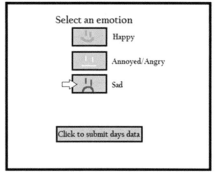

-User selects the emotion they felt during that day by clicking on a button with the image of a face and description to the side of it. The user then submits the data

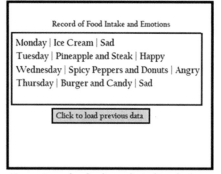

-User reviews what they have submited in a display textbox for future analysis

Fig. 2 Storyboard design

button, a message box displays a message that the data was saved, as shown in Fig. 5. The select day dropdown menu will display only the first 4 days and the user will need to scroll down if they wish to select another day as shown in Fig. 6. The user has the option to display all stored food intake options. To do so, they need to click on the retrieve button located near that drop box as shown in Fig. 7. The user can also retrieve stored data by clicking on the Read Data button. It displays "Saved options are now available in the display box!", as shown in Fig. 8.

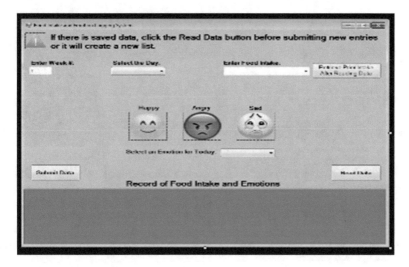

Fig. 3 Application interface without data

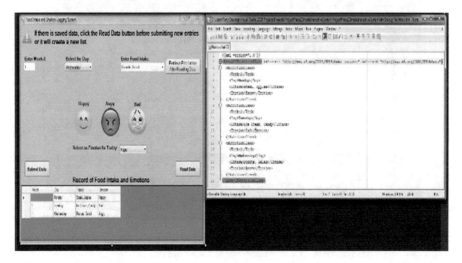

Fig. 4 XML Document with data (file exists only with data)

5 Data Analysis and Evaluation of the Simple Interactive System

5.1 Goals

The main goal of the evaluation is to answer the question: Does the prototype application meet the requirements of allowing the user to record his/her food intake

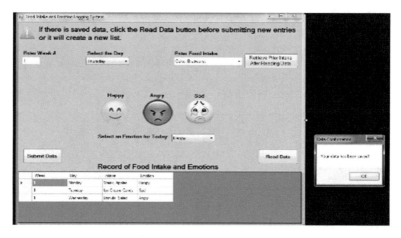

Fig. 5 Interface with data. Message box popup when Submit Data button is clicked. It displays "Your data has been saved!"

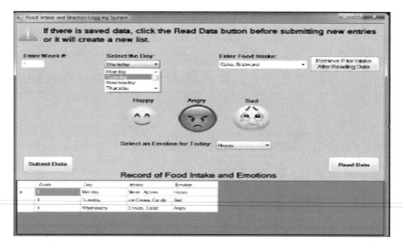

Fig. 6 Day selection Combo box showing how each day of the week is displayed in the dropdown. Shows the first 4 days of the week and a scroll bar to shows the rest

and emotions over time in a fun and simple way? The DECIDE framework is used to guide evaluation [11] and to see if the prototype meets the requirements of the project. It will evaluate how well the application functions when users test it and will help show how it could be improved.

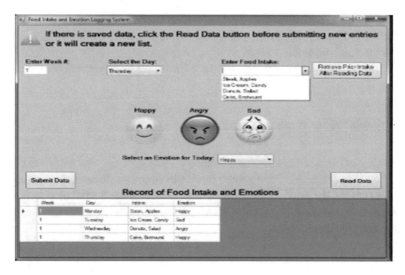

Fig. 7 Food intake Combo box shows all saved intake options when the dropdown arrow is clicked after clicking on the Retrieve button beside it

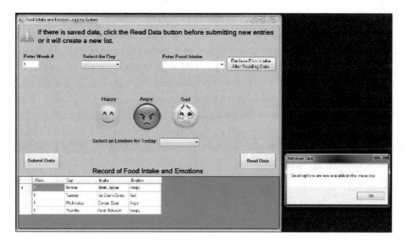

Fig. 8 Retrieve stored data by clicking on the Read Data button. It displays "Saved options are now available in the display box!"

5.2 Exploring the Questions

The following questions will be used to answer and guide the evaluation study and meet the goal of the evaluation.

- How does the user feel toward paper-based forms that they usually have to fill out?
- Would the user prefer to use electronic versions over a paper-based version?

- Did the application allow the user to log food intake and emotions over time?
- Was the GUI of the application simple to use and easy to follow?
- Was the application fun to use?
- What kind of recommendations are there for future improvements of the app?

5.3   Evaluation Method Choice

The evaluation method chosen was usability testing. Since the main goal of the evaluation is to determine if the prototype application was successful at meeting the requirements for the project, usability testing would be the best choice as it allows establishing predetermined questions and tasks that selected users can answer in order to provide crucial data.

5.4   Practical Issues Identified

There were some practical issues that needed to be identified. Time constraints were one of the issues that were faced. Having 10 users test out the application and provide feedback would need to be done in a timely manner to allow each user enough time to use the application but to not take up too much of anyone's time. There was also an issue of accessing the appropriate participants for the usability testing. Participants who would represent the target audience for a food intake and emotion logging system had to be identified and selected. The chosen participants were the ones who dieted, did a ton of activities, or were just curious about their eating patterns. Asking these participants was the best way to get a representative and accurate form of data.

5.5   Collected Data

This section lists users' answer to the usability testing questions. These questions helped collect the necessary data for the evaluation. Because of time limitation and accessibility to users, only 10 subjects participated in the evaluation.

For question 1 that stated, "What do you think of paper-based forms that you often have to fill out?", the users answered:

- User1: "Most of the time they are tedious and not fun to fill out at all."
- User2: "Paper forms are not too bad to use. I usually calculate calories all the time, so I'm used to having to write things down on paper."
- User3: "Since I am a middle-aged adult, I grew up with paper forms, so I don't mind them."

- User4: "Personally if they have an option to fill out forms online, I will do that because it is more convenient for me. I don't enjoy filling out paper forms as much as electronic ones."
- User5: "I don't really mind paper forms as long as they are detailed and don't use technical jargon."
- User6: "Paper forms are oftentimes difficult to follow, most of the time electronic versions are much easier to use and follow along with."
- User7: "Whenever I fill out surveys or forms, I always opt for online ones because I am always on my phone or computer and find it simpler."
- User8: "I don't like paper forms because I would rather type my responses than write them down, and paper forms are oftentimes too wordy."
- User9: "Paper forms oftentimes take too much time to fill out, electronic forms seem to be quicker to me."
- User10: "I like paper-based forms because it is easier to read everything compared to online ones."

For question 2 that stated, "Would you rather prefer to use an electronic version of the form over a paper version? (Y/N)", Fig. 9 shows that 6 users said yes they prefer using an electronic version over a paper version. However, 40% of the users still preferred the paper version.

For question 3 that stated, "Did the application allow you to log your food intake and emotions over time? (Y/N)", Fig. 10 shows that the application did allow them to log their information.

For question 4 that stated, "Was the interface or design of the application simple and easy to follow?", Fig. 11 shows that all users said yes. In addition, some users added the following comments: "Yes, the design was simple and there wasn't much clutter compared to some apps that I have used"; "Yes, it was easy to follow along and fill out the data on the app"; "Yes, it was simple to follow but some buttons should be renamed to make things clearer"; "Yes, the application had a very simple design that was short and straight to the point with many forms of options"; and

Fig. 9 Number of users preferring the use of an electronic version of the form over a paper version

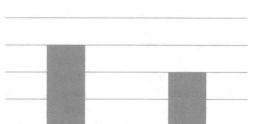

Prefer electronic over paper

■ Prefer electronic over paper

Fig. 10 Number of users
able to log their food intake
and emotions over time

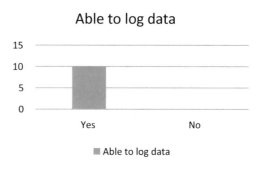

Fig. 11 Number of users indicating that the interface or design of the application was simple and easy to follow

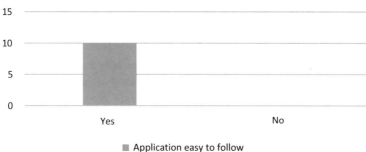

"Yes, it was simple and easy to follow but more explanations would be useful in the long run."

For question 5 that stated, "Was the application fun to use?", Fig. 12 shows that all users indicated that the application was fun to use. Some users had some additional comments such as: "Yes, the app was fun and quick to use"; "Yes, it was interactive and helped me see what a logging system like this could be used for"; "Yes, the app is fun and useful for keeping track of important data"; "Yes, it wasn't anything special but it's still better than if this was in a paper form"; "Yes, it provided useful data that I would personally use for myself"; and "Yes, it was interesting to see the use of tracking food intake with your emotions."

For question 6 that stated, "What type of recommendations if any would you give for future improvements of the application?", the responses of some of the users included: "Make the deletion of specific submitted data available for users to get rid of mistyped information more easily"; "Maybe add a counter for each emotion so the user can easily see how often they felt a certain emotion"; "For an updated version you could try and add more descriptions to each section to make things clearer to follow"; "Rather than typing an emotion you could make them buttons with pictures on them so that there is less typing"; "If possible, you could add a calendar to choose the date rather than selection a week number and day of

Fig. 12 Number of users indicating that the application was fun to use

the week"; and "Improve readability and provide a few more explanations so that everything makes sense."

5.6 Evaluation of Data

After collecting the results, the data shows that the users preferred an electronic form rather than a paper form to fill out if given the choice. The users all responded that the application allowed them to record food intake and emotions over time. They also said that the GUI was simple and easy to follow for the most part with some recommendations for improvements. All the users also agreed that it was fun to use the application since it allowed them to see how something like this could be used in their daily lives. While performing the usability testing, the users enjoyed testing out the system and provided positive feedback and many recommendations for future improvements. The model had a high reliability and showed a positive validity. There was not much bias in the responses since the participants were chosen to each have a slightly different view and lifestyle.

5.7 Data Analysis

The usability testing provided data for my evaluation. Most of the users preferred to use electronic or online forms rather than paper forms. The majority said that the application did what it was supposed to and was easy to follow. They had fun using the application since the reasoning for its creation was unique. They suggested that a future version of the application should consider changing text of the buttons to something more understandable, provide a calendar for them to select the date, and to add a counter to each emotion so that users can summarize how they felt in a specific time frame.

5.8 Interpretation of Data

These questions were chosen to help see how well the application followed the project requirements as well as how useful electronic versions of forms, surveys, etc. can be to acquire data instead of using paper. The data gathered indicated that requirements set by the instructor have been achieved. The recommendations of the users in the feedback question can be considered for future improvements and versions.

5.9 Ethical Issues

There were no ethical issues identified with the use of the system because it does not ask users to submit private information such as their name, address, phone number, etc. A consent form is not needed in this case but if a future update to the application was made that asked them to submit such data then the consent form should be created to protect their privacy. In addition, security should be built in the system to make sure such information is not shared without their consent.

6 Conclusion

When developing the system, the student chose the card-based prototype and implemented the application using Visual Basic C#, which provides easy to use GUI options and a clean design. He enjoyed working on this logging system application. He created a Windows Form Application using Visual Studio C#, which he had previous experience in from a prerequisite course. He was able to meet the requirements for the project without having a cluttered GUI. However, there were some issues that he encountered and had to work on in order to get a fully functional application. One issue was determining how to save the data for future use after the application closes. He set up an XML document that was created to store all the data that could then be displayed when the user needs or chooses to. Another issue was figuring out how to get all the controls to display in one field. He used a dataGridView control that allowed him to display all the input data using a separate class (NutritionClass), which used get/set assessors to retrieve and set the data.

After sampling 10 participants using the usability testing method, he was able to gather positive feedback on the application. The questions that each user answered allowed him to answer his goal of creating a functional, easy to use, and fun food intake and emotion logging system. The data that was gathered revealed that the majority preferred to use electronic versions of forms rather than paper-based forms but there were a few that still preferred to use paper versions. Every user agreed that the application was easy to follow, allowed them to log their food intake and emotions over time, and was fun to use. The users provided plenty of recommendations for improvements. This project helped him apply the core concepts that were learned during the semester to develop the application as well as utilize his C# programming skills from previous semesters.

References

1. World Health Organization. (2011, October) Obesity http://www.who.int/mediacentre/factsheets/fs311/en/index.html
2. W. Jia, R. Zhao, N. Yao, J.D. Fernstrom, M.H. Fernstrom, R.J. Sclabassi, M. Sun Jia, *A Food Portion Size Measurement System for Image-Based Dietary Assessment*, in Bioengineering Conference, IEEE, pp. 3–5, (April 2009)
3. C. Bouchard, G.A. Bray, *Handbook of Obesity*, 2nd edn. (Ennington Biomedical Research Center, Louisiana, 2004)
4. P. Pouladzadeh, S. Shirmohammadi, T. Arici, *Intelligent SVM Based Food Intake Measurement System*, in *Computational Intelligence and Virtual Environments for Measurement Systems and Applications (CIVEMSA) 2013 IEEE International Conference on*, (2013), pp. 87–92
5. N. Hezarjaribi, C.A. Reynolds, D.T. Miller, N. Chaytor, H. Ghasemzadeh, *S2NI: A Mobile Platform for Nutrition Monitoring from Spoken Data*, in *Engineering in Medicine and Biology Society (EMBC) 2016 IEEE 38th Annual International Conference of the, pp. 1991–1994*, (2016)
6. M. Mittal, G. Dhingra, V. Kumar, *Machine Learning Methods Analysis For Calories Measuremnt of Fruits and Vegetables*, in *Signal Processing Computing and Control (ISPCC) 2019 5th international conference on*, (2019), pp. 112–119
7. A.H. Andrew, G. Borriello, J. Fogarty, *Simplifying Mobile Phone Food Diaries*, in *Pervasive Computing Technologies for Healthcare (PervasiveHealth) 2013 7th International Conference on*, (2013), pp. 260–263
8. S. Holmes, A. Moorhead, R. Bond, H. Zheng, V. Coates, M. McTear, *WeightMentor, Bespoke Chatbot for Weight Loss Maintenance: Needs Assessment & Development*, in 2019 IEEE International Conference on Bioinformatics and Biomedicine (BIBM), San Diego, CA, USA, pp. 2845–2851 (2019)
9. E.L. Donaldson, S. Fallows, M. Morris, A text message-based weight management intervention for overweight adults. J. Hum. Nutr. Diet. **27**(12), 90–97 (2014)
10. W.S. Holmes, S.A. Moorhead, V.E. Coates, H. Zheng, R.R. Bond, Impact of digital technologies for communicating messages on weight loss maintenance: a systematic literature review. *Eur J Public Health* **29**(2), 320–328 (2018)
11. Y. Rogers, H. Sharp, J. Preece, INTERACTION DESIGN: beyond human-computer interaction. 3rd Edition (2011)

Geographical Labeling of Web Objects Through Maximum Marginal Classification

K. N. Anjan Kumar, T. Satish Kumar, and J. Reshma

1 Introduction

1.1 Overview on Web Object Search Engine

Web search engines focus on retrieving relevant Web documents for the submitted user query. However, the Web documents can include information about various events, concepts, objects, etc. If the user is specifically requesting information about certain Web object such as automobiles, authors, political parties, countries, etc., the required user information might be confined to a limited section inside Web documents; the user is forced to invest some effort in locating the required information – if the Web document is large and contains multiple hyperlinks.

Web object search engines can be termed as *Vertical Search Engines*, because they directly retrieve the required information about Web objects instead of retrieving the Web documents.

Web object search engines utilize object repository, which is created by extracting information about different Web objects from Web documents and integrating the extracted information. The Web object data is stored in either unstructured or structured format in the Web object repository. However, there are significant

K. N. Anjan Kumar
Department of Computer Science and Engineering, RNS Institute of Technology, Bangalore, Karnataka, India

T. Satish Kumar (✉)
Department of Computer Science and Engineering, BMS Institute of Technology and Management, Bangalore, Karnataka, India

J. Reshma
Department of Computer Science and Engineering, BNM Institute of Technology, Bangalore, Karnataka, India

© Springer Nature Switzerland AG 2021   713
R. Stahlbock et al. (eds.), *Advances in Data Science and Information Engineering*,
Transactions on Computational Science and Computational Intelligence,
https://doi.org/10.1007/978-3-030-71704-9_52

issues in effectively extracting object information from Web documents: (1) Noise might get injected inside the data during extraction process. (2) The extracted object information might contain redundant information.

The *Windows Live Product Search* [1] and *Libra Academic Search* [2] are popular examples of Web object search engines that provide object search facility on products and academic publications, respectively. The Web object search engine presented in [3] provides Web object search facility on various Windows-related products, which were initially distributed in different websites, thus providing effective Web object search using a single application. Similarly, the Web object search engine presented in [2] provides Web object search facility regarding academic publications. In fact, searching for suitable academic publications requires accessing multiple publisher websites. However, Libra Academic Search provided unifying application to reduce the user effort in querying academic publication information.

1.2   Research Issues

Location-based search [4] on Web objects provides a facility to retrieve Web objects confined to a particular geographical location. For example, the query *Pneumonia Brazil* requires Web objects corresponding to pneumonia and confined to Brazil to be retrieved. The Web objects stored present in the Web object repository have to be tagged by location identifier in order to provide effective location-based search. However, many Web objects might not contain explicit location identifier, and in such scenario, suitable location identifiers have to be assigned.

There are multiple issues in providing accurate location labels to Web objects: (1) multiple location names might occur inside Web object information, which can lead to ambiguity in deciding the exact location identifier; (2) presence of noise inside Web object information can lead to poor accuracy in identifying the exact location identifier.

The initial work on assigning location identifiers to Web objects through Gaussian Mixture Model was presented in [5]. However, the empirical result with regard to labeling accuracy is not encouraging. There is considerable opportunity to introduce alternate technique to obtain better results regarding labeling accuracy.

1.3   Contributions

The following contributions are made in this work:

1. The location names present in the stored information of each Web object are extracted to create a corresponding feature vector, which will be subjected to

labeling. Maximum Marginal Classifier (MMC) is utilized to design geographical labeling technique for Web objects.
2. The proposed technique is compared against the contemporary technique. The empirical results are obtained on a real-world data set. The proposed technique outperforms the contemporary technique by at least 40% and by two orders of magnitude in execution efficiency.

This chapter is organized as follows: Sect. 2 outlines the related work in the area of Web object search engines. Section 3 outlines the proposed Web object geographical labeling technique. Section 4 describes the empirical results. Finally, Sect. 5 outlines the chapter conclusion with future directions.

2   Related Work

The initial work on Web object search engines was proposed in [3]. The challenges in building an efficient Web object search engine were highlighted. One of the significant challenges that were described is the addition of noise during the object extraction process. Two functions were proposed to rank the result set of a Web object query, which ranked the attributes and tuples of the Web objects, respectively.

In [6], the problem of assigning geographical labels to Web documents was addressed. The location information present in the Web document, such as physical address, IP address, etc., was utilized to assign labels. However, the problem of assigning labels when explicit location information is not present in Web documents is not addressed.

The Web search engine proposed in [7] attempts to provide query results that cover most of the geographical locations. The document repository was built by utilizing documents that collectively covered all the possible geographical locations.

In [8–12], topic level Web search facility was provided. The goal of these systems is to retrieve those Web pages that have the requested topic information. In [8], each document was divided into number of topics, and spatial labels that correspond to a specific state or country were assigned to these topics. The temporal and spatial distributions of topics were utilized to build this location-based topic level search facility. Another location-based topic search engine was proposed in [13]. Here, the geographical bias present inside Web documents was utilized to build this system. For example, in [13] it was discovered that there was uniform distribution of all website locations for those websites, which were linked to *New York Times*.

The integration of Geographical Information Systems (GIS) with Web search engines was performed and proposed in [14]. The search effectiveness of the Web search engine was increased by utilizing the recorded personal experiences of users in Weblogs.

Geographical labeling technique of documents is called Geo-coding. Geographical Information Systems (GIS) [15] utilize Geo-coding to provide relevant results to the user. Techniques such as feature vector models are used by some GIS to perform

Geo-coding [16, 17]. One of the important issues in Geo-coding is ambiguity in deciding accurate location labels. The problem arises due to the presence of multiple location names inside the documents. Many ambiguities-resolving techniques have been proposed for GIS [18, 19].

3 Geographical Labeling Technique for Web Objects

3.1 Problem Statement

Let, $(o_1.o_2,....o_N)$ indicates a set of Web objects; wherein, each Web object $o_k(1 \leq k \leq N)$ has to be labeled with suitable location identifier/label. The feature vector for a Web object represented in Eq. 1, is constructed using the location identifier that appears in the corresponding Web object information. Here, $\mathbf{d_j}$ is the feature vector for Web object o_j, l indicates the utilized maximum number of location names for the labeling procedure, $p_i(1 \leq i \leq l)$ indicates a specific location identifier/name and $tf(p_i)$ indicates the term frequency of p_i inside o_j. The goal is to assign a suitable location label/name to $\mathbf{d_j}$.

$$
d_f = \begin{bmatrix} tf\,(p_1) \\ tf\,(p_2) \\ . \\ . \\ tf\,(p_t) \end{bmatrix} \tag{1}
$$

3.2 Maximum Marginal Classification Model

Consider a data point $\mathbf{d_j}$, which is modeled through a random vector having the dimensions $k \times 1$.

So, $\mathbf{d_j} = [d_{j1}, d_{j2},...,d_{jk}]$ is the component representation of $\mathbf{d_j}$ and $-a \leq d_{ji} \leq a(1 \leq i \leq k)$. Consider the $2 - class$ classification problem. Assume that there are r classifier functions that can perform $2 - class$ classification on the data point $\mathbf{d_j}$. Let these classifiers be indicated as $F_1(x), F_2(x),....F_r(x)$. If the parameters of these classifiers are unknown, then they can be estimated through the training set. But, the issue is to decide which classifier provides the best classification performance.

$$
R\,(F_c(x)) = \int_{d_{j1}=-a}^{a} \int_{d_{j2}=-a}^{a} \cdots \int_{d_{jk}=-a}^{a} Q\left(F_c\left(\mathbf{d_j}\right), \mathbf{d_j}\right) f\left(\mathbf{d_j}\right) \tag{2}
$$

Risk minimization technique is used to identify the most suitable classifier function. The Equation 2 exhibits the risk functional which decides the optimal

Fig. 1 Hyperplane for first classifier

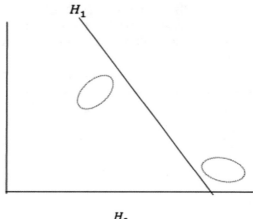

Fig. 2 Hyperplane for second classifier

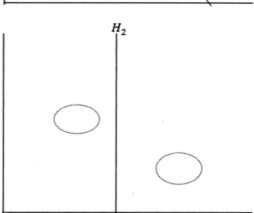

classifier. Here, R is the risk functional, Q is called as the loss function, $1 \leq c \leq r$ and f is the density function of $\mathbf{d_j}$. The classifier function that has the lowest risk functional value is selected as the optimal classifier.

The MMC technique is based on the principle of risk minimization. It provides the most suitable classifier to perform $2 - class$ classification. The main concept of MMC is to detect the hyperplane that provides the maximum separation between the two classes, and this case is illustrated in Figs. 1 and 2. Here, the ovals indicate a specific class data points. Each data point is assumed to have two components, H_1 and H_2 indicate the hyperplanes of two different classifiers. It is apparent that the classifier corresponding to the hyperplane H_2 provides the maximum separation between the two classes.

$$x_j = w\varphi\left(d_j\right) + b \tag{3}$$

$$\varphi\left(a_j\right) = aa_j \tag{4}$$

The classifier function used by MMC is represented in Eq. 3. Here, x_j is the class label of $\mathbf{d_j}$, which can take only two values, $\mathbf{w} \in 1 \times k$ is the weight vector, $\varphi(\mathbf{d_i})$ is a transformation function, which is represented in Equation 4, and a and b are suitable constants.

The training set used in calculating the optimal hyperplane is represented in Eq. 5. The training set contains n ordered pairs; in which, every ordered pair contains data point and its corresponding class label. The optimal hyperplane is obtained by solving the optimization function shown in Eq. 6. Here, $1 \leq m \leq n$ and the magnitude of weight vector is represented in Eq. 7.

$$trinning\ set = [(d_1, x_1), (d_2, x_2), \cdots (d_n, x_n)] \tag{5}$$

$$arg\ \max_{w,b} \frac{1}{\|w\|} \min_m \left[x_m \left(w^T \phi(d_m) \right) + \right] \tag{6}$$

$$\|w\| = \sqrt{w_1^2 + w_2^2 + \cdots + w_k^2} \tag{7}$$

3.3 Algorithm

The Algorithm 1 describes the proposed geographical labeling technique, and which is illustrated in Fig. 3. Initially, parsing of the Web object information is performed to detect explicit location identifier. If such location identifier is not present, Algorithm 1 is executed to assign suitable location identifier.

The details of Algorithm 1 are as follows: Every distinct pair of locations is considered such that, both the locations in the pair should be different. For each

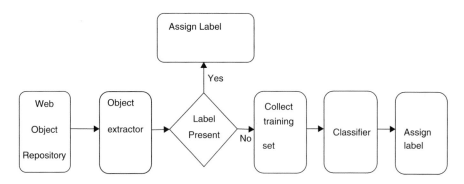

Fig. 3 Web object labeling process

location pair, MMC parameters **w** and *b* are estimated by utilizing the training set (indicated by *Tr set*) shown in Eq. 5, and the function *MMC model(Tr set)* achieves the required optimization represented in Eq. 6.

Here, $\hat{w}$ and $\hat{b}$ are the estimated parameters. The feature vector is classified w.rt considered location pair, and by using the classifier function shown in Eq. 3. The feature vector $\mathbf{d_j}$ is labeled with the location name to which $\mathbf{d_j}$ gets assigned maximum number of times. Here, *label* ($\mathbf{d_j}$) is the assigned location label for $\mathbf{d_j}$.

Algorithm 1 Technique for Web Object Geographical Labeling

---

for $i = 1$ *to* l do
$count(p_i) = 0$
end for
for $Each(p_i, p_k)(i \neq k)do$

$\left(\hat{w}, \hat{b}\right) = MMC model\ (Tr set)$

If $p_i == \left(\hat{w} \oslash \left(d_j\right) + \hat{b}\right)$ then
$count(p_i) = count(p_i) + 1$
else
$count(p_k) = count(p_k) + 1$
end if
end for
$label\left(d_j\right) = \max_r count\left(p_r\right)(1 \leq r \leq l)$

---

4 Results and Discussion

4.1 System Design

An empirical analysis was performed on Web object repository created by utilizing Wikipedia data set [20]. Figure 4 illustrates the utilized object extraction procedure. The *Information Extraction Component* extracts all the object information from various Wikipedia relations. The *Information Filtering* component is responsible for structuring information and its integration for relevant objects, which is then stored in object repository.

The proposed Web object labeling technique is empirically compared against the contemporary technique; for the ease of reference, these techniques will be referred to as *new-label* and *old-label*, respectively. The training set used for training the MMC classifier and test set contained 600 and 1000 Web objects, respectively. The parameter l varied between 50 and 10. The empirical analysis is performed through the metric *labeling accuracy*, which indicates the ratio of accurately labeled objects in the test set to the number of objects in the test set. The value of this metric is between [0−1]; wherein, 1 indicates that accurate labeling of all the Web objects

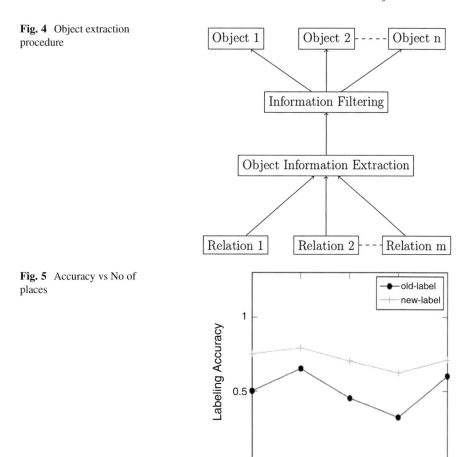

Fig. 4 Object extraction procedure

Fig. 5 Accuracy vs No of places

in the test set was achieved and 0 indicates none of the test set Web objects were accurately labeled.

4.2 Empirical Results Discussions

The first experiment is used for analyzing the performance of new-label and old-label with regard to labeling accuracy and total labeling time by varying the parameter l. Figure 5 illustrates the result analysis regarding labeling accuracy. The new-label outperforms old-label due to the superior effectiveness of its designed classification scheme. The analysis result of the same experiment regarding test set total labeling time is illustrated in Fig. 6. The Gaussian Mixture Model (GMM)

Fig. 6 No of places vs Time

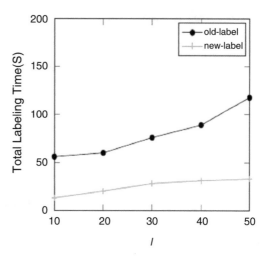

Fig. 7 Latency for training

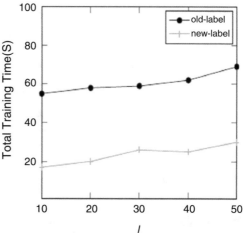

utilized by old-label for classification is computationally expensive, and with the increase of GMM mixing components, execution cost increases considerably. Hence, new-label exhibits superior execution efficiency compared to old-label.

The second experiment analyzes the latency for training the classifiers of old-label and new-label by varying the parameter l. The analysis result of this experiment is illustrated in Fig. 7. There is an increase in the feature vector components of both new-label and old-label as l increases, which increases the training latency of both old-label and new-label. The old-label exhibits increased training latency due to the complexity of GMM.

The effect of noise on the labeling accuracy and total labeling time of new-label and old-label is analyzed through a third experiment. The analysis result regarding labeling accuracy is illustrated in Fig. 8. The labeling accuracy of both new-label

Fig. 8 labeling accuracy

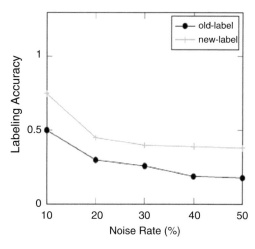

Fig. 9 Labeling Time vs
Noise Rate

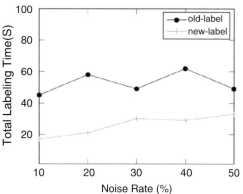

and old-label decreases due to increase in noise rate, because of data manipulation
through noise. Old-label exhibits limited performance due to the same reasons
explained before. The analysis result of the third experiment with regard to total
labeling time is illustrated in Fig. 9; again, new-label outperforms old-label for the
same reasons explained before.

The last experiment is used for analyzing the labeling accuracy of new-label and
old-label on the test set by varying the training set size. Figure 10 illustrates the
analysis result of the last experiment. As the size of the training set increases, it
is evident that both techniques increase their labeling accuracy. Again, new-label
outperforms old-label for the same reasons explained above.

Fig. 10 Training Set Size vs
Accuracy

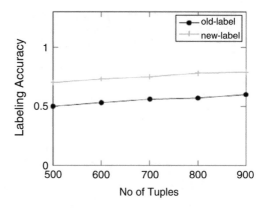

5   Conclusion

The contributions presented in this work are:

1. The importance of providing geography-specific Web object search is outlined. The mathematical framework for Web object geographical labeling was presented, which proposed a feature vector that contained the term frequencies of all the location identifiers found in the Web object information. The MMC model was formally described and it was applied to achieve Web object geographical labeling.
2. The proposed technique was evaluated over real-world data set and it was compared against contemporary technique proposed in [5]. The proposed technique provided around 40% better labeling accuracy and 2 order magnitude improvement in execution efficiency than the contemporary technique.

The future work in this area is summarized below:

1. The labeling accuracy of this proposed technique may improve through parameter tuning, which involves adding few more effective parameters in the classifier function. So, discovery of effective additional parameters can be considered in future.
2. Ranking of Web object results to answer a user query is extremely essential to increase the effectiveness regarding user relevance. Until now, ranking functions have not considered geography-based search. It is important to develop ranking functions that cater to this requirement.

References

1. http://products.live.com
2. http://academic.research.microsoft.com/

3. ZaiqingNie, Yunxiao Ma, Shuming Shi, Ji-Rong Wen, Wei-Ying Ma *Web Object Retrieval* WWW 2007, May8-12-2007, Banff, Alberta, Canada
4. J. Raper, *Geographic relevance*. Journal of Documentation **63**(6), 836–852 (2007)
5. T. Tezuka, H. Kondo, K. Tanaka, *Estimation of Geographic Relevance for Web objects Using Probabilistic Models* (Springer, Berlin/Heidelberg, 2008)
6. K.S. McCurley, *Geospatial mapping and navigation of the Web,* in Proceedings of the 10th international world wide web conference, Hong kong, China, pp. 221–229 (2001)
7. W. Gao, H.C. Lee, Y. Miao, *Geographically focused collaborative crawling,* in Proceedings of the 15th international world wide web conference, Edinburgh, Scotland, pp. 287–296 (2006)
8. Q. Mei, C. Liu, H. Su, C. Zhai, *A probabilistic approach to spatiotemporal theme pattern mining on weblogs,* in Proceedings of the 15th international world wide web conference, Edinburgh, Scotland, pp. 533–542 (2006)
9. L. Gravano, V. Hatzivassiloglou, R. Litchenstein, *Categorizing Web queries according to geographical locality,* in Proceedings of the 12th international conference on information and knowledge management, New Orleans, Lousiana, pp. 325–333 (2003)
10. L. Chen, L. Zhang, F. Jing, K. Deng, W.Y. Ma, *Ranking Web objects from multiple communities*, in Proceedings of the international conference on information and knowledge management, Arlington, Virginia, pp. 377–386 (2006)
11. Z. Nie, Y. Ma, S. Shi, J.R. Wen, W.Y. Ma, *Web object retrieval,* in Proceedings of the 16th international world wide web conference, Banff, Canada, pp. 81–90 (2007)
12. Z. Nie, J.R. Wen, W.Y. Ma, *Object-level vertical search,* in Proceedings of the 3rd biennial conference on innovative data systems research, Asilomar, California, pp. 235–246 (2007)
13. O. Buyukkokten, J. Cho, H. Garcia-Molina, L. Gravano, N. Shivakumar, *Exploiting geographical location information of Web pages,* in proceedings of the ACM SIGMOD workshop on the web and databases, Philadelphia, Pennsylvania (1999)
14. T. Tezuka, T. Kurashima, K. Tanaka, *Toward tighter integration of Web search with a geographic information system,* in Proceedings of the 15th world wide Web conference, Edinburgh, Scotland, pp. 277–286 (2006)
15. C.A. Davis, F.T. Fonseca, *Assessing the certainty of locations produced by an address geo coding system*. Geoinformatica **11**(1), 103–129 (2007)
16. E. Amitay, N. Har El, R. Sivan, A. Soffer, *Geotagging web content*, in Proceedings of the 27th annual international ACM SIGIR conference on research and development in information retrieval, Sheffield, United Kingdom, pp. 273–280
17. M.D. Lieberman, J. Sperling, *STEWARD Architecture of a Spatio-textual Search Engine,* in Proceedings of the 15th annual ACM international symposium on advances in geo-graphic information systems, Seattle, Washington, Article No.25 (2007)
18. M. Schneider, *Geographic Data Modeling: Fuzzy Topological Predicates, their Properties and their Integration into Query Languages,* in Proceedings of the 9th ACM international symposium on advances in geographic information systems, Atlanta, Georgia, (2001), pp. 9–14
19. J. Coffman, A.C. Weaver, *A Framework for Evaluating Database Keyword Search Strategies*, in Proceedings of the 19th ACM International Conference on Information and Knowledge Management (2010). 978-1-4503-0099-5

Automatic Brand Name Translation Based on Hexagonal Pyramid Model

Yangli Jia, Zhenling Zhang, Haitao Wang, and Xinyu Cao

1 Introduction

With the rapid development of the integration of the global economic exchange, a lot of foreign products have entered or are ready to enter the Chinese market. It is very wise and necessary for international corporations to translate their foreign brand names into Chinese when entering the Chinese market considering the diversity of its languages, nationalism, and cultural factors.

At present, there are a number of successfully translated brand names on the Chinese market, such as Coca-Cola(可口可乐/ke-kou-ke-le/), Pepsi Cola(百事可乐/bai-shi-ke-le/), Goldlion(金利来/jin-li-lai/), Nike(耐克/nai-ke/), Pantene(潘婷/pan-ting/), Sprite(雪碧/xue-bi/), Safeguard(舒肤佳/shu-fu-jia/), Colgate(高露洁/gao-lu-jie/), Robust(乐百氏/le-bai-shi/) and so on. These brand names, translated in accordance with Chinese cultural and aesthetic norms, have succeeded in finding resonance with Chinese consumers.

A successfully translated brand name itself is a good advertising while an unsuccessful one would lead to economic loss to the company [1]. Thus, it is necessary and valuable to research on the translation strategies and translation methods.

Some strategies for translating foreign brand names to Chinese have been presented, and most of these translation strategies have been developed from two aspects in linguistics: phoneme and semantics. For example, as Schmitt and Zhang

Y. Jia · Z. Zhang (✉)
School of Computer Science & Technology, Liaocheng University, Liaocheng, China
e-mail: jiayangli@lcu.edu.cn; zhangzhenling@lcu.edu.cn

H. Wang · X. Cao
China National Institute of Standardization, Beijing, China
e-mail: wanght@cnis.ac.cn; caoxy@cnis.ac.cn

© Springer Nature Switzerland AG 2021
R. Stahlbock et al. (eds.), *Advances in Data Science and Information Engineering*,
Transactions on Computational Science and Computational Intelligence,
https://doi.org/10.1007/978-3-030-71704-9_53

point out, phonemes and semantics, as two linguistic dimensions of brand name translation, yield four possible types of name translations from English to Chinese: a translation based on sound; a translation based on meaning; a translation based on sound and meaning; creative translation [2, 3]. These strategies and methods are easy to understand and manually implement. However, brand name translation is a special kind of translation, involving business and products. An effective translated brand name needs to be translated in accordance with the communicative purpose to the targeted market segments [4].

Some business researchers have proposed other translation strategies that consider the combination of the basic translation methods and other strategies such as culture, psychology, and so on. For example, Meilun Gou proposed a creative translation method that is based on sociocultural adaptation and consumer acceptance, but discarded its combination of the other two basic strategies: phoneme and semantics [5]. On the other hand, [6] pays more attention to the aesthetic sentiments when translating brand names for Chinese cosmetic market. These translation methods are not fully summarized and the classification of some examples cited in this chapter is a little indistinct and need to be further discussed.

In this chapter, we put forward a hexagonal pyramid with mid-perpendicular model for brand name translation, and a lot of examples are used to introduce and analyze the model in detail. This model provides a comprehensive summary of brand name translation methods and makes the classification of some translated brand names from vague to clear. In the second section of this chapter, a hexagonal pyramid with mid-perpendicular model for brand name translation is firstly proposed and discussed with some related cases in details based on the comparative linguistic study of loanwords translation. In the third part, an approach supporting all the strategies of the hexagonal pyramid model has been proposed to help find an adequately translated word in Chinese. And an experiment has been done by way of a dedicated program with results of a cluster of recommended Chinese brand words with a good potential to be used.

2   Translation Strategies and Hexagonal Pyramid Model

Based on the research of the existing brand name translations, a model for brand name translation, named hexagonal pyramid model, is proposed. The corresponding algorithm is designed as well. As shown in Fig. 1, in the hexagonal pyramid model, the point (B) refers to a brand name, and three basic points on the bottom, marked as (P), (S), (C), refer to the pure phonetic strategy, pure semantic strategy, and pure commercial strategy separately. The strategies in the three dimensions can be combined with each other. And the points (P, C), (P, S), (S, C) represent the corresponding blending strategies separately. Among them, (P, C) represents the phono-commercial strategy, and (P, S) represents the phono-semantic strategy, and (S, C) represents the semantic-commercial strategy, and the midpoint on the bottom

Fig. 1 Hexagonal pyramid model

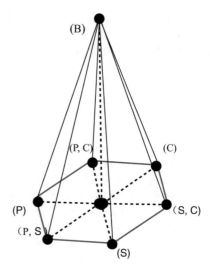

Table 1 Translation strategies based on the hexagonal pyramid model

Translation strategy	Examples
Pure phonetic strategy (P)	星巴克(xing-ba-ke, Starbucks), 派克(Pike), 西门子(xi-men-zi, Siemens)
Pure semantic strategy (S)	先锋(xian-feng, Pioneer), 空客(kong-ke, Airbus)
Commercial strategy (C)	雪碧(xue-bi, Spirit), 飘柔(piao-rou, Rejoice), 汇丰(hui-feng,HSBC)
Phono-commercial strategy (P, C)	兰蔻(lan-kou, Lancome), 必应(bi-ying, Bing), 雅芳(ya-fang, Avon), 宜家(yi-jia, IKEA)
Semantic-commercial strategy (S, C)	七喜(qi-xi, Seven up), 劲量(jin-liang, Energizer)
Phono-semantic strategy (P, S)	台风(tai-feng, typhoon)

(P, S, C) represents the phono-semantic-commercial strategy. Six edges: (B)<->(P), (B)<->(C), (B)<->(S), (B)<->(P, C), (B)<->(P, S), (B)<->(S, C).

On the one hand, this model can be used to classify the existing brand name translations clearly and easily. Some brand name translation examples corresponding to the edges and mid-perpendicular in the model are listed in Table 1. On the other hand, different strategies can be selected according to this model to construct a new brand name translation. The translation types and corresponding translation detail are as follows.

2.1 Phonetic Strategy (P)

In linguistics, phonetic translation or phonetic borrowing is when we accept a term with its pronunciation from foreign languages. Phonetic loanwords are written in Chinese characters based on syllables. One Chinese character is one syllable, so one character or a compound of multiple characters can approximate the phonetic form of a foreign word.

Western brand names can be translated into Chinese based on pronunciation only, which refers to use of Chinese characters to transcribe phonetically the sound of names foreign to the Chinese language. That is to say when a Western brand name is converted into a new Chinese name, its pronunciation is similar to the original one. For example, the converted Chinese brand name "飞利浦"(/fei-li-pu/, Philips) uses the characters "飞"(/fēi/, fly), "利"(/lì/, sharp, benefit), 浦"(/pǔ/, riverine), which are the phonetically imperfect rendering of the English initial syllables.

For example, as shown in Fig. 2, "Philips" (brand name for jeans) has consonant clusters "p" and "h." Firstly, a series of similar syllables written in Chinese Pinyin "*fei-li-pu*" are obtained based on the segment and conversion of the pronunciation of "Philips." The consonant cluster "ph" has been converted to Pinyin "f" and the consonant suffix "s" has been abandoned in this process. And then a sequence of characters "飞利浦,"whose pronunciation is fitted to the modified Pinyin, is chosen as the phonetic loans of "Philips."

As shown in Fig. 3, the brand name Lancome is taken as an example. It is segmented to three syllables (lan, co, me), and then each syllable is converted to

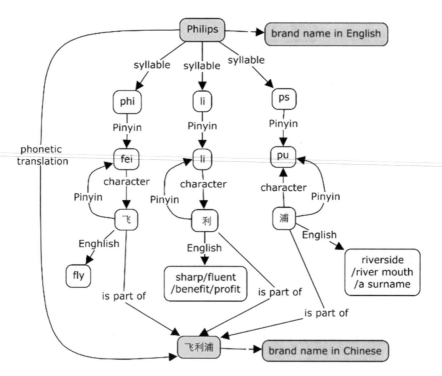

Fig. 2 Word segmentation

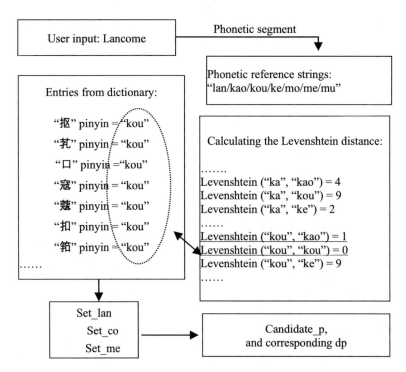

Fig. 3 Phonetic similarity calculation

a Pinyin sequence. We can see that the English syllables can be converted to "lan," "kao," "kou" or "ke," "mo," "me," "mu" in PINYIN, respectively.

A distance between each PINYIN sequence and each Chinese entry in the dictionary will be calculated using a distance algorithm, for example, the Levenshtein algorithm which is a string metric for measuring the difference between two sequences [7, 8]. As shown in Fig. 3, the Levenshtein distance of each Chinese character is marked as dp value.

2.2 Semantic Strategy (S)

Direct semantic translation can be classified to two types: purely semantic translation and loan translation. In purely semantic translation, the formation rules and morphemes of terms are all derived from Chinese, but only introducing the corresponding concept from the foreign language. Such as Apple (苹果/ping-guo/), *Pioneer* (先锋/xian-feng/), Plover (啄木鸟/zhuo-mu-niao/), Volkswagen (大众/da-zhong/), Shell (壳牌/ke-pai/), and Crocodile (鳄鱼/e-yu/), etc. are all translated with purely semantic translation method. In loan translation, a word or phrase is borrowed from another language by word-for-word or root-for-root translation.

The term's structure and semantic meaning come from the foreign language, but the pronunciation and grapheme are Chinese. For example, the brand names Microsoft (微软/wei-ruan/), Western Digital (西部数据/xi-bu-shu-ju/), Airbus(空客/kong-ke/), Dream Works (梦工厂/meng-gong-chang/), Facebook (脸书,lian-shu), General Electric (通用, tong-yong), Playboy (花花公子/hua-hua-gong-zi/), etc. are all translated with loan translation method. Loan translation is usually used in the translation of brand name in compound forms as shown in the above examples. In general, the original and the version, representing the same concept, have the same meaning and give customers the same feeling in both purely semantic translation and loan translation.

2.3   Phono-semantic Strategy (P, S)

This strategy is applied to brand name translation considering both phonetic and semantic dimensions. There are some phono-semantic blending translated brand names on the Chinese market, which use phonetically and semantically similar words from the Chinese language, such as "台风"(/tai-feng/, typhoon).

The similarity is calculated in the phonetic dimension, which has been introduced above. And then the candidates in this dimension, named Candidate_p, are obtained. Each word in the Candidate_p is compared with the original brand name in semantic similarity, and the semantic similarity distance ds between them is obtained then.

Note that brand-name translations based on the phonetic dimension and semantic dimension are rare according to our statistics, although the phono-semantic translation is popular for general words [9].

2.4   Commercial Strategy (C)

Commercial strategy, here referring to the purely commercial strategy focusing on the Chinese market, is a kind of pragmatic translation for Chinese market independent of phoneme or semantics [10]. Commercial translation can obtain good pragmatic effects where phonetic or semantic translation is not suitable. Two kinds of brand name translation can be included in this commercial strategy.

2.4.1   Re-creation

Re-creation is parallel with the renaming translation method, which is usually used in the translation of the names of literary works or commodity. In the process of this kind of translation, the consumer's characteristics including psychology, linguistic level, subject knowledge, aesthetic sentiments and social customs, etc. and the products' characteristics including core characteristics, such as taste, safety, rewords, speed, etc. and the scope of business, such as beverage, food, fashion, automobile, banking, cosmetics, etc. all should be taken into account.

For example, "Rejoice" is the brand name for hair beauty. Its original semantics "happy" is abandoned during the translation process. The re-created brand name is "飘柔"/piao-rou/, which means "gone with the wind and looks supple" literally. It is very reminiscent of the hair fluttering and supple appearance. It is obvious that the commercially translated brand name "飘柔"/piao-rou/ highlights the function of the goods very well. Other examples, such as "辉瑞"/hui-ri/(Pfizer), "汇丰"/hui-feng/(HSBC), "太古"/tai-gu/(Swire) are all well re-created brand name.

This kind of translation is more flexible without the limitation of the similarity of phoneme and semantics. Usually, some Chinese words that have beautiful meanings and can match the aesthetic demands of Chinese consumers are selected to give consumers a positive message. For example, some auspicious words such as "瑞"/rui/, "吉"/ji/, "利"/li/, "达"/da/, "辉"/hui/, "丰"/feng/, "隆"/long/, "福"/fu/, "宝"/bao/, "财"/cai/, "聚"/ju/, "德"/de/, etc. are always selected to meet the consumers' psychology well. And some words such as "靓"/liang/, "倩"/qian/, "雪"/xue/, "莲"/lian/, "碧"/bi/, "飘"/piao/, "海"/hai/, "露"/lu/, "香"/xiang/, "新"/xin/, etc., which are related to or capable of showing the characteristics of products can let consumers have a good association of the products.

Furthermore, this kind of translation, independent of phoneme or semantics of original brand names, reflects more clearly the subjectivity of the translator and adapts more easily to Chinese culture and market. However, it loses some benefits in the consistency of original phoneme and semantics. For example, it is difficult for Chinese consumers to get the exotic flavor from this kind of translation.

2.4.2 No-Translation

Another kind of commercial strategy is the method of no-translation which means brands keep the intrinsic names. This method is suitable for brands that have a simple, catchy and abbreviated name like MAC, SKII, H&M, etc. Its commercial strategy is mainly reflected in the psychology of some Chinese consumers preferring foreign brands. This method can save time and cost of translation, and avoid the risk of conflict from exotic culture. However, there will be a brand gap for Chinese consumers to perceive the brand based on the foreign brand name [11].

2.5 Phono-Commercial Strategy (P, C)

Phonetic-commercial strategy translation tries to keep the balance between phonetic equivalence in linguistics and commercial strategy. There are some good examples, such as Benz – 奔驰(/ben-chi/), Pentium – 奔腾(/ben-teng/), etc. The most famous example is the makeup brand name Lancome translated as "兰蔻"(/lan-cou/). The original one comes from a castle name Lancome in central France, and it has no corresponding semantic translation in the dictionary. The translated Chinese name "兰蔻"(lan-kou) of Lancome is a phono-commercial translated word, wherein "兰"(lan) refers to the flower orchid and "蔻"(cou) refers to cardamom, which is often used as a metaphor for young girls. These brand names are well concerned about the characteristics of products and consumers' psychology and other commercial strategies at the same time as phonetic translation.

Phonetic-commercial strategy, focusing on phonetic similarity and the commercial strategies for the Chinese market, generally discards the original semantic in order to adapt to Chinese culture and Chinese consumers' language habits and shopping psychology. So it is difficult for phonetic-commercial strategy to keep the consistency of brand semantics.

However, extra meaning expected to be perceived easily by Chinese customers is added to the brand names intentionally in accordance with Chinese characteristics during the phonetic-commercial translation process. And, subsequently, positive association related to the product characteristics stands a good chance to promote the purchase desire of Chinese consumers. Lecture [12] argues that when translating a brand name into Chinese there is a golden opportunity to add some extra meanings and benefits.

Phonetic-commercial strategy is regarded as the best way of introducing foreign brand names to the Chinese. It is a complex and difficult work to create a phonetic-commercial translation word, in which ingenious conception and consideration are needed. It will be a more difficult work if taking the original semantics into account as well.

3   Application of Hexagonal Pyramid Model

In this chapter, an automatic translation algorithm for brand names and corresponding experiments are designed, based on the hexagon pyramid model. They aim to find the adequate translated words to the brand name with short distances on phonetic, semantic, commercial, or combined dimensions.

We give the translation algorithm as follows.

Algorithm input: the original brand name in Western language, such as LAN-COME, and three weight values (wp, wc, ws).

Algorithm output: the result of translation.

```
Brandname_translation() {
  User_input (brandname);
  User_input (wp, ws, wc);
  if (wp == 0 && wc == 0 &&ws == 0) return error;
  elif   (wp == 0 && wc == 0 &&ws != 0):
              s_simical&trans();User_output(semantic translated word);
  elif (wp == 0 && wc != 0 &&ws == 0):
              c_simical&trans(); User_output (commercial translated word);
  elif (wp != 0 && wc == 0 &&ws == 0):
              p_simical&trans(); User_output (phonetic translated word);
  elif (wp == 0 && wc != 0 &&ws != 0):
              s_c_simical&trans(), User_output(semantic_commercial translated
              word);
  elif (wp != 0 && wc == 0 &&ws != 0):
              p_s_simical&trans();User_output(phonetic semantic translated word);
  elif (wp != 0 && wc == 0 &&ws != 0):
              p_c_simical&trans();User_output(phonetic_commercial
              translated word);
}
```

Obviously, the algorithm gives the translation result according to the original brand name and the three weight values. Based on the translation algorithm, we can automatically translate the brand names.

In the algorithm, the input data include the original brand name in a Western language, such as Lancome, and three weight values (Wp, Wc, Ws) for three dimensions that meet the following two conditions.

- $0 \leq$ Wp $\leq 1, 0 \leq$ Ws $\leq 1, 0 \leq$ Wc ≤ 1
- Wp + Ws + Wc = 1

As shown in Table 2, three weight values (Wp, Wc, Ws) can let users choose flexibly strategy (P), (C), (S), (P, C), (P, S), and (C, S). For example, if purely phonetic translation is a unique choice, Ws and Wc can be set to zero. When self-created brand names, such as Lancome, are encountered, Ws should be set to zero while Wp or Wc should not.

In the algorithm, a whole similar distance is set as Eq. 1, wherein dp value conveys the similarity distance in the phonetic dimension, the dc value the similarity distance in the semantic dimension, and the dc value the similarity distance in the commercial dimension. The d value is a measure of automatic translation results. All the candidate words are sorted based on the d value, and the words with the lowest whole distance value are selected to be the final candidates.

$$d = \sqrt{w_p \cdot d_p^2 + w_c \cdot d_c^2 + w_s \cdot d_s^2} \tag{1}$$

Table 2 Relations between weight values of three dimensions and translation strategies

Type	W_p	W_s	W_c	Phonetic strategy	Semantic strategy	Commercial strategy
1	1	0	0	+	-	-
2	0	1	0	-	+	-
3	0	0	1	-	-	+
4	1	1	0	+	+	-
5	1	0	1	+	-	+
6	0	1	1	-	+	+

We have experimentally translated some brand names based on the hexagon pyramid model and the automatic translation algorithm, and the model and the algorithm worked effectively.

4 Conclusion

In this chapter, we summarized the strategies and methods of brand name translation from Western languages to Chinese in the context of the boost in global commodity exchange and propose a hexagonal pyramid brand name translation model to provide a comprehensive summary of brand name translation methods. The combination of the strategies based on the model produces an efficient automatic translation method, which can provide help in finding adequately translated words in Chinese. The empirical experiments indicate that our translation model and method are indeed able to improve brand name translation.

As future work, we intend to utilize our approach to provide objective and effective automatic translation for brand names, as well as some efficient and effective candidates for translators.

Acknowledgments This work is supported by the National Natural Science Foundation of China under Grant No. 81973695, Soft Scientific Research Project of Shandong Province under Grant No. 2018RKB01080.

References

1. D. Kum, Yih Hwai Lee, Cheng Qiu, testing to prevent bad translation: Brand name conversions in Chinese–English contexts. Journal of Business Research **64**(6), 594–600 (2011)
2. B. Schmitt, S. Zhang, Selecting the right brand name: An examination of tacit and explicit linguistic knowledge in name translations. Journal of Brand Management **19**(8), 655–665 (2012). https://doi.org/10.1057/bm.2011.62
3. J. Liu, Name selection in international branding: Translating brand culture. International Journal of Business and Management **10**(4), 187–192 (2015)

4. F. Wang, An approach to the translation of brand names. Theory & Practice in Language Studies **2**(9) (2012)
5. M. Gou, *Brand Name Translation Effects: A study about Chinese Consumers' Perception of Alphabetic Language Brands* (Faculty of Economics and Business, University of Amsterdam, Amsterdam, 2011)
6. V. Alleton, Chinese terminologies: On preconceptions, in *New Terms for New Ideas*, ed. by Lackner et al., (Koninklijke Brill, Leiden), pp. 15–34
7. A. Niewiarowski, M. Stanuszek, Paralleliation of the levenshtein distance algorithm. Advances in Physics **20**(86), 493–550 (2015)
8. P. E. Black (ed.), *Levenshtein distance* (U.S. National Institute of Standards and Technology, Gaithersburg, 2008)
9. S. Chaofen, *Chinese: A Linguistic Introduction* (Cambridge University Press, Cambridge, 2006)
10. P. Newmark, Pragmatic translation and literalism. TTR: traduction, terminologie, rédaction **1**(2), 133–145 (1988). http://id.erudit.org/iderudit/037027ar
11. Y.H. Lee, K.S. Ang, Brand name suggestiveness: a Chinese language perspective. International Journal of Research in Marketing **20**(4), 323 (2003)
12. F.C. Hong, A. Pecotich, C.J. Schultz, Brand name translation: Language constraints, product attributes, and consumer perceptions in East and Southeast Asia. Journal of International Marketing **10**(2), 29–45

A Human Resources Competence Actualization Approach for Expert Networks

Mikhail Petrov

1 Introduction

Human resources are one of the most valuable assets of any organization and they play a crucial role in its success. Effective human resource management allows to achieve both project objectives and employees' needs [1]. Moreover, an efficient management of human resource competencies prevents imposition of exorbitant costs, improves the quality of products and services and facilitates better workforce planning. Efficient human resource management needs accurate assessment and representation of available competencies as well as effective mapping of required competencies for specific jobs and positions. The use of expert systems in competence management provides opportunities for this [2].

It follows that information about employees' competencies is important for effective company management. In addition, this information must be relevant, otherwise project management loses effectiveness. If the wrong project team is assigned due to irrelevant experts' competencies data, the project may be unsuccessful. Therefore, an approach is needed to keep competencies in expert networks actualized.

Analysis of the project implementation results is one of the possible solutions to this problem. Whether the project was successful or not provides information on whether the competencies of the experts in the project team were relevant. This information should be used to actualize the competencies of project participants.

This paper presents an approach to human resources competence actualization in expert networks. It includes a reference model of human resources competence actualization and the competence actualization algorithm, which allow to keep experts' competencies information relevant. The approach analyses project implementation

M. Petrov (✉)
Computer Aided Integrated Systems, ITMO University, Saint Petersburg, Russia
e-mail: mikhail.petrov@iias.spb.su

© Springer Nature Switzerland AG 2021   737
R. Stahlbock et al. (eds.), *Advances in Data Science and Information Engineering*,
Transactions on Computational Science and Computational Intelligence,
https://doi.org/10.1007/978-3-030-71704-9_54

results to increase or decrease the competencies of its participants depending on the success or failure of the project. This paper is an extension of work on the competency management systems [3–6] and method of expert group formation for task performing [7, 8].

The rest of the paper is organized as follows. Section 2 considers the related works in the area of human resource management and the impact of competencies on project implementation. The third section is devoted to proposed reference model of human resource competence actualization. The competence actualization algorithm is represented in detail in Sect. 4. The conclusion summarizes the paper.

2   Related Work

Authors of paper [9] collected data from several construction projects to identify relationship between project competencies and project performance. They identified project competencies and project KPIs and evaluated them using prioritized fuzzy aggregation and factor analysis. The results were used to calculate inputs for the fuzzy neural networks, which identified and quantified the relationship between the different project competencies and project key performance indicators.

Paper [10] explores which combinations of competencies contribute to successfully carrying out all the activities in the project's requirements phase. Professionals from the largest Portuguese companies that held different roles in the requirements phase participated in the study. These data were used in a fuzzy-set qualitative comparative analysis to explore which types of competencies (emotional, intellectual, and managerial) are the most relevant for each activity in the requirements phase. The results of this study can be used to determine the influence of degree of competencies and other factors on the project outcome. This can provide a more accurate adjustment of the project participants' competencies.

Paper [11] focuses on examining the mechanisms used to manage competencies in project-based organizations. Based on a multiple case study in different sectors, the paper's results detail the three conceptual dimensions of project management mechanisms (knowledge management, human resource management and strategy), and emphasize the links between these mechanisms and the three levels of project management (individual, collective and organizational). Authors develop an integrative multilevel analysis of mechanisms for managing competencies. Thus, these results allow to take into account not only the contribution of each individual project participant and his or her competencies but also the overall group work.

Paper [12] describes a research project to develop an organizational knowledge architecture that is being specified and developed to support collaboration tasks as well as design and model predictive data analysis and insights for organizational development. The designed architecture and functionalities aim to create coherent web data layers for intranet learning and predictive analysis, defining the vocabulary and semantics for knowledge sharing and reuse projects. Thus, using models based

Table 1 Comparison of the considered approaches' features

Features	[9]	[10]	[11]	[12]	[13]	[14]	[15]
Impact of competencies on project implementation	✓	✓			✓	✓	✓
Impact of project implementation on competencies						✓	
Various degrees of impact	✓	✓			✓		✓
Core competencies	✓				✓		
Core KPI	✓						
Different types of competencies		✓			✓		
Competence management			✓	✓		✓	
Group analysis			✓				✓
Knowledge architecture				✓			

on this architecture to store experts' competencies allows their effective analysis and actualization.

Paper [13] presents results of factor and regression analysis of determinants affecting cognitive skills. Five determinants were considered: age, gender, education, life and professional experience. The results of this analysis can be useful for experts' competencies actualization.

Paper [14] uses theoretical literature analysis to identify four main approaches to study professional competence. The outstanding characteristics and the factors considered are described in the paper for each of the approaches.

Paper [15] studies the effects of team competencies and team processes on the project performance. The results show which types of competencies most influence the outcome. Thus, these results will be useful to consider for an algorithm for experts' competencies actualization.

The work described above was analyzed for general features of the approaches used. The results of the analysis are presented in Table 1.

Related works analysis shows that the relationship of the project performing results and competencies of the participants are well studied. However, a methodology that takes these relationships into account to actualize experts' competencies has not yet been developed. Results of the related works analysis are used to develop a conceptual model of human resources competence actualization. This model can be used to analyze the relationship of the project performing results and competencies of the participants and to suggest competencies changes.

3   Reference Model of Human Resources Competence Actualization

The proposed reference model of human resources competence actualization is shown in Fig. 1. Experts in an expert network are represented in the form of experts. Their profiles contain their preferences and competencies. Expert's preferences

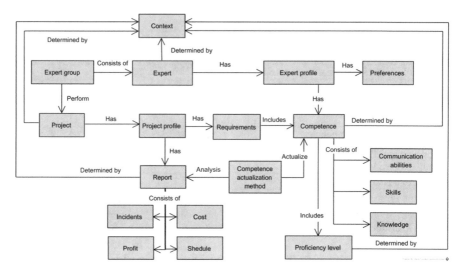

Fig. 1 Reference model of human resources competence actualization

represent individual specifications and can be taken into account while assigning the project. Competence represents expert qualifications and involves the possession of a certain professional skill at a certain level.

Expert groups which consist of one or several experts perform projects by completing different tasks. The project's requirements and experts' competencies are contained in their profiles and define the project participants. Expert group formation for task performing is described in more detail in paper [8].

Project performing results are represented in reports which can contain different determinants such as project's cost and profit, incidents during performing and whether the project was completed on time. These results can be analyzed to identify how the competencies indicated in the participants' profiles correspond to reality. Thus, successful and unsuccessful results of project performing will be used to actualize the participants' competencies. A more detailed description of the actualization procedure is described below.

Additional information about experts, projects, results, competencies and proficiency levels is represented as context. It can be related to a specific time or place, give an informal description of entities. The context is used by managers to better understand the tasks included in project profiles, and the capabilities of potential performers of these tasks. It is also used to evaluate a project and its results under various circumstances.

4 Human Resources Competence Actualization

The description of the competence actualization algorithm is shown in Fig. 2.

The algorithm starts when there are one or several projects whose results have not been analyzed. For each of such projects, two stages are involved. The first stage

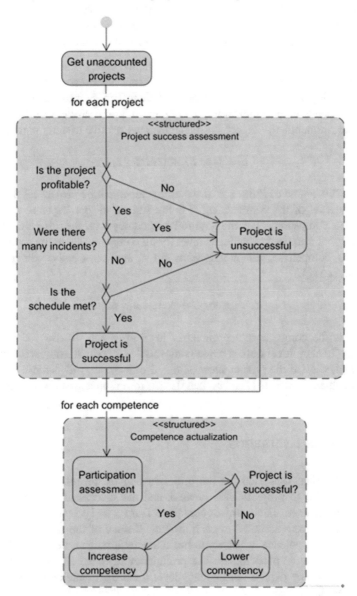

Fig. 2 Competence actualization algorithm

is project success assessment. At this stage, the project results are analyzed to find out if the project is successful. To do this, several determining factors are compared with acceptable values defined by the project manager.

Thus, a project is successful if the following conditions are met:

- the difference between profit and costs is greater than the acceptable value;
- the number of incidents does not exceed the acceptable value; in this paper 'incidents' means unpleasant event that required additional work;
- The project was completed on time or late within the acceptable value.

The second stage is experts' competence actualization. It starts for each competence of the project participants after the project success assessment.

First, the degree of influence of this competence on the project result is determined. It depends on two indicators, see (1): the required proficiency level relative to other requirements, the expert's proficiency level relative to other participants.

$$d = ((p/P) + (r/R))/2, \tag{1}$$

where d is the degree of influence of the competence on the project result; p is the proficiency level of the competence; P is the sum of all participants' proficiency levels in this competence; r is the required proficiency level for this competence; R is the sum of all required proficiency level for all competencies in the project.

Expert's competence change depends on d and on whether the project is successful, see (2).

$$C' = \min\{C + M * d * s, M\} \tag{2}$$

where C' is the competence's proficiency level after actualization; C is competence's proficiency level before actualization; M is maximal competence's proficiency level; s is 1 if the project successful, -1 otherwise. If C' is more than 0 and less than 1, then $C' = 1$. If C' is less than 0, then the competence is removed.

5   Algorithm Evaluation

The synthetic data was used for the initial testing of the model and algorithm. For this purpose, 15 projects were created, and the success of each of them was randomly determined. Each project contained from 5 to 10 requirements. From 2 to 7 performers were assigned to each project. If none of the project participants possessed the competence necessary to fulfill any requirement, the competence was added to one or more participants. The proficiency level for all competencies was set from 1 to 6. The required proficiency level for the generated requirements was set from 1 to 3.

Thus, 6 "successful" and 9 "unsuccessful" projects were created. The average number of requirements was 7.53. The average number of project participants was

5.4. The average proficiency level among competencies was 4.4; average proficiency level among the requirements was 1.94.

After generating the projects, the developed algorithm was applied to each of them. Participants' competencies changes were analyzed. The average competencies were changed to 1.08. In order to analyze the degree of task parameters' influence on competencies changing, the average value was calculated for different amounts of project requirements and participants. These data are presented in Figs. 3 and 4.

The horizontal axis in Fig. 3 shows the number of requirements. The horizontal axis in Fig. 4 shows the number of participants. In both charts, the vertical axis shows the average competence change. As the charts show, the competence changes do not depend on the number of project requirements and are inversely proportional to the number of participants. This demonstrates the distribution of

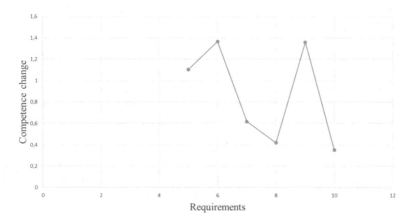

Fig. 3 Competence change for projects with a different number of requirements

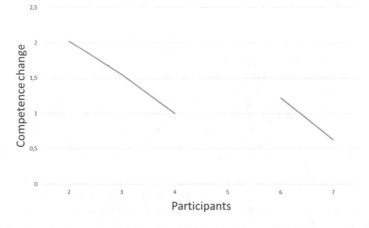

Fig. 4 Competence change for projects with a different number of participants

responsibility among the participants. However, this may be due to the difference in the average proficiency level among competencies and requirements. Further testing and research should clarify this issue.

6 Conclusion

The approach to human resources competence actualization in expert networks presented in this paper allows to keep experts' competencies information relevant by analysis of project implementation results. The reference model used in the approach contains the concepts and relations needed for human resources management and competencies actualization. It also allows to easily add new determining factors for project implementation results, if necessary. The algorithm used in the approach is able to analyze each project in expert network. It takes into account all the factors and participants' competencies.

Further work involves the implementation of the proposed approach, and its testing on real and generated data. At the same time, adjustments and additions to the model and the algorithm are possible. In addition, the degree of project success should be taken into account for a more precise competence change. Project success may vary depending on various factors. Therefore, instead of a binary indicator (successful or unsuccessful project), a coefficient that takes these indicators into account should be used.

Acknowledgment The presented results are part of the research carried out within the project funded by grant #19-37-90094 of the Russian Foundation for Basic Research. I. Algorithm evaluation has been partly supported by Russian State Research # 0073-2019-0005.

References

1. W. Zaouga, L.B.A. Rabai, W.R. Alalyani, Towards an ontology based-approach for human resource management. Procedia Comput. Sci. **151**, 417–424 (2019)
2. M. Bohlouli, N. Mittas, G. Kakarontzas, T. Theodosiou, L. Angelis, M. Fathi, Competence assessment as an expert system for human resource management: A mathematical approach. Expert Syst. Appl. **70**, 83–102 (2017)
3. A. Smirnov, A. Kashevnik, S. Balandin, O. Baraniuc, V. Parfenov, Competency management system for Technopark residents: Smart space-based approach, in *Internet of Things, Smart Spaces, and Next Generation Networks and Systems*, vol. 9870, 2016, pp. 15–24.
4. V. Stepanenko, A. Kashevnik, Competence management systems in organisations: A literature review, in *Proceedings of the 20th Conference of Open Innovations Association FRUCT*, vol. 2017, pp. 427–433.
5. A. Smirnov, A. Kashevnik, M. Petrov, N. Shilov, T. Schäfer, T. Jung, D. Barsch-Harjau, G. Peter, Competence-based language expert network for translation business process management, in *Proceedings of the 25th Conference of Open Innovations Association FRUCT*, vol. 2019, pp. 279–284.

6. В. Степаненко, А. Кашевник, А. Гуртов, Контекстно-ориентированное управление компетенциями в экспертных сетях. Труды СПИИРАН **4**(59), 164–191 (2018)
7. M. Petrov, A. Kashevnik, V. Stepanenko, Competence-based method of human community forming in expert network for joint task solving. *Digital Transformation and Global Society* **858**, 24–38 (2018)
8. M. Petrov, A. Kashevnik, Expert group formation for task performing: Competence-based method and implementation, in *Proceedings of the 23rd Conference of Open Innovations Association FRUCT*, vol. 2018, pp. 315–320
9. M.N. Omar, A.R. Fayek, Modeling and evaluating construction project competencies and their relationship to project performance. Autom Constr **69**, 115–130 (2016)
10. F.P. da Silva, H.M. JerΓinimo, P.R. Vieira, Leadership competencies revisited: A causal configuration analysis of success in the requirements phase of information systems projects. J Bus Res **101**, 688–696 (2019)
11. S. Loufrani-Fedida, L. Saglietto, Mechanisms for managing competencies in project-based organizations: An integrative multilevel analysis. Long Range Plan **49**, 72–89 (2016)
12. A. Barao, J. Vasconcelos, A. Rocha, R. Pereira, A knowledge management approach to capture organizational learning networks. Int J Inf Manag **37**, 735–740 (2017)
13. M. Bruhanov, S. Polyachenko, Determinants of cognitive skills and competencies: Preliminary statistical analysis of PIAAC data. Educ Stud **1**, 214–233 (2015)
14. A. Altunin, Approaches to study competences of employees of scientific medical organizations: A brief literature review. *Social Aspects of Population Health***5**, 18 (2014)
15. P. Klarner, M. Sarstedt, M. Hoeck, C.M. Ringle, Disentangling the effects of team competences, team adaptability, and client communication on the performance of management consulting teams. Long Range Plan. **46**, 258–286 (2013)

Smart Health Emotions Tracking System

Geetika Koneru and Suhair Amer

1 Introductions

Many diseases such as mental health disorders, bipolar disorder, and borderline personality disorder can be linked to variable mood [1] and the dysregulation of the autonomic nervous system [2, 3].

Mood changes and stress can also lead to disease and obesity. Advances in mobile computing allow us to deliver better solutions to those issues. Mobile health interventions in the area of nutrition have been a great asset in changing the lives of many. They can used be as nutrition tracking tools that help improve people's lives and well-being [4]. It has also been noted that the awareness of diet/nutrition tracking methods has gained vast ground in the recent years [5].

There have been several attempts to implement emotional intelligence in Human–Computer Interaction (HCI). Emotion classification is one of the core processes to do that [6].

More users are using smartphone applications for diet/nutrition management and searching for health-related information. More research is still required in the area of user interaction that will help identify features that can promote sustainable lifestyle change [7]. Some researchers state that it is important to identify the potential of HCI when defining the core objective of mobile nutrition tracking. It is important to note that there is growing interest in identifying practical models and theories of how HCI can be mapped onto a mobile nutrition landscape [8].

Advances in IOT are starting to play a major role in the field of health care. It empowers people to connect their health and wealth in a smart way using

G. Koneru · S. Amer (✉)

Department of Computer Science, Southeast Missouri State University, Cape Girardeau, MO, USA

e-mail: samer@semo.edu

© Springer Nature Switzerland AG 2021

R. Stahlbock et al. (eds.), *Advances in Data Science and Information Engineering*,
Transactions on Computational Science and Computational Intelligence,
https://doi.org/10.1007/978-3-030-71704-9_55

technology, apps, smart devices, and wearable gadgets. Smart health is an important application that allows patients with abnormal health conditions to be monitored and provided rapid solution [9, 10].

This paper identifies the importance of technology, IOT, expanding network capabilities, and what roles they can play for monitoring and improving health-related concerns. Coupled with HCI, systems are becoming more attractive and easier to use for people. Here, we are going to explore the development of a system that will track students' emotions regarding their school activities.

2　Requirements

This paper discusses the development of a Smart Health Emotions tracking system. Three phases were completed: (1) establishing requirements and designing a simple interactive system, (2) implementing a simple interactive system, (3) conducting data analysis and evaluation of the simple interactive system.

The system aims at maintaining user records such as school scheduled activities and user's emotions over a period and allow the user to find out how such activities correlated with their emotions at different times.

The user must register as a member and the system should continue to motivate the student to keep updating his/her data. The student developing the system needs to come up with a simple joyful design and graphical user interface that is easy to navigate and use for a system that will collect/display information, such as adding an activity, view list of scheduled active activities, view history. If the user chooses to add an activity, the system will collect information, such as activity type (assignment, quiz, exam, homework, project, club meeting, discussion, paper, tutoring, addNewActivityType), Course name /number/section, due date and due time, grade worth (20% , 30% , service, N/A), Today's date and time (automatic), and Initial emotions regarding completing this activity. If the user chooses to view the list of scheduled active activities, the system will list the following information (in a format easy to read): For each activity, course name /number/section, due date and due time, grade worth, and then asks for emotions and automatically generates date and time of when this emotion was recorded. The system will ask the user if he/she wants to archive activity. If so, it will no longer be listed as active. If the user chooses to view history, the system will list (in a format easy to read) for each activity the following: Course name /number/section, due date and due time, grade worth, and all the dates /times when the emotions were recoded for this activity (preferably in a tabular format).

3 Establish Requirements and Resign a Simple Interactive System

3.1 Usability Goals with Respect to Design Goals

The design goals of the system include the following:

- The usability of the system.
- The user interface of the system should be easy to learn and use.
- Provide interactive webpages that are easy to navigate.
- Record and retrieve activities and emotions efficiently and in a timely manner.
- The content on the website should be easily readable and appealing to the user.
- The website functionality should be error-free and should follow esthetic and minimalistic design principles.

  The user experience goals are the following:

- Since the smart health system tracks the emotions, the observations printed on the history page should be easy to learn and should be appealing to the user.
- User should feel at ease when navigating through the site.
- It should educate the user.
- Allow user to give feedback on an activity he/she completed.
- The website must be colorful and attractive to encourage the student to use it.

3.2 Developer Concerned Questions

To achieve the usability and user experience goals, next are the developer concerned questions and answers:

- Will the system be able to track the emotions of the student in time ?

  Yes, the system will track the time at which the activity is added to the systems and the time at which the emotion is captured. The student will be able to note the emotions on a daily basis and the history of the user lists the emotions along with their recorded times.

- Will the system be able to archive the history of already completed activities?

  Yes, the system will be able to archive the completed activities in terms of both student perspective and in terms of admin perspective. The closed events and completed activities can be archived by cloning the details with respect to the system time.

- Can the images of emotions be captured?

This is an extensive task and is possible if we choose to setup and utilize Microsoft Azure Face API and run the face detection and analysis for emotion analysis.

- Does the system maintain a log of events?

Yes, by logging the times the user logged into the system.

3.3 Use Case Diagram for the Smart Health Tracking System

Figures 1, 2, and 3 describe the use of case diagrams for the landing, admin, and user pages.

3.4 Conceptual Model for Smart Health Tracking System

The system is intended to serve the students and allow them to view the events and register their activities and keep track of their emotions regarding such events. Once the student provides his emotion, the system logs the registration and the time it

Landing page:

Requirement #: 1	Requirement type #: 2	Event/BUC/PUC#4
1		

Description: Have an appealing landing page with scale of emotion
Rationale: To have an aesthetic landing page that attract the student users to view and register for activities while logging their emotions.
Originator: Developer, School authorities
Fit Criterion: The system should be welcoming
Customer satisfaction: 5   **Customer dissatisfaction:** 1
Priority: 1   **Conflicts:** None
Supporting Material: User feeling survey

Fig. 1 User case diagram of landing page

Admin page:

Requirement #: 2	Requirement type #: 3	Event/BUC/PUC#3
1		

Description: Webpage admin should be able to insert the activities on to the system along with its details. Once the activity is added, the time at which the record is entered into the system should be logged. It should also add initial default emotion for the activity.
Rationale: To add the new activities into the system.
Originator: Developer, Activities authority administration
Fit Criterion: The system should accurately log the activities along with the time of activity log.
Customer satisfaction: 5   **Customer dissatisfaction:** 1
Priority: 1   **Conflicts:** None
Supporting Material: Activities authority administration

Fig. 2 User case diagram of admin page

<u>User page:</u>

Requirement #: 3	Requirement type #: 2	Event/BUC/PUC#2
1		

Description: The user i.e. student here should be able to view the available activities and register him/her into one while recording his emotion. At the same time, when his emotion with respect to the activity changes, he should be able to change the emotion. The emotion recording date and time should be logged into the system and archive when needed.

Rationale: To help students register for activities and log the emotions

Originator: Developer, User and Emotions analyst.

Fit Criterion: The system should let users register, log emotions, view history and archive.

Customer satisfaction: 5	**Customer dissatisfaction:** 1
Priority: 1	**Conflicts:** None

Supporting Material: User requirements survey

Fig. 3 User case diagram of user page

occurred. User can record his/her emotion regarding same event on later days and times. The student can also view the history of past events and can archive them if he/she wishes so. Initially the admin sets up the system with several types of activities allowing the student to choose from. The admin also sets up the emotions scale.

In summary, the final product requirements are the following:

- Have a landing page with a scale of emotions.
- Admin should be able to add activities.
- Have a user page where student can login and view available events, choose event by reading its details and benefits, log his/her emotion on registering to the activity, be able to change emotions over time, be able to retrieve history, and archive if needed.
- Have an announcement board for any changes in the activities.
- Have a continuous feedback page.
- Have a leader board on main screen to encourage students and keep them motivated.
- Have an inbuilt emotional analysis page.

3.5 Mental Model from People

From a user perspective, the mental model for the system lists the following as primary concerns for the user:

- Students need a leaderboard that keeps them motivated. A scrolling leader board showing the highest achievers on the web landing page encourages the students to register for activities and keep them motivated.
- Continuous feedback lets the students view their activities and their scales changing, which can allow them to improve.
- An announcement board can help them stay up to date.
- Inbuilt emotional analysis page where students can track emotions over time.

3.6   Interface Design Issues

Some of the interface design issues are the following:

- The user's memory load should be controlled to allow faster and efficient retrieval of data.
- Interface consistency should be maintained and the listed titles and content on the application interface should be accurate.
- The learnability of the application should be given priority.
- The interface should be ordered and should be menu-driven.
- It should be adaptable to all kinds of users, namely, novice, specialized, and sophisticated
- users.
- The graphs and images must be consistent and clear.
- Visibility and color schemas adopted must be consistent and the display should be organized.

4   Implement a Simple Interactive System

The basic idea of the system is to maintain the scheduled student activities along with logging their emotions over time to understand how they reacted on different occasions using a scale of 0 to 10 using emotions scale [11].

There are two types of users using the system, namely, admin and the user. They have three major functionalities.

The admin will be able to add new activities. He/she will be asked to select the activity type such as Assignment, Quiz, Exam, Homework, Project, Club Meeting, Discussion, Paper, Tutoring, and then select Bonus or Retake. Then he/she will be asked to enter Course name/number/section statically, mark grade worth. On a scale of emotions, he/she assigns initial rating for the activity based on the feedback received from previous offerings. Date and time will be automatically read from the system and all the information will be submitted for storage and retrieval.

The admin will be able to view the list of scheduled activities. A table of activities will be displayed along with initial emotion rating, due date and time, and event creation date and time. The admin will be able to view history and can select the type of activity to delete. Once confirmed, the selected activity types can be purged from the table. Finally, the admin can click on Emotional Log (emotional analyzer page) and can view the current state of emotions that the users changed on the activities.

The users can view statistics using stats and view the emotional log of activities. They can also login to register for activities. In this page, the user can select the activity type as in Assignment, Quiz, Exam, Homework, Project, Club Meeting, Discussion, Paper, Tutoring, and clicks on search to view the activities. They then enter the name of the course that they want to register for and change the emotion.

They then enter the emotional scale value and submit it to update the emotional rating. They can view the existing schedule of activities and view the history and purge if needed. This is usually done when the activity deadline is passed. The schedule and history are viewed in the tabular format.

Emotional analyzer page: It is used to view the emotions statistics that the user reports for different activities. This helps in understanding the user's views on the activities in the form of graphical bar charts. These charts can help analyze the stress levels that the students face to prevent them from major chronic diseases like cardiovascular, metabolic, and kidney diseases and cancer, which are the major causes of increased stress.

To implement this system, the following programming languages and software were used:

- Front end Programming: HTLM, CSS, php
- Database: MySQL
- Server: Apache
- Software: Xampp, which is a cross platform open source software for web server solution consisting of Apache HTTP server for interpretation of php.

Figures 4, 5, 6, 7, 8 and 9 are examples of interface screenshots and test case scenarios.

Upon navigating to ADMIN AND STUDENT LOGIN, the page shown in Fig. 10 appears.

After entering the id and password, the user will click on "Admin Login" button that will open the main activity admin page shown in Fig. 11 allowing him/her to add an activity, view schedule details, and view history.

If admin wants to add an activity, he/she will click on the "add data" button. The site will navigate to the page of adding an activity. The user can choose the activity from a drop-down menu like in Fig. 12. He/she can enter course name/number, select date and time from calendar like in Fig. 13. He/she can enter grade worth like

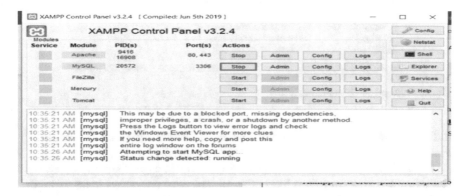

Fig. 4 Xampp servers running status

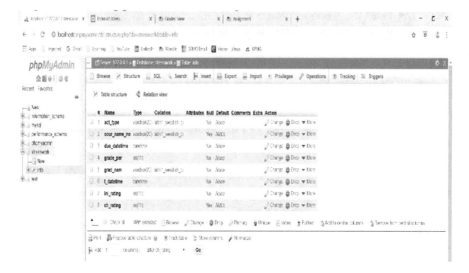

Fig. 5 Table structure in phpMyAdmin Table Name: stresswork

Index of /stress

Name	Last modified	Size	Description
Parent Directory		-	
add.html	2019-06-18 21:49	2.9K	
combo.html	2019-06-18 19:25	4.4K	
dbcon.php	2019-06-18 12:27	1.4K	
delete.php	2019-06-18 12:56	2.3K	
graph.php	2019-06-18 12:58	3.3K	
history.php	2019-06-18 11:32	2.6K	
login.html	2019-06-19 08:29	2.3K	
login.php	2019-06-18 12:55	120	
menu.html	2019-06-18 22:11	1.3K	
reg.html	2019-06-18 22:16	1.9K	
reg.php	2019-06-18 11:26	2.5K	
schedule.php	2019-06-18 13:00	2.2K	
sub1.php	2019-06-19 08:37	1.3K	
three.html	2019-06-18 22:26	1.2K	
u_menu.html	2019-06-18 22:19	1.2K	
user.html	2019-06-18 22:22	2.2K	
user.php	2019-06-18 08:10	122	

Apache/2.4.39 (Win64) OpenSSL/1.0.2s PHP/7.1.30 Server at localhost Port 80

Fig. 6 Html and php files in local host

in Fig. 14. The user can set emotions rating and submits the data like in Fig. 15. Finally, date and time of creation is logged automatically like in Fig. 16.

When clicking on the schedule option from the main page of admin, the schedule is displayed in a tabular format like in Fig. 18. The user can go back to the main page using the "Back to Home" button.

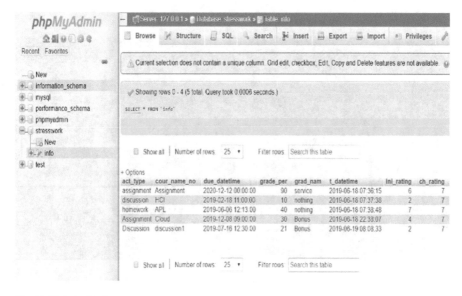

Fig. 7 Data in database

Fig. 8 Landing page

Upon submission of data, a success page will be displayed allowing the user to go back to main menu like in Fig. 17.

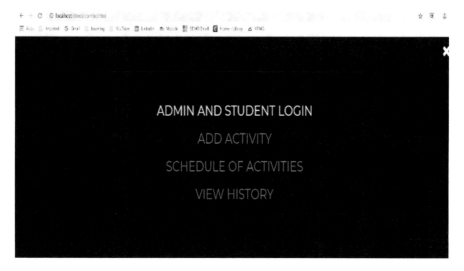

Fig. 9 Main Menu

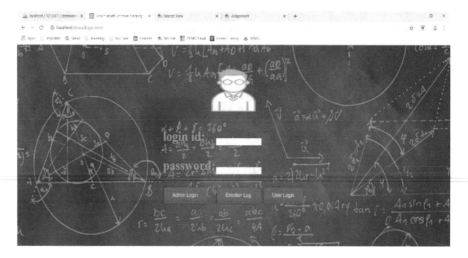

Fig. 10 Login page

Clicking on the "History" button from the main admin page will display the information in a tabular format. The user has the option to delete an activity by first selecting the type of activity that he/she wishes to delete and clicks on delete like Fig. 19. Upon the success of deletion, a page is displayed, which allows the user to go back to main page like Fig. 20, which completes the administration part.

In the main menu, if the admin chooses to view the emotions analyzer chart, he/she clicks on the Emotion Log, which will display a chart like Fig. 21, based on the student ratings.

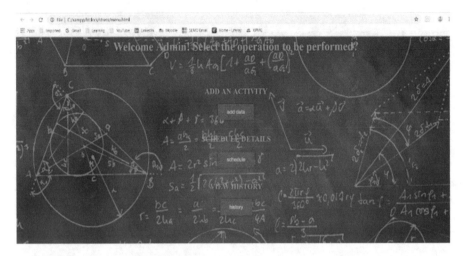

Fig. 11 Main activity admin page

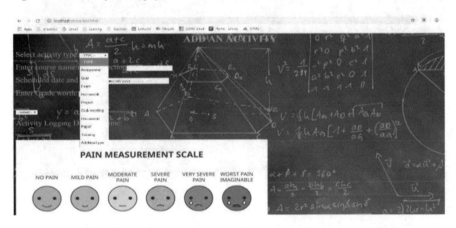

Fig. 12 Choose activity

Like the admin, a user will click the student / User Login button. Student will enter id and password. This will take them to the welcome screen that will ask the user to register for an activity, rate an assignment, view the schedule of activities, and view and delete history. For registration, the user chooses the activity type and clicks search. This displays a table with the selected activity type and then he/she enters the course name for which he/she wishes to change the rating and register. On the emotional scale, the student then sets the value and clicks submit. The student can also view the emotional log page to choose and view stress levels on an activity. The student can also view history and schedule. Examples of such pages can be found in Figs. 22, 23, 24, 25, 26, 27 and 28.

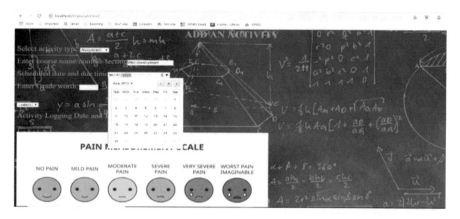

Fig. 13 Enter course name and select date and time of activity

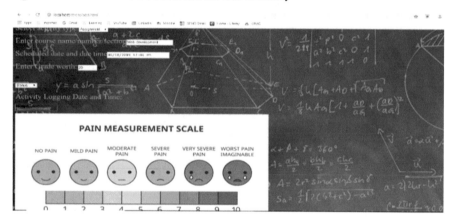

Fig. 14 Enter grade worth

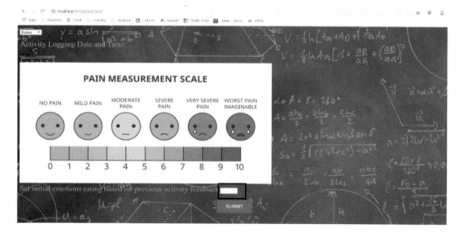

Fig. 15 Initial emotion setting by administrator based on the feedback

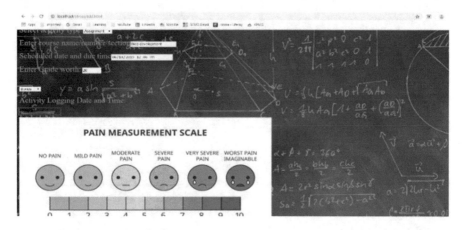

Fig. 16 Dynamic date and time of activity addition

Fig. 17 Success screen

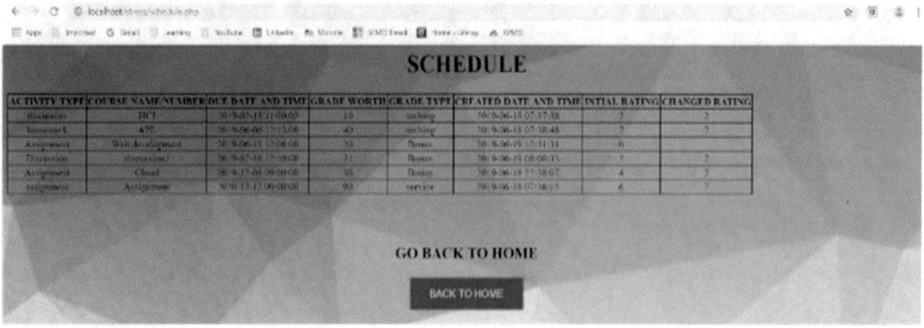

Fig. 18 Displaying schedule

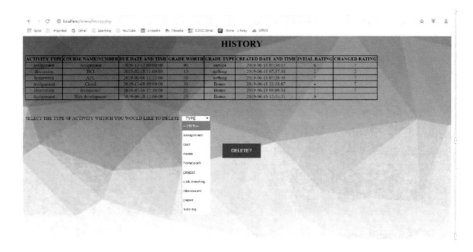

Fig. 19 Display history and be able to delete an activity

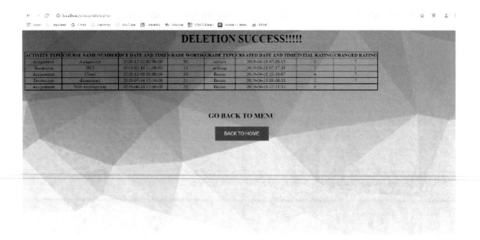

Fig. 20 Deletion success page

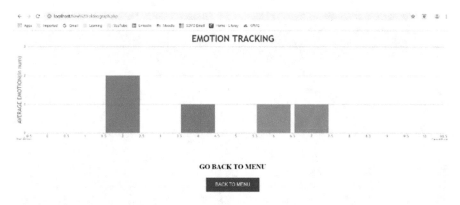

Fig. 21 Analyzed emotions of student

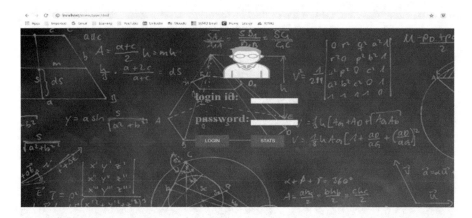

Fig. 22 Student login page

Fig. 23 Student main menu

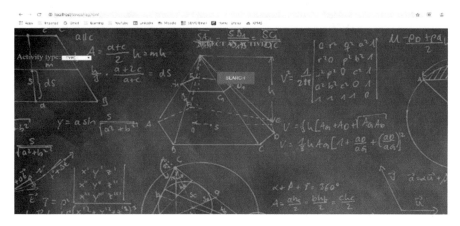

Fig. 24 Selecting /searching for an activity page

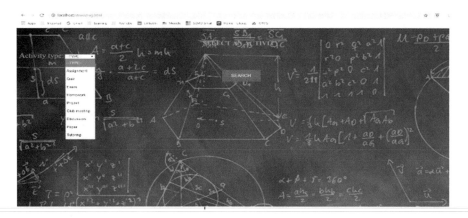

Fig. 25 Searching for an activity using the drop-down menu

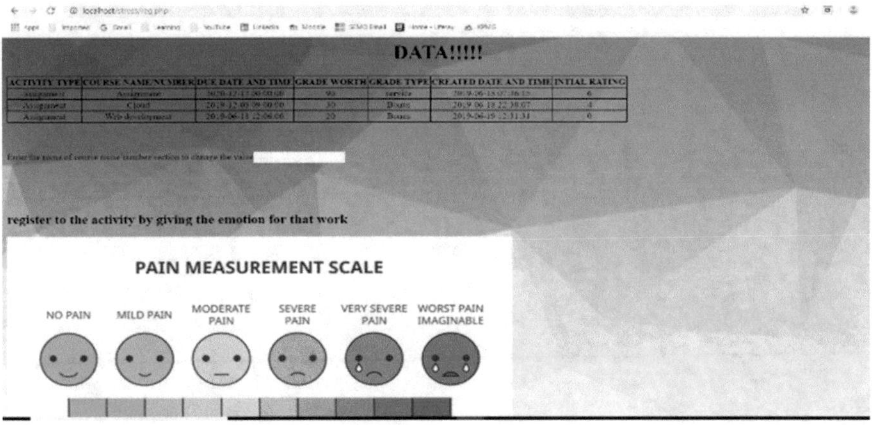

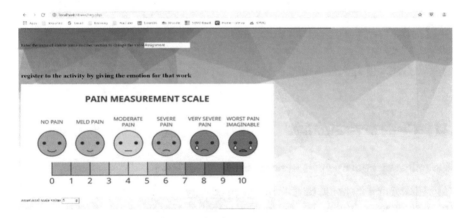

Fig. 26 Displaying stored activity type, course name, due date, grade type, and initial rating

Fig. 27 Enter new emotion to a chosen activity

Several error message windows / pages are displayed when data is missing, or incorrect information is entered. For example, an error page is displayed when null values are passed as the login credentials. For empty data and time, a pop-up appears with the error message. When course name or grade worth are kept nulls, an insertion failure screen appears.

Fig. 28 Update success of emotions page and ability to return to main page

5 Data Analysis and Evaluation

This section will elaborate on the analysis and evaluation of the system.

5.1 Goals to Be Attained

The following primary goals were achieved:

(a) Administrator should be able to add an activity
(b) Administrator should be able to view activities existing in the system
(c) Administrator should be able to set the initial emotion.
(d) User should be able to choose an activity changing emotional rating.
(e) Record the initial date and time and every time an emotion change is made.
(f) History of the scheduled activities along with the archive option should be provided.
(g) Emotional analyzer page helps in providing plot with the recorded ratings.
(h) Appropriate authentication for admin and user through user and admin login pages.

5.2 Evaluation Method

The evaluation method used was in a control setting where the users' activities are controlled in order to test the initial measures and hypothesis. In a controlled setting, the methodology used was usability testing and experiments.

The data was collected using questionnaires and interviews conducted for 10 subjects in a natural setting. The primary goal of this evaluation is to test the usability of the system where the users carry out some tasks. While conducting the evaluation, the number and kinds of errors were recorded. In general, it was not an easy task observing users and collecting data about the systems usability.

5.3 Practical Issues Identified

Some of the practical issues identified after conducting the evaluation:

- For this prototype, static users and admin profiles were used. The system needs a dynamic user and admin registration.
- Instead of asking the user to select the activity type and then add the course name and then giving the rating, a drop-down menu associated with each activity with rating would be better.
- Dynamic delete option can be added to each column of history page to make the delete preference of the admin and the user dynamic.
- Redirection and login user profile can be made automatic.

5.4 Ethical Issues

One of the ethical issues is that the data stored must be secured and should ensure that the user consent is attained before sharing any user-related data.

The data collected during the evaluation process should be secured and confidential data should be protected and not shared.

The developers were ethically required to provide a quality solution to the users and follow standard protocols and frameworks in their design.

5.5 Collection of Data

10 Subjects were interviewed and observed. The data has been collected from 10 subjects that were an undergraduate college student, 8th grader, 4th grader, 6th grader, freshman, a parent, an administrator, a manager, a master's student, and a 3rd grader. They were asked to complete a questionnaire consisting of nine questions. Figure 29 shows an example of the questionnaire where the subject will give his overall feedback as a comment and a satisfaction score for each question.

Questions on Application Functionality	Overall	Satisfaction (Scale 10)
1) How did you feel about application aesthetics?	Satisfactory	5
2) How was the navigation?	Good and Easy	9
3) Authentication working	Can be improved	4
4) Activity List functionality	Good	8
5) Emotional Analyzer page	Good and nice	9
6) Delete operations	Works fine	8
7) Add activity	Working	8
8) Emotion logging	Nice and useful	9
9) Grades allocation is genuine	Yes	10

Fig. 29 Sample evaluation sheet of one subject

5.6 Evaluation of Data:

The questionnaire was conducted over a 10-point scale and it provided high-level insights of the subjects, i.e., users of different levels. The factors evaluated the three dimensions of usability of the smart health emotion tracking system, namely, usefulness, satisfaction, and ease of use.

Next is a summary of their comments:

(a) Usefulness:

- Many users indicated that the system is useful.
- Many users indicated that it met the students' needs.
- Many users indicated that the application functionality is fine.
- Many users indicated that it was useful to view the schedule table with the list of activities.

(b) Satisfaction:

- Many of the users were not satisfied with the esthetics of the application and have said that it can be improved more.
- Most of the users were not satisfied with the authentication functionality and have reported that it needs improvement.
- Some of the users felt it is a good system.
- Many users have reported that the emotional analyzer was effective.
- Most users felt that they would recommend using the system.

(c) Ease of use:

- The students mentioned that the application is easy to use.
- They were able to navigate the system easily.
- All users said that they can use it without written directions.
- They felt the application is user friendly and effortless to use.

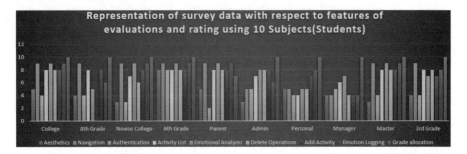

Fig. 30 Representation of survived data with respect to the features evaluated and rated. It shows the rating of 10 subjects

In addition to completing a questionnaire, the subjects were observed in a controlled setting while testing the system and then interviewed. They indicated that the Smart Health Emotions Tracking system needs better authentication. Some indicated that although the application was designed keeping the student perspective in mind, the logging system can be designed much more effectively. One mentioned that one registration page can be enough, allowing the user to set up his/her account (as admin) and still use the system (as user). No need for two separate parts. This can help also with security and maintaining privacy. Some mentioned that a better graphical user interface would make the system more appealing. One mentioned that the report needs to be more vigilant. All users indicated that navigation and learnability of the application were satisfactory.

From the data collected, and as shown in Fig. 30, the attributes of the application are ordered from highly ranked to poorly ranked features as follows:

- Navigation
- Grade allocation
- Activity list functionality
- Emotional analyzer page
- Delete operations
- Add activity functionality
- Emotional logging
- Esthetics
- Authentication working

6   Conclusion

In a Human–Computer Interaction course, the students were asked to develop a Smart Health Emotions tracking system by completing three phases within 2 to 3 weeks. In phase 1, they needed to establish requirements and design a simple interactive system. In phase 2, they needed to implement a simple interactive system. In phase 3, they needed to perform data analysis and evaluation of the simple interactive system. The paper explains the work of one of the students. The students thought that it was good working on the projects. The students indicated that while implementing this system, concentration was on tracking emotions, creating activities, registration, and viewing history. Therefore, less time was spent on authentication and login.

In general, the student had a wonderful learning experience in implementing the system since it helped her gain experience with severs, HTML and php. She developed a simple application that can be developed further into a very useful system that would help students.

References

1. K.C. McDonald, S. Kea, J.R. Geddes, Sleep problems and suicide associated with mood instability in the Adult Psychiatric Morbidity Survey 2007. Aust. New Zeal. J. Psychiatry, 1–7 (2017)
2. I.M. Anderson, P.M. Haddad, J. Scott, Bipolar disorder. BMJ **345**, 8508 (2012)
3. K. Lieb, M.C. Zanarini, C. Schmahl, M.M. Linehan, M. Bohus, Borderline personality disorder. Lancet **364**, 453–461 (2004)
4. S. Scerri, L. Garg, C. Scerri, R. Garg, Human-computer interaction patterns within the mobile nutrition landscape: A review of literature. in *2014 International Conference on Future Internet of Things and Cloud*.
5. J. Kjeldsko, J. Paay, A longitudinal review of Mobile HCI research methods, in *Proceeding of the 15th international conference on Human-computer interaction with mobile devices and services*, ACM, 2012, pp. 69–78.
6. J. Wagner, J. Kim, E. Andre, From physiological signals to emotions: Implementing and comparing selected methods for feature extraction and classification, in *2005 IEEE International Conference on Multimedia and Expo*, Amsterdam, July 2005, pp. 940–943.
7. K.M.J. Azar, L.I. Lesser, B.Y. Laing, J. Stephens, M.S. Aurora, L.E. Burke, L.P. Palaniappan, Mobile applications for weight management: Theory-based content analysis. Am. J. Prev. Med. **45**(5), 583–589 (2013)
8. E.S. Poole, HCI and mobile health interventions. J. Trans. Behav. Med. **3**(4), 402–405 (2013)
9. S. Ananth, P. Sathya, P. Madhan Mohan, Smart Health Monitoring System through IOT, in *International Conference on Communications and Signal Processing* (2019)
10. S. Pradeep Kumar, V. Richard Ranjan Samson, U. Bharath Sai, P. L. S. D. Malleswara Rao, K. Kedar Eswar, From smart health monitoring system of patient through IoT, in *International conference on I-SMAC*, 2017, pp. 551–556.
11. Emotions Scale. Retrieved 4/7/2020 from: https://www.123rf.com/photo_55960284_stock-vector-pain-scale-chart-vertical-cartoon-faces-emotions-scale-doctors-pain-assessment-scale-pain-rating-too.html

Part VII
Video Processing, Imaging Science, and Applications

Content-Based Image Retrieval Using Deep Learning

Tristan Jordan and Heba Elgazzar

1 Introduction

With the exponential increase of information in the modern age of the internet, the need to manipulate these vast data banks grows in parallel. To solve this issue, strategies should be employed which are scalable towards conserving a balance of resources and accuracy. Technology aiming to solve these issues goes by the name content-based image retrieval (CBIR). The content mentioned in CBI could mean any feature which would aid in returning similar images, such features include shape, color, or texture [1]. Many uses can be made of CBIR, which takes advantage of flexibility of data as they use images as opposed to text.

The phrase "a picture's worth a thousand words" may be an exaggeration; however, it can be said that the features which best describe the image may not be easily transferable into words. For instance, if the task was to store images of homes and allows users to query for types of land it may be difficult to put into words the exact type of property that the user would be searching for. By using the features of the image, a user's comment could be of a house that they had seen previously, then the search algorithm would do its best to retrieve matching examples. The elimination of the need to tag will free resources such as human or computing time, and entropy generated from the information within the images itself being lost to generic tags.

With the explosion of the internet and the devices it connects, there are many sources which are creating data at rates that outpace a human's ability to support. Thus, it becomes important to create methods that perform these tasks without the need of human intervention. The primary issues that must be solved pertaining to

T. Jordan · H. Elgazzar (✉)
School of Engineering and Computer Science, Morehead State University, Morehead, KY, USA
e-mail: Tajordan2@moreheadstate.edu; h.elgazzar@moreheadstate.edu

© Springer Nature Switzerland AG 2021   771
R. Stahlbock et al. (eds.), *Advances in Data Science and Information Engineering*,
Transactions on Computational Science and Computational Intelligence,
https://doi.org/10.1007/978-3-030-71704-9_56

CBIR include image representation, image organization, and similarity measurement [2]. These issues will affect the speed that the algorithm runs, accuracy, and/or resources. For example, an image with a resolution of 1920 by 1080 pixels with each pixel holding three separate colors, said image would have 6,220,800 separate color values. This problem amplifies as every stored item would have a similar resolution, giving monumental computational tasks given a poorly optimized comparison. To solve the problem, the object is sent through data reduction which aims to give a best representation with the smallest amount of data. After proper representation of the features, one must store them in such a way as to allow for the goals of the retrieved image to be performed with the least additional steps. Finally, a similarity must be chosen which fairly compares the features, while maintaining a low computation cost with the result being the retrieval of a similar image.

Deep learning is the use of many nonlinear transformations, coming together to model abstract ideas [3]. The name deep learning is derived from the action of adding more layers to the middle of a traditional (shallow) neural network to provide additional transformations to the flowing data. A neural network, often called an artificial neural network, is an algorithm that attempts to model biological brains in a connected series or neurons or nodes. Each neuron is connected to every neuron before and after it in an abstract structure of layers. The data comes into a node, is processed with a weight and an activation function, and then is funneled into the next array of neurons, ending in the output layer. Adding additional layers to the middle of the network provides additional parameters for the data to go through which increase the computational cost with the hopes of better accuracy.

In this paper the discussion is held over utilizing deep learning techniques to classify images, then use a layer of this classification as the features within the images. The process involves taking a predefined number of classes and training a deep learning model to classify images. Then taking this model, each image from the database is run through the predictor and the outputs from a layer within is saved as the feature for the input. Finally, the query image will be input into the deep learning model to retrieve its features and is compared to the outputs from the database to find and return the best match. The usefulness of the proposed algorithm comes from the ability to pull images that are "similar" to classes that would be predefined from the user, using a variety of features within the image. This algorithm hopes to retrieve images that are from the same class, preferably sharing many fine details.

This paper is organized as follows: Sect. 2 describes a discussion of research work related to features within the image and how these descriptors influence the query. In Sect. 3 the details for the functions within the algorithm are explained. Setup of the experiment and a discussion of the results are presented in Sect. 4. A conclusion and directions for future research are provided at the end of the paper in Sect. 5.

2 Related Work

Many researches are being done on the areas related to CBIR. Primary research has led focus into efficiently extracting basic features from images which place them into distinct groupings. Some of the most popular extracted features include shape, texture, and color [4]. These image descriptors are then sent through analysis to reduce the dimensionality of the data creating feature vectors which numerically describe the focus. The algorithms which perform the feature extraction are suited for each feature and can be run together to reach expected results.

Color as a feature is extremely useful due to its efficiency and effectiveness [5]. The goal when using color for image comparison is to boil an image down to a count of the colors or various pixel brightness. These color counts are often referred to as color histograms, a structure of data where the continuous number system is transcribed to an easy-to-digest discrete system. It does so by creating a discrete number of buckets, which are groupings of various colors (can be separated in multiple ways). These buckets contain the count of values that fall within the bounds and allows for easy comparison of color within the images. Using color allows a model to ignore orientation of an image; however, it can fall short in various ways due to losing relational information in positions of the pixels. Color features are useful when a larger identification is already found such as retrieving a blue article of clothing versus a red, this algorithm is often added to others for this reason.

Shape is often seen as the initial starting point when doing image comparisons. This school of thought can however often be complex due to corruptions that may come with the shapes due to noise, irregularities, or other mishaps which may affect how a machine would see the image [6]. There are many steps that have been taken to reduce the control of these problems that are built into the algorithms. One of the methods of collecting this feature data includes counting edges in an image, this data can then be related with a degree of error. Shape can be advantageous in identifying objects from an image; however, without preparation falls short when multiple objects are in the query image, structurally similar objects, data size.

Texture unlike many other features is difficult to describe and often leads to subjective labeling from humans. However, when a machine transcribes texture, it is better with a scientific definition to the texture of an object, with an accurate definition comparison can be performed. Regardless of the exact definitions of the textures that is being studied, it can be said that the study of texture is understanding the relationships of the gray levels (or brightness) [7]. These levels of gray can often signify roughness and other textures based upon a reflection of count and position of said values. Texture as an identifier of its own is niche in what you are trying to retrieve as all furred animals would have a similar texture and the texture of similar objects could have random effects that may not be accounted for (cats without fur) , when used in combination with other features it can create highly detailed searches.

CBIR is useful in real-world applications due to the ability to make quick inferences on large datasets. Some applications where CBIR is currently being applied include medical applications, scientific studies, and digital libraries [8].

These applications tend to deal with the ability to produce a sample that needs to be compared to previously seen examples to make informed decisions on cases that have previously been treated. One example of this being used is an analysis of magnetic images of the brain in the diagnosis of disorders [9].

3 Proposed Method

The method's that are being discussed to perform the image retrieval revolve around a deep learning classification model. Deep learning is used to extract the features from an image automatically as opposed to needing a time-consuming tagging process for incoming images [10]. Using a classifier as the base allows for implicit relational structures the possibility to be noticed. The classifier is a neural network which is a traditionally supervised learning model, meaning that the training data requires both an x value and a y value or result. A visualization of this model can be seen in Fig. 1. These methods allow for all the previously discussed features to be considered.

The environment that the project has been designed and ran in is Python 3 with advantages towards scientific programming due to it being designed as a high-level interpreted language. Many external libraries were used in the project to prevent rewriting algorithms inefficiently. The library Keras is the backbone in the operations dealing with the neural network classifier, allowing easier use of the TensorFlow backend [11, 12]. Using Keras allows the model to be perceived as layers accelerating the drafting process, along with making all other aspects of TensorFlow easier to use. Storing and manipulating large volumes of data requires a modifiable and powerful data structure allowing for simplified matrix manipulation for tasks like images, for this task NumPy was chosen [13].

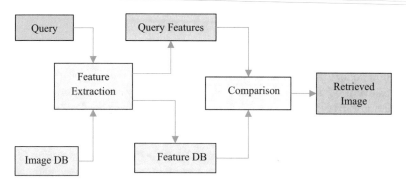

Fig. 1 Diagram illustrating the proposed process of content-based image retrieval

3.1 Preprocessing

Preprocessing is an important task when performing machine learning algorithms. The act of preprocessing is taking your data, both training and testing, to perform any changes that will allow for usability in the algorithm. For the tasks at hand, there are few operations being done to the images. Some operations can include changes to the size, removing backgrounds or denoising.

The training images start as their initial resolution is stored within a folder containing all the images within a class. Each folder or class is given a corresponding class number which will be used as the output value in the neural model. This class number is then encoded into an array that is the same length as the total amount of classes, each object then has a 1 placed in the slot corresponding to their class and a 0 in all other indices. The program then recursively goes through each folder recursively applying preprocessing to the image, then storing the image into one NumPy array and a label in another. Neural networks have fixed inputs, this poses a problem when working with images as they often vary in size and resolution. Ways to solve this problem include adding nulls to fill in space or using traditional algorithms to resize the image. The way that was employed was resizing the image using the built-in Keras function to resize the images to $120\times120\times3$, higher numbers allow for more information to be kept while lower numbers allow for faster computation. Finally, the images expand the x axis by one to match the input shape of the neural network at $1\times120\times120\times3$.

With small image datasets it can be difficult to train image classifiers. Neural networks are often literal in the way that they solve problems; if a model was shown only right side-up images, then it would not have the capacity to understand one that was rotated. To solve this problem, images can be manipulated during the training process. Care must be taken when adding fluff to the dataset as if too much manipulation occurs, the image could be invalid to the original intent. Keras implements a function that does a variety of effects within a range on every epoch, definable by the user. The augments that have been selected include:

- Shear (as shown in Fig. 2)
- Zoom
- Flip (vertical and horizontal)

Fig. 2 A tuning fork and a sheared tuning fork, showing an example of augment [14]

- Rotation
- Brightness

These "new" images make it so that different views of the same object are seen properly.

3.2 Feature Extraction for Image Retrieval

To retrieve images from a database, all images must have the features extracted, then comparison between the feature vectors can take place. For the proposed model, the features are related to the classification of images. Extracting the features involves taking a pretrained machine learning model and extracting statistical data that leads to the output. The benefit to choosing this data is the ability to set the size of the feature vector, choosing the amount of information on this step based on the size of the database. The images that were stored will then be preprocessed and sent through the predictor, saving the feature data to a formatted array. This formatted array is then saved to remove the need to extract the features again. The input to the classification model is a three-color channel image with a 120×120 resolution.

The features that are being extracted in the discussed methods are the relations between the input image and a select group of classes. The classes would ultimately be defined based upon the problem that is being solved. As related to the experiment the ten test classes were used as the output. The features that are pulled are taken from the layer prior to the final classification, this is because there are more nodes on this layer allowing for a more diverse feature set. The width of the data on this layer is definable based upon the expected size of the dataset, growing proportionally to the number of images to be stored (however must be defined prior to training).

Extracted features allow for retrieval to occur. To query an image, the first step would be to extract the features from the query image. These features are then fed into a comparison or distance algorithm which compares the features mined from the database images to the query, storing them in a single dimensional array. Finally, a minimum value is chosen from these distances and data is retrieved from this smallest value (ID, Image, etc.) which can then be used as needed. The comparison algorithm used is as follows:

$$\text{Distance}\left(\vec{X}, \vec{Y}\right) = \sqrt{\sum_{i=0}^{n} (Xi - Yi)^2} \tag{1}$$

Where $\vec{X}$ is the feature vector of the query image, $\vec{Y}$ is the feature vector of the current database image, Xi is the current index within $\vec{X}$, Yi is the current index within $\vec{Y}$, n is the total length of a feature vector.

$$\text{Retrieved Image} = \min\left(\vec{X}\right) \tag{2}$$

Where $\vec{X}$ is an array of the distances between every image.

In some cases, it may be important to know more about a query than a single image. In such event the retrieved image would be popped, and the second smallest distance would be returned. This process is repeatable as many times as needed, allowing for grids of desired information to be easily crafted, stack ranking the importance that is model assigned.

3.3 Deep Neural Networks

The defining characteristic of supervised learning involves using training data with predefined outputs. This form of learning is useful in many applications due to the ability to fit the model to training data. Some of the applications where supervised learning is used include speech recognition, spam detection, and object recognition [15]. Supervised learning is used by the proposed method through a deep neural network (DNN), this model is constructed with the purpose of attempting to classify the data. The training data allows the DNN to understand relationships between the images and their classes.

Classifiers are algorithms that take data and partition it into groups or classes. In the 1990s a revolution in image classification occurred which led to deep neural networks to be used [16]. DNNs allow for complex nonlinear relationships to be worked out by a machine. There have been many successes in using neural networks in image classification including LeNet-5 being able to correctly classify the MNIST (handwritten digits) dataset with a 99.05% accuracy [17]. Table 1 shows the construction of the classifier DNN used in the paper.

Layers The input layer is a relatively simple yet required layer in any neural network. The input layer is the introduction of data into the model and performs no changes on the data. When defining this layer an input size must be defined that all objects must follow.

Convolutional layers are the heart of modern-day image classification. A convolution layer starts with a kernel which is a filtering matrix representing how the network is "seeing" the current layer of data. The kernel scans the matrix of pixels and creates a new array of data; the kernel scan allows for the positional information to be expressed as values are passed through the model. The scan is responsible for understanding the important features of an image that were mentioned before including the shape, texture, and color of the image. The padding in a convolutional scan is related to the ability to expand a data matrix with extra zeros to make the output of the layer keep the same dimensions as the input (same padding) otherwise the output would be the size of the kernel (valid padding). Within every

Table 1 Neural Network
Layers

Layer Number	Layer
1	Input
2	Convolutional 2D
3	Convolutional 2D
4	Max Pooling 2D
5	Convolutional 2D
6	Convolutional 2D
7	Max Pooling 2D
8	Convolutional 2D
9	Convolutional 2D
10	Max Pooling 2D
11	Flatten
12	Dense
13	Dense
14	Output (Dense)

Fig. 3 Example kernel in a
convolutional layer [16]

1	0	1
0	1	0
1	0	1

convolutional layer of the used network the padding value is set to same while the
size of the kernel is 3×3, similar to Fig. 3.

1	0	1
0	1	0
1	0	1

Pooling layers are essential to a fast-convolutional neural network. The image
quality when entering the model begins at $120 \times 120 \times 3$, this means that if left alone,
there would be at least 43,200 neurons in every layer. On a single run this number is
not terribly large; however, these computations build up over many runs (especially
during training). Pooling layers shrink the size of data within the network, with
the goal of preserving important features and speeding up future computation. The
Max Pooling 2D layer operating in the network scans the original and selects the
max value from grouped cells to place in a respective cell in the new smaller
matrix. Within a pooling layer, padding can also be added in the same way as the
convolutional layers, but it is not used in this model as the goal of the pooling layers
is to cut the size of the data. Pooling can also help in denoising data with either of
the pool types. Strides in a pooling layer refer to how far the scanning matrix moves
after the last scan, for example, if you have a stride of 1 it scans every combination
and a stride of 2 skips every other combination and row. The pooling layers that are
being used in this network all use a 2×2 scan with a stride of 2. Fig. 4 shows an

Fig. 4 Example 2×2 max pooling layer with 2 strides, input(left) and output(right) matrices shown

Fig. 5 Visualization of a 4-node dense layer

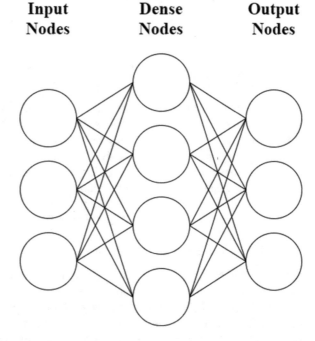

| Input Nodes | Dense Nodes | Output Nodes |

example of a max pooling layer in use, shrinking the cells within the first matrix into a more digestible size.

A flattening layer changes the output from the layer before, in this case a pooling layer, into a single dimension while not affecting the batch size. This information is then passed into an array of fully connected layers. A fully connected layer is one of the earliest designs in a neural network sometimes referred to as a regular layer. Fully connected or dense layers work by pulling the output of every node in the previous layer, summing the data, applying an activation function then sending this output to every neuron in the next layer. A visual representation of a basic form of these layers can be seen in Fig. 5. Learning is performed by training weights that are applied to every output, adjusting their values to better represent the connection of input and output in relation to training data. The number of nodes in these layers are defined by the user with more nodes requiring more memory and time but retaining more information.

Activation Functions Activation functions, also called transfer functions, perform operations on weighted sums to produce an output [18]. These functions can represent linear transformations or nonlinear transformations depending on application. The output of these functions determines the effect of a node on the result of the network, giving the ability to turn off nodes which should not have an effect. The equation which each node in a connected layer follows is denoted as

$$\text{Output} = \text{AF}\left(\sum W_i * X_i\right) \qquad (3)$$

Where AF is the chosen activation function, n is the current node, Wi is the weight at the ith node and Xi is the current output at the ith node.

Rectified Linear Unit (ReLU) The ReLU function is one of the most popular activation functions that sees reliable success through its trials [19]. The primary benefits of the ReLU function is its simplicity, the simplicity of the function makes it quick to run compared to other activation functions. The setup of the ReLU is to pass the contents of the node through if greater than zero, else pass a zero value. The ReLU activation function can be mathematically modeled as the following:

$$\text{ReLU}(x) = \max(0, x) \qquad (4)$$

Where x is the current summed value in a node.

SoftMax The expected output of the classification model is an array with a 1 in the classification. When the classifier takes a guess at what an image would be, it is better to give a list of probabilities of each class, giving the ability to compare the expected output and the prediction. The SoftMax function is responsible for taking the information passed from the previous layer into the class probabilities. The range of available output for the SoftMax function is between 0 and 1 with all possible outputs summed to be 1, the position with the highest value is the target class. The SoftMax equation can be written as:

$$\text{SoftMax}\left(X_j\right) = \frac{e^{X_j}}{\sum_i e^{X_i}} \qquad (5)$$

Where X_j is the current node's sum and X_i is the ith node's sum.

Training A neural network without training is a blank canvas which holds little to no information. The feed forward network that is being discussed in this paper performs this training through comparing the expected output to the prediction with the categorical cross entropy loss function. The network of nodes then self corrects to improve the loss after every epoch (single iteration over all training data). Two hundred fifty epochs are performed over the training data, with every epoch performing the earlier mentioned image tweaks. To ease memory use on the network and further efficiency, an implementation of the Adam optimizer was added [20]. The details of the model are then saved into respective h5 files for quick access during testing and other runs.

4 Experimental Results

4.1 Dataset

The dataset used for this project is a modified version of the Caltech-256 dataset [14]. The original dataset contains 256 categories with a grand total of 30,607 images. The images contain either a uniform or noisy background. The average image size is 351 pixels with an aspect ratio of 1.17; however, the images are not uniform in size and vary greatly. The dataset was compiled by requesting images from a community of dataset users that was then sorted to ensure that all images follow a quality standard.

The size of the dataset was reduced greatly during testing to allow for training speeds to be reasonable. This was completed through hand picking ten categories. The categories were chosen based on trying to find a balance between conflicting classes and distinct classes. These classes also aimed to have a similar number of images contained within, this is to try to remove the bias of a large image set within our classifier.

4.2 Discussion of Results

Three images from each class were pulled to test the image retrieval algorithm, being correctly classified if the first image retrieved was within the same class. Three was the chosen value to keep all categories with equal amount tested and to hold a larger training set. These accuracies are displayed in Table 2, the percentage meaning correct guesses out of total guesses for each category. Poor performance of some categories can reflect issues in category size and variance.

Within the database of images there was close to an average of 90 images per group. With the number of images being this low, the chance that the query image would have a very close image is similarly low. This poor bias towards categories like duck and smokestack where the variance within the category can be high is illustrated in Fig. 6. In categories with very similar images, it can be observed that the algorithm is able to retrieve the images well. This is reflected in tomato and tuning fork where the visual identity of the object is important as demonstrated in Fig. 7.

Table 2 Accuracy per category

Category	Accuracy
Cellphone	66%
Duck	33%
Harp	33%
Smokestack	0%
Toaster	66%
Tomato	100%
Trilobite	66%
Tuning-fork	66%
Waterfall	33%

Fig. 6 Displaying differences in smokestacks

Fig. 7 Input image (left) and retrieved image (right), borders are present during query

To improve the results, one would need a larger image database for retrieval to occur on. The neural network at the base of the algorithm can only learn based on the information that it has and will make few inferences. Care must also be taken to provide a wide enough category to encompass the use case of the algorithm. Given the problem has the niche for a similarity measurement based upon likeness, the algorithm could fit the need.

The type of retrieval that is being discussed within this paper is related to the likeness of objects and their categorical representations. People who may see the use of this type of retrieval would be those who need more information related to the categories that an image may represent. For instance, someone lacking information on a specific species of bird would be able to query a system to retrieve more information regarding their finding. In real-world situations the system would not be locked down to only the best fitting image but could choose any amount from the top to perform the needed analysis upon.

One use that could see success from this style of algorithm is the medical field. An engineer would collect a large collection of images related to scans of a patient to train the network. These scans would then be transcribed into feature data and later used to compare with trial data. The returned image would be useful upon an expert's eye to give a more proper diagnosis of the issue.

5 Conclusion

Content-based image retrieval is important to understand, as it carries many possibilities in modern computing. For instance, medicine and facial recognition are experiencing the effects of CBIR already. The ability to index images eases the load on databases, allowing them to hold more information and in the end convey it properly and concisely to the user.

Classical methods and machine learning algorithms have proven to work on this problem. The issue with traditional ideas lies with combining the features within the images to get a better retrieval in the end. This is where the proposal of convolutional networks come in as the weights train the model to be able to understand the images with little to no human contact after setting up the algorithms and labels.

This research was conducted to be able to retrieve images based on key features within the image as opposed to human labeled text data. How the feature extraction and comparison is setup ultimately decides what images will be retrieved when an input is given. The approach taken within this paper uses convolutional neural networks to do a likeness measure to return images that have categorical similarities and be advantageous for finding images in predefined classes. These images that are returned will theoretically share features defining to their category with the query image and give insight to parallels that may be present within the resulting images.

Explaining the various layers of the neural network at the heart of the algorithm was focused on. Importance within the convolutional layers of the model were especially highlighted as responsible for modern day image learning. Activation functions and their roles in the flow of data was also touched upon. Lastly, the distance function which allows for retrieval to occur was mentioned and explained, giving the formula for scoring which image to return.

Discussion over speed and accuracy is still an important topic in image indexing in the modern era. It is important to know what forces are affecting the two values so that the models are built with the correct balance in mind. The ability to search for information is a powerful tool; however, when working in images it may be difficult to find a way to transcribe them to a computer. The primary research on CBIR into the future is the scalability within these CBIR algorithms.

References

1. J. Eakins, M. Graham, "Content-based image retrieval", a report to the JISC Technology Application Programme, Technical report, Institute for Image Data Research, University of Northumbria at Newcastle, UK (1999).
2. W. Zhou et al., Recent advance in content-based image retrieval: A literature survey. ArXiv:1706.06064 [Cs], pp. 1–22 (2017).
3. R. Saritha, V. Paul, P. Kumar, Content based image retrieval using deep learning process. Clust. Comput., 4187–4200 (2018)
4. R. Torres, A. Falcão, Content-based image retrieval: Theory and applications. RITA, pp. 161–185 (2006).
5. A. Rao, R. Srihari, Z. Zhang, Spatial color histograms for content-based image retrieval, in *Proceedings of the International Conference on Tools with Artificial Intelligence*, pp. 183–186 (1999).
6. S. Arivazhagan, L. Ganesan, S. Selvanidhyananthan, Image retrieval using shape feature. Int. J. Imaging Sci. Eng **1**(3), 101–103 (2007)
7. M. Singha, K. Hemachandran, Content based image retrieval using color and texture. Sig. Image Process. **3**, 39–53 (2012)
8. J. Júnior, R. Marçal, M. Batista, Image retrieval: Importance and applications. Work. Visao. Comput. **20**, 311–316 (2014)
9. Q. Rizvi, Analysis of human brain by magnetic resonance imaging using content-based image retrieval. Int. J. Health Sci. **14**, 3–9 (2020)
10. F. Özyurt, T. Tuncer, E. Avci, et al., A novel liver image classification method using perceptual hash-based convolutional neural network. Arab. J. Sci. Eng. **44**, 3173–3182 (2019)
11. M. Abadi et al., TensorFlow: Large-scale machine learning on heterogeneous systems, (2015). Software available from tensorflow.org
12. F. Chollet et al., Keras (2015). Software available from keras.io.
13. S. Walt, S. Colbert, G. Varoquaux, The NumPy array: A structure for efficient numerical computation. Comput. Sci. Eng. **13**, 22–30 (2011)
14. G. Griffin, A. Holub, P. Perona, Caltech-256 object category dataset, California Institute of Technology (2007).
15. F. Musumeci, C. Rottondi, A. Nag, I. Macaluso, D. Zibar, M. Ruffini, M. Tornatore, An overview on application of machine learning techniques in optical networks. IEEE Commun. Surveys Tuts **21**(2), 1383–1408., 2nd Quart (2019)
16. F. Sultana, et al., Advancements in Image Classification Using Convolutional Neural Network. in *2018 Fourth International Conference on Research in Computational Intelligence and Communication Networks (ICRCICN)*, pp. 122–29 (2018).
17. Y. Lecun, L. Bottou, Y. Bengio, P. Haffner, Gradient-based learning applied to document recognition, Proc. IEEE, pp. 2278–2324 (1998).

18. C. Nwankpa, W. Ijomah, A. Gachagan, S. Marshall, Activation functions: Comparison of trends in practice and research for deep learning. pp. 1–20 (2018).
19. P. Ramachandran, B. Zoph, Q.V. Le, Searching for activation functions. ArXiv **1710**(05941), 1–13 (2017)
20. D. Kingma, J. Ba, Adam: A method for stochastic optimization. ArXiv **1412**(6980), 1–15 (2014)

Human–Computer Interaction Interface for Driver Suspicious Action Analysis in Vehicle Cabin

Igor Lashkov and Alexey Kashevnik

1 Introduction

In recent years, there is a significant increase in the number of vehicles in the whole world. Meanwhile, a great number of research studies is being conducted in the field of development of autonomous intelligent transportation systems. For instance, global research community is concentrated on scientific studies in the fields of smart cities, big data, machine learning, and intelligent transportation systems.

According to the statistics of European Commission, more than 25,000 people are injured and died in traffic accidents on the roads of EU annually [1]. The development of intelligent driver monitoring and driver assistance systems is becoming popular nowadays, as it reaches the high level of functionality and performance. It is a promising approach for improving the in-vehicle cabin driver safety as the early signs of drowsiness or detection can be detected before the emergency situation arises and, in particular, relevant context-based recommendations can be generated for a driver to avoid a possible occurrence of traffic accident, e.g., pull over and take a short nap of 20 min. Thus, for example, the EU proposed to make vehicle safety systems, focused on recognizing the situations when a driver is distracted or drowsy, mandatory for all new produced cars [2]. The authors of the study highlight that 90% of all fatalities and injuries on the roads are due to human error. Moreover, according to the report of National Highway Traffic Safety Administration, 94% of serious crashes are due to human error [3]. Vehicles equipped with intelligent active safety systems have the potential to remove human error from the crash equation, which will help protect drivers and passengers, as well as pedestrians, and other road participants.

I. Lashkov (✉) · A. Kashevnik
SPIIRAS, 39, 14 Line, St. Petersburg, Russia
e-mail: igla@iias.spb.su; alexey@iias.spb.su

© Springer Nature Switzerland AG 2021
R. Stahlbock et al. (eds.), *Advances in Data Science and Information Engineering*,
Transactions on Computational Science and Computational Intelligence,
https://doi.org/10.1007/978-3-030-71704-9_57

There are certain situations in which the driver can intentionally or unintentionally reduce the system performance while operating without even knowing it. Potentially, driver's actions potentially directed at reducing driver safety should be marked by the system as suspicious and, in particular, emergency warnings and recommendations should be given for a driver to stabilize system state. We distinguish explicit driving actions and implicit one. The former kind of actions refers to human attitude placed at a conscious level and are easy to self-report. While, the latter are human attitudes, which are at the unconscious level, are involuntarily formed, and generally unknown.

There is a research and technical gap in researching and employing the video-based solutions aimed at analyzing suspicious human actions and providing human safety while driving. The main purpose of the paper is to present the approach for monitoring in-vehicle driver behavior focused on finding vulnerabilities in interaction interfaces between humans and systems built up with artificial intelligence in transport environment.

Our experiments showed the feasibility of our system in terms of suspicious action analysis in driver behavior along the trip for the purpose of warning the driver to act immediately. Driver actions performed inside the vehicle cabin are subject to thorough interpretation and investigation for the presence of dangerous states observed from recorded video stream. In general, these driving states relate to drowsiness, distraction, alcohol/drug intoxication, aggressive driving, high pulse rate, belt presence, camera sabotage, traffic violations. This paper builds upon our following previous works [4] and [5].

The rest of the paper is organized as follows. Section 2 presents a comprehensive related work in the area of solutions aimed at monitoring driver behavior, analyzing its suspicious actions, recognizing whether the person is drowsy, distracted at the moment, and compares the existing projects and solutions. Section 3 describes in detail the reference model of the proposed approach for monitoring driver behavior. The case study and the details of implementation is presented in Sect. 4. Finally, the main results and potential future work are summarized in Conclusion (Sect. 5).

2   Related Work

This section discusses the research made in driver monitoring systems aimed at analyzing suspicious actions and alerting emergency situations inside the vehicle cabin, that can be a threat to the whole system and decrease its performance.

Measurements obtained from smartphone embedded sensors may extensively describe the environment of the current situation, and, thereby, contribute to the recognition of different kinds of in-cabin real driving situations. One of the recent studies [6] proposed a smartphone-based drowsy driving detection system in real time to detect drowsy driving at an early stage. This system utilizes the Doppler shift of audio signals, the human body produces, to capture the unique patterns of drowsy driving. It leverages smartphone's built-in microphone and speakers to extract particular driving actions, including nodding, yawning, and operating a steering wheel, using long short-term memory networks. The driving statistics collected during real driving environments confirms that different drivers have similar patterns. The authors of the paper perform the evaluation of system performance by analyzing the impact of smartphone location in the car, the structure of the deep network, training dataset size, and other factors, including listening to the radio or driving at different speeds.

Another research study [7] introduces an approach that intensively employs the use of smartphone's built-in accelerometer sensor for monitoring and recognizing different types of aggressive driving maneuvers. Changes of longitudinal acceleration may represent sudden braking and sudden acceleration events. Likewise, another smartphone's built-in gyroscope is actively used to detect safe and unsafe left/right sharp turns driving patterns. The data extracted from mobile sensors aid to classify aggressive and nonaggressive driving events based on patterns. Such smartphone sensors may not always provide accurate measurements related to the surrounding environment and can be error-prone. To tackle this downside, the researchers apply a high pass filter for accelerometer data to remove the high frequency noise and, afterwards, apply the low pass filter to smooth signal values. According to the processed driving statistics, the thresholds for each observed driving pattern (safe/sharp acceleration, deceleration, left and right turns) indicate the severeness of occurred event.

In recent years, video surveillance systems are becoming popular in monitoring crowded areas and alerting authorities on the occurrence of suspicious actions. The research [8] presents an approach based on locating humans in images obtained from camera, evaluating its pose, appearance of body parts (head, arms, and torso) and image data analysis. Image analysis involves classification of human suspicious action using the linear Hough transform technique to compute the score for a particular body part from image frames, and conditional random field statistical modeling method for each image to estimate person pose, extract the exact orientation of each limb, and classify whether the pose is suspicious or not. If a match is found, the human is marked in red rectangle. These extracted orientations act as features in comparison with the suspicious ground truth action dataset, and the evaluated head pose is flagged with action that matches best. The authors tested its approach using an autonomous unmanned aerial vehicle visual surveillance system. The conducted experiments showed a detection accuracy of 71% with no prior knowledge of background, lightness, or scale and location of the human in the images.

Video fragments outputted by video-based surveillance systems must be recorded with correct, not blocked, field of view, with good quality that is critical for further analysis by operator. Violators may forge the recorded video, or the physical device itself, in order to disguise their suspicious activities. Automatic detection of camera sabotage events is critically important for immediate operator intervention. The authors [9] presented a video-based approach for detecting camera sabotage situations, including occlusion, where field of view is partially or fully obstructed; defocus, related with reduced visibility; and displacement, caused by tilt or movement up or down from the intended field of view. The detection of camera occlusion state is based on the use of the foreground objects' area and position, camera defocus state – edge information, and camera displacement – through the use of frame count per panning sweep. The effectiveness of the proposed method is tested using public dataset for camera sabotage detection on panning surveillance systems by comparing it with an existing method. The authors of the paper highlight that the proposed method can be applied in recognizing natural tampering events like careless maintenance by the cleaning staff, dirt on camera lens, fog, and smoke.

Another method is presented in the paper [10] and discusses a convolutional neural network architecture intended for both human pose estimation and seat belt detection. Underneath, the proposed approach utilizes a feature pyramid network backbone with multibranch detection heads, namely, a key point detection head, a part-affinity field detection head, and a seat belt segmentation head. The main goals of the study are to evaluate 2D body posture of the driver and the front-row passenger (if available), and to segment image pixels that relate to seat belts. The neural network developed was tested over 50 driving sessions in different lightness conditions. The seat belt algorithm involves a probability density function over the image showing the likelihood of a pixel being seat belt segmentation. The binary seat belt segmentation model was evaluated with following metrics, including sensitivity, specificity, precision, F1 score, and the intersection of the union. These metrics are shown to be poor to solve this problem since sensitivity, precision, and F1-score were 63.51%, 63.58%, and 63.55%, respectively, and the intersection of the union was only 47%.

Drivers do not always realize that they are in a drowsy state that can lead to a road emergency situation. Researchers [11] propose a method aimed at analyzing a short video sequence for a certain time period to detect sleep deprivation from the driver's face. Developed system utilizes Haar feature-based cascades to locate the driver's face on the image and then applies classification to group it into one of two classes: "sleep derived" and "rested." This approach involves the use of the trained model formed by the convolutional neural networks MobileNets [12], applicable for highly efficient use on mobile and embedded vision-based applications. The deep learning architecture of MobileNet aids to analyze the camera frame and estimate the probability of the frame to be treated as a "sleep deprived" class. The driver is classified as "sleep deprived" if the probability of this class is greater than 0.5. Natural experiments are performed using a prototype mobile application for Android platform. The Android Face API is used to crop and normalize the driver's face, and the TensorFlow framework to compile the MobileNet model previously trained on a standalone laptop. Authors of the study selected Multi-modality Drowsiness Database to provide the ground truth of sleep deprivation. The proposed system showed good results in experiments of classifying rested drivers, but struggles in recognizing the tired ones, providing the accuracy score of 68.7% for image classification on more than 9000 frames.

Personal voice-based assistants are becoming popular nowadays. They can be found quite useful for supporting driver activities inside the vehicle cabin. These virtual assistants mostly rely on the voice-based communication with their users (drivers), that can be potentially vulnerable, lacking proper authentication. Rapid growth of virtual assistants is emerged with third-party functions (skills), developed by various developers. In the paper [13], the authors highlight two types of attacks: *voice squatting* in which the violator exploits the way a skill is invoked, using a malicious skill with a similarly pronounced name or a paraphrased name to hijack the voice command meant for a legitimate skill, and *voice masquerading* in which a malicious skill pretends the personal assistant service or a legitimate skill during the user's conversation with the service to steal its personal information. In order to recognize voice masquerading attack, researchers built a context-sensitive detector upon the private assistant infrastructure, that takes a skill's response and/or the user's utterance as its input to determine whether an impersonation risk presents, and alerts the user once detected. In other case, to detect voice attack, the authors proposed a scanner that runs two steps to capture the competitive invocation names (names with similar pronunciation as that of a target skill or using different variations) for a given invocation name: utterance paraphrasing and pronunciation comparison. The first step is to identify suspicious variations of a given invocation name, and the second one is dedicated to finding the similarity in pronunciation between two different names.

Driver distraction is one of the main challenges in road traffic and cause of road accidents. In this study [14], the authors proposed an image-based approach for detecting driver's distraction state with the following two steps. The first step is to predict the bounding boxes of the driver's right hand and right ear from RGB images utilizing the deep neural network You Only Look Once (YOLO) model [15]. The second module takes the bounding boxes of the ear and hand as input and predicts the type of distraction using a multilayer perceptron to infer the driver's status from the regions of interest. These modules were evaluated with a set of different metrics, including mean accuracy precision, average accuracy, error rate, micro-averaging, and macro-averaging. During the experiments with driving simulator, the participants were involved in five types of distraction tasks, which are talking on a cell phone, texting, drinking water, using the touchscreen, and placing a marker into the cup holder. In terms of classification of driver's distraction state, the results show that the proposed algorithm is able to detect normal driving, using the touchscreen, and, holding a cell phone to her/his ear with F1-score equal to 0.84, 0.69, 0.82, respectively.

It should be noted that driver behavior discussed in observed research studies is critical for providing road safety. To increase safety for public roads, it is important to analyze driver suspicious actions, comprising explicit intentional driving behavior and implicit unintentional one. The presented studies has its own specific pros and cons in considered approaches for recognizing suspicious driving actions. Addressed solutions are limited in its capabilities by considering abnormal user behavior [16] in single cases and do not combine several types of suspicious human actions. The joint use of different kinds of sensor data, including camera-based video recordings, GPS, accelerometer, gyroscope readings, and audio-based signal from microphone, is promising to apply these types of human activities for a driver simultaneously inside the vehicle cabin in order to end up warning him/her about unsafe road situation. The utilization of cloud-based technologies may significantly increase the robustness and performance of our approach for suspicious driving behavior determination.

3   Reference Model of Driver Monitoring System

We propose a reference model (Fig. 1) focused on analyzing in-vehicle driving events and recognizing sequences of driver suspicious actions that can lead to road emergency situations and become a cause of a traffic accident of various severity. In our study, we distinguish two types of driver behavior: the intentional (explicit) driving actions and the unintentional (implicit) ones. The former kind of actions refers to human attitude placed at a conscious level and are easy to self-report. This kind of actions are to be carefully hidden or camouflaged by the driver to trick the system and modify its behavior. Thus, in its case, the driver can disrupt system and, potentially, reduce its performance and accuracy of the operations. While, the latter are human attitudes, which are at the unconscious level, are involuntarily formed, and generally unknown. In this case, the driver may not properly understand the current situation and make judgments due to driver impairment caused by different impacting factors (inattention on the road, strong desire to sleep, etc.) inside the vehicle cabin. This kind of suspicious actions are bound to driver dangerous states, like drowsiness, distraction, fatigue, alcohol intoxication states, and others.

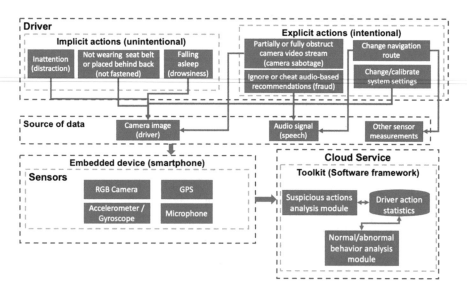

Fig. 1   Reference model of the monitoring system for recognizing driver suspicious actions

We consider driver as the main source of data for the reference model. Camera is used to record video of the driver inside the vehicle cabin, continuously track driver behavior along the entire trip and analyze it for any suspicious implicit or explicit actions performed by the driver. Typical examples of implicit actions are inattention, not wearing seat belt or belt placed behind back, and falling asleep. Whereas, the explicit actions may be caused by partial or full obstruction of the camera, changes in navigation route, modifying the system settings, or cheating the audio-based communication (recommendations generated by the system).

Toolkit is a software framework unit responsible for recognizing suspicious driver actions analysis based on monitoring human behavior for the purpose of warning the driver. Driver actions performed inside the vehicle cabin are subject to thorough interpretation and investigation for the presence of dangerous states observed from recorded video stream. Toolkit is placed in cloud service for making heavy time-consuming computations and uses driver statistics storage to accumulate various types of driver events occurred along the trip; driver patterns analysis module to capture different sensor measurements, process it, analyze driver state in a period of time; and suspicious actions analysis module to make decisions whether the driver behavior is abnormal at the moment of time.

We use RGB camera, GPS, accelerometer, and gyroscope to monitor visual signals of abnormal driver behavior and recognize drowsiness, distraction, and the presence of seat belt. To accomplish this task, we consider a set of distinct parameters, including PERCLOS [17], eye-blink rate, number of yawns per minute, left/right head turn, or tilting the head up/down. In order to detect whether the driver is properly fastened or not, deep neural network engine is used to find the presence of a seat belt on the camera image. Neural network may give information about the belt state by showing the belt coordinates if it is detected and, along with that, estimating the probability confidences that the belt state is like to be properly recognized. Nevertheless, information extracted from camera video stream and sensor readings from GPS, accelerometer, gyroscope, and microphone are efficiently utilized to recognize other types of abnormal driving behavior, including alcohol and drug intoxication, high pulse rate, mobile phone usage, eating, drinking, smoking, aggressive driving, belt presence, camera sabotage, and traffic violations.

Along with that, in order to increase overall system performance, we propose a cloud service, responsible for accumulating driving statistics in a single space together, and processing and comparing information about normal and abnormal behavior of users of the system.

4   Case Study Using a Smartphone

Considering the proposed approach, in order to test and demonstrate the usage of toolkit (software framework), we designed and developed a prototype mobile application intended for Android platform. It is focused on recognizing suspicious driver actions comprising drowsiness, distraction, and seat belt presence, by analyzing human behavior and making computations directly on the device. The current version of the software prototype is implemented in Java, Kotlin, and C++ programming languages in Android Studio IDE.

The scheme of application work is the following. A user chooses a recorded video fragment from gallery or camera that he/she wants to check for the presence of suspicious driving behavior in it. In the beginning, the video fragment is decoded and parsed as a fixed set of distinct frames comprising the whole media file. Afterwards, the output camera frames extracted from the previous step act as a source data for image recognition module, entirely written in C++. In its turn this module employs computer vision and machine learning libraries, OpenCV and Dlib, to locate and obtain the position of the driver's facial landmarks (positions and sizes of head, eyes, mouth). Moreover, it estimates additional visual parameters like the head rotation/tilt angle and the ratio of eye and mouth openness to recognize drowsiness, distraction, and seat belt presence (Fig. 2). Finally, as soon as the analysis process of the whole video frame sequence is finished, the results, describing found suspicious actions in human behavior, are saved physically on the disk in a format of two CSV files. One of them has the detailed statistics describing human behavior, where each row in a file represents information about facial features of the person on each video frame, extracted at some moment of processing time. The other file has the information about concrete suspicious driver actions occurred for some period of time based on the analysis of certain set of video frames. This list of actions provided by the latter file may be eventually used to make a conclusion about the presence of suspicious human behavior.

Although the proposed system defines an initial high level of efficiency, there is a set of different parameters effecting the dangerous state detection and generating alerts for a driver. System preferences include the audio level for generation alerts and recommendations; screen brightness intended to set display brightness while using application. The driver can accurately configure preferences related to dangerous states determination (Fig. 3b, c), including the interval denoting the minimum time between dangerous states recognition, allowing the system not to annoy the driver; the time required to recognize a certain dangerous state. There are

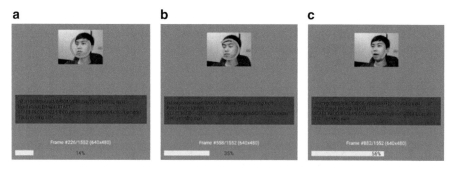

Fig. 2 Screenshots of main screen of Toolkit application. Person behavior is normal (**a**), person closes his eyes (drowsiness) (**b**), and person yawns (drowsiness) (**c**)

Fig. 3 Screenshots of application settings: general (**a**), distraction (**b**), and drowsiness (**c**)

multiple preferences referring to the driver's drowsiness, distraction, and seat belt fastened states, making possible the tuning of parameters affecting the dangerous state determination. Some of them include head turn left/right, head tilt up/down, ratio of left/right eye openness, mouth open ratio to detect yawning. The driver may modify all these parameters at any time of the ongoing trip, and, thereby, affect system operation. These kind of driver actions can be considered suspicious, disruptive for a system, and lead to improper behavior and in turn errors. Along with that, the application is responsible for recording different driver activities as log events, which are then used to identify potentially suspicious driver activities that do not match with the normal behavior of the person. It should be noted that presented parameters need to be manually configured by a driver before starting a trip. Afterwards, this kind of parameters should be automatically determined by the application itself and, further, continuously calibrated during the ongoing trip to consider the driver's profile and context and provide the driver a better overall experience.

We conduct experiments related to drowsiness suspicious action detection using the driver drowsiness detection dataset collected by Computer Vision Lab at National Tsing Hua University [18]. We use the developed toolkit and load the provided video to the dataset and analyze the results that show the average accuracy of the developed system is around 80%. To compare the drowsiness data calculated by the Toolkit with the data provided with the dataset, we developed the Excel-based model.

5 Conclusion

In our study, we proposed the approach for monitoring in-vehicle driver behavior focused on finding vulnerabilities in interaction interfaces between humans and systems built up with artificial intelligence in transportation environment. We presented a reference model focused on analyzing in-vehicle driving events and recognizing sequences of driver suspicious actions that can lead to road emergency situations and become a cause of traffic accidents of various severities. In our study, we highlighted two types of driver behaviors: the intentional (explicit) and unintentional (implicit) driving actions. The results of the study indicate that our solution may be successfully applied in the fast development category of intelligent driver assistance systems, improving dangerous state determination by utilizing sensor measurements by identifying driver suspicious actions leading to system destabilization, error-prone situations, and performance reduction.

The use of data obtained from extra sensors and information about driver's emotions, psychophysiological behavioral state, may potentially increase system robustness and performance in monitoring driver behavior and discovering both explicit and implicit suspicious actions.

Acknowledgements The part of research related to the driver face analysis for dangerous state detection is done by the Russian Foundation for Basic Research project # 19-29-06099.

References

1. 2018 road safety statistics: what is behind the figures?, https://ec.europa.eu/commission/presscorner/detail/en/MEMO_19_1990. Last accessed 20 March 2020.
2. Road safety: Commission welcomes agreement on new EU rules to help save lives, https://ec.europa.eu/commission/presscorner/detail/en/IP_19_1793. Last accessed 20 March 2020.
3. National Highway Traffic Safety Administration, https://www.nhtsa.gov/technology-innovation/automated-vehicles-safety. Last accessed 20 March 2020.
4. A. Kashevnik, I. Lashkov, A. Gurtov, Methodology and mobile application for driver behavior analysis and accident prevention. IEEE Trans. Intell. Transp. Syst., 1–10 (2019). https://doi.org/10.1109/TITS.2019.2918328
5. A. Kashevnik, I. Lashkov, A. Ponomarev, N. Teslya, A. Gurtov, Cloud-based driver monitoring system using a smartphone mounted on a vehicle windshield. IEEE Sens. J. 2020, https://doi.org/10.1109/JSEN.2020.2975382.
6. Y. Xie, F. Li, Y. Wu, S. Yang, Y. Wang, D3-Guard: Acoustic-based drowsy driving detection using smartphones, in *IEEE INFOCOM 2019 - IEEE Conference on Computer Communications*, Paris, France, 2019, pp. 1225–1233. https://doi.org/10.1109/INFOCOM.2019.8737470.
7. R. Chhabra, S. Verma, R. Challa, Detecting aggressive driving behavior using mobile smartphone, in *Proceedings of 2nd International Conference on Communication, Computing and Networking*, 2019, pp. 513–521, https://doi.org/10.1007/978-981-13-1217-5_49.
8. S. Penmetsa, F. Minhuj, A. Singh, S.N. Omkar, Autonomous UAV for suspicious action detection using pictorial human pose estimation and classification. ELCVIA **13**(1), 18–32 (2014)

9. S. Kk, B. Mehtre, Automated camera sabotage detection for enhancing video surveillance systems. Multimed. Tools Appl. **78**(11), 5819 (2018). https://doi.org/10.1007/s11042-018-6165-4

10. S. Chun, N. Ghalehjegh, J. Choi, C. Schwarz, J. Gaspar, D. McGehee, S. Baek, A nimble architecture for driver and seat belt detection via convolutional neural networks, in *The IEEE International Conference on Computer Vision (ICCV)*, 2019.

11. M. García-García, A. Caplier, M. Rombaut, Sleep deprivation detection for real-time driver monitoring using deep learning, in *Image Analysis and Recognition, Lecture Notes in Computer Science*, vol. 10882, (2018), pp. 435–442. https://doi.org/10.1007/978-3-319-93000-8_49

12. W. Wang, Y. Li, T. Zou, X. Wang, J. You, Y. Luo, A novel image classification approach via Dense-MobileNet Models. Mob. Inf. Syst, 1–8 (2020). https://doi.org/10.1155/2020/7602384

13. N. Zhang, X. Mi, X. Feng, X. Wang, Y. Tian, F. Qian, Dangerous skills: Understanding and mitigating security risks of voice-controlled third-party functions on virtual personal assistant systems, in 2019 IEEE Symposium on Security and Privacy (SP), San Francisco, CA, USA, 2019, pp. 1381–1396.

14. L. Li, B. Zhong, C. Hutmacher, Y. Liang, W.J. Horrey, X. Xu, Detection of driver manual distraction via image-based hand and ear recognition. Accid. Anal. Prev. **137**, 105432 (2020)

15. Z.-Q. Zhao, P. Zheng, S.-t. Xu, and X. Wu, Object detection with deep learning: A review, Computer Vision and Pattern Recognition, ArXiv, April 2019.

16. M. Ussath, D. Jaeger, F. Cheng, C. Meinel, Identifying suspicious user behavior with neural networks. in *2017 IEEE 4th International Conference on Cyber Security and Cloud Computing*, 2017, https://doi.org/10.1109/cscloud.2017.10.

17. J. Yan, H. Kuo, Y. Lin, T. Liao, Real-time driver drowsiness detection system based on PERCLOS and grayscale image processing, in *2016 International Symposium on Computer, Consumer and Control (IS3C)*, Xi'an, 2016, pp. 243-246.

18. C. Weng, Y. Lai, S. Lai, Driver drowsiness detection via a hierarchical temporal deep belief network, in *Asian Conference on Computer Vision Workshop on Driver Drowsiness Detection from Video, Taiwan*, 2016, pp. 117-133.

Image Resizing in DCT Domain

Hsi-Chin Hsin, Cheng-Ying Yang, and Chien-Kun Su

1 Introduction

As various displays are abundantly present in our daily lives, there is an increasing demand for effective image resizing. To keep the region of interest (ROI) intact as much as possible, content aware image resizing (CAIR) has drawn a lot of attention in the recent years [1]. State-of-the-art CAIR algorithms can be classified into two categories: the discrete approach [2–5] and the continuous approach [6–9]. Avidan et al. defined a seam as a path of connected pixels from one side of an image to the opposite side, and proposed seam carving (SC) for CAIR [2]. To improve the performance of SC, Shamir et al. used more features related to the ROI [3]. To speed up SC, Patel et al. proposed multiple pixel wide SC in spatial domain [4]. In [5], seams of wavelet trees were constructed, based on which, a scale-recursive SC algorithm was proposed in the wavelet domain. In [6], resizing of images was formulated as a convex quadratic problem, which can be solved by quadratic programming. Wang et al. rescaled salient objects uniformly and concealed non-uniform distortions in the regions of non-interest (RON) [7]. We used the saliency histogram (SH) of an image to evaluate salient regions, and proposed saliency histogram equalization (SHE) [8] and the SHE-SC algorithm [9] to construct a pair of non-uniform meshes for image warping.

H.-C. Hsin (✉)
Department of Computer Science and Information Engineering, National United University, Miaoli City, Taiwan

C.-Y. Yang
Department of Computer Science, University of Taipei, Taipei City, Taiwan

C.-K. Su
Department of Electrical Engineering, Chung Hua University, Hsinchu, Taiwan

© Springer Nature Switzerland AG 2021   799
R. Stahlbock et al. (eds.), *Advances in Data Science and Information Engineering*,
Transactions on Computational Science and Computational Intelligence,
https://doi.org/10.1007/978-3-030-71704-9_58

In many cases, especially for mobile applications, images are compressed to save memory space and transmission bandwidth [10]. Tanaka et al. incorporated image coding with the SC-based concentration and dilution to improve the compression quality at low bit rates [11]. Fang et al. proposed a discrete cosine transform (DCT)-based saliency detection model for image resizing in the joint photographic experts group (JPEG) compressed domain [12]. In [13], a practical system was proposed for transcoding the original video bitstream to the retargeted one in the H.264/advance video coding (AVC) compressed domain. This paper presents an efficient method for obtaining the DCT blocks of the resized image form the DCT blocks of the original image in order to avoid unnecessary computations of inverse and forward DCTs. The rest of the paper is organized as follows. The SHE and SHE-SC algorithms are reviewed in Sect. 2. The proposed algorithm is detailed in Sect. 3. Experimental results are given in Sect. 4. Conclusion can be found in Sect. 5.

2 Review of SHE and SHE-SC

Saliency histogram (SH) provides useful information about the ROI [8]. In the framework of image resizing with mesh warping, a pair of meshes are needed to map the original image onto the resized one. For effective CAIR, the ROI with high SH values should be well preserved at the cost of ignoring the regions of non-interest (RON) with low SH values. It is noted that contrast enhancement can be attained by image histogram equalization (IHE); saliency histogram equalization (SHE) can enhance the ROI likewise. Let $p(x, y)$ be the SH of an $M \times N$ image $\mathbf{I}$, the marginal SH (MSH):

$$p_y(y) = \frac{\sum_{x=1}^{M} p(x, y)}{\sum_{x=1}^{M} \sum_{y=1}^{N} p(x, y)}; y \in [1, N] \tag{1}$$

can be used to partition $\mathbf{I}$ into L vertical strips with equal accumulated amount of SH values as follows

$$v_n = \arg \max_k \frac{n-1}{L} < \sum_{y=1}^{k} p_y(y) \le \frac{n}{L} \tag{2}$$

where v_n is the coordinate of the lower right vertex of the nth vertical strip in the y-direction, $n = 1, \cdots, L$, with the boundary condition: $v_L = N$. Similarly, the partition of K horizontal strips with equal accumulated amount of SH values can be obtained by equalizing the MSH in the x-direction.

SHE provides an efficient way to represent images with non-uniform meshes: the higher the SH value, the finer the mesh cell. To represent the resized image with an adaptive mesh, the SHE-SC algorithm [9] modifies the SC algorithm [2] by taking mesh quads as super-pixels. The significance of a quad: $q_{m,n}$ (super-pixel) is given by

$$\text{Sig}\left(q_{m,n}\right) = \frac{1}{\left|q_{m,n}\right|} \sum_{(i,j) \in q_{m,n}} p\ (i,\ j) \tag{3}$$

where $p\,(i,j)$ is the SH value at coordinates $(i,\ j)$, and $|q_{m,n}|$ is the size of $q_{m,n}$. As each quad has the same accumulated amount of SH values via SHE, $\text{Sig}(q_{m,n})$ is inversely proportional to $|q_{m,n}|$, and the least significant seam (LSS) of quads is the one with maximum sum of quad sizes. To gradually adjust the width of a mesh, a sequence of increasingly significant vertical seams of quads is first identified, and each seam is resized by 1 pixel width. The height of a mesh can also be gradually resized in a similar manner. Based on the pair of meshes, one is obtained by SHE [8] to represent the original image, and the other with the desired size is obtained by SHE-SC [9], the resized image can be obtained by simple interpolation.

3   Proposed Algorithm

In the proposed algorithm, the SH of an image is obtained based on the features extracted from DCT blocks directly. Next, SH-based mesh warping is used to convert the DCT blocks of the original image into the DCT blocks of the resized image. Recent studies have shown that luminance, color, and texture features are suitable for saliency detection [12]. This paper takes the DC coefficients of DCT blocks as a down-sampled image [13], and uses the discriminative regional feature integration (DRFI) algorithm [14] followed by simple interpolation for construction of luminance-color SM (LCSM). In addition, the AC coefficients of DCT blocks are important features for images with textures. The AC-based likelihood of texture [13] has been used for construction of texture SM (TSM). The final SM is obtained by combining LCSM and TSM via sum of normalized maps.

The continuous approach to CAIR is typically formulated as a nonlinear warping problem, which takes a pair of meshes, namely the input mesh representing the original image, and the output mesh representing the resized image. Existing mesh construction algorithms usually take a uniform mesh as the input mesh and deform it into a non-uniform mesh used as the output mesh. In contrast, the proposed algorithm constructs the input mesh by SHE and constrains the output mesh with the same quad size as that of DCT blocks using SHE-SC. To convert the DCT blocks of the compressed image into the DCT blocks of the resized image, the arbitrary downsizing (AD) algorithm has been used [15].

Fig. 1 (**a**) Test image; (**b**) and (**c**) saliency maps (SMs) obtained by DRFI in the spatial domain and the proposed algorithm in the DCT domain; (**d**) and (**e**) resized images obtained by using SHE-SC with SM (**b**) and the proposed algorithm with SM (**c**), respectively

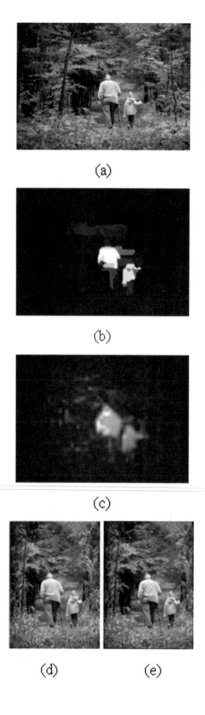

(a)

(b)

(c)

(d)   (e)

Fig. 2 (**a**) Test image; (**b**) and (**c**) saliency maps (SMs) obtained by DRFI in the spatial domain and the proposed algorithm in the DCT domain; (**d**) and (**e**) resized images obtained by using SHE-SC with SM (**b**) and the proposed algorithm with SM (**c**), respectively

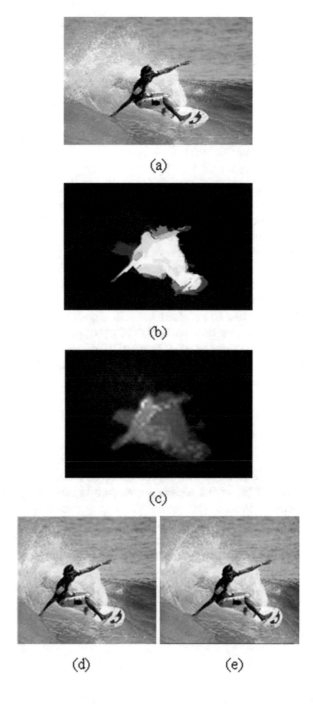

Table 1 Average running time for resizing images at 70% with respect to the maximum dimension of the original image using SC, GVFP and the proposed algorithm

Average running time	Algorithm
21.59 s	SC
25.14 s	GVFP
0.16 s	Proposed algorithm

4 Experimental Results

The proposed algorithm has been implemented in Matlab [®]. Figures 1 and 2, show the performance of the proposed CAIR algorithm in the DCT domain compared with that of the spatial domain SHE-SC algorithm [9]. The saliency maps (SMs) shown in Figs. 1b and 2b, and Figs. 1c and 2c, are obtained by using the DRFI algorithm [14] and the proposed algorithm, respectively. Figures 1d and 2d, and Figs. 1e and 2e are the resized images obtained by using the SHE-SC algorithm and the proposed algorithm, respectively. Though they are visually comparable, the proposed algorithm is preferable in regard to saving of inverse and forward DCTs. Other existing spatial domain CAIR algorithms, namely seam carving (SC) [2] and gradient vector flow paths (GVFP) [16] have also been compared. Table 1 gives the average running time for resizing images at 70% with respect to the maximum dimension of the original image. It shows that the proposed compressed domain CAIR algorithm is superior to SC and GVFP.

5 Conclusions

The joint photographic experts group (JPEG) standard takes discrete cosine transform (DCT) as the underlying method to effectively represent images in spectral domain. Most of the content aware image resizing (CAIR) algorithms are formulated in spatial domain, which are not suitable for coding applications. In this paper, a fast compressed domain CAIR algorithm has been proposed such that various decoded images with different resolutions and aspect ratios can be obtained directly form a JPEG code stream. Compared to other existing CAIR algorithms, the proposed algorithm is preferable, especially in terms of running time. Furthermore, there is no need to perform inverse and forward DCTs.

References

1. J. Kiess, S. Kopf, B. Guthier, W. Effelsberg, A survey on content-aware image and video retargeting. ACM Trans. Multimed. Comput. Commun. Appl **14**(3), article no. 76 (2018) 28 pages
2. S. Avidan, A. Shamir, Seam carving for content-aware image resizing. ACM Trans. Graph. SIGGRAPH **26**, 1–10 (2007)
3. A. Shamir, O. Sorkine, Visual media retargeting, in *Proceeding of ACM SIGGRAPH ASIA,* 2009, pp. 1–13.
4. D. Patel, S. Raman, Accelerated seam carving for image retargeting. IET Image Process **13**(6), 885–895 (2019)
5. H.-C. Hsin, Y.-C. Chan, C.-K. Su, Wavelet saliency map and its application to image resizing, in *APCEAS* 2015, Osaka, Japan, pp. 146–152.
6. R. Chen, D. Freedman, Z. Karni, C. Gotsman, L. Liu, Content-aware image resizing by quadratic programming, in *Computer Vision and Pattern Recognition Workshops (CVPRW),* 2010, pp. 13–18.
7. Y.S. Wang, C.L. Tai, O. Sorkine, T.Y. Lee, Optimized scale and stretch for image resizing. ACM Trans. Graph. **27**, 118–125 (2008)
8. H.-C. Hsin, Saliency histogram equalization and its application to image resizing. IET Image Process **10**(10), 787–798 (2016)
9. H.-C. Hsin, *Combination of saliency histogram equalization and seam carving for image resizing* (IET J Eng, 2017), 7 pages
10. Y. Zhou, L. Zhang, C. Zhang, P. Li, X. Li, Perceptually aware image retargeting for mobile devices. IEEE Trans. Image Process **27**(5), 2301–2313 (2018)
11. Y. Tanaka, M. Hasegawa, S. Kato, Image coding using concentration and dilution based on seam carving with hierarchical search, in *International Conference on Acoustics Speech and Signal Processing*, 2010, pp. 1322–1325.
12. Y. Fang, Z. Chen, W. Lin, C.-W. Lin, Saliency detection in the compressed domain for adaptive image retargeting. IEEE Trans. Image Process. **21**(9), 3888–3901 (2012)
13. J. Zhang, S. Li, C.-C.J. Kuo, Compressed-domain video retargeting. IEEE Trans. Image Process **23**(2), 797–809 (2014)
14. H. Jiang, J. Wang, Z. Yuan, Y. Wu, N. Zheng, S. Li, Salient object detection: A discriminative regional feature integration approach, in *2013 IEEE Conference on Computer Vision and Pattern Recognition*, 2013, pp. 2083–2090.
15. H. Shu, L.-P. Chau, An efficient arbitrary downsizing algorithm for video transcoding. IEEE Trans. Circuits Syst. Video Technol. **14**(6), 887–891 (2004)
16. S. Battiato, G.M. Farinella, G. Puglisi, D. Ravi, Saliency-based selection of gradient vector flow paths for content aware image resizing. IEEE Trans. Image Process **23**(5), 2081–2095 (2014)

Part VIII
Data Science and Information & Knowledge Engineering

Comparative Analysis of Sampling Methods for Data Quality Assessment

Sameer Karali, Hong Liu, and Jongyeop Kim

1 Introduction

The datasets originate from various resources, including websites, social networks, audio and video sources, log files, and other databases. Detecting data defects and inconsistencies through data quality assessment is essential to the improvement of data quality. Quality assurance is the process used to verify that collected data conform to defined standards and includes many rules and methods used to portray data. Applying a data quality assessment to a dataset will help us ensure the reliability of the data regardless of the results. This will allow the data quality assessments to reduce the additional cost of retesting, replacing, and reselling bad data. When results are not satisfied with the collected data, the backlash can damage the entire dataset and negatively impact future results. Marston et al. [11], Elmagarmid et al. [3], and Maria et al. (2014) state that most datasets have several problems, including duplicate and missing data that can make the analysis and maintenance of the data difficult. Moore [13] states that poor quality data costs businesses an average of $15 million per year. The data quality also negatively impacts efficiency, productivity, and credibility. According to researchers Jesmeen

S. Karali
School of Sciences and Informatics, Indiana University Bloomington, Bloomington, USA
e-mail: skarali@iu.edu

H. Liu (✉)
School of Sciences, Indiana University Kokomo, Kokomo, USA
e-mail: hlius@iu.edu

J. Kim
Math & Computer Science, Southern Arkansas University, Magnolia, AR, USA
e-mail: jkim@saumag.edu

© Springer Nature Switzerland AG 2021   809
R. Stahlbock et al. (eds.), *Advances in Data Science and Information Engineering*,
Transactions on Computational Science and Computational Intelligence,
https://doi.org/10.1007/978-3-030-71704-9_59

et al. [6] and Laranjeiro et al. [7], poor quality data costs approximately $13.3 million per organization and $3 trillion per year for the entire US economy.

Big data cleansing is a field that works to examine and process data. This field deals with datasets that are of an extensive scale or too complex for normal approaches since big data have properties of a wide variety and volume. Liu et al. [8, 9] state data cleaning is a process of cleaning erroneous data from primary sources of data; this process will remove errors from the dataset and can leave more time and resources for data analysis. Data are considered reliable if the methods used to collect and analyze the data remain stable over time.

This paper uses the data quality assessment model to reduce the risk of data errors using large scale data for US students. The sample size is determined by calculating the original population, confidence interval, and confidence level. There are many dimensions that can be used for data quality, but this paper will analyze the three most important dimensions: completeness, timeliness, and accuracy. Using these three dimensions, we create controller datasets with various percentages. Samples can be taken from this dataset using four sampling methods, and the extracted data will be scored to determine the most accurate sample.

The paper is organized as follows: data quality assessment, data quality dimensions, and sampling techniques are reviewed in the next section. Section 3 describes the methodology used, and Sect. 4 describes the data collection, experimental settings, and experimental results. Section 5 provides conclusions and future work.

2   Literature Review

2.1   Data Quality Assessment

The quality of data, information collection, usage, assessment, and evaluation of the entire health population are critical issues that the paper discusses. High-quality data are the prerequisite for better population health, decision-making, and overall information. Hazen et al. [5] state how modern age supply chain specialists are provided with a lot of data and are tasked with processing these large amounts of data. There is a need to determine new ways to produce, organize, and analyze this data. While resolving quality issues, any data that are used by the supply chain must be standardized universally. Woodall et al. [17] show that both practical and high-quality data analysis are necessary for an accurate evaluation of the impact of public health interventions and measuring public health outcomes. Woodall et al. [18] state the quality of public health data is not something that should be compromised. All data must be regulated and administered by the appropriate governing body to ensure that it is not compromised. Poor quality data can be detrimental, and even fatal, for an organization, and proper data validity, completeness, comprehensiveness, understandability, and data test coverage protocols must be followed to ensure that poor quality data is eliminated.

Based on these metrics, the quality of data can be conclusively defined and determined. The data used in the hybrid approach must be carefully selected and evaluated to ensure that the data are of decent quality. This hybrid approach should also have a provision for filtering data to ensure quality. Gudivada et al. [4] propose how quality can be assessed, defined, and measured. The syntactic quality criteria are meant to achieve syntactic correctness; they require that all statements in the underlying conceptual model with the implementation must be syntactically correct.

Ryu et al. [15] discuss how data quality has been a significant concern since automated data processing. An effective method of controlling data quality is business rules. The most effective techniques for quality management include a probability-based approach and knowledge-based techniques. All the data being used for business purposes should be filtered, analyzed for congruency, and deemed useful for the business agenda and goals. Knap and Mlýnková [12] state that data quality assessment and the improvement of larger information systems would not be feasible without the appropriate data quality methods. These data quality methods are essential algorithms that can be automatically executed in computer systems to detect and then correct any problems in databases.

The column analysis method can automatically compute various statistics, such as the total number of missing values in a given column of data. Lexical analysis is commonly applied to columns with string values to discover the subcomponents of unstructured content. Pipino et al. [14] state that information quality, software tools, automated data quality tools, and automated data quality software are essential factors that could improve quality and better address data quality issues. International data assessment and improvement frameworks are reviewed to guide and stipulate the required techniques that should be employed to improve the quality of the data being used in a corporation and in organizational contexts, including the techniques of cross-validation and bootstrapping.

Cross-validation is recommended when the data available for model building and testing are limited. Bootstrapping is also used in cases where there is limited data. Both techniques are essential in measuring and establishing data quality levels that should be adopted.

The assessment of social networks is done in a more in-depth manner to ensure that all social network domains can be analyzed in a successful and comprehensive manner [1, 2]. Peer-to-peer applications, question and answer portals, review portals, weblogs, wikis, and forums are all considered social network domains that can create an effective platform where people can exchange ideas and information. It is essential to regulate the activity occurring on various platforms to ensure that the quality of information is not compromised.

Overall, the above studies point out that poor quality data can cause problems, and the cost of solving these problems is high. At the same time, these studies aim to assess the quality of the data and determine a solution to make the data easier to analyze and use.

2.2   Data Quality Dimensions

There are multiple dimensions in data quality assessment. Researchers have been discussing which dimensions are critical to data quality and how each dimension is critical. Wand et al. [16], Jour et al. (2017), and Hassany et al. (2013) state that the following four dimensions are generally considered the most important.

- Completeness
- Timeliness
- Accuracy
- Consistency

To measure data quality, each dimension must be examined and evaluated to obtain a score that evaluates the quality assessment [10]. Each dimension represents a different aspect of the data and has different approaches for obtaining quality assessment scores. The score can also be used to verify the success or unsuccess of the data cleaning process. To understand each dimension of data quality, it is important to understand the role of each dimension. Completeness deals with missing values in the dataset, while timeliness deals with the validity of data. Accuracy is about the authenticity of the data, and consistency is about the form of the data. In this work, we follow the definitions of accuracy, completeness, and timeliness from Liu et al. [10].

2.3   Sample Techniques

This section introduces sampling techniques and explains these terms: simple random samples, stratified samples, clustered samples, and systematic samples. A simple random sample is the most accessible type of probability sample. This sampling range includes the entire dataset. In a simple random sample, each element in the dataset has an equal probability of being selected as a sample. One way to obtain a random sample is to give each element in the dataset a unique ID, and then use a random number generator to determine the elements to include.

In a stratified sample, based on feature attributes such as gender, age, and location, this technique divides the dataset into small subgroups called strata. It then randomly selects elements from each stratum to form a sample, thereby improving the representativeness of each feature in the sample.

In a clustering sample, it also involves dividing the dataset into subgroups, the difference is that each subgroup has similar characteristics to the entire dataset. After grouping, the entire subgroup is randomly selected, instead of sampling from each subgroup.

The selection of systematic sampling elements is systematic; only the first element is randomly selected, and the other elements are selected at fixed intervals. First, all the elements are put together in order, where each element has an equal chance of being selected. Next, the interval is defined, and the first element is randomly selected to obtain the sample.

3   Methodology

In this work, we first determine the sample size based on the original dataset, confidence interval, and confidence level. Next, four sampling techniques (simple random sample, stratified sample, cluster sample, and systematic sample) are applied to get samples. In addition, control datasets are generated based on the original dataset to verify the sampling results. Then, the samples are evaluated in three dimensions to obtain a quality score. Finally, sampling methods are compared based on quality scores. This process is shown in Fig. 1. We focused on data quality using the dimensions of timeliness, completeness, and accuracy. The programming language Java was used to implement this approach and calculate the three dimensions. Each dimension was evaluated by a separate program, and the sample size was determined by examining the size of the original file.

The first dimension is accuracy, which refers to the conformity of the recorded value with the actual value. The program reads the original dataset and creates four control datasets. Each control dataset has a different percentage (D1: 10%, D2: 20%, and D3: 30%, D4: 40%) of updated entities. To update values in the dataset, we counted the number of data units in the file by multiplying the numbers of rows and columns. A data unit is considered correct if the data unit maps back to the original dataset. If the data unit does not map back to the original dataset, the data unit is considered incorrect. The degree of accuracy of a data unit is 1 when the unit is correct.

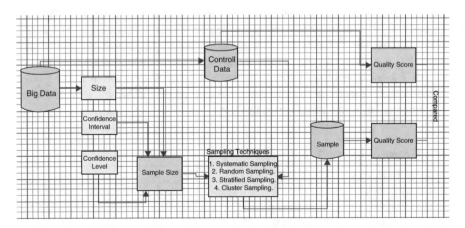

Fig. 1 The process of sampling methods for data quality assessment

The second dimension is completeness. A data unit is complete if the data unit has been assigned a value conforming to the data definition. Otherwise, the data unit is incomplete. The degree of completeness of a data unit is 1 when the unit is complete. The program read the original file and created four datasets. Each dataset has a different percentage (D1: 15%, D2: 25%, D3: 35%, and D4: 45%) of missing entities. The removed data units are randomly selected.

The third dimension is timeliness, which refers to the recorded value being within the relevant date range. The program read the original file and added two new attributes, insertion time and update time. Both times are randomly generated with the condition that the update time is later than the insertion time. Storage time is the duration from when the data are updated to the present.

After the datasets are created for the three dimensions, a method was used to obtain the samples for each dimension. The four sampling techniques are as follows: systematic sampling, sample random sampling, stratified sampling, and cluster sampling. Systematic sampling starts from a predefined line and jumps to a specific line until it reaches the size of the chosen sample. The sample random sampling is randomly selected from the dataset. The third sample is stratified sampling starting from the predefined line and jumping to the descent line. A random line is chosen until the list size is obtained. The fourth type is the cluster sample, which divides the dataset into groups and chooses data randomly from each group until it reaches the target sample size.

4   Experiment and Results

4.1   Experimental Setting and Datasets

The dataset used is student information, a dataset that has approximately 200,000 records with the following five attributes: *idAssessment, idStudent, dateSubmitted, isBanked,* and *score*. Using the four sampling techniques, a sample size of 200 and 2,000 were used for this work. Each technique generates three samples per dataset. Table 1 shows the samples' information. In order to assess different quality dimensions, we randomly deleted and updated some data units, and then evaluated and compared different sampling techniques.

Table 1 Samples and sampling information

Sampling Techniques	Sample Size	D1	D2	D3	D4
Systematic Sampling	200 and 2000	S1;S2;S3	S1;S2;S3	S1;S2;S3	S1;S2;S3
Random Sampling	200 and 2000	S1;S2;S3	S1;S2;S3	S1;S2;S3	S1;S2;S3
Stratified Sampling	200 and 2000	S1;S2;S3	S1;S2;S3	S1;S2;S3	S1;S2;S3
Cluster Sampling	200 and 2000	S1;S2;S3	S1;S2;S3	S1;S2;S3	S1;S2;S3

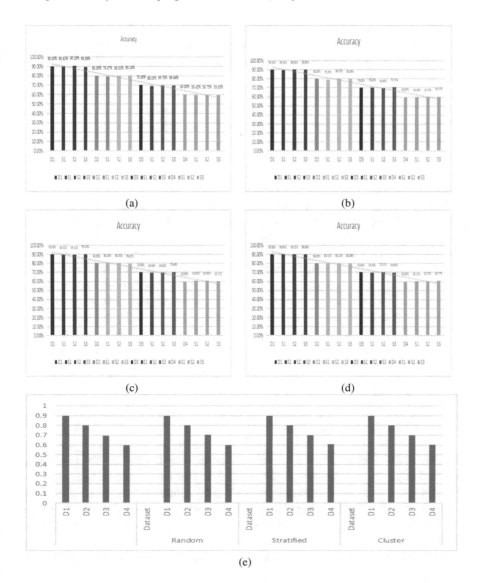

Fig. 2 Comparison of sampling methods for accuracy

4.2 Experimental Results

Fig. 2 shows the comparison results of four sampling methods for accuracy. Fig. 2a–d are the results of systematic, random, stratified, and cluster sampling, respectively. Fig. 2e shows the average accuracy of each sampling method. The x-axis represents the ID of the dataset/sample. There are three datasets, D1, D2, and D3. Each dataset has three samples, S1, S2, and S3. The blue bars are based on D1, the green bars are based on D2, and the red bars are based on D3. The y-axis represents the degree of accuracy. In order to compare the samples with the dataset, the number of updated entities in the samples are counted. Based on the counted number and sample size, the degree of accuracy can be calculated. In order to compare the different sampling methods, we calculated the average accuracy of the samples and then compared the different sample techniques to determine which one is closest to the dataset.

Figure 3 shows the comparison results of four sampling methods for completeness. Figure 3a–d are the results of stratified, random, systematic, and cluster sampling, respectively. Figure 3e shows the average completeness of each sampling method. The x-axis represents the ID of the dataset/sample. To compare the samples with the dataset, some data units are randomly deleted. There are three datasets, D1, D2, and D3. Their completeness is 98%, 95%, and 93%, respectively. Each dataset has three samples, S1, S2, and S3. The blue bars are samples based on D1, the green bars are samples based on D2, and the red bars are samples based on D3. The y-axis represents the degree of completeness. To compare the different sampling methods, we calculated the average completeness of the samples, as shown in Fig. 3d. Then, we compared the different sample techniques to see which is closest to the dataset.

To calculate timeliness, we added two columns to the dataset. One column is for insertion time and the second column is for the time of the update. Both columns were randomly generated. The update time will always be more recent than the insertion time. Figure 4a is the stratified sampling, Fig. 4b is the random sampling, Fig. 4c is the systematic sampling, Fig. 4d is the cluster sampling, and Fig. 4e shows the average timeliness of each sampling method. In these charts, we see the difference between the techniques used to determine what method is the most accurate. The absolute value is used to determine which technique is the best.

Tables 2 and 3 show the results of using one sample T-test for sampling methods: cluster, random, stratified, and systemic based on sample sizes of 200 and 2,000, respectively. The null hypothesis (Ho) addresses that the sample mean is equal to the dataset mean; the alternative hypothesis (Ha) addresses that the sample mean does not equal to dataset mean. This research uses .05 as a significant level. When the P-value is less than .05, the conclusion is that there is enough evidence to infer that the sample mean is different from the dataset mean; otherwise, it concludes that there is not enough evidence to infer that the sample mean is different from the dataset mean.

The P-values lower than .05 are highlighted in the tables. From the results of one sample T-tests in this research, it concludes that when the sample size is 200 or 2000, the sample means of using clustering, random, and systematic sampling methods are equal to the dataset mean; however, the results show that the P-values using stratified sampling method are less than .05, which infers that the sample mean is different from the dataset mean.

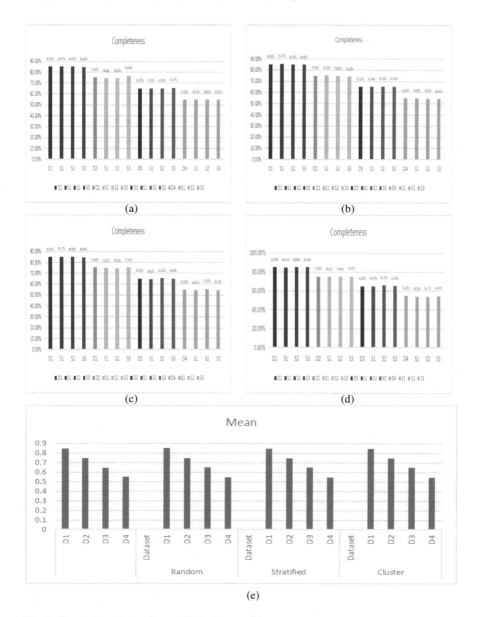

Fig. 3 Comparison of sampling methods for completeness

5 Conclusions and Future Work

Quality assessment is an essential part of ensuring accurate data analysis results and high-quality data. This type of assessment acts as a filter to eliminate data defects so

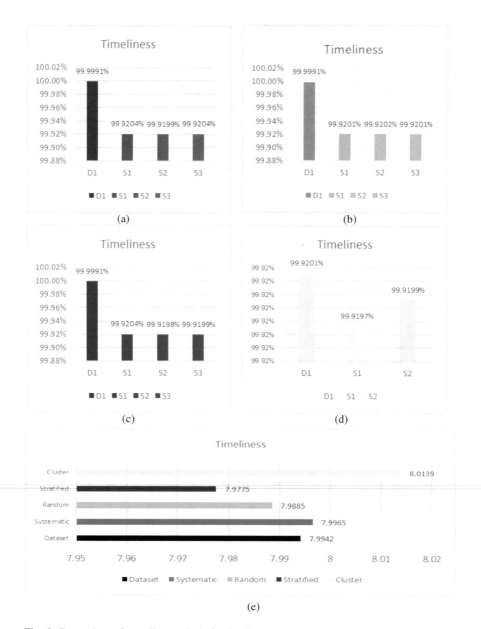

Fig. 4 Comparison of sampling methods for timeliness

that the data is comprehensive and easy to understand before analysis. Determining the quality of data can help improve data defect detection while reducing the time and resources allocated to ensure high quality in data.

Table 2 Results of one sample T-test for sample size 200

Sample Method	Sample	Sample Mean	Standard Deviation	Standard Error	T Statistics	P-Value
Cluster	S1	74.8315	19.2956	1.4463	0.6694	0.504121588
	S2	75.5583	18.8277	1.4747	0.1636	0.870234097
	S3	76.8723	18.4468	1.5535	0.6905	0.490992122
	S4	74.7815	19.1033	1.7512	0.5814	0.562111992
Random	S1	73.9667	18.8564	1.4055	1.3041	0.193866546
	S2	74.9880	20.0578	1.5568	0.5213	0.602826289
	S3	76.8106	19.5163	1.6987	0.5952	0.552744743
	S4	75.5902	18.2525	1.6525	0.1267	0.899370454
Stratified	S1	69.4239	12.6021	0.9290	6.8626	**1.01291E-10**
	S2	69.0061	12.7667	0.9969	6.8145	**1.7408E-10**
	S3	70.3309	12.3102	1.0441	5.2375	**5.94411E-07**
	S4	69.2500	12.8914	1.1395	5.7480	**6.35009E-08**
Systematic	S1	74.2346	21.3807	1.5981	0.9793	0.328777759
	S2	74.2438	20.6485	1.6324	0.9531	0.341994236
	S3	74.2980	20.8834	1.6995	0.8835	0.378354948
	S4	73.5299	20.2372	1.8709	1.2131	0.227549359

Table 3 Results of one sample T-test for sample size 2000

Sample Method	Sample	Sample Mean	Standard Deviation	Standard Error	T Statistics	P-Value
Cluster	S1	75.6400	19.2941	0.4544	0.3511	0.72556647
	S2	75.6765	18.1946	0.4516	0.2725	0.785308336
	S3	76.2233	18.6303	0.4984	0.8502	0.395381051
	S4	76.5498	18.5585	0.5344	1.4038	0.1606451
Random	S1	76.5196	18.3510	0.4347	1.6564	0.097815355
	S2	75.9968	18.5016	0.4660	0.4232	0.672171544
	S3	75.5871	19.1629	0.5100	0.4166	0.677020288
	S4	76.3905	18.8169	0.5430	1.0883	0.276664606
Stratified	S1	71.9346	15.5313	0.3673	10.5227	**3.63405E-25**
	S2	72.0107	15.5673	0.3914	9.6804	**1.42952E-21**
	S3	72.3751	15.1844	0.4086	8.3810	**1.27715E-16**
	S4	72.6245	15.3002	0.4386	7.2394	**7.96759E-13**
Systematic	S1	75.7756	18.6187	0.4393	0.0545	0.956512752
	S2	75.5722	18.8825	0.4753	0.4782	0.632542751
	S3	75.6257	18.6660	0.5003	0.3475	0.728268121
	S4	75.8448	17.9386	0.5224	0.0865	0.931052883

Our contribution includes the following four aspects. First, use four different sampling techniques to obtain an approximate quality assessment. Second, assess the timeliness, completeness, and accuracy of the samples. Third, analyze and compare the evaluation results. Experimental results verify the effectiveness and accuracy of the provided approach. Even if the quality of the original dataset is low, the sample-based assessment approach can effectively reflect the quality of the dataset. We infer that the quality of the dataset itself will not affect the quality assessment of the sample. Besides, by comparing different sampling methods, the experiment provides sample processing suggestions for future quality assessment.

There are some limitations to this work. We only focus on timeliness, completeness, and accuracy. Other dimensions of quality are not involved. The four sampling methods used in this paper belong to probability sampling. Non-probability sampling is not used. The dataset is structured data, and thus further evaluation is required for unstructured data. For future work, we will use more varied datasets to verify this proposed approach. In addition, we will apply this approach to data cleaning to improve data quality.

Acknowledgments Dr. Hong Liu appreciates the support provided by the IUK Summer Faculty Fellowship and Grant-in-Aid program.

References

1. P. Alpar, S. Winkelsträter, Assessment of data quality in accounting data with association rules. Expert Syst. Appl. **41**, 2259–2268 (2014)
2. H. Chen, D. Hailey, N. Wang, P. Yu, A review of data quality assessment methods for public health information systems. Int. J. Environ. Res. Public Health **11**, 5170–5207 (2014)
3. A.K. Elmagarmid, P.G. Ipeirotis, V.S. Verykios, Duplicate record detection: a survey. IEEE Trans. Knowl. Data Eng. **19**, 1–16 (2006)
4. V. Gudivada, A. Apon, J. Ding, Data quality considerations for big data and machine learning: going beyond data cleaning and transformations. Int. J. Adv. Softw **10**, 1–20 (2017)
5. B.T. Hazen, C.A. Boone, J.D. Ezell, L.A. Jones-Farmer, Data quality for data science, predictive analytics, and big data in supply chain management: An introduction to the problem and suggestions for research and applications. Int. J. Prod. Econ **154**, 72–80 (2014)
6. M. Jesmeen, J. Hossen, S. Sayeed, C. Ho, K. Tawsif, A. Rahman, E. Arif, A survey on cleaning dirty data using machine learning paradigm for big data analytics. Indonesian J. Electr. Eng. Comput. Sci. **10**, 1234–1243 (2018)
7. N. Laranjeiro, S. N. Soydemir, J Bernardino, A survey on data quality: Classifying poor data. in *2015 IEEE 21st Pacific rim International Symposium on Dependable Computing (PRDC)*, IEEE, 2015, pp. 179–188.
8. H. Liu, T. A. Kumar, J. P. Thomas, Cleaning framework for big data-object identification and linkage. in *2015 IEEE International Congress on Big Data*, IEEE, 2015, pp. 215–221.
9. H. Liu, A. K. Tk, J. P. Thomas, X Hou, Cleaning framework for bigdata: An interactive approach for data cleaning. in *2016 IEEE Second International Conference on Big Data Computing Service and Applications (BigDataService)*, IEEE, 2016, pp. 174–181.
10. H. Liu, Z. Sang, S Larali, Approximate quality assessment with sampling approaches. in *2019 International Conference on Computational Science and Computational Intelligence (CSCI)*, IEEE, 2019, pp. 1306–1311.
11. L. Marston, J.R. Carpenter, K.R. Walters, R.W. Morris, I. Nazareth, I. Petersen, Issues in multiple imputation of missing data for large general practice clinical databases. Pharmacoepidemiol. Drug Saf. **19**, 618–626 (2010)

12. I. Mlýnková, Quality assessment social networks: A novel approach for assessing the quality of information on the web. Proc. 8th ACM Int' l Workshop Quality in Databases (QDB 10), in *Proc. 36th ACM Int' l Conf. Very Large Data Bases (VLDB 10)*, 2010. pp. 1–10.
13. Moore, S, How to Create a Business Case for Data Quality Improvement. Retrieved July, 19, 2017 (2017).
14. L.L. Pipino, Y.W. Lee, R.Y. Wang, Data quality assessment. Commun. ACM **45**, 211–218 (2002)
15. K.S. Ryu, J.S. Park, J.H. Park, A data quality management maturity model. ETRI J. **28**, 191–204 (2006)
16. Y. Wand, R.Y. Wang, Anchoring data quality dimensions in ontological foundations. Commun. ACM **39**, 86–95 (1996)
17. P.M. Woodall, M. Oberhofer, A. Borek, A classification of data quality assessment and improvement methods. Int. J. Inf. Qual. **3** (2014). https://doi.org/10.1504/IJIQ.2014.068656
18. P. Woodall, A. Borek, A. Parlikad, Evaluation criteria for information quality research. Int. J. Inf. Qual. **4**(2), 124 (2016)

A Resampling Based Semi-supervised Learning Analysis for Identifying School Needs of Backpack Programs

Tahir Bashir, Seong-Tae Kim, Liping Liu, and Lauren Davis

1 Introduction

Globally, food security is one of the major concerns of many governments and nongovernmental organizations. Food security is defined as a situation when people always have physical, social, and economic access to adequate, safe, and nutritious food that meets their dietary needs and food preferences for an active and healthy life [1]. If people cannot have such access occasionally or consistently, then they are in a certain level of food insecurity. A survey [2] showed that 11.1% of U.S. households were food insecure in 2018, and the highest prevalence rate of food insecurity occurred in 2011 due to the financial crisis in 2008. Food insecure households have difficulty providing adequate, nutritious food for all their members including children due to lack of resources. The same survey also characterized several salient findings. Single mother households and households with incomes below poverty level suffered the most from food insecurity. One in seven households with children were food insecure in 2018. The prevalence rate of food insecurity is heterogeneous across the country as some Midwest states and Sunbelt states had higher rates than the rest of states.

Children with food insecurity often experience poor health conditions and poor academic performance. This is especially noticeable during the weekends and

T. Bashir · S.-T. Kim (✉) · L. Liu
Department of Mathematics and Statistics, North Carolina Agricultural & Technical State University, Greensboro, NC, USA
e-mail: tmbashir@aggies.ncat.edu; skim@ncat.edu; lliu@ncat.edu

L. Davis
Department of Industrial and Systems Engineering, North Carolina Agricultural & Technical State University, Greensboro, NC, USA
e-mail: lbdavis@ncat.edu

© Springer Nature Switzerland AG 2021 823
R. Stahlbock et al. (eds.), *Advances in Data Science and Information Engineering*,
Transactions on Computational Science and Computational Intelligence,
https://doi.org/10.1007/978-3-030-71704-9_60

holiday breaks for the time outside school, when children arrive at school after the break with behavioral problems, absentees, altered performance in school activities, etc. [3]. Food insecurity among African-American and Hispanic/LatinX American children is more common, as prevalence rates among these minorities constantly exceed the national average [4]. In response to child food insecurity and hunger, many formal and informal programs have been provided in schools, including the food backpack programs (BPP). The BPPs provide children (with the written permission from their parents) backpacks filled with food to take home on weekends and holiday breaks [5, 6] when school cafeterias are unavailable. Since the first documented school-based food BPP began in Arkansas in 1994, bringing together private donors, faith communities, and public schools, the BPPs have come into national prominence as a response to a perceived crisis of child hunger in America, reaching more than 800,000 children across more than 45 states.

The positive effects of the BPPs can be seen by reported improvements in test scores, positive behavioral gains, decrease in the number of unexcused absences, and increased recognition of potential career paths. However, a recent study [7] assessing the BPP models concludes that some BPPs fit poorly with the needs of most food-insecure children in America and that there are risks of detrimental effects related to worry, shame, and family functioning disarray.

To avoid the possible negative effects that BPPs may bring, a careful study is needed to provide the organizations with some prior information about the true needs of the students before they contact the schools. A previous study [8] utilized multiple linear regression analysis on the data from a backpack program organization and the schools it served in 2 years. The study considered five variables: absentees, rural/urban school locations, minority status, Title I, and low-income students, all in percentages. It reveals that the percentage of low-income students is a significant factor. Through various feature selection methods, a simple regression prediction model is obtained with only one or two predictors.

A limitation of the previous study [8] comes from the limited available data with about 20 schools served by the organization. In linear regression, the sample size rule of thumb is that the regression analysis requires at least 20 cases per independent variable. The data for the 20 schools served by the BPP are so-called labeled data, while the data for the other 100 schools in the same county not served by the organization are unlabeled. With the 20 labeled data, the traditional supervised learning analysis such as the linear regression may not provide us strong statistically supported insights. When there are far more unlabeled data than the labeled data, unsupervised learning is usually employed.

The goal of this study is to propose a novel method to provide a local BPP with a selection guideline for choosing the next schools when the program expands its service. This study is an extension of the previous study [8]. This study included additional schools and one more year of data from the BPP organization as well as more variables (total 12). Most importantly, to efficiently utilize the limited data from the served 21 schools while considering the information from the remaining 104 schools at the same time, a novel resampling based semi-supervised learning (RSSL) method is developed and employed in this study. Our proposed RSSL builds a classifier using the training data combining the currently served schools and a portion of unserved schools and predicts the serving status of the remainders of the unserved schools. This step allows us to obtain the likelihood of each unserved school to be a served school via a resampling method. The classification models include logistic regression (LG), Naïve Bayes (NB), decision tree (DT), random forest (RF), and support vector machines (SVM).

2 Methodology

In Guilford County of North Carolina, there are 125 K-12 schools including 65 elementary, 20 middle schools, 24 high schools, and 16 alternatives in 2019 where this classification was based on the school names. Among 125 schools, 21 schools were served by the BPP, and 104 schools were not. For these unserved schools, a resampling based SSL is developed and employed that combines a small number of served schools with a proportion of unserved schools obtained through a random sampling during training. In the following, after a brief description of the traditional classification models and the general understanding of the supervised/unsupervised learning, our resampling-based SSL algorithm is introduced.

2.1 Supervised Learning and Selected Algorithms

Supervised learning operates on a training sample with labeled data. The following predictive/classification models [14] are supervised learning methods, utilized in this SSL analysis. In the following, the data of the n observations are denoted as $\{(X_i, Y_i)\}_{i=1}^{n}$, where $X_i = (x_{i1}, x_{i2}, \ldots, x_{ip})'$ are for the p predictor measurements and Y_i is for the response measurement, which is Yes ($Y_i = 1$) when the school is served by the BPP and No ($Y_i = 0$) when the school is not served by the BPP in our case.

2.1.1 Logistic Regression (LG)

Logistic regression is the appropriate regression analysis to conduct when the dependent variable is categorical with binary or multiclass labels. The probability that $\mathbf{Y} = \mathbf{y}$ given $\mathbf{X}$ is determined by the logistic function as follows:

$$P(Y = y|x) = \frac{e^{\beta_0 + \beta_1 x_1 + \beta_2 x_2 + \cdots + \beta_p x_p}}{1 + e^{\beta_0 + \beta_1 x_1 + \beta_2 x_2 + \cdots + \beta_p x_p}}, \tag{1}$$

which is a nonlinear sigmoid function. In equation (1), $P(Y = y|x) = P(y|x)$ and y is $1 = Yes$ or $0 = No$. A simple manipulation results in the following linear form called the logistic regression,

$$\text{logit} \; (\pi) := \ln \left(\frac{P \left(Y = y | X \right)}{1 - P \left(Y = y | X \right)} \right) = \beta_0 + \beta_1 x_1 + \beta_2 x_2 + \cdots + \beta_p x_p,$$

where $\pi = P(Y = 1 | X)$. The prediction of Y is made based on the predicted probability, $\hat{\pi}$.

2.1.2 Naïve Bayes (NB)

A Naïve Bayes (NB) classifier applies Bayes' theorem with the assumption of conditional independence among the predictors given the status of the response variable. Conceptually, the NB probability model is derived as

$$\text{posterior} = \frac{\text{prior} \times \text{likelihood}}{\text{evidence}}.$$

Mathematically, the conditional probability of Y given X is defined as

$$P \left(Y = y | x \right) = \frac{P \left(Y = y \right) \prod_{i=1}^{n} P \left(x_i | y \right)}{P \left(x_1, x_2, \cdots x_n \right)}, \tag{2}$$

where $P(Y = y)$ is the prior probability, $\prod_{i=1}^{n} P \left(x_i | y \right)$ is the posterior conditional probability, and $P(x_1, x_2, \cdots x_n)$ is the total probability, the joint density of the predictors. The maximization of the numerator concerning y in equation (2) leads to the prediction of Y.

2.1.3 Decision Tree (DT)

A tree-based classifier involves a set of splitting rules in decision-making process. The decision tree (DT) method is very popular because it is simple to implement and easy to interpret. Decision tree classifier uses a decision tree (as a predictive model), which grows from observations about an item (represented in the branches) to conclusions about the item's target value (represented in the leaves). For the classification tree, leaves represent class labels and branches represent conjunctions of input features that lead to those class labels. Individual leaves contain a single class label which is determined by a majority rule. For the classification, each node is split based on the Gini index or entropy, and the terminal nodes provide the prediction of Y based on the majority rule.

2.1.4 Random Forest (RF)

Although the decision tree is a popular classifier, it is very sensitive to overfitting, which may not be fully solved with pruning. Random forest (RF) is an ensemble learning method for classification and regression tasks that are operated by growing and aggregating multiple decision trees in training. An RF method builds many decorrelated trees on the bootstrapped training samples where each tree uses different samples and predictors, which are in general smaller than the original sample size and the number of predictors, respectively. The final prediction of Y is made based on the average of the generated models. The RF method is known to alleviate the overfitting issue.

2.1.5 Support Vector Machines (SVM)

The support vector machine (SVM) is a popular supervised learning algorithm for both regression and classification like DT and RF. The SVM classifies the categorical response variable using a flexible hyperplane with support vectors where the hyperplane could be the linear, polynomial, and nonparametric kernel. Hence, the SVM is used in a wide range of applications including linear, nonlinear, and high-dimensional classification as well as clustering. More details about the SVM definition and implementation can be found in [9].

2.2 Unsupervised Learning

Unsupervised learning is a type of machine learning algorithm used to draw inferences from datasets consisting of input data without predetermined labeled responses (Y), called the unlabeled data. Since there is no response in the data, unsupervised learning is more challenging compared to supervised learning. Unsupervised learning is often utilized as an exploratory data analysis. Therefore, unsupervised learning techniques usually provide tools to visualize the data in a reduced structure and/or discover subgroups among the variables or the observations which are closely related to principal components analysis and clustering.

2.3 Resampling and Semi-supervised Learning

With the limited available observations, a resampling technique needs to be employed. As an indispensable tool in modern statistical and machine learning, resampling techniques repeatedly draw samples from a training set and refit a model of interest on each sample to obtain additional information about the fitted model. This technique helps us obtain information that would not be available from fitting the model only once using the original training sample. There are two commonly used resampling methods: cross-validation and bootstrap [10]. Both methods are important tools in the practical application of many statistical learning procedures. In this study, the bootstrap is employed, multiple sampling of the training set with replacement.

In practice, it is not easy to obtain labeled data because they are usually associated with cost and time. Semi-supervised learning (SSL) is the machine learning approach concerned with building better classifiers and repressors with utilizing unlabeled data [11]. SSL requires several assumptions: (1) continuity – similar labels (0 or 1) are close to each other; (2) cluster – similar labels have discrete clusters separable, and (3) manifold – data lie on a manifold of much lower dimension than the input space [12]. In this study, some schools are not clear whether they should receive BPP service or not. They need to be defined clearly which group they belong to. Therefore, semi-supervised learning is an appropriate machine learning tool for this kind of unclear labeled data or the data that has hidden characteristics. When used in conjunction with a small amount of labeled data, the SSL can considerably improve the model's performance in terms of accuracy and precision [12, 13].

2.4   Proposed RSSL Algorithm

For the SSL method, it is important to appropriately create the training dataset, which is composed of a part of unlabeled data and labeled data. Our data does not consist of a labeled data set and an unlabeled data set. Instead, we have the 21 served schools and the 104 unserved schools for the BPP. Our goal is to predict which of the unserved schools should have a priority for the BPP service. Hence, we slightly modified the conventional SSL. We define the training set by combining the served schools and the k portion of the unserved schools. The $1 - k$ portion of the unserved schools is used as the test dataset. Next, we can estimate the likelihood of each unserved school predicted as served using the bootstrap method, which allows us to determine the rank of the BPP service. In a nutshell, our proposed method combines the resampling technique and the semi-supervised learning method, called the *resampling-based semi-supervised learning* (RSSL). The following explains the RSSL developed for the school needs for BPP program:

Step 1: Split the schools unserved into two groups with k and (1−k), 0<k<1

Step 2: Create a training data set using the schools served and the k portion of the schools unserved

Step 3: Build predictive models in the training data using the following selected models/algorithms: Logistic Regression, Naïve Bayes, Decision Tree, Random Forest, and Support Vector Machine.

Step 4: Apply the prediction models to the test data (the 1−k portion of the schools unserved) to obtain the predicted values (served = 1or unserved = 0)

Step 5: Repeat N times of Steps 1–4. At each iteration, the k portion of schools is changed due to the random sampling.

Step 6: After the N times iteration, for each school (of the schools unserved) and each classifier, the prediction measure (PM) is calculated:

$$PM = \frac{\text{the number of 1's (predicted as the served school)}}{\text{the number of times selected for the test data in N iterations}} \tag{3}$$

2.5 The Ensemble of the SSL Results

In general, supervised learning performance is evaluated using various assessment metrics such as accuracy, recall, precision, specificity, F1, and Mathews correlation coefficients [14, 15]. This study, however, does not calculate these metrics; instead it focuses on the evaluation of likelihood of being served to determine the priority. The priority shall be given to the schools with relatively high PM values from the RSSL analysis aforementioned. However, the RSSL results from using different predictive models may vary and sometimes vary largely, which makes it difficult for us to draw a reasonable conclusion for the recommendation of the schools for BPP. To quantify the ranking for the school needs for the BPP, an ensemble technique may be employed to combine the RSSL results from all predictive models. Various weights may be imposed on the RSSL results from various predictive models. The following two formulas are adopted to calculate the overall ranking for the schools.

$$P(S) = \sum_{j=1}^{5} PM(S, j) * w_j \tag{4}$$

and

$$P(S) = \sum_{j=1}^{5} PM(S, j) * w_r, \tag{5}$$

where the P is for the overall score, S is for the school (total 104), the j is for the predictive model (total 5), the w is for the preset weight, and the r is for the ranking based on the PM values in the descending order of the PM values. The overall ranking for the schools can thus be obtained from the schools' overall scores P-values in the descending order.

3 Data Description

3.1 Data Collection

The data sets for this study were collected from a backpack program organization in North Carolina (NC), the district offices of the Guilford County Schools (GCS) public school system, and the State of North Carolina Department of Public Instruction.

The backpack program organization in NC provided (internal communication) a complete data set for their food bags distribution in three years: 2016–2017, 2017–2018, and 2018–2019. They served for 20 schools in 2016–2017, and 21 schools in the following two years. The food bags data for the earlier years were incomplete. Therefore, this study focuses on these three years. In this study, the response variable is defined as 1 for being served and 0 for not being served by the BPP.

The raw data for each school were collected from the public data on the GCS district websites [16, 17], the School Information Dashboard data. The collected data include race/ethnicity percentages, chronic absence rates, and lists of Title I schools, etc. Also, the State of NC Department of Public Instruction provided data with the rural/urban location of the schools, percentages of economically disadvantaged and low-income students, etc. After some organizing and cleaning up the data consists of 376 observations, 10 variables remain complete for all the 125 schools in the county. These 10 variables are listed in the following:

- Title I (TI), is indicated as Y and N for yes Title I and no Title I, respectively. The Title I define as the fund that school receives if it has a high percentage of children from a low-income family.
- Levels of Schools, elementary school, middle school, high school, or other schools
- School Performance (SP), is the overall school performance.
- Enrollment (EM)
- Academic Assessment (AA)
- Science Score (SS)
- Reading Overall (RO)
- Reading Achievement (RA)
- Reading Growth Score (RGS)
- Mathematics Achievement (MA)
- The number of Low-Income Students (NLI)
- The Percentage of Low-Income Students at the schools (PLI).

3.2 Descriptive Analysis

In this study, the exploratory data analysis of the 2018–2019 data is presented in detail. For the other years, the results are similar and thus skipped here. As can be seen in Table 1, we summarize two categorical variables, Title I status and level of school along with the BPP status. This result shows that Title I status is statistically significantly associated with the serving status of the BPP (P-value = 0.0075) while the level of school is not associated with the BPP. This result indicates that the backpack program provided its service for Title I schools with some priority while there was no difference across school levels in selecting schools

In Table 2, we summarize the numerical variables with the sample mean and the sample standard deviation in each status of the backpack program. The variables of School Performance, Enrollment, and Mathematics Achievement are not statistically significant between the two groups of the BPP. The socioeconomic status measured by the number and the percentage of the Low-Income Student is significantly different between the two groups. Other academic performance measures based on reading, mathematics, and science scores presented a significant difference between the two groups.

4   Results and Discussions

Before implementing our proposed RSSL algorithm, we performed data preprocessing. First, we checked missing values, which were not observed in the 2018–2019 data. Second, we standardized the numerical predictors to avoid possible impact of different units. Third, to reduce the overfitting issue in the training model, we removed the predictors which are not statistically significant between the two categories of the response variable as indicated in Tables 1 and 2.

Table 1 Characteristics of the categorical variables for the backpack program

Variable	Category	Backpack program				P-value
		Yes		No		
		Count	Percent	Count	Percent	
Total		21	16.8	104	83.2	
TI	Yes	17	19	48	46.2	
	No	4	81	56	53.8	0.0075[a]
School	Elementary	15	71.4	50	48.1	
Level	Middle	2	9.52	18	17.3	
	High	2	9.52	22	21.2	
	Other	2	9.52	14	13.5	0.3410[b]

[a]The P-value was obtained from the chi-square test with Yates correction
[b]The P-value was obtained from the Fisher exact test

Table 2 Characteristics of
the numerical variables for
the backpack program

| Variable | Backpack program | | | | P-value |
| | Yes | | No | | |
	Mean	SD	Mean	SD	
AA	42.9	13.5	56.5	14	0.0004
SP	77.8	7.3	79.9	6.6	0.221
EM	591.6	410.3	586.5	383	0.958
NLI	286.4	108.8	214.9	158.9	0.0161
PLI	57.9	19.1	38.5	19.3	0.0002
RO	52.5	14	60.4	12.9	0.0241
RA	53.1	12.7	64.2	14.4	0.0011
RGS	46.9	14.7	61.6	17.1	0.0003
MA	78	10	75.4	10.7	0.297
SS	63	13.5	73.8	14	0.0023

P-values were obtained using a two-sample t-test

4.1 RSSL Result and Impacts of Selected Classifiers

The preprocessed data were separated into the two subsets: 1) the training dataset that consists of entire schools participating in the backpack program and the k = 0.8 portion of the schools that do not participate in the BPP and 2) the test dataset which contains the remaining portion schools that do not participate in the BPP. The RSSL built five different predictive models using the training data and predicted the BPP service status in the test data. We iterated these steps N = 10,000 times as stated in the aforementioned RSSL algorithm. The proposed RSSL was implemented in R 3.62 in R-Studio with packages, *tidyverse, ggplot2,* and *GGally* for the data preprocessing and visualization and packages *tree, randomForest, e1071,* and *naivebayes* for the classifiers where the *e1071* package was used for SVM.

Table 3 presents the analytical results of the resampling based semi-supervised learning to predict the backpack program status. We deidentified the school names on the table for privacy issues. Table 3 displays the top 10 schools selected by the five different machine learning algorithms in the proposed algorithm. The proportion was calculated using the PM of the proposed algorithm in Sect. 2. There are 29 distinct schools consisting of 16 elementary schools, 4 middle schools, 5 high schools, and 4 other schools. Interestingly, several schools were selected by multiple classifiers. School E41 was selected four times except the DT classifier. Schools E38, E36, E23, and E16 were selected by the three algorithms, and 13 schools were selected by two classifiers. As a result, we recommend the Backpack Program organization to consider these schools as a top priority.

The RF algorithm provides a clear cutoff in the proportion at .5 separating the schools into the two groups. The NB and RF provided relatively low proportions overall, and the LG and SVM reported relatively high portions. Although several schools are selected commonly across different machine learning algorithms, the

Table 3 Prediction of school needs of backpack programs using resampling-based semi-supervised learning

| | Machine learning method | | | | | | | | | |
| | LG | | NB | | DT | | RF | | SVM | |
Rank	School	Portion	School	Portion	School	Portion	School	Portion	School	Portion
1	O4	1	M1	0.709	O10	0.929	E29	0.998	O4	0.976
2	H11	0.988	E41	0.655	E23	0.887	E23	0.827	H11	0.874
3	E41	0.976	E23	0.629	E16	0.76	M13	0.793	E36	0.845
4	H6	0.947	E16	0.617	O14	0.741	M12	0.789	E26	0.841
5	E26	0.937	H16	0.61	M20	0.7	O1	0.682	H6	0.831
6	E38	0.926	E36	0.604	E24	0.631	E2	0.594	H9	0.808
7	E36	0.892	E15	0.588	O6	0.615	E21	0.525	E41	0.758
8	E9	0.889	E27	0.584	H17	0.598	M20	0.386	E9	0.731
9	H1	0.835	E3	0.583	E8	0.572	E27	0.366	E38	0.724
10	H9	0.828	E38	0.582	M16	0.558	E41	0.348	M13	0.719

selected schools are diverse. This result warrants further fine-tuning for hyperparameters in the classifiers. For example, LG used all the predictors without variable selection. The DT did not apply a pruning to mitigate overfitting. The RF used 4 predictors for each tree and grew 500 trees. The SVM used the linear kernel at the cost of 10. All these values need to be optimally selected via grid search. Also, it is not easy to directly compare the performance of the five selected classifiers in the real data. The choice of k may affect the result, especially in the small sample size. Simulation studies, carefully designed, will help to address these issues. One should also consider different machine learning techniques such as artificial neural network and deep learning.

4.2 Ensemble Results

An overall score is obtained for each school using the formula provided in equation (5) of Section II.E. The top list of the schools for the BPP needs is then generated accordingly, which is displayed in Table 4.

In Table 4, we report the top 10 schools in the average of the five ranks and the average of the five portions using the formula in equation (5). For the ranking, the lower the ranking number, the higher the priority for the school. For the portion, the higher the portion value, the higher the priority for the school. The average rank selects schools E41, M13, and E23 as the top 3 schools, while the average portion selects schools O4, M13, and H6 as the top 3 schools. School M13 is selected by both methods in the top 3 schools. In this table, we simply used the average value of ranks and the average value of portions to determine the weight rank, where the ranks in Table 3 used equal weights. However, we can consider a couple of different options, which could be considered in future study. First, one can choose the best

Table 4 Weighted ranking of the schools using the rank and portion

Weighted rank	School	Average rank	School	Average portion
1	E41	8.8	O4	0.469
2	M13	10.6	M13	0.444
3	E23	13.2	H6	0.369
4	H9	15.2	E41	0.364
5	E2	15.6	E29	0.361
6	O1	17.2	H12	0.358
7	O4	17.4	E23	0.329
8	M12	17.8	H9	0.269
9	E9	20.6	E26	0.269
10	M6	21.6	E36	0.26

classifier among the five methods aforementioned. Second, one can choose different types of weight instead of equal weight.

5 Conclusion and Future Research

In this paper, the resampling-based semi-supervised learning (RSSL) analysis is proposed to identify the unlabeled observations which are probabilistically closed to the interesting label in the labeled data. After extensive numerical simulations, the proposed RSSL algorithm with the selected classifiers, SVM, DT, LR, NB, and RF, identified potential schools that could be served by the backpack program preferentially. Some classifiers such as the random forest reported a clear separation of the proportion which could aid decision making. The procedure can be easily implemented to analyze the schools in any other areas for their needs for the BPPs with some slight or no modification and limited available data.

Semi-supervised learning is considered a relatively new machine learning approach to utilize unlabeled data and data scientists are trying to be familiar with this sound method. Semi-supervised learning possesses its own mathematical assumptions which necessitates more and deeper study on the assumptions and other aspects that can make SSL more flexible and useful. Last but not least, SSL can work effectively when the data contains labeled and unlabeled ones. Thus, this issue needs further research and study to make it more approachable in broad concepts with the flexibility of dealing with all types of data. Our proposed RSSL has potential to be applied to other application areas that need accurate labeling in the unlabeled data, which is an important alternative to supervised learning.

References

1. Food and Agriculture Organization of the United Nations. *Trade Reforms and Food Security: Conceptualizing the Linkages*. (Food & Agriculture Organization, 2005)
2. Food Security and Nutrition Assistance. (2020, July 20). Retrieved from https://www.ers.usda.gov/data-products/ag-and-food-statistics-charting-the-essentials/food-security-and-nutrition-assistance/
3. A. Coleman-Jensen, Food insecurity among children has declined overall but remains high for some groups. *Amber Waves: The Economics of Food, Farming, Natural Resources, and Rural America, 2019*(1490-2020-857). (2019).
4. J. Bernal, E.A. Frongillo, H.A. Herrera, J.A. Rivera, Food insecurity in children but not in their mothers is associated with altered activities, school absenteeism, and stunting. J. Nutr. **144**(10), 1619–1626 (2014)
5. B.H. Fiese, *Backpack program evaluation* (The University of Illinois at Urbana–Champaign, Urbana, 2013)
6. N. Cotugna, S. Forbes, A backpack program provides help for weekend child hunger. J. Hunger Environ. Nutr. **2**(4), 39–45 (2008)
7. M.E. Ecker, S.K. Sifers, The BackPack Food Program's effects on US elementary students' hunger and on-task behavior. J. Child Nutr. Manag **37**(2), n2 (2013)
8. D. Black, L. Liu, S. Kim, L. Davis, *Proceedings of the 2019 International Conference on Data Science. A prediction model for backpack programs* (2019).
9. B. Schölkopf, A.J. Smola, R.C. Williamson, P.L. Bartlett, New support vector algorithms. Neural Comput. **12**(5), 1207–1245 (2000)
10. G. James, D. Witten, T. Hastie, R. Tibshirani, *An Introduction to Statistical Learning*, vol 112 (Springer, New York, 2013)
11. Matthias, S. (2001). Learning with labeled and unlabeled data. Inst. Adapt. Neural Comput.
12. X. Zhu, A.B. Goldberg, Introduction to semi-supervised learning, in *Synthesis Lectures on Artificial Intelligence and Machine Learning*, vol. 3(1), (2009), pp. 1–130
13. E. Protopapadakis, Decision making via semi-supervised machine learning techniques. *arXiv preprint arXiv:1606.09022*. (2016).
14. D. Chicco, G. Jurman, The advantages of the Matthews correlation coefficient (MCC) over F1 score and accuracy in binary classification evaluation. BMC Genomics **21**(1), 6 (2020)
15. C. Goutte, E. Gaussier, A probabilistic interpretation of precision, recall, and F-score, with implication for evaluation, in *European Conference on Information Retrieval*, (Springer, Berlin/Heidelberg, 2005, March), pp. 345–359
16. Guilford County Schools Data Dashboard (2020, July 20). Retrieved from https://www.gcsnc.com/Page/44123
17. Guilford County Schools Title I (2020, July 20). Retrieved from https://www.gcsnc.com/domain/5042

Data-Driven Environmental Management: A Digital Prototype Dashboard to Analyze and Monitor the Precipitation on Susquehanna River Basin

Siamak Aram, Maria H. Rivero, Nikesh K. Pahuja, Roozbeh Sadeghian, Joshua L. Ramirez Paulino, Michael Meyer, and James Shallenberger

1 Introduction

The Susquehanna River is the nation's sixteenth largest river that flows into the Atlantic Ocean. The Susquehanna drains 27,510 square miles, covering half the land area, the states of New York, Pennsylvania, and Maryland. The river wanders 444 miles from its origin at Otsego Lake near Cooperstown, New York, until it empties into the Chesapeake Bay at Havre de Grace, Maryland. The Susquehanna contributes one-half of the freshwater flow into the bay [1]. Since 1971, the Susquehanna River Basin Commission (SRBC) leads the conservation, development, and administration of the Susquehanna River Basin (SRB) resources, reinforcing its mission of providing public welfare through comprehensive planning, water supply allocation, and management of the water resources [1].

The SRBC presents a long history of data-driven approaches, measuring key parameters that contribute to the monitoring and conservation of the basin's resources. The model on this study seeks to develop a predictive tool that can automate the supervision of the river flow and water quality of the SRB. A data-driven approach used several Machine Learning (ML) techniques that permit the interpretation of data in real-time. Mosavi et al. [2] affirms that ML techniques provide improved performance for environmental modeling by introducing a cost-effective solution to the management of water resources [2]. Marçais et al. [3],

S. Aram (✉) · M. H. Rivero · N. K. Pahuja · R. Sadeghian · J. L. Ramirez Paulino · M. Meyer
Harrisburg University School of Science and Technology, Harrisburg, PA, USA
e-mail: SAram@HarrisburgU.edu; MRivero@my.harrisburgu.edu; NPahuja@my.harrisburgu.edu; RSadeghian@harrisburgu.edu; JLRamirez@my.harrisburgu.edu; MMeyer@HarrisburgU.edu

J. Shallenberger
Susquehanna River Basin Commission, Harrisburg, PA, USA
e-mail: jshallenberger@srbc.net

© Springer Nature Switzerland AG 2021
R. Stahlbock et al. (eds.), *Advances in Data Science and Information Engineering*,
Transactions on Computational Science and Computational Intelligence,
https://doi.org/10.1007/978-3-030-71704-9_61

refers to a study during the 1990s, where Artificial Neural Networks (ANN) became popular in this field, considering the prediction of rainfall-runoff processes. The author affirms that interest in Machine Learning methods such as Neural Networks (NNs) for hydrological sciences are being explored in this field [3].

Our proposed model aims to build an environmental monitoring digital dashboard for SRBC using Machine Learning methods such as Linear Regression (LR), Decision Tree (DT), Random Forest (RF), Nonlinear Regression (NLR), Lasso Model (LM), and Neural Networks (NNs) model applied on the Pine creek watershed of the SRB. Furthermore, this study developed a tool using *ArcGIS* (Desktop Help 10.2 Geostatistical Analyst), developed by the Environmental Systems Research Institute (ESRI), that connects and illustrates relationships between river flow and precipitation [4]. The model outcomes will improve the existing approaches into a robust predictive model. The resulting decision-support tool's mission is to continuously add more environmental parameters and deploy the application nationally for government agencies and businesses. This unprecedented study validated the collaboration between data science and the existing environmental policy, providing a powerful tool to make better informed, pragmatic, and more effective decisions to preserve the environment and the water quality of SRB.

2   Machine Learning Approaches

Several studies have used Machine Learning approaches for measuring the rate of river flow in a particular time frame. Akhtar et al. [5], used an Artificial Neural Networks (ANNs) model adapted on a geographical information system that displays the discharge dataset to predict the river flow of the large-scale Ganges river basin located in Asia. In this study, the ANNs model forecasted 3 days of discharges considering as inputs the river flow length, travel time, and the local streamflow measurements. This study intended to expand the forecasting horizon from 3 days to a period of 7–10 days with different combinations of input from the rainfall dataset. As a result, the discharge data along with rainfall information variable performed better for a 7–10 days forecasting [5]. Another study uses Machine Learning for monitoring the flood protection systems. Pyayt et al. [6], integrates the proposed model into an Early Warning Systems (EWS) platform to predict the dike's abnormal behavior characteristics and the probability of the flood in a location. For this methodology, four different types of Machine Learning models were developed using linear regression and neural clouds. Also, a web-based dashboard was built to visualize the detection of abnormal behavior scheme for the EWS [6]. In more recent studies, Neugebauer et al. [7] describe the Artificial Neural Networks (ANNs) as a Machine Learning algorithm suited for modeling complex physical phenomena occurring in time. This study evaluates the hydrological data from the Lyna River located in the northeastern region of Poland [7]. The primary data parameters involved are the water flow and the amount of precipitation from the period of 2000 until 2016. As a result, the ANNs suit the modeling properly predicting the size of the precipitation and the river streamflow.

The impact of Machine Learning approaches has influenced the prediction of water quality management. Lu et al. [8] describe the practice of managing water resources vital for the community's life and economic development. Usually, the water resource management departments employ monitoring points for observing water quality changes instead of predicting its quality. Although the physical methods to predict water quality are labor-intensive and time-consuming, the authors introduce Machine Learning techniques as an intelligent model to deal with the instability of water and the nonlinearity in time series. The methodology of this study proposes two hybrid models: Extreme Gradient Boosting (XGBoost) and Random Forest (RF). The model will predict six water quality indicators: water temperature, dissolved oxygen, pH value, specific conductance, turbidity, and fluorescent dissolved organic matter. This Machine Learning model uses the techniques of hybrid decision-tree to obtain a more accurate and short-term water quality of prediction results [8]. As a result, the prediction stability is higher than current benchmark. This study will implement Machine Learning process, train the model, and evaluate the best performance of the set. In addition, the interpretations and conclusions will assess the feasibility of implementing the study's goals.

2.1 Precipitation and Streamflow

The United States Environmental Protection Agency (EPA) [9] published a survey to inform modelers about the types, resolutions, and sources of precipitation datasets available to the public. It supports environmental modeling that involves observing precipitation, to understand the erosion, transportation of contaminants, and water quality. It also provides information about the past and current conditions of the water, which supports the predictive model that will simulate data to obtain the future state of the water as well as re-create historical conditions. According to the EPA, the study presents simulated data of the precipitation dataset classified into rain gauge, radar, and satellite-based measurements. The survey concludes that selecting the appropriate precipitation dataset has an impact on the effectiveness of hydrological model performance [9]. Modelers must choose datasets carefully and leverage the uncertainty within the model. As a result, modelers have the challenge to find the appropriate dataset that suits the objectives of an environmental model. Previous studies in Machine Learning referred to ungauged streamflow prediction, which is another challenge that modelers can encounter within a study. Besaw et al. (2010) explore the use of Artificial Neural Networks (ANNs) to determine the gauged and/or ungauged streamflow, applying a data-driven technique that is valuable for real-time applications. This model includes two data-driven ANNs that register the time-lagged records of precipitation and temperature from basins located in Northern Vermont. This Machine Learning model predicts the flow from one basin to the nearest one that has a more representative climate input published by the US Geological Survey (USGS) streamflow records [10]. The overall goal of this study demonstrated that the ANNs were capable of predicting streamflow in a nearby basin as accurately as in the basin on which it trained previously. Furthering this application will provide the watershed and water resources stakeholders the information required to make strategic and informed decisions.

Choubin et al. [11] consider that precipitation forecasting is one of the most critical issues of the hydrological cycle, and is also essential to water resource development, planning, and management of droughts. The study's results show that the model of ANN is more accurate than linear regression. Modelers need to research and compare different Machine Learning models to assess the most accurate result in the study river flow prediction [11]. Recent studies in the past year strengthen the need and benefits of applying Machine Learning to make predictions of the river flows and impact the existing hydrological models. For instance, Sidrane et al. [12] highlight the issue that communities are threatened by climate change and that flooding could become a dangerous hazard to the region. The study proposes a multi-basin model of river flooding susceptibility using the geographically distributed data from the U.S. Geographical Survey (USGS). The Machine Learning approach considered a supervised framework to predict measurements of flood susceptibility from a mix of river basin attributes, and the historical records of rainfall and stream height [12]. The results of this model outline the shortcomings of physics-based flood prediction models. The main purpose of this study is to make accessible the flood prediction to all at-risk communities. Moreover, there is a need to introduce real-time data visualization tools for effective decision-making in the field of hydrology. Maguire D. (2008) expands the notion of a Geographic Information System (GIS). A GIS is crucial for the understanding of geography, monitoring, and making decisions based on a specific location [13]. GIS is a technology that captures, manages, manipulates, and visualizes geographic information. Kurakina [14] highlights the GIS significance during environmental impact occurring in water bodies. A GIS can contribute to the preservation of natural resources due to the advanced spatial visualization and capability analysis. Also, the GIS interface runs with the *ArcView* software that enables the user to make real-time and space analysis of the water quality in different measuring points [14].

Many studies have tested the reliability of the GIS. The flood of the Red River in 1997 made a tremendous impact on the citizens of the United States and Canada. Unfortunately, when these natural disasters happen, the individual agencies in floodplain management have the responsibility to react and achieve their function collaboratively across different geographical locations. Simonovic [15] published a report written for the International Joint Commission (IJC) and the Red River Basin Task Force. The IJC proposed a framework titled the Red River Basin Decision Support System (REDES) to provide stakeholders the tools to estimate flood management strategies that will enhance the planning and response area recovery of river floods. This conceptual framework operates with a Development of Decision Support System (DSS), an application technology that includes databases, tools, and flood strategies. This system emphasizes the flood prediction, emergency response, and public involvement [15].

A GIS Dashboard follows standardized guidelines. The U.S. Environmental Protection Agency (EPA) [16] published a framework to develop a Water Quality Surveillance and Response System (SRS) to monitor and manage the distribution of system water quality of a location. The primary goal of the EPA is to detect and respond to water quality incidents quickly. Thus, this system identifies components in both areas of surveillance and response [16]. The EPA designed this guide to help utilities and stakeholders that would like to plan and implement a multicomponent SRS in a location. The EPA presents the general guidelines to plan and build a dashboard that can potentially incorporate data-streams as well as an interactive interface to display the computed data. Chen et al. [17], built a Machine Learning model to evaluate the flood risk at the Yangtze River Delta in China. The methodology included the Random Forest (RF) algorithm, which screened significant indexes of flood risk. Also, a risk assessment model was built based on the Radial Basis Function (RBF) neural network to evaluate the flood risk level in this region from 2009 to 2018 presenting the levels of the flood risk via GIS [17]. The resulting study suggests that the flood risk has been increasing in the area during the past 10 years following the urbanization background. Some of the regions have already started to implement measures to control the flood risk-level sustaining the importance of integrating a real-time dashboard and its stakeholders.

3   Methodology and Datasets

This paper analyzes the Susquehanna River Basin data collected at different sites using several Machine Learning techniques. The data comprises a monthly time-series of three variables from 1980 to 2018. The daily river flow rate of Susquehanna River Basin was predicted using supervised learning. We used several Machine Learning techniques such as Linear Regression (LR), Decision Tree (DT), Random Forest (RF), Nonlinear Regression (NLR), Lasso Model (LM), and Neural Network (NN) model.

The following three variables act as the input to the model: the previous day's flow, temperature, and precipitation. Using the inputs, the model predicted future flows of the river presented in Table 1.

Table 1 lists the dataset used in the prediction. We divided the dataset into training and validation sets and then used the Machine Learning model for training. The model performance was then evaluated on the validation dataset using the performance metrics such as Root Mean Square Error (RMSE) and R-Square.

Table 1 Data sources and parameters

Data from sites	Units	Source	Remarks
Precipitation	Inches	Prism[a]	4 × 4 km raster resolution daily values
Flow	Cubic ft. per sec	USGS[b]	All available historical data
Ambient and water temperature	Deg F	PRISM/USGS[c]	Proxy for evaporation

[a]Publisher: PRISM Climate Group at Oregon State University
[b]U.S. Geological Survey
[c]Publisher: PRISM Climate Group at Oregon State University & U.S. Geological Survey

Table 2 Model results

Method	Fit descriptor	Value (analysis on original data)	Value (analysis on processed data)
Linear regression	RMSE	508.95	5.78
Decision tree	RMSE	209.97	0.53
Random Forest	RMSE	238.27	0.58
NonlinearRegression	RMSE	350.69	0.78
Lasso model	RMSE	366.47	0.81
NN model	RMSE	238.27	0.58

4 Data Exploratory Analysis

Performance metrics of various models are indicated in Table 2. This pilot study provides information on whether it is feasible to predict the Susquehanna River Basin (SRB) river flow rate using a Machine Learning model.

The performance of different models is compared against each other. Figure 1 shows the actual vs. predicted daily flow by the Decision Tree (DT) model between the period of November 2013 and June 2017. The actual flow is plotted on the y-axis in blue color, while the predicted flow is plotted on the y-axis in orange color, and the dates are plotted on the x-axis. The best performance was recorded by the Decision Tree (DT) model. Both Neural Network (NN) and Random Forest (RF) models tied for second best performance. Nonlinear regression (NLR), though captured nonlinearity in data, performed worse than DT and it was followed by the Lasso model (LM). The worst performance belonged to the linear regression model (LR). Further research considers Neural Networks (NN) and applying Deep Learning models to improve its performance. However, a low amount of training data could be an issue.

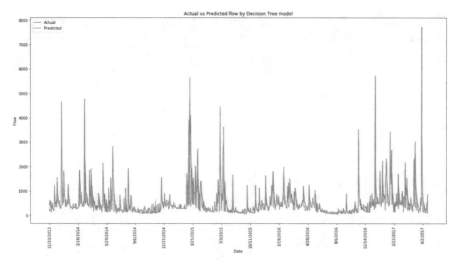

Fig. 1 Actual vs. predicted flow by Decision Tree model

5 Operational Dashboard

The operational dashboard is a versatile and powerful geospatial tool that can be used to guide organizational decisions across numerous datatypes and formats. Key data sets can be focused in one entire screen without losing context. The dashboard's central feature is a map element that includes layer of precipitation focused on the Susquehanna River Basin (SRB). A map is a base form of visualizing data for environmental scientists, both in the field and electronically. Hence, due to their specialized background, this operational dashboard feature allows them as users to access the content in a very streamlined and intuitive way. The layers provide access to the information through the element actions, while further navigation can be directed by layer-based feature queries.

The proposed dashboard modeled data for a robust prediction on the Susquehanna River Basin conditions. Existing water quality portals for the SRB are overly simple, nonnavigable, and hold little real-time data. This model of the operational dashboard illustrated in Fig. 2 displays a group of elements that interact with one another when a feature is selected. The map's element contains layers that show some of the Susquehanna River Basin Commission (SRBC) model sites, the sub-basin boundaries, water trails (rivers, streams, creeks, etc.), the soils, and bedrock geology of the SRB. There is an indicator that displays three features: average discharge, minimum discharge, and maximum discharge of the sites overall. A list element is under the indicator which shows each site, and displays the city and state that it resides in. To avoid pop-ups on the map element, a detail element is below the map, configured to filter the detail depending on what is selected on the list element and on the map element as well. Lastly, there are serial charts on the

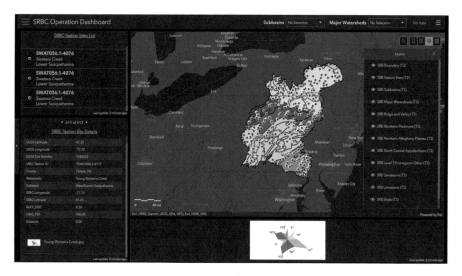

Fig. 2 The GIS Dashboard incorporated modeling data for a robust prediction on the Susquehanna River Basin (SRB) conditions. SRBC boundary (black outline), station sites (dots), sub-basins (red lines), and ecoregions (colored areas) are highlighted. Maucha diagrams are shown in site info side bar and in lower bar for certain sites

left. The mission of this operational dashboard is to integrate the prediction of the river flow for SRBC sites using Machine Learning techniques. This GIS model will continue to evolve when adding more environmental parameters and elements to the selected Machine Learning model. For instance, the initial phase will incorporate flow status into predictions, adding value to the monitoring of water quality and biologic integrity interests.

6 Conclusion

The field of hydrology is expanding its forecasting methodology due to the implementation of Machine Learning models and its integration into a Geographic Information System digital dashboard. Selected past and current environmental monitoring studies have enhanced a data-driven methodology for accurate predictions and comprehensive decision-making for environmental organizations.

The proposed study conducted a Machine Learning approach and data exploration to build a predictive model that monitors the precipitation and streamflow on the Pine Creek of the Susquehanna River Basin. The results described a significant output for accurate forecasting using the DT model that will be integrated into a digital dashboard that will serve as a robust predictive model and decision-making tool for the Susquehanna River Basin Commission (SRBC). Furthermore, the *ArcGIS* software connects datasets with the relationships between the river flow and precipitation contributing to the streamlined access to data that can make forecasting, monitoring, and timely action simpler to organizations [18].

Our proposed model will evolve continuously when adding more environmental parameters and demonstrate its utility when making the deployment of this tool nationwide within government agencies and businesses. The river flow rate is one of the primary variables for assessing the range of aquatic resources that environmental scientists are interested in the river basin. Therefore, the SRBC and the Basin's stakeholders will benefit when this prototype evolves the prediction of river flow and other hydrologic variables (actual & predicted) into reliable water quality and biologic condition predictions, especially at unmeasured settings. Further benefits will be realized when the models evolve to incorporate features of the watershed, particularly, human-induced items.

The SRBC will be able to integrate its environmental policy with data science tools to innovate and improve their analytical decision-making towards a more efficient and accurate model and continue to monitor and protect the Susquehanna River Basin. The recommendations for future studies suggest implementing a Neural Networks (NN) approach, using deep learning techniques which can lead to possible performance improvement. However, modelers should be aware of maintaining a low amount of training data, since it can become an issue in the methodology of the study.

Acknowledgement This research was granted by Harrisburg University Presidential Grant 2020. Also, we would like to thank John Quigley, Director of the Center for Environment, Energy and Economy at Harrisburg University. We would like to thank students Aditya. V. Singh, Sridhar Ravula, Anshuman Chakravarty, Debosruti Dutta, Himangshu Pal, Pranita Patil, and Alyson Marshal who helped with the different steps of this research project.

References

1. The Susquehanna River Basin Commission. *Susquehanna River Basin Facts* (2016). https://www.srbc.net/our-work/fact-sheets/docs/river-basin-facts.pdf. Accessed May 20 2020
2. A. Mosavi, P. Ozturk, K. Chau, Flood prediction using machine learning models. Water **10**, 1–40 (2018) (in press)
3. J. Marçais, J.R. De Dreuzy. Prospective interest of deep learning for hydrological inference. HAL Arch, 688–692 (2018)
4. Environmental Systems Research Institute (ESRI). *ArcGIS Desktop Help 10.2 Geostatistical Analyst* (2014). http://resources.arcgis.com/en/help/main/10.2/index.html
5. M.K. Akhtar, G.A. Corzo, S.J. Van Andel, A. Jonoski, River flow forecasting with artificial neural networks using satellite observed precipitation preprocessed with flow length and travel time information: A case study of the Ganges river basin. Hydrol. Earth Syst. Sci. **13**, 1607–1618 (2009) in press
6. A.L. Pyayt, I.I. Mokhov, B. Lang, V.V. Krzhizhanovskaya, R.J. Meijer, Machine learning methods for environmental monitoring and flood protection. World Acad. Sci. Eng. Technol **54**, 118–123 (2011) in press
7. M. Neugebauer, R. Augustyniak, P. Solowiej, K. Parszuoto, J. Halacz. Hydrological modeling of water flow in the river using artificial neural network, in *18th International Multidisciplinary*

Scientific GeoConference (SGEM), Section Hydrology and Water Resources,. 365–370 (in press) (2018)

8. L. H, X. Ma, Hybrid decision tree-based machine learning models for short-term water quality prediction. Chemosphere **249**, 1–12 (2020)

9. The United States Environmental Protection Agency (EPA), A survey of precipitation data for environmental modeling Office of Research and Development national exposure research laboratory Athens. Georgia, 1–32 (2017) (in press)

10. L.E. Besaw, D.M. Rizzo, P.R. Bierman, W.R. Hackett. Advances in ungauged streamflow prediction using artificial neural network. J. Hydrol. **36**, 27–37 (in press) (2010)

11. B. Choubin, S. Khalighi-Sigaroodi, A. Malekian, O. Kisi, Multiple linear regression, multi-layer perceptron network and adaptive neuro-fuzzy inference system for forecasting precapitation based on large-scale climate signals. Hydrol. Sci. J. **6**, 1001–1009 (2016)

12. C. Sidrane, D. Fitzpatrick, A. Annex, D. O'Donoghue, Y. Gal, P. Bilinksi. *Machine Learning for Generalizable Prediction of Flood Susceptability*, in 33rd Conference on Neural Information Processing Systems, NeurIPS, Vancouver, Canada 1–7 (2019)

13. D. Maguire. GIS and science. *ArcNews Magazine*, 3–8 (2007/2008)

14. N. Kurakina. River pollutants monitored with GIS: analyzing the environmental impact of water bodies in russia. *ArcNews Magazine,* 39–42 (2007/2008)

15. S.P. Simonovic. *Decision Support System for Flood Management in the Red River Basin.* Slobodan P. Simonovic Consulting Engineer Ltd. Winnipeg, Canada 1–41 (1998)

16. The United States Environmental Protection Agency (EPA). Guidance for Developing Integrated Water Quality Surveillance and Response Systems the Water Security Division of the Office of Ground Water and Drinking Water, 1–66 (2015)

17. J. Chen, L. Qian, W. Huimin, M. Deng. A Machine Learning Ensemble Approach Based on Random Forest and Radial Basis Function Neural Network for Risk Evaluation of Regional Flood Disaster: A Case Study of the Yangtze River Delta, China. Int. J. Environ. Res. Pub. Health **17**, 1–21 (in press) (2019)

18. Esri. *ArcGIS: A Cloud-based GIS Mapping Software* (2020). https://www.esri.com/en-us/arcgis/products/arcgis-online/overview Accessed Jun 10 2020

Viability of Water Making from Air in Jazan, Saudi Arabia

Fathe Jeribi ⓘ **and Sungchul Hong**

1 Introduction

Jazan is a region located in the southwest of Saudi Arabia with a border of Yemen [1]. The climate in Jazan is classified as desert [2]. Usually, the temperature in Jazan varies from 73 degrees Fahrenheit to 98 degrees Fahrenheit [3]. However, Jazan area has relatively high humidity (annual average 66%) and constant steady wind (6.8 to 8 miles per hour or 3 to 3.57 meters per second) [3, 4]. This paper proposes a water making process, which can generate water from air. This water generation from air needs a water making device, and this device is operated by electricity. This operational electricity can be generated by solar energy and wind. The viability of this water making method was tested by a computer simulation.

1.1 Solar Energy

Solar energy is a power that results from the sun. It benefits by providing electricity without burning fuels [5]. Solar panel is an equipment that can convert sunlight into electricity [6, 7]. It consists of many photovoltaic cells. Grouping cells together can make a solar panel [8]. The sun is considered one of the main renewable energy sources. Solar energy is indicated as sustainable energy due to its availability when

F. Jeribi (✉)
Jazan University, Jazan, Saudi Arabia
e-mail: fjeribi@jazanu.edu.sa; fjerib1@students.towson.edu

S. Hong
Towson University, Towson, MD, USA
e-mail: shong@towson.edu

© Springer Nature Switzerland AG 2021 847
R. Stahlbock et al. (eds.), *Advances in Data Science and Information Engineering*,
Transactions on Computational Science and Computational Intelligence,
https://doi.org/10.1007/978-3-030-71704-9_62

Table 1 Advantages of solar energy

No.	Advantages	Explanation
1	Renewable	Compared to nonrenewable energy sources, such as coal and fossil fuel, renewable energy sources such as solar, cannot be depleted. In other words, it will be available as long as there is sun.
2	Abundant	The volume of sunlight cannot be imagined. The earth can receive 120,000 TW of sunlight. In other words, it receives 20,000 times more power than that is required to supply the whole world.
3	Environmentally friendly	Using solar energy cannot result in pollution.
4	Availability	Solar energy can be available all over the world.
5	Reducing electricity bills	Solar energy can help people to reduce the electricity bills when they apply it to their homes.
6	Lowering maintenance	Systems of solar energy do not need a lot of maintenance.

Table 2 Sun hours in Jazan, Saudi Arabia, in 2017

Month	Sun Hours
January	235.5
February	244
March	295
April	348.5
May	387.5
June	373.5
July	374
August	365.5
September	365
October	232.5
November	196.5
December	214.5

the sun shines [9]. There are many advantages of solar energy. Table 1 below lists and explains some of these advantages [10].

This paper uses SolarWorld SW 250 Poly as a sample device, and it can generate 250 watts per hours in a good condition [11]. Table 2, below, shows the sun hours in the Jazan region in 2017 [12]. Table 3, below, shows the total of solar energy that can be generated using SolarWorld SW 250 Poly for all months in the Jazan region in 2017.

1.2 Wind Energy

Wind energy, i.e. wind power, is the procedure of making electricity through wind. Contemporary wind turbines are utilized to gather kinetic energy and create electricity [13]. There are many advantages of using wind energy [9]:

Table 3 The total of solar energy generated using Solarworld Sw 250 Poly for all months in the Jazan Region in 2017

Month	Sun Hours	Sun Hours × 250 watts	60 Solar Panels
January	235.5	58,875 watt-hour (Wh)	3,532,500 Wh
February	244	61,000 Wh	3,660,000 Wh
March	295	73,750 Wh	4,425,000 Wh
April	348.5	87,125 Wh	5,227,500 Wh
May	387.5	96,875 Wh	5,812,500 Wh
June	373.5	93,375 Wh	5,602,500 Wh
July	374	93,500 Wh	5,610,000 Wh
August	365.5	91,375 Wh	5,482,500 Wh
September	365	91,250 Wh	5,475,000 Wh
October	232.5	58,125 Wh	3,487,500 Wh
November	196.5	49,125 Wh	2,947,500 Wh
December	214.5	53,625 Wh	3,217,500 Wh

- Wind energy does not produce air or water emissions [9].
- Wind energy does not burn gas or oil. Consequently, it does not affect the environment via transportation and extraction of resources [9].

Wind turbines can be classified into three categories: utility-scale wind, distributed wind, and offshore wind [13].

1. Utility-scale wind.

The scope of this wind turbine is from 100 kw to many mw. In this type, electricity is transmitted to the grid and then distributed to the user, either through operators of power systems or power utilities [13].

2. Distributed wind.

The scope of this wind turbine is less than 100 kw. In this type, electricity is utilized at home, in small businesses, etc. In addition, it is not connected to the grid [13].

3. Offshore wind.

This turbine is built in water. Compared to turbines that are on land, offshore turbines are larger. Consequently, they can generate more electricity [13].

Using 100 kw wind turbine can produce 100 kw per hour at its maximum speed (10 m/s) [14]. For a day (24 h), it can produce 2400 kwh as maximum. Usually, for a whole month, it can generate 72,000 kwh as maximum. In Jazan, it can generate 3 kw at 3.5 meters per second and 100 kw at 10 meters per second [15]. The average wind speed is between 6.8 to 8 mph (miles per hour) [3]. The cost of 100 kw wind turbine is $300,000 to $800,000 [16]. We assumed a small-scale electricity production; if we apply utility scale windmill, it will generate a lot more electricity.

1.3   Water Making Device

An Atmospheric Water Generator is a device that uses the technology of condensing or dehumidification. The goal of this device is extracting water from the humidity. The extracted water is passes through many filters to make it pure drinking water [17]. An Atmospheric Water Generator does the same procedure of refrigerators and air conditioners, or in other words, the procedure of cooling through vaporization. It uses a compressor with the goal of transforming atmospheric air into pressurized air. After that, the resultant air goes through pipes of a condenser. The function of these pipes is reducing the temperature of air to dew point [17]. There are many advantages of using Atmospheric Water Generator. Some of these advantages are off-grid living, water to go, cheap and easy to maintain, and emergency water solutions [18].

- Off-grid Living.
  Occasionally, off-grid living is an outcome of inappropriate circumstances. However, sometimes, it is simply by choices. Many people decide to utilize alternative solutions of water collection and energy to avoid the costs of living in public utilities [18].
- Water to Go.
  When resources have low operational costs, government sectors that travel, like army, can utilize an atmospheric water generator to make water. Atmospheric Water Generator could be very useful in regions or deserts that have little safe drinking water [18].
- Cheap and Easy to Maintain.
  Cost-efficiency is one of the main benefits of Atmospheric Water Generators. In the beginning, you will see that the cost of purchasing the device is more than the cost of buying water from a retailer. However, the whole value enhances within a short time. Another advantage is that no extra costs are required to perform the water making. In addition, the electrical bills are eliminated because of the autonomy of solar and wind electricity generation [18].
- Emergency Water Solutions.
  Using an atmospheric water generator is useful for emergency situations. It can also help people who live in remote regions to get water [18].

2   Water Generation from Air

An Atmospheric Water Generator is a machine that utilizes the technology of condensing or dehumidification. Extracting water from the humidity is the goal of this machine. The extracted water is going over through various filters with the goal to produce pure drinking water [17]. Figure 1, below, shows how water can be collected from the air. People can use solar panels and wind turbines to collect

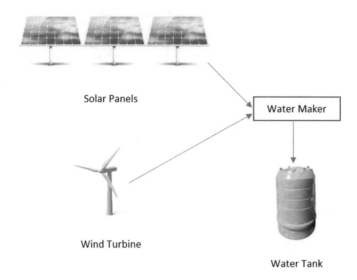

Fig. 1 Water from air (WFA) structure

electricity, which can be used to run water maker device. After collecting water, they can store it in the water tank.

3 Comparison between Water Desalination and Water from Air

To compare the two water making methods, this paper considers two variables: location and price.

- For the location variable, the water generation location could be near a seashore or an inland place that is far away from a seashore. For near seashore locations, desalination plants have an advantage of making water. For locations far from seashore, Water from air (WFA) has an advantage of making water. For an inland location, people can get water from two different ways:

  (a) People can get water from WFA, i.e. atmosphere water generator.
  (b) People can get water from a desalination plant by a water tank truck.

- For price variable, a market considers two factors for price: desalination plant and water from air.

  Desalination plant is an expensive process. It uses electricity and oil burning. In this factor, there are two methods of desalination water:

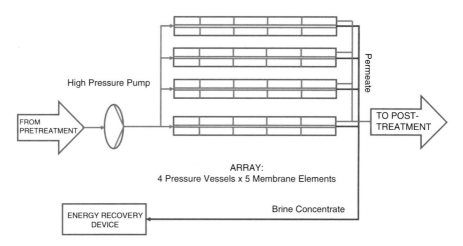

Fig. 2 Reverse osmosis desalination process

(a) Boiling and Cooling Method.

   This method is a traditional way and it has salt disposal problem. When desalination plant finishes desalinating water, they throw the byproduct back to sea. This can cause environmental problems by increasing salt in sea if there are too many desalination plants located in small area. There is a concern about salinity in Persian Gulf because they are too many desalination plants in the area.

(b) Reverse Osmosis Desalination Process Method.

   This method is a different process of desalination and it is made of a high-pressure pump. This pump is followed by an energy recovery device as well as elements of the reverse osmosis membranes (Fig. 2) [19]. This method can convert about 40% of sea water into fresh water. The remaining water has higher salinity and too much of this can cause some environmental issues.

There are many types of Atmospheric Water Generators. One example is AKVO atmospheric water generator. AKVO is an abbreviation for Advanced Knowledge Ventures & Opportunities. It is an Indian organization established in 1999. The goal of this organization is to supply pure and secure drinking water through the technology of air to water. This technology works by reproducing the condensation procedure through simulation of dew points. Even if humidity is low, the technology can make water constantly [20]. The AKVO Atmospheric Water Generator works using the following steps [21]:

– A high-speed fan draws air from atmosphere.
– There are three layers of air filter that can help to take of impurities by 99.97%.
– Moisture condenses through cold coils.
– Water drops into the tank.

Table 4 Capacities per day and nominal power consumptions for all models of AVKO atmospheric water generator

Model No.	Capacity Per Day	Nominal Power Consumption by Hour
AVKO 36K	100 liters	1.8 kw/hour
AVKO 55K	150 liters	2.1 kW/hour
AVKO 110K	300 liters	3.3 kW/hour
AVKO 180K	500 liters	4 kW/hour
AVKO 365K	1000 liters	8.2 kW/hour

– There is another filtering for additional filtration, which results in getting drinking water.

The ideal conditions to make WFA are relative humidity between 40% and 100% and a temperature between 21 and 32 degree Celsius [21]. In the Jazan area, the annual average of humidity is 66% and the temperature is between 23 ~ 37 degree Celsius [3, 4]. There are many models of devices from AVKO: AVKO 36K, AVKO 55K, AVKO 110K, AVKO 180K, and AVKO 365K. Every model has a different capacity and power consumption. Table 4 below summarizes the capacities per day and nominal power consumptions by hour for all models [22].

Using AVKO 365K, the nominal power consumption is 8.2 kw/hour. There is a need for 196.8 kwh to get 1000 liters per day. If there are 10 AVKO 365K devices, the total capacity of water will be 10,000 liters per day. To run these 10 devices, there is a need for 82 kw/hour per hour. In addition, there is a need for 1968 kwh (approximately 2 MWh) per day.

In 2017, in the Jazan region, the water consumption was 59 mcm (million cubic meters), and the population was about 1.6 million inhabitants [23, 24]. Water consumption per person per year was 36.875 cm. In other words, water consumption per person per day is 0.10 cm.

Assume that we use the AVKO 365K model, which can generate 1000 liters per day (1 cm per day). This model needs 8.2 kw/h (196.8 kwh per day). If people use 60 panels, where every panel can generate 250 watts, the total electricity generated will be 15,000 watts [25]. For 10 h, the total of electricity will be 150 kwh, which can supply electricity for AVKO 365K atmospheric water generator.

4 Analysis and Simulation

4.1 Analysis

The transportation cost of water considers truck driver's wage, the depreciation of truck, and fuel. For the truck driver, assume that the driver gets a payment of $73.000 per year ($6083.3 per month and $202.8 per day). For truck depreciation, assume that the cost of the truck's water is $200,000, which usually can be used for 20 years.

Table 5 The estimated transportation costs

Destination	30 Km	60 Km	120 Km	240 Km
Truck diver cost	$25.35	$50.7	$101.4	$202.8
Truck depreciation cost	$3.48	$6.95	$13.9	$27.8
Fuel	$5	$10	$20	$40
Original Price	$34	$34	$34	$34
Total (34,068.71 Liter) for one truck	$67.83	$101.65	$169.2	$304.6
Total (12,000 Liter)	$23.89	$35.8	$59.63	$107.29

The cost for a year is $10,000, and a day is $27.8. For fuel, assume that the cost of fuel for 8 h is $40. Usually, the truck can hold about 34,0000 liters. The original cost of 1000 liter is $1; however, it costs more money to deliver it to mountains or remote areas outside of the city. The total cost of delivering water out of city is calculated based on the sum of truck driver cost, truck depreciation cost, fuel cost, and the original price of water. Table 5 below shows the estimated transportation costs of water based on distance in kilometers.

4.2 Simulation

This simulation system is built using Java programming language. The goal of this simulation is to show the amount of solar energy and wind energy. This collected energy can help to run water making devices. This simulation does the following:

- For the 12 months (January, February, March, April, May, June, July, August, September, October, November, December), it calculates the daily sunlight hours in the Jazan region, which is based on the average sun light hours for the whole month. It uses the normal distribution approach, i.e. Gaussian to generate random numbers for sunlight hours.
- Sum up the daily sunlight hours of the whole months. The formula used to calculate the total of sunlight hours is:

$$\sum_{j=1}^{i} SunHour_j \tag{1}$$

Where i is days of a month.

- To show the solar energy per month, it multiplies the sum of sunlight hours for the whole month in 60 and 250 (Eq. 2). Sixty means that they use 60 solar panels and 250 means that every solar panel generates 250 watts. The resulted sun hours per month will be divided by 1000 to make it in kwh.

$$Total\ of\ Solar\ Energy = \sum_{j=1}^{i} \left(SunHour_j \times 30 \times 250 \right) \tag{2}$$

- Calculates the daily wind energy per hour, which is based on the average of wind energy per hour for the whole month. Also, it uses normal distribution approach, i.e. Gaussian to generate random numbers for wind energy per hour. Then multiply the wind energy per hour in 3 and 24 to show wind energy per day. Three means that they use 3 wind turbines.
- Sum up the wind energy for the whole months.

$$\sum_{j=1}^{i} W_j \tag{3}$$

Where W is wind energy.

- Sum up the total of solar energy and wind energy. Then divide this total by 5904 kwh (AVKO 365 K, 1 month) to show the number of devices that can be run per day.

$$Total\ of\ Solar\ and\ Wind = \sum_{j=1}^{i} \left(SunHour_j \times 30 \times 250 \right) + \sum_{j=1}^{i} W_j \tag{4}$$

4.3 Simulation Results

This paper calculates the sunlight hours per month for the Jazan region, Saudi Arabia. In addition, it calculates the total of solar energy and wind energy. The goal of this paper is to show how many water making devices can be run per day. The number of devices can help to know the amount of water that can be collected per month. The result of the simulation system is summarized in Table 6 below.

Table 6 The results of single run of the simulation system

Month	Total Sunlight Hours	Solar Energy	Wind Energy	Solar & Wind Energy	Per Day Devices	Water Amount Per Month
JAN	239.16	3587.4	989	6555.24	1	31,000 Liters
FEB	243.24	3648.6	754	5910.84	1	28,000 Liters
MAR	296.84	4452.6	759	6728.52	1	31,000 Liters
APR	340.11	5101.65	864	7695.09	1	30,000 Liters
MAY	392.15	5882.25	928	8665.05	1	31,000 Liters
JUNE	375.48	5632.2	2551	13284.36	2	60,000 Liters
JULY	378.24	5673.6	2852	14230.8	2	62,000 Liters
AUG	358.72	5380.8	2768	13684.56	2	62,000 Liters
SEP	359.9	5398.5	2330	12387.54	2	60,000 Liters
OCT	235.04	3525.6	2691	11598.96	1	31,000 Liters
NOV	209.51	3142.65	2709	11268.57	1	30,000 Liters
DEC	212.92	3193.8	2719	11351.4	1	31,000 Liters

In the Jazan region, one person needs 0.10 cubic meters (1000 liters per day), or in other words, 3000 liters per month. For February, there is one device that can be run per day. This means that the total amount of water that can be generated for the whole month is 28,000 Liters, which can support approximately nine people. For April and November, there is one device that can be run per day. This means that the total amount of water that can be generated for the whole month is 30,000 Liters, which can support 10 people. For January, March, May, October, and December, there is one device that can be run per day. This means that the total amount of water that can be generated for the whole month is 31,000 Liters, which can support approximately 10 people. For June and September, there are two devices that can be run per day. This means that the total amount of water that can be generated for the whole month is 60,000 Liters, which can support 20 people. For July and August, there are two devices that can be run per day. This means that the total amount of water that can be generated for the whole month is 62,000 Liters, which can support approximately 20 people.

In conclusion, this simulation shows that the most generated water can be for July and August. However, people can generate enough water in other months and avoid the transportation cost of water. If you have utility level of electricity, you can generate a lot more water. This paper simulates a small-scale water generation.

5 Conclusion

A WFA system is proposed in this paper. This paper shows the viability of generating water from air. The required electricity can be generated by solar panels and windmill. Weather condition near Jazan area is classified as desert, but its humidity and wind can support necessary electricity. This region has high humidity, steady wind, and great amount of sunlight but no rain. Two systems are compared: desalination plant and WFA. Desalination plants costs almost $32 million to build a 2.5 million gallons per day [26]. However, water from the desalination plant needs transportation cost to deliver to remote area. WFA can be built in this remote area and need a lot less transportation cost, or possibly none. By using small-scale wind turbines and 60 solar panels, the WFA system can generate enough water for 9 to 20 people in the Jazan region, Saudi Arabia.

References

1. Jazan Project Overview. http://www.airproducts.com/Microsites/jazan-project/project-overview.aspx. Last accessed 2018/11/12
2. Jazan Region Climate. https://en.climate-data.org/asia/saudi-arabia/jazan-region-1998/. Last accessed 2018/11/12
3. Average Weather in Jizan. https://weatherspark.com/y/102295/Average-Weather-in-Jizan-Saudi-Arabia-Year-Round. Last accessed 2018/11/13

4. Average humidity in Jazan (Jazan Province). https://weather-and-climate.com/average-monthly-Humidity-perc,j-z-n,Saudi-Arabia. Last accessed 2018/12/5

5. A. Silbajoris, *A Brief Summary of Solar Energy*. https://sciencing.com/brief-summary-solar-energy-5806468.html. (2018)

6. Solar panel brief history and overview. https://www.energymatters.com.au/panels-modules/. Last accessed 2018/11/15

7. H. Terzioglu, F.A. Kazan, M. Arslan, A new approach to the installation of solar panels, in *Proceedings of the 2015 2nd International Conference on Information Science and Control Engineering*, (IEEE, Shanghai, 2015). https://doi.org/10.1109/icisce.2015.133

8. M. Dhar, *How Do Solar Panels Work?*https://www.livescience.com/41995-how-do-solar-panels-work.html. (2017)

9. A.R. Prasad, S. Singh, H. Nagar, Importance of solar energy technologies for development of rural area in India. Int. J. Sci. Res. Sci. Technol. **3**(6), 585–599 (2017)

10. Solar Energy Pros and Cons, Energy Informative. https://energyinformative.org/solar-energy-pros-and-cons/. Last accessed 2018/11/20

11. SolarWorld SW 250 Poly, 250 Watt Solar Panel, SoW. www.gogreensolar.com/products/solarworld-sw-250-poly-250-watt-solar-panel. Last accessed 2018/11/15

12. Jazan Historical Weather. https://www.worldweatheronline.com/lang/en-us/jazan-weather-history/jizan/sa.aspx. Last accessed 2018/11/12

13. Basics of Wind Energy. https://www.awea.org/wind-101/basics-of-wind-energy. Last accessed 2018/11/15

14. 100kW Ø25m Variable Pitch. http://www.polarisamerica.com/turbines/100kw-wind-turbines/. Last accessed 2018/11/18

15. T100 100kW Wind Turbine. http://www.argolabe.es/pdf/Argolabe(en).pdf. Last accessed 2018/12/7

16. Windustry. http://www.windustry.org/. Last accessed 2018/12/10

17. A. Tripathi, S. Tushar, S. Pal, S. Lodh, S. Tiwari, P.R.S. Desai, Atmospheric water generator. Int. J. Enhanc. Res. Sci. Technol. Eng. **5**(4), 69–72 (2016)

18. 4 Main Benefits of Atmospheric Water Generators (2018). http://akvosphere.com/4-main-benefits-of-atmospheric-water-generators/. Last accessed 2018/11/25

19. Water Treatment Solutions. https://www.lenntech.com/processes/desalination/reverse-osmosis/general/reverse-osmosis-desalination-process.htm. Last accessed 2018/2/19

20. About AVKO, Company Profile. http://akvosphere.com/about-us/. Last accessed 2018/3/10

21. Water from Air. http://akvosphere.com/air-to-water-technology/. Last accessed 2018/3/12

22. Akvo 36K. https://akvosphere.com/akvo-atmospheric-water-generators/. Last accessed 2018/3/15

23. A. Puri-Mirza, *Saudi Arabia: Water Consumption by Region 2018*. https://www.statista.com/statistics/627146/saudi-arabia-water-consumption-by-region/. Last accessed 2018/10/1

24. Population in Saudi Arabia's Jazan region by gender and nationality 2018. https://www.statista.com/statistics/617269/saudi-arabia-population-gender-and-nationality-in-jazan-region/. Last accessed 2018/11/7

25. B. Zientara, *How Much Electricity Does a Solar Panel Produce?*, Solar Power Rocks. www.solarpowerrocks.com/solar-basics/how-much-electricity-does-a-solar-panel-produce/. (2012)

26. General FAQs. http://www.twdb.texas.gov/innovativewater/desal/faq.asp. Last accessed 2018/12/2

A Dynamic Data and Information Processing Model for Unmanned Aircraft Systems

Mikaela D. Dimaapi, Ryan D. L. Engle, Brent T. Langhals,
Michael R. Grimaila, and Douglas D. Hodson

1 Introduction

Dynamic Data and Information Processing (DDIP) involves feedback through sensor reconfiguration to integrate real-time data into a predictive method for system behavior. Fundamentally, DDIP presents opportunities to advance understanding and analysis of activities, operations, and transformations that contribute to system performance, thereby aiding in decision-making and event prediction. Previously examined DDIP application domains include weather monitoring and forecasting, supply chain system analysis, power system and energy analysis, and structural health monitoring. DDIP is formerly known as *Dynamic Data Driven Application Systems*. [1]. Currently, there is limited existing work that implements DDIP in support of Unmanned Aircraft Systems (UAS).

UAS, often called *drones*, have exploded in popularity over the last decade providing value to business, governmental, and research interests [2]. Larger-scale UAS typically have three components: a control system, an Unmanned Aerial Vehicle (UAV), and a command and control system to link the two. While UAS are available for commercial use in security and photography, they are also integral to intelligence, surveillance, and reconnaissance (ISR) efforts in a military context. UAS also support military training and exercises when sustained endurance efforts are required and physical infrastructure imposes limitations and restrictions on the use of manned aircraft [3].

To this $70 billion industry [2], any reduction in the operating or maintenance costs may be desirable. As a model, DDIP has already demonstrated its utility for

M. D. Dimaapi · R. D. L. Engle (✉) · B. T. Langhals · M. R. Grimaila · D. D. Hodson
Air Force Institute of Technology, Wright-Patterson AFB, Dayton, OH, USA
e-mail: mikaela.dimaapi@afit.edu; ryan.engle@afit.edu; michael@grimaila.com;
michael.grimaila@afit.edu; doug@openeaagles.org; douglas.hodson@afit.edu

© Springer Nature Switzerland AG 2021
R. Stahlbock et al. (eds.), *Advances in Data Science and Information Engineering*,
Transactions on Computational Science and Computational Intelligence,
https://doi.org/10.1007/978-3-030-71704-9_63

improving a preventative maintenance schedule for a semiconductor supply chain [4]. In this example, Koyuncu et. al. improved preventative maintenance schedules to reduce unexpected down time by reacting to anomalous conditions. This research seeks to develop a similar DDIP model for a UAS maintenance scenario.

Specifically, this research will consider an unspecified, commercially produced, large-scale UAS employing a fixed preventative maintenance schedule. The authors believe that indicators, such as engine temperature and pressure, can be monitored to drive a DDIP application. The application's output should propose a revised maintenance schedule for a particular UAS when anomalous events suggest a mechanical failure may occur ahead of a fixed preventative maintenance activity.

2 Background

DDIP applications include symbiotic control feedback via sensor reconfiguration to integrate real-time data into a predictive method for system behavior. Key elements within the DDIP feedback loop are a physical system, a sensor reconfiguration loop, and a data assimilation loop. These systems involve the ability of a running application to incorporate real-time data into the decision process, and conversely, the ability of an application to respond to those inputs [5] [6]. This conceptual framework seeks to implement adaptive state estimation by reconfiguring sensors during operation [7].

Dynamic data adapts intelligently to evolving conditions and infers knowledge in ways that are not predetermined by start-up parameters [8]. Through dynamic integration of computational and measurement aspects of an application, DDIP incorporates a symbiotic feedback control system [8]. For example, in space weather monitoring, Cubesats provide data to dynamically monitor solar phenomena within the different layers of the Earth's upper atmosphere. The DDIP phenomenon applies as components within each Cubesat. Data assimilation must be performed at a nearly real-time rate to determine significance of a given atmospheric disruption. Sensors on the Cubesats affect resolution of data collection and transmission, and the Cubesats must determine an efficient means for transmitting their data to a ground station for modeling and prediction [9]. In essence, within applications such as Cubesat surveillance and beyond, DDIP offers potential to augment analysis and prediction capabilities of simulations [1].

DDIP implements computational feedback rather than traditional state estimation techniques by seeking to reconfigure system sensors during operation. Sensor errors correct the simulation of the physical system rather than the physical system itself. The model guides the sensor data collection, and in turn, the sensor data improves model accuracy [7]. Integrating data assimilation and sensor reconfiguration are central to methods and applications using DDIP. In addition to advancing sensing capabilities in data-driven Cubesats, DDIP can be used to control preventative maintenance scheduling within supply chain systems. Large quantities of data exist in supply chain systems. Within supply chain models, data is analyzed to prevent unnecessary usage of already-constrained computing resources [4]. The proposed DDIP construct consists of physical systems and real-time simulations. The process involves an ongoing, continuous loop comprising sensor data from each physical machine and algorithms in the real-time simulation that use that sensor data. Updates are sent back to the physical systems and databases via a communication server to optimize the preventative maintenance schedules of different supply chain stages.

3   Methodology

Despite development within the aforementioned domains, limited research has been conducted using DDIP in the UAS domain of applications. Challenges remain in defining parameters needed for system control and determining the necessary sampling rates for sensors to monitor the structural health of the UAS [7]. Within the UAS domain, DDIP could support dynamic response and control necessary for optimization of maintenance scheduling. A DDIP model for UAS application would require representation of physical and estimated states, estimation algorithms, and data collection and storage mechanisms. Upon identifying the role of DDIP within this framework, this model could be applied to lean maintenance in UAS.

Finding an optimized preventative maintenance schedule for a UAS could demonstrate the effectiveness of applying DDIP to this domain. Building DDIP into a UAS-specific model could have the consequence of improving current maintenance timing and saving money. By creating a more accurate, on-demand preventative maintenance schedule rather than defaulting to a fixed schedule dictated by standard procedure, this approach could save manhours and ultimately, incurred costs overtime.

Figure 1 presents a notional configuration depicting the data flow with a DDIP UAS model using SysML. Within this model, the *Unmanned Aircraft System* is the physical system of interest. While in flight, the *UAS Sensors* capture the state of the system by monitoring and collecting data about the UAS conditions. System characteristics, such as oil pressure and engine temperature, can provide valuable information relevant to events that precede engine failure. Following the model downward from UAS Sensors, a node routes *Sensor Data* to both the *Simulation* and a database for longer-term storage. Specifically, this data is used to update a *Sensor Database* containing flight data while also directly providing inputs for the *Simulation*. Together, the input sensors and the simulation comprise the data assimilation loop within the DDIP pattern.

The *Simulation* integrates the historical and real-time data inputs to predict engine failure. This is accomplished using algorithms defining the expected life of an engine, while considering safe operating ranges and thresholds of severity with respect to the sampled variables of interest from the *Sensor Data*. The simulation processes the input data to determine where it falls within those predefined ranges. When the measurements of *Parameters of Interest* approach or exceed acceptable

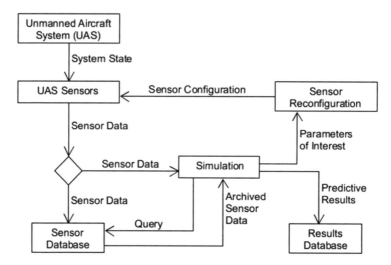

Fig. 1 Data Driven Information Processing (DDIP) Unmanned Aircraft System (UAS) Data Flow Model

limits, the simulation will flag them as anomalous. The resulting information would provide input back to the *UAS Sensors* via *Sensor Reconfiguration* controls to adjust sampling rates for variables approaching or outside of the safe zones. In this manner, the simulation outputs will guide the UAS sensor sampling rates. The influence of the simulation on the UAS sensors constitutes the sensor reconfiguration loop within the DDIP model.

Lastly, the *Predictive Results* of the *Simulation* will be stored in a results database. This information may be queried by end-user applications to aid decisions regarding predictive maintenance scheduling.

4 Anticipated Outcomes

This problem presents an opportunity model and approach to alternate between two sampling rates for monitored indicators of interest. If an event occurs, the system will increase the sampling rate. The simulation could select data from these heterogeneous samples where appropriate. Ultimately, the output would be a recommendation for early, additional, or future scheduled maintenance.

This work could be generalized refining the larger preventative maintenance and cost model or applying DDIP to other systems. Cost savings within the domain of UAS is only one of several applications for which DDIP can be used.

References

1. F. Darema, *Dynamic Data Driven Application Systems: New Capabilities for Application Simulations and Measurements* (National Science Foundation, Arlington, 2005)
2. D. Joshi, *Drone technology uses and applications for commercial, industrial and military drones in 2020 and the future*, 18 12 2019. [Online]. Available: https://www.businessinsider.com/drone-technology-uses-applications
3. U.S. Department of Defense, Unmanned Aircraft Systems (UAS), 03 2020. [Online]. Available: dod.defense.gov/UAS/
4. N. Koyuncu, S. Lee, K.K. Vasudevan, Y.-J. Son, P. Sarfare, DDDAS-Based Multi-fidelity Simulation for Online Preventive Maintenance Scheduling in Semiconductor Supply Chain, in *Winter Simulation Conference*, (2007)
5. National Science Foundation, DDDAS: Dynamic Data Driven Application Systems, (2005). [Online]. Available: https://www.nsf.gov/pubs/2005/nsf05570/nsf05570.htm
6. F. Darema, *Dynamic Data Driven Application Systems*, (4 April 2003). [Online]. Available: https://pswscience.org/meeting/dynamic-data-driven-application-systems/
7. E. Blasch, D. Bernstein, M. Rangaswamy, Introduction to dynamic data driven applications systems, in *Handbook of Dynamic Data Driven Applications Systems*, (Springer, 2019)
8. R. Bahsoon, *Defining Dependable Dynamic Data-Driven Software Architectures* (Aston University Birmingham, Birmingham, 2007)
9. D. Bernstein, *Transformative Advances in DDDAS with Application to Space Weather Monitoring* (AF Office of Scientific Research, Arlington, 2015)

Utilizing Economic Activity and Data Science to Predict and Mediate Global Conflict

Kaylee-Anna Jayaweera, Caitlin Garcia, Quinn Vinlove, and Jens Mache

1 Introduction

Were there economic pointers attached to Covid-19? Could taking a step back from relying on a country's hearsay when it comes to solving and containing natural, civil, or international disasters have allowed more lives to be saved, more hospitals to be prepped, and more civilians to be informed?

Living through a global pandemic in a divided world that is routinely being flooded with misinformation, one begins to wonder if the tools needed to mediate global and local catastrophes are available. From this, the question very quickly becomes not only whether there are predictive tools but also whether they are readily available for citizens as well as countries.

Are there factors global citizens can look further into that may correlate with, and hopefully predict, conflict? And if they are available, how can we utilize Data Science, Machine Learning, and Artificial Intelligence to aid in keeping global societies honest?

In pursuit of the answer to these very dire questions, the goal became truly clear: creating a system or method of taking in public data and simplifying its points into concise conflict predictions. The task of understanding what causes conflict soon followed.

It should be stated and understood that there are an infinite number of possible triggers to conflict. Acknowledging this, it also is clear that our small team, nor any team, could ever account for every aspect, so we needed to zoom out and get a bigger picture before we could narrow down our options. The main field acknowledged

K.-A. Jayaweera (✉) · C. Garcia · Q. Vinlove · J. Mache
Lewis & Clark College, Portland, OR, USA
e-mail: kjayaweera@lclark.edu; garciac@lclark.edu; quinnvinlove@lclark.edu; jmache@lclark.edu

© Springer Nature Switzerland AG 2021
R. Stahlbock et al. (eds.), *Advances in Data Science and Information Engineering*,
Transactions on Computational Science and Computational Intelligence,
https://doi.org/10.1007/978-3-030-71704-9_64

measurable factors are climate change, large-scale human migration, education, new technologies, economic motive, regional motive, political motive, and social factors [1, 2]. Each of these dimensions has a sure effect on human action, but measuring and accounting for each seemed still too large. So, we looked at what others in the field had done.

2 Related Work

We are certainly not the first to try and take on the challenge of predicting and simplifying human aggressions into a system that global citizens could access in order to prepare and hold each other accountable. Two significant preexisting systems are ViEWS [3], created by Uppsala University for the Department of Peace and Conflict Research located in Sweden, and GUARD [4], created by the United Kingdom's Alan Turing Institute for its Defense and Security programs.

Both are impressive programs; they highlight areas of likely conflict, and seem to take in different dimensions of data. However, they also seem to be focused solely on a specific area of conflict (instead of the entire world and its territories), and the data they were taking in required a large amount of human input. It seems that without gross amounts of human resource and input, the output data would not be easily obtainable by any global citizen.

3 Moving Towards Efficiency

Our team took advantage of our position as a small but passionate group and realized that our size meant zero stakes in the matter of producing accurate predictions. We are not funded or relaying details to larger organizations that rely on accurate data – at least not yet. Quickly we realized that because of this, we can experiment wildly with new concepts that in theory could prove to be effective, if not leaping steps in the right direction.

Could there be merit to studying one portion of the contributors of conflict?

To further this revelation of the freedom by our project size, we started digging deeper into something that we could have access to all the time, the economy. What if we took public and government trading, stock, and hoarding into account and found small changes in investment that tip whether a country or its citizens are preparing for some sort of conflict that the rest of the world is not privy to yet?

4   Thought Process

Our thought process is as follows:

Imagine you are the leader of a country, and you have just realized that there is a virus within your borders that has hazardous trends likely to cause a global pandemic if it were to remain unchecked.

Thinking as a strong leader, you want to make sure that your own country has the supplies and necessities in order to survive the potential wave of pandemic to follow. To do this, you now invest heavily in Personal Protective Equipment (PPE), hoard food and take investments from the transportation industry and move them into your own domestic no-contact shipping.

In the name of self-preservation and protection from the chaos to follow, ideally this would all happen before you tipped off, and likely downplayed, the severity of the situation to the rest of the international community.

The flip side:

Now, take this scenario and imagine that there was a widely available and easily readable system that could point to potential areas of conflict or unrest by observing economic activity that historically was paired with international and/or domestic conflict. It lets the entire world know that you, the strong leader, and your country are acting and investing in a way that would point to domestic issues.

In this new scenario, as a response to this system there would be widespread accountability, resource management, crisis mitigation, global security and knowledge accessibility potentially saving thousands, if not millions, of lives.

5   Real Application

Some version of this scenario has turned into reality within the past 12 months. The unfortunate reality is that the real time consequence has been the unnecessary loss of life.

With history and this modern reality comes the added value in the potential to track the economy in a way that pairs slight changes with historical conflict to prevent human and economic loss.

6   First Approach

To begin with, our team hoped to use an existing tool that would enable more ability on our end to make conflict predictions.

To create this system, we started by experimenting with the Amazon Web Services (AWS) product Amazon Forecast. AWS is a cloud platform that offers various products in Machine Learning, game technology, security, and the like.

Forecast is well known for its time-series forecasting and is popular with companies that hope to predict supply chain demand and business metrics [5].

We believed that the tool's ability to easily account for irregularities such as holidays and adapt to the nature of the data sets made it an appealing product given its forgiving nature.

After reformatting the product placement data sheet to Forecast standards, we started off by testing the product on soybean imports. We quickly encountered many walls that were preventing us from being as effective as possible.

The product was unable to accommodate multiple data sets with varying time series. We realized that Forecast easily accommodates more stagnant datasets, such as metadata, to be included. We ultimately need one that can support many time series data sets. Thus, we abandoned Amazon Forecast and decided it would be best to continue with a model of our own that we could tailor to our own research.

7   Current Approach

The first two figures, Figs. 1 and 2, provide an example of the preliminary correlation between economic activity and conflict we were able to produce by simply graphing USDA export data [6] and pairing it with global conflict data from the Armed Conflict Location & Event Data Project (ACLED) [7]. Figure 1 displays US exports of corn to Saudi Arabia in 2015, while Fig. 2 shows actual conflict in Saudi Arabia per month in 2015. Significant spikes in correlation can be seen between both Figs. 1 and 2 between April and July in each graph.

Within Figs. 3 and 4, a more striking correlation is very evident. Figures 3 and 4 respectively show US exports of soy to Indonesia and conflict within Indonesia in 2015. Notice the clear alignment of pattern between spikes in each graph from February to May. Our team was very pleasantly surprised with these preliminary findings, especially with the current simplicity of our code and the clear correlation between its outputs.

Fig. 1 US exports of corn to Saudi Arabia

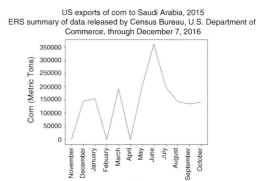

US exports of corn to Saudi Arabia, 2015
ERS summary of data released by Census Bureau, U.S. Department of Commerce, through December 7, 2016

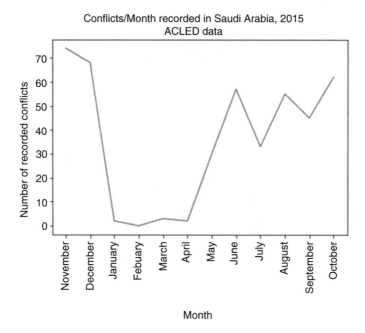

Fig. 2 Conflicts per month in Saudi Arabia

Fig. 3 US Exports of Soy to Indonesia

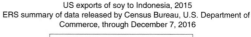

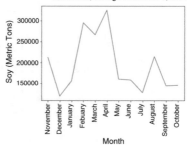

The input values can be seen below the two sets of figures within Tables 1 and 2. Table 1 displays the USDA data which outlines a specific good and the export of that good from country to country. Table 2 is a sample of the Armed Conflict Location & Event Data Project (ACED) data produced. Within it are conflict locations, their description, their severity, and their conflict type.

These significant and similar spike areas lead us to believe that there may be further correlation that could become more evident as more parameters are put into place. With further research, we will test methods of comparing and quantifying this degree of similarity between conflict and product data. The goal is to be able to use Artificial Intelligence and Machine Learning to produce predictions on where

Table 1 Sample USDA export data from November 2015 to October 2015

Region	Nov. 2015	Dec. 2015	Jan. 2015	Feb. 2015	Mar. 2015	Apr. 2015	May 2015	Jun. 2015	Jul. 2015	Aug. 2015	Sep. 2015	Oct. 2015
0 Japan	418341	342761	560135	656818	1063283	972244	1248961	1249743	1440594	1218643	1643916	385733
1 Mexico	753709	1029277	890251	1098896	1249991	1372285	1344127	1227860	1114138	1216459	1251920	1001596
2 Colombia	176314	474488	374044	530066	460149	1084779	132383	10708	246717	282138	345006	429814
3 Taiwan	58624	65845	29829	60114	92629	387312	162273	379944	303823	386162	311092	57308
4 Peru	121806	237184	228035	154869	232496	260310	189307	256474	132569	340905	286504	187155
5 Saudi Arabia	0	143703	153497	0	190963	225	199586	360281	197624	142763	133777	139930
6 Guatemala	73875	30040	48835	116842	70746	85080	77979	83112	73127	95095	85592	79187

Showing quantity of product moved from each country

Table 2 Sample Armed Conflict Location & Event Data Project (ACED) data for conflict showing December and March dates with varying types of conflict (shelling/artillery/missile attack, armed clash, attack, etc.)

	data_id	iso	event_id_cnty	event_id_no_cnty	event_date	year	time_precision	event_type	sub_event_type	actor1	...	location
31	6439313	682	SAU5708	5708	31 December 2015	2015	1	Explosions/Remote violence	Shelling/artillery/missile attack	Military Forces of Yemen (2015-2016) Supreme R…	...	Jizan
84	7084802	682	SAU5701	5701	30 December 2015	2015	1	Explosions/Remote violence	Shelling/artillery/missile attack	Military Forces of Yemen (2015-2016) Supreme R…	...	Abha Regional Airport
146	6439519	682	SAU5706	5706	30 December 2015	2015	1	Explosions/Remote violence	Shelling/artillery/missile attack	Military Forces of Yemen (2015-2016) Supreme R…	...	Jizan
147	6439520	682	SAU5707	5707	30 December 2015	2015	1	Explosions/Remote violence	Shelling/artillery/missile attack	Military Forces of Yemen (2015-2016) Supreme R…	...	Wadi Aleeb
176	6439548	682	SAU5704	5704	30 December 2015	2015	1	Battles	Armed clash	Military Forces of Yemen (2015-2016) Supreme R…	...	Al Khobh
…	...	...	...	...	...	...	...	...	...	...	...	
27328	6446985	682	SAU5065	5065	29 March 2015	2015	1	Battles	Armed clash	Unidentified Armed Group (Saudi Arabia)	...	Riyadh
28483	6447214	682	SAU5064	5064	16 March 2015	2015	1	Violence against civilians	Attack	Unidentified Armed Group (Saudi Arabia)	...	Medina

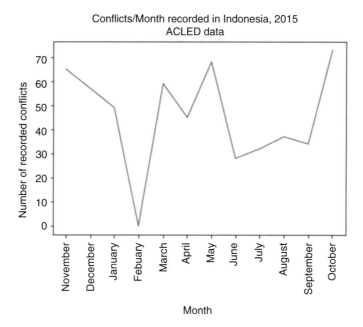

Fig. 4 Conflicts per month in Indonesia

conflicts may arise by utilizing historical data that outline conflict in certain regions as well as types of products whose trade data correlates to the said regional conflict data.

8 Discussion and Conclusion

Though this is all a preliminary study, we believe that this is certainly a route that should be further explored. We want to be sure not to overstep and assume that this seeming correlation is foundation enough to point to consistent and reliable conflict prediction, only that it is certainly a sure step in the right direction.

With further research and training data, hopefully we will be able to discover and follow trends within the global economy that will point to the probability of domestic or international conflict happening all over the world in real time. To be sure the demand will still remain present however far into history this exploration takes us. To quote the United Nations, "early warning is an essential component of prevention, and the United Nations will continue to carefully monitor developments around the world to detect threats to international peace and security" [8].

Now remains the task of weeding out effective economic market pointers. So far, soybean and other soy products prove promising; however, we do not want to close ourselves off from the idea that there may be certain types of products that are better predictors than others depending on the global location of conflicts.

In any case, the discovery continues at the horizon of efficient conflict prediction and global cooperation, preparedness, and growth.

Acknowledgements This material is bases upon work partially supported by the National Science Foundation grant 1723714, and the John S. Rogers Science research program at Lewis & Clark College.

Special thanks to Professor Cyrus Partovi of Lewis & Clark College's International Affairs Department, Dr. Alain Kägi of Lewis & Clark College's Mathematical Science department, and fellow peer Kris Gado for their consistent enthusiastic encouragement and curiosity along the way.

References

1. Peace Research Institute Oslo (PRIO). "Conflict Prediction". Accessed 8 June 2020. https://www.prio.org/Projects/Project/?x=1401
2. World Resources Institute. "UN Security Council Examines the Connection Between Water Risk and Political Conflict", November 5, 2018. https://www.wri.org/blog/2018/11/un-security-council-examines-connection-between-water-risk-and-political-conflict
3. Allansson, Marie. "ViEWS - Department of Peace and Conflict Research - Uppsala University, Sweden". Uppsala University, Sweden. Accessed 8 June 2020. https://www.pcr.uu.se/research/views/
4. The Alan Turing Institute. "Predicting Conflict – a Year in Advance". Accessed 8 June 2020. https://www.turing.ac.uk/research/impact-stories/predicting-conflict-year-advance
5. "What Is Amazon Forecast? - Amazon Forecast". Accessed 8 June 2020. https://docs.aws.amazon.com/forecast/latest/dg/what-is-forecast.html
6. "USDA ERS - Soybeans & Oil Crops". Accessed 8 June 2020. https://www.ers.usda.gov/topics/crops/soybeans-oil-crops/
7. "Armed Conflict Location & Event Data Project". Accessed 20 July 2020. https://reliefweb.int/organization/acled
8. "Peace and Security", August 30, 2016. https://www.un.org/en/sections/issues-depth/peace-and-security/

Part IX
Machine Learning, Information & Knowledge Engineering, and Pattern Recognition

A Brief Review of Domain Adaptation

Abolfazl Farahani, Sahar Voghoei, Khaled Rasheed, and Hamid R. Arabnia

1 Introduction

Traditional machine learning aims to learn a model on a set of training samples to find an objective function with minimum risk on unseen test data. To train such a generalized model, however, it assumes that both training and test data are drawn from the same distribution and share similar joint probability distributions. This constraint can be easily violated in real-world applications since training and test sets can originate from different feature spaces or distributions. The difficulty of collecting new instances with the same property, dimension, and distribution, as we have in the training data, may happen due to various reasons, e.g., the statistical properties of a domain can evolve in time or new samples can be collected from different sources, causing domain shift. Besides, if possible, it is preferred to utilize a related publicly available annotated data as a training dataset instead of creating a new labeled dataset, which is a laborious and time-consuming task. However, when the training data is not an accurate reflection of test data distribution, the trained model will likely experience degradation in performance when applying on the test data. To tackle the above problem, researchers proposed a new research area in machine learning called domain adaptation. In this setting, training and test sets are termed as the source and the target domains, respectively. Domain adaptation generally seeks to learn a model from a source labeled data that can be generalized to a target domain by minimizing the difference between domain distributions.

Domain adaptation is a special case of transfer learning [41]. These two closely related problem settings are subdiscipline of machine learning which aim to improve the performance of a target model with insufficient or lack of annotated data by using

A. Farahani (✉) · S. Voghoei · K. Rasheed · H. R. Arabnia
University of Georgia, Athens, GA, USA
e-mail: a.farahani@uga.edu; Voghoei@uga.edu; Khaled@uga.edu; hra@uga.edu

© Springer Nature Switzerland AG 2021 877
R. Stahlbock et al. (eds.), *Advances in Data Science and Information Engineering*,
Transactions on Computational Science and Computational Intelligence,
https://doi.org/10.1007/978-3-030-71704-9_65

the knowledge from another related domain with adequate labeled data. We first briefly review transfer learning and its categories and then address some existing related works to transfer learning and domain adaptation. Transfer learning refers to a class of machine learning problems where either the tasks or domains may change between the source and target, while in domain adaptations, only domains differ and tasks remain unchanged. In transfer learning, a domain consists of feature space and marginal probability distribution, and a task includes a label space and an objective predictive function. Thus, various possible scenarios in domains and tasks create different transfer learning settings. Pan et al. [41] categorize transfer learning into three main categories: inductive, transductive, and unsupervised transfer learning. Inductive transfer learning refers to the situation where the source and target tasks differ, no matter whether or not domains are different. In this setting, the source domain may or may not include annotated data, but a few labeled data in the target domain are required as training data. In transductive transfer learning, tasks remain unchanged while domains differ, and labeled data are available only in the source domain. However, part of the unlabeled data in the target domain is required at training time to obtain its marginal probability distribution. Finally, unsupervised transfer learning refers to the scenario where the tasks are different similar to inductive transfer learning; however, both source and target domains include unlabeled data.

Similar to transfer learning and domain adaptation, semi-supervised classification addresses the problem of having insufficient labeled data. This problem setting employs abundant unlabeled samples and a small amount of annotated samples to train a model. In this approach, both labeled and unlabeled data samples are assumed to be drawn from the equivalent distributions. In contrast, transfer learning and domain adaptation relax this assumption to allow domains to be from distinct distributions [45, 75].

Multitask learning [11] is another related task that aims to improve generalization performance in multiple related tasks by simultaneously training them. Since related tasks are assumed to utilize common information, multi-task learning tends to learn the underlying structure of data and share the common representations across all tasks. There are some similarities and differences between transfer learning and multitask learning. Both of these learning techniques use similar procedures, such as parameter sharing and feature transformation, to leverage the transferred knowledge to improve learners' performance. However, transfer learning aims to boosts the target learner performance by first training a model on a source domain and then transferring the related knowledge to the target learner, while multitask learning aims to improve the performance of multiple related tasks by jointly training them.

Similarly, multi-view learning aims to learn from multi-view data or multiple sets of distinctive features such as audio+video, image+text, text+text, etc. For example, a web page can be described by the page and hyperlink contents, which is an example of describing data by two different sources of text. The intuition behind this type of learning is that the multi-view data contains complementary information, and a model can learn more comprehensive and compact representation to improve the generalization performance. Some real-world application examples of multi-view learning are video analysis [60?], speaker recognition [3, 35], natural language processing [13, 47, 49], and recommender system [65, 66]. Canonical correlation analysis (CCA) [28] and co-training [6] are the first representative techniques introduced in the concept of multi-view learning which are also used in transfer learning [7].

Domain generalization [39] also tends to train a model on multiple annotated source domains which can be generalized into an unseen target domain. In domain generalization, target samples are not available at the training time. However, domain adaptation requires the target data during training to align the shift across domains.

2 Notations and Definitions

In domain adaptation, domains can be considered as three main parts: input or feature space $\mathcal{X}$, output or label space $\mathcal{Y}$, and an associated probability distribution $p(x, y)$, i.e., $\mathcal{D} = \{\mathcal{X}, \mathcal{Y}, p(x, y)\}$. Feature space $\mathcal{X}$ is a subset of a d-dimensional space; $\mathcal{X} \subset \mathbb{R}^d$, $\mathcal{Y}$ refers to either a space of binary $\{-1, +1\}$ or multi-class $\{1, \ldots K\}$, where K is the number of classes; and $p(x, y)$ is a joint probability distribution over the feature-label space pair $\mathcal{X} \times \mathcal{Y}$. We can decompose the joint probability distribution as $p(x, y) = p(y)p(x|y)$ or $p(x, y) = p(x)p(y|x)$, where $p(.)$ is a marginal distribution and $p(.|.)$ is a conditional distribution.

Given a source domain $\mathcal{S}$ and a target domain $\mathcal{T}$, the source dataset samples drawn from the source domain consist of feature-label pairs $\{(x_i, y_i)\}_{i=1}^n$, where n is the number of samples in the source dataset. Similarly, the target dataset can be denoted as $\{z_i, u_i\}_{i=1}^m$, where (z_i, u_i) refers to the target samples and their associated labels. In unsupervised domain adaptation where the labels are not available in the target domain, u is unknown. When the source and target domains are related but from different distributions, naively extending the underlying knowledge contained in one domain into another might negatively affect the learner's performance in the target domain. Therefore, domain adaptation was proposed to tackle this problem by reducing the disparity across domains and further training a model that performs well on the target samples. In other words, domain adaptation aims to learn a generalized classifier in the presence of a shift between source and target domain distributions. Classification is a machine learning task that aims to learn a function from labeled training data to map input samples to real numbers $h : \mathcal{X} \rightarrow \mathcal{Y}$, where h is a function or an element of a hypothesis space $\mathcal{H}$ and $\mathcal{H}$ refers to a set of all possible functions. For example, in image classification, a classifier assigns each input image a category such as a dog or a cat. Generally, to obtain the best predictive function, we learn a model on a given source dataset by minimizing the expected risk of the source labeled data:

$$R_S(h) = \mathbb{E}_{(x,y) \sim P_S(x,y)} \left[\ell(h(x), y) \right]$$

$$= \sum_{y \in \mathcal{Y}} \int_{\mathcal{X}} \ell(h(x), y) p_S(x, y) dx, \qquad (1)$$

where the expectation is taken with respect to the source distribution P_S, $\ell(h(x), y)$ is a loss function that denotes the disagreement between the label predicted by the classifier h and the true label, and $R_S(h)$ is the sum of all the disagreements or misclassified samples in the source domain. However, in the supervised learning, the goal is to learn a model with the lowest risk when applying on the target domain. Thus, we can rewrite the above equation as follows:

$$\begin{aligned}
R_T(h) &= \mathbb{E}_{(x,y) \sim P_T} \left[\ell(h(x), y) \right] \\
&= \sum_{y \in \mathcal{Y}} \int_{\mathcal{X}} \ell(h(x), y) p_T(x, y) dx \\
&= \sum_{y \in \mathcal{Y}} \int_{\mathcal{X}} \ell(h(x), y) p_T(x, y) \frac{p_S(x,y)}{p_S(x,y)} dx \\
&= \sum_{y \in \mathcal{Y}} \int_{\mathcal{X}} \ell(h(x), y) \frac{p_T(x,y)}{p_S(x,y)} p_S(x, y) dx \\
&= \mathbb{E}_{(x,y) \sim P_S} \left[\frac{p_T(x,y)}{p_S(x,y)} \ell(h(x), y) \right],
\end{aligned} \qquad (2)$$

where $P_S(x, y)$ and $P_T(x, y)$ are joint probability distributions of source and target domains. For more information about the risk minimization and supervised learning, see [38, 63]. In classical machine learning, $\frac{P_T(x,y)}{P_S(x,y)} = 1$ since the assumption is that both training and test data are drawn from the same distribution. However, domain adaptation relaxes this assumption, which will be discussed in the following sections.

3 Categorization of Domain Adaptation

Conventional domain adaptation assumes that feature and label spaces remain unchanged while their probability distributions may vary between domains. However, finding a source and target domains with equivalent label space is usually arduous or even impossible. When the source and target label spaces are not identical, matching the whole source and target distributions will create a representation space containing the features of the data belonging to the source classes which do not exist in the target classes. The target domain sees these classes as outlier classes, and extending their knowledge into the target domain will cause negative transfer, which significantly harms the model performance. Thus, in addition to marginal distribution disparity, we need to consider different scenarios where label spaces differ across domains. Different marginal distributions and different label spaces across domains are termed as domain gap and category gap, respectively. Based on the category gap, domain adaptation can be divided into four main categories: closed set, open set, partial, and universal domain adaptation.

- Closed set domain adaptation refers to the situation where both source and target domains share the same classes while there still exists a domain gap between domains. Traditional domain adaptation falls into this category.
- In open set domain adaptation, related domains share some labels in the common label set, and also they may have private labels [43]. Saito et al. [48] introduced new open set domain adaptation in which the data in the source private classes is removed. In the modified open set, the source label set is considered to be a subset of the target label set. Open set domain adaptation is suitable when there are multiple source domains where each includes a subset of target classes. Domain adaptation techniques aim to utilize all the source domain information contained in the shared classes to boost the model's performance in the target domain.
- In contrast to open set, partial domain adaptation refers to the situation where the target label set is a subset of the source label set [10, 70]. In this setting, the available source domain can be considered as a generic domain that consists of an abundant number of classes, and the target is only a subset of the source label set with fewer classes.
- Universal domain adaptation (UDA) [67] generalizes the above scenarios. In contrast to the above settings, which require prior knowledge about the source and target label sets, universal domain adaptation is not restricted to any prior knowledge. In this setting, source and target domains may share common label sets, and also each domain may have a private label set or outlier classes. Universal domain adaptation first tends to find the shared label space across domains and then similar to open set, and partial domain adaptation aligns the data distributions in the common label set. Ultimately, a classifier will be trained on the matched source labeled data to be applied safely to the unlabeled target data. In the testing phase, the trained classifier in both open set and universal domain adaptation is expected to assign accurate labels to the target samples belonging to the shared label space and mark the samples in the outlier classes as unknown.

This overview focuses on closed set unsupervised domain adaptation. This type of domain adaptation aims to utilize the labeled source data and the unlabeled target data to learn an objective predictive function that can perform well on the target domain where there is a shift between domains. Based on the definition of a domain (see Sect. 2), two domains can be different if at least one of input space, output space, or the probability density function changes between domains. Closed set domain adaptation considers the situation where the feature and label spaces are identical between domains, while the joint probability distributions may differ. Domain shift mainly can be categorized into three classes: prior shift, covariate shift, and concept shift.

- Prior shift or class imbalance considers the situation where posterior distributions are equivalent, $p_s(y|x) = p_t(y|x)$, and prior distributions of classes are different between domains, $p_s(y) \neq p_t(y)$. To solve a domain adaptation problem with a prior shift, we need labeled data in both source and target domains.
- Covariate shift refers to a situation where marginal probability distributions differ, $p_s(x) \neq p_t(x)$, while conditional probability distributions remain constant across domains, $p_s(y|x) = p_t(y|x)$. Sample selection bias and missing data are two causes for the covariate shift. Most of the proposed domain adaptation techniques aim to solve this class of domain gap.
- Concept shift, also known as data drift, is a scenario where data distributions remain unchanged, $p_s(x) = p_t(x)$, while conditional distributions differ between domains, $p_x(y|x) \neq p_t(y|x)$. Concept shift also requires labeled data in both domains to estimate the ratio of conditional distributions.

4 Approaches

Existing domain adaptation approaches can be broadly categorized into methods with shallow and deep architectures. Shallow domain adaptation approaches [23, 25, 31, 42] mainly utilize instance-based and feature-based techniques to align the domain distributions. One way of aligning the distributions is by minimizing the distance between domains. The mostly used distance measures in domain adaptation are maximum mean discrepancy (MMD) [24], Wasserstein metric, correlation alignment (CORAL)[57], Kullback-Leibler (KL) divergence [33], and contrastive domain discrepancy (CDD) [32]. Deep domain adaptation approaches [1, 18, 20, 36], on the other hand, utilize neural networks. This class of approaches usually uses convolutional, autoencoder, or adversarial based networks to diminish the domain gap. Some of the approaches in this category may also utilize a distance metric in one or multiple layers of two networks, one for source data and one for target data, to measure the discrepancy between the feature representations on the corresponding layers.

4.1 Instance-Based Adaptation

Instance-based domain adaptation approaches aim to deal with the shift between data distributions by minimizing the target risk based on the source labeled data. As mentioned in Sect. 3, domain shift is mainly categorized into covariate shift, prior shift, and concept shift. In unsupervised domain adaptation where the labeled data is available only in the source domain, if the source and target distributions are different, $p_T(x, y) \neq p_S(x, y)$, and the source and the target posteriors are arbitrary, $p_T(y|x) \neq p_S(y|x)$, the problem in Eq. (2) becomes intractable without labeled data in the target domain. Hence, we need to assume that both distributions are different only in marginal distribution, while the posteriors remain unchanged. This setting is called covariate shift in which the target risk can be simplified as follows:

$$R_T(h) = \mathbb{E}_{(x,y)\sim P_S}\left[\frac{p_T(x)\,\cancel{p_T(y|x)}}{p_S(x)\,\cancel{p_S(y|x)}}\ell(h(x), y)\right],\tag{3}$$

where the ratio of two density functions is considered as importance weight, i.e., $w(x) = \frac{p_T(x)}{p_S(x)}$. When domains consist of prior shift, the assumption is that the conditional distributions remain equal while prior distributions of classes differ across domains. Therefore, we can simplify the target risk as follows:

$$R_T(h) = \mathbb{E}_{(x,y)\sim P_S}\left[\frac{p_T(y)\,\cancel{p_T(x|y)}}{p_S(y)\,\cancel{p_S(x|y)}}\ell(h(x), y)\right],\tag{4}$$

where $w(y) = \frac{p_T(y)}{p_S(y)}$ is known as class weights. To solve the prior shift, we need labeled data in both source and the target domains which is out of our scope.

A typical solution to the covariate shift problem is to use importance weighting approaches to compensate for the bias by re-weighting the samples in the source domain based on the ratio of target and source domain densities, $w(x) = p_T(x)/p_S(x)$, where $w(x)$ is re-weighting factor for the samples in the source domain. [51, 68] proved that using the density ratio could remove sample selection bias by re-weighting the source instances. Thus, this problem can be viewed as a density ratio estimation (DRE). In this approach, the key idea is to find appropriate weights for the source samples through an optimization problem, such that the discrepancy between the re-weighted source data distribution and the actual target data distribution can be minimized.

To estimate the importance, one can indirectly estimate the marginal data distributions of each domain separately and then estimate the ratio. However, indirect density estimation is usually ineffective and a very challenging task, as the importance is usually unknown in reality, especially when having high-dimensional features [26]. A solution to this problem is to estimate the weights directly in an optimization procedure in which the model minimizes the discrepancy between the weighted source and target distributions. The disparity between domains can be reduced using Kernel Mean Matching (KMM) [25, 30]. KMM is a nonparametric method that directly estimates the weights by minimizing the discrepancy between domains using maximum mean discrepancy (MMD) in the reproducing kernel Hilbert space (RKHS). KMM first maps the samples in both domains into RKHS using a nonlinear kernel function such as Gaussian and then obtains the sample weights by minimizing the means of target data and the weighted source data:

$$D_{MMD}[w, p_S(x), p_T(x)] = \min_{w} \| \mathbb{E}_S[w(x)\phi(x)] - \mathbb{E}_T[\phi(x)] \|_{\mathcal{H}}$$
$$s.t. \quad w(x) \in [0, W], \quad \mathbb{E}_S[w(x)] = 1,\tag{5}$$

where $\phi(x)$ is a nonlinear kernel function that maps the data samples into RKHS and both constraints on $w(x)$ ensure that the variance of sample weights is bounded to be low and the weighted source data distribution is close to the probability distribution. Similar to many kernel-based approaches, KMM uses quadratic program in the optimization process which restricts the model to work well only on small datasets. Kullback-Leibler Importance Estimation Procedure (KLIEP) [53, 55, 56] directly estimates the density ratio using KL-divergence between target distribution and weighted source distribution. In this setting, KL-divergence can be simplified as follows:

$$
D_{KL}[w(x), p_S(x), p_T(x)] = \int_{\mathcal{X}} p_T(x) log \frac{p_T}{p_S w(x)} dx
$$

$$
= \int_{\mathcal{X}} p_T(x) log \frac{p_T}{p_S} dx - \int_{\mathcal{X}} p_T(x) \log w(x) dx
$$

$$(6)$$

$$
\propto \quad m^{-1} \sum_{j=1}^{m} \log w(z_j).
$$

In the second line of above equation, the first term does not depend on $w(x)$ which means that it is constant and can be removed from the objective function since we are optimizing it w.r.t $w(x)$. The above objective function is based on the weighted samples in the target domain that makes it computationally expensive for large-scale problems since we need to compute new weights for each new target data. To solve this problem, [54] proposed a new model that uses a linear model $w(x) = \phi(x)\alpha$, where $\phi(x)$ is a basis function and α is a set of parameters to be learned. Therefore, the objective function can be written as:

$$
D_{KL}[\phi(x), p_S(x), p_T(x)] = m^{-1} \sum_{j=1}^{m} \log \phi(z_j)\alpha. \tag{7}
$$

KLIEP minimizes the KL-divergence between the actual target distribution and the importance-weighted source distribution in a nonparametric manner to find the instance weights.

4.2 Feature-Based Adaptation

Feature-based adaptation approaches aim to map the source data into the target data by learning a transformation that extracts invariant feature representation across domains. They usually create a new feature representation by transforming the original features into a new feature space and then minimize the gap between domains in the new representation space in an optimization procedure while preserving the underlying structure of the original data. Subspace-based, transformation-based, and construction-based are some of the main feature-based adaptation methods. Below, we elaborate on each category and address some related approaches.

Subspace-Based Subspace-based adaptation aims to discover a common intermediate representation that is shared between domains. Many techniques have been proposed to construct this representation from the low-dimensional representation of the source and target data. Most of the adaptation approaches in this category first create a low-dimensional representation of original data in the form of a linear subspace for each domain and then reduce the discrepancy between the subspaces to construct the intermediate representation. A dimensionality reduction technique such as principal component analysis (PCA) can be used to construct the subspaces as two points, one for each domain, in a low-dimensional Grassmann manifold. The distance between the points in Grassmann manifold indicates the domain shift which can be reduced by applying different methods. Goplan et al. [23] proposed sampling geodesic flow (SGF) that first finds a geodesic path between the source and target points on a Grassmann manifold and then samples a set of points, subspaces, including the source and the target points along this path. In the next step, the data from both domains are projected onto all sampled subspaces along the geodesic path and will be concatenated to create a high-dimensional vector. Finally, a discriminative classifier can learn from the source-projected data to classify the unlabeled samples. Sampling more points from the geodesic path would help to map the source subspace into the target subspace more precisely. However, sampling more subspaces extends the dimensionality of the feature vector, which makes this technique computationally expensive. Geodesic flow kernel [21] was proposed to extend and improve SGF. GFK is a kernel-based domain adaptation method that deals with shift across domains. It aims to represent the smoothness of transition from a source to a target domain by integrating an infinite number of subspaces to find a geodesic line between domains in a low-dimensional manifold space. Fernando et al. [17] proposed a subspace alignment (SA) technique to directly reduce the discrepancy between domains by learning a linear mapping function that projects the source point directly into the target point in the Grassmann manifold. The projection matrix M can be learned by minimizing the divergence in the Bergman matrix:

$$M = \underset{M}{\arg\min} \quad ||X_S M - X_T||_F^2 = X_S^T X_T, \tag{8}$$

where X_S, X_T are the low-dimensional representation, basis vectors, of the source and the target data in the Grassmann manifold, respectively, and $||.||_F^2$ is the Frobenius norm. SA only aligns the subspace bases and ignores the difference between subspace distributions. Subspace distribution alignment (SDA) [58] extends the work in SA by aligning both subspace distributions and the bases at the same time. In SDA the projection matrix M can be formulated as $M = X_S^T X_T Q$, where Q is a matrix to align the discrepancy between distributions.

Transformation-Based Feature transformation transforms the original features into a new feature representation to minimize the discrepancy between the marginal and the conditional distributions while preserving the original data's underlying structure and characteristics. To reduce the domain discrepancy, we need some metrics such as maximum mean discrepancy (MMD), Kullback-Leibler divergence (KL-divergence), or Bregman divergence to measure the dissimilarity across domains. Transfer component adaptation (TCA) [42] was proposed to learn a domain-invariant feature transformation in which the marginal distributions between the source and target domains are minimized in RKHS using maximum mean discrepancy (MMD) criterion. After finding the domain-invariant features, we can utilize any classical machine learning technique to train the final target classifier. Joint domain adaptation (JDA) [37] extends TCA by simultaneously matching both marginal and conditional distributions between domains. JDA utilizes PCA as a dimensionality reduction technique to extract more robust features. Low-dimensional features can then be embedded into a high-dimensional feature space where the difference between marginal distributions is minimized using MMD. To align conditional distributions, we need labeled data in both domains. When labels are unavailable in the target domain, we can use pseudo labels, which can be estimated by the classifier trained on the labeled source data. After obtaining the pseudo labels, the model minimizes the distance between conditional distributions by modifying MMD. All the above steps are jointly and iteratively performed to find the best mapping function that aligns both marginal and conditional distributions across domains. The final target classifier can be trained on domain-invariant features discovered by the algorithm.

Reconstruction-Based The feature reconstruction-based methods aim to reduce the disparity between domain distributions by sample reconstruction in an intermediate feature representation. Jhuo et al. [31] proposed a Robust visual Domain Adaptation with Low-rank Reconstruction (RDALR) method to reduce the domain discrepancy. RDALR learns a linear projection matrix W that transforms the source samples into an intermediate representation where they can be linearly represented by the samples in the target domain to align the domain shift. The domain adaptation problem can be addressed by minimizing the following objective function:

$$\min_{W,Z,E} rank(Z) + \alpha||E||_{2,1},$$
$$s.t. \quad WX_S = X_T Z + E, \quad WW^T = I, \tag{9}$$

where X_S and X_T are sets of samples in the source and target domains, respectively, WX_S denotes the transformed source samples, Z is the reconstruction coefficient matrix including a set of coefficient vectors corresponding to the projected source samples, E is a matrix of noise and outlier information of source domain, and α is the regularization parameter. By minimizing the rank of the coefficient matrix Z, the method tends to reconstruct the different source samples together and find the underlying structure of source samples. $WW^T = I$ ensures to have a basis transformation matrix by enforcing it to be orthogonal. Besides, the noises and outliers in the source domain are decomposed into E. By minimizing E, the noises and outliers will be removed from the projected source data. However, RDALR is restricted to the rotation only, and the discriminative source domain information may not be transferred to the target domain, causing an unreliable alignment. Shao et al. [50] proposed Low-Rank Transfer Subspace Learning (LTSL) to resolve the RDALR problem. LTSL intends to discover a common low-rank subspace between domains where the source samples can reconstruct the target samples. Learning the common subspace makes LTSL more flexible on data representation. LTSL performs the adaptation by minimizing the following objective function:

$$\min_{W,Z,E} F(W, X_S) + \lambda_1 rank(Z) + \lambda_2 ||E||_{2,1},$$
$$s.t. \quad W^T X_T = W^T X_S Z + E, \quad W^T U_2 W = I, \tag{10}$$

where F(.) is a generalized subspace learning function. In contrast to RDALR that considers all target samples to represent the projected source samples, LTSL assumes that each sample in the target domain is more related to its neighborhood. Thus, each datum in the target domain can be reconstructed by only a set of source samples. In this way, the method can transfer both locality and discriminative properties of the source domain into the target domain.

4.3 Deep Domain Adaptation

In recent years, deep neural networks have been widely employed for representation learning and achieved remarkable results in many machine learning tasks such as image classification [29], sentiment analysis [71], speech recognition [14, 15], object detection [40, 73], and object recognition [2, 16].

Deep neural networks are very powerful techniques to extract the generalized feature representation of data. However they require an abundant labeled data for training, while annotating large amount of data is laborious, costly, and sometimes impossible. Besides, neural networks assume that both source and target domains are sampled from the same distribution and domain shift can greatly degrade the performance. Hence, deep domain adaptation was proposed to address the lack of sufficient labeled data while boosting the model's performance by deploying deep network properties along with adaptation techniques. Deep network adaptation techniques are mainly categorized into discrepancy-based, reconstruction-based, and adversarial-based adaptation.

Discrepancy-Based For the first time, Long et al. [36] propose Deep Adaptation Network (DAN) which utilizes the deep neural networks in the domain adaptation setting to learn transferable features across domains. DAN assumes that there is a shift between marginal distributions, while the conditional distributions remain unchanged. Therefore, it tends to match the marginal distributions across domains by adding multiple adaptation layers for the task-specific representations. Adaptation layers utilize multiple kernel variant of MMD (MK-MMD) to embed all the task-specific representations into RKHS and align the shift between marginal distributions. DAN only aligns the marginal distributions and does not consider the conditional distribution disparity across domains. Deep Transfer Network (DTN) [72] was proposed to match both marginal and conditional distributions simultaneously. DTN is a MMD-based distribution matching technique and composed of two types of layers. The shared feature extraction layer learns a subspace to align the marginal distributions across domains, while discrimination layer matches the conditional distributions using classifier transduction. To reduce the difference in the conditional distributions, the source labels and the target pseudo labels are used to project the data points into different hyperplanes for different classes. To align the conditional distributions, the data points are first projected into different hyperplanes using the source labels and the target pseudo labels, and then the discrepancy between conditional distributions is reduced by measuring and minimizing the conditional MMD-based objective function.

Reconstruction-Based Another category of deep adaptation networks utilizes autoencoder to align the discrepancy between domains by minimizing the reconstruction error and learning invariant and transferable representation across domains. The purpose of using autoencoder in domain adaptation is to learn the parameters of the encoder based on the samples in one domain (source) and adapt the decoder to reconstruct the samples in another domain (target). Glorot et al. [20] proposed a deep domain adaptation network based on Stacked Denoising Autoencoders (SDAs) [64], to extract high-level representation to represent both source and target domain data. The proposed model is assumed to have access to various domains with unlabeled data and only one source labeled data. In the first step, the higher-level feature extraction can be obtained by learning from all the available domains in an unsupervised manner. In the first step, the model obtains high-level representations from data in all domains by minimizing the reconstruction error. Next, a linear classifier, such as linear SVM, is trained on the extracted features of the source labeled data by minimizing a squared hinge loss. Finally, this classifier can be used on the target domains. The model shows remarkable results; however, using SDAs makes it computationally expensive and unscalable, especially when having high-dimensional features. Marginalized SDA (mSDA) [12] was proposed to extend the work in SDAs and address its limitations. mSDA marginalizes noise with linear denoisers to induce the model to learn the parameters in a closed-form solution without using stochastic gradient descent (SGD) [8]. Ghifary et al. [19] proposed a Deep Reconstruction-Classification Network (DRCN) that uses an encoder-decoder network for unsupervised domain adaptation in object recognition. DRCN consists of a standard convolutional network (encoder) to predict the source labels and a deconvolutional network [69] (decoder) to reconstruct the target samples. The model jointly utilizes supervised and unsupervised learning strategies to learn the encoder parameters to predict the source labels and the parameters of the decoder to reconstruct the target data. The encoder parameters are shared across both label prediction and reconstruction tasks, while the decoder parameters are private only for the reconstruction task.

Adversarial-Based The success of adversarial learning as a powerful domain-invariant feature extractor has motivated many researchers to embed it into deep networks. Adversarial domain adaptation approaches tend to minimize the distribution discrepancy between domains to obtain transferable and domain-invariant features. The main idea of adversarial domain adaptation was inspired by the generative adversarial networks (GANs) [22], which tend to minimize the cross-domain discrepancy through an adversarial objective. GANs are deep learning-based generative models composed of a two-player game, a generator model G and a discriminator model D. The generator aims to produce the samples similar to the domain of interest from the source data and confuse the discriminator to make a wrong decision. The discriminator then tends to discriminate between the true data in the domain of interest and the counterfeits generated by model G.

Ganin et al. [18] proposed a deep adversarial-based domain adaptation approach to match the domain gap by adding an effective gradient reversal layer (GRL) to the model. During the forward propagation, GRU acts like an identity function that leaves the input unchanged. However, it multiplies the gradient by a negative scalar to reverse it during the backpropagation phase. The proposed model can be trained on a massive amount of source labeled data and a large amount of unlabeled target data. The model is generally decomposed into three parts, feature extractor G_f, label predictor G_y, and domain classifier G_d. G_f produces the features for both label predictor and domain classifier. The GRL is inserted between G_f and G_d. The three parts of the model can learn their corresponding parameters $\theta_f,$, θ_y, and θ_d based on the source labeled data and the unlabeled target data. During the learning stage, the model learns the parameters θ_f and θ_y by minimizing the label predictor loss. This process enforces the feature extractor to produce the discriminative features for a good label prediction on the source domain. At the same time, the model learns the parameters θ_f and θ_d by maximizing and minimizing the domain classifier loss, respectively, to obtain domain-invariant features. Pei et al. [44] argued that using a single domain discriminator can only match the marginal distributions and ignores the discrepancy between class-conditional distributions causing negative transfer and falsely aligning the classes across domains. Multi-adversarial domain adaptation (MADA) [44] was proposed to reduce the shift in the joint distribution between domains and enable fine-grained alignment of different class distributions by constructing multiple class-wise domain discriminators. Such that each discriminator only matches the source and target data samples belonging to the same class.

The generator G and discriminator D also can be implemented through a minimax optimization of an adversarial objective for a domain confusion. This approach was inspired by theory on domain adaptation [4, 5], suggesting that the transferable representation across domains does not include discriminative information and, thus, an algorithm is not able to learn the origin of an input sample based on this representation. In other words, the minimax optimization process aims to match the domain disparity by learning deep invariant representation across domains. Tzeng et al. [61] and Ajakan et al. [1] constructed the minimax optimization-based adversarial domain adaptation in deep neural networks for the first time. Tzeng et al. [61] introduced an adversarial CNN-based architecture that tends to align both marginal and conditional distributions across domains by minimizing the classification loss, soft label loss, domain classifier loss, and domain confusion loss. The model first computes the average probability distribution, soft label, for the source labeled data in each category. It can be later fine-tuned over the target labeled data to match the class-conditional distributions to the soft labels across domains. Unlike hard labels, the soft labels provide more transferable and useful information about each sample's category and also the relationship between categories, e.g., *bookshelves* are more similar to *filling cabinets* than to *bicycles*. The domain confusion loss is the cross-entropy between the uniform distribution over domain labels and the output predicted domain labels. Minimizing this loss enforces the algorithm to extract invariant representation across domains that maximally confuses the discriminator.

Visual adversarial domain adaptation also utilizes generative adversarial networks (GANs) to reduce the shift between domains. In generative adversarial domain adaptation, generator model G aims to synthesize implausible images, while discriminator model D seeks to identify between synthesized and the real samples. Visual domain adaptation techniques using generative adversarial networks adopt representations in pixel-level, feature-level, or both. The pixel-level approaches perform adaptation in the raw pixel space to directly translate the images in one domain into the style images of another domain. Bousmal et al. [9] proposed an unsupervised pixel-level domain adaptation method (PixelDA) that utilizes generative adversarial networks to adopt one domain to another. The model changes source domain images to appear as if drawn from the target domain distribution while preserving their original content. Simulated+Unsupervised learning (SimGAN) [52] aims to generate synthetic images that are realistic and similar to those in the domain of interest. A simulator first generates synthetic images from a noise vector. A refiner network then refines them to resemble the target images by optimizing an adversarial loss. In contrast to the pixel-level approaches, the feature-level methods modify the representation in the discriminative feature space to alleviate the discrepancy across domains. The Domain Transfer Network (DAN) [59] is a feature-level generative-based adaptation that enforces the consistency in the embedding space to transform a source image into a target image. Hoffman et al. [27] proposed Cycle-Consistent Adversarial Domain Adaptation (CyCADA) that fuses both pixel-level [9, 34, 52] and feature-level [36, 62] adversarial domain adaptation methods with cycle-consistent image-to-image translation techniques [74] to direct the mapping from one domain to another. The model simultaneously adopts representations at both pixel-level and feature-level, by minimizing several losses, including pixel-level, feature-level, semantic, and cycle consistency losses in both domains.

References

1. H. Ajakan, P. Germain, H. Larochelle, F. Laviolette, M. Marchand, Domain-adversarial neural networks (2014). arXiv preprint arXiv:1412.4446
2. S. Amirian, Z. Wang, T.R. Taha, H.R. Arabnia, Dissection of deep learning with applications in image recognition, In *Computational Science and Computational Intelligence; "Artificial Intelligence" (CSCI-ISAI); 2018 International Conference on. IEEE* (2018), pp. 1132–1138
3. E. Asali, F. Shenavarmasouleh, F.G. Mohammadi, P.S. Suresh, H.R. Arabnia, Deepmsrf: a novel deep multimodal speaker recognition framework with feature selection (2020). arXiv preprint arXiv:2007.06809
4. S. Ben-David, J. Blitzer, K. Crammer, A. Kulesza, F. Pereira, J.W. Vaughan, A theory of learning from different domains. Mach. Learn. **79**(1–2), 151–175 (2010)
5. S. Ben-David, J. Blitzer, K. Crammer, F. Pereira, Analysis of representations for domain adaptation, in *Advances in Neural Information Processing Systems* (2007), pp. 137–144
6. A. Blum, T. Mitchell, Combining labeled and unlabeled data with co-training, in *Proceedings of the Eleventh Annual Conference on Computational Learning Theory* (ACM, 1998), pp. 92–100
7. Z. Bo, S. Zhong-Zhi, Z. Xiao-Fei, Z. Jian-hua, A transfer learning based on canonical correlation analysis across different domains Chinese. Chin. J. Comput. **38**(7), 1326–1336 (2015)
8. L. Bottou, Stochastic gradient descent tricks, in *Neural Networks: Tricks of the Trade* (Springer, Berlin, 2012), pp. 421–436
9. K. Bousmalis, N. Silberman, D. Dohan, D. Erhan, D. Krishnan, Unsupervised pixel-level domain adaptation with generative adversarial networks, in *Proceedings of the IEEE Conference on Computer Vision and Pattern Recognition* (2017), pp. 3722–3731
10. Z. Cao, M. Long, J. Wang, M.I. Jordan, Partial transfer learning with selective adversarial networks, in *Proceedings of the IEEE Conference on Computer Vision and Pattern Recognition* (2018), pp. 2724–2732
11. R. Caruana, Multitask learning. Machine Learn. **28**(1), 41–75 (1997)
12. M. Chen, Z. Xu, K. Weinberger, F. Sha, Marginalized denoising autoencoders for domain adaptation (2012). arXiv preprint arXiv:1206.4683
13. K. Cho, B. Van Merriënboer, C. Gulcehre, D. Bahdanau, F. Bougares, H. Schwenk, Y. Bengio, Learning phrase representations using RNN encoder-decoder for statistical machine translation (2014). arXiv preprint arXiv:1406.1078
14. G.E. Dahl, D. Yu, L. Deng, A. Acero, Context-dependent pre-trained deep neural networks for large-vocabulary speech recognition. IEEE Trans. Audio Speech Lang. Process. **20**(1), 30–42 (2011)
15. L. Deng, M.L. Seltzer, D. Yu, A. Acero, A.-R. Mohamed, G. Hinton, Binary coding of speech spectrograms using a deep auto-encoder, in *Eleventh Annual Conference of the International Speech Communication Association* (2010)
16. A. Eitel, J.T. Springenberg, L. Spinello, M. Riedmiller, W. Burgard, Multimodal deep learning for robust RGB-D object recognition, in *2015 IEEE/RSJ International Conference on Intelligent Robots and Systems (IROS)* (IEEE, Piscataway, 2015), pp. 681–687
17. B. Fernando, A. Habrard, M. Sebban, T. Tuytelaars, Unsupervised visual domain adaptation using subspace alignment, in *Proceedings of the IEEE International Conference on Computer Vision* (2013), pp. 2960–2967
18. Y. Ganin, V. Lempitsky, Unsupervised domain adaptation by backpropagation, in *International Conference on Machine Learning* (2015), pp. 1180–1189
19. M. Ghifary, W.B. Kleijn, M. Zhang, D. Balduzzi, W. Li, Deep reconstruction-classification networks for unsupervised domain adaptation, in *European Conference on Computer Vision* (Springer, Berlin, 2016), pp. 597–613
20. X. Glorot, A. Bordes, Y. Bengio, Domain adaptation for large-scale sentiment classification: A deep learning approach, in *ICML* (2011)

21. B. Gong, Y. Shi, F. Sha, K. Grauman, Geodesic flow kernel for unsupervised domain adaptation, in *2012 IEEE Conference on Computer Vision and Pattern Recognition* (IEEE, Piscataway, 2012), pp. 2066–2073

22. I. Goodfellow, J. Pouget-Abadie, M. Mirza, B. Xu, D. Warde-Farley, S. Ozair, A. Courville, Y. Bengio, Generative adversarial nets, in *Advances in Neural Information Processing Systems* (2014), pp. 2672–2680

23. R. Gopalan, R. Li, R. Chellappa, Domain adaptation for object recognition: an unsupervised approach, in *2011 International Conference on Computer Vision* (IEEE, Piscataway, 2011), pp. 999–1006

24. A. Gretton, K. Borgwardt, M. Rasch, B. Schölkopf, A.J. Smola, A kernel method for the two-sample-problem, in *Advances in Neural Information Processing Systems* (2007), pp. 513–520

25. A. Gretton, A. Smola, J. Huang, M. Schmittfull, K. Borgwardt, B. Schölkopf, Covariate shift by kernel mean matching. Dataset Shift Mach. Learn. **3**(4), 5 (2009)

26. W.K. Härdle, M. Müller, S. Sperlich, A. Werwatz, *Nonparametric and Semiparametric Models* (Springer, Berlin, 2012)

27. J. Hoffman, E. Tzeng, T. Park, J.-Y. Zhu, P. Isola, K. Saenko, A.A. Efros, T. Darrell, Cycada: cycle-consistent adversarial domain adaptation (2017). arXiv preprint arXiv:1711.03213

28. H. Hotelling, Relations between two sets of variates, in *Breakthroughs in Statistics* (Springer, Berlin, 1992), pp. 162–190

29. J. Hu, L. Shen, G. Sun, Squeeze-and-excitation networks, in *Proceedings of the IEEE Conference on Computer Vision and Pattern Recognition* (2018), pp. 7132–7141

30. J. Huang, A. Gretton, K. Borgwardt, B. Schölkopf, A.J. Smola, Correcting sample selection bias by unlabeled data, in *Advances in Neural Information Processing Systems* (2007), pp. 601–608

31. I.-H. Jhuo, D. Liu, D. Lee, S.-F. Chang, Robust visual domain adaptation with low-rank reconstruction, in *2012 IEEE Conference on Computer Vision and Pattern Recognition* (IEEE, Piscataway, 2012), pp. 2168–2175

32. G. Kang, L. Jiang, Y. Yang, A.G. Hauptmann, Contrastive adaptation network for unsupervised domain adaptation, in *Proceedings of the IEEE Conference on Computer Vision and Pattern Recognition* (2019), pp. 4893–4902

33. S. Kullback, R.A. Leibler, On information and sufficiency. Ann. Math. Stat. **22**(1), 79–86 (1951)

34. M.-Y. Liu, O. Tuzel, Coupled generative adversarial networks, in *Advances in Neural Information Processing Systems* (2016), pp. 469–477

35. K. Livescu, M. Stoehr, Multi-view learning of acoustic features for speaker recognition, in *2009 IEEE Workshop on Automatic Speech Recognition & Understanding* (IEEE, Piscataway, 2009), pp. 82–86

36. M. Long, Y. Cao, J. Wang, M.I. Jordan, Learning transferable features with deep adaptation networks (2015). arXiv preprint arXiv:1502.02791

37. M. Long, J. Wang, G. Ding, J. Sun, P.S. Yu, Transfer feature learning with joint distribution adaptation, in *Proceedings of the IEEE International Conference on Computer Vision* (2013), pp. 2200–2207

38. M. Mohri, A. Rostamizadeh, A. Talwalkar, *Foundations of Machine Learning* (MIT Press, Cambridge, 2018)

39. K. Muandet, D. Balduzzi, B. Schölkopf, Domain generalization via invariant feature representation, in *International Conference on Machine Learning* (2013), pp. 10–18

40. W. Ouyang, X. Wang, X. Zeng, S. Qiu, P. Luo, Y. Tian, H. Li, S. Yang, Z. Wang, C.-C. Loy, et al., Deepid-net: deformable deep convolutional neural networks for object detection, in *Proceedings of the IEEE Conference on Computer Vision and Pattern Recognition* (2015), pp. 2403–2412

41. S. Pan, A survey on transfer learning. IEEE Trans. Knowl. Data Eng. **22**(10), 1345–1359 (2010)

42. S.J. Pan, I.W. Tsang, J.T. Kwok, Q. Yang, Domain adaptation via transfer component analysis. IEEE Trans. Neural Netw. **22**(2), 199–210 (2010)

43. P. Panareda Busto, J. Gall, Open set domain adaptation, in *Proceedings of the IEEE International Conference on Computer Vision* (2017), pp. 754–763
44. Z. Pei, Z. Cao, M. Long, J. Wang, Multi-adversarial domain adaptation, in *Thirty-Second AAAI Conference on Artificial Intelligence* (2018)
45. N.N. Pise, P. Kulkarni, A survey of semi-supervised learning methods, in *2008 International Conference on Computational Intelligence and Security*, vol. 2 (IEEE, Piscataway, 2008), pp. 30–34
46. V. Ramanishka, A. Das, D.H. Park, S. Venugopalan, L.A. Hendricks, M. Rohrbach, K. Saenko, Multimodal video description, in *Proceedings of the 24th ACM International Conference on Multimedia* (2016), pp. 1092–1096
47. A.M. Rush, S. Chopra, J. Weston, A neural attention model for abstractive sentence summarization (2015). arXiv preprint arXiv:1509.00685
48. K. Saito, S. Yamamoto, Y. Ushiku, T. Harada, Open set domain adaptation by backpropagation, in *Proceedings of the European Conference on Computer Vision (ECCV)* (2018), pp. 153–168
49. I.V. Serban, A. Sordoni, Y. Bengio, A. Courville, J. Pineau, Building end-to-end dialogue systems using generative hierarchical neural network models, in *Thirtieth AAAI Conference on Artificial Intelligence* (2016)
50. M. Shao, D. Kit, Y. Fu, Generalized transfer subspace learning through low-rank constraint. Int. J. Comput. Vis. **109**(1–2), 74–93 (2014)
51. H. Shimodaira, Improving predictive inference under covariate shift by weighting the log-likelihood function. J. Stat. Plan. Inference **90**(2), 227–244 (2000)
52. A. Shrivastava, T. Pfister, O. Tuzel, J. Susskind, W. Wang, R. Webb, Learning from simulated and unsupervised images through adversarial training, in *Proceedings of the IEEE Conference on Computer Vision and Pattern Recognition* (2017), pp. 2107–2116
53. M. Sugiyama, M. Krauledat, K.-R. MÃžller, Covariate shift adaptation by importance weighted cross validation. J. Mach. Learn. Res. **8**, 985–1005 (2007)
54. M. Sugiyama, K.-R. Müller, Input-dependent estimation of generalization error under covariate shift. Stat. Decis. **23**(4/2005), 249–279 (2005)
55. M. Sugiyama, K.-R. Müller, Model selection under covariate shift, in *International Conference on Artificial Neural Networks* (Springer, 2005), pp. 235–240
56. M. Sugiyama, S. Nakajima, H. Kashima, P.V. Buenau, M. Kawanabe, Direct importance estimation with model selection and its application to covariate shift adaptation, in *Advances in Neural Information Processing Systems* (2008), pp. 1433–1440
57. B. Sun, J. Feng, K. Saenko, Correlation alignment for unsupervised domain adaptation, in *Domain Adaptation in Computer Vision Applications* (Springer, Berlin, 2017), pp. 153–171
58. B. Sun, K. Saenko, Subspace distribution alignment for unsupervised domain adaptation, in *BMVC*, vol. 4 (2015), pp. 24–1
59. Y. Taigman, A. Polyak, L. Wolf, Unsupervised cross-domain image generation (2016). arXiv preprint arXiv:1611.02200
60. D. Tran, L. Bourdev, R. Fergus, L. Torresani, M. Paluri, Learning spatiotemporal features with 3d convolutional networks, in *Proceedings of the IEEE International Conference on Computer Vision* (2015), pp. 4489–4497
61. E. Tzeng, J. Hoffman, T. Darrell, K. Saenko, Simultaneous deep transfer across domains and tasks, in *Proceedings of the IEEE International Conference on Computer Vision* (2015), pp. 4068–4076
62. E. Tzeng, J. Hoffman, K. Saenko, T. Darrell, Adversarial discriminative domain adaptation, in *Proceedings of the IEEE Conference on Computer Vision and Pattern Recognition* (2017), pp. 7167–7176
63. V. Vapnik, Principles of risk minimization for learning theory, in *Advances in Neural Information Processing Systems* (1992), pp. 831–838
64. P. Vincent, H. Larochelle, I. Lajoie, Y. Bengio, P.-A. Manzagol, L. Bottou, Stacked denoising autoencoders: learning useful representations in a deep network with a local denoising criterion. J. Mach. Learn. Res. **11**, 3371–3408 (2010)

65. C. Wang, D.M. Blei, Collaborative topic modeling for recommending scientific articles, in *Proceedings of the 17th ACM SIGKDD International Conference on Knowledge Discovery and Data Mining* (2011), pp. 448–456
66. H. Wang, N. Wang, D.-Y. Yeung, Collaborative deep learning for recommender systems, in *Proceedings of the 21th ACM SIGKDD International Conference on Knowledge Discovery and Data Mining* (2015), pp. 1235–1244
67. K. You, M. Long, Z. Cao, J. Wang, M.I. Jordan, Universal domain adaptation, in *Proceedings of the IEEE Conference on Computer Vision and Pattern Recognition* (2019), pp. 2720–2729
68. B. Zadrozny, Learning and evaluating classifiers under sample selection bias, in *Proceedings of the Twenty-First International Conference on Machine Learning* (2004), p. 114
69. M.D. Zeiler, D. Krishnan, G.W. Taylor, R. Fergus, Deconvolutional networks, in *2010 IEEE Computer Society Conference on Computer Vision and Pattern Recognition* (IEEE, Piscataway, 2010), pp. 2528–2535
70. J. Zhang, Z. Ding, W. Li, P. Ogunbona, Importance weighted adversarial nets for partial domain adaptation, in *Proceedings of the IEEE Conference on Computer Vision and Pattern Recognition* (2018), pp. 8156–8164
71. L. Zhang, S. Wang, B. Liu, Deep learning for sentiment analysis: a survey. Wiley Interdiscip. Rev. Data Mining Knowl. Discovery **8**(4), e1253 (2018)
72. X. Zhang, F.X. Yu, S.-F. Chang, S. Wang, Deep transfer network: unsupervised domain adaptation (2015). arXiv preprint arXiv:1503.00591
73. Z.-Q. Zhao, P. Zheng, S.-T. Xu, X. Wu, Object detection with deep learning: a review. IEEE Trans. Neural Netw. Learn. Syst. **30**(11), 3212–3232 (2019)
74. J.-Y. Zhu, T. Park, P. A.A. Isola, Efros, Unpaired image-to-image translation using cycle-consistent adversarial networks, in *Proceedings of the IEEE International Conference on Computer Vision* (2017), pp. 2223–2232
75. X.J. Zhu, Semi-supervised learning literature survey. Tech. rep., University of Wisconsin-Madison Department of Computer Sciences, 2005

Fake News Detection Through Topic Modeling and Optimized Deep Learning with Multi-Domain Knowledge Sources

Vian Sabeeh, Mohammed Zohdy, and Rasha Al Bashaireh

1 Introduction

Fake news has become the most debated topic since 2016. Also, the lack of cognitive capacity of an individual to discern fake news from the massive amount of disseminated information in social media platforms leads to the increased propagation and fake belief towards the false information. Moreover, the fast-paced developments and collaborative information sharing on social media give an additional dimension for false information. With the exponential development of technology, the proliferation of fake news began to explode. The prevalence of fake news over social media not only declines the trustworthiness of news sources but also misleads the public to believe the deceptive information. The rapid prolifer-ation and dissemination of false information can have far-reaching and disastrous outcomes in several fields, such as political events [1] and financial markets [2]. Furthermore, fake news often emerges during natural disasters, including Hurricane Sandy and Japan earthquake that made disorder and increased the panic [3]. The term false information has not only attracted individual attention but has also received considerable attention from research communities. Accordingly, a majority of research on the fake news detection exploits machine learning techniques with hand-engineered features. However, it requires domain expertise for analysis of the inherent characteristics of the news. In recent years, deep learning methods

V. Sabeeh (✉) · R. Al Bashaireh
Computer Science and Engineering Department, Oakland University, Rochester, MI, USA
e-mail: viansabeeh@oakland.edu; ralbashaireh@oakland.edu

M. Zohdy
Faculty of Electrical and Computer Engineering Department, Oakland University, Rochester, MI, USA
e-mail: zohdyma@oakland.edu

© Springer Nature Switzerland AG 2021   895
R. Stahlbock et al. (eds.), *Advances in Data Science and Information Engineering*,
Transactions on Computational Science and Computational Intelligence,
https://doi.org/10.1007/978-3-030-71704-9_66

have been identified to be a highly promising decision-making algorithm. Accurate and automatic identification of fake news enables the removal of deliberately false information. As a consequence, there has been a rising interest in the development of autonomous systems to discover the fake news. With the ability of rich feature extraction, deep learning has been extensively used in innovative fields in a real-time environment. The deep learning method transfers the data onto the multiple hidden layers to assist the learning of complex data [4]. In the recent years, researchers have begun to employ deep learning methods for acquiring promising results in a fake news detection system. However, the deep learning method only accomplishes the promising performance with the abundance of labeled data, which obscures the reliability of the detection system. Even though there have been considerable research efforts made to solve issues in the fake news detection systems using deep learning, many problems are still open. Making decisions only from the labeled data without fundamental knowledge becomes computationally ineffective. The automatic fact-checking system is critical for timely identification and mitigation of fake news. Fake news detection and mitigation methods need to be capable of detecting unseen and newly emerging events. Thus, it is necessary to apply the pre-trained deep learning models for improving the identification process. Most of the current work does not focus on the optimal hidden structure while decision making, which has a negative impact on detection accuracy and complexity of the system. Therefore, it is necessary to select the optimal number of a hidden node for acquiring consistent results.

The key contributions of the Fake news Identification using BERT with optimal Neurons and Domain knowledge (FIND) are summarized as follows:

- It enhances the fake news detection model with the incorporation of the pre-trained model, optimal hidden neuron selection, and intelligent decision-making.
- Initially, the FIND approach significantly pre-trains the BERT model on a massive unsupervised corpus to perform the two-step classification for fake news detection with the help of the LDA model.
- It employs the LDA method to discover the underlying topic of the headline and body of the news articles individually as well as collaboratively to detect the fake news.
- In the first-step classification, FIND selects the optimal number of hidden nodes and presents the classified results as generic four-classes with the assistance of the BERT model.
- In the second-step classification, FIND applies the intelligent decision-making involving LDA and web source-based generic multi-class to specific two-class classification that is fake news detection.

2   Related Work

The exponential growth of misinformation in social networking platforms has become a potential threat to implicitly influencing the opinion of the public and threatening the news industry. Thus, in recent years, fake news detection has become a surging research area. From the Natural Language Processing (NLP) perspective, several research efforts have been presented to employ stylistic [5] and linguistic features [6]. Other research efforts involved word embeddings, latent semantic analysis features [7], and a bag of words (BOW). In some instances, the researchers exploited the retrieved evidence and external features from the internet. Most of the existing fake news detection researchers have employed the deep learning algorithms to detect and classify the fake news information reviewed as follows.

2.1   Fake News Detection and Classification Approaches

The work [8] employs Bing and Google search engines accordingly to explore the factuality of claims related to the politics. The deep learning-based models have outperformed conventional machine learning techniques owing to the ability of feature extraction. In other attempts, some research works employed deep learning methods to evaluate fake news in a system. The deep learning-based hybrid fake news detection model [9] employs text-Convolutional Neural Network (CNN) for capturing linguistic features of the statement. It involves the extraction of the relation between the contextual information for accurate fake news detection. It also employs a self-attention mechanism to discover the global representation of the element in a sequence. By using the heterogeneous data, including news propagation, content, user activity, demographic information, and social network structure, the work [10] detects the fake information. It also considers the propagation pattern of news for detecting false information. A sifted multitask learning model [11] employs filtering and stance detection mechanism for discovering fake news. By utilizing attention and gate mechanisms in the sharing layer, it accurately captures the potential shared features. Also, it applies the transformer model for an encoding task, which improves performance. By integrating the Markov Random Field (MRF) and deep neural networks, the work [12] detects the fake news. It effectively analyzes the correlations between the news articles based upon the MRF method. In addition, by using deep neural networks, it learns the high-level representations. Fake news detection based on the multiple Bidirectional LSTM method in [13] classifies the fake news headlines for stance detection. It resolves the overfitting issue by tuning the parameters of the multilayer perceptron predictor layer. However, it did not generalize well for input pairs. The Deep Neural Networks-based fake news classification model [14] relies on content-based factors, authenticity of the source, perceived cognitive authority, and natural language features for discovering the fake claims. The claim classification model incorporates two sub-modules, whereas the first one exploits the word-level features and claim for extracting the relevant article among the knowledge sources. It helps to evaluate the trustworthiness of the claim. Subsequently, the second sub-module employs a deep neural network to learn the style of a fake claim.

2.2 Stance Detection-Based Fake News Assessment Approaches

In the fake news detection model, stance detection is significant for evaluating the veracity of the news because it supports the fact-checking by detecting a stance from heterogeneous information sources. As a consequence, several stance detection research works have been proposed to combat the detection and disseminating of fake news [15]. The rumor stance detection model [16] exploits the word embeddings, a combination of n-grams, and lexical cue words to detect the stance towards the headline of the article. It focuses on cue words for enhancing the accuracy of detection. However, it often faces a problem when the length ratio difference between the titles and articles is massive. A stance detection-based fact-checking model [17] employs a deep bidirectional transformer for encoding the article-claim pairs. It improves the performance of the stance detection model by pre-training a transformer model, which helps to reduce the complexity of the system. The automatic stance detection model [18] applies the concepts of conditional encoding and neural attention LSTM model for enhancing the accuracy of stance detection. The LSTM model with the neural attention mechanism helps to understand the textual entailment. The exBAKE fake news detection approach [19] employs the BERT model to categorize the data based on weighted cross-entropy (WCE). Additionally, it incorporates Daily Mail news and CNN news data for pre-training the BERT model, which helps to identify the fake news accurately.

3 Proposed Methodology

With the target of effectively detecting the fake news, the proposed approach develops the enhanced deep learning model with the pre-training, LDA-based topic knowledge incorporation, optimal neuron selection, and two-step classification. Initially, the FIND approach pre-trains the BERT model and fine-tunes the model for the deceptive information classification. The proposed fake news detection system involves two main phases, such as the BERT model pre-training phase and the fine-tuning phase. Fig. 1 describes the overall process of the FIND approach. In the pre-training phase, the FIND approach utilizes the massively unlabeled data sources such as news databases to pre-train the BERT model. In the fine-tuning phase, the proposed approach is responsible for extracting the topic of both the headlines and body text using the LDA, fake news data-based parameter tuning of the BERT

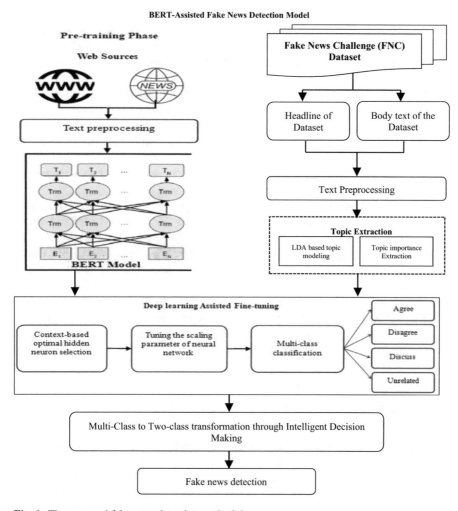

Fig. 1 The proposed fake news detection methodology

model, and multi-class to two-class transformation through intelligent decision-making. In essence, the parameter tuning involves either the optimal hidden node or neuron selection in the hidden layer, which enforces the learning model to avoid the data overfitting and extended training time.

3.1   Pre-training the BERT Model

In the recent years, fake news detection research has gained popular attention in different real-world applications due to the increased amount of fake news generation. Despite this, developing an accurate fact-checking detection model is critical due to the need for a massive amount of training data, which is inefficient for the emerging news content. With the increased advantage of the BERT model in text processing, the proposed FIND approach exploits the BERT model as the unsupervised pre-training model to further help the fake news detection model. The BERT model considers the sentence context in terms of utilizing the context of the previous and next sentences while analyzing each sentence. Moreover, it enables transform learning on the previous pre-trained models to provide custom training for the corresponding data. In contrast to other deep learning methods, the BERT model jointly analyzes the given text with the consideration of both the right and left contexts over the layers, which contributes to higher performance to the text processing applications.

In the FIND approach, the pre-trained process assists the restriction of inaccurate fake news detection. By utilizing the BERT model, the proposed FIND approach significantly understands and discovers the relationship among the keywords in each sentence. Accordingly, the proposed FIND approach pre-trains the BERT model using web information sources and news corpus to accomplish accurate detection in the fact-checking applications. Moreover, with the assistance of the BERT pre-training model, the proposed FIND approach alleviates the process of training the learning model from scratch and avoids the misclassification. After the pre-training process, the FIND approach leverages the BERT model for the fine-tuning phase to adopt the fake news detection task.

3.2 Topic Extraction from News Using LDA

To facilitate intelligent decision-making, the FIND approach employs the LDA model to extract the topic of both the headlines and the body text of the FNC dataset. In the FIND approach, intelligent decision-making involves the transformation of multi-class data to two-class data based on the classification results and LDA results of the fake news data. Hence, by applying the LDA method, the proposed methodology inherently extracts the topic from the headline as well as the body of the text. LDA is a generative probabilistic model that is designed for discovering underlying hidden topics of the discrete data. Also, it explores the hidden semantic information of the document without the use of knowledge sources. Initially, the LDA model randomly creates a summary of topics for each word of the document for a discrete probability distribution. In the proposed deceptive detection system, the LDA-based topic extraction model discovers the latent factors of the headline and the body of the text to recognize the underlying topic of the text significantly. Let, a fake news challenge dataset consists of the headline and the body of the text referring a set of 'n' headlines represented by $H = \{H_1, \ldots, H_n\}$, and set of 'n' body of the text denoted by$\{B_1, \ldots, B_n\}$. Before applying the LDA model, the FIND approach applies different text preprocessing methods to generate the most informative keywords in each sentence. By applying the LDA algorithm, the proposed approach obtains 'm' number of topics for headline and body of the text $T = \{T1, \ldots, Tm\}$. The LDA model applies probability distribution for each 'w' words in a headline, $Tm = \{w1, \ldots, wk\}$, whereas, the term wk represents the probabilistic score of the kth word in a headline that related to the topic 'm.' Similarly, the proposed FIND approach measures the probability distribution over 'w' words in a body of the text, $Tm = \{w1, \ldots, wl\}$, in which wl denotes the lth word of the body of the text. Accordingly, the proposed approach discovers the topic for the headline as well as the body of the text. As a result, the FIND approach utilizes the extracted topic information of the news content in the second-step classification that is deep learning-assisted fake news classification from the multi-class data.

3.3 Deep Learning-Assisted Fine-Tuning

The proposed FIND approach performs a two-step classification such as deep learning-assisted multi-class classification and intelligent decision-making based fake news classification to determine the fake news from the FNC dataset. Instead of only determining the relationship between the headline and body text in the FNC dataset, the FIND approach determines the fake news by individually and collaboratively analyzing the headlines and body text. In the two-step classification, deep learning-assisted fine-tuning involves the multi-class classification with the optimal number of neurons in the BERT model. Fine-tuning of the proposed approach is concerned with the training of the pre-trained BERT model on the target task along with the optimal neurons. Accordingly, the proposed BERT-based fake news detection model is pre-trained in web information sources and fine-tuned to identify fake news with the highest degree of accuracy. In other words, the proposed FIND approach transfers the gained recognition knowledge from web information sources to the target domain of fake news detection using the attention model. Notably, the great success of the proposed fake news detection model relies on a valid selection of several hidden nodes in the BERT model. Therefore, it is necessary to discover a sufficient number of hidden nodes based on the input data to adopt the BERT model accurately.

$$N_H = \begin{cases} N_{H(k)} & H_k = H_1 \\ N_{H(k+1)} & H_k \neq H_1 \end{cases} \tag{1}$$

In the proposed FIND approach, the pre-trained model does not focus on the optimal selection of the number of neurons in the hidden layer of the BERT model, which leads to the overfitting of data and results in unstable detection results. To overcome this constraint, the proposed approach selects the optimal hidden nodes on BERT pre-trained model without compromising the accuracy. The FIND approach computes the optimal number of hidden nodes (NH) for the BERT learning model in two different ways with respect to the number of the hidden layer, which is mentioned in Eq. 1. According to Eq. 2, the proposed approach computes the number of neurons for the first hidden layer (k), i.e., k=1. Wherein, the terms Ni and No represents the total number of input and output hidden nodes in the hidden layer. If the hidden layer is not the first hidden layer, the proposed approach selects the optimal number of neurons using Eq. 3 for remaining (k+1) layers.

$$N_{H(k)} = \frac{N_i + N_o}{2} \tag{2}$$

To improve the performance of the fake news detection system, the proposed FIND approach additionally considers the length of sentences (Ls) along with the number of input samples (Ns) while deciding the optimal number of neurons for the remaining k+1 layers. In Eq. 3, $\alpha 1$ and $\alpha 2$ indicate the scaling parameters, which are computed using Eq. 4.

$$N_{H(k+1)} = \left(\frac{\exp^{(N_s)}}{\alpha_1 * \log (N_i + N_o)} + \frac{\exp^{(L_s)}}{\alpha_2 * \log (N_i + N_o)} \right) \tag{3}$$

$$\alpha_1 = \log 2 (N_s) + 2 \quad ; \quad \alpha_2 = \log 2 (L_s) + 2 \tag{4}$$

In the FIND approach, the selection of an optimal number of hidden nodes has a tremendous impact on the fake news detection results even it does not directly influence the structure of the pre-training of the learning model. In the deep learning model, the improper selection of hidden nodes in the hidden layer leads the overfitting. Hence, the proposed approach significantly prevents the fake news detection system from the overfitting issue by fixing the scaling parameter, which is shown in Eq. 4. In consequence, the FIND approach classifies the incoming FNC dataset into multiple classes such as agree, disagree, unrelated, and discuss with the help of the optimally tuned BERT model. From the result of the first-step classification, the proposed approach retains the agree, disagree, unrelated, and discuss classes of the FNC dataset.

3.4 Fake News Detection Through Intelligent Decision-Making

In the two-step classification, fake news detection is the second-level classification, i.e., the transformation of multi-classes into two classes based on intelligent decision-making. During the multi-class classification, the proposed approach considers only the trained data in the FNC dataset and the pre-trained information, which does not result from the data as the fake news or real news due to the lack of the consideration of factual and non-factual information. Moreover, the fake news classification emphasizes not only the syntactic structure but also on validation with the real facts residing in the web source. Hence, to evaluate the trustworthiness of the news, the proposed FIND methodology utilizes the knowledge of the pre-trained model for decision making. By utilizing the knowledge of news corpus and web information sources along with the LDA information, the proposed FIND approach transforms the four-classes of agree, disagree, unrelated, and discuss into two-classes of fake or real news. The proposed intelligent decision-making model utilizes the LDA topic model score and vector matching score with the web source during the final decision making.

$$FNS = \left(\sum_{T} TMS_{HB} \times \sum_{D} VMS \right) + exp^{(class\ priority)} \tag{5}$$

By applying Eq. 5, the proposed approach accurately detects the fake news. In this Equation, TMSHB represents the Topic Match Score (TMS) of each topic 'T,' which is obtained from the LDA for both the body and headline of the text. The term VMS denotes the Vector Match Score (VMS) of the text, which is obtained from matching the text with the factual web sources. From the analysis of the FNC dataset, the proposed FIND approach models the priority for four-classes derived from the first-step classification in the fake news detection model. In essence, it assigns a priority value '0' for unrelated class, '1' for discussing class, and '2' for both agree and disagree classes. Thus, the FIND approach decides that the news is fake or real using Eq. 5 for decision-making in the second step of classification.

4 Experimental Evaluation

To demonstrate the performance of the proposed FIND approach, the experimental framework compares the FIND approach with the existing exBAKE approach [19] using the test FNC data set.

4.1 Dataset

The experimental model employs the Fake News Challenge (FNC) dataset [20]. The fake news challenge dataset comprises 2,595 headlines and 300 claims. In this dataset, the body of the text, i.e., claims are gathered from Twitter accounts and rumor sites, including @Hoaxalizer and Snopes.com. The journalist also summarizes the article into a headline.

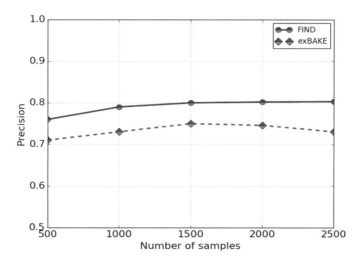

Fig. 2 Number of samples vs. precision

4.2 Evaluation Metric

The experimental framework evaluates the performance of the proposed algorithm FIND with the existing algorithm exBAKE using three different performance metrics: precision, recall, accuracy, and error rate while varying the number of samples. Fig. 2 shows the precision results of both the proposed FIND and existing exBAKE approach while varying the number of samples from 500 to 2500. The precision value increases with the increase of the number of samples. When the number of samples is 500, the precision value of the FIND approach attains by 6.57% more than the existing exBAKE approach by evaluating the incongruity between the headline and body text in the FIND approach. By applying the LDA-based topic detection model and performing the two-step classification, the FIND approach determines the influence between the headline and body text and accurately determines the fake news. In contrast to the exBAKE approach, the FIND approach inherently analyzes the hidden content of the news using the LDA model, which helps to enhance the detection accuracy. Hence, the FIND approach maintains the optimal range of precision value even when escalating the number of samples, which is obtained by an effective deep understanding of the unstructured text.

The recall value of both the FIND and the exBAKE approaches are depicted in Fig. 3. The proposed FIND approach yields a higher recall rate, which reveals that it detects the misinformation much more effectively than the existing exBAKE approach. When the number of samples is 2000, the FIND approach accomplishes the recall rate as 80% and maintains the recall rate even when increasing the number of samples from 2000 to 2500. At the same time, the existing exBAKE approach attains only 72% recall value and rapidly drops down its recall value by 0.2% due to the absence of optimal hidden node consideration. In essence, the recall value

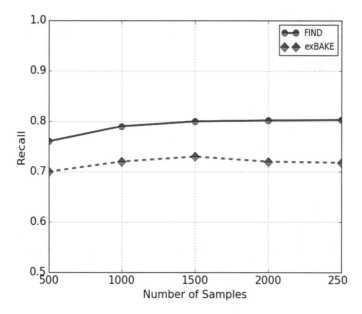

Fig. 3 Number of samples vs. recall

of the proposed fake news detection model relies on an optimal number of hidden nodes in the learning model. The FIND approach applies the optimal hidden neuron selection method on the BERT pre-trained model to improve the performance of the fake news detection model rather than learning the input data with the random number of neurons, regardless of the characteristics of the input data.

Fig. 4 compares the F-score results of both the proposed FIND approach and the existing exBAKE approach with the variation of the number of samples from 500 to 2500. The FIND approach is concerned with the pre-training of the BERT model for further improving the fake news detection accuracy.

It pre-trains the BERT model on a vast amount of unsupervised corpora and fine-tuning on fake news detection tasks with intelligent decision-making. Hence, it discovers the fake news accurately from the generic multi-classes of the FNC dataset. From Fig. 4, it is observed that the FIND approach attains a greater F-score even when processing a large number of samples. Accordingly, when dealing with a large number of samples, i.e., 2500, the FIND approach improves the F-score value by 10.78% when compared to the existing exBAKE approach.

This is because the FIND approach prevents the overfitting of data by selecting the optimal number of the hidden node and also deciding the fake news based on the vector matching and topic matching score, which enhances the detection accuracy. Moreover, rather than analyzing the relationship between the headline and body text alone, the FIND approach examines the influence of the headline as well as the influence of the body text on the factual information using the LDA and web sources.

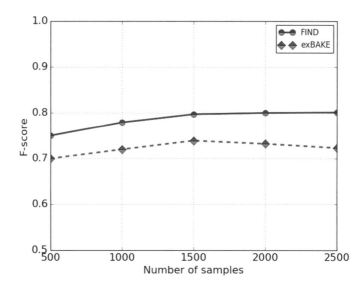

Fig. 4 Number of samples vs. F-score

5   Conclusion

This paper presented the FIND approach, the multi-domain knowledge-based two-step automatic fake news detection model. The proposed FIND approach involves the LDA-based news topic detection for determining the influence between the headlines and body of the news articles individually as well as collaboratively. To effectively detect fake news using pre-trained BERT model, the proposed approach performs the optimal hidden node selection in the fine-tuning stage for avoiding the overfitting, which helps to improve the accuracy of the fake news detection. Thus, the pre-trained model, optimal hidden neuron selection, and the intelligent decision-making model play a significant role in enhancing the fake news detection model. The experimental results on fake news challenge dataset demonstrate the effectiveness of the proposed FIND approach and the significance of an optimal number of hidden node selections along with intelligent decision-making. In essence, the proposed FIND approach enhances F-score value by 10.78% when compared to the existing exBAKE approach, even when there are a large number of samples.

References

1. H. Allcott, M. Gentzkow, Social media and fake news in the 2016 election. J. Econ. Perspect. **31**(2), 211–236 (2017)
2. S. Kogan, T.J. Moskowitz, M. Niessner, *Fake News: Evidence from Financial Markets*. Available at SSRN 3237763, (2018)
3. M. Takayasu, K. Sato, Y. Sano, K. Yamada, W. Miura, H. Takayasu, Rumor diffusion and convergence during the 3.11 earthquake: a Twitter case study. PLoS one **10**(4), e0121443 (2015)
4. Q. Zhang, L.T. Yang, Z. Chen, P. Li, A survey on deep learning for big data. Inf. Fusion **42**, 146–157 (2018)
5. M. Potthast, J. Kiesel, K. Reinartz, J. Bevendorff, B. Stein, A stylometric inquiry into hyperpartisan and fake news, arXiv preprint arXiv:1702.05638, (2017)
6. S. Volkova, K. Shaffer, J.Y. Jang, N. Hodas, Separating facts from fiction: Linguistic models to classify suspicious and trusted news posts on twitter, in *Proceedings of the 55th Annual Meeting of the Association for Computational Linguistics*, vol. 2, pp. 647–653, (2017)
7. G. Karadzhov, P. Gencheva, P. Nakov, I. Koychev, We built a fake news & click-bait filter: what happened next will blow your mind!, arXiv preprint arXiv:1803.03786, (2018)
8. S. Guha, Related Fact Checks: a tool for combating fake news, arXiv preprint arXiv:1711.00715, (2017)
9. Y. Wang, H. Han, Y. Ding, X. Wang, Q. Liao, Learning contextual features with multi-head self-attention for Fake News Detection, in *International Conference on Cognitive Computing*, pp. 132–142, (2019)
10. F. Monti, F. Frasca, D. Eynard, D. Mannion, M.M. Bronstein, Fake News Detection on Social Media using Geometric Deep Learning, arXiv preprint arXiv:1902.06673, (2019)
11. L. Wu, Y. Rao, H. Jin, A. Nazir, L. Sun, Different absorption from the same sharing: Sifted multi-task learning for fake news detection. arXiv preprint arXiv:1909.01720, (2019)
12. D.M. Nguyen, T.H. Do, R. Calderbank, N. Deligiannis, Fake news detection using deep markov random fields, in *Proceedings of the 2019 Conference of the North American Chapter of the Association for Computational Linguistics: Human Language Technologies*, vol. 1, pp.1391–1400, (2019)
13. K. Miller, A. Oswalt, Fake news headline classification using neural networks with attention, tech. rep., California State University, (2017)
14. S. Ghosh, C. Shah, Towards automatic fake news classification, in *Proceedings of the Association for Information Science and Technology*, vol. 55, No.1, pp.805–807, (2018)
15. A.E. Lillie, E.R. Middelboe, Fake news detection using stance classification: a survey. arXiv preprint arXiv:1907.00181 (2019)
16. B. Ghanem, P. Rosso, F. Rangel, Stance detection in fake news a combined feature representation, in *Proceedings of the First Workshop on Fact Extraction and VERification (FEVER)*, pp.66–71, (2018)
17. C. Dulhanty, J.L. Deglint, I.B. Daya, A. Wong, Taking a Stance on Fake News: Towards Automatic Disinformation Assessment via Deep Bidirectional Transformer Language Models for Stance Detection, arXiv preprint arXiv:1911.11951, (2019)
18. O.T. Pfohl, O. Triebe, F. Legros, Stance detection for the fake news challenge with attention and conditional encoding, in *CS224n: Natural Language Processing with Deep Learning*, (2017)
19. H. Jwa, O. Dongsuk, K. Park, J.M. Kang, H. Lim. *exBAKE: Automatic Fake News*
20. Fake News Challenge (FNC) Dataset. Available Online at:https://github.com/FakeNewsChallenge/fnc-1. Accessed On May 2020

Accuracy Evaluation: Applying Different Classification Methods for COVID-19 Data

Sameer Karali and Hong Liu

1   Introduction

Coronaviruses are a cluster of viruses that cause symptoms such as severe fever, extreme acute respiratory syndrome, and other respiratory syndromes. A newly discovered form has resulted in the recent outbreak of a respiratory disease now called COVID-19. In 1968, the word "coronavirus" was coined since the virus had a crown-like surface that replicated the outer shell of the sun—the corona—when viewed through an electron microscope. When an infected individual coughs or sneezes, the new coronavirus spreads through particles released into the air. The droplets frequently move no further than a few feet and fall to a surface within a few seconds. For this reason, social and physical distancing is an efficient method for avoiding spread.

COVID-19 emerged in December 2019 in Wuhan, a city in China [10]. As of April 15, 2020, 153,822 deaths have been attributed to the disease. More than 500,819 people have survived it. The virus is assumed to be transmitted from individual to individual. Respiratory droplets are created and transmitted when an infected individual coughs, sneezes, or speaks. These droplets may land in the mouth or nose of nearby individuals or may be inhaled into the lungs. Avoiding physical contact and isolating yourself if you are sick are the best ways to prevent spread.

S. Karali
School of Sciences and Informatics, Indiana University Bloomington, Bloomington, IN, USA
e-mail: skarali@iu.edu

H. Liu (✉)
School of Sciences, Indiana University Kokomo, Kokomo, IN, USA
e-mail: hlius@iu.edu

© Springer Nature Switzerland AG 2021
R. Stahlbock et al. (eds.), *Advances in Data Science and Information Engineering*,
Transactions on Computational Science and Computational Intelligence,
https://doi.org/10.1007/978-3-030-71704-9_67

2   Literature Review

This section reviews the history, background, and current status of coronaviruses, which is a significant and growing research field. Kahn and McIntosh [5] state that coronaviruses first reported in the 1960s are accountable for a large proportion of upper respiratory tract illnesses in childhood. Since 2003, five new coronaviruses have been reported, including severe acute respiratory coronavirus disorders. Worldwide, NL63 was described as a population of newly found group I coronaviruses such as NL and the New Haven coronavirus. There has been no established global distribution of a group II coronavirus, HKU1. The SARS outbreak focused on animal coronaviruses. Wertheim et al. [9] state that coronaviruses are found in a variety of species of bats and birds. The analysis of coronaviruses by molecular clock suggests that the common ancestor of this type of virus lived approximately 10,000 years ago. Mohd et al. [7] state that the Middle East respiratory syndrome coronavirus is a novel coronavirus that was reported in 2012 and is accountable for extreme human breathing syndrome.

Chinazzi et al. [1] record the proliferation of COVID-19 in mainland China. The authors use an international population disease transmission model to predict the impact of travel limitations on the epidemic's domestic and international spread. The prototype of the design is based on incidents recorded globally and indicates that most Chinese cities had received some infected travelers from Wuhan on January 23, 2020, the beginning of the travel ban. The Wuhan travel quarantine delayed overall disease development in China by approximately 3 to 5 days, but it had a more pronounced impact on an international scale where cases were reduced by almost 80% through mid-February 2020. Model findings also suggest that the sustained travel restrictions of 90% of travels to and from mainland China impacted the disease trajectory moderately when paired with a 50% or higher reduction in population transmission.

Cohen [2] and Peng et al. [8] claim that vaccine development is being aided by actively attempting to infect volunteers receiving "candidate" items from the experiment. Eyal et al. [3] state that human experiments were needed to speed up the authorization of coronavirus vaccines. Controlled human challenge tests of SARS-CoV-2 vaccination targets could improve research and the potential roll-out of effective vaccines. New trials will eliminate several phases from the approval process by eliminating the standard Phase 3 vaccine testing, something that can make good vaccines easier to obtain.

3   Methodology

We proposed a framework to understand the data characteristics of COVID-19, including data transformation, descriptive analysis, and inference analysis. In addition, we evaluated three classification methods, namely KNN algorithm, K-

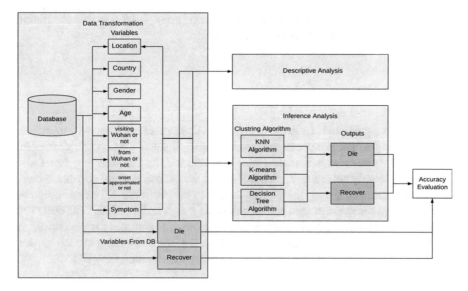

Fig. 1 Framework of accuracy evaluation

means algorithm, and decision tree algorithm. Figure 1 shows the framework of this work.

In the data transformation stage, we selected 10 of the 20 variables to perform the analysis. The "death" attribute and the "recovered" attribute were merged to form a new attribute: "*status.*" We tested the eight variables using "*status*" to assess their importance. Finally, we evaluated the KNN algorithm, the K-means algorithm, and the decision tree algorithm and compared their accuracy.

3.1 KNN

KNN is a supervised machine learning algorithm that classifies a data point into a target class, depending on the features of the point's neighboring data points. The algorithm checks how similar a data point is to its neighbors and classifies the point into the class. KNN is a nonparametric model; therefore, it does not make any assumptions about the dataset. The algorithm uses Euclidean distance to measure the distances.

In this work, we defined an eight-dimensional space, and the distance was evaluated based on the following equation,

$$dist = \sqrt{\sum_{i=1}^{8} \left(X_i^{(a)} - X_i^{(b)} \right)^2}$$

where i is the number of dimensions, a denotes one point, b denotes the other point, and *dist* denotes the distance between a and b.

3.2 K-Means

The K-means algorithm is an iterative algorithm that tries to partition the dataset into K predefined distinct nonoverlapping subgroups (clusters) where each data point belongs to only one group. K-means initializes centroids by first shuffling the dataset and then randomly selecting K data points for the centroids without replacement. The interactions occur until there is no change to the centroids. K means computes the sum of the squared distance between the data points and all of the centroids. The data points are then assigned to the closest centroid. The centroid for the clusters is computed by finding the average of all of the data points that belong to each cluster.

In this analysis, we calculate the distance using the equation below:

$$J(V) = \sum_{i=1}^{C} \sum_{j=1}^{Ci} \left(\| X_i - V_j \| \right)^2$$

$||Xi\text{-}Vj||$ is the Euclidean distance between Xi and Vj. Where X is the set of data points, and V is the set of centers.

Ci is the number of data points in the ith cluster.
C is the number of cluster centers.

3.3 Decision Tree Algorithm

A decision tree is a supervised nonparametric algorithm that can be used for both classification and regression. The decision tree classifies a new object into the target class by going through the attribute branches from the root attribute to the ending target class leaf. The decision tree places the best attribute that most impacts the dependent variable at the root of the tree. The tree will then split branches by creating sub-nodes to ensure that the homogeneity of the resulting sub-node is high. The attribute variable for the next branch that splits from the resulting sub-node is selected by splitting the nodes on all available variables and then selecting the split that creates the most homogeneous sub-nodes. Deciding which attribute to place at the root or different levels of the tree as internal nodes are determined using certain criteria such as the Gini Index or the Gain Ratio for each attribute and then placing the attributes that have the highest value first.

For the decision tree algorithm, we used this equation:

$$Gini = 1 - \sum_{i=1}^{C} (P_i)^2$$

To calculate the *Gini* using the formula above, we used the weighted *Gini* score for each spilled data.

4   Experiment and Results

4.1   Datasets and Preprocessing

The dataset used for the analysis is provided by Kaggle for COVID-19 and named "COVID19_line_list_data" [4]. There are 20 variables with 1085 observations; the variables of the dataset are: id, case_in_country, reporting.date, summary, location, country, gender, age, symptom_onset, if_onset_approximated, hosp_visit_date, exposure_start, exposure_end, visiting.wuhan, death, recovered, symptom, source, and link.

We selected ten variables from the original dataset to run the analysis. The selected variables were as follows: location, country, gender, age, and visiting. Wuhan, from. Wuhan, if_onset_approximated, symptom, death, and recovered. Based on these ten variables, we analyzed, classified, and compared the data. The other variables were deleted because they are not relevant to our analysis. This work focuses on the general classification and does not involve time constraints.

4.2   Experimental Setting

In the experimental stage, the dataset is first loaded, and the variables of interest are extracted. Since some patients do not have "status" information, this can make it impossible to verify the accuracy of classification. Therefore, patient data without this information is removed. We end up with 141 observations and nine variables, including a dependent variable of "status" and eight independent variables. In this work, we used 10-fold validations to split the dataset into subsets, and each split involved taking one subset as a test set and the remaining nine subsets as a training set.

On the folds validations, the left side of Table 1 is the folds, and the right side is observations. In this analysis, we used 10-folds cross-validation as a resampling method. We split our dataset into 10 folds. We held one fold as the test set at each time and trained the model using the remaining nine folds. The reason for doing resampling is to obtain more accuracy for each model since the model's accuracy depends on the training set used. By using different training sets to train the model

Table 1 10-fold validations

Folds	Index
1	102
2	112
3	4
4	55
5	70
6	98
7	135
8	7
9	43
10	51

and using the accuracy for a different set of tests, we are able to obtain the general accuracy of the model.

We used Euclidean distance to measure the distance from the data point to nearest neighbors in the KNN; the variables should be numeric. Therefore, the variables must be converted into numeric variables. Different variables can have different scaling units, and we need to normalize these units. We use age as the actual numeric variables, and other variables are obtained using different factors. The dataset is split into the training and test sets, and then our program is run ten times. Each time we obtained a different accuracy and the average accuracy was calculated for each model.

In this work, we focus on classifying patients based on datasets and models. Therefore, we use the following three algorithms:

KNN Algorithm

By choosing the optimum K value, we fit the KNN for each training set to make the model highly accurate. Then, we use the KNN model with optimum K value to predict the test set and obtain accuracy. We store the accuracy of 10-folds cross-validation in the first column of the accuracy matrix. For the KNN algorithm, we used the *train()* function to automatically assign the K value within the function. The tune length is 10, which means the 10 K value is 1:10, and the best one is selected to provide the highest accuracy. The results show that the values of 5 and 7 have the highest accuracy, so they are the optimum of K.

K-Means Algorithm

For the K-means algorithm, we fit K-means for each training set by choosing K as 2, because we need to classify our data into two categories. Then, we used each created K-means model to predict the test set and obtain accuracy. We store each accuracy of 10-fold cross-validation in the second column of the accuracy matrix.

Decision Tree Algorithm

We fit a decision tree for each training set by choosing default values for specific tree parameters. Then, we use each decision tree to predict the test set and get

Table 2 Average accuracy

Algorithm	Average accuracy
KNN	0.9224
K-means	0.5124
Decision Tree	0.8938

accuracy. We stored each accuracy of 10-fold cross-validation in the third column of the accuracy matrix.

4.3 Experimental Results

In order to understand the characteristics of COVID-19 data, we have done a descriptive analysis and inference analysis. The results of the descriptive analysis are shown in Fig. 2 and are based on data from January and February 2020.

Figure 2 shows the classification results based on different variables. State 0 represents the death category, and state 1 represents the recovery category. Through the analysis of the results, our main findings are the following: (1) Patients over 55 years of age belong to high-risk groups, while patients under the age of 50 have a high recovery rate. (2) Compared with other countries, Singapore has a higher recovery rate from COVID-19. (3) Gender also plays an important role. The results show that the recovery rate of women is higher than that of men.

According to Fig. 3, KNN and decision tree show stability between (0.7, 1). The accuracy of K-means is more unstable than the other two methods. Moreover, the accuracy of K-means is lower than KNN and decision tree. Therefore, KNN and decision tree algorithms are better than the K-means algorithm in predicting the classification of COVID-19 patients (Table 2).

By comparing the average accuracy of each algorithm, the highest accuracy given by KNN is 0.9224. Therefore, for our dataset, KNN is the best model to predict patient classification. We need to find the optimum number of nearest neighbors to fit the model. The below table gives the result of each cross-validation analysis (Tables 3, 4 and 5).

The data shows that when K is 7, the accuracy rate is higher than when K is 5. Based on the dataset, experimental settings, and classification model, the results show that the best model for predicting COVID-19 patient classification is the KNN algorithm with seven nearest neighbors.

5 Conclusions

To better understand the data characteristics of COVID-19 and the accuracy of different classification models, we performed data transformation, descriptive analysis, and inference analysis on the COVID-19 dataset. We used three different

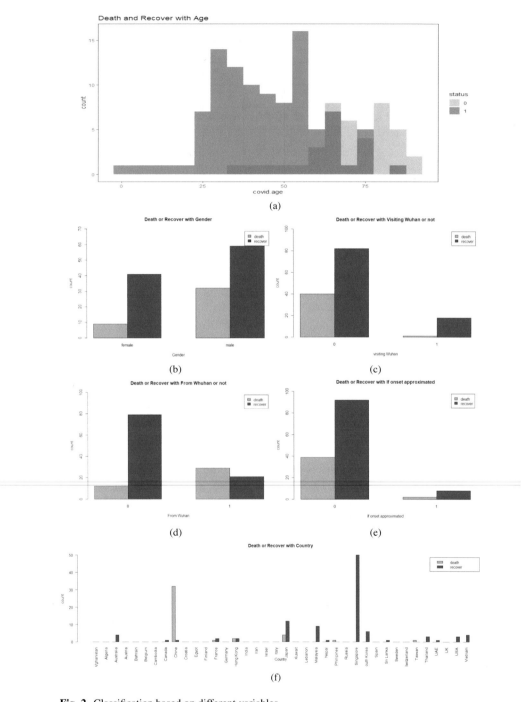

Fig. 2 Classification based on different variables

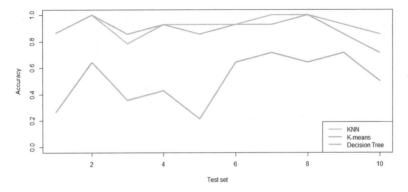

Fig. 3 Accuracy plot

Table 3 Accuracy results of
KNN algorithm

Test Folds	Optimum K neighbors	KNN Accuracy
1	5	0.8667
2	7	1.0000
3	5	0.7857
4	5	0.9285
5	7	0.9285
6	5	0.9285
7	7	1.0000
8	5	1.0000
9	7	0.9285
10	7	0.8571

Table 4 Accuracy results of
K-means algorithm

Test Folds	Optimum K neighbors	K-means Accuracy
1	2	0.2667
2	2	0.6429
3	2	0.3571
4	2	0.4286
5	2	0.2143
6	2	0.6429
7	2	0.7143
8	2	0.6429
9	2	0.7143
10	2	0.5000

algorithms to classify and compare their accuracy rates: KNN algorithm, K-means algorithm, and decision tree algorithm. Our contributions are threefold. First, we proposed a framework for classifying patients based on the existing dataset. Based on the two attributes, "death and recovery," we created a new data attribute as a target attribute to verify and predict the classification of the confirmed patient. Second, the combination of both descriptive and inferential statistical analyses

Table 5 Accuracy results of
decision tree algorithm

Test Folds	Decision Tree Accuracy
1	0.8667
2	1.0000
3	0.8571
4	0.9286
5	0.8571
6	0.9286
7	0.9286
8	1.0000
9	0.7143
10	0.7143

allowed us not only to investigate the features of the data but also to precisely measure the accuracy of the three methods. Third, we summarized the analysis results and gave suggestions. Based on the data and the experimental setup as well as the model used, the results show that the best classification model is the KNN algorithm, with an accuracy of 92.24%.

The current work has several limitations. The scope of the dataset is not comprehensive, and the timeline for the dataset is based on data from January and February 2020. We will collect more extensive data for future work, and comprehensive accuracy analysis, including more algorithms, will be conducted in future work.

Acknowledgments The authors appreciate the support provided by the IUK Summer Faculty Fellowship and Grant-in-Aid program.

References

1. M. Chinazzi, J.T. Davis, M. Ajelli, C. Gioannini, M. Litvinova, S. Merler, A.P. y Piontti, K. Mu, L. Rossi, K. Sun, The effect of travel restrictions on the spread of the 2019 novel coronavirus (COVID-19) outbreak. Science **368**, 395–400 (2020)
2. J. Cohen, *Infect Volunteers to Speed a Coronavirus Vaccine?: American Association for the Advancement of Science* (2020)
3. N. Eyal, M. Lipsitch, P.G. Smith, Human challenge studies to accelerate coronavirus vaccine licensure. J. Infect. Dis. **221**, 1752–1756 (2020)
4. Kaggle COVID-19 data with SIR. https://www.kaggle.com/lisphilar/covid-19-data-with-sir-model/data. Accessed 15 Apr 2020
5. J.S. Kahn, K. Mcintosh, History and recent advances in coronavirus discovery. Pediatr. Infect. Dis. J. **24**, S223–S227 (2005)
6. M. Mcnulty, D. Bryson, G. Allan, E. Logan, Coronavirus infection of the bovine respiratory tract. Vet. Microbiol. **9**, 425–434 (1984)
7. H.A. Mohd, J.A. Al-Tawfiq, Z.A. Memish, Middle East respiratory syndrome coronavirus (MERS-CoV) origin and animal reservoir. Virol. J. **13**, 87 (2016)

8. X. Peng, L. Wang, Y. Qiao, Q. Peng, A joint evaluation of dictionary learning and feature encoding for action recognition, in *2014 22nd International Conference on Pattern Recognition*, (IEEE, 2014), pp. 2607–2612

9. J.O. Wertheim, D.K. Chu, J.S. Peiris, S.L.K. Pond, L.L. Poon, A case for the ancient origin of coronaviruses. J. Virol. **87**, 7039–7045 (2013)

10. Y. Yan, H. Chen, L. Chen, B. Cheng, P. Diao, L. Dong, X. Gao, H. Gu, L. He, C. Ji, Consensus of Chinese experts on protection of skin and mucous membrane barrier for health-care workers fighting against coronavirus disease 2019. Dermatol Ther, e13310 (2020)

Clearview, an Improved Temporal GIS Viewer and Its Use in Discovering Spatiotemporal Patterns

Vitit Kantabutra

1 Introduction

In many fields of study such as epidemiology and the health sciences, history and the digital humanities, environmental science, evolution, archaeology, anthropology, energy development, urban planning, transportation, disaster management, criminology, and more, it is important to have temporal GIS software that enables the user to view, notice, and correct interesting geographic patterns and processes that occur and evolve through time. Unfortunately, the current versions of even the best known GIS software, Google Earth 7, and ESRI ArcGIS 10 are not always capable of performing correctly or reliably, even when presented with simple data files.

One major problem with those software packages is *time linkage error* or *time mismatch,* which means matching a user-specified point in time with displayable data items that belong to a different, incorrect point in time. For example, when the user specifies the year 1754, such software might incorrectly display the map for year 1755 instead of the map for 1754. Another problem with those two extant software packages is the existence of *forbidden time steps,* that is, the user can't freely specify whatever time step he or she wants even though such time step should be within the displayable range. Only certain, unpredictable time steps are permitted as input to the software. For example, the user may be interested in seeing the map for the year 1824, but that year may not be permitted as input. The closest years that are permitted as input may be 1823 and 1827.

This paper will describe the detailed design of a temporal GIS viewer that correctly displays temporal maps efficiently and smoothly and additionally does not have the two problems just mentioned. In particular, this paper will describe

V. Kantabutra (✉)
Department of Electrical and Computer Engineering, Idaho State University, Pocatello, ID, USA
e-mail: vititkantabutra@isu.edu

© Springer Nature Switzerland AG 2021
R. Stahlbock et al. (eds.), *Advances in Data Science and Information Engineering,*
Transactions on Computational Science and Computational Intelligence,
https://doi.org/10.1007/978-3-030-71704-9_68

the author's prototypical temporal GIS viewer, Clearview, that has such properties. Clearview uses a decidedly different data structure from the more conventional, tabular data structures used in Google Earth and ESRI ArcGIS and the data structures in well-known database storage schemes for temporal GIS such as Key Dates (also known as Time Slices) and Space-Time Composites. The main data structure used in Clearview is more similar to the one used in multiresolution cinema [3] than the ones used in other GIS software. The correctness and efficiency of Clearview can partly be attributed to better data structuring techniques, as well as the absence of a need to fudge data to fit an oversimplified data structure as would be required by Time Slices.

To be specific, the data structure used by Clearview provides efficient, no-search linkages from the time dimension to the geographical objects to be displayed, enabling correct, clear display of geographical objects such as point locations, boundaries, and labels at user-specified time steps as well as smooth animation of time-varying maps not possible with the extant schemes for storing temporal GIS data.

This paper will also demonstrate that by using a good GIS temporal viewer on a well-known HGIS (Historical GIS) database, CHGIS (Chinese Historical GIS) [1], the user can visually discover interesting patterns in the database that apparently have not been reported before. For example, using Clearview, one discovers something important about the CHGIS Version 4 time series data, namely, that in the provincial-level data, there is a 22% rate of provincial capitals occurring outside their respective provincial boundaries. This most likely indicates significant database errors. There is more on this and other important observations about CHGIS in Sect. 5.

Though currently hardcoded to display the CHGIS, Clearview can be retailored to handle a wide range of databases. There are no real technical obstacles that prevent the handling of other databases. The hard-codedness of Clearview is more due to a lack of convention in data formats in temporal KML files that allows different input files to have different formats for some detailed fields that should be displayed. It is technically easy to require all data files to adhere to certain input formats, but up until now, there are not enough people creating temporal GIS files for a consistent input formatting standard to be developed. Therefore two different files may have slightly different formats, even when the two files belong to the same database.

Looking ahead, Sect. 2 explains the serious time linkage errors and issues of Google Earth 7 and ESRI ArcGIS 10. Section 3 reviews existing data structures for storing temporal GIS data and ends by explaining why the current work is needed. This should clarify, albeit briefly, the current work's contribution to geographic information science. Section 4 explicates the inner workings of Clearview. Section 5 explains in greater depth the patterns in the CHGIS temporal database uncovered by using Clearview that we just discussed here in the introduction.

2   Extant Temporal GIS Software Packages with Display Capabilities

Back in 2007, the lack of ability of GIS software to handle temporal data was well known [5], from which we quote, "A lack of temporal functionality in GIS software is commonly criticised. Progress in adding it has been limited, as GIS software vendors do not see this as particularly important for their market."

Fast forward to the current time, now the two major GIS vendors, namely, Google and ESRI, have for a few years been claiming to handle temporality. However, as stated previously and as to be detailed shortly, neither vendor's GIS, Google Earth 7, nor ESRI ArcGIS 10 handles temporality correctly.

The following is a list of time linkage and similar errors in ESRI ArcGIS and Google Earth:

1. Time linkage error, as explained in the Introduction. To examine a real example of this error, we use a temporal database that was used by Google itself on its tutorial web site [4] as a sample input to Google Earth as well to ESRI ArcGIS 10. The database in question is a database by Brian Flood showing when each of the 50 US states was admitted to the Union. Google Earth 7.3.3.7699, build date May 7, 2020, running on Mac OSX 10.15.4, shows that the first three states to be admitted were admitted in 12/1786. The time slider display is incorrect since the actual admission year for those three states was 1787.

   ESRI ArcGIS 10 exhibits a more serious problem with time linkage when used on the same database. According to ArcGIS 10's display, all 50 states entered the Union in 1787! In fact all states are displayed at all times, that is, it seems that ArcGIS simply ignores the time interval of validity of each geographic object.
2. The points in time allowed as input to the software packages are limited and are apparently unrelated to the points in time at which events or changes in the database occur. In other words, the designers of the two software packages seemed to have completely disregarded the time points of significance in the temporal GIS database. These software packages are only able to display data that pertain to certain years or points in time rather than being able to display data for any legitimate time step that the user desires. To understand this problem, we use CHGIS' time series data, provincial part. This consists of two files, one for the provincial administrative seats and the other for the provincial borders.

   We first tested Google Earth on this database. The year progression as we slide the time slider from left to right is 187, 195, 202, 209, ..., 771, 778, 786, 793, etc., instead of being keyed to the database's temporal events. This year progression is consistent with a scheme for temporal GIS data storage called Time Slices, which is too simple for real historical GIS applications.

   Now we tested ESRI ArcGIS. When used with the CHGIS database, ArcGIS does display different data for different points in time, unlike in the case of the US states' database, for which ArcGIS displays the same data no matter what point in time is supposed to be current.

ArcGIS's year progression as the time slider input is slid from left to right is best described as bizarre and uncorrelated with the time progression in the database itself.

3. Lack of clarity due to clutter. The base map, which seemingly cannot be omitted in Google Earth, makes the map hard to read. The historian is often only interested in geographic features that pertain to past times. While current information can be useful, it would be nice to be able to omit or deselect such information. This problem is not a technical inadequacy. Instead, it just means that Google as a GIS software provider does not think of historical applications as being a high-priority item.

4. Animation is too fast for historical applications with 1 year time resolution. This observation, stated in [1], still holds true in today's Google Earth version, in Nov. 2016. Again this means the historical applications are not a high-priority item to GIS software providers.

3   Previous Ways to Represent Data in Historical and Temporal GIS Databases

The simplest method of representing temporal data is called Key Dates or Time Slices. This method involves picking important time steps, called key dates, and then representing all spatial objects at every key date. This method results in much data redundancy and inaccuracy due to fudging data to fit the sparsely available key dates [1].

Large projects need better data representation. For example, the GBHGIS (Great Britain Historical GIS) Project uses the Datestamped data model [10] instead. Berman [1] reviewed this data model as well as the closely related Space-Time Composites (STC) data model [7, 8]. He concluded, "... one of the major drawbacks in both developing and maintaining Datestamped and STC GIS systems is spatial fragmentation."

Berman [1] introduces a new data organization for temporal GIS, for use in CHGIS, that solves the spatial fragmentation problem. Berman's main data object is called the *historical instance,* which records placename, feature type, a spatial object (such as a point location or polygon), and a date range. For identification purposes, Berman also added an object id to each historical instance. Whenever any one of these components changes, a new historical instance must be generated. This approach avoids the spatial fragmentation of earlier approaches but instead runs into *fragmentation in time.* Overall, though, this approach works better than the others available thus far.

Kantabutra et al. [6] showed how to use the non-relational ILE (Intentionally Linked Entities) database model to represent temporal GIS data. Using ILE, it is possible to eliminate both fragmentation in space and fragmentation in time.

It is notable that none of the data storage approaches mentioned so far in this section deals with the issue of having to supply data fast enough for a smooth cinematic display on a typical laptop computer. This paper introduces a data structuring technique that solves the fast display problem.

We should also comment on the data structures used by Google Earth and ESRI ArcGIS. While the implementations of these packages appear to be proprietary, some knowledge of the internal data structures can be inferred from the software packages' manuals and from our tests of the packages.

From Google Earth's manual [4], we can see that all geographic elements in KML, Google Earth's language for representing geographic objects, can have time data associated with it in the form of either a time stamp or a time interval. Google Earth takes input chiefly in the form of KML (Keyhole Markup Language) files, and for temporal data, this would be an extension called TE KML (Time-Enabled KML) files.

However, when we experimented with Google Earth, it only permits access to time steps that are spaced approximately regularly, which, in the case of using Google Earth with CHGIS, means every 7 or sometimes 8 years. This behavior makes it strongly appear like Google Earth's proprietary internal data structure is an implementation of the Time Slices scheme with all its problems as we have discussed, rather than the Datestamped or STC scheme which would be more similar to the data model of the KML file itself.

In ESRI ArcGIS, temporal functionality is mainly meant for tracking lightweight geographical objects such as points. However, shapes such as polygons can also change in time. However, the representation of time-varying polygons is awkward because, according to [2], it necessitates having a new object id for each time period during which the shape is fixed.

4 Clearview, a Temporal GIS Viewer for Spatiotemporal History

A temporal GIS display system must be based on a data structure that is fast enough to supply data in real time to the display system to create a smooth video. Both Google Earth and ESRI ArcGIS use table-based databases which may not yield query answers quickly enough to supply data smoothly in real time to a cinematic display system. Relational DBMS and other table-based databases are not meant to supply query answers at cinematic display speeds, even when aided by index structures such as B-trees and other search trees.

We therefore resort to a direct linking scheme where, when presented with a query (a time step), the data structure gives a direct link (or a two-step link) to a graphical object, which then displays itself by calling its display method. No searching is allowed because searching could take too much time to permit smooth video output.

Fig. 1 Activator node
example

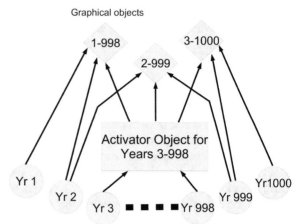

Given the above discussion, it makes sense to examine using the time linkage approach described in [6] as a possible solution. That approach involves having each pointer pointing from a time object directly to a geographic object valid at that particular time step. Therefore, when a time step is specified, no search at all is needed to find all the geographic objects that are active (should be displayed) at that time step.

There is, however, a potential problem with the approach just described. Having a pointer from every time step object pointing directly to geographic, or graphical, objects may result in too many pointers for a database as complex as the time series part of CHGIS. For example, imagine a situation where geographic objects can last hundreds or even thousands of years where the time step is a year. There would be a large number of pointers from the time step objects to each such long-lasting geographic object. For example, suppose the time steps are years, and suppose we have three geographic objects to be displayed, one from year 1 to year 998, another one from year 2 to year 999, and a third geographic object lasting from year 3 to year 1000. Having a pointer from each time step object to each geographic object requires $3 \times 998 = 2994$ pointers, that is, almost 3000 pointers.

We now introduce an auxiliary linkage data structure that will reduce the number of links without sacrificing much performance. We note a redundancy in the pointers in our example: all three geographic objects have pointers from years 3 through 998. We use a new kind of object, one that represents time intervals, as an intermediary between the time step objects and the geographic objects to help reduce the redundancy. We can call these time interval objects *activator objects,* because they can be thought of as an object that is used to "activate" a graphical object and make it draw itself (in OOP parlance).

In this example, as shown in Fig. 1, by inserting an activator object in between the time step objects and the graphical objects, it is not hard to show that we can reduce the number of pointers from 2994 to only 1005, a reduction of almost a factor of 3.

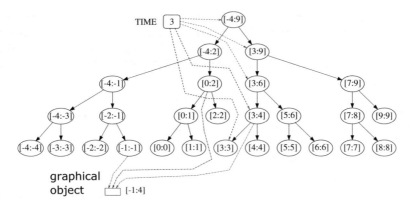

Fig. 2 Tree of Time example

However, a real database such as CHGIS is not as simple as this example, where there are only three graphical objects, and these objects share most of their intervals of existence. This is where a better solution is needed. It turns out that the activator object idea still works, but instead of having a single activator object, we need an entire tree of them. This tree is the *Tree of Time*.

In detail, the Tree of Time is, indeed, a binary tree, as drawn in Fig. 2 with its nodes represented by labeled ovals and its edges represented by solid lines with arrowheads. Each node's label is a time range. In this example, suppose the entire database's range is time $= -4$ to time $= 9$. The root node of the tree, labeled with the entire time range, is notated as $[-4 : 9]$. The children of the root, that is, the two nodes reachable from the root by traversing only one edge, are each labeled with half the root's interval. In particular, the left child is labeled with the left half interval, $[-4 : 2]$, and the right child with the right half interval, $[3 : 9]$. In general, for any node n, n's left and right children are labeled with the left and right halves of n's interval. We continue growing children of nodes, halving the intervals, until we have nodes labeled with single time points. Such nodes are the leaves of the tree.

Having defined the Tree of Time, we now have to consider how to use it to point directly to graphical objects to activate (display) at a given point in time. The main idea is this: instead of activating each graphical object directly with an object representing a point in time (which is what we did before using the Tree of Time), we will now activate each graphical object with nodes of the Tree of Time. In particular, we will activate the graphical object with nodes of the tree that attains maximal subinterval coverage. Here, we define a tree node n to attain maximal subinterval coverage for a time interval $[t1 : t2]$ if n's interval is a subinterval of the graphical object's interval (of validity) and if n's sibling (n's parent's other child) is not a subinterval of the graphical object's interval.

In Fig. 2 it can be seen that the example graphical object, whose validity interval is $[-1 : 4]$, is activated by the nodes labeled $[-1 : -1]$, $[0 : 2]$, and $[3 : 4]$. That's three links from tree nodes to the graphical object.

In addition to the links from tree nodes to the graphical object, we will need links from each time step object itself to certain tree nodes. In particular, let's consider the time step object representing time = 3 as shown in Fig. 2. From that time object, we need a link from it to all tree nodes whose interval includes 4. There are five such nodes.

Adding these five links to the three links we had earlier gives us eight links in all. Using the old scheme, we would need one link from the time object to each time step in the geographical object, which gives us only 6 links! The reader might wonder why we went through all this trouble, only to make things worse. The answer is that things are only worse in this simple example because the time range of the graphical object is very small. In fact, if there are many short-lived geographical objects, then such objects should be linked from the time step objects the old way—directly. However, long-lived objects that live from tens of time steps or more can benefit greatly from the reduction in number of total links using the new linking scheme with the Tree of Time.

Skeptics may wonder if such a complicated tree could be constructed quickly. In fact, on a 2009 model MacBook Pro, it only takes less than a minute to construct the Tree of Time for the entire CHGIS time series database. If this were to be a database with extremely frequent revisions in real time, we could modify the data structure to allow faster revision without complete reconstruction. However, most historical databases are fairly static, and so we have not put thought into the issue of database revisions since a complete reconstruction takes so little time anyway. Also note that our approach seems to be the only known one that works at this point in time, so one should keep that in mind when comparing the costs of this approach with those of Google Earth or ESRI ArcGIS.

There is also the issue of scaling up. Perhaps a better data structure is needed if the database were to increase substantially in size. Clearview operates correctly and comfortably (without delays that would annoy the user) on CHGIS V.4 on a 2009 MacBook Pro, but when all the administrative levels are displayed, it is apparent that handling a much larger database may be a problem because the delays are beginning to creep up. However, at this time, Clearview seems to be the only GIS software capable of even correctly displaying a database the size of CHGIS. Google Earth and ESRI are unable to handle CHGIS or in fact even the much smaller US states' database used as Google Earth's example on Google's own web site. Additionally, while Google Earth is mostly stable, ArcGIS crashes easily and often. Clearview can crash if we intentionally move the time slider or dial back and forth very quickly and repeatedly. However, in real use with the CHGIS database, Clearview does not seem to crash easily at all.

5   Previously Unseen Patterns in the CHGIS V.4 Time Series Database Discovered Using Clearview

Through Clearview's ability to show interesting patterns in the CHGIS V. 4 temporal data, it will hopefully become clear why a good temporal CHGIS viewer such as Clearview can potentially be an essential tool in the creation, refinement, and exposition of all historical and temporal databases. We will examine the following patterns:

1. By simply running Clearview on CHGIS' provincial temporal data through the early years of data validity, it becomes immediately apparent that provincial capitals often lie outside the boundaries of their respective provinces. An example of this anomaly is shown in a screenshot of Clearview in Fig. 3.

    In order to count how frequently provincial capitals lie outside their respective provincial boundaries, we define a *historical instance* for the provincial part of the CHGIS time series database as the ordered quadruple (province name, provincial capital point location, provincial boundary polygon, time interval), where a quadruple is a valid historical instance only if the time interval is the maximal time interval during which the other components of the quadruple remain at the specified fixed value. Under that definition, 22% of all the historical instances in existence in the provincial part of the CHGIS time series database each involves a provincial capital that lies outside the respective provincial boundary. We can immediately infer that many such instances are database

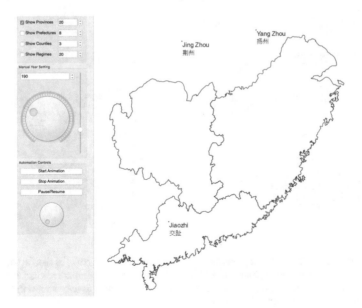

Fig. 3 CHGIS time series database, provincial part, at year 190, as seen on Clearview

errors, since, historically, there have been very few instances of an administrative region anywhere in the world that were administered from outside the region itself.

Bol and Berman of CHGIS, in their speeches and writings [1, 9], discouraged relying too much on boundary data. Yet one hopes that any boundaries that are included in temporal GIS maps, including CHGIS, are good approximations to the true boundaries.

2. We'll now examine what CHGIS' director, Harvard's P. Bol, used as a prime example in talks he gave about how CHGIS can be used to learn from the "ancient bureaucrats' scrupulously kept records" [9]. This example concerns Southern China, in parts of modern-day Guangdong and Guangxi Provinces, where the number of county seats actually decreased significantly from years 750 to 1050, even though the population of China doubled over the same 300-year period. What happened? Did the Huang Chao Rebellion, which massacred foreign merchants in 878–879 and captured the capital Chang'an in 881, decimating the economy and severely weakening the Tang Dynasty, cause the decrease?

To shed light on this question, we ran Clearview's automated time sweep animation, which tells us visually that all the significant reduction in the number of counties in the 300-year period actually occurs during 971–974. See Fig. 4.

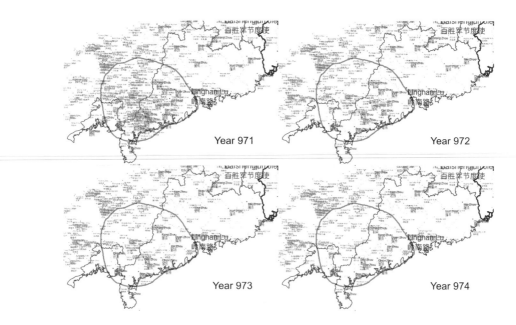

Fig. 4 Coastal area around Guangdong and Guangxi from years 971 to 974 as seen in Clearview—despite the label collision problem, this display is much more informative than the displays available using other software packages in that Clearview's display at least shows the density of the counties being reduced in the Pearl River area

Therefore, using Clearview, we have sharpened the original 300-year period of thinning down of county seats observed initially to a period of just 3 years, a hundredfold increase in the precision of pinpointing the interval of reduction of the county seats. Since the new Song Dynasty came to power in 971, we can reasonably hypothesize that the Song Dynasty reorganized the counties as a cleanup act to match the population decrease in the local area caused by the rebels or whatever other causes there might have been. Further study of this situation by historians and archeologists could yield interesting results.

3. Bol clearly would like CHGIS to be the definitive source for historical geographical information for China, more reliable than printed sources. Quoting from [9]: "Bol makes a bold claim, stating that what he and his colleagues are creating not only will be but already is more authoritative than any print atlas. If you want authority, you have to turn to the GIS. Now, it is generally understood that CHGIS has replaced any and all print atlases. demonstrating the power, potentially of digital history and geography."

However, in order for CHGIS to be a definitive source of information, it must be both as correct and complete as possible. This would require a gigantic amount of work, of course. However, a good viewer like Clearview could make the work easier by enabling the users to visualize what data may be missing. For example, by running quickly through time, it is obvious that most of CHGIS V.4 temporal data is on the southeastern part of China. In fact, there is no data as far north as Beijing at all, for any time period covered by CHGIS's time series database. Since the Beijing area has been prominent in Chinese history for millennia, it is important to correct this missing data problem. Note that P. Bol privately informed me that CHGIS V.6 includes complete prefectural boundaries for the core provinces.

6 Conclusions

This paper took the reader through various important example of flaws in the extant GIS viewers, including time linkage errors and forbidden time steps. Next the paper explains how to design error-free temporal viewing functionality into a GIS. The resulting temporal GIS viewer is called Clearview. Finally, the paper gives examples that show that Clearview can aid discovery of knowledge through geospatial process visualization.

References

1. M.L. Berman, Modeling and visualizing historical GIS data, in *Proceedings of the Spatio-Temporal Workshop* (Harvard University, Cambridge, 2009)

2. ESRI: Arcmap 10.5 manual: How time is supported in spatial data (2017). http://desktop.arcgis.com/en/arcmap/latest/map/time/how-time-is-supported-in-spatial-data.htm
3. A. Finkelstein, C.E. Jacobs, D.H. Salesin, Multiresolution video, in *Proceedings of SIGGRAPH 96* (1996), pp. 281–290
4. Google, Google developers: keyhole markup language: time and animation. https://developers.google.com/kml/documentation/time?hl=en. Accessed May 2020
5. I.N. Gregory, P.S. Ell, *Historical GIS: Technologies, Methodologies, and Scholarship* (Cambridge University Press, Cambridge, 2007)
6. V. Kantabutra, J.B.J. Owens, D.P. Ames, C.N. Burns, B. Stephenson, Using the newly-created ILE DBMS to better represent temporal and historical GIS data. Trans. GIS **14**, 39–58 (2010)
7. G. Langran, *Time in Geographic Information Systems* (Taylor Francis, London, 1992)
8. G. Langran, N.R. Chrisman, A framework for temporal geographic information. Cartographica **25**(3), 1–14 (1988)
9. C. Rasmussen, Peter Bol: Creating the China Historical GIS, a presentation for the Digital History Project, University of Nebraska, Lincoln (2006). https://digitalhistory.wordpress.com/author/crasmus/. Accessed July 2020
10. H. Southall, A. von Lünen, P. Aucott, Great Britain Historical Geographic Information System, in *Proceedings of the Fifth IEEE International Conference on e-Science Workshops* (2009), pp. 162–166

Using Entropy Measures for Evaluating the Quality of Entity Resolution

Awaad Al Sarkhi and John R. Talburt

1 Introduction

Entity resolution (ER), sometimes called record linking, is the process of determining if two references to real-world entities in an information system are referring to the same or different entities [1]. When two references are referring to the same entity they are said to be equivalent references [2]. To assess the accuracy of an ER decision, it is necessary to know a priori whether the references are equivalent or not. Because ER is a classification process, the most typical measure of ER accuracy is the F-measure statistic commonly used to measure classification algorithms in data mining. Given a pair of entity references, there are four ER outcomes

- The references are equivalent and the ER process linked them, true positive (TP) outcome
- The references are equivalent and the ER process did not link them, a false negative (FN) outcome
- The references are not equivalent and the ER process linked them, a false positive (FP) outcome
- The references are not equivalent and the ER process did not link them, a true negative (TN) outcome

The precision of the ER process is the ratio of TP outcomes to all positive outcomes (TP+FP), whereas the recall of the process is the ratio of the TP outcomes to all true outcomes (TP + FP). The F-measure is simply the harmonic mean of the precision and the recall measures.

A. Al Sarkhi (✉) · J. R. Talburt
University of Arkansas, Little Rock, AR, USA
e-mail: aalsarkhi@ualr.edu; jrtalburt@ualr.edu

© Springer Nature Switzerland AG 2021

933

R. Stahlbock et al. (eds.), *Advances in Data Science and Information Engineering*,
Transactions on Computational Science and Computational Intelligence,
https://doi.org/10.1007/978-3-030-71704-9_69

$$F - \text{measure} = \frac{2 \cdot \text{Precison} \cdot \text{Recall}}{\text{Precision} + \text{Recall}}$$

2   Problem Statement

While other alternatives to the F-measure been proposed such as the weighted F-measure [3], and merge distance [4], they all require knowing the true equivalences between references. Datasets of entity references where the true equivalences are known are said to be "fully annotated" reference set. Such annotated reference sets are very valuable for evaluating the efficacy of an ER process. For example, in research to assess the effectiveness of a data matching algorithm, or to gauge the effect of making a change to a matching rule in an industry setting.

However, these annotations are difficult to build and are not always available. The objective of the research described here is to assess using an alternative method for assessing linking accuracy based on the notion of entropy that does not require annotations.

3   Proposed Method for Entropy Evaluation

The method proposed here is applied to the result of an ER process. Most ER processes have three components:

1. Blocking—dividing the references into smaller subsets, called blocks, and only applying the matching rules to pairs of references in a block as a way to improve ER performance by decreasing run time.
2. Pairwise matching—assessing the similarity between two references, and making the decision to link or not link.
3. Transitive closure—labeling all of the mutually linked references into a single subset called a cluster and labeling each cluster with a unique identifier.

The method proposed here is based on Shannon entropy calculation [5] to assess the level of organization (consistency) in each cluster of two or more references. The formula for the calculation the entropy of a cluster is

$$E = -\sum_{j=1}^{N} p\left(t_j\right) \cdot \log_2 p\left(t_j\right),$$

where it is the j-th vertical token group in the cluster, and $p(t_j)$ is the probability of t_j.

A vertical token group is defined to be the same token counted only once in each reference of a cluster. Thinking of the cluster as a matrix where the references are the rows and the columns are the tokens, then a vertical token group is a vertical grouping of the same token across different references. However, each token is only counted once in each reference. This means the maximum size of a vertical token group is equal to the number of references in the cluster. The probability of a vertical token group is the size of the token group divided by the number of references in the cluster. For example, consider the following cluster of three references

R1: JOHN GRANT 123 GRANT ST
R2: MARY GRANT 21 OAK STREET
R3: MARY GRANT 21 OAK ST

The first vertical token group is for the token "JOHN" which only occurs once in R1 forming a vertical token group of size 1 with a group probability of 1/3. The second vertical token group is for "GRANT" which has three tokens, one token each from R1, R2, and R3 giving this group a probability of 1.0 (3/3). The second "GRANT" in R1 is not part of this token group because each token is only counted once in each reference. The token group for "123" has a probability of 1/3, the second "GRANT" group has a probability of 1/3, and "ST" group a probability of 2/3.

After exhausting all of the tokens in R1, there are still four uncounted tokens in R2 forming the "MARY" group with probability 2/3, "21" group probability 2/3, the "OAK" group probability 2/3, and the "STREET" group probability 1/3. Finally, there are no remaining uncounted tokens in R3.

In total, there are nine vertical token groups in the example cluster. The total entropy of the cluster is calculated from Formula (2) by

$$E = - \left(\tfrac{1}{3} \cdot \log\left(\tfrac{1}{3}\right) + 1 \cdot \log(1) + \tfrac{1}{3} \cdot \log\left(\tfrac{1}{3}\right) + \tfrac{1}{3} \cdot \log\left(\tfrac{1}{3}\right) + \tfrac{2}{3} \cdot \log\left(\tfrac{2}{3}\right) \right.$$
$$\left. + \tfrac{2}{3} \cdot \log\left(\tfrac{2}{3}\right) + \tfrac{2}{3} \cdot \log\left(\tfrac{2}{3}\right) + \tfrac{2}{3} \cdot \log\left(\tfrac{2}{3}\right) + \tfrac{1}{3} \cdot \log\left(\tfrac{1}{3}\right) \right) = 3.67318$$

Entropy is a measure of the organization of a cluster in terms of having similar tokens. The entropy of a cluster decreases as references in a cluster have more and more similar tokens. By this measure, a cluster will have an entropy of 0 if, and only if, all of the references have the same set of tokens.

4 Research Method

4.1 Reference Sets

To evaluate the effectiveness of the entropy evaluation, it is necessary to have a fully annotated reference set. For these experiments, samples were drawn from two corpora of fully annotated reference sets. The first is the R-corpus of 800K person references synthetically generated using the R-package "generator" [6] and degraded with data quality errors using the R-package "relErrorGeneratoR" from GitHub.com.

While some reference-level errors such as misspelling, truncation, mixed formatting, and missing values were injected into the data during generation, the individual references in the 800K corpus are of relatively high quality.

The majority of the data quality errors introduced into the R-corpus were data redundancy (duplicate record) errors to make the corpus more useful for entity resolution research. Shown here are two references from Sample S4 with Record Layout A. The only variations between the two references are name truncation (initial) and different formats for telephone numbers and social security numbers.

A926344: ANDREW, AARON, STEPHEN, 2475 SPICEWOOD DR, WINSTON SALEM, NC, 27106, 601-70-6106, (159)-928-5341
A930444: A, AARON, STEPHEN, 2475 SPICEWOOD DR, WINSTON SALEM, NC, 27106, 601706106, (159)9285341

The other source of the fully annotated sample is the SOG corpus [7]. The SOG corpus has approximately 270K references with three different record layouts A, B, and C. The SOG corpus has a much higher level of data quality errors than the R-corpus. Most records exhibit at least one error such as missing value, misspelling, truncation, inconsistent formatting, nicknames, and name changes. Shown here are three equivalent references from Sample S8 exhibiting a number of these data quality issues.

A960175,lucia,r,oster,t20672,southwood,oaks,dr,porter,,tx,77365,,,10896980,,
A966807,lucia,r,wi   son,12006,MOUNTAIN,RIDGE,RD,HOUSTON,,TEXAS, 77043,PO,BOX,280034, houston,,tx,77228,10896980,1917
A971069,LUCIA,R,WILSON,20672,SOUTHWOOD,OAKS,DR,PORTE,,TEXAS, 77365,,,001-89-6980,,

4.2 Blocking and Stop Word Removal

The blocking for the experiments was to use a simple frequency-based technique [8]. For each sample, the references were first preprocessed to

- replace non-word characters with blanks
- uppercase all letters
- split into tokens based on whitespace

All of the resulting tokens were counted to determine the frequency of each token in the sample. The token frequencies were used to determine both blocking and stop word removal [9, 10]. First, a blocking frequency threshold was set. If two references shared a token with a frequency less than the blocking frequency threshold, then the two references were placed in the same block. Also, a stop word frequency threshold was set. If a token in a reference had a frequency higher than the stop-work frequency threshold, then it was removed from the reference.

4.3 Pairwise Matching with the Monge-Elkan Comparator

The scoring matrix used for this research is a variation of the Monge-Elkan method [11] for comparing multi-token values, but with the removal of stop words as described previously. The individual tokens were compared using the normalized Levenshtein edit distance similarity function, which returns a value in the interval [0,1] when comparing two tokens. Because the scoring matrix calculates an overall similarity based on the average of the highest token similarities in each row and column of the matrix, it also returns a value in the interval [0,1].

In addition to the blocking frequency threshold and the stop word frequency threshold, a third threshold set was the match threshold. If two references compared by the scoring matrix produced a score at or above the match score threshold, they were linked, otherwise, they were not linked.

4.4 Comparing the Entropy Value to F-Measure

The experiments were conducted by taking samples from the R-corpus and the SOG-corpus and performing entity resolution using various thresholds for the blocking, stop word, and match score. These produced a variety of clusters. For each cluster, both its true F-measure and entropy were calculated and compared.

5 Results

Sample S4 from the R-corpus contains 1912 synthetic person references of relatively high quality. Table 1 summarizes the analysis of clustering S4 at various match levels. The Match column shows the match threshold as varied from 0.5 to 0.9 in increments of 0.1. For each match threshold, the rows are arranged in ascending order by size of the cluster produced. The column Count shows the number of clusters of that size. For example, the match threshold of 0.5 produced 63 clusters containing four linked references.

The Min column shows the minimum entropy and the Max column shows the maximum entropy for the clusters of a given size. For example, 0.0 was the minimum entropy for the 63 clusters of size 4 at match level 0.5, and 18.0 was the maximum entropy. The 90% column shows the entropy cutoff value at which 90% of the clusters having an entropy below this value are correctly linked. This was

Table 1 Result from Sample S4

Match	Size	Count	Min	Max	90%	80%
0.5	2	192	0.0	11.5	5.0	9.8
0.5	3	92	0.0	13.8	4.1	5.7
0.5	4	63	0.0	18.0	2.8	3.3
0.5	5	36	0.0	15.8	3.5	3.5
0.5	6	21	3.7	17.5	**	**
0.5	7–15				**	**
0.6	2	98	0.0	11.5	4.4	5.0
0.6	3	53	0.0	11.0	4.7	5.3
0.6	4	39	1.0	11.5	4.2	4.3
0.6	5	19	4.4	13.2	**	**
0.6	6–9				**	**
0.7	2	49	0.0	11.5	4.5	11.5
0.7	3	29	0.0	9.0	4.3	4.7
0.7	4	26	0.0	11.5	3.0	4.1
0.7	5	12	0.0	11.6	4.4	4.4
0.7	6	4	4.6	9.6	**	**
0.7	7–9				**	**
0.8	2	46	0.0	9.0	9	9
0.8	3	19	0.0	5.4	4.7	5.1
0.8	4	6	0.0	6.5	4.5	4.5
0.8	5	3	0.9	5.9	4.7	4.7
0.8	6	1	0.0	0.0	1	1
0.9	2	7	0.0	3.0	3	3
0.9	3	5	0.4	4.2	4.3	4.3

done by listing each cluster in a table along with its true F-measure (by annotation) and its calculated entropy value.

For example, if the cutoff value is set at 2.8 for the 63 size 4 clusters at match score 0.5, then 90% of those clusters with an entropy of 2.8 or less will be correctly linked, i.e. have an F-measure of 1.0. Similarly, the 80% column shows that for the same 63 clusters, 80% of the size 4 clusters with an entropy of 3.3 or less are correctly linked.

The entry "**" in Table 1 indicates the 90% or 80% threshold is not possible because none of the clusters for that size are correctly linked. For example, for Match threshold 0.5 cluster size 6, none of the 21 clusters formed at this threshold had an F-measure of 1.0, and the minimum entropy was 3.7.

While lower entropy appears to be an indicator of clusters with high precision, it does not provide insight into the recall. False positive links in a cluster are only half of the linking accuracy. Examining individual clusters cannot detect false negative linking. The purpose of linking the references at different thresholds, especially at lower match thresholds, is to broadly include references into the clusters being evaluated. For example, 0.50 is a very low scoring matrix threshold. It means, that

Table 2 Overall F-measures for linking S4 at different entropy thresholds

Starting match	Final match	Blocking frequency	Stop word frequency	Entropy threshold	Final F-measure	Precision	Recall
0.50	1.00	12	22	3.0	0.9209	0.9930	0.8586
0.50	1.00	12	22	3.5	0.9235	0.9818	0.8717
0.50	1.00	12	22	3.7	0.9272	0.9776	0.8818
0.50	1.00	12	22	3.8	0.9219	0.9657	0.8818
0.50	1.00	12	22	4.2	0.9088	0.9274	0.8909

Table 3 Overall F-measures for linking S8 at different entropy thresholds

Starting match	Final match	Blocking frequency	Stop word frequency	Entropy threshold	Final F-measure	Precision	Recall
0.50	1.00	7	15	6.0	0.2111	0.9462	0.1188
0.50	0.70	7	15	15.0	0.3581	0.7005	0.2405
0.50	0.60	7	15	25.0	0.5066	0.6290	0.4205
0.50	0.60	7	15	35.0	0.5373	0.5301	0.5446
0.50	0.60	7	15	40.0	0.5140	0.4735	0.5621

on average, the tokens between two references were only 50% similar to character strings.

However, by using an entropy threshold as a regulator and monotonically increasing the match threshold, it is possible in some cases to achieve both high recall and high precision. Table 2 shows the results of an iterative process performed on S4. The process starts by forming clusters with the scoring matrix starting at match threshold 0.50. For each cluster formed, the entropy is calculated. If the entropy of the cluster is at or below a fixed threshold, then the cluster is kept. If not, the references are set aside for reprocessing at the next higher match threshold of 0.60. This process is repeated until all until there are no remaining references to process. Table 2 shows a series of trials using this process with varying entropy thresholds.

The same cluster-by-cluster analysis was done for Sample S8 comprising 1000 references drawn from the SOG corpus which exhibits much low data quality. These data quality errors also tend to increase the entropy of a cluster, even if the cluster is correctly linked. Missing values misspelled and truncated words, and other inconsistencies tend to increase the entropy, while match rules can often overcome these issues and still arrive at the correct linking. For example, using a nickname similarity function for first name matching. For Sample S8, the lower entropy values provided high precision, but very low recall as shown in Table 3. While raising the entropy threshold increased the recall, it decreased the precision. The best overall result obtained from the sample was just over 50% linking accuracy.

6   Conclusion and Future Research

The experiments conducted in this research show promise for allowing ER processes to self-regulate linking based on cluster entropy. While the results are very promising for entity references of relatively high quality, using this process for low-quality data needs further improvement.

There are several factors that can be explored to further this research. In particular, an improved method for computing the best entropy threshold for a given reference set. For these experiments, the blocking frequency and stop word frequency thresholds were left constant. The interaction of these thresholds with the entropy threshold remains to be explored. Other factors needing more research are the method for tokenizing the references and other similarity functions used by the scoring matrix.

References

1. J.R. Talburt, *Entity resolution and information quality* (Elsevier, Burlington, 2011)
2. J.R. Talburt, Y. Zhou, *Entity Information Life Cycle for Big Data: Master Data Management and Information Integration* (Elsevier, Waltham, 2015)
3. D. Hand, P. Christen, A note on using the F-measure for evaluating record linking algorithms. Stat. Comput. **28**, 539–547 (2018)
4. D. Menstrina, S. Whang, H. Garcia-Moliina, Evalutation entity resolution results, in *Proceedings of the VLDB Endowment*, (2010)
5. C. E. Shannon, A note on the concept of entropy, Bell Syst. Tech. J., 1948.
6. Y. Ye, J.R. Talburt, Generating synthetic data to support entity resolution education and research. J. Comput. Sci Coll **34**(7), 12–19 (2019)
7. J.R. Talburt, Y. Zhou, S.Y. Shivaiah, SOG: A synthetic occupancy generator to support entity resolution instruction and research. MIT Int. Conf. Inf. Qual., 91–105 (2009)
8. A. Alsarkhi, J. R. Talburt, Optimizing inverted index blocking for the matrix comparator in linking unstandardized references, in *Proceedings of the 2019 International Conference on Scientific Computing*, 2019.
9. A. Alsarkhi, J. Talburt, An analysis of the effect of stop words on the performance of the matrix comparator for entity resolution. J. Comput. Sci. Coll., 67–71 (2019)
10. A. Alsarkhi, R. T. John, A scalable, hybrid entity resolution process for unstandardized entity references, *The Journal of Computing Sciences in Colleges Papers of the 18th Annual CCSC Mid-South Conference*, 2020, pp. 19–29.
11. A. E. Monge, C. P. Elkan, The field matching problem: Algorithms and applications, in *KDD-96 Proceedings*, 1996.

Improving Performance of Machine Learning on Prediction of Breast Cancer Over a Small Sample Dataset

Neetu Sangari and Yanzhen Qu

1 Introduction

1.1 Background

Breast cancer is the second leading cause of mortality in women. In the USA, it is estimated that 276,480 new cases of invasive breast cancer and 48,530 new cases of noninvasive breast cancer are likely to be diagnosed in women [1]. American Cancer Society's estimates 42,170 expected deaths from breast cancer in the US for 2020 [1]. The incidence rates have increased by 0.3% per year [1]. 1 in 8 women (12.5%) are at risk of breast cancer in their lifetime, and 1 in 38 women (2.6%) may die of breast cancer [1]. Approximately 3.5 million US women with a history of breast cancer were alive in September 2019 [2].

Breast cancer screening is critical for early detection and to ensure a higher probability of having a good outcome in treatment. Our research aims to assess how models based on data can be collected in routine blood analyses - notably, glucose, insulin, HOMA, leptin, adiponectin, resistin, MCP-1, age, and body mass index (BMI) [3]. Patricio et al. stated that resistin, glucose, age, and BMI data might be used to predict the presence of breast cancer [4].

Accurate breast cancer prediction models help early diagnosis, which will reduce mortality rates, and help clinicians to develop personalized treatments. Goldstein et al. (2017) mentioned that a lot of emphasis is placed on the development and application of machine learning (ML) algorithms, and not much effort is put on model performance evaluation. For a simple ML algorithm like the linear regres-

---

N. Sangari · Y. Qu (✉)
Colorado Technical University, Colorado Springs, CO, USA
e-mail: neetu.sangari@student.ctuonline.edu; yqu@coloradotech.edu

© Springer Nature Switzerland AG 2021   941
R. Stahlbock et al. (eds.), *Advances in Data Science and Information Engineering*,
Transactions on Computational Science and Computational Intelligence,
https://doi.org/10.1007/978-3-030-71704-9_70

sion, there are well-understood metrics available to evaluate model performance. However, for many other machine learning algorithms like support vector machines, gradient boosting machines, and random forests there are no uniform and acceptable metrics available for the model performance evaluation [5]. Additionally, most of the research in ML-based prognostic tools in oncology has come from technology-focused labs, and there are several concerns about their applicability in the real world. This is also noted in Parikh et al. (2019), where the authors highlight the need to investigate if ML prognostic algorithms improve traditional regression models, especially in oncology settings [6].

1.2   Related Work

In this paper, we will present our research work on the model performance of multiple ML algorithms based on an existing small sample dataset related to breast cancer. The data used in this study is very recent, and to our knowledge, only one systematic study of the application of ML algorithms has been reported in the literature. Patricio et al. (2017) used this dataset to perform exploratory analysis to see if the biomarkers could be the right candidate for predicting the risk of breast cancer. They applied different ML algorithms on the original dataset comprising 116 records and concluded that some of the biomarkers could be used to model for breast cancer screening. Our work is different in two ways: First, in addition to Support Vector Machine (SVM) and Random Forest (RF), we also used another ML algorithm Gradient Boosting (GB) and an ensemble method with different hyper-parameters. Second, since the original dataset is very small and we found strong evidence of massive model overfitting, the final models were fitted on the augmented dataset. For data augmentation, we used Predictive Mean Matching (PMM) with bootstrapping - multiple imputation techniques.

1.3   Paper Organization

This paper is structured as follows. The second section provides the problem statement for the research and presents the hypothesis and research questions. The third section describes the work related to answer the first research question and shows how to effectively augment the original small dataset without destroying the characteristics of the original dataset. The fourth section presents the work related to answer the second research question and discusses how much further model performance improvement for different ML algorithms can be made by using an augmented dataset. The final section presents the conclusion and future work.

2   Problem, Hypothesis, and Research Questions

2.1   Problem Statement

Although many ML algorithm-based prognostic tools have been studied for early breast cancer diagnosis, there is no research on consistently evaluating the performance of broad-range ML algorithms. The limitation on the evaluation is due to the lack of a clear understanding of what good criteria to evaluate the model performance of any ML algorithm-based prognostic tool used for early breast cancer diagnosis.

2.2   Hypothesis Statement

If an ML algorithm-based prognostic tool can improve traditional regression models in oncology settings, then that tool will be helpful to improve the predictions generated from randomized clinical trials and more accurately identifying patients with high breast cancer risk.

2.3   Research Questions

It is common knowledge that the performance of any ML algorithm is heavily dependent on the quality of the dataset used for training and testing. Our research will use a smaller dataset to train and test the relevant ML algorithm used by the prognostic tool, the overfitting problem caused by the smaller dataset will also be addressed. Therefore, very naturally, the first research question for us will be the following.

1st Research Question: How can we resolve the overfitting issue caused by the small dataset to ensure the quality of the dataset used for training and testing an ML algorithm used by a prognostic tool?

After we have had a high-quality dataset, our next question will be if there is any ML prognostic algorithm that indeed can outperform the predictions generated from randomized clinical trials. This leads to the second research question, as shown below.

2nd Research Question: How much further model performance improvement for an ML algorithm can be made by using an augmented dataset?

3   Works That Answer the 1st Research Question

In this section, we will present our work related to answering the first research question. The focus point of the work is the details of how to effectively augment the original small dataset without destroying the characteristics of the original dataset.

3.1   Stratified Sampling

The objective of the proposed study is to apply an ML algorithm for feature selection and prediction. The response variable "Classification" indicates the presence (target) and absence (control) of breast cancer. The proportion of target and control is 55% and 45% respectively in the original data set; we aim to preserve the same composition in the training and testing datasets. For this, we adopted the following procedure:

(a) Partition the dataset into training and testing data. We used stratified sampling to preserve the proportion of control and target variables in the training and testing datasets.
(b) Fit the ML method on training data set using k-fold cross-validation. Analyze the accuracy statistics and the confusion matrix to evaluate the model fit.
(c) Use the fitted model to make predictions using testing and training datasets.
(d) Analyze the model performance using accuracy metrics.

3.2   Data Augmentation

This section aims to provide a brief overview of the data augmentation procedure that was implemented for the machine learning model fitting. The original data with 116 records appears to be a small sample for Support Vector Machine with Radial Basis, the fitting of which highlighted significant overfitting for different training and testing data set compositions. It was, therefore, decided to augment the original dataset and rerun the model fitting. Even though there are numerous data augmentation methods, most of those are not suitable for the type of dataset. The goal was to augment the data set so that original features such as the correlation structure and relationships are preserved.

We explored Expectation-Maximization (EM)-based and Imputation-based methods that could be used for data augmentation. Most of the EM-based methods directly or indirectly assume a multivariate normal distribution, which does not apply to the original dataset. So, it was decided to use the *Predictive Mean Matching with bootstrapping* –multiple imputation techniques – for data augmentation [7]. After some initial analysis and experiments, we employed the following procedure:

(a) Identify the significant variables that could be used for Imputation. For this, we
    studied the correlation structure of the original data and the results of the initial
    model fitting exercise. The cross-correlations of predictor variables are shown
    in Fig. 1. Notice high correlation pairs: Glucose-HOMA (70%), BMI & Leptin
    (57%), Glucose & Insulin (50%). There are different techniques to identify the
    values of the variables, which can be removed for imputation. Since the aim
    is to impute values that can preserve the correlation structure of the predictor
    variables, we decided to remove values of selected variables, one variable from
    each of the high correlation segment.

Besides, cross-correlation, we also cluster analysis, which uses the correlation
distance between the predictor variables to identify the relationships between them.
This is depicted as a *dendrogram* at the top of Fig. 1. By using an appropriate
threshold level of three, we were able to identify three clusters from which we could

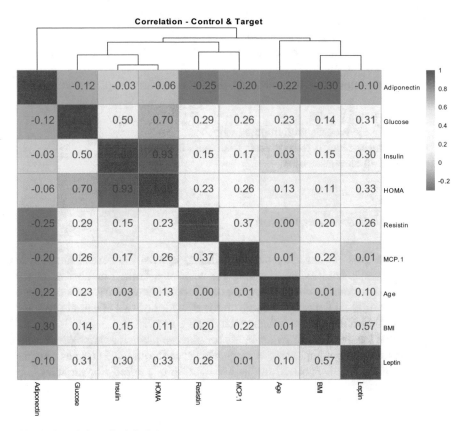

Fig. 1 Correlation of original dataset

select one variable as an imputation candidate[1] This way, the imputation method will be able to impute the removed values using the information available from other variables.

In Fig. 1, we can identify three distinct groups of variables with high correlation among them: Glucose, Insulin, HOMA; BMI, Leptin; MCP1, Resistin.

(b) Remove the extreme valued records of "Glucose," "BMI," and "Resistin" as imputation algorithms are sometimes sensitive to such values. We used the Median Absolute Deviation (MAD) to identify and remove the records for the data set.

(c) Identify a suitable missing value percentage that could be used to remove the data points from the significant predictor variables. After some trial and error, we decided to use 30% as missing percentage value resulting in ~ 30 records per batch with missing values resulting in 10 batch runs to generate around 300 records needed to augment the original data set. For each batch:

- Get a random seed. A new random seed is used for each batch because we would like to delete a different set of observations.
- Use the seed to make missing values of ~30% of the values from three predictor variables.
- Run the imputation algorithm on the dataset with complete and missing values.
- Get imputed results.
- Identify the records with imputed values.
- Store these records to be used for data augmentation.
- Merge all batches results.

3.3 Distribution Comparison

One of the dangers of PMM imputation methods, especially for small samples, is that these methods tend to repeat values for want of a limited number of donor records [7]. Therefore, to check if the imputed records have distribution coverage similar to the original dataset, we compare the densities of two essential predictor variables for which the data were imputed in Figs. 2 and 3.

Notice that the densities of two datasets for glucose predictor variables are very close except for minor bumps at the extreme. However, the imputed data has a higher peak for resistin. We also noticed similar behavior for other predictor variables.

---

[1] The clusters are represented by three different sets of coloring schemes on the main diagonal.

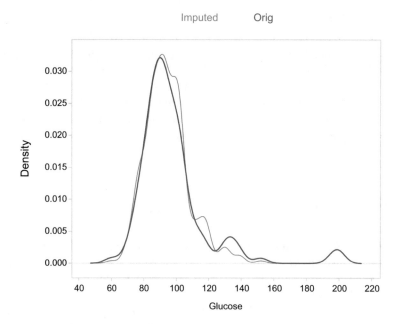

Fig. 2 Glucose – density comparison

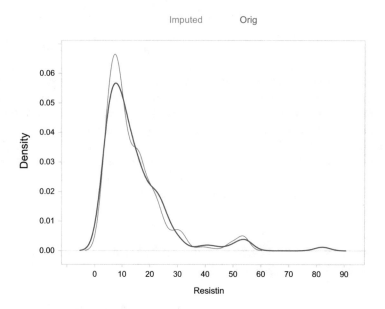

Fig. 3 Resistin – density comparison

3.4   Summary

The correlation distance between predictor variables highlighted three clusters in Fig. 1: glucose, insulin, HOMA; BMI, leptin; MCP1, resistin. The initial analysis highlighted "glucose," "BMI," and "resistin" as significant predictor variables across different models. Thus, these significant predictor variables are used as targets for missing data imputation using the Median Absolute Deviation. The goal is to keep the imputed records distribution coverage as the original dataset and check densities comparison between the imputed and original dataset.

4   Works That Answer the 2nd Research Question

The feature selection process identified the most potential variables "glucose," "BMI," and "resistin" as significant predictor variables, and the correlation analysis showed that glucose, insulin, HOMA; BMI, leptin; MCP1, and resistin are clinically supported by the following:

Ando et al. [8] summarized the epidemiological link between obesity, leptin, and breast cancer, i.e., leptin's circulating levels increase proportionally to total adipose tissue mass, a vital member of the molecular network obesity associated with breast tumor cell growth. Muti et al. [9] reported that glucose, insulin, and insulin-like growth factors (IGFs) are contributors to breast cancer. Ando et al. have reported that adipokines have been found to link obesity to cancer. Leptin and adiponectin, in particular, have come to be recognized for their influence on breast cancer risk and tumor biology. In obesity, fat accumulation, as well as a dysregulation of adipokine production, results in increased leptin and decreased adiponectin levels that strongly contribute to the onset of obesity-associated breast cancers. The clinical investigations suggested that low adiponectin concentrations are associated with an increased risk of breast cancer. Monocyte chemoattractant protein 1 (MCP-1) is one of the inflammatory chemokines implicated in cancer development and progression. In a recent study, Dutta et al. [10] have found elevated levels of MCP.1 in certain categories of patients with triple-negative breast cancer (TNBC).

4.1   Support Vector Machine with Radial Basis (SVMRB): Original vs. Augmented Data

SVMRB is a maximum margin classification algorithm. The generalization ability of the optimal hyperplane can be directly related to the number of support vectors. The support vectors are the data points that lie at the margin borders. These are the data points closest to the decision surface and hence are most difficult to classify. Therefore, support vectors define the location of the decision surface and provide a bound on the expected error rate for the test sample.

Table 1 Number of support vectors

Data points	Original	Augmented
Training	80	296
Support Vectors	54	108
Percentage	68%	36%

Table 2 SVMRB confusion matrix

Augmented (70/30)			Original(70/30)		
Pred.	control	target	Pred.	control	target
control	48	10	control	13	3
target	5	65	target	2	18
Accuracy: 0.8828			Accuracy: 0.8611		
95% CI: (0.814, 0.932)			95% CI: (0.705, 0.953)		
Kappa: 0.7618			Kappa: 0.717		

(a) The number of support vectors in Table 1 provides an indication of the generalization ability of the model – the higher the number of support vectors, the lower the model generalization [11]. The training set has 296 records (134 control and 162 target). The final fitted model has 108 support vectors indicating a low generalization ability of the model and higher expectation of the error rate. This is also reflected in a significant reduction in Kappa when the fitted model was used for predicting the class labels for the training set.

(b) A comparison of the fitted model shows some improvement in the fit for SVMRB with an augmented dataset. The final model with an augmented data set has a lower proportion of the data points as support vectors as compared to the fitted model with the original dataset. Also, the cost function value has increased to 128 (from 16) for the final model. These two indicate some improvement in the SVMRB model with augmented data set in Table 2. 70/30 denotes the proportion of training and testing data.

(c) However, the improvements are not significant to warrant the model to be flexible and generalizable, and overfitting still persists in all the scenarios.

(d) Accuracy comparison of the fitted model with and without augmented data also shows some improvement, which is also supported by the tighter confidence intervals.

4.2 Gradient Boosting (GB): Original vs. Augmented Data

The GB is an ensemble modeling technique. The idea is to use a *weak learner*, a learner who performs better than random guessing, that can be boosted into a strong learner with improved performance. It is an iterative technique, where at each iteration the base model is fitted to a different training dataset, the error is computed as the sum of the misclassified instances, the gradient (residuals) is calculated, and a model is then fit to the residuals to minimize the loss function [12]. This way,

Table 3 GBM confusion matrix

Augmented (70/30)				Original(70/30)			
gbm	class	pred	test	gbm	class	pred	test
control		target		control		target	
control		49	9	control		8	8
target		12	58	target		1	19
Avg. Accuracy: 0.83594				Avg. Accuracy: 0.75			
AUC: 0.8818				AUC: 0.8719			
95% CI: 0.8204–0.9431				95% CI: 0.7494–0.9943			
(DeLong)				(DeLong)			

the algorithm always trains the models using the data samples, which are difficult to learn in the previous iterations. The final model is obtained using the weighted average of all the fitted models.

GBM is an ensemble method in which the final method is obtained as a weighted average of all hierarchically fitted models. Therefore, the performance of model prediction on the original and augmented dataset is analyzed using average accuracy and area under the curve (AUC). In addition, we look at 95% CIs as an indication of the model variance. The results show some improvement in model prediction performance with augmented data in Table 3.

a. The AUC has increased slightly; however, 95% CIs bounds have tightened, indicating lower uncertainty in the predicted model fitted on the augmented dataset.
b. Average accuracy improved significantly from 75% to 84% when the augmented dataset was used to train the model, which was then used for making predictions.

4.3 Random Forest (RF): Original vs. Augmented Data

Random Forest is one of the most popular and powerful ensembles machine learning algorithms based on Bootstrap Aggregation or bagging. Each ensemble model is used to generate a prediction for a new sample, and the average of these predictions provides the forest's prediction. The performance of model prediction on the original and augmented dataset is analyzed using accuracy and 95% CIs. The results show some improvement in model prediction performance with augmented data in Table 4. Accuracy improved significantly from 78% to 91% when the augmented dataset was used to train the model and used for the predictions.

Table 4 Random forest
confusion matrix

Augmented (70/30)				Original(70/30)			
rf	class	pred	test	rf	class	pred	test
control		target		control		target	
control		55	3	control		10	6
target		8	62	target		2	18
Accuracy: 0.9141				Accuracy: 0.77781			
95% CI: 0.8514–0.9563				95% CI: 0.6085–0.8988			
Kappa: 0.7618				Kappa: 0.5385			

4.4   Summary

The machine learning methods SVMRB, GBM, and RF based on augmented data (424 records) show improvement in accuracy and overfitting as compared to original data (116 records). After analysis of the feature selection from different models, the predictor variables can be used for early breast cancer diagnosis to improve and complement randomized clinical trials significantly in the future.

5   Conclusion

Small samples are a very common occurrence in research areas involving human participants or where it is very difficult to collect additional data such as dynamic batch systems. While a lot of effort has been devoted to developing ML algorithms for large datasets, not much effort has been devoted to areas where small datasets are the norm. In this research, we have studied the effects of the small dataset on the model performance of ML algorithms. In particular, we have used Coimbra Breast Cancer data to train and test the Support Vector Machine Radial Basis, Gradient Boosting, and Random Forest three ML algorithms. The analysis SVMRB highlights severe overfitting and low model generalization when fitted with the original dataset with 116 observations. The performance improved with the augmented dataset for all three ML algorithms, although the degree of improvement is different among three ML algorithms.

Nevertheless, some concerns over model generalization remain. The study also highlights that in scenarios where the input dataset is small, we cannot fully rely on the typical model performance measures from the confusion matrix such as accuracy. Instead, we need to evaluate the overall model performance by looking at different measures. For example, for SVM, we need to look at the number of support vectors, the cost parameters, and accuracy measures. ML algorithms can contribute to the early detection of breast cancer and play a major role in randomized clinical trials. The collaboration of researchers, statisticians, clinicians, and industries can expedite drug development and increase the quality of care to our individual patients.

For the future work, we are looking into increasing the size of the augmented dataset, and will also try different augmentation methods such as applying adversarial approach or by constraining the complexity of the model in order to further reduce overfitting.

References

1. American Cancer Society. How Common Is Breast Cancer? Jan. 2020. https://www.cancer.org/cancer/breast-cancer/about/how-common-is-breast-cancer.html
2. American Cancer Society. Cancer Facts & Figures 2020. https://www.cancer.org/content/dam/CRC/PDF/Public/8577.00.pdf
3. UCI Machine Learning Repository Irvine, CA: University of California, School of Information and Computer Science. http://archive.ics.uci.edu/ml
4. M. Patricio, J. Pereira, J. Crisostomo, P. Matafome, M. Gomes, R. Seiça, F. Caramelo, Using Resistin, glucose, age and BMI to predict the presence of breast cancer. BMC Cancer **18**(1), 29 (2018). https://doi.org/10.1186/s12885-017-3877-1.10
5. T. Parr, K. Turgutlu, C. Csiszar, J. Howard, Beware default random forest importances. March **26**, 2018 (2018). https://explained.ai/rf-importance/index.html
6. R.B. Parikh, C. Manz, C. Chivers, S.H. Regli, J. Braun, M.E. Draugelis, L.M. Schuchter, L.N. Shulman, A.S. Navathe, M.S. Patel, N.R. O'Connor, Machine learning approaches to predict 6-month mortality among patients with cancer. JAMA Network Open **2**(10), e1915997–e1915997 (2019). https://doi.org/10.1001/jamanetworkopen.2019.15997
7. F.E. Harrell Jr., *Regression modeling strategies: with applications to linear models, logistic and ordinal regression, and survival analysis* (Springer, 2015). https://link.springer.com/content/pdf/10.1007/978-3-319-19425-7.pdf
8. S. Ando, L. Gelsomino, S. Panza, C. Giordano, D. Bonofiglio, I. Barone, S. Catalano, Obesity, Leptin and breast cancer: epidemiological evidence and proposed mechanisms. Cancers **11**(1), 62 (2019). https://doi.org/10.3390/cancers11010062
9. P. Muti, T. Quattrin, B.J. Grant, V. Krogh, A. Micheli, H.J. Schünemann, et al., Fasting glucose is a risk factor for breast cancer: a prospective study. Cancer Epidemiol. Prev. Biomarkers **11**(11), 1361–1368 (2002). https://cebp.aacrjournals.org/content/11/11/1361.full-text.pdf
10. P. Dutta, M. Sarkissyan, K. Paico, Y. Wu, J.V. Vadgama, MCP-1 is overexpressed in triple-negative breast cancers and drives cancer invasiveness and metastasis. Breast Cancer Res. Treat. **170**(3), 477–486 (2018). https://doi.org/10.1007/s10549-018-4760-8
11. B.A. Goldstein, A.M. Navar, M.J. Pencina, J. Ioannidis, Opportunities and challenges in developing risk prediction models with electronic health records data: a systematic review. J. Am. Med. Inf. Assoc. **24**(1), 198–208 (2017). https://doi.org/10.1093/jamia/ocw042
12. M. Kuhn, K. Johnson, *Applied predictive modeling*, vol 26 (Springer, New York, 2013). https://link.springer.com/content/pdf/10.1007/978-1-4614-6849-3.pdf

Development and Evaluation of a Machine Learning-Based Value Investing Methodology

Jun Yi Derek He and Joseph Ewbank

1 Introduction

The underlying principle behind value investing is to purchase stocks when the investor believes it is undervalued by the market, so that its price will appreciate to its intrinsic value over time [1]. An investor uses the company's fundamental data found in their income statement, balance sheet, and cashflow statements to estimate its intrinsic value [2]. A common method of estimating a company's intrinsic value is using the Discounted Cash Flow calculation, popularized by John Burr Williams [3] in *the Theory of Investment Value*.

The goal of this paper is to utilize sophisticated data pipelines and develop machine learning models to learn value investing.

One related work is by Rasekhschaffe and Jones [4]. The researchers used multiple machine learning algorithms, ensembles, and feature generation to find stocks that underperform and outperform the market. One area for improvement is to discriminate between large outperformance and small outperformance. A similar work was conducted by Quah [5]. The work analyzed soft-computing models on picking stocks among the Dow Jones Industrial Average from 1995 to 2016. The researchers identified optimal training parameters and algorithm evaluation methods that achieved above-average returns. Another work was conducted by Sim et al. [6]. The researchers used 3D Subspace modeling with stock fundamental data to generate models that can identify profitable stocks to purchase. The forecast horizons were defined on an annual basis, which is a value that can be experimented with to produce better results.

J. Y. D. He (✉) · J. Ewbank
The Woodlands College Park High School, Woodlands, TX, USA
e-mail: jewbank@conroeisd.net

© Springer Nature Switzerland AG 2021
R. Stahlbock et al. (eds.), *Advances in Data Science and Information Engineering*,
Transactions on Computational Science and Computational Intelligence,
https://doi.org/10.1007/978-3-030-71704-9_71

2   Experimental Setup

2.1   Dataset Description

The dataset was obtained from gurufocus.com and it comprises 30 years of fundamental data from 1990 to 2019 from the S&P 500, the Russell 2000, and the Russell 1000 indexes [7]. The data consists of approximately 200 companies and 4000 data entries.

The algorithm needed context about each company for each data instance. This was done by adding the previous X-1 years of the company's data to each instance, where X is the number of years of context to be provided to an algorithm. Overlapped data from shifting to create context were removed to avoid data leakage.

Experiment 1 used the train test split method, where 89% of the data were randomly sampled for training, and the rest were used for testing [8].

For experiment 2, the 15 years between 1990 and 2004 were used for in-sample learning, and the remaining 12 years were used for out-of-sample testing. The 15 years for training were treated with train test split to create training and validation.

Based on Quah [5] and Buffett and Clark [9], 11 features were selected: Month End Stock Price, Price to Book Ratio, Price to Earning Ratio, Year over Year Revenue per Share Growth, Gross Margin, Net Margin, Retained Earnings, Return on Equity, Total Current Assets, Current Ratio, and Capital Expenditure.

2.2   Testing Criteria and Algorithms

The most important metric is the return of the portfolio with a forecast horizon of 3 years. The return is equal to the cube root of the percent change over 3 years. The prediction targets for any given company at a certain year were either 0, 1, or 2. A prediction target of 0 was assigned to the bottom-third percentile of returns, 1 was assigned to the middle-third percentile, and 2 was assigned to the top-third percentile. This allows the models to discriminate a large outperformance from a small outperformance.

Volatility was also measured by finding the interquartile range (IQR) of the companies' returns in a given set of data. Standard deviation was not used because it assumes that the return distribution is normally distributed, which is often times not the case [10].

"Precision 1" is the number of stocks that are class 2 over the number of stocks predicted as class 2. "Precision 2" is the number of stocks that are class 1 or 2 over the number of stocks predicted as class 2. Precision 2 calculates the percentage of predictions that have at least average return, while Precision 1 calculates that for above average return.

Different algorithms were used to learn from this data: Support Vector Machines (SVM), Random Forests (RF), AdaBoost (Ada), and an ensemble of the three using simple voting. An algorithm predicts a stock when it outputs a label 2. A benchmark was made to represent a portfolio that does not have any rules when making an investment, which is the market average.

Another method for algorithm decision-making is to utilize algorithm confidence. This involves first assigning a confidence value to each class using the softmax function [11]. Next, the stocks are ranked from greatest to least in those confidence values, and the top 20 stocks are selected.

3   Experimental Results

3.1   Experiment 1: Determining Optimal Context Years

The first experiment involved determining the optimal number of years given as context for each data instance. There exists a trade-off between the amount of data per instance and the number of instances available for training that had to be identified. After experimenting with five combinations ranging from five years per instance to one year, the return on investments (ROI) are reported in Fig 1.

The average ROI values were calculated by finding the mean of the returns of the algorithms on both the in-sample (before 2013) test data and the out-of-sample (post 2013) test data. Figure 1 shows that the average of the algorithms returns is the highest for one year of context and three years of forecast.

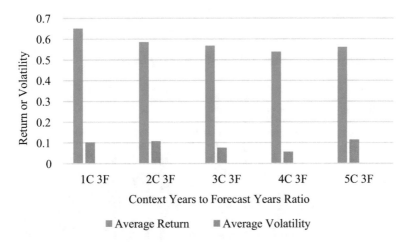

Fig. 1 Average return and volatility by context to forecast year ratio

3.2   Experiment 2: 15 Years (1990–2005), 11 Years (2005–2016)

Figures 2 and 3 showcase the RF returns and volatility obtained from the second experiment. Figures 4 and 5 are the Precision 1 and Precision 2 values obtained during experiment 2 for all of the algorithms.

Each algorithm has three values tracked: the algorithm's precision on the training data, the algorithm's average precision on the out-of-sample test data, and the algorithm's average precision on the out-of-sample test data using confidence predictions. The same values were tracked in Fig. 5 but using Precision 2.

From Fig. 2, it can be seen that the confidence predictions increase the ROI by roughly 35% on average. According to Fig. 4, RF shows the least deviation from the

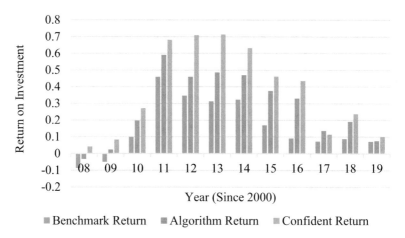

Fig. 2   Annual return on investment RF vs. benchmark (2008–2019)

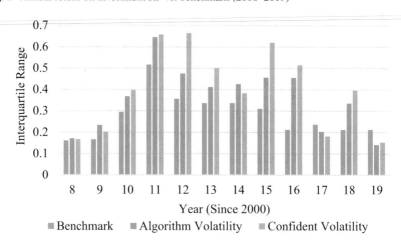

Fig. 3   Annual interquartile range RF vs. benchmark (2008–2019)

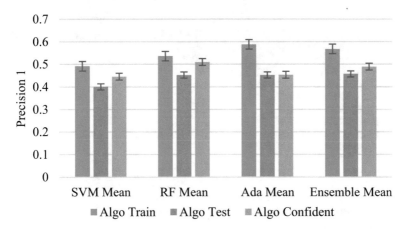

Fig. 4 Precision 1 in-sample vs. out-of-sample

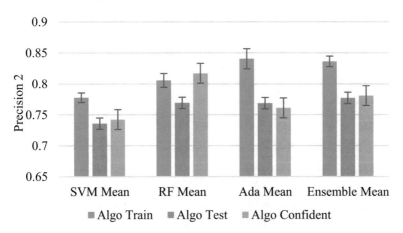

Fig. 5 Precision 2 in-sample vs. out-of-sample

Table 1 Gross asset worth confident predictions from 2005–2019 from all algorithms

Year	Benchmark	SVM	RF	Ada	Ensemble
ROI Total Percent Change	246.6	1063.	1471.	841.9	1161.
ROI Compound Annual Growth %	10.03	20.77	23.6	18.82	21.53
IQR Total Percent Change	1165.	2018.	2291.	1821.	2101.
IQR Compound Annual Growth %	21.55	26.47	27.66	25.52	26.85

training precision and both the in-sample and out-of-sample testing precision. From Fig. 5, a similar conclusion can be drawn.

Table 1 below shows the out-of-sample returns and interquartile range of each algorithm and that RF had the best risk–reward ratio with 28.77% to 23.60%. The benchmark had 21.42% to 10.03%, respectively.

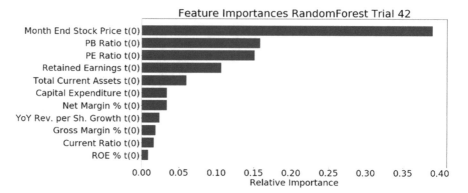

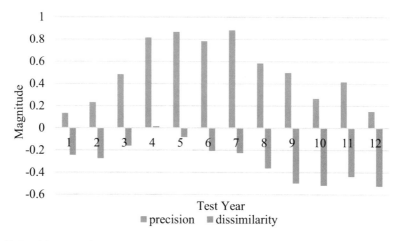

Fig. 6 Feature importance assigned using the random forest algorithm

Fig. 7 Precision vs. weighted dissimilarity for random forest

Figure 6 below shows the relative feature importance of each variable as reported by the random forest algorithm. It can be seen that the most impactful metrics are the "Month End Stock Price," "Price to Book Ratio," and "Price to Earning Ratio."

To gauge the algorithm's reliability in the future years, one can determine the relationship between the algorithm's precision and the comparability of the out-of-sample distribution and training data, which is shown in Fig. 7. For each independent variable, the z-score between the mean of the training distribution and that of each out-of-sample year's distribution was determined. The net z-score for each year was found by adding each feature's z-scores multiplied by the weight value from Fig. 6. The net z-score was graphed with the precision for each year.

From Fig. 7, it can be seen that when there is smaller dissimilarity (signified by a smaller magnitude in dissimilarity), the precision of the model tends to increase.

The converse is true. This inverse relationship can be put into consideration when using the algorithms in out-of-sample data.

4 Conclusion

The researchers concluded that the random forest and ensemble are strong candidates for users who would like to utilize machine learning to help aid in picking stocks using a value investing approach. Evidence was gathered that the results are scalable to future years if the user calculates the dissimilarity between the current year's data distribution with the training data's distribution. The random forest algorithm generally found valuation metrics like the "price to earning ratio" and "price to book value ratio" to be influential in its decision-making process.

Acknowledgment The researcher would like to thank his mentor at the Woodlands College Park High School for offering advice on the research. The project was self-funded.

References

1. E. Riley, *Advisor Today* (Advisor Today, 2010, June).
2. J. Chen, *Investment Analysis: The Key to Sound Portfolio Management Strategy*. (2019, April 23). Retrieved from https://www.investopedia.com/terms/i/investment-analysis.asp
3. J.B. Williams, *The Theory of investment value* (BN Publishing, Miami, 2014)
4. K.C. Rasekhschaffe, R.C. Jones, Machine learning for stock selection. Financ. Anal. J. **75**(3), 70–88 (2019). https://doi.org/10.1080/0015198x.2019.1596678
5. T.-S. Quah, DJIA stock selection assisted by neural network. Expert Syst. Appl. **35**(1-2), 50–58 (2008). https://doi.org/10.1016/j.eswa.2007.06.039
6. K. Sim, V. Gopalkrishnan, C. Phua, G. Cong, 3D subspace clustering for value investing. IEEE Intell. Syst. **29**(2), 52–59 (2014). https://doi.org/10.1109/mis.2012.24
7. C. Tian, *Stock Financials Data Batch Download* (n.d.). Retrieved from https://www.gurufocus.com/download_financials_batch.php
8. Y. Xu, R. Goodacre, On splitting training and validation set: A comparative study of cross-validation, bootstrap and systematic sampling for estimating the generalization performance of supervised learning. J. Anal. Test. **2**, 249–262 (2018). https://doi.org/10.1007/s41664-018-0068-2
9. M. Buffett, D. Clark, *Warren Buffett and the Interpretation of Financial Statements the Search for the Company with a Durable Competitive Advantage* (Simon & Schuster, London, 2011)
10. T. Adkins, Calculating volatility: A simplified approach. in *Investopedia*, Investopedia, 25 Jan. 2020. www.investopedia.com/articles/basics/09/simplified-measuring-interpreting-volatility.asp
11. K. Duan, S.S. Keerthi, W. Chu, S.K. Shevade, A.N. Poo, Multi-category classification by soft-max combination of binary classifiers, in *Multiple Classifier Systems Lecture Notes in Computer Science*, (2003), pp. 125–134. https://doi.org/10.1007/3-540-44938-8_13

Index

Printed in the United States
by Baker & Taylor Publisher Services